Convection in Porous Media

Convection in Porous Media

Third Edition

Donald A. Nield

University of Auckland
Auckland, New Zealand

and

Adrian Bejan

Duke University
Durham, North Carolina, USA

Donald. A. Nield
Associate Professor
Department of Engineering Science
University of Auckland
Private Bag 92019, Auckland
New Zealand
d.nield@auckland.ac.nz

Adrian Bejan
J. A. Jones Professor of Mechanical Engineering
Department of Mechanical Engineering and
Materials Science
Duke University
Durham, North Carolina 27708
USA
abejan@duke.edu

Library of Congress Control Number: 2005932308

10 digit ISBN: 0-387-29096-6
13 digit ISBN: 978-0387-29096-6

Printed in the United States of America (BS/DH)

9 8 7 6 5 4 3 2 1

springer.com

To our wives
Rachel Nield and Mary Bejan

Our children
Cherry, Alexandra, and Peter Nield
Cristina, Teresa, and William Bejan

Our grandchildren
Michael and Rachel van der Mark
Charlotte and Susan Nield
Elizabeth and John Hayman

Preface to the Third Edition

Papers on convection in porous media continue to be published at the rate that is now over 200 per year. The indication of the continued importance of the subject, together with the wide acceptance of the first and second editions of this volume, has encouraged us to prepare an expanded third edition. We have retained the basic structure and most of the text of the second edition. We have been somewhat selective in our choice of references, but nevertheless there are over 1400 new references. Again, we have made an effort to highlight new conceptual developments and engineering applications.

We found that it was possible to fit a lot of the new material under the existing section headings. However, we now have new sections on bidisperse porous media, local thermal nonequilibrium, electrodiffusion, transverse heterogeneity in channels, thermal development of forced convection, effects of temperature-dependent viscosity, constructal multiscale flow structures, optimal spacings for plates separated by porous structures, control of convection using vertical vibration, and bioconvection.

Once again we decided that, except for a brief mention, convection in unsaturated media had to be beyond the scope of this book. Also, we are aware that there are some topics in the area of hydrology that could be regarded as coming under the umbrella of the title of our book but are not treated here.

We are grateful to a large number of people who provided us, prior to publication, with copies of their chapters of books that survey research on various topics. Other colleagues have continued to improve our understanding of the subject of this book in ways too numerous to mention here.

We wish to thank our employers, the University of Auckland and Duke University, for their ongoing support.

Once again we relied on the expertise and hard work of Linda Hayes and Deborah Fraze for the preparation of the electronic version of our manuscript.

D. A. Nield
A. Bejan

Preface to the Second Edition

Papers on convection in porous media continue to be published at the rate of over 100 per year. This indication of the continued importance of the subject, together with the wide acceptance of the first edition, has encouraged us to prepare an expanded second edition. We have retained the basic structure and most of the text of the first edition. With space considerations in mind, we have been selective in our choice of references, but nevertheless there are over 600 new references. We also made an effort to highlight new conceptual developments and engineering applications.

In the introductory material, we judged that Chapters 2 and 3 needed little alteration (though there is a new Section 2.6 on other approaches to the topic), but our improved understanding of the basic modeling of flow through a porous medium has led to a number of changes in Chapter 1, both within the old sections and by the addition of a section on turbulence in porous media and a section on fractured media, deformable media, and complex porous structures.

In Chapter 4, on forced convection, we have added major new sections on compact heat exchangers, on heatlines for visualizing convection, and on constructal tree networks for the geometric minimization of the resistance to volume-to-point flows in heterogeneous porous media.

In Chapter 5 (external natural convection) there is a substantial amount of new material inserted in the existing sections. In Chapters 6 and 7, on internal natural convection, we now have included descriptions of the effects of a magnetic field and rotation, and there are new sections on periodic heating and on sources in confined or partly confined regions; the latter is a reflection of the current interest in the problem of nuclear waste disposal. In Chapter 8, on mixed convection, there are no new sections, but in a new subsection we have given some prominence to the unified theory that has been developed for boundary layer situations. In Chapter 9, on double-diffusive convection (heat and mass transfer) there is a new section on convection produced by inclined gradients, a topic that also has been given wider coverage in the related section in Chapter 7.

In Chapter 10, which deals with convection with change of phase, we have a new subsection on the solidification of binary alloys, a research area that has blossomed in the last decade. We also have a new section on spaces filled with fluid and fibers coated with a phase-change material. In the first edition we had little to say about two-phase flow, despite its importance in geothermal and other contexts. We now have included a substantial discussion on this topic, which we have placed at the end of Chapter 11 (geophysical aspects). Once again we decided that, except for a brief mention, convection in unsaturated media had to be beyond the scope of this book.

D.A.N. again enjoyed the hospitality of the Department of Mechanical Engineering and Materials Science at Duke University while on Research and Study Leave from the University of Auckland, and both of those institutions again provided financial support.

We are grateful for comments from Graham Weir and Roger Young on a draft of Section 11.9, a topic on which we had much to learn. We also are grateful to a large number of people who provided us with preprints of their papers prior to publication. Other colleagues have improved our understanding of the subject of this book in ways too numerous to mention here.

Once again we relied on the expertise and hard work of Linda Hayes for the preparation of the electronic version of our manuscript, and again the staff at the Engineering Library of Duke University made our search of the literature an enjoyable experience.

D. A. Nield
A. Bejan

Preface to the First Edition

In this book we have tried to provide a user-friendly introduction to the topic of convection in porous media. We have assumed that the reader is conversant with the basic elements of fluid mechanics and heat transfer, but otherwise the book is self-contained. Only routine classic mathematics is employed. We hope that the book will be useful both as a review (for reference) and as a tutorial work (suitable as a textbook in a graduate course or seminar).

This book brings into perspective the voluminous research that has been performed during the last two decades. The field recently has exploded because of worldwide concern with issues such as energy self-sufficiency and pollution of the environment. Areas of application include the insulation of buildings and equipment, energy storage and recovery, geothermal reservoirs, nuclear waste disposal, chemical reactor engineering, and the storage of heat-generating materials such as grain and coal. Geophysical applications range from the flow of groundwater around hot intrusions to the stability of snow against avalanches.

We believe that this book is timely because the subject is now mature in the sense that there is a corpus of material that is unlikely to require major revision in the future. As the reader will find, the relations for heat transfer coefficients and flow parameters for the case of saturated media are now known well enough for engineering design purposes. There is a sound basis of underlying theory that has been validated by experiment. At the same time there are outstanding problems in the cases of unsaturated media and multiphase flow in heterogeneous media, which are relevant to such topics as the drying of porous materials and enhanced oil recovery.

The sheer bulk of the available material has limited the scope of this book. It has forced us to omit a discussion of convection in unsaturated media and also of geothermal reservoir modeling; references to reviews of these topics are given. We also have excluded mention of several hundred additional papers, including some of our own. We have emphasized reports of experimental work, which are in relatively short supply (and in some areas are still lacking). We have also emphasized simple analysis where this illuminates the physics involved. The excluded material includes some good early work, which has now been superseded, and some recent numerical work involving complex geometry. Also excluded are papers involving the additional effects of rotation or magnetic fields; we know of no reported experimental work or significant applications of these extensions. We regret that our survey could not be exhaustive, but we believe that this book gives a good picture of the current state of research in this field.

The first three chapters provide the background for the rest of the book. Chapters 4 through 8 form the core material on thermal convection. Our original plan, which was to separate foundational material from applications, proved to be impractical, and these chapters are organized according to geometry and the form of heating. Chapter 9 deals with combined heat and mass transfer and Chapter 10 with convection coupled with change of phase. Geophysical themes involve additional physical processes and have given rise to additional theoretical investigations; these are discussed in Chapter 11.

* * *

This book was written while D.A.N. was enjoying the hospitality of the Department of Mechanical Engineering and Materials Science at Duke University, while on Research and Study Leave from the University of Auckland. Financial support for this leave was provided by the University of Auckland, Duke University, and the United States–New Zealand Cooperative Science Program. We are particularly grateful to Dean Earl H. Dowell and Prof. Robert M. Hochmuth, both of Duke University, for their help in making this book project possible.

Linda Hayes did all the work of converting our rough handwritten notes into the current high-quality version on computer disk. She did this most efficiently and with tremendous understanding (i.e., patience!) for the many instances in which we changed our minds and modified the manuscript.

At various stages in the preparation of the manuscript and the figures we were assisted by Linda Hayes, Kathy Vickers, Jong S. Lim, Jose L. Lage, and Laurens Howle. Eric Smith and his team at the Engineering Library of Duke University went to great lengths to make our literature search easier. We are very grateful for all the assistance we have received.

D. A. Nield
A. Bejan

Contents

Nomenclature

B	transition number for electrodiffusion, Eq. (3.95)
Be	Bejan number, Eq. (4.145)
Br	Brinkman number, Section 2.2.2
C	concentration
c	specific heat
c_a	acceleration coefficient
c_F	Forchheimer coefficient
c_P	specific heat at constant pressure
D	diameter
D	d/dz
D_m	solute diffusivity
D_{CT}	thermodiffusion coefficient (Soret coefficient times D_m)
Da	Darcy number
d_p	particle diameter
Ec	Eckert number, Section 2.2.2
g	gravitational acceleration
Ge	Gebhart number, Section 2.2.2
H	vertical dimension
$\mathbf{i},\mathbf{j},\mathbf{k}$	unit vectors
Ja	Jakob number
K	permeability
k	thermal conductivity
k_m	thermal conductivity of the porous medium
L	horizontal dimension
Le	Lewis number
N	buoyancy ratio
Nu	Nusselt number
P	pressure
Pe	Péclet number
Pr	Prandtl number
q', q'', q'''	heat transfer rate per unit length, area, volume, respectively
r.e.v.	representative elementary volume, Fig. 1.2
Ra	thermal Rayleigh (Rayleigh-Darcy) number
Ra_D	solutal Rayleigh number
Re	Reynolds number
r	radial coordinate

Sc, Sh	Jakob numbers
Ste	Stefan number
s	time constant
T	temperature
t	time
$\mathbf{V}$	intrinsic velocity
$\mathbf{v}$	(u,v,w), seepage velocity
x,y,z	position coordinates
α	nondimensional wavenumber
α_{BJ}	Beavers-Joseph coefficient
α_m	thermal diffusivity of the porous medium
β	thermal expansion coefficient
β_C	concentration expansion coefficient
δ	boundary layer thickness
ζ	inter-phase momentum transfer coefficient
η	similarity variable
θ	angle
θ	temperature perturbation amplitude
λ	exponent in power law variation
μ	dynamic viscosity
	effective viscosity (Brinkman)
ν	kinematic viscosity
ρ	density
σ	heat capacity ratio, $\sigma = \varphi + (1-\varphi)(\rho c)_s/(\rho c_P)_f$
τ	nondimensional time
φ	porosity
φ	angle
ψ	streamfunction
ω	frequency
χ	$c_F K^{1/2}$

Subscripts

b	basic state
b	bulk
C	concentration
c	critical
D	parameter based on length D
e	effective
eff	effective
f	fluid
g	gas
H	horizontal
L	parameter based on length L
l	liquid

m	porous medium
p	particle
ref	reference
s	solid
V	vertical
w	wall
x	parameter based on length x
0	reference
∞	far field

Superscripts

$'$	perturbation
$\hat{}$	nondimensional perturbation

1
Mechanics of Fluid Flow through a Porous Medium

1.1. Introduction

By a porous medium we mean a material consisting of a solid matrix with an interconnected void. We suppose that the solid matrix is either rigid (the usual situation) or it undergoes small deformation. The interconnectedness of the void (the pores) allows the flow of one or more fluids through the material. In the simplest situation ("single-phase flow") the void is saturated by a single fluid. In "two-phase flow" a liquid and a gas share the void space.

In a natural porous medium the distribution of pores with respect to shape and size is irregular. Examples of natural porous media are beach sand, sandstone, limestone, rye bread, wood, and the human lung (Fig. 1.1). On the pore scale (the microscopic scale) the flow quantities (velocity, pressure, etc.) will be clearly irregular. But in typical experiments the quantities of interest are measured over areas that cross many pores, and such space-averaged (macroscopic) quantities change in a regular manner with respect to space and time, and hence are amenable to theoretical treatment.

How we treat a flow through a porous structure is largely a question of distance—the distance between the problem solver and the actual flow structure (Bejan *et al.*, 2004). When the distance is short, the observer sees only one or two channels, or one or two open or closed cavities. In this case it is possible to use conventional fluid mechanics and convective heat transfer to describe what happens at every point of the fluid- and solid-filled spaces. When the distance is large so that there are many channels and cavities in the problem solver's field of vision, the complications of the flow paths rule out the conventional approach. In this limit, volume-averaging and global measurements (e.g., permeability, conductivity) are useful in describing the flow and in simplifying the description. As engineers focus more and more on designed porous media at decreasing pore scales, the problems tend to fall between the extremes noted above. In this intermediate range, the challenge is not only to describe *coarse* porous structures, but also to *optimize* flow elements and to *assemble* them. The resulting flow structures are *designed* porous media (see Bejan *et al.*, 2004; Bejan, 2004b).

The usual way of deriving the laws governing the macroscopic variables is to begin with the standard equations obeyed by the fluid and to obtain the macroscopic equations by averaging over volumes or areas containing many pores. There are

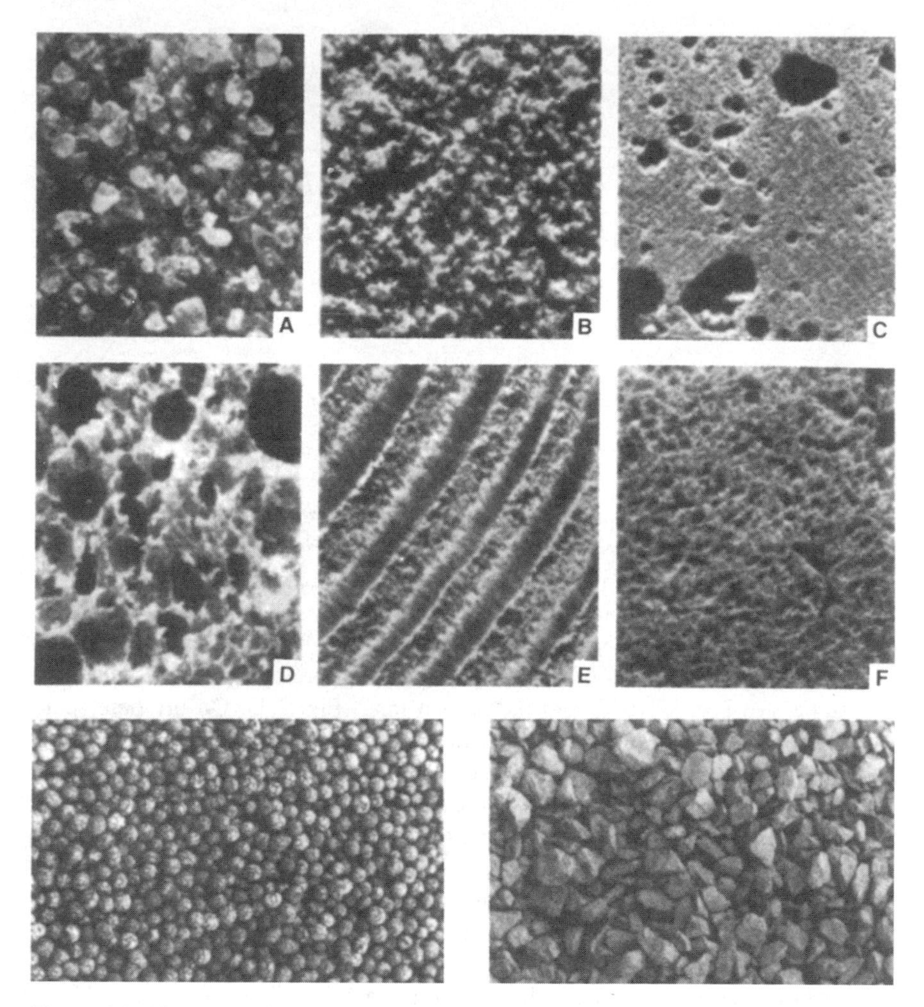

Figure 1.1. Top: Examples of natural porous materials: A) beach sand, B) sandstone, C) limestone, D) rye bread, E) wood, and F) human lung (Collins, 1961, with permission from Van Nostrand Reinhold). Bottom: Granular porous materials used in the construction industry, 0.5-cm-diameter Liapor® spheres (left), and 1-cm-size crushed limestone (right) (Bejan, 1984).

two ways to do the averaging: spatial and statistical. In the spatial approach, a macroscopic variable is defined as an appropriate mean over a sufficiently large *representative elementary volume* (r.e.v.); this operation yields the value of that variable at the centroid of the r.e.v. It is assumed that the result is independent of the size of the representative elementary volume. The length scale of the r.e.v. is much larger than the pore scale, but considerably smaller than the length scale of the macroscopic flow domain (Fig. 1.2).

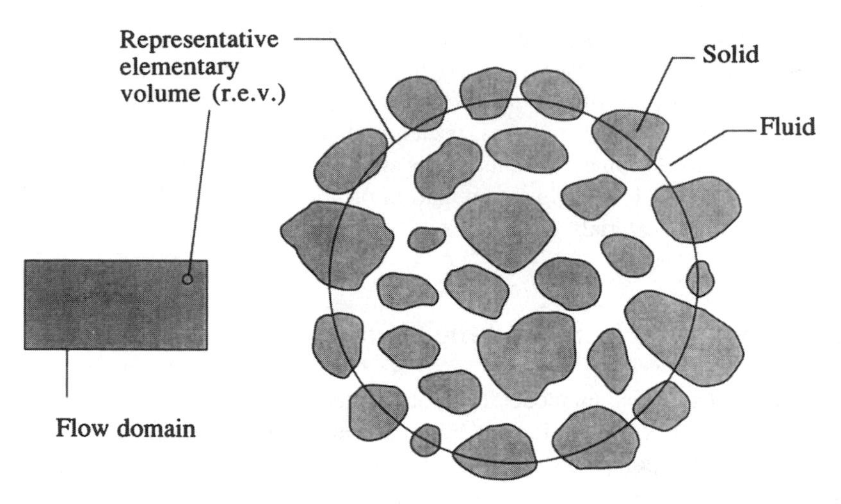

Figure 1.2. The representative elementary volume (r.e.v.): the figure illustrates the intermediate size relative to the sizes of the flow domain and the pores.

In the statistical approach the averaging is over an ensemble of possible pore structures that are macroscopically equivalent. A difficulty is that usually the statistical information about the ensemble has to be based on a single sample, and this is possible only if statistical homogeneity (stationarity) is assumed.

If one is concerned only with deriving relationships between the space-averaged quantities and is not concerned about their fluctuation, then the results obtained by using the two approaches are essentially the same. Thus in this situation one might as well use the simpler approach, namely the one based on the r.e.v. An example of its use is given in Section 3.5. This approach is discussed at length by Bear and Bachmat (1990). In recent years a number of problems have come to the fore which require a statistical approach [see, for example, Georgiadis and Catton (1987, 1988) and Georgiadis (1991)].

1.2. Porosity

The porosity φ of a porous medium is defined as the fraction of the total volume of the medium that is occupied by void space. Thus $1 - \varphi$ is the fraction that is occupied by solid. For an isotropic medium the "surface porosity" (that is, the fraction of void area to total area of a typical cross section) will normally be equal to φ.

In defining φ in this way we are assuming that all the void space is connected. If in fact one has to deal with a medium in which some of the pore space is disconnected from the remainder, then one has to introduce an "effective porosity," defined as the ratio of connected void to total volume.

For natural media, φ does not normally exceed 0.6. For beds of solid spheres of uniform diameter φ can vary between the limits 0.2595 (rhombohedral packing) and 0.4764 (cubic packing). Nonuniformity of grain size tends to lead to smaller porosities than for uniform grains, because smaller grains fill the pores formed by larger grains. For man-made materials such as metallic foams φ can approach the value 1. Table 1.1 shows a compilation of porosities and other properties of common porous materials.

1.3. Seepage Velocity and the Equation of Continuity

We construct a continuum model for a porous medium, based on the r.e.v. concept. We introduce a Cartesian reference frame and consider volume elements that are sufficiently large compared with the pore volumes for reliable volume averages to be obtained. In other words, the averages are not sensitive to the choice of volume element. A distinction is made between an average taken with respect to a volume element V_m of the medium (incorporating both solid and fluid material) and one taken with respect to a volume element V_f consisting of fluid only. For example, we denote the average of the fluid velocity over V_m by $\mathbf{v} = (u, v, w)$. This quantity has been given various names, by different authors, such as seepage velocity, filtration velocity, superficial velocity, Darcy velocity, and volumetric flux density. We prefer the term Darcy velocity since it is short and distinctive. Taking an average of the fluid velocity over a volume V_f we get the intrinsic average velocity $\mathbf{V}$, which is related to $\mathbf{v}$ by the Dupuit-Forchheimer relationship $\mathbf{v} = \varphi\mathbf{V}$.

Once we have a continuum to deal with, we can apply the usual arguments and derive differential equations expressing conservation laws. For example, the conservation of mass is expressed by the continuity equation

$$\varphi\frac{\partial\rho_f}{\partial t} + \nabla\cdot(\rho_f\mathbf{v}) = 0 \tag{1.1}$$

where ρ_f is the fluid density. This equation is derived by considering an elementary unit volume of the medium and equating the rate of increase of the mass of the fluid within that volume, $\partial(\varphi\rho_f)/\partial t$, to the net mass flux into the volume, $-\nabla\cdot(\rho_f\mathbf{v})$, noting that φ is independent of t.

1.4. Momentum Equation: Darcy's Law

We now discuss various forms of the momentum equation which is the porous-medium analog of the Navier-Stokes equation. For the moment we neglect body forces such as gravity; the appropriate terms for these can be added easily at a later stage.

Table 1.1. Properties of common porous materials [based on data compiled by Scheidegger (1974) and Bejan and Lage (1991)]

Material	Porosity φ	Permeability K [cm^2]	Surface per unit volume [cm^{-1}]
Agar-agar		2×10^{-10}–4.4×10^{-9}	
Black slate powder	0.57–0.66	4.9×10^{-10}–1.2×10^{-9}	7×10^3–8.9×10^3
Brick	0.12–0.34	4.8×10^{-11}–2.2×10^{-9}	
Catalyst (Fischer-Tropsch, granules only)	0.45		5.6×10^5
Cigarette		1.1×10^{-5}	
Cigarette filters	0.17–0.49		
Coal	0.02–0.12		
Concrete (ordinary mixes)	~ 0.1		
Concrete (bituminous)		1×10^{-9}–2.3×10^{-7}	
Copper powder (hot-compacted)	0.09–0.34	3.3×10^{-6}–1.5×10^{-5}	
Cork board		2.4×10^{-7}–5.1×10^{-7}	
Fiberglass	0.88–0.93		560–770
Granular crushed rock	0.45		
Hair (on mammals)	0.95–0.99		
Hair left		8.3×10^{-6}–1.2×10^{-5}	
Leather	0.56–0.59	9.5×10^{-10}–1.2×10^{-9}	1.2×10^4–1.6×10^4
Limestone (dolomite)	0.04–0.10	2×10^{-11}–4.5×10^{-10}	
Sand	0.37–0.50	2×10^{-7}–1.8×10^{-6}	150–220
Sandstone ("oil sand")	0.08–0.38	5×10^{-12}–3×10^{-8}	
Silica grains	0.65		
Silica powder	0.37–0.49	1.3×10^{-10}–5.1×10^{-10}	6.8×10^3–8.9×10^3
Soil	0.43–0.54	2.9×10^{-9}–1.4×10^{-7}	
Spherical packings (well shaken)	0.36–0.43		
Wire crimps	0.68–0.76	3.8×10^{-5}–1×10^{-4}	29–40

1.4.1. Darcy's Law: Permeability

Henry Darcy's (1856) investigations into the hydrology of the water supply of Dijon and his experiments on steady-state unidirectional flow in a uniform medium revealed a proportionality between flow rate and the applied pressure difference. In modern notation this is expressed, in refined form, by

$$u = -\frac{K}{\mu}\frac{\partial P}{\partial x} \tag{1.2}$$

Here $\partial P/\partial x$ is the pressure gradient in the flow direction and μ is the dynamic viscosity of the fluid. The coefficient K is independent of the nature of the fluid but it depends on the geometry of the medium. It has dimensions $(\text{length})^2$ and is called the *specific permeability* or *intrinsic permeability* of the medium. In the case of single phase flow we abbreviate this to permeability. The permeabilities of common porous materials are summarized in Table 1.1. It should be noted that in Eq. (1.2) P denotes an intrinsic quantity, and that Darcy's equation is not a balance of forces averaged over a r.e.v. Special care needs to be taken when adding additional terms such as the one expressing a Coriolis force. One needs to take averages over the fluid phase before introducing a Darcy drag term. (See Section 1.5.1 below.)

In three dimensions, Eq. (1.2) generalizes to

$$\mathbf{v} = \mu^{-1}\mathbf{K}\cdot\nabla P, \tag{1.3}$$

where now the permeability $\mathbf{K}$ is in general a second-order tensor. For the case of an isotropic medium the permeability is a scalar and Eq. (1.3) simplifies to

$$\nabla P = -\frac{\mu}{K}\mathbf{v}. \tag{1.4}$$

Values of K for natural materials vary widely. Typical values for soils, in terms of the unit m^2, are: clean gravel 10^{-7}–10^{-9}, clean sand 10^{-9}–10^{-12}, peat 10^{-11}–10^{-13}, stratified clay 10^{-13}–10^{-16}, and unweathered clay 10^{-16}–10^{-20}. Workers concerned with geophysics often use as a unit of permeability the *Darcy*, which equals 0.987×10^{-12} m^2.

Darcy's law has been verified by the results of many experiments. Theoretical backing for it has been obtained in various ways, with the aid of either deterministic or statistical models. It is interesting that Darcy's original data may have been affected by the variation of viscosity with temperature (Lage, 1998). A refined treatment of the mass and momentum conservation equations, based on volume averaging, has been presented by Altevogt *et al.* (2003).

1.4.2. Deterministic Models Leading to Darcy's Law

If K is indeed determined by the geometry of the medium, then clearly it is possible to calculate K in terms of the geometrical parameters, at least for the case of simple geometry. A great deal of effort has been spent on this endeavor, and the results are well presented by Dullien (1992).

For example, in the case of beds of particles or fibers one can introduce an effective average particle or fiber diameter D_p. The hydraulic radius theory of Carman-Kozeny leads to the relationship

$$K = \frac{D_{p_2}^2 \varphi^3}{180(1-\varphi)^2}, \tag{1.5}$$

where

$$D_{p_2} = \int_0^\infty D_p^3 h\left(\mathrm{D_p}\right) dD_p \bigg/ \int_0^\infty D_p^2 h\left(D_p\right) dD_p \tag{1.6}$$

and $h(D_p)$ is the density function for the distribution of diameters D_p. The constant 180 in Eq. (1.5) was obtained by seeking a best fit with experimental results. The Carman-Kozeny equation gives satisfactory results for media that consist of particles of approximately spherical shape and whose diameters fall within a narrow range. The equation is often not valid in the cases of particles that deviate strongly from the spherical shape, broad particle-size distributions, and consolidated media. Nevertheless it is widely used since it seems to be the best simple expression available. A modified Carman-Kozeny theory was proposed by Liu *et al.* (1994). A fibrous porous medium was modeled by Davis and James (1996). For randomly packed monodisperse fibers, the experiments of Rahli *et al.* (1997) showed that the Carman-Kozeny "constant" is dependent on porosity and fiber aspect ratio. The Carman-Kozeny correlation has been applied to compressed expanded natural graphite, an example of a high porosity and anisotropic consolidated medium, by Mauran *et al.* (2001). Li and Park (1998) applied an effective medium approximation to the prediction of the permeability of packed beds with polydisperse spheres.

1.4.3. Statistical Models Leading to Darcy's Law

Many authors have used statistical concepts in the provision of theoretical support for Darcy's law. Most authors have used constitutive assumptions in order to obtain closure of the equations, but Whitaker (1986) has derived Darcy's law, for the case of an incompressible fluid, without making any constitutive assumption. This theoretical development is not restricted to either homogeneous or spatially periodic porous media, but it does assume that there are no abrupt changes in the structure of the medium.

If the medium has periodic structure, then the homogenization method can be used to obtain mathematically rigorous results. The method is explained in detail by Ene and Poliševski (1987), Mei *et al.* (1996), and Ene (2004). The first authors derive Darcy's law without assuming incompressibility, and they go on to prove that the permeability is a symmetric positive-definite tensor.

1.5. Extensions of Darcy's Law

1.5.1. Acceleration and Other Inertial Effects

Following Wooding (1957), many early authors on convection in porous media used an extension of Eq. (1.4) of the form

$$\rho_f\left[\frac{\partial \mathbf{V}}{\partial t}+(\mathbf{V}\cdot\nabla)\mathbf{V}\right]=-\nabla P-\frac{\mu}{K}\mathbf{v} \tag{1.7}$$

which, when the Dupuit-Forchheimer relationship is used, becomes

$$\rho_f\left[\varphi^{-1}\frac{\partial \mathbf{v}}{\partial t}+\varphi^{-2}(\mathbf{v}\cdot\nabla)\mathbf{v}\right]=-\nabla P-\frac{\mu}{K}\mathbf{v}. \tag{1.8}$$

This equation was obtained by analogy with the Navier-Stokes equation. Beck (1972) pointed out that the inclusion of the $(\mathbf{v}\cdot\nabla)\mathbf{v}$ term was inappropriate because it raised the order (with respect to space derivatives) of the differential equation, and this was inconsistent with the slip boundary condition (appropriate when Darcy's law was employed). More importantly, the inclusion of $(\mathbf{v}\cdot\nabla)\mathbf{v}$ is not a satisfactory way of expressing the nonlinear drag, which arises from inertial effects, since $(\mathbf{v}\cdot\nabla)\mathbf{v}$ is identically zero for steady incompressible unidirectional flow no matter how large the fluid velocity, and this is clearly in contradiction to experience.

There is a further fundamental objection. In the case of a viscous fluid a material particle retains its momentum, in the absence of applied forces, when it is displaced from a point A to a neighboring arbitrary point B. But in a porous medium with a fixed solid matrix this is not so, in general, because some solid material impedes the motion and causes a change in momentum. The $(\mathbf{v}\cdot\nabla)\mathbf{v}$ term is generally small in comparison with the quadratic drag term (see Section 1.5.2) and then it seems best to drop it in numerical work. This term needs to be retained in the case of highly porous media. Also, at least the irrotational part of the term needs to be retained in order to account for the phenomenon of choking in high-speed flow of a compressible fluid (Nield, 1994b). Nield suggested that the rotational part, proportional to the intrinsic vorticity, be deleted. His argument is based on the expectation that a medium of low porosity will allow scalar entities like fluid speed to be freely advected, but will inhibit the advection of vector quantities like vorticity. It is now suggested that even when vorticity is being continuously produced (e.g. by buoyancy), one would expect that it would be destroyed by a momentum dispersion process due to the solid obstructions. The claim that the $(\mathbf{v}\cdot\nabla)\mathbf{v}$ term is necessary to account for boundary layer development is not valid; viscous diffusion can account for this. Formal averaging of the Navier-Stokes equation leads to a $(\mathbf{v}\cdot\nabla)\mathbf{v}$ term, but this is deceptive. Averaging methods inevitably involve a loss of information with respect to the effects of geometry on the flow.

With the $(\mathbf{v}\cdot\nabla)\mathbf{v}$ term dropped, Eq. (1.8) becomes

$$\frac{\rho_f}{\varphi}\frac{\partial \mathbf{v}}{\partial t}=-\nabla P-\frac{\mu}{K}\mathbf{v}. \tag{1.9}$$

One can now question whether the remaining inertial term (the left-hand side of this equation) is correct. It has been derived on the assumption that the partial derivative with respect to time permutes with a volume average, but in general this is not valid. The inadequacy of Eq. (1.9) can be illustrated by considering an ideal medium, one in which the pores are identical parallel tubes of uniform circular cross section of radius a. Equation (1.9) leads to the prediction that in the presence of a constant pressure gradient any transient will decay like $\exp[-(\mu\varphi/K\rho_f)t]$, whereas from the exact solution for a circular pipe [see, for example, formula (4.3.19) of Batchelor (1967)] one concludes that the transient should decay approximately like $\exp[-({\lambda_1}^2\mu/a^2\rho_f)t]$, where $\lambda_1 = 2.405$ is the smallest positive root of $J_0(\lambda) = 0$, and where J_0 is the Bessel function of the first kind of order zero. In general, these two exponential decay terms will not be the same. It appears that the best that one can do is to replace Eq. (1.9) by

$$\rho_f \mathbf{c}_a \cdot \frac{\partial \mathbf{v}}{\partial t} = -\nabla P - \frac{\mu}{K}\mathbf{v}, \tag{1.10}$$

where $\mathbf{c}_a$ is a constant tensor that depends sensitively on the geometry of the porous medium and is determined mainly by the nature of the pore tubes of largest cross sections (since in the narrower pore tubes the transients decay more rapidly). We propose that $\mathbf{c}_a$ be called the "acceleration coefficient tensor" of the porous medium. For the special medium introduced above, in which we have unidirectional flow, the acceleration coefficient will be a scalar, $c_a = a^2/{\lambda_1}^2 K$. If the Carman–Kozeny formula (Eq. 1.5) is valid and if D_{p2} can be identified with a/γ where γ is some constant, then

$$\mathrm{c_a} = 180\gamma^2(1-\varphi)^2/\lambda_1^2\varphi^3 = 31.1\gamma^2(1-\varphi)^2/\varphi^3. \tag{1.11}$$

Liu and Masliyah (2005) present an equation, obtained by volumetric averaging, that does indicate a slower decaying speed than that based on the straight passage model. They also say that the decaying speed is expected to be much faster than that for a medium free from solids, and it is this characteristic that makes the flow in a porous medium more hydrodynamically stable than that in an infinitely permeable medium and delayed turbulence is expected.

In any case, one can usually drop the time derivative term completely because in general the transients decay rapidly. An exceptional situation is when the kinematic viscosity $\nu = \mu/\rho_f$ of the fluid is small in comparison with K/t_0 where t_0 is the characteristic time of the process being investigated. This criterion is rarely met in studies of convection. Even for a liquid metal ($\nu \sim 10^{-7}\mathrm{m}^2\,\mathrm{s}^{-1}$) and a material of large permeability ($K \sim 10^{-7}\mathrm{m}^2$) it requires $t_0 \ll 1\mathrm{s}$. However, it is essential to retain the time-derivative term when modeling certain instability problems: see Vadasz (1999).

For a porous medium in a frame rotating with angular velocity Ω with respect to an inertial frame, in Eq. (1.8) P is replaced by $P - \rho_f|\Omega \times \mathbf{x}|^2/2$, where $\mathbf{x}$ is the position vector, and a term $\rho_f\Omega \times \mathbf{v}/\varphi$ is added on the left-hand side.

If the fluid is electrically conducting, then in Eq. (1.8) P is replaced by $P + |\mathbf{B}|^2/2\mu_m$, where $\mathbf{B}$ is the magnetic induction and μ_m is the magnetic permeability, and a term $(\mathbf{B}.\nabla)\mathbf{B}/\varphi\mu_m$ is added to the right-hand side. In most practical cases the effect of a magnetic field on convection will be negligible, for reasons spelled out in Section 6.21.

The solution of the momentum equation and equation of continuity is commonly carried out by using the vector operators div and curl to solve in succession for the rotational and irrotational parts of the velocity field. The accuracy of the numerical solution thus obtained depends on the order of performing the operations. Wooding (2005) showed that taking a certain linear combination of the two solutions produces a solution of optimal accuracy.

1.5.2. Quadratic Drag: Forchheimer's Equation

Darcy's equation (1.3) is linear in the seepage velocity $\mathbf{v}$. It holds when $\mathbf{v}$ is sufficiently small. In practice, "sufficiently small" means that the Reynolds number Re_p of the flow, based on a typical pore or particle diameter, is of order unity or smaller. As $\mathbf{v}$ increases, the transition to nonlinear drag is quite smooth; there is no sudden transition as Re_p is increased in the range 1 to 10. Clearly this transition is not one from laminar to turbulent flow, since at such comparatively small Reynolds numbers the flow in the pores is still laminar. Rather, the breakdown in linearity is due to the fact that the form drag due to solid obstacles is now comparable with the surface drag due to friction. According to Joseph *et al.* (1982) the appropriate modification to Darcy's equation is to replace Eq. (1.4) by

$$\nabla P = -\frac{\mu}{K}\mathbf{v} - c_F K^{-1/2}\rho_f\,|\mathbf{v}|\,\mathbf{v}, \tag{1.12}$$

where c_F is a dimensionless form-drag constant. Equation (1.12) is a modification of an equation associated with the names of Dupuit (1863) and Forchheimer (1901); see Lage (1998). For simplicity, we shall call Eq. (1.12) the Forchheimer equation and refer to the last term as the Forchheimer term, but in fact the dependence on $\rho_f K^{-1/2}$ is a modern discovery (Ward, 1964). Ward thought that c_F might be a universal constant, with a value of approximately 0.55, but later it was found that c_F does vary with the nature of the porous medium and can be as small as 0.1 in the case of foam metal fibers. Beavers *et al.* (1973) showed that the bounding walls could have a substantial effect on the value of c_F, and found that their data correlated fairly well with the expression

$$c_F = 0.55\left(1 - 5.5\frac{d}{D_e}\right), \tag{1.13}$$

where d is the diameter of their spheres and D_e is the equivalent diameter of the bed, defined in terms of the height h and width w of the bed by

$$D_e = \frac{2wh}{w+h}. \tag{1.14}$$

The numerical calculations of Coulaud *et al.* (1988) on flow past circular cylinders suggest that c_F varies as φ^{-1} for φ less than 0.61.

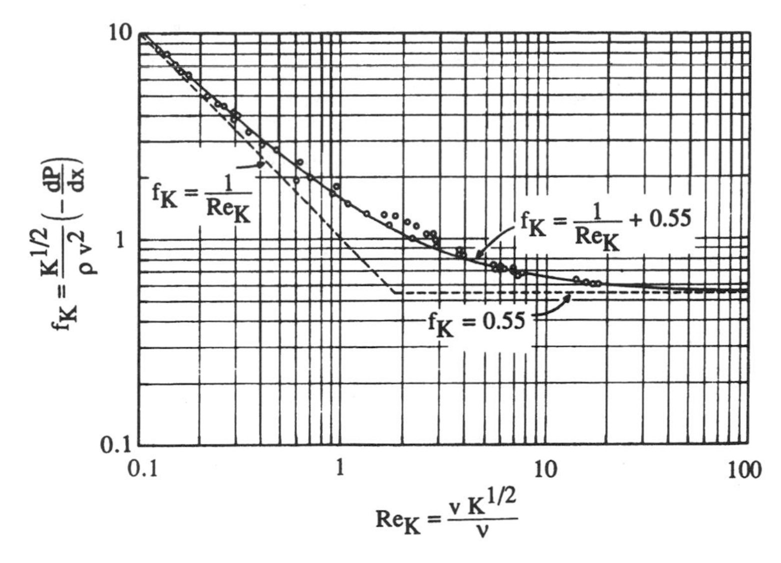

Figure 1.3. The transition from the Darcy regime to the Forchheimer regime in unidirectional flow through an isothermal saturated porous medium (Ward, 1964).

Equation (1.12) is invariant under a rotation of coordinate frame. Kaviany (1995) gives a form for the Forchheimer term [see his Eq. (2.57)], which does not have this property, and he gives no evidence for his claim that his form is more in accordance with the experimental results.

The transition from the Darcy regime to the Forchheimer regime is illustrated in Fig. 1.3. The data refer to unidirectional isothermal flow with the seepage velocity v in the direction x. Plotted on the ordinate is the "friction factor" f_K, which is based on $K^{1/2}$ as length scale. The abscissa belongs to the Reynolds number based on $K^{1/2}$. Figure 1.3 shows that the transition occurs in the Re_K range 1–10. At higher Reynolds numbers, the quadratic drag term dominates on the right-hand side of Eq. (1.12), and f_K becomes the same as c_F.

Associated with the transition to pore-scale turbulence, the coefficient c_F varies with velocity. For a *limited range*, one can take c_F to be linear in velocity. That means that the drag is cubic in velocity. Experiments reported by Lage *et al.* (1997) show this behavior. Extensive experimental data for flow in packed beds were presented by Achenbach (1995). This sort of cubic variation is distinct from that which occurs for small values of the pore-based Reynolds number. Firdaouss *et al.* (1997) showed that, under fairly general assumptions and for periodic porous media whose period is of the same order as that of the inclusion, the nonlinear correction to Darcy's law is cubic with respect to the Darcy number. In this case the quadratic term vanishes. The case of anisotropic media was discussed by Skjetne and Auriault (1999a). However, Lage and Antohe (2000) demonstrated that this mathematically valid cubic extension is irrelevant in practice, and they suggested an alternative parameter, in place of the Reynolds number, to characterize the transition from linearity. A further limit on the applicability of a Forchheimer-type law was noted by Montillet (2004). The validation of Forchheimer's law for flow through porous media with converging boundaries was discussed by Venkataraman and Rao (2000). An extra term, involving $|\mathbf{v}|^{1/2}\mathbf{v}$ [effectively the geometric mean of the two terms on the right-hand side of Eq. (1.12)] was introduced by Hsu and Cheng (1990). They argued that this modification was necessitated by the need to allow for the viscous boundary layer effect at intermediate values of the Reynolds number. The modification is supported by the results of pressure drop experiments reported by Hsu *et al.* (1999). However, for practical thermal convection problems the inclusion of this term in the model leads to relatively little improvement in explanatory power, and so the term is usually neglected.

The transition from Darcy flow (1.4) to Darcy-Forchheimer flow [Eq. (1.12)] occurs when Re_K is of order 10^2. This transition is associated with the occurrence of the first eddies in the fluid flow, for example, the rotating fluid behind an obstacle or a backward facing step. The order of magnitude $\mathrm{Re}_K \sim 10^2$ is one in a long list of constructal theory results that show that the laminar-turbulent transition is associated with a universal *local Reynolds number* of order 10^2 (Bejan, 1984, p. 213).

To derive $\mathrm{Re}_K \sim 10^2$ from turbulence, assume that the porous structure is made of three-dimensional random fibers that are so sparsely distributed that $\varphi \leq 1$. According to Koponen *et al.* (1998), in this limit the permeability of the structure is correlated very well by the expression $K = 1.39D^2/[e^{10.1(1-\varphi)} - 1]$, where D is the fiber diameter. In this limit the volume-averaged velocity $u\varphi$ has the same scale as the velocity of the free stream that bathes every fiber. It is well known that vortex shedding occurs when $\mathrm{Re}_D = uD/\nu \sim 10^2$ (e.g., Bejan, 2000, p. 155). By eliminating D between the above expressions for K and Re_D, we calculate $\mathrm{Re}_K = uK^{1/2}/\nu$ and find that when eddies begin to appear, the Re_K value is in the range 100 to 200 when φ is in the range 0.9 to 0.99.

Equation (1.12) is the form of Forchheimer's equation that we recommend for use, but for reference we note that Irmay (1958) derived an alternate equation, for unidirectional flow, of the form

$$\frac{dP}{dx} = -\frac{\beta\mu(1-\varphi)^2 v}{d_p^2\varphi^3} - \frac{\alpha\rho_f(1-\varphi)v^2}{d_p\varphi^3} \tag{1.15}$$

where d_p is the mean particle diameter and α and β are shape factors that must be determined empirically. With $\alpha = 1.75$ and $\beta = 150$ this equation is known as Ergun's equation. The linear terms in Eq. (1.15) and the unidirectional case of Eq. (1.12) can be made identical by writing

$$K = \frac{d_P^2\varphi^3}{\beta(1-\varphi)^2} \tag{1.16}$$

which is Kozeny's equation, but it is not possible at the same time to make the quadratic terms identical, in general. Some authors have forced them to be identical by taking $c_F = \alpha\beta^{-1/2}\varphi^{-3/2}$, and they have then used this expression in their numerical computations. It should be appreciated that this is an *ad hoc* procedure.

Either Eqs. (1.12) or (1.15) correlates well with available experimental data (see, for example, Macdonald *et al.*, 1979). A correlation slightly different from that of Ergun was presented by Lee and Ogawa (1994). Papathanasiou *et al.* (2001) showed that for fibrous material the Ergun equation overpredicts the observed friction factor when the usual Reynolds number (based on the particle diameter) is greater than unity, and they proposed an alternative correlation, based directly on the Forchheimer equation and a Reynolds number based on the square root of the permeability.

For further discussion of the Forchheimer equation, supporting the viewpoint taken here, see Barak (1987) and Hassanizadeh and Gray (1988). They emphasize that the averaging of microscopic *drag forces* leads to a macroscopic nonlinear theory for flow, but the average of microscopic *inertial terms* is negligible in typical practical circumstances. It seems that the need for fluid to flow around solid particles leads to a reduction in the coherence of the fluid momentum pattern, so that on the macroscopic scale there is negligible net transfer of momentum in a direction transverse to the seepage velocity vector. An analytical development based on form drag was given by du Plessis (1994). An analysis of the way in which microscopic phenomena give rise to macroscopic phenomena was presented by Ma and Ruth (1993).

The ratio of the convective inertia term $\rho\varphi^{-2}(\mathbf{v}\cdot\nabla)\mathbf{v}$ to the quadratic drag term is of order $K^{1/2}/c_F\varphi^2 L$, where L is the characteristic length scale. This ratio is normally small, and hence it is expected that the calculations of the heat transfer which have been made by several authors, who have included both terms in the equation of motion, are not significantly affected by the convective inertia term. This has been confirmed for two cases by Lage (1992) and Manole and Lage (1993).

A momentum equation with a Forchheimer correction was obtained using the method of volume averaging by Whitaker (1996). A generalized Forchheimer equation for two-phase flow based on hybrid mixture theory was proposed by Bennethum and Giorgi (1997). Other derivations have been given by Giorgi (1997) (via matched asymptotic expansions), Chen *et al.* (2001) (via homogenization), and Levy *et al.* (1999) (for the case of a thermoelastic medium). A generalized tensor form applicable to anisotropic permeability was derived by Knupp and Lage (1995). An alternative derivation for anisotropic media was given by Wang *et al.* (1999). An attempt to determine the values of the constants in an Ergun-type equation by numerical simulation for an array of spheres was reported by Nakayama *et al.* (1995). A reformulation of the Forchheimer equation, involving two Reynolds numbers, was made by Teng and Zhao (2000). Lee and Yang (1997) investigated Forchheimer drag for flow across a bank of circular cylinders. The effective inertial coefficient for a heterogenous prorus medium was discussed by Fourar *et al.* (2005).

Lage *et al.* (2005) prefer to work in terms of a form coefficient C related to c_F by $C = c_F L/K^{1/2}$, where L is a global characteristic length such as the length of a channel. They introduce a protocol for the determination of K and C, using Darcy's law for a porous medium and Newton's law of flow round a bluff body as constitutive equations defining K and C, respectively. Their analysis shows that the model equation for measuring C requires the separation between the viscous-drag effect imposed by the porous medium and the viscous effect of the boundary walls on the measured pressure drop when defining K.

1.5.3. Brinkman's Equation

An alternative to Darcy's equation is what is commonly known as Brinkman's equation. With inertial terms omitted this takes the form

$$\nabla P = -\frac{\mu}{K}\mathbf{v} + \tilde{\mu}\nabla^2\mathbf{v}. \tag{1.17}$$

We now have two viscous terms. The first is the usual Darcy term and the second is analogous to the Laplacian term that appears in the Navier-Stokes equation. The coefficient $\tilde{\mu}$ is an effective viscosity. Brinkman set μ and $\tilde{\mu}$ equal to each other, but in general that is not true.

In recent papers Eq. (1.17) has been referred to as "Brinkman's extension of Darcy's law" but this is a misleading expression. Brinkman (1947a,b) did not just add another term. Rather, he obtained a relationship between the permeability K and the porosity φ for an assembly of spheres a "self-consistant" procedure, which is valid only when the porosity is sufficiently large, $\varphi > 0.6$ according to Lundgren (1972). This requirement is highly restrictive since, as we have noted earlier, most naturally occurring porous media have porosities less than 0.6.

When the Brinkman equation is employed as a general momentum equation, the situation is more complicated. In Eq. (1.17) P is the intrinsic fluid pressure, so each term in that equation represents a force per unit volume of the *fluid*. A detailed averaging process leads to the result that, for an isotropic porous medium, $\tilde{\mu}/\mu = 1/\varphi T^*$, where T^* is a quantity called the tortuosity of the medium (Bear and Bachmat, 1990, p. 177). Thus $\tilde{\mu}/\mu$ depends on the geometry of the medium. This result appears to be consistent with the result of Martys *et al.* (1994), who on the basis of a study in which a numerical solution of the Stokes' equation was matched with a solution of Brinkman's equation for a flow near the interface between a clear fluid and a porous medium, concluded that the value of $\tilde{\mu}/\mu$ had to exceed unity, and increased monotonically with decreasing porosity. Liu and Masliyah (2005) summarize the current understanding by saying that the numerical simulations have shown that, depending upon the type of porous media, the effective viscosity may be either smaller or greater then the viscosity of the fluid. On the one hand, straight volume averaging as presented by Ochoa-Tapia and Whitaker (1995a) gives $\tilde{\mu}/\mu = 1/\varphi$, greater than unity. On the other hand, analyses such as that by Sáez *et al.* (1991) give $\tilde{\mu}/\mu$ close to a tortuosity τ, defined as dx/ds where $s(x)$ is the distance along a curve, a quantity that is less than unity. Liu and Masliyah (2005) suggest that one can think of the difference between $\tilde{\mu}$ and μ as being due to momentum dispersion. They say that it has been generally accepted that $\tilde{\mu}$ is strongly dependent on the type of porous medium as well as the strength of flow. They note that there are further complications if the medium is not isotropic. They also note that it is common practice for $\tilde{\mu}$ to be taken as equal for μ for high porosity cases.

Experimental checks of Brinkman's theory have been indirect and few in number. Lundgren refers to measurements of flows through cubic arrays of spherical beads on wires, which agree quite well with the Brinkman formula for permeability as a function of porosity. Givler and Altobelli (1994) matched theoretical and observed velocity profiles for a rigid foam of porosity 0.972 and obtained a value of about 7.5 for $\tilde{\mu}/\mu$. In our opinion the Brinkman model is breaking down when such a large value of $\tilde{\mu}/\mu$ is needed to match theory and experiment. Some preliminary results of a numerical investigation by Gerritsen *et al.* (2005) suggest that the Brinkman equation is indeed not uniformly valid as the porosity tends to unity.

It was pointed out by Tam (1969) that whenever the spatial length scale is much greater than $(\tilde{\mu}K/\mu)^{1/2}$ the $\nabla^2\mathbf{v}$ term in Eq. (1.17) is negligible in comparison with the term proportional to $\mathbf{v}$, so that Brinkman's equation reduces to Darcy's equation. Levy (1981) showed that the Brinkman model holds only for particles whose size is of order η^3, where η ($\ll 1$) is the distance between neighboring particles; for larger particles the fluid filtration is governed by Darcy's law and for smaller particles the flow does not deviate from that for no particles. Durlofsky and Brady (1987), using a Green's function approach, concluded that the Brinkman equation was valid for $\varphi > 0.95$. Rubinstein (1986) introduced a porous medium having a very large number of scales, and concluded that it could be valid for φ as small as 0.8.

We conclude that for many practical purposes there is no need to include the Laplacian term. If it is important that a no-slip boundary condition be satisfied, then the Laplacian term is indeed required; but its effect is significant only in a thin boundary layer whose thickness is of order $(\tilde{\mu}K/\mu)^{1/2}$, the layer being thin since the continuum hypothesis requires that $K^{1/2} \ll L$ where L is a characteristic macroscopic length scale of the problem being considered. When the Brinkman equation is employed, it usually will be necessary to also account for the effects of porosity variation near the wall (see Section 1.7).

There are situations in which some authors have found it convenient to use the Brinkman equation. One such situation is when one wishes to compare flows in porous media with those in clear fluids. The Brinkman equation has a parameter K (the permeability) such that the equation reduces to a form of the Navier-Stokes equation as $K \to \infty$ and to the Darcy equation as $K \to 0$. Another situation is when one wishes to match solutions in a porous medium and in an adjacent viscous fluid. But usage of the Brinkman equation in this way is not without difficulty, as we point out in the following section.

Several recent authors have added a Laplacian term to Eq. (1.12) to form a "Brinkman-Forchheimer" equation. The validity of this is not completely clear. As we have just seen, in order for Brinkman's equation to be valid, the porosity must be large, and there is some uncertainty about the validity of the Forchheimer law at such large porosity. A scale analysis by Lage (1993a) revealed the distinct regimes in which the various terms in the Brinkman-Forchheimer equation were important or not.

It is possible to derive a Brinkman-Forchheimer equation by formal averaging, but only after making a closure that incorporates some empirical material and that inevitably involves loss of information. Clarifying earlier work by Vafai and Tien (1981) and Vafai and Tien (1982), Hsu and Chang (1990) obtained an equation that in our notation can be written

$$\rho_f\left[\frac{1}{\varphi}\frac{\partial \mathbf{v}}{\partial t}+\frac{1}{\varphi}\nabla\left(\frac{\mathbf{v}\cdot\mathbf{v}}{\varphi}\right)\right]=-\nabla P+\frac{\mu}{\varphi\rho_f}\nabla^2\mathbf{v}-\frac{\mu}{K}\mathbf{v}-\frac{c_F\rho_f}{K^{1/2}}|\mathbf{v}|\,\mathbf{v}. \quad (1.18)$$

For an incompressible fluid, $\nabla\cdot\mathbf{v}=0$, and so $\varphi^{-1}\nabla(\varphi^{-1}\mathbf{v}\cdot\mathbf{v})$ reduces to $\varphi^{-1}\mathbf{v}\cdot\nabla(\mathbf{v}/\varphi)$, and then Eq. (1.21) becomes an easily recognizable combination of Eqs. (1.8), (1.12), and (1.17). The position of the factor φ in relation to the spatial derivatives is important if the porous medium is heterogeneous.

If L is the appropriate characteristic length scale, the ratio of the last term in Eq. (1.17) to the previous term is of the order of magnitude of $(\tilde{\mu}/\mu)K/L^2$, the Darcy number. Authors who assume that $\tilde{\mu}=\mu$ define the Darcy number to be K/L^2. The value of Da is normally much less than unity, but Weinert and Lage (1994) reported a sample of a compressed aluminum foam 1-mm thick, for which Da was about 8. Nield and Lage (1997) have proposed the term "hyperporous medium" for such a material. The flow in their sample was normal to the smallest dimension, and so, unlike in Vafai and Kim (1997), the sample was not similar to a thin screen. When the Brinkman term is comparable with the Darcy term throughout the medium, the K which appears in Eq. (1.17) can no longer be determined by a simple Darcy-type experiment.

Further work in the spirit of Brinkman has been carried out. For example, Howells (1998) treated flow through beds of fixed cylindrical fibers. Efforts to produce consistency between the Brinkman equation and the lattice Boltzmann method have been reported by Martys (2001).

In the case when the fluid is a rarefied gas and the Knudsen number (ratio of the mean free path to a characteristic length) has a large value, velocity slip occurs in the fluid at the pore boundaries. This phenomenon is characteristic of a reduction in viscosity. Hence in these circumstances one could expect that the Darcy and Brinkman drag terms (the viscous terms) would become insignificant in comparison with the Forchheimer drag term (the form drag term). At very large values of the Knudsen number a continuum model is not appropriate on the pore scale, but on the REV scale a continuum model may still be useful.

1.5.4. Non-Newtonian Fluid

Shenoy (1994) has reviewed studies of flow in non-Newtonian fluids in porous media, with attention concentrated on power-law fluids. He suggested, on the basis of volumetric averaging, that the Darcy term be replaced by $(\mu^*/K^*)v^{n-1}\mathbf{v}$, the Brinkman term by $(\mu^*/\varphi^n)\nabla\{|[0.5\Delta:\Delta]^{1/2}|^{n-1}\Delta\}$ for an Ostwald-deWaele fluid, and the Forchheimer term be left unchanged (because it is independent of the viscosity). Here n is the power-law index, μ^* reflects the consistency of the fluid, K^* is a modified permeability, and Δ is the deformation tensor. We would replace μ^* in the Brinkman term by an equivalent coefficient, not necessarily the same as that in the Darcy term. A similar momentum equation was obtained by Hayes *et al.* (1996) using volume averaging.

Some wider aspects have been discussed by Shah and Yortsos (1995). Using homogenization theory, they show that the macroscopic power law has the same form as the power law for a single capillary, at low Reynolds numbers (a regime that is reached at low velocities only if $n < 2$). However, the power-law permeability may depend also on the orientation of the pressure gradient. The homogenization method, together with the theory of isotropic tensor function of tensor arguments, was used by Auriault *et al.* (2002b) to treat anisotropic media. An alternative model was proposed by Liu and Masliyah (1998). Numerical modeling of non-Newtonian fluids in a three-dimensional periodic array was reported by Inoue and Nakayama (1998).

1.6. Hydrodynamic Boundary Conditions

In order to be specific, we consider the case where the region $y < 0$ is occupied by a porous medium, and there is a boundary at $y = 0$, relative to Cartesian coordinates (x, y, z). If the boundary is impermeable, then the usual assumption is that the normal component of the seepage velocity $\mathbf{v} = (u, v, w)$ must vanish there, i.e.,

$$v = 0 \text{ at } y = 0. \tag{1.19}$$

If Darcy's law is applicable, then, since that equation is of first-order in the spatial derivatives, only one condition can be applied at a given boundary. Hence the other components of the velocity can have arbitrary values at $y = 0$; i.e., we have slip at the boundary.

If instead of being impermeable the boundary is free (as in the case of a liquid-saturated medium exposed to the atmosphere), then the appropriate condition is that the pressure is constant along the boundary. If Darcy's law is applicable and the fluid is incompressible, this implies that

$$\frac{\partial v}{\partial y} = 0 \text{ at } y = 0. \tag{1.20}$$

This conclusion follows because at $y = 0$ we have $P =$ constant for all x and z, so $\partial P/\partial x = \partial P/\partial z = 0$, and hence $u = w = 0$ for all x and z. Hence $\partial u/\partial x = \partial w/\partial z = 0$ at $y = 0$. Since the equation of continuity

$$\frac{\partial u}{\partial x} + \frac{\partial v}{\partial y} + \frac{\partial w}{\partial z} = 0 \tag{1.21}$$

holds for $y = 0$, we deduce the boundary condition (1.20).

If the porous medium is adjacent to clear fluid identical to that which saturates the porous medium, and if there is unidirectional flow in the x direction (Fig. 1.4), then according to Beavers and Joseph (1967) the appropriate boundary condition

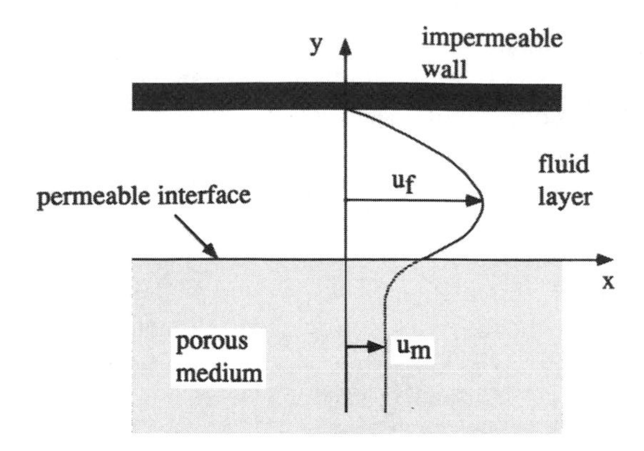

Figure 1.4. Velocity profile for unidirectional flow through a fluid channel bounded by an impermeable wall and a saturated porous medium.

is the empirical relationship

$$\frac{\partial u_f}{\partial y} = \frac{\alpha_{BJ}}{K^{1/2}}(u_f - u_m), \tag{1.22}$$

where u_f is the velocity in the fluid and u_m is the seepage velocity in the porous medium. It is understood that in Eq. (1.22) u_f and $\partial u_f/\partial y$ are evaluated at $y = 0^+$ and u_m is evaluated at some small distance from the plane $y = 0$, so there is a thin layer just inside the medium over which the transition in velocity takes place.

The quantity α_{BJ} is dimensionless and is independent of the viscosity of the fluid, but it depends on the material parameters that characterize the structure of the permeable material within the boundary region. In their experiments Beavers and Joseph found that α_{BJ} had the values 0.78, 1.45, and 4.0 for Foametal having average pore sizes 0.016, 0.034, and 0.045 inches, respectively, and 0.1 for Aloxite with average pore size 0.013 or 0.027 inches. More evidence for the correctness of this boundary condition was produced by Beavers *et al.* (1970, 1974). Sahraoui and Kaviany (1992) have shown that the value of α_{BJ} depends on the flow direction at the interface, the Reynolds number, the extent of the clear fluid, and nonuniformities in the arrangement of solid material at the surface.

Some theoretical support for the Beavers-Joseph condition is provided by the results of Taylor (1971) and Richardson (1971), based on an analogous model of a porous medium, and by the statistical treatment of Saffman (1971). Saffman pointed out that the precise form of the Beavers-Joseph condition was special to the planar geometry considered by Beavers and Joseph, and in general was not in fact correct to order K. Saffman showed that on the boundary

$$u_f = \frac{K^{1/2}}{\alpha_{BJ}} \frac{\partial u_f}{\partial n} + \mathrm{O}(K), \tag{1.23}$$

where n refers to the direction normal to the boundary. In Eq. (1.22) u_m is O(K), and thus may be neglected if one wishes.

Jones (1973) assumed that the Beavers-Joseph condition was essentially a relationship involving shear stress rather than just velocity shear, and on this view Eq. (1.22) would generalize to

$$\frac{\partial u_f}{\partial y} + \frac{\partial v_f}{\partial x} = \frac{\alpha_{BJ}}{K^{1/2}}(u_f - u_m) \tag{1.24}$$

for the situation when v_f was not zero. This seems plausible, but apparently it has not yet been confirmed. However, Straughan (2004b) has argued that one should give consideration to the Jones version, because it and not the original Beavers-Joseph version is properly invariant.

Taylor (1971) observed that the Beavers-Joseph condition can be deduced as a consequence of the Brinkman equation. This idea was developed in detail by Neale and Nader (1974), who showed that in the problem of flow in a channel bounded by a thick porous wall one gets the same solution with the Brinkman equation as one gets with the Darcy equation together with the Beavers-Joseph condition, provided that one identifies α_{BJ} with $(\tilde{\mu}/\mu)^{1/2}$.

Near a rigid boundary the porosity of a bed of particles is often higher than elsewhere in the bed because the particles cannot pack so effectively right at the boundary (see Section 1.7). One way of dealing with the channeling effect that can arise is to model the situation by a thin fluid layer interposed between the boundary and the porous medium, with Darcy's equation applied in the medium and with the Beavers-Joseph condition applied at the interface between the fluid layer and the porous medium. Nield (1983) applied this procedure to the porous-medium analog of the Rayleigh-Bénard problem. Alternatively, the Brinkman equation, together with a formula such as Eq. (1.26), can be employed to model the situation.

Haber and Mauri (1983) proposed that the boundary condition $\mathbf{v} \cdot \mathbf{n} = 0$ at the interface between a porous medium and an impermeable wall should be replaced by

$$\mathbf{v} \cdot \mathbf{n} = K^{1/2} \nabla_t \cdot \mathbf{v}_t, \tag{1.25}$$

where $\mathbf{v}$ is the velocity inside the porous medium and $\mathbf{v}_t$ is its tangential component, and where ∇_t is the tangential component of the operator ∇. Haber and Mauri argue that Eq. (1.25) should be preferred to $\mathbf{v} \cdot \mathbf{n} = 0$, since the former accords better with solutions obtained by solving some model problems using Brinkman's equation. For most practical purposes there is little difference between the two alternatives, since $K^{1/2}$ will be small compared to the characteristic length scale L in most situations.

A difficulty arises when one tries to match the solution of Brinkman's equation for a porous medium with the solution of the usual Navier-Stokes equation for an adjacent clear fluid, as done by Haber and Mauri (1983), Somerton and Catton (1982), and subsequent authors. In implementing the continuity of the tangential component of stress they use equations equivalent to the continuity of $\mu \partial u/\partial y$ across the boundary at $y = 0$. Over the fluid portion of the interface the clear fluid value of $\mu \partial u/\partial y$ matches with the intrinsic value of the same quantity in the porous medium, but over the solid portion of the interface the matching breaks down because there in the clear fluid $\mu \partial u/\partial y$ has some indeterminate nonzero value while the porous medium value has to be zero. Hence the average values of $\mu \partial u / \partial y$ in the clear fluid and in the medium do not match.

Authors who have specified the matching of $\mu \partial u / \partial y$ have overdetermined the system of equations. This leads to overprediction of the extent to which motion induced in the clear fluid is transmitted to the porous medium. The availability of the empirical constant α_{BJ} in the alternative Beavers-Joseph approach enables one to deal with the indeterminancy of the tangential stress requirement.

There is a similar difficulty in expressing the continuity of normal stress, which is the sum of a pressure term and a viscous term. Some authors have argued that the pressure, being an intrinsic quantity, has to be continuous across the interface. Since the total normal stress is continuous, that means that the viscous term must also be continuous. Such authors have overdetermined the system of equations. It is true that the pressure has to be continuous on the microscopic scale, but on the macroscopic scale the interface surface is an idealization of a thin layer in which the pressure can change substantially because of the pressure differential across solid material. In practice the viscous term may be small compared with the pressure, and in this case the continuity of total normal stress does reduce to the approximate continuity of pressure. Also, for an incompressible fluid, the continuity of normal stress does reduce to continuity of pressure if one takes the effective Brinkman viscosity equal to the fluid viscosity, as shown by Chen and Chen (1992). Authors who have formulated a problem in terms of stream function and vorticity have failed to deal properly with the normal stress boundary condition (Nield, 1997a). For a more soundly based procedure for numerical simulation and for a further discussion of this matter, the reader is referred to Gartling *et al.* (1996).

Ochoa-Tapia and Whitaker (1995a,b) have expressly matched the Darcy and Stokes equations using the volume-averaging procedure. This approach produces a jump in the stress (but not in the velocity) and involves a parameter to be fitted experimentally. They also explored the use of a variable porosity model as a substitute for the jump condition and concluded that the latter approach does not lead to a successful representation of all the experimental data, but it provides insight into the complexity of the interface region. Kuznetsov (1996a) applied the jump condition to flows in parallel-plate and cylindrical channels partially filled with a porous medium. Kuznetsov (1997b) reported an analytical solution for flow near an interface. Ochoa-Tapia and Whitaker (1998) included inertia effects in a momentum jump condition. Questions about mathematical continuity were discussed by Payne and Straughan (1998). Homogenization of wall-slip gas flow was treated by Skjetne and Auriault (1999b). Matching using a dissipation function was proposed by Cieszko and Kubik (1999). Modeling of the interface using a transition layer was introduced by Murdoch and Soliman (1999) and Goyeau *et al.* (2003, 2004). Layton *et al.* (2003) introduced a finite-element scheme that allows the simulation of the coupled problem to be uncoupled into steps involving porous media and fluid flow subproblems. (They also proved the existence of weak solutions for the coupled Darcy and Stokes equations.) Numerical treatments of jump conditions include those by Silva and de Lemos (2003a) and Costa *et al.* (2004). The interfacial region was modeled by Stokes flow in a channel partly filled with an array of circular cylinders beside one wall by James and Davis (2001). Their calculations show that the external flow penetrates the porous medium very little, even for sparse arrays, with a velocity u_m about one quarter of that predicted by the Brinkman model.

Shavit *et al.* (2002, 2004) have simulated the interface using a Cantor-Taylor brush configuration to model the porous medium. They also reported the results of particle image velocimetry measurements that showed that the concept of apparent viscosity did not provide a satisfactory agreement. They proposed that the standard Brinkman equation be replaced by a set of three equations.

Salinger *et al.* (1994a) found that a Darcy-slip finite-element formulation produced solutions that were more accurate and more economical to compute than those obtained using a Brinkman formulation. A further study using a finite-element scheme was reported by Nassehi (1998).

Similar considerations apply at the boundary between two porous media. Conservation of mass requires that the normal component of $\rho_f \mathbf{v}$, the product of fluid density and seepage velocity, be continuous across the interface. For media in which Darcy's law is applicable only one other hydrodynamic boundary condition can be imposed and that is that the pressure is continuous across the interface. The fluid mechanics of the interface region between two porous layers, one modeled by the Forchheimer equation and the other by the Brinkman equation, were analyzed by Allan and Hamdan (2002).

A range of hydrodynamic and thermal interfacial conditions between a porous medium and a fluid layer were analyzed by Alazmi and Vafai (2001). In general it is the velocity field that is sensitive to variation in boundary conditions, while the temperature field is less sensitive and the Nusselt number is even less sensitive. Goharzadeh *et al.* (2005) performed experiments and observed that the thickness of the transition zone is order of the grain diameter, and hence much larger than the square root of the permeability as predicted by some previous theoretical studies. Min and Kim (2005) have used the special two-dimensional model of Richardson (1971) as the basis for an extended analysis of thermal convection in a composite channel.

1.7. Effects of Porosity Variation

In a porous bed filling a channel or pipe with rigid impermeable walls there is in general an increase in porosity as one approaches the walls, because the solid particles are unable to pack together as efficiently as elsewhere because of the presence of the wall. Experiments have shown that the porosity is a damped oscillatory function of the distance from the wall, varying from a value near unity at the wall to nearly core value at about five diameters from the wall. These oscillations are illustrated by the experimental data (the circles) plotted in Fig. 1.5.

The notion of volume averaging over a r.e.v. breaks down near the wall, and most investigators have assumed a variation of the form (Fig. 1.5)

$$\varphi = \varphi_\infty \left[1 + C \exp\left(-N \frac{y}{d_p} \right) \right], \tag{1.26}$$

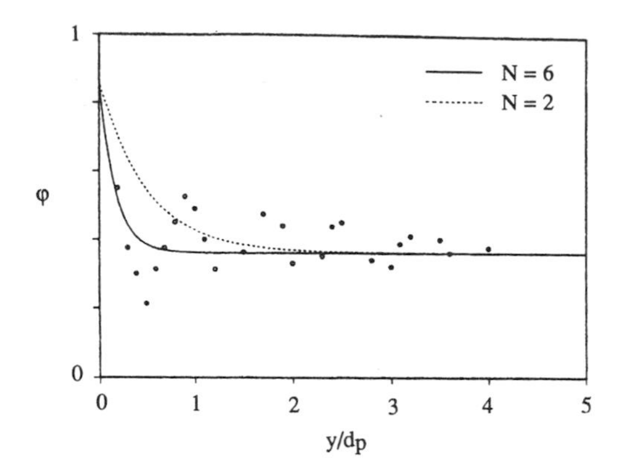

Figure 1.5. The variation of porosity near the wall (Cheng *et al.*, 1991, with permission from Kluwer Academic Publishers).

where y is the distance from the wall, d_p is the particle diameter, and C and N are empirical constants. Recent experiments have indicated that appropriate values are $C = 1.4$ and $N = 5$ or 6 for a medium with $\varphi_\infty = 0.4$.

As a consequence of the porosity increase in the vicinity of the wall, the velocity of a flow parallel to the wall increases as the wall is approached and goes through a maximum before it decreases to zero (to satisfy the no-slip condition). In general, this leads to a net increase in volume flux, i.e., to the phenomenon called the *channeling effect.*

As Georgiadis and Catton (1987) have pointed out, there also is a more general phenomenon that arises because of porosity variation in association with quadratic drag. To illustrate this, consider the steady fully developed two-dimensional flow through a channel. The unidirectional nondimensional velocity profile $q(y)$ for flow parallel to the x axis is the solution of a boundary value problem of the following form (Brinkman-Forchheimer):

$$\frac{d^2q}{dy^2} = \frac{dP}{dx} + Kq + \Lambda\,|q|\,q \text{ with } q(\pm 1) = 0. \tag{1.27}$$

The quantities K and Λ depend on the porosity φ [compare the Irmay-Ergun equation (1.15)]. The solution of Eq. (1.27), with the boundary layer term omitted, is

$$q = \frac{(3\alpha)^{1/2}}{\Lambda} - \frac{K}{2\Lambda}, \tag{1.28}$$

where

$$\alpha = -\frac{dP}{dx}\frac{\Lambda}{3} + \frac{K^2}{12}. \tag{1.29}$$

The mean flow rate over the channel cross section is given by the spatial average of Eq. (1.27), and assuming statistical homogeneity this is equivalent to an ensemble average with φ as the variable. It is easily shown that the function $q(\varphi)$ of the random variable φ is convex in the interval [0,1] if the Ergun relationships hold. This implies that for the same pressure gradient along the channel the mean flux is larger when there is a spatial variation of porosity: $\bar{q}(\bar{\varphi}) > q(\bar{\varphi})$. This means that if we use the average value $\bar{\varphi}$ of the porosity, we obtain only a lower bound for the flow rate through the packed bed. Georgiadis and Catton (1987) found that in one realistic case $q(\bar{\varphi})$ could be 9 percent greater than $q(\bar{\varphi})$. Pressure drop/flow rate measurements therefore would give an effective value for the permeability greater than that otherwise expected. Fu and Huang (1999) showed that random porosity led to a negative correlation between local Nusselt number and nearwall local porosity.

1.8. Turbulence in Porous Media

The nonlinear spectral analysis of Rudraiah (1988) was based on a Brinkman model valid for high porosity only, and so is of questionable use for media in which the solid material inhibits the formation of macroscopic eddies. Masuoka and Takatsu (1996) used a volume-averaging procedure to produce a zero-equation model. Nield (1997c) questioned their basic assumption that the Forchheimer flow resistance and dispersion are caused mainly by turbulent mixing, and that the drag force caused by the molecular stress can be equated to the Darcy term alone. Takatsu and Masuoka (1998) and Masuoka and Takatsu (2002) have further developed their model and have conducted experiments on flow through banks of tubes. They have persisted with their faulty assumption, based on the assumption that the deviation from Darcy's law appears at the same value of the Reynolds number (based on a characteristic particle diameter) as that at which turbulent vortices appear. Nield (1997c) pointed out that the experimental work on which Masuoka and Takatsu relied in fact indicates otherwise. More recent experiments have been conducted by Seguin *et al.* (1998).

Travkin and Catton (1994, 1995, 1998, 1999), Gratton *et al.* (1996), and Catton and Travkin (1996) have developed general models in which the solid-phase morphology is emphasized. They have not related their models to critical experiments, and so it is not clear that this refinement is justified from a practical point of view.

Lee and Howell (1987) performed extensive numerical calculations, of forced convective heat transfer from a heated plate, using a volume-averaged κ-ε model. The κ-ε model of Antohe and Lage (1997b), which is more general than the ones introduced by Lee and Howell (1987) and Prescott and Incropera (1995), is promising from a practical aspect. Their analysis leads to the conclusion that, for a medium of small permeability, the effect of the solid matrix is to damp the turbulence, as one would expect. This analysis was further extended by Getachew *et al.* (2000). Further work with a κ-ε model was reported by L. Chen *et* al. (1998) and by Laakkonen (2003). Modeling with one energy equation was performed by Chung *et al.* (2003). Numerical modeling of composite porous-medium/clear-fluid ducts has been reported by Kuznetsov and Xiong (2003), Kuznetsov (2004a) and Yang and Hwang (2003).

Kuwahara *et al.* (1996) performed numerical modeling of the turbulent flow within the pores of a porous medium using a spatially periodic array, and obtained some macroscopic characteristics of that flow. Note that this is different from turbulence on a macroscopic scale, because the period length in the simulations (something that is representative of the pore scale) provides an artificial upper bound on the size of the turbulent eddies that can be generated. This was pointed out by Nield (2001b). Further numerical modeling using periodic arrays was conducted by Kuwahara and Nakayama (1998), Kuwahara *et al.* (1998), Nakayama and Kuwahara (1999, 2000) and Nakayama *et al.* (2004).

In his discussion of transition to turbulence, Lage (1998) has noted the difference in pressure-drop versus flow-speed relationship between the case of a porous medium that behaves predominantly like an aggregate of conduits (characterized by a balance between pressure drop and viscous diffusion) and the case of a medium that behaves like an aggregate of bluff bodies (characterized by a balance between pressure drop and form drag).

An alternative approach has been extensively developed by de Lemos and co-workers: de Lemos (2004) (review), de Lemos and Braga (2003), de Lemos and Mesquita (2003), de Lemos and Pedras (2000, 2001), Rocamora and de Lemos (2000), de Lemos and Rocamora (2002), de Lemos and Tofaneli (2004), Pedras and de Lemos (2000, 2001a,b,c, 2003) and Silva and de Lemos (2003b). It is based on volume averages and a double decomposition concept involving both spatial deviations and time fluctuations. To a limited extent this approach unifies the work of Masuoka, Takatsu, Nakayama, and Kuwahara (who applied a time average followed by a volume average) and Lage and his co-workers and predecessors (who applied the two averages in the opposite order). Simplified models for turbulence in porous media, or related systems such as vegetation, have been presented by Wang and Takle (1995), Nepf (1999), Macedo *et* al. (2001), Hoffman and van der Meer (2002), Flick *et al.* (2003), and Alvarez *et al.* (2003).

Work on the topic of this section has been reviewed by Lage *et al.* (2002). A related paper is the study of hydrodynamic stability of flow in a channel or duct occupied by a porous medium by Nield (2003). As one would expect from the conclusions of Antohe and Lage (1997b) cited above, for such flows the critical Reynolds number for the onset of linear instability is very high. Darcy drag, Forchheimer drag, and additional momentum dispersion all contribute to a flattening of the velocity profile in a channel, and thus to increased stability. Also contributing to increased stability is the rapid decay with time noted in Section 1.5.1. Work to date indicates that turbulence changes the values of drag coefficients from their laminar flow values but does not qualitatively change convective flows in porous media except when the porosity is high. A further review of turbulence in porous media is that by Vafai *et al.* (2005).

Further numerical modeling using periodic arrays was conducted by Kuwahara and Nakayama (1998), Nakayama and Kuwahara (1999, 2000, 2005), and Nakayama *et al.* (2004). Studies of turbulence in relation to the interface between a porous medium and a clear fluid region have been made by de Lemos (2005b), Assato *et al.* (2005), and Zhu and Kuznetsov (2005).

1.9. Fractured Media, Deformable Media, and Complex Porous Structures

The subject of flow in fractured media is an important one in the geological context. In addition to continuum models, discrete models have been formulated. In these models, Monte Carlo simulations and various statistical methods are employed, and the concepts of percolation processes, universal scaling laws, and fractals are basic tools. These matters are discussed in detail by Barenblatt *et al.* (1990) and Sahimi (1993, 1995). The lattice Boltzmann method is widely employed; see, for example, Maier *et al.* (1998).

Likewise, little research has been done yet on convection with deformable porous media, although some thermoelastic aspects of this subject have been studied. For example, dual-porosity models (involving two overlapping continua) have been developed by Bai and Roegiers (1994) and Bai *et al.* (1994a,b, 1996). Another exception is the discussion of the flow over and through a layer of flexible fibers by Fowler and Bejan (1995). Some flows in media formed by porous blocks separated by fissures have been studied by Levy (1990) and Royer *et al.* (1995), who employed a homogenization method, and also by Lage (1997). There is one published study of convection in a saturated fissured medium, that by Kulacki and Rajen (1991). This paper contains a useful review, an experimental study of heat transfer in an idealized fissured medium, and supporting numerical work. They conclude that one interconnected fissure in every one tenth of the domain is sufficient for an equivalence between a saturated fissure system and a porous medium, and that the assumption that a fissured system can be treated as a porous medium leads to an overestimate (i.e., an upper bound) for the heat transfer.

It is likely that in the future an increasing use of numerical simulation will be used in the study of complex porous structures, such as geological structures. An interesting development is the finite-element program that has been used by Joly *et al.* (1996) to study the onset of free convections and the stability of two-dimensional convective solutions to three-dimensional perturbations. Further numerical studies have been reported by Ghorayeb and Firoozabadi (2000a,b, 2001) and by Saghir *et al.* (2001).

Biological applications have motivated the investigation of other phenomena related to convection in porous media. Khaled and Vafai (2003) surveyed some investigations of diffusion processes within the brain, diffusion during tissue generation, applications of magnetic resonance to the categorization of tissue properties, blood flow in tumors, blood flow in perfusion tissues, bioheat transfer in tissues, and bioconvection. Lage *et al.* (2004a) have used a porous medium model to investigate the red cell distribution effect on alveolar respiration.

1.10. Bidisperse Porous Media

A bidisperse porous medium (BDPM) is composed of clusters of large particles that are agglomerations of small particles (Fig. 1). Thus there are macropores between the clusters and micropores within them. Applications are found in bidisperse adsorbent or bidisperse capillary wicks in a heat pipe. Since the bidisperse wick structure significantly increases the area available for liquid film evaporation, it has been proposed for use in the evaporator of heat pipes.

A BDPM thus may be looked at as a standard porous medium in which the solid phase is replaced by another porous medium. We then can talk about the f-phase (the macropores) and the p-phase (the remainder of the structure). An alternative way of looking at the structure is to regard it as a porous medium in which fractures or tunnels have been introduced. One then can think of the f-phase as being a "fracture phase" and the p-phase as being a "porous phase."

Extending the Brinkman model for a monodisperse porous medium, Nield and Kuznetsov (2005c) proposed to model the steady-state momentum transfer in a BDPM by the following pair of coupled equations for $\mathbf{v}_f$ and $\mathbf{v}_p$.

$$\mathbf{G} = \left(\frac{\mu}{K_f}\right)\mathbf{v}_f + \zeta(\mathbf{v}_f - \mathbf{v}_p) - \tilde{\mu}_f \nabla^2 \mathbf{v}_f \tag{1.30}$$

$$\mathbf{G} = \left(\frac{\mu}{K_p}\right)\mathbf{v}_p + \zeta(\mathbf{v}_p - \mathbf{v}_f) - \tilde{\mu}_p \nabla^2 \mathbf{v}_p. \tag{1.31}$$

Here $\mathbf{G}$ is the negative of the applied pressure gradient, μ is the fluid viscosity, K_f and K_p are the permeabilities of the two phases, and ζ is the coefficient for momentum transfer between the two phases. The quantities $\tilde{\mu}_f$ and $\tilde{\mu}_p$ are the respective effective viscosities.

For the special case of the Darcy limit one obtains

$$\mathbf{v}_f = \frac{(\mu/K_p + 2\zeta)\,\mathbf{G}}{\mu^2/K_f K_p + \zeta\mu(1/K_f + 1/K_p)} \tag{1.32}$$

$$\mathbf{v}_p = \frac{(\mu/K_f + 2\zeta)\,\mathbf{G}}{\mu^2/K_f K_p + \zeta\mu(1/K_f + 1/K_p)}. \tag{1.33}$$

In this case the bulk flow thus is given by

$$\mathbf{G} = (\mu/K)\mathbf{v} \tag{1.34}$$

where

$$\mathbf{v} = \varphi\mathbf{v}_f + (1-\varphi)\mathbf{v}_p \tag{1.35}$$

$$K = \frac{\varphi K_f + (1-\varphi)K_p + 2(\zeta/\mu)K_f K_p}{1 + (\zeta/\mu)(K_f + K_p)}. \tag{1.36}$$

Thus in this case the effect of the coupling parameter ζ is merely to modify the effective permeabilities of the two phases, via the parameter ζ/μ.

2
Heat Transfer through a Porous Medium

2.1. Energy Equation: Simple Case

In this chapter we focus on the equation that expresses the first law of thermodynamics in a porous medium. We start with a simple situation in which the medium is isotropic and where radiative effects, viscous dissipation, and the work done by pressure changes are negligible. Very shortly we shall assume that there is local thermal equilibrium so that $T_s = T_f = T$, where T_s and T_f are the temperatures of the solid and fluid phases, respectively. Here we also assume that heat conduction in the solid and fluid phases takes place in parallel so that there is no net heat transfer from one phase to the other. More complex situations will be considered in Section 6.5. The fundamentals of heat transfer in porous media also are presented in Bejan *et al.* (2004) and Bejan (2004a).

Taking averages over an elemental volume of the medium we have, for the solid phase,

$$(1-\varphi)(\rho c)_s \frac{\partial T_s}{\partial t} = (1-\varphi)\nabla \cdot (k_s \nabla T_s) + (1-\varphi) q_s''' \tag{2.1}$$

and for the fluid phase,

$$\varphi(\rho c_P)_f \frac{\partial T_f}{\partial t} + (\rho c_P)_f \mathbf{v} \cdot \nabla T_f = \varphi \nabla \cdot (k_f \nabla T_f) + \varphi q_f'''. \tag{2.2}$$

Here the subscripts s and f refer to the solid and fluid phases, respectively, c is the specific heat of the solid, c_P is the specific heat at constant pressure of the fluid, k is the thermal conductivity, and q'''[W/m^3] is the heat production per unit volume.

In writing Eqs. (2.1) and (2.2) we have assumed that the surface porosity is equal to the porosity. This is pertinent to the conduction terms. For example, $-k_s \nabla T_s$ is the conductive heat flux through the solid, and thus $\nabla \cdot (k_s \nabla T_s)$ is the net rate of heat conduction into a unit volume of the solid. In Eq. (2.1) this appears multiplied by the factor $(1-\varphi)$, which is the ratio of the cross-sectional area occupied by solid to the total cross-sectional area of the medium. The other two terms in Eq. (2.1) also contain the factor $(1-\varphi)$ because this is the ratio of volume occupied by solid to the total volume of the element. In Eq. (2.2) there also appears a convective term, due to the seepage velocity. We recognize that $\mathbf{V} \cdot \nabla T_f$ is the rate of change of temperature in the elemental volume due to the convection of fluid into it, so this, multiplied by $(\rho c_P)_f$, must be the rate of change of thermal energy, per unit volume of fluid, due to the convection. Note further that in writing Eq. (2.2) use has been made of the Dupuit-Forchheimer relationship $\mathbf{v} = \varphi\mathbf{V}$.

Setting $T_s = T_f = T$ and adding Eqs. (2.1) and (2.2) we have

$$(\rho c)_m \frac{\partial T}{\partial t} + (\rho c)_f \mathbf{v} \cdot \nabla T = \nabla \cdot (k_m \nabla T) + q_m''', \tag{2.3}$$

where

$$(\rho c)_m = (1 - \varphi)(\rho c)_s + \varphi(\rho c_P)_f, \tag{2.4}$$

$$k_m = (1 - \varphi)k_s + \varphi k_f, \tag{2.5}$$

$$q_m''' = (1 - \varphi)q_s''' + \varphi q_f''' \tag{2.6}$$

are, respectively, the overall heat capacity per unit volume, overall thermal conductivity, and overall heat production per unit volume of the medium.

2.2. Energy Equation: Extensions to More Complex Situations

2.2.1. Overall Thermal Conductivity of a Porous Medium

In general, the overall thermal conductivity of a porous medium depends in a complex fashion on the geometry of the medium. As we have just seen, if the heat conduction in the solid and fluid phases occurs in parallel, then the overall conductivity k_A is the weighted arithmetic mean of the conductivities k_s and k_f:

$$k_A = (1 - \varphi)k_s + \varphi k_f. \tag{2.7}$$

On the other hand, if the structure and orientation of the porous medium is such that the heat conduction takes place in series, with all of the heat flux passing through both solid and fluid, then the overall conductivity k_H is the weighted harmonic mean of k_s and k_f:

$$\frac{1}{k_H} = \frac{1 - \varphi}{k_s} + \frac{\varphi}{k_f}. \tag{2.8}$$

In general, k_A and k_H will provide upper and lower bounds, respectively, on the actual overall conductivity k_m. We always have $k_H \leq k_A$, with equality if and only if $k_s = k_f$. For practical purposes, a rough and ready estimate for k_m is provided by k_G, the weighted geometric mean of k_s and k_f, defined by

$$k_G = k_s^{1-\varphi} k_f^{\varphi}. \tag{2.9}$$

This provides a good estimate so long as k_s and k_f are not too different from each other (Nield, 1991b). More complicated correlation formulas for the conductivity of packed beds have been proposed. Experiments by Prasad *et al.* (1989b) showed that these formulas gave reasonably good results provided that k_f was not significantly greater than k_s. The agreement when $k_f \gg k_s$ was not good, the observed conductivity being greater than that predicted. This discrepancy may be due to porosity variation near the walls. Since k_m depends on φ, there is an effect analogous to the hydrodynamic effect already noted in Section 1.7. Some of the discrepancy may be due to the difficulty of measuring a truly stagnant thermal conductivity in this case (Nield, 1991b).

In the case when the fluid is a rarefied gas and the Knudsen number has a large value, temperature slip occurs in the fluid at the pore boundaries. In these circumstances one could expect that the fluid conductivity would tend to zero as the Knudsen number increases. Then in the case of external heating the heat would be conducted almost entirely through the solid matrix. In the case of just internal heating in the fluid the situation is reversed as the fluid phase becomes thermally isolated from the solid phase.

Further models for stagnant thermal conductivity have been put forward by Hsu *et al.* (1994, 1995), Cheng *et al.* (1999) and Cheng and Hsu (1998, 1999). In particular, Cheng *et al.* (1999) and also Hsu (2000) contain comprehensive reviews of the subject. Volume averaging was used by Buonanno and Carotenuto (1997) to calculate the effective conductivity taking into account particle-to-particle contact. Experimental studies have been made by Imadojemu and Porter (1995) and Tavman (1996). The former concluded that the thermal diffusivity and conductivity of the fluid played the major role in determining the effective conductivity of the medium. Hsu (1999) presented a closure model for transient heat conduction, while Hsiao and Advani (1999) included the effect of heat dispersion. Hu *et al.* (2001) discussed unconsolidated porous media, Paek *et al.* (2000) dealt with aluminum foam materials, and Fu *et al.* (1998) studied cellular ceramics. Carson *et al.* (2005) obtained thermal conductivity bounds for isotropic porous materials. A unified closure model for convective heat and mass transfer has been presented by Hsu (2005). He notes that r.e.v. averaging leads to the introduction of new unknowns (dispersion, interfacial tortuosity, and interfacial transfer) whose determination constitutes the closure problem. More experiments are needed to determine some of the coefficients that are involved. His closure relation for the interfacial force contains all the components due to drag, lift, and transient inertia to the first-order approximation. He concludes that the macroscopic energy equations are expected to be valid for all values of the time scale and Reynolds number, for the case of steady flows. Further investigations are needed for unsteady flows.

So far we have been discussing the case of an isotropic medium, for which the conductivity is a scalar. For an anisotropic medium $\mathbf{k}_m$ will be a second-order tensor. Lee and Yang (1998) modeled a heterogeneous anisotropic porous medium.

A fundamental issue has been raised by Merrikh *et al.* (2002, 2005a,b) and Merrikh and Lage (2005). This is the question of how the internal regularity of a solid/fluid physical domain affects global flow and heat transfer. These authors have considered a situation (a regular distribution of rectangular solid obstacles in a rectangular box) that is sufficiently simple for a comparison to be made between the results of numerical modeling involving a treatment of the fluid and solid phases considered separately ("continuum model") and a standard r.e.v.-averaged porous medium ("porous continuum model"). The results for the two models can be substantially different. In other words, the internal regularity can have an important effect. The authors considered situations where the obstacles were separated from the boundary walls, and thus some of the difference is due to a channeling effect.

2.2.2. Effects of Pressure Changes, Viscous Dissipation, and Absence of Local Thermal Equilibrium

If the work done by pressure changes is not negligible [i.e., the condition $\beta T(g\beta/c_{Pf})L \ll 1$ is not met], then a term $-\beta T(\partial P/\partial t + \mathbf{v}\cdot\nabla P)$ needs to be added to the left-hand side of Eq. (2.3). Here β is the coefficient of volumetric thermal expansion, defined by

$$\beta = -\frac{1}{\rho}\left(\frac{\partial\rho}{\partial T}\right)_P. \tag{2.10}$$

Viscous dissipation is negligible in natural convection if $(g\beta/c_{Pf})L \ll 1$, which is usually the case. If it is not negligible, another term must be added to the right-hand side of Eq. (2.3), as noted first by Ene and Sanchez-Palencia (1982). If Darcy's law holds, that term is $(\mu/K)\mathbf{v}\cdot\mathbf{v}$ in the case of an isotropic medium, and $\mu\mathbf{v}\cdot\mathbf{K}^{-1}\cdot\mathbf{v}$ if the medium is anisotropic. To see this, note that the average of the rate of doing work by the pressure, on a unit volume of an r.e.v., is given by the negative of $\text{div}(P\varphi\mathbf{V}) = \text{div}(P\mathbf{v}) = \mathbf{v}.\text{grad}\ P$, since $\text{div}\ \mathbf{v} = 0$. The Forchheimer drag term, dotted with the velocity vector, contributes to the dissipation, despite the fact that the viscosity does not enter explicitly. This apparent paradox was resolved by Nield (2000). The contribution of the Brinkman drag term is currently a controversial topic. Nield (2004b) proposed that the Brinkman term be treated in the same way as the Darcy and Forchheimer terms, so that the total viscous dissipation remains equal to the power of the total drag force. Al-Hadhrami *et al.* (2003) prefer a form that remains positive and reduces to that for a fluid clear of solid material in the case where the Darcy number tends to infinity. Accordingly, they add the clear fluid term to the Darcy term. Nield (2004b) suggested that the Brinkman equation may break down in this limit.

Nield (2000) noted that scale analysis, involving the comparison of the magnitude of the viscous dissipation term to the thermal diffusion term, shows that viscous dissipation is negligible if $N \ll 1$, where $N = \mu U^2L^2/Kc_Pk_m\Delta T = \text{Br/Da}$, where the Brinkman number is defined by $\text{Br} = \mu U^2/c_Pk_m\Delta T = \text{EcPr}$, where the Eckert number Ec is defined by $\text{Ec} = U^2/c_P\Delta T$. For most situations the Darcy number K/L^2 is small, so viscous dissipation is important at even modest values of the Brinkman number. For forced convection the choice of the characteristic velocity is obvious. For natural convection, scale analysis leads to the estimate $U \sim (k_m/\rho c_PL)\text{Ra}^{1/2}$ and the condition that viscous dissipation is negligible becomes $\text{Ge} \ll 1$, where Ge is the Gebhart number defined by $\text{Ge} = g\beta L/c_P$. An exception is the case of a uniformly laterally heated box, studied by Costa (2005a). For this case the net global release of kinetic energy by the buoyancy force is zero, the viscous dissipation is balanced by the pressure work, the characteristic velocity scale is $k_m/\rho c_PL$, and the ratio of the viscous dissipation and conduction terms is of order EcPr/Da, where Ec is now equal to $(k_m/\rho c_PL)^2/c_P\Delta T$. The above comments on forced convection are made on the assumption that the Péclet number $\mathrm{Pe} = \rho c_P U L / k_m$ is not large. If it is large, then the proper comparison is one between the magnitudes of the viscous dissipation term and the convective transport term. This ratio is of order Ec/DaRe, where the Reynolds number $\mathrm{Re} = \rho U L / \mu$. Further aspects of the effects of viscous dissipation on the flow in porous media are discussed in the survey by Magyari *et al.* (2005b).

If one wishes to allow for heat transfer between solid and fluid (that is, one no longer has local thermal equilibrium), then one can replace Eqs. (2.1) and (2.2) by

$$(1-\varphi)(\rho c)_s \frac{\partial T_s}{\partial t} = (1-\varphi)\nabla \cdot (k_s \nabla T_s) + (1-\varphi) q_s''' + h(T_f - T_s), \quad (2.11)$$

$$\varphi(\rho c_P)_f \frac{\partial T_f}{\partial t} + (\rho c_P)\mathbf{v} \cdot \nabla T_f = \varphi \nabla (k_f \nabla T_f) + (1-\varphi) q_f''' + h(T_s - T_f), \quad (2.12)$$

where h is a heat transfer coefficient. See also Eqs. (2.11a) and (2.12a) later in this section. A critical aspect of using this approach lies in the determination of the appropriate value of h. Experimental values of h are found in an indirect manner; see, e.g., Polyaev *et al.* (1996). According to correlations for a porous bed of particle established in Dixon and Cresswell (1979),

$$h = a_{fs} h^*, \quad (2.13)$$

where the specific surface area (surface per unit volume) a_{fs} is given by

$$a_{fs} = 6(1-\varphi)/d_p, \quad (2.14)$$

and

$$\frac{1}{h^*} = \frac{d_p}{\mathrm{Nu}_{fs} k_f} + \frac{d_p}{\beta k_s} \quad (2.15)$$

where d_p is the particle diameter and $\beta = 10$ if the porous bed particles are of spherical form. The fluid-to-solid Nusselt number Nu_{fs} is, for Reynolds numbers (based on d_p) $\mathrm{Re}_p > 100$, well correlated by the expression presented in Handley and Heggs (1968):

$$\mathrm{Nu}_{fs} = (0.255/\varphi)\,\mathrm{Pr}^{1/3}\,\mathrm{Re}_p^{2/3}, \quad (2.16)$$

while for low values of Re_p the estimates of Nu_{fs} vary between 0.1 and 12.4, these being based on Miyauchi *et al.* (1976) and Wakao *et al.* (1976, 1979). Other authors have used alternative expressions for h^* and a_{fs} and some of these were considered by Alazmi and Vafai (2000), who found that the various models give closely similar results for forced convection channel flow when the porosity is high or the pore Reynolds number is large or the particle diameters are small. Theoretical and experimental results reported by Grangeot *et al.* (1994) indicate that h^* depends weakly on the Péclet number of the flow. This subject is discussed further in Sections 6.5 and 6.9.2. The topic in the context of turbulence has been discussed by Saito and de Lemos (2005). An experimental study for a metallic packed bed was reported by Carrillo (2005). A discussion of further aspects of the two-medium approach to heat transfer in porous media is given by Quintard *et al.*

(1997) and Quintard and Whitaker (2000). Nield (2002a) noted that Eqs. (2.11) and (2.12) are based on the implicit assumption that the thermal resistances of the fluid and solid phases are in series. For the case of a layered medium in a parallel plate channel with fluid/solid interfaces parallel to the x-direction, he suggested that the appropriate equations in the absence of internal heating are

$$(1-\varphi)(\rho c)_s \frac{\partial T_s}{\partial t} = (1-\varphi)\left[\frac{\partial}{\partial x}\left(k_s' \frac{\partial T_s}{\partial x}\right) + \frac{\partial}{\partial y}\left(k_s \frac{\partial T_s}{\partial y}\right)\right] + h(T_f - T_s), \tag{2.11a}$$

$$\varphi(\rho c_P)_f \frac{\partial T_f}{\partial t} + (\rho c_P)\mathbf{v}\cdot\nabla T_f = \varphi\left[\frac{\partial}{\partial x}\left(k_f' \frac{\partial T_f}{\partial x}\right) + \frac{\partial}{\partial y}\left(k_f \frac{\partial T_f}{\partial y}\right)\right] + h(T_s - T_f), \tag{2.12a}$$

where $k_f' = k_s' = k_H$ with k_H given by Eq. (2.8). Equations (2.11) and (2.12) have to be solved subject to certain applied thermal boundary conditions. If a boundary is at uniform temperature, then one has $T_f = T_s$ on the boundary. If uniform heat flux is imposed on the boundary, then there is some ambiguity about the distribution of flux between the two phases. Nield and Kuznetsov (1999) argued that if the flux is truly uniform, then it has to be uniform with respect to the two phases, and hence the flux on the r.e.v. scale has to be distributed between the fluid and solid phases in the ratio of the surface fractions; for a homogeneous medium that means in the ratio of the volume fractions, that is in the ratio $\varphi : (1-\varphi)$. This distribution allows the conjugate problem considered by them to be treated in a consistent manner. The consequences of other choices for the distribution were explored by Kim and Kim (2001) and Alazmi and Vafai (2002). The Nield and Kuznetsov (1999) approach is equivalent to Model 1D in Alazmi and Vafai (2002) and is not equivalent to either approach used in Kim and Kim (2001).

The particular case of local thermal nonequilibrium in a steady process is discussed by Nield (1998a). Petit *et al.* (1999) have proposed a local nonequilibrium model for two-phase flow. A numerical study of the interfacial convective heat transfer coefficient was reported by Kuwahara *et al.* (2001). An application of the method of volume averaging to the analysis of heat and mass transfer in tubes was made by Golfier *et al.* (2002). An alternative two-equation model for conduction only was presented by Fourie and Du Plessis (2003a,b). Vadasz (2005) demonstrated that, for heat conduction problems, local thermal equilibrium applies for any conditions that are a combination of constant temperature and insulation. He also questioned whether a linear relationship between the average temperature difference between the phases and the heat transferred over the fluid–solid surface was appropriate in connection with conditions of local thermal nonequilibrium. Rees and Pop (2005) surveyed studies of local thermal nonequilibrium with special attention to natural and forced convection boundary layers and on internal natural convection. Their survey complements that by Kuznetsov (1998e) for internal forced convection.

2.2.3. Thermal Dispersion

A further complication arises in forced convection or in vigorous natural convection in a porous medium. There may be significant thermal dispersion, i.e., heat transfer due to hydrodynamic mixing of the interstitial fluid at the pore scale. In addition to the molecular diffusion of heat, there is mixing due to the nature of the porous medium. Some mixing is due to the obstructions; the fact that the flow channels are tortuous means that fluid elements starting a given distance from each other and proceeding at the same velocity will not remain at the same distance apart. Further mixing can arise from the fact that all pores in a porous medium may not be accessible to a fluid element after it has entered a particular flow path.

Mixing can also be caused by recirculation caused by local regions of reduced pressure arising from flow restrictions. Within a flow channel mixing occurs because fluid particles at different distances from a wall move relative to one another. Mixing also results from the eddies that form if the flow becomes turbulent. Diffusion in and out of dead-end pores modifies the nature of molecular diffusion. For details, see Greenkorn (1983, p. 190).

Dispersion is thus a complex phenomenon. Rubin (1974) took dispersion into account by generalizing Eq. (2.3) so that the term $\nabla \cdot (\alpha_m \nabla T)$, where $\alpha_m = k_m/(\rho c)_m$ is the thermal diffusivity of the medium, is replaced by $\nabla \cdot \mathbf{E} \cdot \nabla T$ where $\mathbf{E}$ is a second-order tensor (the dispersion tensor). In an isotropic medium the dispersion tensor is axisymmetric and its components can be expressed in the form

$$E_{ij} = F_1 \delta_{ij} + F_2 V_i V_j, \tag{2.17}$$

where $V_i (= v_i/\varphi)$ is the ith component of the barycentric (intrinsic) velocity vector, and F_1 and F_2 are functions of the pore size and the Péclet and Reynolds numbers of the flow.

At any point in the flow field it is possible to express $\mathbf{E}$ with reference to a coordinate system in which the first axis coincides with the flow direction; when this is done we have

$$\begin{aligned} E_{11} &= \eta_1 U + \alpha_m, \\ E_{22} &= E_{33} = \eta_2 U + \alpha_m, \\ E_{ij} &= 0 \text{ for } i \neq j, \end{aligned} \tag{2.18}$$

where E_{11} is the longitudinal dispersion coefficient, E_{22} and E_{33} are the lateral dispersion coefficients, and U is the absolute magnitude of the velocity vector.

If the Péclet number of the flow is small, then η_1 and η_2 are small and the molecular thermal diffusivity α_m is dominant. If the Péclet number of the flow is large, then η_1 and η_2 are large and almost constant. It is found experimentally that $\eta_2 = \eta_1/30$, approximately.

For an account of the treatment of dispersion in anisotropic media in the context of convection, the reader is referred to Tyvand (1977). In the particular case when heat conduction is in parallel, Catton *et al.* (1988) conclude on the basis of their statistical analysis that the effective thermal conductivity k^*_{zz}, for mass and thermal transport in the z-direction through a bed of uniform spherical beads, is given by

$$k_{zz}^{*} = (1-\varphi)k_s + \varphi\left(\frac{2B}{\pi}\right)\mathrm{Pe}\,k_f \tag{2.19}$$

In this expression B is a constant introduced by Ergun (empirically, $B = 1.75$) and Pe is the Péclet number defined by $\mathrm{Pe} = vd_p/\alpha_f(1-\varphi)$, where d_p is the spherical particle diameter and α_f is the thermal diffusivity of the fluid, defined by $\alpha_f = k_f/(\rho c_P)_f$.

Thermal dispersion plays a particularly important role in forced convection in packed columns. The steep radial temperature gradients that exist near the heated or cooled wall were formerly attributed to channeling effects, but more recent work has indicated that thermal dispersion is also involved. For a nearly parallel flow at high Reynolds numbers, the thermal dispersivity tensor reduces to a scalar, the transverse thermal dispersivity. Cheng and his colleagues [see Hsu and Cheng (1990) and the references given in Section 4.9] assumed that the local transverse thermal dispersion conductivity k'_T is given by

$$\frac{k'_T}{k_f} = D_T \mathrm{Pe}_d \ell \frac{u}{u_m}. \tag{2.20}$$

In this equation Pe_d is a Péclet number defined by $\mathrm{Pe}_d = u_m d_p/\alpha_f$, in terms of the mean seepage velocity u_m, the particle diameter d_p, and fluid thermal diffusivity α_f, while D_T is a constant and ℓ is a dimensionless dispersive length normalized with respect to d_p. In recent work the dispersive length is modeled by a wall function of the Van Driest type:

$$\ell = 1 - \exp(-y/\omega d_p). \tag{2.21}$$

The empirical constants ω and D_T depend on the coefficients N and C in the wall porosity variation formula [Eq. (1.28)]. The best match with experiments is given by $D_T = 0.12$ and $\omega = 1$, if $N = 5$ and $C = 1.4$. The theoretical results based on this *ad hoc* approach agree with a number of experimental results.

A theoretical backing for this approach has been given by Hsu and Cheng (1990). This is based on volume averaging of the velocity and temperature deviations in the pores in a dilute array of spheres, together with a scale analysis. The thermal diffusivity tensor **D** is introduced as a multiplying constant which accounts for the interaction of spheres. For the case of high pore Reynolds number flow, Hsu and Cheng (1990) found the thermal dispersion conductivity tensor $\mathbf{k}'$ to be given by

$$\mathbf{k}' = \mathbf{D}k_f \frac{1-\varphi}{\varphi}\mathrm{Pe}_d \tag{2.22}$$

The linear variation with Pe_d is consistent with most of the existing experimental correlations for high pore Reynolds number flow. At low pore Reynolds number flow they found

$$\mathbf{k}' = \mathbf{D}^* k_f \frac{1-\varphi}{\varphi^2} \mathrm{Pe}_d^2 \tag{2.23}$$

where $\mathbf{D}^*$ is another constant tensor. The quadratic dependence on Pe_d has not yet been confirmed by experiment.

Kuwahara *et al.* (1996) and Kuwahara and Nakayama (1999) have studied numerically thermal diffusion for a two-dimensional periodic model. A limitation of their correlation formulas as the porosity tends to unity was discussed by Yu (2004) and Nakayama and Kuwahara (2004). A similar model was examined by Souto and Moyne (1997a,b). The frequency response model was employed by Muralidhar and Misra (1997) in an experimental study of dispersion coefficients. The role of thermal dispersion in the thermally developing region of a channel with a sintered porous metal was studied by Hsieh and Lu (2000). Kuwahara and Nakayama (2005) have extended their earlier numerical studies to the case of three-dimensional flow in highly anisotropic porous media. For further information about dispersion in porous media the reader is referred to the review by Liu and Masliyah (2005), which deals with the dispersion of mass, heat and momentum.

2.3. Oberbeck-Boussinesq Approximation

In studies of natural convection we add the gravitational term $\rho_f\,\mathbf{g}$ to the right-hand side of the Darcy equation (1.4) or its appropriate extension. [Note that in Eq. (1.4) the term ∇P denotes an *intrinsic* quantity, so we add the gravitational force per unit volume of the *fluid*.] For thermal convection to occur, the density of the fluid must be a function of the temperature, and hence we need an equation of state to complement the equations of mass, momentum, and energy. The simplest equation of state is

$$\rho_f = \rho_0\,[1 - \beta\,(\mathrm{T} - \mathrm{T}_0)]\,, \tag{2.24}$$

where ρ_0 is the fluid density at some reference temperature T_0 and β is the coefficient of thermal expansion.

In order to simplify the subsequent analysis, one employs the Boussinesq approximation whenever it is valid. Strictly speaking, one should call this the *Oberbeck-Boussinesq approximation*, since Oberbeck (1879) has priority over Boussinesq (1903), as documented by Joseph (1976). The approximation consists of setting constant all the properties of the medium, except that the vital buoyancy term involving β is retained in the momentum equation. As a consequence the equation of continuity reduces to $\nabla \cdot \mathbf{v} = 0$, just as for an incompressible fluid. The Boussinesq approximation is valid provided that density changes $\Delta\rho$ remain small in comparison with ρ_0 throughout the flow region and provided that temperature variations are insufficient to cause the various properties of the medium (fluid and solid) to vary significantly from their mean values. Johannsen (2003) discussed the validity of the Boussinesq approximation in the case of a bench mark problem known as the Elder problem.

2.4. Thermal Boundary Conditions

Once the thermal conductivity in the porous medium has been determined, the application of thermal boundary conditions is usually straightforward. At the interface between two porous media, or between a porous medium and a clear fluid, we can impose continuity of the temperature (on the assumption that we have local thermodynamic equilibrium) and continuity of the normal component of the heat flux. We note that two conditions are required because the equation of energy (2.3) contains second-order derivatives.

The heat flux vector is the sum of two terms: a convective term $(\rho c_P)_f T\mathbf{v}$ and a conductive term $-k\nabla T$. The normal component of the former is continuous because both T and the normal component of $\rho_f\mathbf{v}$ are continuous. It follows that the normal component of $k\nabla T$ also must be continuous. At an impermeable boundary the usual thermal condition appropriate to the external environment can be applied, e.g., one can prescribe either the temperature or the heat flux, or one can prescribe a heat transfer coefficient.

Sahraoui and Kaviany (1993, 1994) have discussed the errors arising from the use of approximations of the effective conductivity near a boundary, due to nonuniformity of the distributions of the solid and fluid phases there. They have introduced a slip coefficient into the thermal boundary condition to adjust for this, for the case of two-dimensional media.

Ochoa-Tapia and Whitaker (1997, 1998) have developed flux jump conditions applicable at the boundary of a porous medium and a clear fluid. These are based on a nonlocal form of the volume-averaged thermal energy equations for fluid and solid. The conditions involve excess surface thermal energy and an excess nonequilibrium thermal source. Min and Kim (2005) have used the special two-dimensional model of Richardson (1971) in order to obtain estimates of the coefficients that occur in the thermal and hydrodynamic jump conditions.

2.5. Hele-Shaw Analogy

The space between two plane walls a small distance apart constitutes a Hele-Shaw cell. If the gap is of thickness h and the walls each of thickness d, then the governing equations for gap-averaged velocity components (parallel to the plane walls) are identical with those for two-dimensional flow in a porous medium whose permeability K is equal to $h^3/[12(h + 2d)]$, for the case where the heat flow is parallel to the plane walls (Hartline and Lister, 1977). The Hele-Shaw cell thus provides a means of modeling thermal convection in a porous medium, as in the experiments by Elder (1967a).

For the analogy to hold, the three quantities h/δ, $Uh^2/\nu\delta$, and $Uh^2/\alpha_f\delta$ must all be small compared with unity. Here U is the velocity scale and δ the smallest length scale of the motion being modeled, while ν and $/\alpha_f$ are the kinematic viscosity and thermal diffusivity of the fluid. These conditions ensure that there is negligible advection of vorticity and rapid diffusion of vorticity and heat across the flow.

The experimental temperature profiles found by Vorontsov *et al.* (1991) were in good agreement with the theory. Schöpf (1992) has extended the comparison to the case of a binary mixture. Specific studies of convection in a Hele-Shaw cell were reported by Cooper *et al.* (1997), Goldstein *et al.* (1998), and Gorin *et al.* (1998).

The Hele-Shaw cell experiments are especially useful for revealing streamline patterns when the walls are made of transparent material. The analogy has obvious limitations. For example, it cannot deal with the effects of lateral dispersion or instabilities associated with three-dimensional disturbances. The discrepancies associated with these effects have been examined by Kvernvold (1979) and Kvernvold and Tyvand (1981).

Hsu (2005) has compared the governing equations for the averaged flows and heat transfer in Hele-Shaw cells with those of porous media and he observed the following differences: (a) the averaged Hele-Shaw cell is two-dimensional, (b) the interfacial force in the averaged Hele-Shaw flows is contributed entirely from the shear force, and (c) there exists no thermal tortuosity for the averaged Hele-Shaw flows. Thus the Hele-Shaw analogy is good for viscous dominated two-dimensional flow with negligible thermal tortuosity. However, these simplifications help in the verification of closure modeling. Furthermore, a three-dimensional numerical simulation of the convection heat transfer in Hele-Shaw cells may reveal some detailed physics of heat transfer in porous media that are impossible to tackle due to the randomness and the complexity of the microscopic solid geometry. Hsu (2005) illustrates this with results for the case of oscillating flows past a heated circular cylinder.

2.6. Other Approaches

Direct numerical simulation of heat and fluid flow, using the full Navier-Stokes equations at the pore scale, for regularly spaced square or circular rods or spheres has been conducted by Kuwahara *et al.* (1994). A direct numerical simulation was applied by He and Georgiadis (1992) to the study of the effect of randomness on one-dimensional heat conduction. Lattice gas cellular automata simulations were performed by McCarthy (1994) for flow through arrays of cylinders, and by Yoshino and Inamura (2003) for flow in a three-dimensional structure. Buikis and Ulanova (1996) have modeled nonisothermal gas flow through a heterogeneous medium using a two-media approach. A diffuse approximation has been applied by Prax *et al.* (1996) to natural convection. Martins-Costa *et al.* (1992, 1994), Martins-Costa and Saldanhar da Gama (1994), and Martins-Costa (1996) have applied the continuous theory of mixtures to the modeling and simulation of heat transfer in various contexts. Modeling of convection in reservoirs having fractal geometry has been conducted by Fomin *et al.* (1998). Spaid and Phelan (1997) applied lattice Boltzmann methods to model microscale flow in fibrous porous media. Some aspects relevant to biological tissues were discussed by Khanafer *et al.* (2003) and Khaled and Vafai (2003). A general discussion of the dynamic modeling of convective heat transfer in porous media was provided by Hsu (2005). Further simulation studies with a lattice Boltzmann model, with the viscosity independent or dependent on the temperature, have been reported by Guo and Zhao (2005a,b).

Radiative heat transfer is beyond the scope of this book, but we mention that a review of this subject was made by Howell (2000) and a combined radiation and convection problem was studied by Talukdar *et al.* (2004).

3
Mass Transfer in a Porous Medium: Multicomponent and Multiphase Flows

3.1. Multicomponent Flow: Basic Concepts

The term "mass transfer" is used here in a specialized sense, namely the transport of a substance that is involved as a component (constituent, species) in a fluid mixture. An example is the transport of salt in saline water. As we shall see below, convective mass transfer is analogous to convective heat transfer.

Consider a batch of fluid mixture of volume V and mass m. Let the subscript i refer to the ith component (component i) of the mixture. The total mass is equal to the sum of the individual masses m_i so $m = \Sigma m_i$. Hence if the concentration of component i is defined as

$$C_i = \frac{m_i}{V}, \tag{3.1}$$

then the aggregate density ρ of the mixture must be the sum of all the individual concentrations,

$$\rho = \Sigma C_i. \tag{3.2}$$

Clearly the unit of concentration is kg m^{-3}. Instead of C_i the alternative notation ρ_i is appropriate if one thinks of each component spread out over the total volume V.

When chemical reactions are of interest it is convenient to work in terms of an alternative description, one involving the concept of *mole*. By definition, a mole is the amount of substance that contains as many molecules as there are in 12 grams of carbon 12. That number of entities is 6.022×10^{23} (Avogadro's constant). The molar mass of a substance is the mass of one mole of that substance. Hence if there are n moles in a mixture of molar mass M and mass m, then

$$n = \frac{m}{M}. \tag{3.3}$$

Similarly the number of moles n_i of component i in a mixture is the mass of that component divided by its molar mass M_i,

$$n_i = \frac{m_i}{M_i}. \tag{3.4}$$

The *mass fraction* of component i is

$$\Phi_i = \frac{m_i}{m} \tag{3.5}$$

so clearly $\Sigma\Phi_i = 1$. Similarly the *mole fraction* of component i is

$$x_i = \frac{n_i}{n} \tag{3.6}$$

and $\Sigma x_i = 1$.

To summarize, we have three alternative ways to deal with composition: a dimensional concept (concentration) and two dimensionless ratios (mass fraction and mole fraction). These quantities are related by

$$C_i = \rho\,\Phi_i = \rho\,\frac{M_i}{M} x_i, \tag{3.7}$$

where the equivalent molar mass (M) of the mixture is given by

$$M = \Sigma M_i x_i. \tag{3.8}$$

If, for example, the mixture can be modeled as an *ideal gas*, then its equation of state is

$$PV = mR_m T \quad \text{or} \quad PV = nRT, \tag{3.9}$$

where the gas constant of the mixture (R_m) and the universal gas constant (R) are related by

$$R_m = \frac{n}{m} R = \frac{R}{M}. \tag{3.10}$$

The *partial pressure* P_i of component i is the pressure one would measure if component i alone were to fill the mixture volume V at the same temperature T as the mixture. Thus

$$P_i V = m_i R_m T \quad \text{or} \quad P_i V = n_i RT. \tag{3.11}$$

Summing these equations over i, we obtain Dalton's law,

$$P = \Sigma P_i, \tag{3.12}$$

which states that the pressure of a mixture of gases at a specified volume and temperature is equal to the sum of the partial pressures of the components. Note that $P_i/P = x_i$, and so using Eqs. (3.7) and (3.8) we can relate C_i to P_i.

The nomenclature we have used in this section applies to a mixture in *equilibrium*, that is, to a fluid batch whose composition, pressure, and temperature do not vary from point to point. In a convection study we are (out of necessity) involved with a *nonequilibrium* mixture which we view as a patchwork of small equilibrium batches: the equilibrium state of each of these batches is assumed to vary only slightly as one moves from one batch to its neighbors.

3.2. Mass Conservation in a Mixture

We apply the principle of mass conservation to each component in the mixture. For the moment we use the notation ρ_i instead of C_i for the concentration of component i. In the absence of component generation we must have

$$\frac{\partial \rho_i}{\partial t} + \nabla \cdot (\rho_i \mathbf{V}_i) = 0, \tag{3.13}$$

where $\mathbf{V}_i$ is the (intrinsic) velocity of particles of component i. Summing over i, we obtain

$$\frac{\partial \rho}{\partial t} + \nabla \cdot (\Sigma \rho_i \mathbf{V}_i) = 0. \tag{3.14}$$

This is the same as

$$\frac{\partial \rho}{\partial t} + \nabla \cdot (\rho \mathbf{V}) = 0 \tag{3.15}$$

provided that we identify $\mathbf{V}$ with the mass-averaged velocity,

$$\mathbf{V} = \frac{1}{\rho} \Sigma \rho_i \mathbf{V}_i. \tag{3.16}$$

Motion of a component relative to this mass-averaged velocity is called *diffusion*. Thus $\mathbf{V}_i - \mathbf{V}$ is the diffusion velocity of component i and

$$\mathbf{j}_i = \rho_i (\mathbf{V}_i - \mathbf{V}) \tag{3.17}$$

is the *diffusive flux* of component i. Equation (3.13) now gives

$$\frac{\partial \rho_i}{\partial t} + \nabla \cdot (\rho_i \mathbf{V}) = -\nabla \cdot \mathbf{j}_i. \tag{3.18}$$

Reverting to the notation C_i for concentration, and assuming that the mixture is incompressible, we have

$$\frac{DC_i}{Dt} = -\nabla \cdot \mathbf{j}_i, \tag{3.19}$$

where $D/Dt = \partial/\partial t + \mathbf{V} \cdot \nabla$.

For the case of a two-component mixture, Fick's law of mass diffusion is

$$\mathbf{j}_1 = -D_{12} \nabla C_1, \tag{3.20}$$

where D_{12} is the mass diffusivity of component 1 into component 2, and similarly for $\mathbf{j}_2$. In fact, $D_{12} = D_{21} = D$. The diffusivity D, whose units are $m^2 s^{-1}$, has a numerical value which in general depends on the mixture pressure, temperature, and composition. From Eq. (3.19) and (3.20) we have

$$\frac{DC_1}{Dt} = \nabla \cdot (D \nabla C_1). \tag{3.21}$$

If the migration of the first component is the only one of interest, then the subscript can be dropped. For a homogeneous situation we have

$$\frac{DC}{Dt} = D\nabla^2 C. \tag{3.22}$$

The analogy between this equation and the corresponding energy equation (temperature T, thermal diffusivity α_m)

$$\frac{DT}{Dt} = \alpha_m \nabla^2 T \tag{3.23}$$

is obvious. Fourier's law of thermal diffusion $\mathbf{q} = -k\nabla T$, where $\mathbf{q}$ is the heat flux and k is the thermal conductivity, is analogous to Fick's law of mass diffusion $\mathbf{j} = -D\nabla C$.

So far in this chapter we have been concerned with the fluid only, but now we consider a porous solid matrix saturated by fluid mixture. Within the solid there is of course neither flow nor any component of the mixture. Multiplying Eq. (3.21) (with the suffix dropped) by the porosity φ we have

$$\varphi\frac{\partial C}{\partial t} + \varphi\mathbf{V}\cdot\nabla C = \varphi\nabla\cdot(D\nabla C).$$

Recalling the Dupuit-Forchheimer relationship $\mathbf{v} = \varphi\mathbf{V}$, we see that this equation can be written, if φ is constant, as

$$\varphi\frac{\partial C}{\partial t} + \mathbf{v}\cdot\nabla C = \nabla\cdot(D_m\nabla C), \tag{3.24}$$

where $D_m = \varphi D$ is the mass diffusivity of the porous medium. Some authors invoke tortuosity and produce a more complicated relationship between D_m and D. The diffusive mass flux in the porous medium (rate of flow of mass across unit cross-sectional area of the medium) is

$$\mathbf{j}_m = -D_m\nabla C = \varphi\mathbf{j} \tag{3.25}$$

This is consistent with the surface porosity of the medium being equal to φ. Equation (3.24) also may be derived directly by using as control volume an element of the medium. If the mass of the substance whose concentration is C is being generated at a rate $\dot{m}'''$ per unit volume of the medium, then the term $\dot{m}'''$ must be added to the right-hand side of Eq. (3.24). The result may be compared with Eq. (2.3).

3.3. Combined Heat and Mass Transfer

In the most commonly occurring circumstances the transport of heat and mass (e.g., salt) are not directly coupled, and both Eqs. (2.3) and (3.24) (which clearly are uncoupled) hold without change. In double-diffusive (e.g., thermohaline) convection the coupling takes place because the density ρ of the fluid mixture depends on both temperature T and concentration C (and also, in general, on the pressure P). For sufficiently small isobaric changes in temperature and concentration the mixture density ρ depends linearly on both T and C, and we have approximately

$$\rho = \rho_0[1 - \beta(T - T_0) - \beta_C(C - C_0)], \tag{3.26}$$

where the subscript zero refers to a reference state, β is the volumetric thermal expansion coefficient,

$$\beta = -\frac{1}{\rho}\left(\frac{\partial \rho}{\partial T}\right)_{P,C}, \tag{3.27}$$

and β_C is the volumetric concentration expansion coefficient,

$$\beta_C = -\frac{1}{\rho}\left(\frac{\partial \rho}{\partial C}\right)_{T,P}. \tag{3.28}$$

Both β and β_C are evaluated at the reference state.

In some circumstances there is direct coupling. This is when cross-diffusion (Soret and Dufour effects) is not negligible. The Soret effect refers to mass flux produced by a temperature gradient and the Dufour effect refers to heat flux produced by a concentration gradient. For the case of no heat and mass sources we have, in place of Eqs. (2.3) and (3.24),

$$\frac{(\rho c)_m}{(\rho c)_f}\frac{\partial T}{\partial t} + \mathbf{v}\cdot\nabla T = \nabla\cdot(D_T\nabla T + D_{TC}\nabla C), \tag{3.29}$$

$$\varphi\frac{\partial C}{\partial t} + \mathbf{v}\cdot\nabla C = \nabla\cdot(D_C\nabla C + D_{CT}\nabla T), \tag{3.30}$$

where D_T $(= k_m/(\rho c)_f)$ is the thermal diffusivity, D_C $(= D_m)$ is the mass diffusivity, D_{TC}/D_T is the Dufour coefficient, and D_{CT}/D_C is the Soret coefficient of the porous medium.

The variation of density with temperature and concentration gives rise to a combined buoyancy force, proportional to $\beta(T - T_0) + \beta_C(C - C_0)$. The fact that the coefficients of Eq. (3.29) differ from those of Eq. (3.30) leads to interesting effects, such as flows oscillating in time in the presence of steady boundary conditions.

The Soret and Dufour effects are usually minor and can be neglected in simple models of coupled heat and mass transfer. According to Platten and Legros (1984), the mass fraction gradient established under the effect of thermal diffusion is very small. However, it has a disproportionately large influence on hydrodynamic stability relative to its contribution to the buoyancy of the fluid. They also state that in most liquid mixtures the Dufour effect is inoperative, but that this may not be the case in gases. Mojtabi and Charrier-Mojtabi (2000) confirm this by noting that in liquids the Dufour coefficient is an order of magnitude smaller than the Soret effect. They conclude that for saturated porous media, the phenomenon of cross diffusion is further complicated because of the interaction between the fluid and the porous matrix and because accurate values of the cross-diffusion coefficients are not available.

The thermodiffusion coefficient D_{TC} and the isothermal diffusion coefficient D_T were separately measured by Platten and Costeseque (2004) for both a porous medium and the corresponding liquid clear of solid material. They found that the measured value of the ratio of these two quantities (what they call the Soret coefficient) was the same for the clear fluid as for the porous medium to within experimental error.

The thermodynamic irreversibility of coupled heat and mass transfer in saturated porous media is treated based on the method of irreversible thermodynamics in Bejan *et al.* (2004). Viskanta (2005) has reviewed studies of combustion and heat transfer in inert porous media.

3.4. Effects of a Chemical Reaction

In recent years it has been realized that it is not always permissible to neglect the effects of convection in chemical reactors of porous construction. Suppose that we have a solution of a reagent whose concentration C is defined as above. If m is the molar mass of the reagent, then its concentration in moles per unit volume of the fluid mixture is $C_m = C/m$. Suppose that the rate equation for the reaction is

$$\frac{dC_m}{dt} = -kC_m^n. \tag{3.31}$$

The integer power n is the order of the reaction. The rate coefficient k is a function of the absolute temperature T given by the Arrhenius relationship

$$k = A\exp\left(-\frac{E}{RT}\right), \tag{3.32}$$

where E is the activation energy of the reaction (energy per mole), R is the universal gas constant, and A is a constant called the preexponential factor.

Assume further that the solid material of the porous medium is inert, that the reaction produces a product whose mass can be ignored, and that there is negligible change in volume. Then the rate of increase of C due to the reaction is $m\,dC_m/dt$. It follows that Eq. (3.24) is to be replaced by

$$\varphi\frac{\partial C}{\partial t} + \mathbf{v}\cdot\nabla C = \nabla\cdot(D_m\nabla C) - \varphi A m^{1-n}C^n\exp\left(-\frac{E}{RT}\right). \tag{3.33}$$

If the consumption of one mole of reagent causes the heat energy to increase by an amount $-\Delta H$ due to the reaction, then the increase in energy per unit volume of the fluid mixture is $(\Delta H)dC_m/dt$. Thus in place of Eq. (2.3) we have

$$(\rho c)_m\frac{\partial T}{\partial t} + (\rho c)_f\mathbf{v}\cdot\nabla T$$
$$= \nabla\cdot(k_m\nabla T) + \dot{m}''' - \varphi A(\Delta H)m^{-n}C^n\exp\left(-\frac{E}{RT}\right). \tag{3.34}$$

Equation (3.33), for the case of a first-order reaction ($n = 1$), is in accord with the formulation of Kolesnikov (1979). We note that for a zero-order reaction ($n = 0$) the thermal equation (3.34) is decoupled from Eq. (3.33) in the sense that Eq. (3.34) does not depend explicitly on C [though C and T are still related by Eq. (3.33)].

These equations are appropriate if the reaction is occurring entirely within the fluid. Now suppose that we have a catalytic reaction taking place only on the solid surface of the porous matrix. If the surface porosity is equal to the (volume) porosity φ, and if the reaction rate is proportional to the mass of the solid material, then Eqs. (3.33) and (3.34) should be altered by replacing φA by $(1-\varphi)\rho_s A'$ where A' is a new constant preexponential factor (compare Gatica *et al.*, 1989).

Recent papers on the effects of chemical reactions include those by Balakotaiah and Portalet (1990a,b), Stroh and Balakotaiah (1991, 1992, 1993), Farr *et al.* (1991), Gabito and Balakotaiah (1991), Nandakumar and Weinitschke (1992), Salinger *et al.* (1994b), Nguyen and Balakotaiah (1995), Subramanian and Balakotaiah (1995, 1997), Vafai *et al.* (1993), Kuznetsov and Vafai (1995b), and Chao *et al.* (1996).

3.5. Multiphase Flow

If two or more miscible fluids occupy the void space in a porous medium, then even if they occupy different regions initially they mix because of diffusive and other dispersive effects, leading ultimately to a multicomponent mixture such as what we just have been considering. If immiscible fluids are involved, the situation is more complicated. Indeed the complexities are such that, insofar as convection studies are concerned, only the simplest situations have been treated. It invariably has been assumed that Darcy's law is valid. Consequently our discussion of the momentum and energy equations in this section will be comparatively brief. This will enable us to present a derivation of the basic equations using formal averages. We follow the presentation of Cheng (1978).

We consider "two-phase" fluid flow in a porous medium. This means that we actually have three phases: two fluids and the solid matrix. The fluids could well both be liquids, but to simplify the discussion we suppose that we have a liquid phase (which we can label by the suffix l) and a gas phase (suffix g). As in previous chapters the suffix s refers to the solid matrix, which in this section is not necessarily fixed.

We take a representative elementary volume V occupied by the liquid, gas, and solid, whose interfaces may move with time, so

$$V = V_l(t) + V_g(t) + V_s(t). \tag{3.35}$$

We define the phase average of some quantity ψ_α as

$$\langle \psi_\alpha \rangle \equiv V^{-1} \int_V \psi_\alpha dV, \tag{3.36}$$

where ψ_α is the value of ψ in the α phase ($\alpha = l, g, s$) and is taken to be zero in the other phases. The intrinsic phase average of ψ_α is defined as

$$\langle \psi_\alpha \rangle^\alpha \equiv V_\alpha^{-1} \int_{V_\alpha} \psi_\alpha dV, \tag{3.37}$$

that is, the integration is carried out over only the α phase. Since ψ_α is zero in the other phases, Eq. (3.37) can be rewritten as

$$\langle\psi_\alpha\rangle^\alpha \equiv V_\alpha^{-1}\int_V \psi_\alpha dV. \tag{3.38}$$

Comparing Eqs. (3.36) and (3.38) we see that

$$\langle\psi_\alpha\rangle = \varepsilon_\alpha\langle\psi_\alpha\rangle^\alpha \tag{3.39}$$

where

$$\varepsilon_\alpha = \frac{V_\alpha}{V} \tag{3.40}$$

is the fraction of the total volume occupied by the α phase. In terms of the porosity φ of the medium we have

$$\varepsilon_l + \varepsilon_g = \varphi, \qquad \varepsilon_s = 1 - \varphi. \tag{3.41}$$

We define deviations (from the respective average values, for the α phase)

$$\tilde{\psi}_\alpha \equiv \psi_\alpha - \langle\psi_\alpha\rangle^\alpha, \quad \tilde{\chi}_\alpha \equiv \chi_\alpha - \langle\chi_\alpha\rangle^\alpha \tag{3.42}$$

and note that in the other phases $\tilde{\psi}_\alpha$ and $\tilde{\chi}_\alpha$ are zero. It is easily shown that

$$\langle\psi_\alpha\chi_\alpha\rangle^\alpha = \langle\psi_\alpha\rangle^\alpha\langle\chi_\alpha\rangle^\alpha + \langle\tilde{\psi}_\alpha\tilde{\chi}_\alpha\rangle^\alpha \tag{3.43}$$

and

$$\langle\psi_\alpha\chi_\alpha\rangle = \varepsilon_\alpha\langle\psi_\alpha\rangle^\alpha\langle\chi_\alpha\rangle^\alpha + \langle\tilde{\psi}_\alpha\tilde{\chi}_\alpha\rangle. \tag{3.44}$$

The following theorems are established by integration over an elementary volume.

Averaging theorem:

$$\langle\nabla\psi_\alpha\rangle = \nabla\langle\psi_\alpha\rangle + V^{-1}\int_{A_\alpha}\psi_\alpha \mathbf{n}_\alpha dS. \tag{3.45}$$

Modified averaging theorem:

$$\langle\nabla\psi_\alpha\rangle = \varepsilon_\alpha\nabla\langle\psi_\alpha\rangle^\alpha + V^{-1}\int_{A_\alpha}\tilde{\psi}_\alpha \mathbf{n}_\alpha dS. \tag{3.46}$$

Transport theorem:

$$\langle\frac{\partial\psi_\alpha}{\partial t}\rangle = \frac{\partial}{\partial t}\langle\psi_\alpha\rangle - V^{-1}\int_{A_\alpha}\psi\mathbf{w}_\alpha\cdot\mathbf{n}_\alpha dS \tag{3.47}$$

where A_α denotes the interfaces between the α phase and the other phases, $\mathbf{w}_\alpha$ is the velocity vector of the interface, and $\mathbf{n}_\alpha$ is the unit normal to the interface pointing outward from the α phase.

3.5.1. Conservation of Mass

The microscopic continuity equation for the liquid phase is

$$\frac{\partial \rho_l}{\partial t} + \nabla \cdot (\rho_l \mathbf{V}_l) = 0, \tag{3.48}$$

which can be integrated over an elementary volume to give

$$\left\langle \frac{\partial \rho_l}{\partial t} \right\rangle + \langle \nabla \cdot (\rho_l \mathbf{V}_l) \rangle = 0, \tag{3.49}$$

where ρ_l and $\mathbf{V}_l$ are the density and velocity of the liquid. Application of the transport theorem to the first term and the averaging theorem to the second term of this equation, with the aid of Eq. (3.44), leads to

$$\frac{\partial}{\partial t}(\varepsilon_l \langle \rho_l \rangle^l) + \nabla \cdot (\langle \rho_l \rangle^l \langle \mathbf{V}_l \rangle + \langle \tilde{\rho}_l \tilde{\mathbf{V}}_l \rangle \\ + V^{-1} \int_{A_{lg}} \rho_l (\mathbf{V}_l - \mathbf{w}_{lg}) \cdot \mathbf{n}_l dS + V^{-1} \int_{A_{ls}} \rho_l (\mathbf{V}_l - \mathbf{w}_{ls}) \cdot \mathbf{n}_l dS = 0 \tag{3.50}$$

where A_{lg} and A_{ls} are the liquid-gas and liquid-solid interfaces that move with velocities $\mathbf{w}_{lg}$ and $\mathbf{w}_{ls}$. The first integral in Eq. (3.50) represents mass transfer due to a change of phase from liquid to gas, and in general this is nonzero; but the second integral vanishes, since there is no mass transfer across the liquid-solid interface. The dispersive term $\langle \tilde{\rho}_l \tilde{\mathbf{V}}_l \rangle$ is generally small and we suppose that it can be neglected. Accordingly, Eq. (3.50) reduces to

$$\frac{\partial}{\partial t}(\varepsilon_l \langle \rho_l \rangle^l) + \nabla \cdot (\langle \rho_l \rangle^l \langle \mathbf{V}_l \rangle) + V^{-1} \int_{A_{lg}} \rho_l (\mathbf{V}_l - \mathbf{w}_{lg}) \cdot \mathbf{n}_l dS = 0 \tag{3.51}$$

Similarly the macroscopic continuity equations for the gas and for the solid are

$$\frac{\partial}{\partial t}(\varepsilon_g \langle \rho_g \rangle^g) + \nabla \cdot (\langle \rho_g \rangle^g \langle \mathbf{V}_g \rangle) + V^{-1} \int_{A_{gl}} \rho_g (\mathbf{V}_g - \mathbf{w}_{gl}) \cdot \mathbf{n}_g dS = 0 \tag{3.52}$$

and

$$\frac{\partial}{\partial t}(\varepsilon_s \langle \rho_s \rangle^s) + \nabla \cdot (\langle \rho_s \rangle^s \langle \mathbf{V}_s \rangle) = 0. \tag{3.53}$$

The mass gained by change of phase from liquid to gas is equal to the mass lost by change of phase from gas to liquid. Thus the surface integrals in Eqs. (3.51) and (3.52) are equal in magnitude but opposite in sign. The integrals thus cancel each other when Eqs. (3.51)–(3.53) are added to give

$$\frac{\partial}{\partial t}\left[\varepsilon_l \langle \rho_l \rangle^l + \varepsilon_g \langle \rho_g \rangle^g + \varepsilon_s \langle \rho_s \rangle^s\right] \\ + \nabla \cdot \left(\langle \rho_l \rangle^l \langle \mathbf{V}_l \rangle^l + \langle \rho_g \rangle^g \langle \mathbf{V}_g \rangle^g + \langle \rho_s \rangle^s \langle \mathbf{V}_s \rangle^s\right) = 0 \tag{3.54}$$

Note that, for example, $\langle \mathbf{V}_l \rangle = \varepsilon_l \langle \mathbf{V}_l \rangle^l$ since $\mathbf{V}_l$ is taken to be zero in the gas and solid phases. If the volumetric liquid and gas saturation, S_l and S_g, are defined by

$$S_l = \frac{V_l}{V_l + V_g}, \quad S_g = \frac{V_g}{V_l + V_g} \tag{3.55}$$

so that

$$S_l + S_g = 1, \ \varepsilon_l = \varphi S_l, \ \varepsilon_g = \varphi S_g, \ \text{and } \varepsilon_s = 1 - \varphi, \tag{3.56}$$

then Eq. (3.54) can be rewritten as

$$\begin{aligned} &\frac{\partial}{\partial t}\left[\varphi S_l \langle \rho_l \rangle^l + \varphi S_g \langle \rho_g \rangle^g + (1-\varphi)\langle \rho_s \rangle^s\right] \\ &\quad + \nabla \cdot \left(\langle \rho_l \rangle^l \langle \mathbf{V}_l \rangle^l + \langle \rho_g \rangle^g \langle \mathbf{V}_g \rangle^g + \langle \rho_s \rangle^s \langle \mathbf{V}_s \rangle^s\right) = 0. \end{aligned} \tag{3.57}$$

3.5.2. Conservation of Momentum

The microscopic momentum equation for the liquid phase is

$$\frac{\partial}{\partial t}(\rho_l \mathbf{V}_l) + \nabla \cdot (\rho_l \mathbf{V}_l \mathbf{V}_l) + \nabla P_l - \nabla \cdot \tau_l - \rho_l \mathbf{f} = 0, \tag{3.58}$$

where P_l, τ_l, and $\mathbf{f}$ are, respectively, the pressure, the viscous stress tensor, and the body force per unit mass of the liquid. If the body force is entirely gravitational, then

$$\mathbf{f} = \mathbf{g} = -\nabla \Phi, \tag{3.59}$$

where Φ is the gravitational potential. We substitute Eq. (3.59) into Eq. (3.58), integrate the resulting equation over an elementary volume, apply the transport theorem to the first term and the averaging theorem to the second, third, and fourth terms, and use Eq. (3.44). We also make use of the equation of continuity (3.57) and replace $\nabla \cdot \tau_l$ by $\mu_l \nabla^2 \langle \mathbf{V}_l \rangle$. (See Gray and O'Neill, 1976.) We get

$$\begin{aligned} &\left[\varepsilon_l \langle \rho_l \rangle^l \frac{\partial}{\partial t}\langle \mathbf{V}_l \rangle^l + \varepsilon_l \langle \rho_l \rangle^l \langle \mathbf{V}_l \rangle \cdot \nabla \langle \mathbf{V}_l \rangle \right. \\ &\quad \left. + V^{-1}\int_{A_{lg}} \rho_l \mathbf{V}_l (\mathbf{V}_l - \mathbf{w}_{lg}) \cdot \mathbf{n}_l dS + \nabla \cdot \left(\langle \rho_l \rangle^l \langle \tilde{V}_l \tilde{V}_l \rangle\right)\right] \\ &\quad + \varepsilon_l \nabla \langle P_l \rangle^l + \varepsilon_l \langle \rho_l \rangle^l \nabla \langle \Phi_l \rangle^l \\ &\quad + V^{-1}\int_{A_{lg}} \left(\tilde{P}_l + \langle \rho_l \rangle^l \tilde{\Phi}_l\right) \mathbf{n}_l dS + V^{-1}\int_{A_{ls}} \left(\tilde{P}_l + \langle \rho_l \rangle^l \tilde{\Phi}_l\right) \mathbf{n}_l dS \\ &\quad - \mu_l \nabla^2 \langle \mathbf{V}_l \rangle - V^{-1}\int_{A_{lg}} \mathbf{n}_l \cdot \tau_l dS - V^{-1}\int_{A_{ls}} \mathbf{n}_l \cdot \tau_l dS = 0, \end{aligned} \tag{3.60}$$

where density gradients at the microscopic level have been assumed to be small compared to the corresponding velocity gradients.

For an isotropic medium, Gray and O'Neill (1976) argued that

$$V^{-1}\int_{A_{lg}} \mathbf{n}_l \cdot \tau_l dS + V^{-1}\int_{A_{ls}} \mathbf{n}_l \cdot \tau_l dS = \mu\varepsilon_l B(\langle \mathbf{V}_s\rangle^s - \langle \mathbf{V}_l\rangle^l) \tag{3.61}$$

and

$$V^{-1}\int_{A_{lg}} (\tilde{P}_t + \langle\rho_l\rangle^l \tilde{\Phi}_l)\mathbf{n}_l dS + V_l^{-1}\int_{A_{ls}} (\tilde{P}_l + \langle\rho_l\rangle^l \tilde{\Phi}_l)\mathbf{n}_l dS$$
$$= F(\nabla\langle P_l\rangle^l + \langle\rho_l\rangle^l \nabla\langle\Phi_l\rangle^l), \tag{3.62}$$

where B and F are constants that depend on the nature of the isotropic medium. Substituting Eqs. (3.61) and (3.62) into Eq. (3.60) and neglecting the inertia terms in the square brackets and the term $\mu\nabla^2\langle\mathbf{V}_l\rangle$ (compare the discussion in Section 1.5) yields

$$\langle \mathbf{V}_l\rangle^l - \langle \mathbf{V}_s\rangle^S = -\frac{k_{sl}K}{\varepsilon_l \mu_l}(\nabla\langle P_l\rangle^l + \langle\rho_l\rangle^l \nabla\langle\Phi_l\rangle^l), \tag{3.63}$$

where $k_{sl}K \equiv \varepsilon_l(1+F)/B$. Here K denotes the intrinsic permeability of the porous medium, as defined for one-phase flow. The new quantity k_{sl} is the relative permeability of the porous medium saturated with liquid. It is a dimensionless quantity.

Similarly, when inertia terms and the term $\mu_g\nabla^2\langle\mathbf{V}_g\rangle$ are neglected, the momentum equation for the gas phase is

$$\langle \mathbf{V}_g\rangle^g - \langle \mathbf{V}_s\rangle^s = -\frac{k_{sg}K}{\varepsilon_g \mu_g}(\nabla\langle P_g\rangle^g + \langle\rho_g\rangle^g \nabla\langle\Phi_g\rangle^g), \tag{3.64}$$

where k_{sg} denotes the relative permeability of the porous medium saturated with gas. Equations (3.63) and (3.64) are the Darcy equations for a liquid-gas combination in an isotropic porous medium. A similar expression for an anisotropic medium has been developed by Gray and O'Neill (1976). A permeability tensor is involved. They also obtain an expression for flow in an isotropic medium with nonnegligible inertial effects.

3.5.3. Conservation of Energy

The microscopic energy equation, in terms of enthalpy for the liquid phase, is

$$\frac{\partial}{\partial t}(\rho_l h_l) + \nabla\cdot(\rho_l h_l \mathbf{V}_l - k_l\nabla T_l) - \left(\frac{\partial P_l}{\partial t} + \mathbf{V}_l \cdot \nabla P_l\right) = 0, \tag{3.65}$$

where h_l and k_l are the enthalpy and thermal conductivity of the liquid. In writing this equation we have neglected the viscous dissipation, thermal radiation, and any internal energy generation. Integrating this equation over a representative elementary volume and applying the transport equations to the first and fourth terms, Eqs. (3.44) and (3.45) to the second term, Eq. (3.46) to the third term, and Eq. (3.44) to the fifth term yields

$$\frac{\partial}{\partial t}\left(\varepsilon_l\langle\rho_l\rangle^l\langle h_l\rangle_l\right)+\nabla\cdot\left(\langle\rho_l\rangle_l\langle h_l\rangle_l\langle\mathbf{V}_l\rangle\right)-\nabla\cdot\left(\varepsilon_l k_l^*\nabla\langle T_l\rangle^l\right)$$
$$-\left[\varepsilon l\frac{\partial}{\partial t}\left(\langle\mathbf{P}_t\rangle^l\right)+\langle V_l\rangle\cdot\nabla\langle P_l\rangle l\right]+Q_{lg}+Q'_{lg}+Q'_{ls}=0, \quad (3.66)$$

where k_l^* is the effective thermal conductivity of the liquid in the presence of the solid matrix. This k_l^* is the sum of the stagnant thermal conductivity k_l' (due to molecular diffusion) and the thermal dispersion coefficient k_l'' (due to mechanical dispersion), which in turn are defined by

$$-\varepsilon_l k_l'\nabla\langle T_l\rangle^l=-\langle k_l\rangle^l\left(\varepsilon_l\nabla\langle T_l\rangle^l+V^{-1}\int_{A_{lg}}\tilde{T}_l\mathbf{n}_l dS+V^{-1}\int_{A_{ls}}\tilde{T}_l\mathbf{n}_l dS\right) \quad (3.67a)$$

and

$$-\nabla\cdot\left(\varepsilon_l k_l''\nabla\langle T_l\rangle^l\right)=\nabla\cdot\left(\rho_l\tilde{h}_l\tilde{\mathbf{V}}_l\right)-\langle\tilde{\mathbf{V}}_l\cdot\nabla\tilde{P}_l\rangle$$
$$+V^{-1}\int_{A_{lg}}\tilde{P}_l\tilde{\mathbf{V}}_l\cdot\mathbf{n}_l dS+V^{-1}\int_{A_{ls}}\tilde{P}_l\tilde{\mathbf{V}}_l\cdot\mathbf{n}_l dS. \quad (3.67b)$$

The integrals in Eq. (3.67a) account for the change in thermal diffusion due to the microstructure of the solid matrix. The terms Q_{lg}, Q'_{lg}, and Q'_{ls} are given, respectively, by

$$Q_{lg}=V^{-1}\int_{A_{lg}}(\rho_l h_l-\tilde{P}_l)(\mathbf{V}_l-\mathbf{w}_{lg})\cdot\mathbf{n}_l dS\approx V^{-1}\int_{A_{lg}}\rho_l h_l(\mathbf{V}_l-\mathbf{w}_{lg})\cdot\mathbf{n}_l dS, \quad (3.68a)$$

$$Q'_{lg}=V^{-1}\int_{A_{lg}}\mathbf{q}\cdot\mathbf{n}_l dS \quad (3.68b)$$

$$Q'_{ls}=V^{-1}\int_{A_{ls}}\mathbf{q}\cdot\mathbf{n}_l dS=A_{ls}h_l V^{-1}(T_s-T_l) \quad (3.68c)$$

where $\mathbf{q}$ in Eqs. (3.68b) and (3.68c) is the conduction heat flux across the interface, and h_l in Eq. (3.68c) is defined as the local volume averaged heat transfer coefficient at the liquid-solid interface, which depends on the physical properties of the liquid and its flow rate.

Similarly, the energy equation for the gas phase and for the solid-matrix phase are, respectively,

$$\frac{\partial}{\partial t}\left(\varepsilon_g\langle\rho_g\rangle^g\langle h_g\rangle^g\right)+\nabla\cdot\left(\langle\rho_g\rangle^g\langle h_g\rangle^g\langle\mathbf{V}_g\rangle\right)-\nabla\cdot\left(\varepsilon_g k_g^*\nabla\langle T_g\rangle^g\right)$$
$$-\left(\varepsilon_g\frac{\partial}{\partial t}\langle\rho_g\rangle^g+\langle\mathbf{V}_g\rangle\cdot\nabla\langle P_g\rangle^g\right)+Q_{gl}+Q'_{gl}+Q'_{gs}=0 \quad (3.69)$$

and

$$\frac{\partial}{\partial t}\left(\varepsilon_s\langle\rho_s\rangle^s\langle h_s\rangle^s\right)+\nabla\cdot\left(\langle\rho_s\rangle^s\langle h_s\rangle^s\langle\mathbf{V}_s\rangle\right)-\nabla\cdot\left(\varepsilon_s k_s^*\nabla\langle T_s\rangle^s\right)$$
$$-\left(\varepsilon_s\frac{\partial}{\partial t}\langle P_s\rangle^s+\langle\mathbf{V}_s\rangle\cdot\nabla\langle P_s\rangle^s\right)+Q'_{sl}+Q'_{sg}=0, \tag{3.70}$$

where k_g^* and k_s^* are defined analogously to k_l^*, and similarly for the various Q terms. Note that

$$Q_{gl}=-Q_{lg},\quad Q'_{gl}=-Q'_{lg},\quad Q'_{sl}=-Q'_{ls} \tag{3.71}$$

and

$$Q'_{gs}=V^{-1}\int_{A_{gs}}\mathbf{q}\cdot\mathbf{n}_g dS=A_{gs}h_g V^{-1}\left(T_s-T_g\right)=-Q'_{sg} \tag{3.72}$$

where h_g is the heat transfer coefficient at the gas-solid interface.

The difference between P_g and P_l is called the capillary pressure. In many circumstances, including most geophysical situations, the capillary pressure can be neglected, so in this case we have

$$\langle P_l\rangle^l=\langle P_g\rangle^g=\langle P_s\rangle^s=\langle P\rangle. \tag{3.73}$$

Furthermore, we can usually assume local thermodynamic equilibrium and so

$$\langle T_l\rangle^l=\langle T_g\rangle^g=\langle T_s\rangle^s=\langle T\rangle. \tag{3.74}$$

Adding Eqs. (3.66), (3.69), and (3.70) in this case, we get

$$\frac{\partial}{\partial t}\left[\varphi S_l\langle\rho_l\rangle^l\langle h_l\rangle^l+\varphi S_g\langle\rho_g\rangle^g\langle h_g\rangle^g+(1-\varphi)\langle\rho_s\rangle^s\langle h_s\rangle^s\right]$$
$$+\nabla\cdot\left[\langle\rho_l\rangle^l\langle h_l\rangle^l\langle\mathbf{V}_l\rangle+\langle\rho_g\rangle^g\langle h_g\rangle^g\langle\mathbf{V}_g\rangle+\langle\rho_s\rangle^s\langle h_s\rangle^s\langle\mathbf{V}_s\rangle\right]$$
$$-\nabla\cdot(\mathrm{k}\nabla\langle\mathrm{T}\rangle)-\left[\frac{\partial}{\partial \mathrm{t}}\langle\mathrm{P}\rangle+(\langle\mathbf{V}_l\rangle+\langle\mathbf{V}_g\rangle+\mathbf{V}_s\rangle)\cdot\nabla\langle\mathrm{P}\rangle\right]=0, \tag{3.75}$$

where $k=\varphi(S_l k_l^*+S_g k_g^*)+(1-\varphi)k_s^*$ is the effective thermal conductivity of the porous medium saturated with liquid and gas at local thermal equilibrium, with the heat conduction assumed to be in parallel. (See Section 2.2.1.)

3.5.4. Summary: Relative Permeabilities

The governing equations for two-phase flow, for the case of negligible capillary pressure and local thermal equilibrium, are Eqs. (3.57), (3.63), (3.64), and (3.75). Since P and T are independent of phase we can drop the angle brackets in $\langle P\rangle$ and $\langle T\rangle$. Also we note that $\langle\mathbf{V}_l\rangle$ is just $\mathbf{v}_l$, the seepage velocity for the liquid phase, etc. Also, in Eq. (3.57), $\langle\rho_l\rangle\langle\mathbf{V}_l\rangle^l=\varepsilon_l^{-1}\langle\rho_l\rangle\langle\mathbf{V}_l\rangle=\langle\rho_l\rangle^l\langle\mathbf{V}_l\rangle$, etc. For a gravitational body force we have $\nabla\Phi_l=\nabla\Phi_g=-\mathbf{g}$. Thus we can rewrite the four governing equations, with the angle brackets for intrinsic averages dropped, as

$$\frac{\partial}{\partial t}\left[\varphi S_l\rho_l+\varphi S_g\rho_g+(1-\varphi)\rho_s\right]+\nabla\cdot(\rho_l\mathbf{v}_l+\rho_g\mathbf{v}_g+\rho_s\mathbf{v}_s)=0, \tag{3.76}$$

$$\mathbf{v}_l-\frac{\varepsilon_l}{\varepsilon_s}\mathbf{v}_s=-\frac{k_{sl}K}{\mu_l}(\nabla P-\rho_l\mathbf{g}), \tag{3.77}$$

$$\mathbf{v}_g - \frac{\varepsilon_g}{\varepsilon_s}\mathbf{v}_s = -\frac{k_{sg}K}{\mu_g}(\nabla P - \rho_g \mathbf{g}), \tag{3.78}$$

$$\frac{\partial}{\partial t}\left[\varphi S_l \rho_l h_l + \varphi S_g \rho_g h_g + (1-\varphi)\rho_s h_s\right] + \nabla \cdot (\rho_l h_l \mathbf{v}_l + \rho_g h_g \mathbf{v}_g + \rho_s h_s \mathbf{v}_s)$$
$$-\nabla \cdot (k \nabla T) - \left[\frac{\partial P}{\partial t} + (\mathbf{v}_l + \mathbf{v}_g + \mathbf{v}_g) \cdot \nabla P\right] = 0. \tag{3.79}$$

We can now extend Eqs. (3.76) and (3.79) by allowing for source terms q_M''' (rate of increase of mass per unit volume of the medium) and q_E''' (rate of increase of energy per unit volume of the medium). At the same time we can introduce A_M and A_E, respectively, the mass and energy per unit volume of the medium, and $\mathbf{F}_M$ and $\mathbf{F}_E$, respectively, the mass flux and energy flux in the medium. These are given by

$$A_M = \varphi S_l \rho_l + \varphi S_g \rho_g + (1-\varphi)\rho_s, \tag{3.80}$$

$$A_E = \varphi S_l \rho_l h_l + \varphi S_g \rho_g h_g + (1-\varphi)\rho_s h_s, \tag{3.81}$$

$$\mathbf{F}_M = \rho_l \mathbf{v}_l + \rho_g \mathbf{v}_g + \rho_s \mathbf{v}_s, \tag{3.82}$$

$$\mathbf{F}_E = \rho_l h_l \mathbf{v}_l + \rho_g h_g \mathbf{v}_g + \rho_s h_s \mathbf{v}_s - k\nabla T. \tag{3.83}$$

We also write

$$\frac{D^* P}{Dt} = \frac{\partial P}{\partial t} + (\mathbf{v}_l + \mathbf{v}_g + \mathbf{v}_s) \cdot \nabla P. \tag{3.84}$$

Thus D^*/Dt is a material derivative based on the sum of $\mathbf{v}_l$, $\mathbf{v}_g$, and $\mathbf{v}_s$, rather than the mass-weighted average of the velocities. The extended forms of the mass equation (3.76) and the energy equation (3.79) are

$$\frac{\partial A_M}{\partial t} + \nabla \cdot \mathbf{F}_M = q_M''' \tag{3.85}$$

and

$$\frac{\partial A_E}{\partial t} + \nabla \cdot \mathbf{F}_E - \frac{D^* P}{Dt} = q_E'''. \tag{3.86}$$

We are now confronted with the task of solving the Darcy equations (3.77) and (3.78), the mass equation (3.85), and the energy equation (3.85) subject to appropriate initial and boundary conditions. In many practical situations there will be no source terms ($q_M''' = q_E''' = 0$), the solid matrix will be fixed ($\mathbf{v}_s = 0$), and the pressure term D^*P/Dt will be negligible. Even then the task is not straightforward, because the relative permeabilities are not constant.

It is observed experimentally that in general the relative permeability for the liquid phase k_{sl} increases in a nonlinear fashion from 0 to 1 as the liquid saturation S_l increases from 0 to 1, and the functional relationship is not single valued. The value observed as S_l increases differs from that observed as S_l decreases, i.e., one has hysteresis. Also, k_{sl} may not differ from zero until S_l exceeds some nonzero critical value S_{l0}. This last behavior is illustrated in Fig. 3.1.

The complications arise because usually one fluid "wets" the solid and adheres to its surfaces, and each fluid can establish its own channels of flow through the

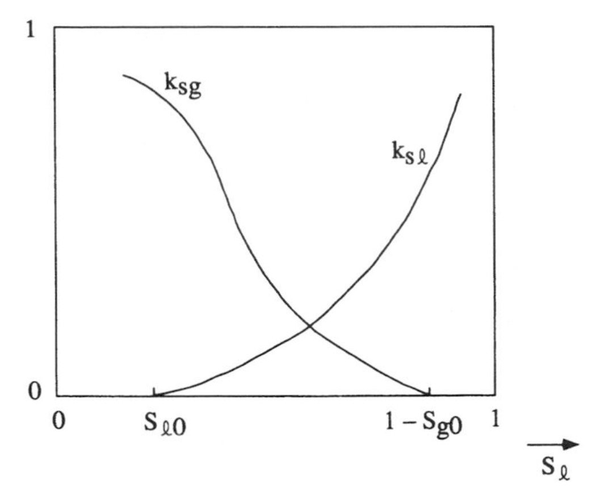

Figure 3.1. The general form of the relative permeability curves for two-phase flow through a porous medium.

medium only to a limited extent. Further, the flow of one fluid can destroy the connectivity of the pores available for the flow of the other fluid. Another factor affecting permeability is the difference in viscosity between fluids; one fluid can act as a lubricant for the other. Also, permeabilities tend to be slightly higher at higher pressure gradients.

In view of this complexity it is fortunate that experience has shown that the main qualitative features of convection flows are not sensitive to the precise form of the relative permeability versus saturation relationship. For several situations, satisfactory results have been reported when use has been made of a simple linear relationship, namely,

$$k_{sl} = S_l, \qquad k_{sg} = S_g \, (= 1 - S_l). \tag{3.87}$$

For the case when the liquid is oil, Corey *et al.* (1956) proposed the use of the semiempirical formulas

$$k_{sl} = \hat{S}_l^4 \text{ and } k_{sg} = (1 - \hat{S}_l)^2 (1 - \hat{S}_l^2), \tag{3.88a}$$

where

$$\hat{S}_l = \frac{S_l - S_{l0}}{1 - S_{l0} - S_{g0}}. \tag{3.88b}$$

The Corey formulas also have been used with water and steam.

A general alternative description of two-phase flow has been proposed by Hassanizadeh and Gray (1993). An experimental study of relative permeabilities and the various flow regimes that arise during steady-state two-phase flow was reported by Avroam and Payatakes (1995). A new model for multiphase, multicomponent transport in capillary porous media, in which the multiple phases are considered as constituents of a multiphase mixture, has been developed by Wang and Cheng (1996). This model is mathematically equivalent to the traditional model but involves a reduced number of model equations. An experimental and theoretical study of two-phase flow and heat transfer was conducted by Jamialahmadi *et al.* (2005). Some specific situations involving two-phase flows are discussed in Section 11.9.

3.6. Unsaturated Porous Media

Here we provide introductory references to an important topic that we have not discussed because of lack of space. The modeling of convection in unsaturated porous media, with and without boiling or condensation, has been discussed by Plumb (1991a). The particular topic of drying of porous media has been surveyed by Bories (1991) and Plumb (1991b, 2000). Some additional references to convection in unsaturated porous media are given in the general review by Tien and Vafai (1990a). The subject of multiphase flow and heat transfer in porous media has been reviewed by Wang and Cheng (1997) and Chang and Wang (2002). These papers reveal that convection in unsaturated media is a difficult problem.

One difficulty is that because of instabilities the interface between phases is on the macroscopic scale often far from being a well-defined smooth surface. A second difficulty is caused by the effects of surface tension. This produces a pressure difference that is proportional to the interface curvature on the *pore scale,* something that is completely different from the interface curvature on the macroscopic scale. Since the local pressure difference is affected by contact angle, and this is dependent on a number of things, there is a fundamental difficulty in calculating the appropriate average pressure difference on the macroscopic scale. A third difficulty is that hysteresis is commonly associated with the advance and recession of a phase interface.

Some recent papers involving the drying of porous media include those by Francis and Wepfer (1996), Daurelle *et al.*(1998), Lin *et al.*(1998), Oliveira and Haghighi (1998), Mhimid *et al.* (1999, 2000), Zili and Ben Nasrallah (1999), Coussot (2000), Landman *et al.* (2001), Natale and Santillan Marcus (2003), Plourde and Prat (2003), Salagnac *et al.* (2004), Nganhou (2004), Dayan *et al.* (2004), Frei *et al.* (2004), and Tao *et al.* (2005).

Recent papers of other aspects of convection in unsaturated media include those of Yu *et al.* (1993), Hanamura and Kaviany (1995), Larbi *et al.*(1995), Zhu and Vafai (1996), Dickey and Peterson (1997), Gibson and Charmchi (1997), Bouddour *et al.* (1998), H. Chen *et al.* (1998), Figus *et al.* (1998), Yan *et al.* (1998), Wang and Cheng (1998), Moya *et al.* (1999), Peng *et al.* (2000), Zhao and Liao (2000), Liu *et al.* (2002), Kacur and Van Keer (2003), Shen *et al.* (2003), Zili-Ghedira *et al.* (2003), and Jadhav and Pillai (2003).

3.7. Electrodiffusion through Porous Media

Diffusion is a slow process. When the diffusing species are electrically charged, diffusion can be accelerated by applying externally an electric current or by imposing a gradient of electrical potential. There are many applications at several scales, for example, the delivery of drugs by iontophoresis through the human body and the dechlorination of concrete structures such as bridges contaminated and corroded by sea water.

The basics of diffusion of ionic species through nonreactive and reactive porous media were reviewed most recently in the book by Bejan *et al.* (2004), based on the work of Frizon *et al.* (2003) and others. This section is based on the simplest presentation of electrodiffusion through nonreactive porous media, which was made based on scale analysis by Lorente and Ollivier (2005).

Instead of the classic Fick diffusion equation (3.22), the presence of electrical forces requires the use of the more general Nernst-Planck equation

$$\varphi \frac{\partial C_i}{\partial t} = D_i \frac{\partial}{\partial x}\left(\frac{\partial C_i}{\partial x} + z_i \frac{F}{RT} C_i \frac{\partial \psi}{\partial x}\right) \tag{3.89}$$

The subscript i indicates the ionic species that diffuses through the porous medium, z_i is the charge number, F is the Faraday constant, A is the ideal gas constant, T is the absolute temperature, and ψ is the electric potential created by the ionic species. In the same equation, C_i is the ionic species concentration and D_i is the effective diffusion coefficient of the species. For simplicity, we consider time-dependent diffusion in one direction (x).

The problem is closed by solving Eq. (3.89) in conjunction with the current conservation equation,

$$F \Sigma_i z_i \; j_i = j \tag{3.90}$$

where j_i is the ionic flux through the porous medium,

$$j_i = -D_i \left(\frac{\partial C_i}{\partial x} + z_i \frac{F}{RT} C_i \frac{\partial \psi}{\partial x}\right) \tag{3.91}$$

and j is the constant current density applied from the outside. The electric potential gradient follows from Eqs. (3.90) and (3.91):

$$\frac{\partial \psi}{\partial x} = -\frac{RT}{F} \frac{\dfrac{j}{F} + \Sigma_i z_i D_i \dfrac{\partial C_i}{\partial x}}{\Sigma_i z_i^2 D_i C_i}. \tag{3.92}$$

As an example, consider a one-dimensional porous medium (a slab) of thickness L. Initially the species of interest has $C_i = 0$ throughout the porous medium ($0 < x < L$). At the time $t = 0$, a new concentration level is imposed on one face, $C_i = \Delta C_i$ at $x = 0$, while the $x = L$ face is maintained at $C_i = 0$.

Lorente and Ollivier (2005) established the scales of diffusion in two limits. When the dominant driving force is the concentration gradient, the scales are those of classic diffusion, and the time of diffusion penetration over the distance L is

$$t_{diff} \sim \varphi \frac{L^2}{D_i}. \tag{3.93}$$

When electrical effects dominate, the time of diffusion over L is

$$t_{el} \sim \varphi \frac{LF}{j} \Delta C_i. \tag{3.94}$$

The transition between the two regimes is described by the new dimensionless group

$$B = \frac{FD\Delta C_i}{Lj} \tag{3.95}$$

which is the ratio of the two characteristic time scales,

$$B \sim \frac{t_{el}}{t_{diff}}. \tag{3.96}$$

Lorente and Ollivier (2005) modeled the same one-dimensional time-dependent electrodiffusion numerically, in a nondimensionalization based on the correct scales revealed by scale analysis. Numerical simulations conducted for practical examples (e.g., the extraction of an ionic species from a contaminated block) validated the predictions based on scale analysis and confirmed the correctness of both methods.

4
Forced Convection

The fundamental question in heat transfer engineering is to determine the relationship between the heat transfer rate and the driving temperature difference. In nature, many saturated porous media interact thermally with one another and with solid surfaces that confine them or are embedded in them. In this chapter we analyze the basic heat transfer question by looking only at *forced convection* situations, in which the fluid flow is caused (forced) by an external agent unrelated to the heating effect. First we discuss the results that have been developed based on the Darcy flow model and later we address the more recent work on the non-Darcy effects. We end this chapter with a review of current engineering applications of the method of forced convection through porous media. Some fundamental aspects of the subject have been discussed by Lage and Narasimhan (2000) and the topic has been reviewed by Lauriat and Ghafir (2000).

4.1. Plane Wall with Prescribed Temperature

Perhaps the simplest and most common heat transfer arrangement is the flow parallel to a flat surface that borders the fluid-saturated porous medium. With reference to the two-dimensional geometry defined in Fig. 4.1, we recognize the equations governing the conservation of mass, momentum (Darcy flow), and energy in the flow region of thickness δ_T:

$$\frac{\partial u}{\partial x} + \frac{\partial v}{\partial y} = 0, \tag{4.1}$$

$$u = -\frac{K}{\mu}\frac{\partial P}{\partial x}, \qquad v = -\frac{K}{\mu}\frac{\partial P}{\partial y}, \tag{4.2}$$

$$u\frac{\partial T}{\partial x} + v\frac{\partial T}{\partial y} = \alpha_m \frac{\partial^2 T}{\partial y^2}. \tag{4.3}$$

Note the boundary layer-approximated right-hand side of Eq. (4.3), which is based on the assumption that the region of thickness δ_T and length x is slender ($\delta_T \ll x$). The fluid mechanics part of the problem statement [namely, Eqs. (4.1) and (4.2)] is satisfied by the uniform parallel flow

$$u = U, \qquad v = 0, \tag{4.4}$$

The constant pressure gradient that drives this flow ($-dP/dx = \mu U_\infty/K$) is assumed known.

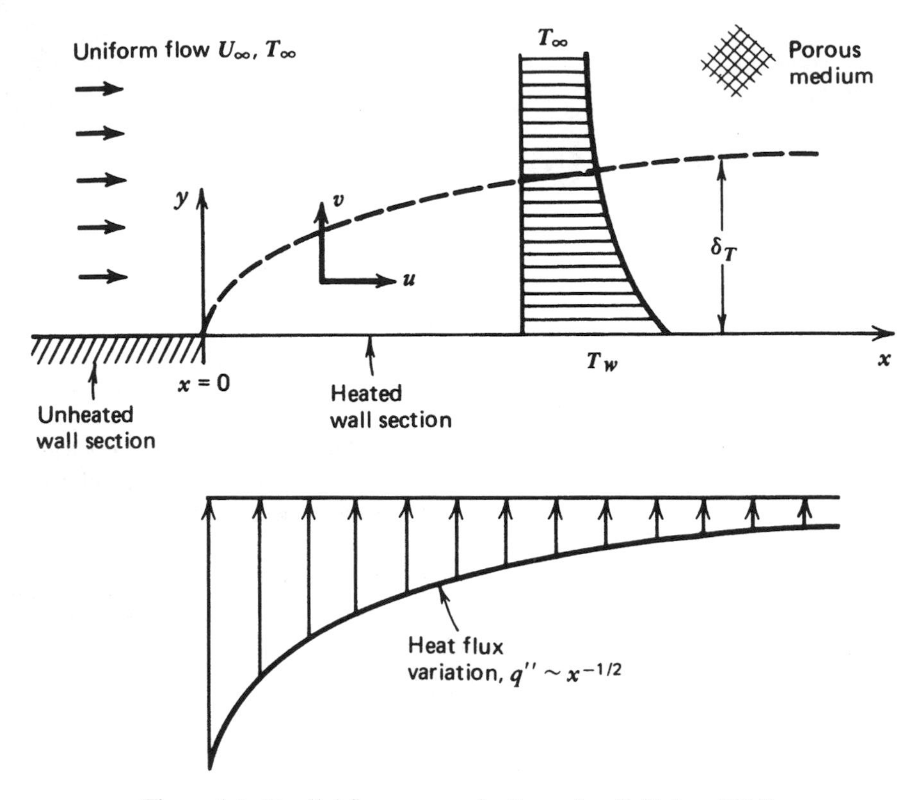

Figure 4.1. Parallel flow near an isothermal wall (Bejan, 1984).

The heat transfer rate between the surface at temperature T_w and the saturated porous medium at far-field temperature T_∞ can be determined in several ways. The scale analysis begins with writing $\Delta T = T_w - T_\infty$, so that the order-of-magnitude counterpart of Eq. (4.3) becomes

$$U_\infty \frac{\Delta T}{x} \sim \alpha_m \frac{\Delta T}{\delta_T^2}. \tag{4.5}$$

From this we can determine the thickness of the thermal boundary layer

$$\delta_T \sim x\, Pe_x{}^{-1/2}, \tag{4.6}$$

in which Pe_x is the Péclet number based on U_∞ and x:

$$\mathrm{Pe}_x = \frac{U_\infty x}{\alpha_m}. \tag{4.7}$$

For the local heat flux q'' we note the scale $q'' \sim k_m\, \Delta T/\delta_T$, or the corresponding local Nusselt number

$$\mathrm{Nu}_x = \frac{q''}{\Delta T}\frac{x}{k_m} \sim \mathrm{Pe}_x{}^{1/2}. \tag{4.8}$$

Figure 4.1 qualitatively illustrates the main characteristics of the heat transfer region, namely, the boundary layer thickness that increases as $x^{1/2}$ and the heat flux that decays as $x^{-1/2}$. The exact analytical solution for the same problem can be derived in closed form by introducing the similarity variables recommended by the scale analysis presented above:

$$\eta = \frac{y}{x}\,\mathrm{Pe}_x{}^{1/2}, \qquad \theta(\eta) = \frac{T - T_w}{T_\infty - T_w}. \tag{4.9}$$

In this notation, the energy equation (4.3) and the boundary conditions of Fig. 4.1 become

$$\theta'' + \frac{1}{2}\eta\theta' = 0, \tag{4.10}$$

$$\theta(0) = 0, \qquad \theta(\infty) = 1. \tag{4.11}$$

Equation (4.10) can be integrated by separation of variables, and the resulting expressions for the similarity temperature profile and the surface heat flux are (Bejan, 1984):

$$\theta = \mathrm{erf}\left(\frac{\eta}{2}\right), \tag{4.12}$$

$$\mathrm{Nu}_x = \frac{q''}{T_w - T_\infty}\frac{x}{k_m} = 0.564\,\mathrm{Pe}_x^{1/2}, \tag{4.13}$$

The overall Nusselt number based on the heat flux $\overline{q}''$ averaged from $x = 0$ to a given plate length $x = L$ is

$$\overline{\mathrm{Nu}_L} = \frac{\overline{q''}}{T_w - T}\frac{L}{k_m} = 1.128\,\mathrm{Pe}_L^{1/2}. \tag{4.14}$$

Cheng (1977c) found the same Nu_x result by integrating numerically the equivalent of Eqs. (4.10) and (4.11) for a wider class of problems. The similarity temperature profile (4.12) has been plotted as $(1 - \theta)$ versus η in Fig. 4.2. The effect of viscous dissipation has been included in the analysis by Magyari *et al.* (2003b). An experimental study of forced convection over a horizontal plate in a porous medium by Afifi and Berbish (1999). Magyari *et al.* (2001a) presented some exact analytical solutions for forced convection past a plane or axisymmetric body having a power-law surface distribution.

4.2. Plane Wall with Constant Heat Flux

When the surface heat flux q'' is independent of x the temperature difference $T_w - T_\infty$ increases as x in the downstream direction. This can be seen by combining the heat flux scale $q'' \sim k_m(T_w - T_\infty)/\delta_T$ with the δ_T scale (4.6), which applies to the constant q'' configuration as well. The similarity solution for the temperature distribution along and near the $y = 0$ surface was determined numerically by

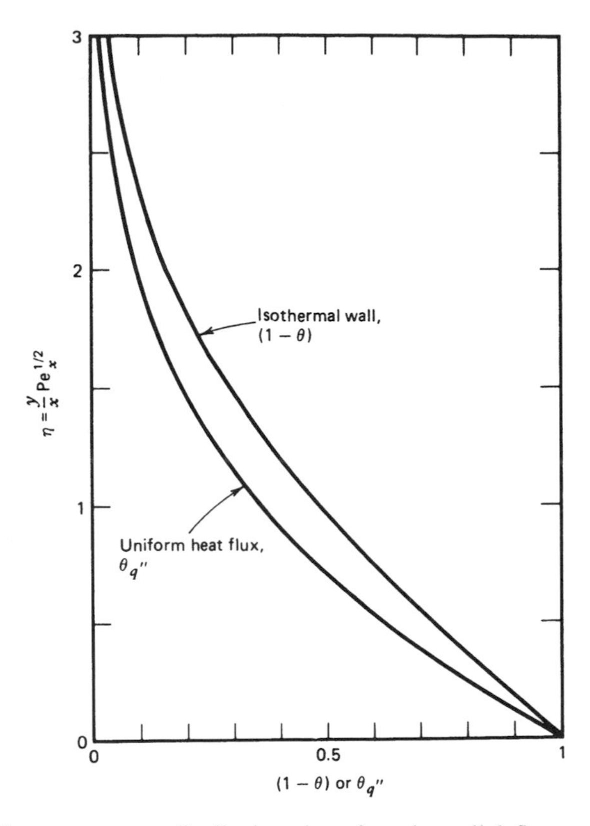

Figure 4.2. The temperature distributions in a forced parallel flow near walls with constant temperature and constant heat flux (Bejan, 1984).

Bejan (1984),

$$T(x, y) - T_{\infty} = \frac{q''/k_m}{(-d\theta_{q''}/d\eta)_{\eta=0}} \left(\frac{\alpha_m x}{U}\right)^{1/2} \theta_{q''}(\eta), \tag{4.15}$$

in which $\theta_{q''}(\eta)$ is the similarity temperature profile displayed in Fig. 4.2. The similarity variable η is defined on the ordinate of the figure. Since the calculated slope of the $\theta_{q''}$ profile at the wall is $(-d\,\theta_{q''}/d\eta)_{\eta=0} = 0.886$, the inverse of the local temperature difference can be nondimensionalized as the local Nusselt number

$$\mathrm{Nu_x} = \frac{q''}{T_w(x) - T} \frac{x}{k_m} = 0.886\,\mathrm{Pe}_x^{1/2}. \tag{4.16}$$

The overall Nusselt number that is based on the average wall temperature $\overline{T}_w$ (specifically, the temperature averaged from $x = 0$ to $x = L$) is

$$\overline{\mathrm{Nu_L}} = \frac{q''}{\overline{T}_w - T} \frac{L}{k_m} = 1.329\,\mathrm{Pe}_L^{1/2}. \tag{4.17}$$

We use this opportunity to communicate the exact solution for the problem of heat transfer from an embedded wall with uniform heat flux. The closed-form analytical alternative to the numerical solution (4.15) shown in Fig. 4.2 is

$$\frac{T(x,y)-T_\infty}{q''\,x/k_m}\,\mathrm{Pe}_x^{1/2} = 2\pi^{-1/2}\exp\left(-\frac{\eta^2}{4}\right) - \eta\,\mathrm{erfc}\left(\frac{\eta}{2}\right). \tag{4.18}$$

The right-hand side of Eq. (4.18) now replaces the function $\theta_{q''}/(-d\theta_{q''}/d\eta)_{\eta=0}$ used earlier in (4.15). This exact solution also reveals the exact values of the numerical coefficients that appear in Eqs. (4.16) and (4.17), namely $0.886 = \pi^{1/2}/2$ and $1.329 = (3/4)\pi^{1/2}$.

It is worth reviewing the Nusselt number results (4.13), (4.16), and (4.17), in order to rediscover the order-of-magnitude trend anticipated in Eq. (4.8). All these results are valid if $\delta_T \ll x$, i.e., when the Péclet number is sufficiently large so that $\mathrm{Pe}_x^{1/2} \gg 1$. The effect of variation of viscosity with temperature was studied by Ramirez and Saez (1990) and Ling and Dybbs (1992).

4.3. Sphere and Cylinder: Boundary Layers

A conceptually similar forced-convection boundary layer develops over any other body that is imbedded in a porous medium with uniform flow. Sketched in Fig. 4.3 is the thermal boundary layer region around a sphere, or around a circular cylinder that is perpendicular to the uniform flow with volume averaged velocity u. The sphere or cylinder radius is r_0 and the surface temperature is T_w.

The distributions of heat flux around the sphere and cylinder were determined by Cheng (1982), who assumed that the flow obeys Darcy's law. With reference to the angular coordinate θ defined in Fig. 4.3, Cheng obtained the following expressions for the local peripheral Nusselt number:

Sphere:

$$\mathrm{Nu}_\theta = 0.564\left(\frac{u r_0 \theta}{\alpha_m}\right)^{1/2}\left(\frac{3}{2}\theta\right)^{1/2}\sin^2\theta\left(\frac{1}{3}\cos^3\theta - \cos\theta + \frac{2}{3}\right)^{1/2}. \tag{4.19}$$

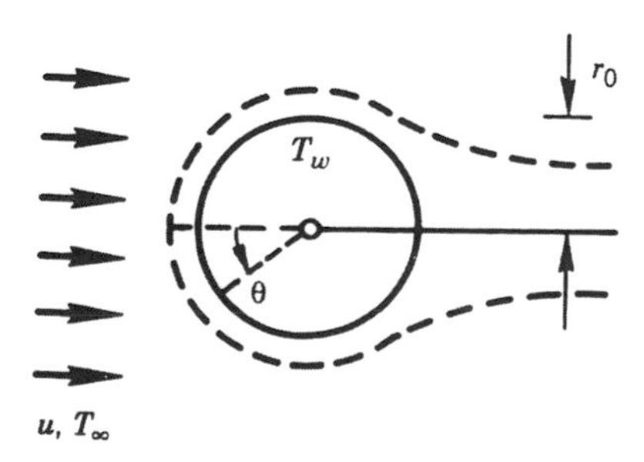

Figure 4.3. The forced-convection thermal boundary layer around a sphere or perpendicular cylinder embedded in a porous medium.

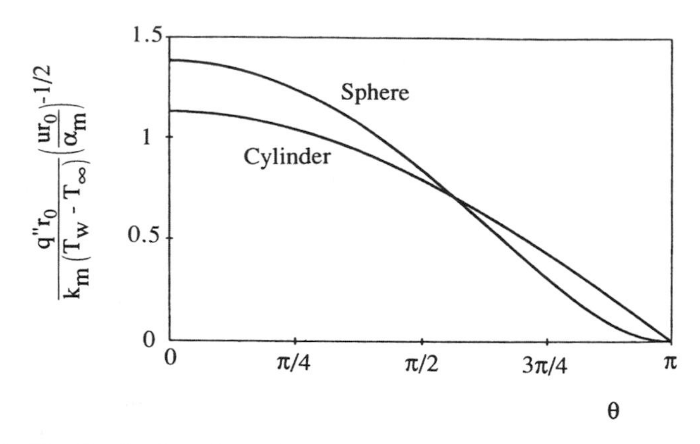

Figure 4.4. The distribution of heat flux over a cylinder or sphere with forced-convection boundary layer.

Cylinder:

$$\text{Nu}_\theta = 0.564 \left(\frac{u\, r_0\, \theta}{\alpha_m} \right)^{1/2} (2\,\theta)^{1/2} \sin\theta\, (1 - \cos\theta)^{1/2}. \tag{4.20}$$

Worth noting in these expressions is the Péclet number based on the swept arc $r_0\theta$, namely $\text{Pe}_\theta = u r_0 \theta / \alpha_m$. The local Nusselt number is defined as

$$\text{Nu}_\theta = \frac{q''}{T_w - T_\infty} \frac{r_0\, \theta}{k_m}. \tag{4.21}$$

The variation of the local heat flux over the cylinder or sphere circumference is illustrated in terms of $[q'' r_0 / k_m (T_w - T_\infty)](u r_0/\alpha_m)^{-1/2}$ versus θ in Fig. 4.4.

Equations (4.19) and (4.20) are valid when the boundary layers are distinct (thin), i.e., when the boundary layer thickness $r_0\ \text{Pe}_\theta^{1/2}$ is smaller than the radius r_0. This requirement can also be written as $\text{Pe}_\theta^{1/2} \gg 1$, or $\text{Nu}_\theta \gg 1$.

The conceptual similarity between the thermal boundary layers of the cylinder and the sphere (Fig. 4.3) and that of the flat wall (Fig. 4.1) is illustrated further by the following attempt to correlate the heat transfer results for these three configurations. The heat flux averaged over the area of the cylinder and sphere, $\overline{q}''$, can be calculated by averaging the local heat flux q'' expressed by Eqs. (4.19)–(4.21). We have done this on this occasion, and the results are:

Sphere: $$\overline{\text{Nu}_D} = 1.128\, \text{Pe}_D^{1/2}, \tag{4.22}$$

Cylinder: $$\overline{\text{Nu}_D} = 1.015\, \text{Pe}_D^{1/2}. \tag{4.23}$$

In these expressions, the Nusselt and Péclet numbers are based on the diameter $D = 2r_0$,

$$\overline{\text{Nu}_D} = \frac{\overline{q''}}{T_w - T} \frac{D}{k_m}, \qquad \text{Pe}_D = \frac{u\, D}{\alpha_m}. \tag{4.24}$$

Remarkable at this stage is the similarity between the $\overline{\mathrm{Nu}_D}$ expressions (4.22) and (4.23), and between this set and the corresponding $\overline{\mathrm{Nu}_L}$ formula for the isothermal flat wall, Eq. (4.14). The correlation of these three results is very successful because in each case the length scale used in the definition of the overall Nusselt number and the Péclet number is the dimension that is aligned with the direction of flow, the diameter in Fig. 4.3, and the length L in Fig. 4.1.

In an earlier attempt to correlate the overall heat transfer rates for these three configurations, as length scale we used Lienhard's (1973) "swept" length l, namely $l = L$ for the flat wall and $l = \pi r_0$ for the cylinder and sphere. We found that this length scale does not work nearly as well; in other words, the resulting $\overline{\mathrm{Nu}_l} \sim \mathrm{Pe}_l$ expressions change appreciably from one configuration to the next. In defense of Lienhard's length scale, however, it must be said that it was originally proposed for natural convection boundary layers, not forced convection.

The heat transfer by forced convection from a cylinder with elliptic cross section to the surrounding saturated porous medium was analyzed by Kimura (1988a). This geometry bridges the gap between the circular cylinder and the plane wall discussed in Section 4.1. The elliptic cylinder in cross-flow is in itself relevant as a model for the interaction between a uniform flow and a circular cylinder that is not perpendicular to the flow direction. The extreme case in which the circular cylinder is parallel to the flow direction was also analyzed by Kimura (1988b).

Murty *et al.* (1990) investigated non-Darcy effects and found that heat transfer from a cylinder was only weakly dependent on Darcy and Forchheimer numbers for $\mathrm{Da} < 10^{-4}$, $\mathrm{Re} < 200$.

An experimental study of heat transfer from a cylinder embedded in a bed of spherical particles, with cross-flow of air, was made by Nasr *et al.* (1994). Agreement with theory based on Darcy's law and boundary layer approximations was found to be moderately successful in predicting the data, but improved correlations were obtained with an equation modified to better account for particle diameter and conductivity variations.

For axial flow past a cylinder, an experimental study, with water and glass beads, was carried out by Kimura and Nigorinuma (1991). Their experimental results agreed well with an analysis, similar to that for the flat plate problem but with the curvature taken into account.

Heat transfer from a large sphere imbedded in a bed of spherical glass beads was studied experimentally by Tung and Dhir (1993). They concluded that the total rate of heat transfer could be predicted from the equation

$$\mathrm{Nu} = \mathrm{Nu}_{\mathrm{conduction}} + \mathrm{Nu}_{\mathrm{radiation}} + \left(\mathrm{Nu}^3_{\mathrm{natural}} + \mathrm{Nu}^3_{\mathrm{forced}}\right)^{1/3}, \tag{4.25}$$

where

$$\mathrm{Nu}_{\mathrm{forced}} = 0.29\,\mathrm{Re}^{0.8}\mathrm{Pr}^{1/2}, \qquad 0.7 \le \mathrm{Pr} \le 5, \mathrm{Re} \le 2400. \tag{4.26}$$

where Re is the Reynolds number based on the diameter of the large sphere.

Asymptotic solutions, valid for high or low (respectively) Pe, for the case of a sphere with either prescribed temperature or prescribed flux, were obtained by

Romero (1994, 1995a). Analytical solutions for large Péclet numbers for flow about a cylinder or sphere were reported by Pop and Yan (1998). Numerical simulation of forced convection past a parabolic cylinder was carried out by Haddad *et al.* (2002). MHD and viscous dissipation effects for flow past a cylinder were studied by El-Amin (2003a).

4.4. Point Source and Line Source: Thermal Wakes

In the region downstream from the hot sphere or cylinder of Fig. 4.3, the heated fluid forms a thermal wake whose thickness increases as $x^{1/2}$. This behavior is illustrated in Fig. 4.5, in which x measures the distance downstream from the heat source. Seen from the distant wake region, the imbedded sphere appears as a point source (Fig. 4.5, left), while the cylinder perpendicular to the uniform flow (u, T_∞) looks like a line source (Fig. 4.5, right).

Consider the two-dimensional frame attached to the line source q' in Fig. 4.5, right. The temperature distribution in the wake region, $T(x, y)$, must satisfy the energy conservation equation

$$u\frac{\partial T}{\partial x} = \alpha_m \frac{\partial^2 T}{\partial y^2}, \tag{4.27}$$

the boundary conditions $T \rightarrow T_\infty$ as $y \rightarrow \pm\infty$, and the integral condition

$$q' = \int_{-\infty}^{\infty} (\rho c_P)_f \, u \, (T - T_\infty) dy. \tag{4.28}$$

Restated in terms of the similarity variable η and the similarity temperature profile θ,

$$\eta = \frac{y}{x}\,\mathrm{Pe}_x^{1/2}, \qquad \theta(\eta) = \frac{T - T_\infty}{q/k_m}\mathrm{Pe}_x^{1/2}, \tag{4.29}$$

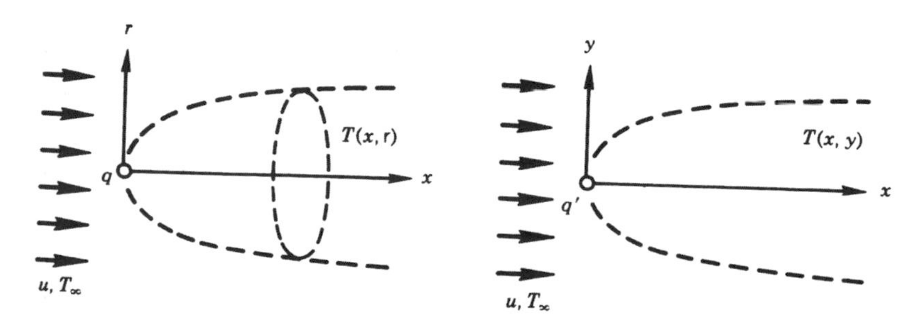

Figure 4.5. The thermal wakes behind a point source (left), and behind a line source perpendicular to the uniform flow (right).

in which $\mathrm{Pe}_x = ux/\alpha_m$, the problem statement becomes

$$-\frac{1}{2}(\theta + \eta\theta') = \theta'', \tag{4.30}$$

$$\theta \to 0 \quad \text{as} \quad \eta \to \pm\infty \tag{4.31}$$

$$\int_{-\infty}^{\infty} \theta d\eta = 1. \tag{4.32}$$

The solution can be determined analytically,

$$\theta = \frac{1}{2\pi^{1/2}} \exp\left(-\frac{\eta^2}{4}\right). \tag{4.33}$$

In terms of the physical variables, the solution is

$$T - T_\infty = 0.282 \frac{q'}{k_m} \left(\frac{\alpha_m}{ux}\right)^{1/2} \exp\left(-\frac{uy^2}{4\alpha_m x}\right). \tag{4.34}$$

In conclusion, the wake temperature distribution has a Gaussian profile in y. The width of the wake increases as $x^{1/2}$, while the temperature excess on the centerline $[T(x, 0) - T_\infty]$ decreases as $x^{-1/2}$.

The corresponding solution for the temperature distribution $T(x, r)$ in the round wake behind the point source q of Fig. 4.5, left is

$$T - T_\infty = \frac{q}{4\pi k_m x} \exp\left(-\frac{ur^2}{4\alpha_m x}\right), \tag{4.35}$$

In this case, the excess temperature on the wake centerline decreases as x^{-1}, that is more rapidly than on the centerline of the two-dimensional wake.

Both solutions, Eqs. (4.34) and (4.35), are valid when the wake region is slender, in other words when $\mathrm{Pe}_x \gg 1$. When this Péclet number condition is not satisfied, the temperature field around the source is dominated by the effect of thermal diffusion, not convection. In such cases, the effect of the heat source is felt in all directions, not only downstream.

In the limit where the flow (u, T_∞) is so slow that the convection effect can be neglected, the temperature distribution can be derived by the classic methods of pure conduction. A steady-state temperature field can exist only around the point source,

$$T(r) - T = \frac{q}{4\pi k_m r}. \tag{4.36}$$

The pure-conduction temperature distribution around the line source remains time-dependent (all the temperatures rise; e.g., Bejan, 1993, p. 181). When the time t is sufficiently long so that $(x^2 + y^2)/(4\alpha_m t) \ll 1$, the excess temperature around the line source is well approximated by

$$T(r, t) - T_\infty \cong \frac{q'}{4\pi k_m} \left[\ln\left(\frac{4\alpha_m t}{\sigma r^2}\right) - 0.5772\right]. \tag{4.37}$$

In this expression, r^2 is shorthand for $(x^2 + y^2)$. We will return to the subject of buried heat sources in Sections 5.10 and 5.11.

4.5. Confined Flow

We now consider the forced convection heat transfer in a channel or duct packed with a porous material, Fig. 4.6. In the Darcy flow regime the longitudinal volume-averaged velocity u is uniform over the channel cross section. For this reason, when the temperature field is fully developed, the relationship between the wall heat flux q'' and the local temperature difference $(T_w - T_b)$ is analogous to the formula for fully developed heat transfer to "slug flow" through a channel without a porous matrix. The temperature T_b is the mean or bulk temperature of the stream that flows through the channel (e.g., Bejan 1984, p. 83). The T_b definition for slug flow reduces to

$$T_b = \frac{1}{A} \int_A T dA, \tag{4.38}$$

in which A is the area of the channel cross section.

In cases where the confining wall is a tube with the internal diameter D, the relation for fully developed heat transfer can be expressed as a constant Nusselt number (Rohsenow and Choi, 1961):

$$\mathrm{Nu_D} = \frac{q''(x)}{T_w - T_b(x)} \frac{D}{k_m} = 5.78 \quad (\text{tube}, T_w = \text{constant}), \tag{4.39}$$

$$\mathrm{Nu}_D = \frac{q''}{T_w(x) - T_b(x)} \frac{D}{k_m} = 8 \quad (\text{tube}, q'' = \text{constant}). \tag{4.40}$$

When the porous matrix is sandwiched between two parallel plates with the spacing D, the corresponding Nusselt numbers are (Rohsenow and Hartnett, 1973)

$$\mathrm{Nu}_D = \frac{q''(x)}{T_w - T_b(x)} \frac{D}{k_m} = 4.93 \quad (\text{parallel plates}, \ T_w = \text{constant}), \tag{4.41}$$

$$\mathrm{Nu}_D = \frac{q''}{T_w(x) - T_b(x)} \frac{D}{k_m} = 6 \quad (\text{parallel plates}, \ q'' = \text{constant}). \tag{4.42}$$

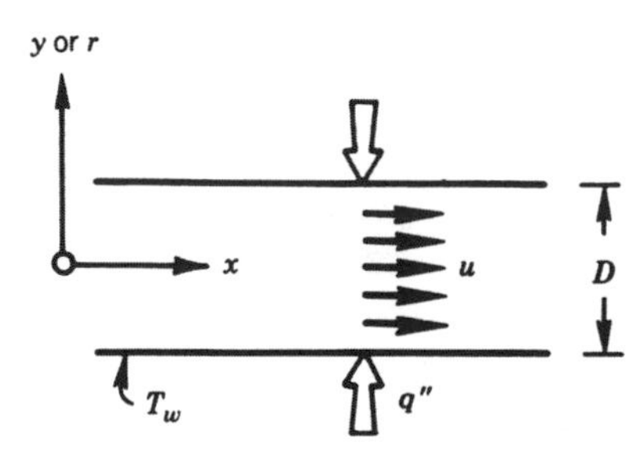

Figure 4.6. Heat transfer to the Darcy flow forced through the porous medium confined by the walls of a channel or duct.

The forced-convection results [Eqs. (4.39)–(4.42)] are valid when the temperature profile across the channel is fully developed, i.e., sufficiently far from the entrance $x = 0$ (Fig. 4.6). The entrance length, or the length needed for the temperature profile to become fully developed, can be estimated by recalling from Eq. (4.6) that the thermal boundary layer thickness scales is $(\alpha_m x/u)^{1/2}$. By setting $(\alpha_m x/u)^{1/2} \sim D$ we obtain the thermal entrance length $x_T \sim D^2 u/\alpha_m$. Inside the entrance region $0 < x < x_T$, the heat transfer is impeded by the forced-convection thermal boundary layers that line the channel walls, and can be calculated approximately with the formulas presented in Sections 4.1 and 4.2.

One important application of the results for a channel packed with a porous material is in the area of heat transfer augmentation. The Nusselt numbers for fully developed heat transfer in a channel without a porous matrix are given by expressions similar to Eqs. (4.39)–(4.42), except that the saturated porous medium conductivity k_m is replaced by the thermal conductivity of the fluid alone, k_f. The relative heat transfer augmentation effect is indicated approximately by the ratio

$$\frac{h_x \text{ (with porous matrix)}}{h_x \text{ (without porous matrix)}} \sim \frac{k_m}{k_f}, \tag{4.43}$$

in which h_x is the local heat transfer coefficient $q''/(T_w - T_b)$. In conclusion, a significant heat transfer augmentation effect can be achieved by using a high-conductivity matrix material, so that k_m is considerably greater than k_f.

An experimental study of forced convection through microporous enhanced heat sinks was reported by Lage *et al.* (2004b). An experimental study of flow of CO_2 at supercritical pressure was carried out by Jiang *et al.* (2004i,j). Correlations for forced convection between two parallel plates or in a circular pipe were obtained by Haji-Sheikh (2004). A numerical study, using a Green's function solution method and dealing with the effects due to a temperature change at the wall and the contributions of frictional heating, was conducted by Haji-Sheikh *et al.* (2004a). The role of longitudinal diffusion in fully developed forced convection slug flow in a channel was studied by Nield and Lage (1998). Forced convection in a helical pipe was analyzed by Nield and Kuznetsov (2004b). Curvature of the pipe induces a secondary flow at first order and increases the Nusselt number at second order, while torsion affects the velocity at second order and does not affect the Nusslet number at second order. A numerical study of this problem was made by Cheng and Kuznetzov (2005).

4.6. Transient Effects

Most of the existing work on forced convection in fluid-saturated porous media is concerned with steady-state conditions. Notable exceptions are the papers on time-dependent forced convection heat transfer from an isothermal cylinder (Kimura, 1989a) and from a cylinder with uniform heat flux (Kimura, 1988c). Nakayama and Ebinuma (1990) studied the forced convection heat transfer between a suddenly heated plate and a non-Darcy flow that starts initially from rest.

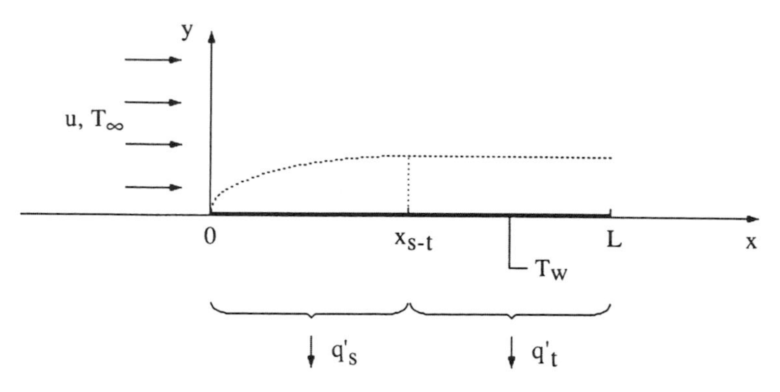

Figure 4.7. Forced-convection thermal boundary layer near a plate embedded in a porous medium with steady, parallel, and uniform flow.

These three papers show that the simplest and perhaps most important forced convection configuration has been overlooked. In that configuration, the flow through the saturated porous medium is steady, parallel, and uniform (Bejan and Nield, 1991). The flow is driven by a pressure difference that is applied in the x direction in Fig. 4.7, and can be either a Darcy flow or a non-Darcy flow in which the quadratic drag (Forchheimer effect) plays a role in the overall flow resistance. What distinguishes the Bejan and Nield (1991) configuration from the one analyzed by Nakayama and Ebinuma (1990) is that the flow is and remains steady as the embedded plate is suddenly heated or cooled to a different temperature.

4.6.1. Scale Analysis

Consider the uniform flow, with volume-averaged velocity u, which is parallel to the wall $y = 0$ shown in Fig. 4.7. The initial temperature of the fluid-saturated porous medium is T_∞. Beginning at time $t = 0$, the temperature of the wall section $0 < x < L$ is maintained at a different constant temperature, T_w. In time, the flow in the fluid-saturated porous medium adjusts to this change by developing a near-wall region wherein the variation from T_w to T_∞ is smoothed.

We can develop a feel for the size and history of the near-wall region by examining the order of magnitude implications of the energy equation for that region,

$$\sigma \frac{\partial T}{\partial t} + u \frac{\partial T}{\partial x} = \alpha_m \frac{\partial^2 T}{\partial y^2}. \tag{4.44}$$

The temperature boundary conditions are as indicated in Fig. 4.7, specifically

$$T = T_w \quad \text{at} \quad y = 0 \tag{4.45}$$

$$T \to T_\infty \quad \text{as} \quad y \to \infty \tag{4.46}$$

Implicit in the writing of the energy equation (4.42) is the assumption that the near-wall region is slender, or boundary layer-like. To this assumption we will return in Eqs. (4.62)–(4.65).

One way to perform the scale analysis is by considering the entire boundary layer region of length L. The thickness of this thermal boundary layer is denoted by δ. If we further write $\Delta T = T_\infty - T_w$, we find the following scales for the three terms of Eq. (4.42):

$$\begin{array}{ccc} \sigma\dfrac{\Delta T}{t}, & u\dfrac{\Delta T}{L}, & \alpha_m\dfrac{\Delta T}{\delta^2}. \\ \text{thermal} & \text{longitdinal} & \text{transverse} \\ \text{inertia} & \text{convection} & \text{conduction} \end{array} \tag{4.47}$$

At sufficiently short times t, the transverse heating effect is balanced by the thermal inertia of the saturated porous medium. This balance yields the time-dependent thickness

$$\delta_t \sim \left(\frac{\alpha_m t}{\sigma}\right)^{1/2}. \tag{4.48}$$

As t increases, the thermal inertia scale decreases relative to the longitudinal convection scale, and the energy equation becomes ruled by a balance between transverse conduction and longitudinal convection. The steady-state boundary layer thickness scale in this second regime is

$$\delta_s \sim \left(\frac{\alpha_m L}{u}\right)^{1/2}. \tag{4.49}$$

The time of transition t_c, when the boundary layer region becomes convective, can be estimated by setting $\delta_t \sim \delta_s$:

$$t_c \sim \frac{\sigma L}{u}. \tag{4.50}$$

Not all of the L-long boundary layer is ruled by the balance between conduction and inertia when t is shorter than T_c. When t is finite, there is always a short enough leading section of length x in which the energy balance is between transverse conduction and longitudinal convection. In that section of length x and thickness δ_x, the scales of the three terms of Eq. (4.44) are

$$\sigma\frac{\Delta T}{t}, \quad u\frac{\Delta T}{x}, \quad \alpha_m\frac{\Delta T}{\delta_x^2}, \tag{4.51}$$

showing that $u\Delta T/x \sim \alpha_m \Delta T/\delta_x^2$, or

$$\delta_x \sim \left(\frac{\alpha_m x}{u}\right)^{1/2} \tag{4.52}$$

when $\sigma\Delta T/t < u\Delta T/x$, i.e., when

$$x < \frac{u\,t}{\sigma}. \tag{4.53}$$

The boundary layer changes from the convective (steady) section represented by Eq. (4.52) to the conductive (time-dependent) trailing section of Eq. (4.48). The change occurs at $x = x_{s-t}$ where

$$x_{s-t} \sim \frac{u\,t}{\sigma}. \tag{4.54}$$

4.6.2. Wall with Constant Temperature

The two-section structure of the thermal boundary layer is indicated in Fig. 4.7. Its existence was also recognized by Ebinuma and Nakayama (1990b) in the context of transient film condensation on a vertical surface in a porous medium. The chief benefit of this insight is that it enables us to delineate the regions in which two analytical solutions are known to apply: first the steady leading section where according to Eqs. (4.9)–(4.12)

$$\frac{T - T_w}{T_\infty - T_w} = \operatorname{erf}\left[\frac{y}{2}\left(\frac{u}{\alpha_m x}\right)^{1/2}\right] \quad (x < x_{s-t}) \tag{4.55}$$

and farther downstream the time-dependent section where

$$\frac{T - T_w}{T_\infty - T_w} = \operatorname{erf}\left[\frac{y}{2}\left(\frac{\sigma}{\alpha_m t}\right)^{1/2}\right] \quad (x > x_{s-t}). \tag{4.56}$$

The time-dependent section is no longer present when $x_{s-t} \sim L$, i.e., when $t \sim \sigma L/u$, in accordance with Eq. (4.50).

We see from the condition (4.52) that the temperature distributions (4.55) and (4.56) match at $x = x_{s-t}$. The longitudinal temperature gradient $\partial T/\partial x$ experiences a discontinuity across the $x = x_{s-t}$ cut, but this discontinuity becomes less pronounced as t increases, i.e., as the x_{s-t} cut travels downstream. It also must be said that neither Eq. (4.55) nor (4.56) is exact at $x = x_{s-t}$, because at that location none of the three effects competing in Eq. (4.45) can be neglected.

The instantaneous heat transfer rate (W/m) through the surface of length L can be deduced by taking the heat transfer rate through the leading (steady-state) section $0 < x < x_{s-t}$, cf. Eq. (4.14),

$$q_s' = k_m (T_\infty - T_w)\frac{2}{\pi^{1/2}}\left(\frac{u}{\alpha_m} x_{s-t}\right)^{1/2} \tag{4.57}$$

and adding to it the contribution made by the time-dependent trailing section $x_{s-t} < x < L$:

$$q_t' = (L - x_{s-t})\frac{k_m(T_\infty - T_w)}{(\pi\alpha_m t/\sigma)^{1/2}}. \tag{4.58}$$

The total heat transfer rate $q' = q_s' + q_t'$ can be compared with the long-time (steady-state) heat transfer rate of the L-long plate,

$$q'_{final} = k(T_\infty - T_w)\frac{2}{\pi^{1/2}}\left(\frac{u}{\alpha_m} L\right)^{1/2} \tag{4.59}$$

and the resulting expression is

$$\frac{q'}{q'_{final}} = 1 + \frac{1 - \tau}{2\tau^{1/2}}. \tag{4.60}$$

In this expression τ is the dimensionless time

$$\tau = \frac{ut}{\sigma L}. \tag{4.61}$$

According to Eq. (4.50), $\tau = 1$ marks the end of the time interval in which Eq. (4.60) holds. The beginning of that time interval is dictated by the validity of the assumption that the leading (steady-state) section of the boundary layer is always slender, cf. Eq. (4.49),

$$\left(\frac{\alpha_m x_{s-t}}{u}\right)^{1/2} < x_{s-t}. \tag{4.62}$$

This requirement translates into

$$\frac{u x_{s-t}}{\alpha_m} > 1 \tag{4.63}$$

or, in view of Eqs. (4.54) and (4.61),

$$\tau > \frac{1}{\mathrm{Pe}_L}, \tag{4.64}$$

where Pe_L is the Péclet number based on L,

$$\mathrm{Pe}_L = \frac{uL}{\alpha_m}. \tag{4.65}$$

At times τ shorter than $1/\mathrm{Pe}_L$, the leading section is not a forced convection boundary layer, and the entire L length produces a time-dependent heat transfer rate of type (4.58):

$$q' = L \frac{k_m (T_\infty - T_w)}{(\pi \alpha_m t/\sigma)^{1/2}}. \tag{4.66}$$

The dimensionless counterpart of this estimate is

$$\frac{q'}{q'_{final}} = \frac{1}{2\,\tau^{1/2}}. \tag{4.67}$$

In summary, the total heat transfer rate is given by three successive expressions, each for one regime in the evolution of the temperature field near the suddenly heated plate:

$$\frac{q'}{q'_{final}} = \begin{cases} \dfrac{1}{2\tau^{1/2}}, & 0 < \tau < \mathrm{Pe}_L^{-1} \\[2ex] 1 + \dfrac{1-\tau}{2\tau^{1/2}}, & \mathrm{Pe}_L^{-1} < \tau < 1 \\[2ex] 1, & \tau > 1. \end{cases} \tag{4.68}$$

The domain occupied by each regime is indicated on the (Pe_L, τ) plane of Fig. 4.8. The approximate solution (4.66) shows that relative to the long-time result (4.59), the transient heat transfer rate depends on two additional dimensionless groups, τ and Pe_L.

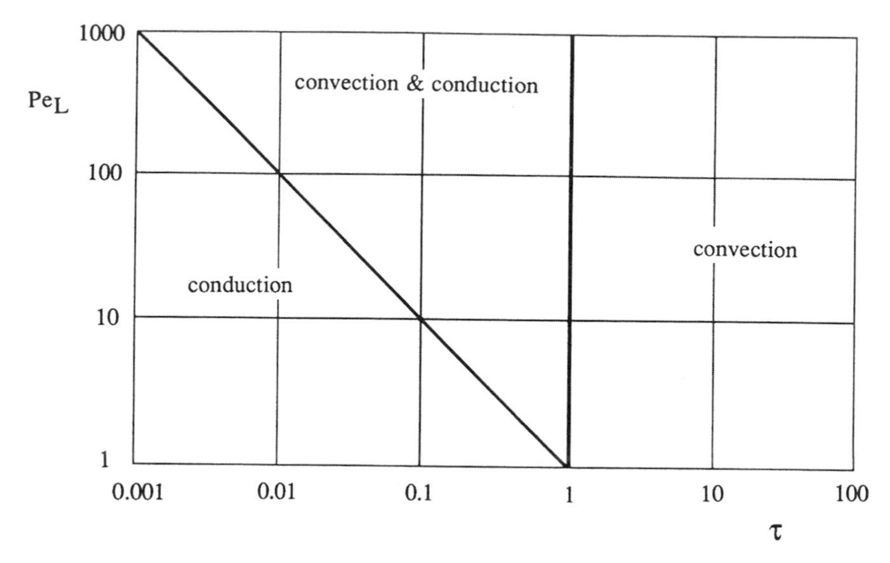

Figure 4.8. The τ-Pe_L ranges in which the three parts of the solutions (4.66) and (4.68) are applicable.

4.6.3. Wall with Constant Heat Flux

The thermal boundary layer formed in the vicinity of a plate with sudden heat flux q'' can be described in a way that is analogous to the analysis presented between Eqs. (4.55) and (4.68). The structure shown in Fig. 4.7 is present here as well, and Eqs. (4.54) and (4.61) continue to hold. The upstream portion $0 < x < x_{s-t}$ closely approximates the steady forced convection boundary layer with uniform heat flux (Section 4.2). The downstream section $x_{s-t} < x < L$ is dominated by time-dependent conduction into a semi-infinite medium with uniform heat flux at the surface.

The history of the L-averaged temperature of the wall or of the L-averaged wall-medium temperature difference $\overline{\Delta T}$ approaches [cf. Eq. (4.17)] the value

$$\overline{\Delta T}_{final} = \frac{4}{3\pi^{1/2}} \frac{q''L}{k_m} \left(\frac{uL}{\alpha_m}\right)^{1/2}. \tag{4.69}$$

Expressed in dimensionless form, the L-averaged temperature difference is

$$\frac{\overline{\Delta T}}{\overline{\Delta T}_{final}} \cong \begin{cases} \frac{3}{2}\tau^{1/2}, & 0 < Pe_L^{-1} \\ \left(\frac{3}{3} - \frac{\tau}{2}\right)\tau^{1/2}, & Pe_L^{-1} < \tau < 1 \\ 1, & \tau > 1. \end{cases} \tag{4.70}$$

The solutions (4.66) and (4.68) are based on the assumption that $\mathrm{Pe}_L \gg 1$. For example, Eq. (4.66) shows that the heat transfer ratio q'/q'_{final} experiences a change of relative magnitude O (Pe_L^{-1}) at $\tau = \mathrm{Pe}_L^{-1}$. The same observation applies to the $\Delta T/\Delta T_{final}$ ratio of Eq.(4.68).

Unsteady forced convection on a flat plate, with the effect of inertia and thermal dispersion accounted for, was analyzed by Cheng and Lin (2002). The dispersion accelerates the rate of unsteady heat transfer but does not affect the response time to reach a steady state.

4.6.4. Other Configurations

Kimura (1989b) has studied transient forced convection about a vertical cylinder. He obtained analytic solutions for small time (conduction solution) and large time (boundary layer solution) and numerical results for the general time situation. Thevenin (1995) performed other calculations.

Al-Nimr *et al.* (1994a,b) have investigated numerically convection in the entrance region of either a tube or an annulus, when a timewise step change of wall temperature is imposed, for Darcy and non-Darcy models. A conjugate problem involving concentric annuli was studied numerically by El-Shaarawi *et al.* (1999). Alkam and Al-Nimr (1998) performed a numerical simulation of transient forced convection in a circular pipe partly filled with a porous substrate. Unsteady forced convection about a sphere was studied numerically by Yan and Pop (1998). H. L. Fu *et al.* (2001) studied experimentally heat transfer in a channel subject to oscillating flow, while Mohamad and Karim (2001) reported experiments in a pipe with core and sheath occupied by different porous materials.

In a series of papers, Kuznetsov (1994, 1995a,b, 1996b-f, 1998e) has investigated the effect of local thermal non-equilibrium on heat transfer, for the problem when a porous bed is initially at a uniform temperature and then suddenly subjected to a step increase of fluid inlet temperature. The locally averaged fluid velocity v is assumed to be uniform in space and constant in time. The analytical solution obtained by Kuznetsov, using a perturbation method based on the assumption that the fluid-to-solid heat transfer coefficient is large, shows that the temperature of the fluid (T_f) or solid (T_s) phase takes the form of an advancing front, while the temperature difference $T_f - T_s$ takes the form of an advancing pulse. The amplitude of that pulse decreases as the pulse propagates downstream. Kuznetsov treated in turn a one-dimensional semi-infinite region, a one-dimensional finite region, a two-dimensional rectangular region, a circular tube, a concentric tube annulus, and a three-dimensional rectangular box. In the one-dimensional semi-infinite case the wave speed v_{wave} is related to the fluid flow speed v by

$$v_{wave} = \frac{(\rho c)_f}{\varphi(\rho c)_f + (1-\varphi)(\rho c)_s} v. \tag{4.71}$$

In the two-dimensional and three-dimensional cases the amplitude of the pulse also decreases from the central flow region to the walls of the packed bed. Kuznetsov's (1996c) paper deals with a one-dimensional slab with a fluid-to-solid heat transfer coefficient (something whose value is difficult to determine experimentally) that varies about a mean value in a random fashion. He calculated the mean and standard deviation of $T_f - T_s$.

The effects of thermal nonequilibrium have been included in numerical simulations by Sözen and Vafai (1990, 1993), Vafai and Sözen (1990a,b), Amiri and Vafai (1994), and Amiri *et al.* (1995), e.g., in connection with the condensing flow of a gas or longitudinal heat dispersion in a gas flow in a porous bed. They found that the local thermal equilibrium condition was very sensitive to particle Reynolds number and Darcy number, but not to thermophysical properties. Amiri and Vafai (1998) and Wu and Hwang (1998) performed further numerical simulations.

4.7. Effects of Inertia and Thermal Dispersion: External Flow

When quadratic drag is taken into account, the Darcy equations (4.2) are replaced by the approximate equations

$$u + \frac{\chi}{v} u^2 = -\frac{K}{\mu}\frac{\partial P}{\partial x}, \quad v = -\frac{K}{\mu}\frac{\partial P}{\partial y} \tag{4.72}$$

for the case when the primary flow is in the x direction, so $v/u \ll 1$. Here $\chi = c_F K^{1/2}$, where c_F was introduced in Eq. (1.12). Eliminating P from these equations and introducing the stream-function ψ defined by $u = \partial\psi/\partial y$, $v = -\partial\psi/\partial x$ so that Eq. (4.1) is satisfied, we obtain

$$\frac{\partial^2\psi}{\partial y^2} + \frac{\chi}{v}\frac{\partial}{\partial y}\left[\left(\frac{\partial\psi}{\partial y}\right)^2\right] = 0, \tag{4.73}$$

and Eq. (4.3) becomes

$$\frac{\partial\psi}{\partial y}\frac{\partial T}{\partial x} - \frac{\partial\psi}{\partial x}\frac{\partial T}{\partial y} = \alpha_m \frac{\partial^2 T}{\partial y^2}. \tag{4.74}$$

If one considers the case where $T_w = T_\infty + Ax^\lambda$, $U_\infty = Bx^m$, where A, B, λ, and m are constants, one finds that a similarity solution is possible if and only if $m = 0$ and $\lambda = 1/2$. One can check that the similarity solution is given by

$$\psi = (\alpha_m U_\infty x)^{1/2} f(\eta), \tag{4.75}$$

$$T - T_\infty = (T_w - T)\,\theta(\eta), \tag{4.76}$$

$$\eta = \left(\frac{U_\infty x}{\alpha_m}\right)^{1/2} \frac{y}{x}, \tag{4.77}$$

provided that f and η satisfy the differential equations

$$f'' + R^* [(f')^2] = 0, \tag{4.78}$$

$$\theta'' = \frac{1}{2}(f'\theta - f\theta'), \tag{4.79}$$

where

$$R^* = \frac{\chi U_\infty}{\nu}. \tag{4.80}$$

The boundary conditions

$$y = 0: T = T_w, \quad v = 0, \tag{4.81}$$

$$y \to \infty: T = T_w, \quad u = U_\infty, \tag{4.82}$$

lead to

$$\theta(0) = 1, f(0) = 0, \theta(\infty) = 0, f'(\infty) = 1. \tag{4.83}$$

The local wall heat flux is

$$q'' = -k_m \left(\frac{\partial T}{\partial y}\right)_{y=0} = -k_m A \left(\frac{B}{\alpha_m}\right)^{1/2} \theta'(0), \tag{4.84}$$

where $\theta'(0) = -0.886$. We recognize that this is the case of constant wall heat flux. In nondimensional form this result is precisely the same as Eq. (4.16) and is independent of the value of R^*. Thus in this case quadratic drag has no effect on the wall heat flux (for fixed U_∞), but it does have the effect of flattening the dimensionless velocity profile (Lai and Kulacki, 1987).

The effect of thermal dispersion in the same case was discussed by Lai and Kulacki (1989a). In the present context it is the transverse component that is important. If one allows for thermal dispersion by adding a term Cud_p (where d_p is the mean particle or pore diameter and C is a numerical constant) to α_m in the term $\alpha_m \partial^2 T/\partial y^2$ in Eq.(4.3), then Eq.(4.16) is replaced by

$$\mathrm{Nu}_x = 0.886\,(1 + C\,\mathrm{Pe}_d)\,\mathrm{Pe}_x^{1/2}, \tag{4.85}$$

where $\mathrm{Pe}_d = U_\infty\, d_p/\alpha_m$. Thus thermal dispersion increases the heat transfer because it increases the effective thermal conductivity in the y direction.

The effect of quadratic drag in the transient situation for the case of constant wall temperature was examined by Nakayama and Ebinuma (1990), who found that it had the effect of slowing the rate at which a steady-state solution is approached. One can deduce from their steady-state formulas that (as for the constant flux situation) quadratic drag does not affect the Nu_x (Pe_x) relationship, in this book the formula (4.13).

4.8. Effects of Boundary Friction and Porosity Variation: Exterior Flow

When one introduces the Brinkman equation in order to satisfy the no-slip condition on a rigid boundary, one runs into a complex problem. The momentum equation no longer has a simple solution, and a momentum boundary layer problem must be treated. For the purposes of this discussion, we follow Lauriat and Vafai (1991) and take the boundary layer form of the momentum equation

$$\frac{1}{\varphi^2}\left(u\frac{\partial u}{\partial x}+v\frac{\partial u}{\partial y}\right)=\frac{v}{K}(U-u)+\frac{c_F}{K^{1/2}}(U^2-u^2)+\frac{v}{\varphi}\frac{\partial^2 u}{\partial y^2}. \quad (4.86)$$

For the reasons pointed out in Section 1.5, we drop the left-hand side of this equation at the outset, and in the last term we replace φ^{-1} by $\widetilde{\mu}/\mu$. The condition on a plane wall is now

$$u=v=0,\ T=T_w \quad \text{for } x>0,\ y=0. \quad (4.87)$$

The remaining equations and boundary conditions are unaltered.

The integral method, as used by Kaviany (1987), provides an approximate solution of the system. If the velocity profile is approximated by

$$u=U_\infty\left[\frac{3}{2}\frac{y}{\delta}-\frac{1}{2}\left(\frac{y}{\delta}\right)^3\right], \quad (4.88)$$

one finds that the momentum boundary layer thickness δ is given by

$$\frac{\delta^2}{K/\varphi}=\frac{140}{(35+48c_F\,\mathrm{Re}_p)}(1-e^{-\gamma x^*}) \quad (4.89)$$

where

$$\mathrm{Re}_p=U_\infty K^{1/2}/v \quad (4.90)$$

is the pore Reynolds number

$$\gamma=\left(\frac{70}{13}\frac{1}{\mathrm{Re}_p}+\frac{96}{13}c_F\right)\varphi^{3/2}, \quad (4.91)$$

and

$$x^*=\frac{x}{(K/\varphi)^{1/2}}. \quad (4.92)$$

The momentum boundary layer thickness δ is almost constant when $x^*>5/\gamma$. Thus the hydrodynamic development length can be taken as

$$x_e=\frac{5}{\gamma}\left(\frac{K}{\varphi}\right)^{1/2} \quad (4.93)$$

and the developed momentum boundary layer thickness is given by

$$\delta=\left[\left(\frac{140}{35+48\,c_F\,\mathrm{Re}_p}\right)\frac{K}{\varphi}\right]^{1/2}. \quad (4.94)$$

For the developed region, exact solutions have been obtained by Cheng (1987), Beckermann and Viskanta (1987), and Vafai and Thiyagaraja (1987). They show that the velocity is constant outside a boundary layer whose thickness decreases as c_F and/or Re_p increases, in accordance with Eq. (4.86).

Wall effects caused by nonuniform porosity (Section 1.7) have been investigated experimentally by a number of investigators and theoretically by Vafai (1984, 1986), Vafai *et al.* (1985), and Cheng (1987). The degree to which hydrodynamic wall effects influence the heat transfer from a heated wall depends on the Prandtl number Pr of the fluid. The ratio of the thermal boundary layer thickness δ_T to the momentum boundary layer thickness δ is of order Pr^{-1}. For low Prandtl number fluids ($\mathrm{Pr} \to 0$), $\delta \ll \delta_T$ and the temperature distribution, and hence the heat transfer, is given by the Darcy theory of Sections 4.1 and 4.2. For a more general case where the inertial effects are taken into account and for a variable wall temperature in the form $T_w = T_\infty + Ax^p$, an exact solution was obtained by Vafai and Thiyagaraja (1987) for low Prandtl number fluids in terms of gamma and parabolic cylindrical functions. They found the temperature distribution to be

$$\begin{aligned} T = T_\infty &+ A\Gamma(p+1) \\ &\times \left\{2^{p+1/2}\pi^{-1/2}x^p \exp(-\alpha y^2/x)\, D_{-(2p+1)}[(4\alpha y^2/x)^{1/2}]\right\}, \end{aligned} \tag{4.95}$$

where $\alpha = U_\infty/8\alpha_m$. The corresponding local Nusselt number is

$$\mathrm{Nu_x} = \frac{\Gamma(\mathrm{p}+1)}{\Gamma(\mathrm{p}+1/2)}(\mathrm{Re_p\,Pr_e})^{1/2}, \qquad \mathrm{Da_x^{-1/4}} = \frac{\Gamma(\mathrm{p}+1)}{\Gamma(\mathrm{p}+1/2)}\mathrm{Pe_x^{1/2}}, \tag{4.96}$$

which reduces to Eq. (4.13) when $p = 0$.

When the Prandtl number is very large, $\delta_T \ll \delta$ and so the thermal boundary layer lies completely inside the momentum boundary layer. As $\mathrm{Pr} \to \infty$ one can assume that the velocity distribution within the thermal boundary layer is linear and given by

$$u = \frac{\tau_w y}{\mu_f}, \tag{4.97}$$

where τ_w is the wall stress which is given by

$$\tau_w = \frac{\mu_f U_\infty}{(K/\varphi)^{1/2}}\left(1 + \frac{4}{3}c_F\,\mathrm{Re}_p\right)^{1/2}. \tag{4.98}$$

This means that the energy equation can be approximated by

$$y\frac{\partial T}{\partial x} = \frac{\alpha_m \mu_f}{\tau_w}\frac{\partial^2 T}{\partial y^2}. \tag{4.99}$$

We now introduce the similarity variables

$$\eta = y\left(\frac{1}{9\xi x}\right)^{1/3}, \qquad \theta(\eta) = \frac{T - T_w}{T_\infty - T_w}, \tag{4.100}$$

where

$$\xi = \frac{\alpha_m \mu_f}{\tau_w} = \frac{K}{\mathrm{Re}_p\,\mathrm{Pr}_e}\left[\varphi\left(1 + \frac{4}{3}c_F\,\mathrm{Re}_p\right)\right]^{-1/2} \tag{4.101}$$

and where the *effective* Prandtl number Pr_e is defined as

$$\mathrm{Pr}_e = \frac{\nu}{\alpha_m}. \tag{4.102}$$

We then have the differential equation system

$$\theta'' + 3\eta^2\theta' = 0, \tag{4.103}$$

$$\theta(0) = 0, \quad \theta(\infty) = 1, \tag{4.104}$$

which has the solution (Beckermann and Viskanta, 1987)

$$\theta = \frac{1}{\Gamma(4/3)} \int_0^\eta e^{-\xi^3}\, d\xi. \tag{4.105}$$

Hence the local Nusselt number is

$$\begin{aligned} \mathrm{Nu}_x &= \frac{k_m (\partial T/\partial y)_{y=0}}{k_m (T_w - T_\infty)/x} = 1.12 \left(\frac{x^2}{9\xi}\right)^{1/3} \\ &= 0.538 \left[\varphi\left(1 + \frac{4}{3} c_F \mathrm{Re}_p\right)\right]^{1/6} \left(\frac{\mathrm{Re}_p \mathrm{Pr}_e}{\mathrm{Da}_x}\right)^{1/3} \end{aligned} \tag{4.106}$$

and the overall Nusselt number over a length L from the leading edge becomes

$$\overline{\mathrm{Nu}} = 1.68 \left(\frac{L^2}{9\xi}\right)^{1/3}. \tag{4.107}$$

Vafai and Thiyagaraja (1987) have compared these analytical results with numerical solutions. They found that the low Prandtl number analytical solution accurately predicts the temperature distribution for a Prandtl number Pr_e as high as 8, while the high-Pr_e analytical solution is valid for Pr_e as low as 100 and possibly for somewhat lower values.

The combined effects of inertia and boundary friction were considered by Kaviany (1987). He expressed his results in terms of a parameter Γ_x defined as the total flow resistance per unit volume (Darcy plus Forchheimer drag) due to the solid matrix, scaled in terms of $8\rho U_\infty^2/3\varphi x$. He concluded that the "Darcian regime" where Nu_x varies as $\mathrm{Pr}_e^{1/2}$ holds when $\Gamma_x > 0.6\,\mathrm{Pr}_e$ and the "non-Darcian regime" where Nu_x varies as $\mathrm{Pr}_e^{1/3}$ holds when $0.07 < \Gamma_x < 0.6\,\mathrm{Pr}_e$. When $\Gamma_x = 0.07$ the presence of the solid matrix is not significant. Another study is that by Kumari *et al.* (1990c).

Vafai *et al.* (1985) experimentally and numerically investigated the effects of boundary friction and variable porosity. Their experimental bed consisted of glass beads of 5 mm and 8 mm diameter saturated with water. They found good agreement between observation of the average Nusselt number and numerical predictions when the effect of variable porosity was included (but not otherwise). Cheng (1987) noted that since their experiments were conducted in the range $100 < \mathrm{Re}_p < 900$, thermal dispersion effects should have been important, and in fact they neglected these. He pointed out that in their numerical work Vafai *et al.* (1985) used a value of thermal conductivity about three times larger than was warranted, and by doing so they had fortuitously approximated the effect of transverse thermal dispersion.

Further experimental work was undertaken by Renken and Poulikakos (1989). They reported details of thermal boundary layer thickness, temperature field, and local Nusselt number. Good agreement was found with the numerical results of Vafai *et al.* (1985) with the effects of flow inertia and porosity variation accounted for.

Some further details on the content of this section can be found in the review by Lauriat and Vafai (1991). Nakayama *et al.* (1990a) used novel transformed variables to produce a local similarity solution for flow over a plate. Vafai and Kim (1990) have analyzed flow in a composite medium consisting of a fluid layer overlaying a porous substrate that is attached to the surface of a plate. Luna and Mendez (2005) have used a Brinkman model to study analytically and numerically the conjugate problem of forced convection on a plate with finite thermal conductivity and with constant heat flux at the extreme boundary.

For the case of cross flow across a cylinder, Fand *et al.* (1993) obtained empirical correlation expressions for the Nusselt number. For the same geometry, a numerical study was made by Nasr *et al.* (1995). They reported that the effect of decreasing Da was an increase in Nu, but Lage and Nield (1997) pointed out that this is true only if the Reynolds number Re is held constant. If the pressure gradient is kept constant, Nu increases with Da. Nasr *et al.* (1995) also noted that Nu increased with increase of either Re or effective Prandtl number, and that the effect of quadratic drag on Nu is via the product DaRe.

Heat transfer around a periodically-heated cylinder was studied experimentally (with water and glass beads) and numerically by Fujii *et al.* (1994). They also modeled the effects of thermal dispersion and thermal nonequilibrium.

Unsteady forced convection, produced by small amplitude variations in the wall temperature and free stream velocity, along a flat plate was studied by Hossain *et al.* (1996)

4.9. Effects of Boundary Friction, Inertia, Porosity Variation, and Thermal Dispersion: Confined Flow

In porous channels the velocity field generally develops to its steady-state form in a short distance from the entrance. To see this, let t_c be a characteristic time for development and u_c a characteristic velocity. During development the acceleration term is of the same order of magnitude as the Darcy resistance term, so $u_c/t_c \sim \nu u_c/K$, and so the development length $\sim t_c u_c \sim K u_c/\nu$, which is normally small. [Note that, in contrast with the argument used by Vafai and Tien (1981), the present argument holds whether or not the convective inertial term is negligible.] Further, the numerical results of Kaviany (1985) for flow between two parallel plates show that the entrance length decreases linearly as the Darcy number decreases. In this section we assume that the flow is also fully developed thermally.

We start by considering a channel between two plane parallel walls a distance $2H$ apart, the boundaries being at $y = H$ and $y = -H$. For fully developed flow the velocity is $u(y)$ in the x-direction. We suppose that the governing equations are

$$G = \frac{\mu u^*}{K} + \frac{c_F \rho u^{*^2}}{K^{1/2}} - \tilde{\mu}\frac{d^2 u^*}{dy^{*^2}}, \tag{4.108}$$

$$u^* \frac{\partial T^*}{\partial x^*} = \frac{k_m}{(\rho c_P)_f} \frac{\partial^2 T^*}{\partial y^{*^2}}. \tag{4.109}$$

Here the asterisks denote dimensional variables, and G is the applied pressure gradient. Local thermal equilibrium has been assumed, dispersion is neglected, and it is assumed that the Péclet number is sufficiently large for the axial thermal conduction to be insignificant. We define the dimensionless variables

$$x = \frac{x^*}{H}, \quad y = \frac{y^*}{H}, \quad u = \frac{\tilde{\mu} u^*}{GH^2}, \tag{4.110}$$

and write

$$M = \frac{\tilde{\mu}}{\mu}, \ \mathrm{Da} = \frac{K}{H^2}, \ F = \frac{c_F \rho G H^4}{K^{1/2}\mu^2}. \tag{4.111}$$

Thus M is a viscosity ratio, Da is a Darcy number, and F is a Forchheimer number. Then Eq. (4.108) becomes

$$M\frac{d^2 u}{dy^2} - \frac{u}{\mathrm{Da}} - Fu^2 + 1 = 0. \tag{4.112}$$

This equation is to be solved subject to the boundary/symmetry conditions

$$u = 0 \text{ at } y = 1, \quad \frac{du}{dy} = 0 \ \text{ at } \ y = 0. \tag{4.113}$$

When F is not zero, the solution can be expressed in terms of standard elliptic functions (Nield *et al.*, 1996). When $F = 0$, the solution is

$$u = \mathrm{Da}\left(1 - \frac{\cosh Sy}{\cosh S}\right), \tag{4.114}$$

where for convenience we introduce

$$S = \frac{1}{(M\mathrm{Da})^{1/2}}. \tag{4.115}$$

We also introduce the mean velocity U^* and the bulk mean temperature $T_m{}^*$ defined by

$$U^* = \frac{1}{H}\int_0^H u^* dy^*, \quad T_m{}^* = \frac{1}{HU^*}\int_0^H u^* T^* dy^*. \tag{4.116}$$

We then define further dimensionless variables defined by

$$\hat{u} = \frac{u^*}{U^*}, \quad \hat{T} = \frac{T^* - T_w{}^*}{T_m{}^* - T_w{}^*}, \tag{4.117}$$

and the Nusselt number

$$\mathrm{Nu} = \frac{2Hq''}{k_m(T_m{}^* - T_w{}^*)}. \tag{4.118}$$

Here $T_w{}^*$ and q'' are the temperature and heat flux on the wall.

For the case of uniform heat flux on the boundary, the first law of thermodynamics leads to

$$\frac{\partial T^*}{\partial x^*} = \frac{dT_m{}^*}{dx^*} = \frac{q''}{(\rho c_P)_f H U^*} = \text{constant}. \tag{4.119}$$

In this case Eq. (4.109) becomes

$$\frac{d^2\hat{T}}{dy^2} = -\frac{1}{2}\mathrm{Nu}\,\hat{u}. \tag{4.120}$$

The boundary conditions for this equation are

$$\hat{T} = 0 \text{ at } y = 1, \quad \frac{d\hat{T}}{dy} = 0 \text{ at } y = 0. \tag{4.121}$$

For the Brinkman model, with u given by Eq. (4.114), we have

$$\hat{u} = \frac{S}{S - \tanh S}\left(1 - \frac{\cosh Sy}{\cosh S}\right), \tag{4.122}$$

$$\hat{T} = \frac{S\mathrm{Nu}}{S - \tanh S}\left[\frac{1}{4}(1 - y^2) - \frac{\cosh S - \cosh Sy}{2S^2 \cosh S}\right]. \tag{4.123}$$

The definition of the dimensionless temperature leads to an identity that we call the integral compatibility condition (Nield and Kuznetsov, 2000), namely

$$\int_0^1 \hat{u}\hat{T}dy = 1. \tag{4.124}$$

Substitution from Eqs. (4.122) and (4.123) then leads to

$$\mathrm{Nu} = \frac{12S(S - \tanh S)^2}{2S^3 - 15S + 15\tanh S + 3S\tanh^2 S}, \tag{4.125}$$

in agreement with an expression obtained by Lauriat and Vafai (1991). As the Darcy number Da increases from 0 to ∞, i.e., as S decreases from ∞ to 0, the Nusselt number Nu decreases from the Darcy value 6 [agreeing with Eq. (4.42)] to the clear fluid value $210/51 = 4.12$. Thus the effect of boundary friction is to decrease the heat transfer by reducing the temperature gradient at the boundary.

For $F \neq 0$, Vafai and Kim (1989) used a boundary-layer approximation in obtaining a closed form solution. This solution becomes inaccurate for hyperporous media, those for which $\mathrm{Da} > 0.1$. For such media, the Brinkman term is comparable with the Darcy term throughout the flow (and not just near the walls) and K can no longer be determined by a simple Darcy-type experiment. A closed form solution of the Brinkmann-Forchheimer equation, valid for all

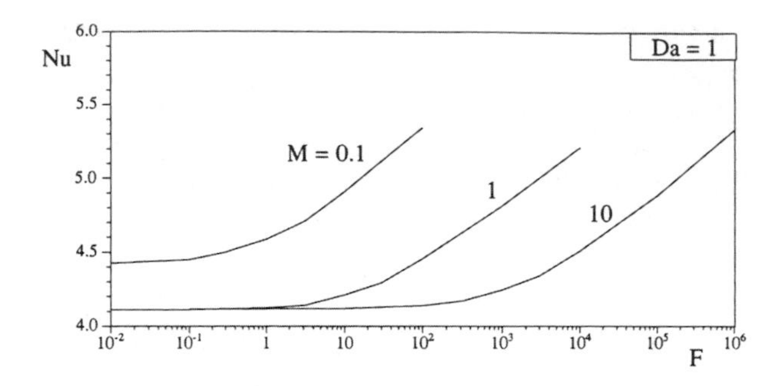

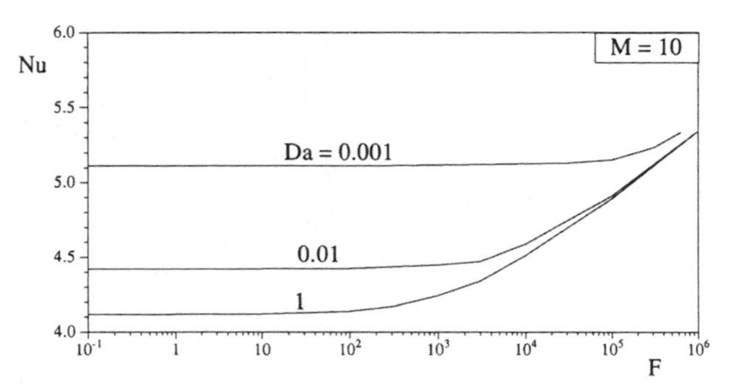

Figure 4.9. Effect of the Forchheimer number, F, on the Nusselt number Nu, for a channel with isoflux boundaries (Nield *et al.* 1996).

values of Da, was obtained by Nield *et al.* (1996). Some typical results are given in Fig. 4.9.

The results of Nield *et al.* (1996) may be summarized as follows. For each type of thermal boundary condition, the temperature profile is little changed as a result of variation of M, Da or F. It is slightly more peaked when Da is small or when F is large. On the other hand, the Nusselt number is significantly altered, primarily as a result of the change in velocity profile. The effect of an increase in F is to produce a more slug-like flow; and because of the way the mean velocity is defined this decreases $(T_w - T_m)$, and hence increases Nu. In particular, for the case of isoflux boundaries, the following holds. When simultaneously Da is large and F is small, the velocity profile is approximately parabolic and the Nusselt number is near 70/17 (a lower bound). When either Da is sufficiently small or F is sufficiently large, the velocity profile is approximately uniform (apart from a thin boundary layer) and the Nusselt number is near 6 (an upper bound). For the case of isothermal surfaces the story is similar, but the Nusselt numbers are smaller [the reason for this is spelled out in Nield *et al.* (1996, p. 211)].

For the case of a circular tube, with H replaced by the radius R of the tube in the scaling, one finds (Nield *et al.*, 2003b) that the solution can be expressed in terms of modified Bessel functions:

$$\hat{u} = \frac{S\,[I_0(S) - I_0(Sr)]}{SI_0(S) - 2I_1(S)}, \tag{4.126}$$

$$\hat{T} = \frac{S\,\mathrm{Nu}}{SI_0(S) - 2I_1(s)}\left[\frac{I_0(S)}{4}(1 - r^2) - \frac{I_0(S) - I_0(Sr)}{S^2}\right], \tag{4.127}$$

$$\mathrm{Nu} = \frac{8S[SI_0(S) - 2I_1(S)]^2}{(S^3 - 24S)[I_0(S)]^2 + 48I_0(S)I_1(S) + 8S[I_1(S)]^2}. \tag{4.128}$$

When the uniform flux boundary condition is replaced by the uniform temperature condition, one finds that Eq. (4.120) is replaced by

$$\frac{d^2\hat{T}}{dy^2} = -\frac{1}{2}\mathrm{Nu}\,\hat{u}\hat{T}. \tag{4.129}$$

The boundary condition given by Eq. (4.121) still applies. We see that we now have an eigenvalue problem with Nu as the eigenvalue. Now Eq. (4.124) is satisfied trivially, and instead of this compatibility condition one uses an integral compatibility condition (previously satisfied trivially), namely

$$\mathrm{Nu} = -2\frac{d\hat{T}}{dy}(1). \tag{4.130}$$

Equation (4.130) enables the amplitude of the eigenfunction to be determined. For the case of Darcy flow (Da $= 0$) we have $\hat{u} = 1$, $\hat{T} = (\pi/2)\cos(\pi y/2)$ and $\mathrm{Nu} = \pi^2/2 = 4.93$. For other values of Da the value of Nu can be found numerically, most readily by expressing the second-order differential equation as two first-order equations and then using a shooting method. Details of the method may be found in Nield and Kuznetsov (2000).

The above results for symmetric heating can be extended to the case of asymmetric heating, using a result established by Nield (2004c). The result applies when the heat flux along each boundary is uniform, or the temperature along each boundary is uniform. With the Nusselt number defined in terms of the mean wall temperature and the mean wall heat flux, the value of the Nusselt number is independent of the asymmetry whenever the velocity profile is symmetric with respect to the midline of the channel. This means that the above results also apply to the case of heating asymmetric with respect to the midline.

In the case of a circular tube, Eqs. (4.129), (4.130) are replaced by

$$\frac{d^2\hat{T}}{dr^2} + \frac{1}{r}\frac{d\hat{T}}{dr} = -\mathrm{Nu}\,\hat{u}\hat{T}, \tag{4.131}$$

$$\mathrm{Nu} = -2\frac{d\hat{T}}{dr}(1). \tag{4.132}$$

For the case Da $= 0$ one finds that $\mathrm{Nu} = \lambda^2$ where $\lambda = 2.40483$ is the smallest positive root of the Bessel function $J_0(x)$, so that $\mathrm{Nu} = (2.40483)^2 = 5.783$, and $\hat{T} = \lambda J_0(\lambda r)/2J_1(\lambda)$.

Variable porosity effects in a channel bounded by two isothermal parallel plates and in a circular pipe were examined numerically by Poulikakos and Renken (1987), for the case of a fully developed velocity field. They assumed that the porosity variation had negligible effects on the thermal conductivity, an assumption that breaks down when there is a large difference between the thermal conductivities of the two phases (David *et al.*, 1991). Poulikakos and Renken (1987) found that in the fully developed region the effect of channeling was to produce a Nusselt number increase (above the value based on the Darcy model) of 12 percent for a parallel plate channel and 22 percent for a circular pipe.

Renken and Poulikakos (1988) performed an experimental investigation for the parallel plate configuration with the walls maintained at constant temperature, with particular emphasis on the thermally developing region. They also performed numerical simulations incorporating the effects of inertia, boundary friction, and variable porosity. Their experimental and numerical findings agreed on predicting an enhanced heat transfer over that predicted using the Darcy model.

Poulikakos and Kazmierczak (1987) obtained closed form analytical solutions of the Brinkman equation for parallel plates and a circular pipe with constant heat flux on the walls for the case where there is a layer of porous medium adjacent to the walls and clear fluid interior. They also obtained numerical results when the walls were at constant temperature. For all values of Da the Nusselt number Nu goes through a minimum as the relative thickness of the porous region s varies from 0 to 1. The minimum deepens and is attained at a smaller value of s as Da increases. A general discussion of Brinkman, Forchheimer, and dispersion effects was presented by Tien and Hunt (1987). For the Brinkman model and uniform heat flux boundaries, Nakayama *et al.* (1988) obtained exact and approximate solutions. Analytical studies giving results for small or large Darcy numbers for convection in a circular tube were reported by Hooman and Ranbar-Kani (2003, 2004).

Hunt and Tien (1988a) have performed experiments that document explicitly the effects of thermal dispersion in fibrous media. They were able to correlate their Nusselt number data, for high Reynolds number flows, in terms of a parameter $u_a L^{1/2} K^{1/4}/\alpha_m$, where u_a is the average streamwise Darcy velocity and L is a characteristic length. Since this parameter does not depend explicitly on the thermal conductivity, they concluded that dispersion overwhelmed transport from solid conduction. They were able to explain this behavior using a dispersion conductivity of the form

$$k_d = \rho\, c_P \gamma\, K^{1/2} u, \tag{4.133}$$

where γ is a numerical dispersion coefficient, having the empirically determined value of 0.025. An analytical study of the effect of transverse thermal dispersion was reported by Kuznetsov (2000c).

Hunt and Tien (1988b) modeled heat transfer in cylindrical packed beds such as chemical reactors by employing a Forchheimer-Brinkman equation. They allowed the diffusivity to vary across the bed. Marpu (1993) found that the inclusion of axial conduction leads to a significant increase in Nusselt number in the thermally developing region of pipes for Péclet number less than 100. In similar circumstances, the effect of axial dispersion was found by Adnani *et al.* (1995) to be important for Péclet number less than 10.

Cheng *et al.* (1991) have reviewed methods for the determination of effective radial thermal conductivity and Nusselt number for convection in packed tubes and channels and have reanalyzed some of the previous experimental data in the light of their own recent contributions to thermal dispersion theory with variable porosity effects taken into account. They found that for forced convection in a packed column the average Nusselt number depends not only on the Reynolds number but also on the dimensionless particle diameter, the dimensionless length of the tube, the thermal conductivity ratio of the fluid phase to the solid phase, and the Prandtl number of the fluid. They summarized their conclusions by noting that in their recent work [Cheng *et al.* (1988), Cheng and Hsu (1986a,b), Cheng and Zhu (1987), Cheng and Vortmeyer (1988), and Hsu and Cheng (1988, 1990)] they had developed a consistent theory for the study of forced convection in a packed column taking into consideration the wall effects on porosity, permeability, stagnant thermal conductivity, and thermal dispersion. These effects become important as the particle/tube diameter ratio is increased. Various empirical parameters in the theory can be estimated by comparison of theoretical and experimental results for the pressure drop and heat transfer, but there is at present a need to perform more experiments on forced convection in packed columns where both temperature distribution and heat flux are measured to enable a more accurate determination of the transverse thermal dispersivity.

Chou *et al.* (1994) performed new experiments and simulations for convection in cylindrical beds. They concluded that discrepancies in some previous models could be accounted for by the effect of channeling for the case of low Péclet number and the effect of thermal dispersion in the case of high Péclet number. Chou *et al.* (1992b,c) had reported similar conclusions, on the basis of experiments, for convection in a square channel.

The effect of suction at permeable walls was investigated by Lan and Khodadadi (1993). An experimental study of convection with asymmetric heating was reported by Hwang *et al.* (1992). Bartlett and Viskanta (1996) obtained analytical solutions for thermally developing convection in an asymmetrically heated duct filled with a medium of high thermal conductivity.

Lage *et al.* (1996) performed a numerical study for a device (designed to provide uniform operating temperatures) consisting of a microporous layer placed between two sections of a cold plate. The simulation was based on two-dimensional equations derived from three-dimensional equations by integration over the small dimension of the layer.

For convection in cylindrical beds, Kamiuto and Saitoh (1994) investigated Nu_P, κ, and Γ, where Nu_P and Re_P are Nusselt and Reynolds numbers based on the particle diameter, while κ is the ratio of thermal conductivity of solid to that of fluid and Γ is the ratio of bed radius to particle diameter. They found that as $Re_P Pr$ tends to zero, Nu_P tends to a constant value depending on both κ and Γ, while for large $Re_P Pr$ the value of Nu_P depends on both Re_P Pr and Pr but only to a small extend on κ.

For pipes packed with spheres, Varahasamy and Fand (1996) have presented empirical correlation equations representing a body of new experimental data. Experimental studies involving metal foams have been reported by Calmidi and Mahajan (2000), Hwang *et al.* (2002), and Zhao *et al.* (2004b). Further experimental and theoretical studies of convection in a circular pipe were conducted by Izadpanah *et al.* (1998). Extending previous experimental work by Jiang *et al.* (1999b), Jiang *et al.* (2004e,f,h) studied numerically and experimentally the wall porosity effect for a sintered porous medium. A similar study of nonsintered material was reported by Jiang *et al.* (2004g). Sintered materials also were discussed by Kim and Kim (2000). Forced convection in microstructures was discussed by Kim and Kim (1999). Another numerical study in a metallic fibrous material was reported by Angirasa (2002a), and that was followed with an experimental study by Angirasa (2000b). An experimental study with aluminum foam in an asymmetrically heated channel was made by S. J. Kim *et al.* (2001).

Entropy generation in a rectangular duct was studied by Demirel and Kahraman (1999). For a square duct, a numerical study of three-dimensional flow was reported by Chen and Hadim (1999b).

The effect of viscous dissipation has been studied numerically by Zhang *et al.* (1999) for a parallel plate channel and by Yih and Kamioto for a circular pipe. An analytical study of the effects of both viscous dissipation and flow work in a channel, for boundary conditions of uniform temperature or uniform heat flux, was reported by Nield *et al.* (2004b). These authors specifically satisfied the first law of thermodynamics when treating the fully developed flow. They also considered various models for the contribution from the Brinkman term to the viscous dissipation. Some general matters related to the possibility of fully developed convection were discussed by Nield (2006). An analytical study of heat transfer in Couette flow was made by Kuznetsov (1998c). An analytical study of a conjugate problem, with conduction heat transfer inside the channel walls accounted for, was made by Mahmud and Fraser (2004). Entropy generation in a channel was studied analytically and numerically by Mahmud and Fraser (2005b). Vafai and Amiri (1998) briefly surveyed some of the work done on the topics that here are discussed mainly in Sections 4.9 and 4.10.

Convection in a hyperporous medium saturated by a rarefied gas, with both velocity-slip and temperature-slip at the boundaries of a parallel-plate channel or a circular duct, was analyzed by Nield and Kuznetsov (2006). They found that temperature slip leads to decreased transfer, while the effect of velocity-slip depends on the geometry and the Darcy number.

4.10. Local Thermal Nonequilibrium

It is now commonplace to employ a two-temperature model to treat forced convection with local thermal nonequilibrium (LTNE). Authors who have done this include Vafai and Tien (1989), Jiang *et al.* (1998, 1999, 2001, 2002), You and Song (1999), Kim *et al.* (2000), Kim and Jang (2002), Muralidhar and Suzuki (2001), and Nakayama *et al.* (2001). Transient and time-periodic convection in a channel has been treated analytically by Al-Nimr and Abu-Hijleh (2002), Al-Nimr and Kiwan (2002), Abu-Hijleh *et al.* (2004), and Khashan *et al.* (2005). A further study of transient convection was conducted by Spiga and Morini (1999). An analysis involving a perturbation solution was presented by Kuznetsov (1997d). The specific aspect LTNE involving steady convective processes was analyzed by Nield (1998a). The modeling of local nonequilibrium in a structured medium was discussed by Nield (2002), and a conjugate problem was analyzed by Nield and Kuznetsov (1999). A problem in a channel with one wall heated was analyzed by Zhang and Huang (2001); see also the note by Magyari and Keller (2002). The departure from local thermal equilibrium due to a rapidly changing heat source was analyzed by Minkowycz *et al.* (1999). Further analysis was carried out by Lee and Vafai (1999) and Marafie and Vafai (2001). The particular case of various models for constant wall heat flux boundary conditions was discussed by Alazmi and Vafai (2002). The present authors think that the best model is the one where there is uniform flux over the two phases, as employed by Nield and Kuznetsov (1999). Alazmi and Vafai (2004) showed that thermal dispersion has the effect of increasing the sensitivity of LTNE between the two phases. The case of non-Newtonian fluid was treated numerically by Khashan and Al-Nimr (2005). Most work on LTNE has been done for confined flows, but Wong *et al.* (2004) treated finite Péclet number effects in forced convection past a heated cylinder.

4.11. Partly Porous Configurations

For complicated geometries numerical studies are needed. The use of porous bodies to enhance heat exchange motivated the early studies of Koh and Colony (1974) and Koh and Stevans (1975). Huang and Vafai (1993, 1994a–d) and Vafai and Huang (1994), using a Brinkman-Forchheimer model, have performed studies of a composite system made of multiple porous blocks adjacent to an external wall (either protruding or embedded) or along a wall with a surface substrate. Khanafer and Vafai (2001, 2005) investigated isothermal surface production and regulation for high heat flux applications using porous inserts. Cui *et al.* (2001) conducted an experimental study involving a channel with discrete heat sources.

Convection in a parallel-plate channel partially filled with a porous layer was studied by Jang and Chen (1992). They found that the Nusselt number is sensitive to the open space ratio and that the Nusselt number is a minimum at a certain porous layer thickness, dependent on Darcy number. A similar study was reported by Tong *et al.* (1993). Srinivasan *et al.* (1994) analyzed convection in a spirally fluted tube using a porous substrate approach. Hadim and Bethancourt (1995) simulated convection in a channel partly filled with a porous medium and with discrete heat sources on one wall. Chikh *et al.* (1995b, 1998) studied convection in an annulus partly filled with porous material on the inner heated wall and in a channel with intermittent heated porous disks, while Rachedi and Chikh (2001) studied a similar problem. Ould-Amer *et al* (1998) studied numerically the cooling of heat generating blocks mounted on a wall in a parallel plate channel. Fu *et al.* (1996) and Fu and Chen (2002) dealt with the case of a single porous block on a heated wall in a channel. Sözen and Kuzay (1996) studied round tubes with porous inserts. Zhang and Zhao (2000) treated a porous block behind a step in a channel. Masuoka *et al.* (2004) studied experimentally and numerically, with alternative interface conditions considered, the case of a permeable cylinder placed in a wind tunnel of rectangular cross section. Layeghi and Nouri-Borujerdi (2004) discussed forced convection from a cylinder or an array of cylinders in the presence or absence of a porous medium. Huang *et al.* (2004b) studied numerically the enhancement of heat transfer from multiple heated blocks in a channel using porous covers.

Abu-Hijleh (1997, 2000, 2001b, 2002) numerically simulated forced in various geometries with orthotropic porous inserts, while Abu-Hijleh (2003) treated a cylinder with permeable fins. A transient problem involving partly filled channels was studied by Abu-Hijleh and Al-Nimr (2001).

Analytical solutions for some flows through channels with composite materials were obtained by Al-Hadrami *et al.* (2001a,b). Pipes with porous substrates were treated numerically by Alkam and Al-Nimr (1999a,b, 2001), while parallel-plate channels were similarly treated by Alkam *et al.* (2001, 2002). A tubeless solar collector and an unsteady problem involving an annulus were likewise treated by Al-Nimr and Alkam (1997a, 1998a). Hamdan *et al.* (2000) treated a parallel-plate channel with a porous core. W. T. Kim *et al.* (2003c) studied both a porous core and a porous sheath in a circular pipe. A Green's function method was used by Al-Nimr and Alkam (1998b) to obtain analytical solution for transient flows in parallel-plate channel. Experimental and numerical investigations of forced convection in channels containing obstacles were conducted by Young and Vafai (1998, 1999) and Pavel and Mohamad (2004a–c).

The limitation of the single-domain approach for the computation of convection in composite channels was exposed by Kuznetsov and Xiong (1999). The effect of thermal dispersion in a channel was analyzed by Kuznetsov (2001). Kuznetsov and Xiong (2000) numerically simulated the effect of thermal dispersion in a composite circular duct.

Kuznetsov (2000a) reviewed a number of analytical studies, including those by Kuznetsov (1998b, 1999a,c, 2001) for flow induced by pressure gradients, and by Kuznetsov (1998d, 2000b) and Xiong and Kuznetsov (2000) for Couette flow. The effect of turbulence on forced convection in a composite tube was discussed by Kuznetsov *et al.* (2002, 2003b), Kuznetsov (2004a), and Kuznetsov and Becker (2004). A numerical study of turbulent heat transfer above a porous wall was conducted by Stalio *et al.* (2004). Convection past a circular cylinder sheathed with a porous annulus, placed perpendicular to a tutbulent air flow, was studied numerically and experimentally by Sobera *et al.* (2003). Hydrodynamically and thermally developing convection in a partly filled square duct was studied numerically using the Brinkman model by Jen and Yan (2005).

A boundary-layer analysis of unconfined forced convection with a plate and a porous substrate was presented by Nield and Kuznetsov (2003d). A more general analytical investigation of this situation had been presented earlier by Kuznetsov (1999b).

4.12. Transversely Heterogeneous Channels and Pipes

Analytical studies on the effect on forced convection, in channels and ducts, of the variation in the transverse direction of permeability and thermal conductivity were initiated by Nield and Kuznetsov (2000), who used the Darcy model for local thermal equilibrium. Both parallel-plate channels and circular ducts were considered, and walls at uniform temperature and uniform heat flux, applied symmetrically, were treated in turn. Both continuous variation and stepwise variation of permeability and conductivity were treated. For the parallel plate channel, this work was extended to the Brinkman model by Nield and Kuznetsov (2003d). For the case of a parallel-plate channel with uniform heat flux boundaries, Sundaravadivelu and Tso (2003) extended the basic analysis to allow for the effect of viscosity variations. Asymmetric property variation and asymmetric heating in a parallel-plate channel were considered by Nield and Kuznetsov (2001a). A conjugate problem, with either a parallel-plate channel or a circular duct, was treated by Kuznetsov and Nield (2001). The interaction of thermal nonequilibrium and heterogeneous conductivity was studied by Nield and Kuznetsov (2001b). With application to the experimental results reported by Paek *et al.* (1999b) in mind, Nield and Kuznetsov (2003a) treated a case of gross heterogeneity and anisotropy using a layered medium analysis. A conjugate problem, involving the Brinkman model and with temperature-dependent volumetric heat inside the solid wall, was treated analytically and numerically by Mahmud and Fraser (2005a).

For illustration, we present the results obtained by Nield and Kuznetsov (2000) for the effect of heterogeneity on Nusselt number. We first consider the case where the permeability and thermal conductivity distributions are given by

$$K = K_0\left\{1 + \varepsilon_K\left(\frac{|y^*|}{H} - \frac{1}{2}\right)\right\},$$

$$k = k_0\left\{1 + \varepsilon_k\left(\frac{|y^*|}{H} - \frac{1}{2}\right)\right\}. \tag{4.134a,b}$$

Here the boundaries are at $y^* = -H$ and $y^* = H$. The mean values of the permeability and conductivity are K_0 and k_0, respectively. The coefficients ε_K and ε_k are each assumed to be small compared with unity. To first order, one finds that for the case of uniform flux boundaries

$$\mathrm{Nu} = 6\left(1 + \frac{1}{4}\varepsilon_K - \frac{1}{8}\varepsilon_k\right). \tag{4.135}$$

and for the case of uniform temperature boundaries,

$$\mathrm{Nu} = \frac{\pi^2}{2}\left\{1 + \frac{2}{\pi^2}(\varepsilon_K - \varepsilon_k)\right\}. \tag{4.136}$$

4.13. Thermal Development

In forced convection in a porous medium, hydrodynamic development is not normally of importance. This is because the hydrodynamic development length is readily shown to be of order of magnitude $(K/\varphi)^{1/2}$ and usually this is very small compared with the channel width. In contrast, the thermal development length can be much greater.

For the Darcy model one has slug flow, and for the case of walls at uniform temperature the classical Graetz solution for thermal development is applicable. An analysis based on the Brinkman model was reported by Nield *et al.* (2004a), for both a parallel-plate channel and a circular tube. The additional effect of a Forchheimer term has not yet been treated, but one would anticipate that since an increase in Forchheimer number would produce a more slug-like flow, the effect of quadratic drag would be similar to that produced by a reduction in Darcy number. The corresponding case where the walls are at uniform heat flux was treated by Nield *et al.* (2003b). The effect of local thermal nonequilibrium was examined by Nield *et al.* (2002), and the additional effects of transverse heterogeneity were studied by Nield and Zuznetsosv (2004a). Thermal development in a channel occupied by a non-Newtonian power-law fluid was studied by Nield and Kuznetsov (2005a). In the standard analysis of the Graetz type the axial conduction and viscous dissipation effects are neglected, but in the studies by Nield *et al.* (2003a) and Kuznetsov *et al.* (2003c) these effects were included, for the cases of a parallel-plate channel and a circular duct, respectively. For the case of a circular duct, axial conduction effects and viscous dissipation effects and were studied numerically by Hooman *et al.* (2003) and Ranjbar-Kani and Hooman (2004), respectively.

A numerical study of heat transfer in the thermally developing region in an annulus was reported by Hsieh and Lu (1998). Thermal developing forced convection inside ducts of various shapes (including elliptical passages) were analyzed by Haji-Sheikh and Vafai (2004). Haji-Sheikh *et al.* (2005) illustrated the use of a combination a Green's function solution and an extended weighted residuals method to the study of isosceles triangular passages. They noted that their methodology is equally applicable when the boundary conditions are of the first, second, or third kind.

The general feature of thermal development is that the Nusselt number increases as one moves from the fully developed region toward the entrance region. It is found that the rate of increase decreases as the Darcy number increases.

4.14. Surfaces Covered with Porous Layers

The hair growth on the skin of a mammal is an example of a saturated porous medium where, locally, the solid matrix (hair) is *not* in thermal equilibrium with the permeating fluid (air). A theory for the heat transfer by forced convection through a surface covered with hair has been developed by Bejan (1990a). It was tested subsequently in the numerical experiments of Lage and Bejan (1990). This entire body of work was reviewed by Bejan and Lage (1991) and Bejan (1992b).

The most essential features of the geometry of an actual surface covered with hair are retained in the model presented in Fig. 4.9. The skin surface is connected to a large number of perpendicular strands of hair, the density of which is assumed constant,

$$n = \frac{\text{number of strands of hair}}{\text{unit area of skin surface}}. \tag{4.137}$$

The hair population density n is related to the porosity of the "hair + air" medium that resides above the skin,

$$\varphi = \frac{\text{air volume}}{\text{total volume}} = 1 - n\,A_s. \tag{4.138}$$

Each strand of hair is modeled as a cylinder with the cross section A_s.

Parallel to the skin surface and through the porous structure formed by the parallel hair strands flows a uniform stream of air of velocity U. This stream is driven longitudinally by the dynamic pressure rise formed over that portion of the animal's body against which the ambient breeze stagnates. The longitudinal length L swept by the air flow is a measure of the linear size of the animal. The constant air velocity U is a quantity averaged over the volume occupied by air. It is assumed that the strand-to-strand distances are small enough so that the air flow behaves according to the Darcy law, with apparent slip at the skin surface.

At every point in the two-dimensional (x, y) space occupied by the porous medium described above, we distinguish two temperatures: the temperature of the solid structure (the local hair strand), T_s, and the temperature of air that surrounds the strand, T_a. Both T_s and T_a are functions of x and y. The transfer of heat from the skin to the atmosphere is driven by the overall temperature difference $(T_w - T_\infty)$, where T_w is the skin temperature and T_∞ the uniform temperature of the ambient air that enters the porous structure. The temperature of the interstitial air, T_a, is equal to the constant temperature T_∞ in the entry plane $x = 0$.

For the solid structure, the appropriate energy equation is the classic conduction equation for a fin (in this case, single strand of hair),

$$k_s A_s \frac{\partial^2 T_s}{\partial y^2} - h\,p_s\,(T_s - T_a) = 0, \tag{4.139}$$

where p_s is the perimeter of a strand cross section. The thermal conductivity of the strand, k_s, and the perimeter-averaged heat transfer coefficient, h, are both constant. The constancy of h is a result of the assumed low Reynolds number of the air flow that seeps through the hair strands.

The second energy conservation statement refers to the air space alone, in which (ρc_P) and k_a are the heat capacity and thermal conductivity of air:

$$\rho c_P U \frac{\partial^2 T_a}{\partial x^2} = k_a \frac{\partial^2 T_a}{\partial y^2} + n\,h\,p_s\,(T_s - T_a). \tag{4.140}$$

On the left-hand side of this equation, we see only one convection term because the air-space-averaged velocity U points strictly in the x direction. The first term on the right-hand side of the equation accounts for air conduction in the transversal direction (y). By not writing the longitudinal conduction term $k_a \partial^2 T_a / \partial x^2$, we are

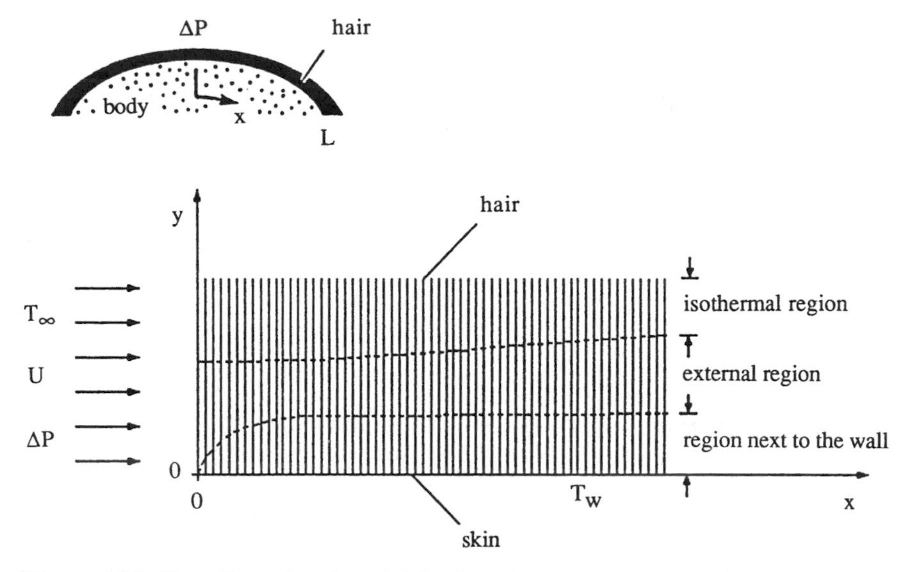

Figure 4.10. Two-dimensional model for forced convection through the hair growth near the skin (after Bejan, 1990a).

assuming that the flow region in which the effect of transversal air conduction is important is thin.

The last term in Eq. (4.140) accounts for the "volumetric heat source" effect that is due to the contact between the air stream and the local (warmer) hair strand. Note the multiplicative role of the strand density n in the makeup of this term: the product (np_s) represents the total contact area between hair and air, expressed per unit of air volume. The heat source term of Eq. (4.140) is the air-side reflection of the heat sink term (the second term) encountered in the fin conduction equation (4.139).

In an air region that is sufficiently close to the skin, the air stream is warmed up mainly by contact with the skin, i.e., not by the contact with the near-skin area of the hair strands. Consequently, for this region, in Eq. (4.140) the heat source term $nhp_s(T_s - T_a)$ can be neglected. On the other hand, sufficiently far from the skin most of the heating of the air stream is effected by the hair strands that impede the flow. In the energy balance of this external flow the vertical conduction term can be neglected in Eq. (4.140).

For the details of the heat transfer analysis of the two-temperatures porous medium of Fig. 4.10 the reader is referred to the original paper (Bejan, 1990a). One interesting conclusion is that the total heat transfer rate through a skin portion of length L is minimized when the hair strand diameter assumes the optimal value D_{opt} given by

$$\frac{D_{opt}}{\nu}\left(\frac{\Delta P}{\rho}\right)^{1/2} = \left(\frac{k_z^2 k_s}{2k_a}\right)^{1/4}\left(\frac{1-\varphi}{\varphi}\right)^{5/4}\left[\frac{L}{\nu}\left(\frac{\Delta P}{\rho}\right)^{1/2}\right]^{1/2}. \quad (4.141)$$

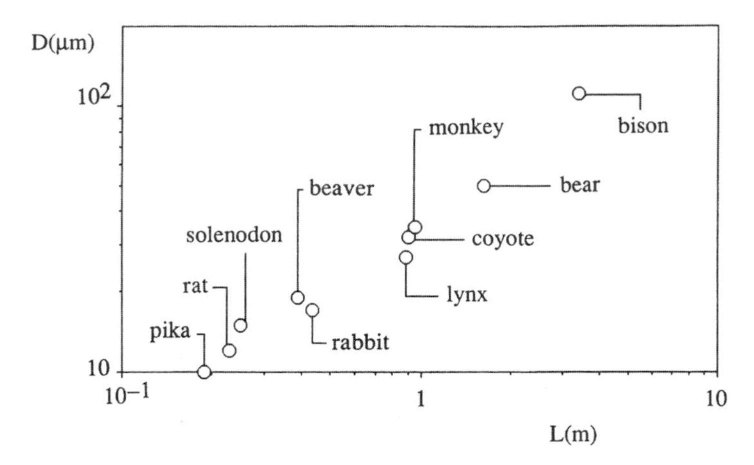

Figure 4.11. The hair strand diameters and body lengths of ten mammals (Bejan and Lage, 1991).

That lowest heat transfer rate is

$$\frac{q'_{min}}{k_a\,(T_w - T_\infty)} = \left(32\frac{k_s}{k_a}\right)^{1/4} \varphi^{3/4}\,(1-\varphi)^{1/4} \left[\frac{L}{\nu}\left(\frac{\Delta P}{\rho}\right)^{1/2}\right]^{1/2}. \tag{4.142}$$

These results are based on several additional assumptions, which include a model of type (1.5) for the permeability of the hair matrix

$$K \cong \frac{D^2\,\varphi^3}{k_z\,(1-\varphi)^2}, \tag{4.143}$$

where the constant k_z is a number of order 10^2.

Equation (4.130) shows that the minimum heat transfer rate increases with the square root of the linear size of the body covered with hair, $L^{1/2}$. The optimal hair strand diameter is also proportional to $L^{1/2}$. This last trend agrees qualitatively with measurements of the hair sizes of mammals compiled by Sokolov (1982). Figure 4.11 shows the natural hair strand diameters (D) of ten mammals, with the length scale of the body of the animal plotted on the abscissa.

More recent studies of surfaces covered with fibers have focused on the generation of reliable pressure drop and heat transfer information for low Reynolds number flow through a bundle of perpendicular or inclined cylindrical fibers (Fowler and Bejan, 1994). There is a general need for data in the low Reynolds number range, as most of the existing results refer to heat exchanger applications (i.e., higher Reynolds numbers). Fowler and Bejan (1995) studied numerically the heat transfer from a surface covered with flexible fibers, which bend under the influence of the interstitial flow. Another study showed that when the effect of radiation is taken into account, it is possible to anticipate analytically the existence of an optimal packing density (or porosity) for minimal heat transfer across the porous cover (Bejan, 1992b).

Vafai and Kim (1990) and Huang and Vafai (1993, 1994) have shown that a porous coating can alter dramatically the friction and heat transfer characteristics of a surface. This effect was also documented by Fowler and Bejan (1995). Depending on its properties and dimensions, the porous layer can act either as an insulator or as a heat transfer augmentation device. The engineering value of this work is that it makes it possible to "design" porous coatings such that they control the performance of the solid substrate.

4.15. Designed Porous Media

A potentially revolutionary application of the formalism of forced convection in porous media is in the field of heat exchanger simulation and design. Heat exchangers are a century-old technology based on information and concepts stimulated by the development of large-scale devices (see, for example, Bejan, 1993, chapter 9). The modern emphasis on heat transfer augmentation, and the more recent push toward miniaturization in the cooling of electronics, have led to the development of compact devices with much smaller features than in the past. These devices operate at lower Reynolds numbers, where their compactness and small dimensions ("pores") make them candidates for modeling as saturated porous media.

Such modeling promises to revolutionize the nomenclature and numerical simulation of the flow and heat transfer through heat exchangers. Decreasing dimensions, increasing compactness, and constructal design (Section 4.18) make these devices appear and function as *designed porous media* (Bejan, 2004b). This emerging field is outlined in two new books (Bejan, 2004a; Bejan *et al.*, 2004).

To illustrate this change, consider Zukauskas' (1987) classic chart for the pressure drop in cross-flow through arrays of staggered cylinders (e.g., Fig. 9.38 in Bejan, 1993). The four curves drawn on this chart for the transverse pitch/cylinder diameter ratios 1.25, 1.5, 2, and 2.5 can be made to collapse into a single curve, as shown in Fig. 4.12 (Bejan and Morega, 1993). The technique consists of treating the bundle as a fluid saturated porous medium and using the volume-averaged velocity U, the pore Reynolds number $UK^{1/2}/\nu$ on the abscissa, and the dimensionless pressure gradient group $(\Delta P/L)K^{1/2}/\rho U^2$ on the ordinate.

The similarities between Figs. 4.12 and 4.3 are worth noting. The effective permeability of the bundle of cylinders was estimated using Eq. (4.131) with $k_z = 100$, and Zukauskas' chart. Figure 4.12 shows very clearly the transition between Darcy flow (slope -1) and Forchheimer flow (slope 0). The porous medium presentation of the array of cylinders leads to a very tight collapse of the curves taken from Zukauskas' chart. The figure also shows the pressure drop curve for turbulent flow through a heat exchanger core formed by a stack of parallel plates. An added benefit of Fig. 4.12 is that it extends the curves reliably into the low Reynolds number limit (Darcy flow), where classic heat exchanger data are not available.

This method of presentation (Fig. 4.12) deserves to be extended to other heat exchanger geometries. Another reason for pursuing this direction is that the heat and fluid flow process can be simulated numerically more easily if the heat exchanger is replaced at every point by a porous medium with volume averaged properties. An example is presented in Fig. 4.13 (Morega *et al.*, 1995). Air flows from left to

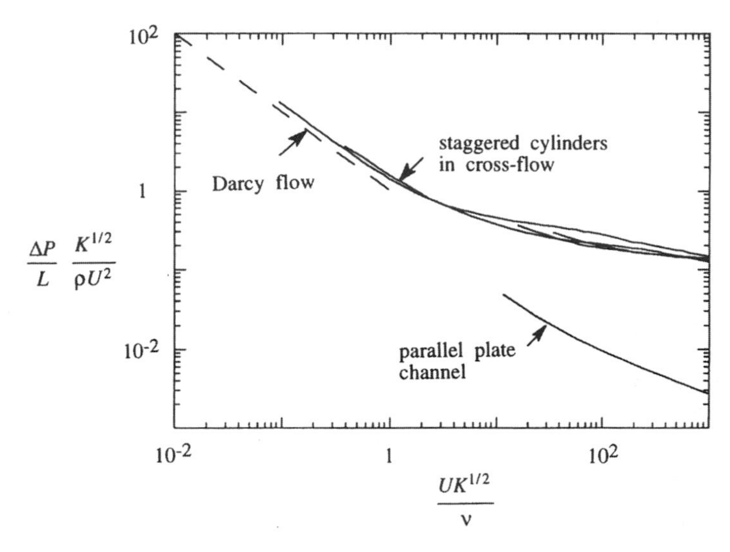

Figure 4.12. Porous medium representation of the classic pressure-drop data for flow through staggered cylinders and stacks of parallel plates (Bejan and Morega, 1993).

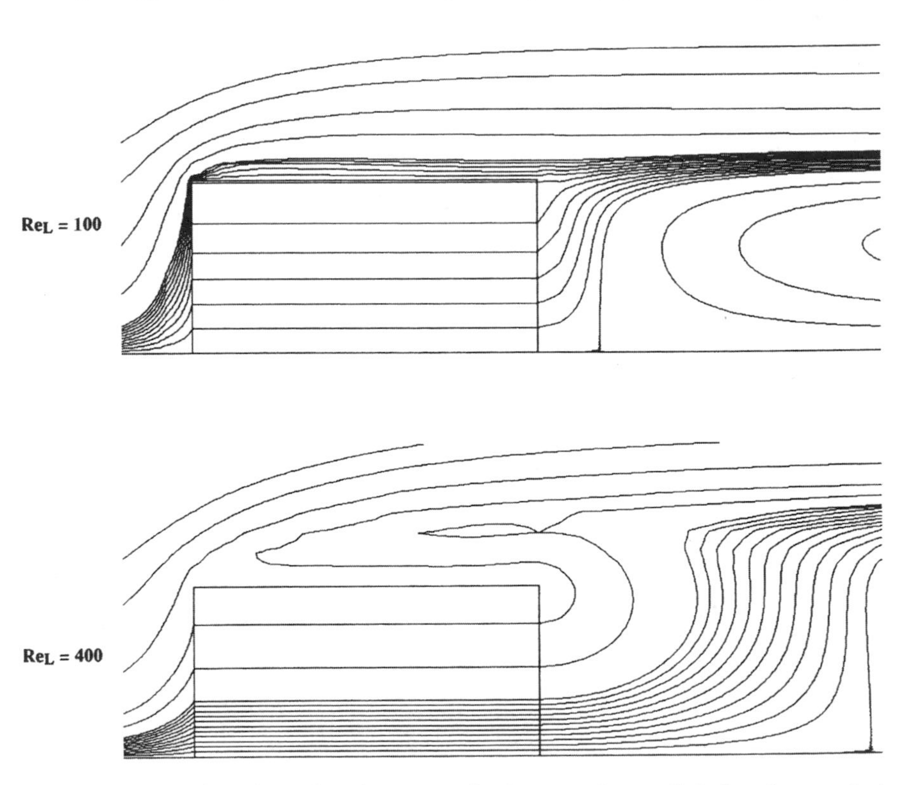

Figure 4.13. The flow through and over a stack of rectangular parallel-plate fins attached to a base, and modeled as a porous medium (Morega *et al.*, 1995).

right along a hot horizontal surface (the electronics module) and through an array of parallel plate fins of rectangular profile (the heat sink). The plate thickness and plate-to-plate spacing are $t/L = 0.05$ and $d/L = 0.069$, where L is the length of the plate in the flow direction. The Reynolds number Re_L is based on L and the approach velocity. The air flows through and over the heat sink. The corresponding temperature field and the effect of changing the Reynolds number are illustrated in Morega *et al.* (1995). One advantage of the numerical model is that it accounts in a volume-averaged sense for the conduction heat transfer through each plate, longitudinally and transversally. Another advantage comes from the relative simplicity and high computational speed, because in the thermal design and optimization of cooling techniques it is necessary to simulate a large number of geometric configurations such as Fig. 4.13.

Another important application of porous media concepts in engineering is in the optimization of the internal spacings of heat exchangers subjected to overall volume constraints (see Sections 4.19 and 4.20). Packages of electronics cooled by forced convection are examples of heat exchangers that must function in fixed volumes. The design objective is to install as many components (i.e., heat generation rate) as possible, while the maximum temperature that occurs at a point (hot spot) inside the given volume does not exceed a specified limit. Bejan and Sciubba (1992) showed that a very basic trade-off exists with respect to the number of installed components, i.e., regarding the size of the pores through which the coolant flows. This trade-off is evident if we imagine the two extremes: numerous components (small pores) and few components (large spacings).

When the components and pores are numerous and small, the package functions as a heat-generating porous medium. When the installed heat generation rate is fixed, the hot spot temperature increases as the spacings become smaller, because in this limit the coolant flow is being shut off gradually. In the opposite limit, the hot spot temperature increases again because the heat transfer contact area decreases as the component size and spacing become larger. At the intersection of these two asymptotes we find an optimal spacing (pore size) where the hot spot temperature is minimal when the heat generation rate and volume are fixed. The same spacing represents the design with maximal heat generation rate and fixed hot spot temperature and volume. Bejan and Sciubba (1992), Bejan (1993), and Morega *et al.* (1995) developed analytical and numerical results for optimal spacings in applications with solid components shaped as parallel plates. Optimal spacings for cylinders in cross-flow were determined analytically and experimentally by Bejan (1995) and Stanescu *et al.* (1996). The spacings of heat sinks with square pin fins and impinging flow were optimized numerically and experimentally by Ledezma *et al.* (1996). The latest conceptual developments are outlined in Section 4.19.

The dimensionless results developed for optimal spacings (S_{opt}) have generally the form

$$\frac{S_{opt}}{L} \sim \mathrm{Be}_L^{-n} \tag{4.144}$$

where L is the dimension of the given volume in the flow direction, and Be_L is the dimensionless pressure drop that Bhattacharjee and Grosshandler (1988) termed the Bejan number,

$$\mathrm{Be}_L = \frac{\Delta P \cdot L^2}{\mu_f \alpha_f}. \tag{4.145}$$

In this definition ΔP is the pressure difference maintained across the fixed volume. For example, the exponent n in Eq. (4.144) is equal to 1/4 in the case of laminar flow through stacks of parallel-plate channels. The Bejan number serves as the forced-convection analog of the Rayleigh number used in natural convection (Petrescu, 1994).

The optimization of heat transfer processes in porous media, which we just illustrated, is an important new trend in the wider and rapidly growing field of thermodynamic optimization (Bejan, 1996a). Noteworthy are two optimal-control papers of Kuznetsov (1997a,c), in which the heat transfer is maximized during the forced-convection transient cooling of a saturated porous medium. For example, Kuznetsov (1997a) achieved heat transfer maximization by optimizing the initial temperature of the porous medium subject to a fixed amount of energy stored initially in the system, and a fixed duration of the cooling process.

Other work on heat exchangers as porous media has been reported by Lu *et al.* (1998), Jiang *et al.* (2001), Boomsma *et al.* (2003) and Mohamad (2003).

4.16. Other Configurations or Effects

4.16.1. Effect of Temperature-dependent Viscosity

The study of the effect of a temperature-dependent viscosity on forced convection in a parallel-plate channel was initiated by Nield *et al.* (1999). The original analysis was restricted to small changes of viscosity, carried out to first order in Nield *et al.* (1999) and to second order in Narasimhan *et al.* (2001b), but the layered medium analysis of Nield and Kuznetsov (2003b) removed this restriction. For the case of a fluid whose viscosity decreases as the temperature increases (the usual situation) it is found that the effect of the variation is to reduce/increase the Nusselt number for cooled/heated walls. The analysis predicts that for the case of small Darcy number the effect of viscosity variation is almost independent of the Forchheimer number, while for the case of large Darcy number the effect of viscosity variation is reduced as the Forchheimer number increases. Within the limitations of the assumptions made in the theory, experimental verification was provided by Nield *et al.* (1999) and Narasimhan *et al.* (2001a).

For example, in the case of uniform flux boundaries and Darcy's law, Nield *et al.* (1999) showed that the mean velocity is altered by a factor $(1 + N/3)$ and the Nusselt number is altered by a factor $(1 - 2N/15)$, where the viscosity variation number N is defined as

$$N = \frac{q''H}{k}\frac{1}{\mu_0}\left(\frac{d\mu}{dT}\right)_0, \tag{4.146}$$

where the suffix 0 indicates evaluation at the reference temperature T_0.

The extension to the case where there is a substantial interaction between the temperature-dependence of viscosity and the quadratic drag effect was carried out in a sequence of papers by Narasimhan and Lage (2001a,b, 2002, 2003, 2004a). The effect on pump power gain for channel flows was studied by Narasimhan and Lage (2004b). In these papers the authors developed what they call a Modified Hazen-Dupuit-Darcy model which they then validated with experiments with PAO as the convecting liquid and compressed aluminum-alloy porous foam as the porous matrix. This work on temperature-dependent viscosity was reviewed by Narasimhan and Lage (2005).

The effects of a magnetic field and temperature-dependent viscosity on forced convection past a flat plate, with a variable wall temperature and in the presence of suction or blowing, was studied numerically by Seddeek (2002, 2005).

4.16.2. Other Flows

Non-Darcy boundary-layer flow over a wedge was studied using three numerical methods by Hossain *et al.* (1994). An application to the design of small nuclear reactors was discussed by Aithal *et al.* (1994). Convection with Darcy flow past a slender body was analyzed by Romero (1995b), while Sattar (1993) analyzed boundary-layer flow with large suction. The effect of blowing or suction on forced convection about a flat plate was also treated by Yih (1998d,e). The interaction with radiation in a boundary layer over a flat plate was studied by Mansour (1997). A porous medium heated by a permeable wall perpendicular to the flow direction was studied experimentally by Zhao and Song (2001). The boundary layer at a continuously moving surface was analyzed by Nakayama and Pop (1993). The effect of liquid evaporation on forced convection was studied numerically by Shih and Huang (2002).

Convection in an asymmetrically heated sintered porous channel was investigated by Hwang *et al.* (1995). Various types of sintered and unsintered heat sinks were compared experimentally by Tzeng and Ma (2004). Convection in a sintered porous channel with inlet and outlet slots was studied numerically by Hadim and North (2005). Sung *et al.* (1995) investigated flow with an isolated heat source in a partly filled channel. Conjugate forced convection in cross flow over a cylinder array with volumetric heating in the cylinders was simulated by Wang and Georgiadis (1996). Heat transfer for flow perpendicular to arrays of cylinders was examined by Wang and Sangani (1997). An internally finned tube was treated as a porous medium by Shim *et al.* (2002). Forced convection in a system of wire screen meshes was examined experimentally by Ozdemir and Ozguc (1997). The effect of anisotropy was examined experimentally by Yang and Lee (1999); numerically by S. Y. Kim *et al.* (2001), Nakayama *et al.* (2002), and Kim and Kuznetsov (2003); and analytically by Degan *et al.* (2002) The effect of fins in a heat exchanger was studied numerically by S. J. Kim *et al.* (2000, 2002) and Kim and Hyun (2005). An experimental study involving finned metal foam heat sinks was reported by Bhattachrya and Mahajan (2002). Forced convection in a channel with a localized heat source using fibrous materials was studied numerically by Angirasa and Peterson (1999). A numerical investigation with a random porosity model was made by W. S. Fu *et al.* (2001). Experimental studies involving a rectangular duct heated only from the top wall were conducted by Demirel *et al.* (1999, 2000). A thermodynamic analysis of heat transfer in an asymmetrically heated annular packed bed was reported by Demirel and Kahraman (2000). A laboratory investigation of the cooling effect of a coarse rock layer and a fine rock layer in permafrost regions was reported by Yu *et al.* (2004). Forced convection in a rotating channel was examined experimentally by Tzeng *et al.* (2004).

4.16.3. Non-Newtonian Fluids

Boundary-layer flow of a power-law fluid on an isothermal semi-infinite plate was studied by Wang and Tu (1989). The same problem for an elastic fluid of constant viscosity was treated by Shenoy (1992). These authors used a modified Darcy model. A non-Darcy model for a power-law fluid was employed by Shenoy (1993a) and Hady and Ibrahim (1997) for flow past a flat plate, by Alkam *et al.* (1998) for flow in concentric annuli, and by Nakayama and Shenoy (1993b) and Chen and Hadim (1995, 1998a,b, 1999a) for flow in a channel. These studies showed that in the non-Darcy regime the effect of increase of power-law index n is to increase the thermal boundary-layer thickness and the wall temperature and to decrease the Nusselt number; in the Darcy regime the changes are small. As the Prandtl number increases, the Nusselt number increases, especially for shear-thinning fluids ($n < 1$). As n decreases, the pressure drop decreases.

An elastic fluid was treated by Shenoy (1993b). A viscoelastic fluid flow over a nonisothermal stretching sheet was analysed by Prasad *et al.* (2002). An experimental study for heat transfer to power-law fluids under flow with uniform heat flux boundary conditions was reported by Rao (2001, 2002).

4.16.4. Bidisperse Porous Media

A bidisperse (or bidispersed—we have opted for the shorter and more commonly used form) porous medium (BDPM), as defined by Z. Q. Chen *et al.* (2000), is composed of clusters of large particles that are agglomerations of small particles. Thus there are macropores between the clusters and micropores within them. Applications are found in bidisperse adsorbent or bidisperse capillary wicks in a heat pipe. Since the bidisperse wick structure significantly increases the area available for liquid film evaporation, it has been proposed for use in the evaporator of heat pipes.

A BDPM thus may be looked at as a standard porous medium in which the solid phase is replaced by another porous medium, whose temperature may be denoted by T_p if local thermal equilibrium is assumed within each cluster. We can then talk about the f-phase (the macropores) and the p-phase (the remainder of the structure). An alternative way of looking at the structure is to regard it as a porous medium in which fractures or tunnels have been introduced. One can then think of the f-phase as being a "fracture phase" and the p-phase as being a "porous phase."

Questions of interest are how one can determine the effective permeability and the effective thermal conductivity of a bidisperse porous medium. Fractal models for each of these have been formulated by Yu and Cheng (2002a,b). In the first paper, the authors developed two models for the effective thermal conductivity based on fractal geometry and the electrical analogy. Theoretical predictions based on these models were compared with those from a previous lumped-parameter model and with experimental data for the stagnant thermal conductivity reported by Z. Q. Chen *et al.* (2000). In this paper a three-dimensional model of touching spatially periodic cubes, which are approximated by touching porous cubes, was used; Cheng and Hsu (1999) had previously used a two-dimensional model. On the basis of their experiments, Z. Q. Chen *et al.* (2000) concluded that, when the ratio of solid/fluid thermal conductivity is greater than 100, the effective thermal conductivity of a bidisperse porous medium is smaller than that of a monodisperse porous medium saturated with the same fluid, because of the contact resistance at the microscale and the higher porosity for the bidisperse medium in comparison with the monodisperse one.

Extending the Brinkman model for a monodisperse porous medium, we propose to model the steady-state momentum transfer in a BDPM by the following pair of coupled equations for $\mathbf{v}_f^*$ and $\mathbf{v}_p^*$, where the asterisks denote dimensional variables,

$$\mathbf{G} = \left(\frac{\mu}{K_f}\right)\mathbf{v}^*{}_f + \zeta(\mathbf{v}^*{}_f - \mathbf{v}^*{}_p) - \tilde{\mu}_f \nabla^{*2}\mathbf{v}^*{}_f \tag{4.147}$$

$$\mathbf{G} = \left(\frac{\mu}{K_p}\right)\mathbf{v}^*{}_p + \zeta(\mathbf{v}^*{}_p - \mathbf{v}^*{}_f) - \tilde{\mu}_p \nabla^{*2}\mathbf{v}^*{}_p. \tag{4.148}$$

Here $\mathbf{G}$ is the negative of the applied pressure gradient, μ is the fluid viscosity, K_f and K_p are the permeabilities of the two phases, and ζ is the coefficient for momentum transfer between the two phases. The quantities $\tilde{\mu}_f$ and $\tilde{\mu}_p$ are the respective effective viscosities. From Eqs. (4.147) and (4.148), $\mathbf{v}^*{}_p$ can be eliminated to give

$$\begin{aligned} &\tilde{\mu}_f\tilde{\mu}_p\nabla^{*4}\mathbf{v}_f^* - [\tilde{\mu}_f(\zeta + \mu/K_p) + \tilde{\mu}_p(\zeta + \mu/K_f)]\nabla^{*2}\mathbf{v}^*{}_f \\ &\quad + [\zeta\mu(1/K_f + 1/K_p) + \mu^2/K_fK_p]\mathbf{v}^*{}_f = \mathbf{G}(2 + \mu/K_p) \end{aligned} \tag{4.149}$$

and $\mathbf{v}^*{}_p$ is given by the same equation with subscripts swapped. For the special case of the Darcy limit one obtains

$$\mathbf{v}^*{}_f = \frac{(\mu/K_p + 2\zeta)\mathbf{G}}{\mu^2/K_fK_p + \zeta\mu(1/K_f + 1/K_p)}, \tag{4.150}$$

$$\mathbf{v}^*{}_p = \frac{(\mu/K_f + 2\zeta)\mathbf{G}}{\mu^2/K_fK_p + \zeta\mu(1/K_f + 1/K_p)}. \tag{4.151}$$

These equations were obtained by Nield and Kuznetsov (2005a). The bulk flow thus is given by

$$\mathbf{G} = (\mu/K)\mathbf{v}^*, \tag{4.152}$$

where

$$\mathbf{v}^* = \varphi \mathbf{v}^*{}_f + (1 - \varphi)\mathbf{v}^*{}_p, \tag{4.153}$$

$$K = \frac{\varphi K_f + (1 - \varphi)K_p + 2(\zeta/\mu)K_f K_p}{1 + (\zeta/\mu)(K_f + K_p)}. \tag{4.154}$$

Thus, in this case, the effect of the coupling parameter ζ merely is to modify the effective permeabilities of the two phases, via the parameter ζ/μ.

Nield and Kuznetsov (2005b) treated forced convection in a parallel-plate channel occupied by a BDPM, using a two-temperature model similar to Eqs. (6.54) and (6.55) in this book. Nield and Kuznetsov (2004c) extended the analysis to the case of a conjugate problem with plane solid slabs bounding the channel. They found that the effect of the finite thermal resistance due to the slabs is to reduce both the heat transfer to the porous medium and the degree of local thermal non-equilibrium. An increase in the value of the Péclet number leads to decrease in the rate of exponential decay in the downstream direction, but does not affect the value of a suitably defined Nusselt number. The case of thermally developing convection in a BDPM was treated by Kuznetsov and Nield (2005). Heat transfer in a BDPM has been reviewed by Nield and Kuznetsov (2005).

4.16.5. Oscillatory Flows

For an annulus and a pipe, Guo *et al.* (1997a,b) treated pulsating flow. For a completely filled channel, Kim *et al.* (1994) studied a pulsating flow numerically. Soundalgekhar *et al.* (1991) studied flow between two parallel plates, one stationary and the other oscillating in its own plane. Hadim (1994a) simulated convection in a channel with localized heat sources.

Sözen and Vafai (1991) analyzed compressible flow through a packed bed with the inlet temperature or pressure oscillating with time about a nonzero mean. They found that the oscillation had little effect on the heat storage capacity of the bed. Paek *et al.* (1999a) studied the transient cool down of a porous medium by a pulsating flow. Experiments involving steady and oscillating flows were conducted by Leong and Jin (2004, 2005).

4.17. Heatlines for Visualizing Convection

The concepts of heatfunction and heatlines were introduced for the purpose of visualizing the true path of the flow of energy through a convective medium (Kimura and Bejan, 1983; Bejan, 1984). The heatfunction accounts simultaneously for the transfer of heat by conduction and convection at every point in the medium. The heatlines are a generalization of the flux lines used routinely in the field of conduction. The concept of heatfunction is a spatial generalization of the concept of Nusselt number, i.e., a way of indicating the magnitude of the heat transfer rate through any unit surface drawn through any point on the convective medium.

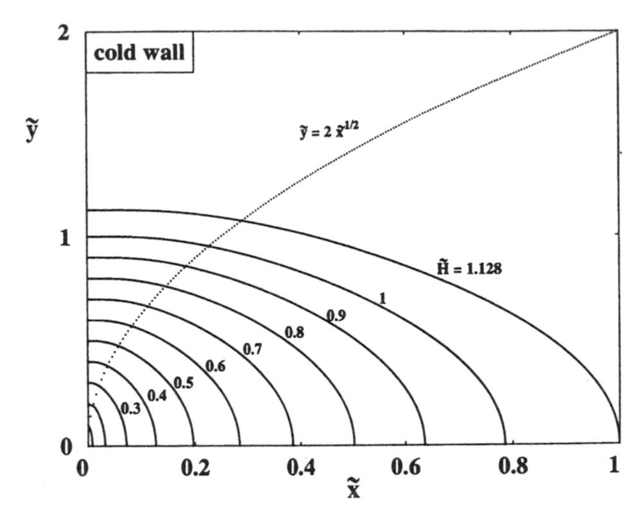

Figure 4.14. The heatlines of the boundary layer near a cold isothermal wall (Morega and Bejan, 1994).

The heatline method was extended to several configurations of convection through fluid-saturated porous media (Morega and Bejan, 1994). To illustrate the method, consider the uniform flow with thermal boundary layer, which is shown in Fig. 4.1. The heatfunction $H(x, y)$ is defined such that it satisfies identically the energy equation for the thermal boundary layer, Eq. (4.3). The H definition is in this case

$$\frac{\partial H}{\partial y} = (\rho c_P) u (T - T_{ref}), \tag{4.155}$$

$$-\frac{\partial H}{\partial x} = (\rho c_P) v (T - T_{ref}) - k_m \frac{\partial T}{\partial y}, \tag{4.156}$$

where the reference temperature T_{ref} is a constant. The flow field (u, v) and the temperature field (T) are furnished by the solutions to the convective heat transfer problem. It was pointed out in Trevisan and Bejan (1987a) that T_{ref} can have any value and that a heatline pattern can be drawn for each T_{ref} value. The most instructive pattern is obtained when T_{ref} is set equal to the lowest temperature that occurs in the convective medium that is being visualized. This choice was made in the construction of Figs. 4.14 and 4.15. In both cases the heatfunction can be obtained analytically. When the wall is colder (T_w) than the approaching flow (T_∞) (Fig. 4.14), the nondimensionalized heatfunction is

$$\widetilde{H}(\widetilde{x}, \widetilde{y}) = \widetilde{x}^{1/2} \left[\eta \operatorname{erf}\left(\frac{\eta}{2}\right) + \frac{2}{\pi^{1/2}} \exp\left(-\frac{\eta^2}{4}\right) \right], \tag{4.157}$$

where $\widetilde{H} = H/[k_m(T_\infty - T_w)\mathrm{Pe}_l^{1/2}$, $\mathrm{Pe}_L = U_\infty L/\alpha_m$, $\widetilde{x} = x/L$, and $\eta = y(U_\infty/\alpha_m x)^{1/2}$. In these expressions L is the length of the $y = 0$ boundary. Figure

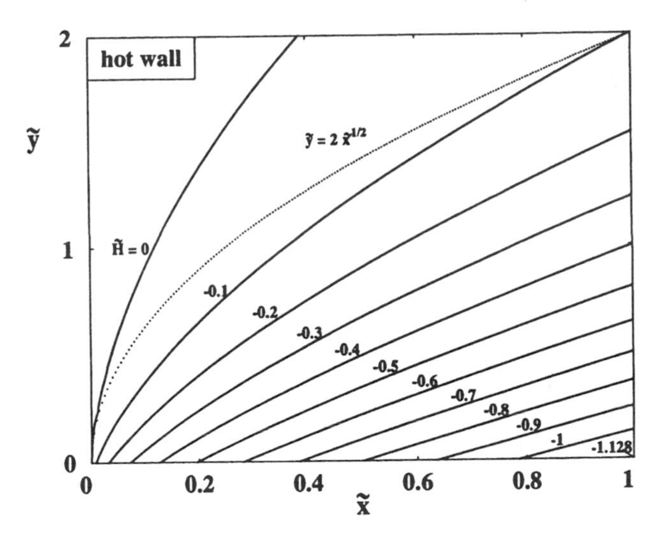

Figure 4.15. The heatlines of the boundary layer near a hot isothermal wall (Morega and Bejan, 1994).

4.14 shows that the H = constant curves visualize several features of convection near a cold wall. The energy that is eventually absorbed by the wall is brought into the boundary layer ($\widetilde{y} \cong 2\widetilde{x}^{1/2}$) by fluid from upstream of the cold section of the wall. The heatlines that enter the wall are denser near $\widetilde{x} = 0$, i.e., the heat flux is more intense. Finally, the value of the heatfunction increases along the wall, because the wall absorbs the heat released by the fluid. The trailing-edge $\widetilde{H}$ value matches the total heat transfer rate through the wall, Eq. (4.14).

Figure 4.15 shows the corresponding pattern of heatlines when the wall is warmer than the approaching fluid,

$$\widetilde{H}(\widetilde{x}, \widetilde{y}) = \widetilde{x}^{1/2}\left[\eta \operatorname{erfc}\left(\frac{\eta}{2}\right) - \frac{2}{\pi^{1/2}} \exp\left(-\frac{\eta^2}{4}\right)\right]. \tag{4.158}$$

The heatlines come out of the wall at an angle because, unlike in Fig. 4.14, the gradient $\partial H/\partial y$ is not zero at the wall. Above the wall, the heatlines are bent even more by the flow because the effect of transversal conduction becomes weaker. The higher density of heatlines near $\widetilde{x} = 0$ indicates once again higher heat fluxes. The $\widetilde{H}$ value at the wall decreases in the downstream direction because the wall loses heat to the boundary layer.

Morega and Bejan (1994) displayed the heatlines for two additional configurations: boundary layers with uniform heat flux and flow through a porous layer held between parallel isothermal plates. As in Figs. 4.14 and 4.15, the heatlines for cold walls are unlike the heatlines for configurations with hot walls. In other words, unlike the patterns of isotherms that are used routinely in convection heat transfer (e.g., Fig. 7.4), the heatline patterns indicate the true direction of heat flow and distinguish between cold walls and hot walls.

Costa (2003) has reported a study of unified streamline, heatline, and massline methods of visualization of two-dimensional heat and mass transfer in anisotropic media. His illustrations include a problem involving natural convection in a porous medium.

Heatlines and masslines are now spreading throughout convection research as the proper way to visualize heat flow and mass flow. This method of visualization is particularly well suited for computational work, and should be included in commercial computational packages. The growing activity based on the heatlines method is reviewed in Bejan (2004a) and Costa (2005).

4.18. Constructal Tree Networks: Minimal Resistance in Volume-to-Point Flows

It was discovered recently that by minimizing geometrically the thermal resistance between one point and a finite-size volume (an infinity of points) it is possible to predict a most common natural structure that previously was considered nondeterministic: the tree network (Bejan, 1996b, 1997a,b; Ledezma *et al.*, 1997). Tree network patterns abound in nature, in both animate and inanimate systems (e.g., botanical trees, lightning, neural dendrites, dendritic crystals). The key to solving this famous problem was the optimization of the shape of each finite-size element of the flow volume, such that the flow resistance of the element is minimal. The optimal structure of the flow—the tree network—then was *constructed* by putting together the shape-optimized building blocks. This construction of multiscale, hierarchical geometry became the starting point of the *constructal theory* of self-optimization and self-organization in Nature (Bejan, 1997c, 2000).

The deterministic power of constructal theory is an invitation to new theoretical work on natural flow structures that have evaded determinism in the past. This section is about one such structure: the dendritic shape of the low-resistance channels that develop in natural fluid flows between a volume and one point in heterogeneous media (Bejan, 1997b,c; Bejan *et al.*, 2004). Examples of volume-to-point fluid flows are the bronchial trees, the capillary vessels, and the river drainage basins and deltas.

The deterministic approach outlined in this section is based on the proposition that a naturally occurring flow structure—its geometric form—is the end result of a process of geometric optimization. The objective of the optimization process is to construct the path (or assembly or paths) that provides minimal resistance to flow, or, in an isolated system, maximizes the rate of approach to equilibrium.

4.18.1. The Fundamental Volume-to-Point Flow Problem

Consider the fundamental problem of minimizing the resistance to fluid flow between one point and a finite-size volume (an infinity of points). For simplicity

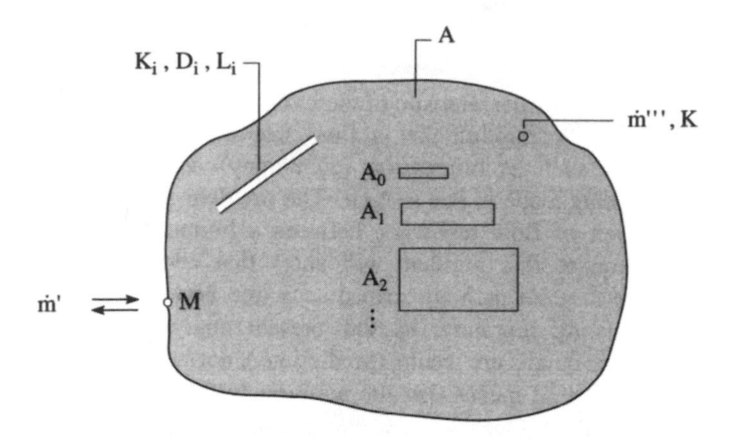

Figure 4.16. The two-dimensional flow between one point (M) and a finite-size volume (A).

we assume that the volume is two-dimensional and represented by the area A (Fig. 4.16). The total mass flow rate $\dot{m}'$ (kg/sm) flows through the point M and reaches (or originates from) every point that belongs to A. We also assume that the volumetric mass flow rate $\dot{m}'''$(kg/sm^3) that reaches all the points of A is distributed uniformly in space, hence $\dot{m}' = \dot{m}''' A$.

The space A is filled by a porous medium saturated with a single-phase fluid with constant properties. The flow is in the Darcy regime. If the permeability of the porous medium is uniform throughout A, then the pressure field $P(x, y)$ and the flow pattern can be determined uniquely by solving the Poisson-type problem associated with the point sink or point source configuration of Fig. 4.16. This classic problem is not the subject of this section.

Instead, we consider the more general situation where the space A is occupied by a nonhomogeneous porous medium composed of a material of low permeability K and a number of layers (e.g., cracks, filled or open) of much higher permeabilities ($K_1, K_2, \ldots$). The thicknesses ($D_1, D_2, \ldots$) and lengths ($L_1, L_2, \ldots$) of these layers are not specified.

For simplicity we assume that the volume fraction occupied by the high-permeability layers is small relative to the volume represented by the K material. There is a very large number of ways in which these layers can be sized, connected, and distributed in order to collect and channel $\dot{m}'$ to the point M. In other words, there are many designs of composite materials ($K, K_1, K_2, \ldots$) that can be installed in A: our objective is to find not only the internal architecture of the composite that minimizes the overall fluid-flow resistance, but also a *strategy* for the geometric optimization of volume-to-point flows in general.

The approach we have chosen is illustrated in Fig. 4.16. We regard A as a patchwork of rectangular elements of several sizes ($A_0, A_1, A_2, \ldots$). We will show that the shape (aspect ratio) of each such element can be optimized for minimal flow resistance. The smallest element (A_0) contains only low-permeability material and

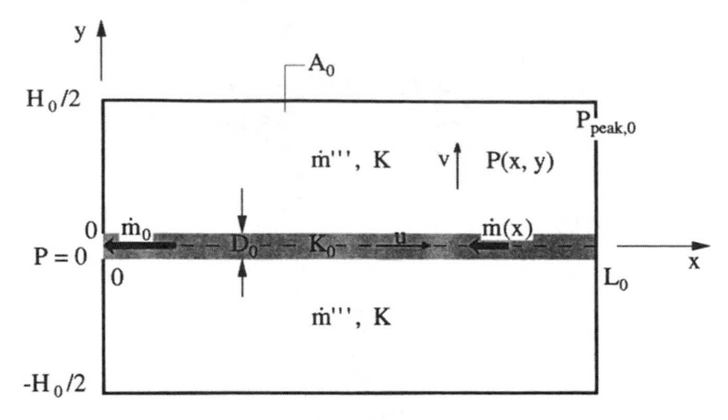

Figure 4.17. The smallest volume element, with volumetric flow through the K porous medium and "channel" flow along a high-permeability layer (K_0).

one high-permeability layer (K_0, D_0), Fig. 4.17. Each successively larger volume element (A_i) is an assembly of elements of the preceding size (A_{i-1}), which act as tributaries to the collecting layer (K_i, D_i, L_i) that defines the assembly. We will show that the optimally shaped assemblies can be arranged like building blocks to collect the volumetric flow $\dot{m}'''$ and transform it into the single stream $\dot{m}'$ at the point M.

Before presenting the analysis, it is worth commenting on the reasons for doing it and how it fits next to the vast amount of work that has been done in the same field. A general characteristic of the exiting studies is that they begin with the often tacit assumption that a fluid tree network exists. Geometric details such as bifurcation (dichotomy) are assumed. No such assumptions are being made in this section. The problem solved in this section is the minimization of flow resistance between a finite-size volume and one point. The solution to this problem will show that certain portions of the optimized volume-to-point path are shaped as a tree network. In other words, unlike in the existing literature, in the present analysis the tree and its geometric details are results (predictions), not assumptions. This is a fundamental difference. It means that the solution to the volume-to-point flow problem sheds light on the universal design principle that serves as origin for the formation of fluid tree networks in nature.

4.18.2. The Elemental Volume

In Fig. 4.17 the smallest volume $A_0 = H_0 L_0$ is fixed, but its shape H_0/L_0 may vary. The flow, $\dot{m}'_0 = \dot{m}''' A_0$, A_0 is collected from the K medium by a layer of much higher permeability K_0 and thickness D_0. The flow is driven toward the origin (0, 0) by the pressure field $P(x, y)$. The rest of the rectangular boundary $H_0 \times L_0$ is impermeable. Since the flow rate $\dot{m}'_0$ is fixed, to minimize the flow resistance means to minimize the peak pressure (P_{peak}) that occurs at a point inside A_0. The pressure at the origin is zero.

The analysis is greatly simplified by the assumptions that were mentioned already ($K \ll K_0$, $D_0 \ll H_0$), which, as we will show in Eq. (4.156), also mean that the optimized A_0 shape is such that H_0 is considerably smaller than L_0. According to these assumptions the flow through the K domain is practically parallel to the y direction,

$$P(x, y) \cong P(y) \quad \text{for} \quad H_0/2 > |y| > D_0/2 \tag{4.159}$$

while the flow through the K_0 layer is aligned with the layer itself $P(x, y) \cong P(x)$ for $|y| < D_0/2$. Symmetry and the requirement that P_{peak} be minimum dictate that the A_0 element be oriented such that the K_0 layer is aligned with the x axis. The mass flow rate through this layer is $\dot{m}'(x)$, with $\dot{m}'(0) = \dot{m}'_0$ at the origin (0, 0), and $\dot{m}'(L_0) = 0$. The K material is an isotropic porous medium with flow in the Darcy regime,

$$v = \frac{K}{\mu}\left(-\frac{\partial P}{\partial y}\right) \tag{4.160}$$

In this equation v is the volume-averaged velocity in the y direction (Fig. 4.17). The actual flow is oriented in the opposite direction. The pressure field $P(x, y)$ can be determined by eliminating v between Eq. (4.151) and the local mass continuity condition

$$\frac{\partial v}{\partial y} = \frac{\dot{m}'''}{\rho} \tag{4.161}$$

and applying the boundary conditions $\partial P/\partial y = 0$ at $y = H_0/2$ and $P = P(x, 0)$ at $y \cong 0$ (recall that $D_0 \ll H_0$):

$$P(x, y) = \frac{\dot{m}'''v}{2K}(H_0 y - y^2) + P(x, 0). \tag{4.162}$$

Equation (4.162) holds only for $y \gtrsim 0$. The corresponding expression for $y \lesssim 0$ is obtained by replacing H_0 with $-H_0$ in Eq. (4.162).

The pressure distribution in the K_0 material, namely $P(x, 0)$, is obtained similarly by assuming Darcy flow along a D_0-thin path near $y = 0$,

$$u = \frac{K_0}{\mu}\left(-\frac{\partial P}{\partial x}\right), \tag{4.163}$$

where u is the average velocity in the x direction. The flow proceeds toward the origin, as shown in Fig. 4.17. The mass flow rate channeled through the K_0 material is $\dot{m}'(x) = -r D_0 u$. Furthermore, mass conservation requires that the mass generated in the infinitesimal volume slice ($H_0 dx$) contributes to the $\dot{m}'(x)$ stream: $\dot{m}''' H_0 dx = -d\dot{m}'$. Integrating this equation away from the impermeable plane $x = L_0$ (where $\dot{m}' = 0$), and recalling that $\dot{m}'_0 = \dot{m}''' H_0 L_0$, we obtain

$$\dot{m}(x) = \dot{m}''' H_0(L_0 - x) = \dot{m}_0\left(1 - \frac{x}{L_0}\right). \tag{4.164}$$

Combining these equations we find the pressure distribution along the x axis

$$P(x,0) = \frac{\dot{m}_0' \nu}{D_0 K_0}\left(x - \frac{x^2}{2L_0}\right). \tag{4.165}$$

Equations (4.162) and (4.165) provide a complete description of the $P(x, y)$ field. The peak pressure occurs in the farthest corner ($x = L_0$, $y = H_0/2$):

$$P_{peak,0} = \dot{m}_0' \nu \left(\frac{H_0}{8KL_0} + \frac{L_0}{2K_0 D_0}\right). \tag{4.166}$$

This pressure can be minimized with respect to the shape of the element (H_0/L_0) by noting that $L_0 = A_0/H_0$ and $\phi_0 = D_0/H_0 \ll 1$. The number ϕ_0 is carried in the analysis as an unspecified parameter. For example, if the D_0 layer was originally a crack caused by the volumetric shrinking (e.g., cooling, drying) of the K medium, then D_0 must be proportional to the thickness H_0 of the K medium. The resulting geometric optimum is described by

$$\frac{H_0}{L_0} = 2(\widetilde{K}_0\phi_0)^{-1/2} \qquad \widetilde{L}_0 = 2^{-1/2}(\widetilde{K}_0\phi_0)^{1/4} \tag{4.167}$$

$$\widetilde{H}_0 = 2^{1/2}(\widetilde{K}_0\phi_0)^{-1/4} \qquad \Delta\widetilde{P}_0 = \frac{1}{2}(\widetilde{K}_0\phi_0)^{-1/2} \tag{4.168}$$

The nondimensionalization used in Eqs. (4.146)–(4.147) and retained throughout this section is based on using $A_0^{1/2}$ as length scale and K as permeability scale:

$$(\widetilde{H}_i, \widetilde{L}_i) = \frac{(H_i, L_i)}{A_0^{1/2}}, \qquad \widetilde{K}_i = \frac{K_i}{K}, \tag{4.169}$$

$$\Delta\widetilde{P}_i = \frac{P_{peak,i}}{\dot{m}''' A_i \nu / K}, \qquad \phi_i = \frac{D_i}{H_i}. \tag{4.170}$$

At the optimum, the two terms on the right side of Eq. (4.166) are equal. The shape of the A_0 element is such that the pressure drop due to flow through the K material is equal to the pressure drop due to the flow along the K_0 layer. Note also that the first of Eq. (4.168) confirms the assumptions made about the D_0 layer at the start of this section: high permeability ($\widetilde{K}_0 \gg 1$) and small volume fraction ($\phi_0 \ll 1$) mean that the optimized A_0 shape is slender, $H_0 \ll L_0$, provided that $\widetilde{K}_0 \gg \phi_0^{-1}$.

4.18.3. The First Construct

Consider next the immediately larger volume $A_1 = H_1 L_1$ (Fig. 4.18), which can contain only elements of the type optimized in the preceding section. The streams $\dot{m}_0'$ collected by the D_0-thin layers are now united into a larger stream $\dot{m}_1'$ that connects A_1 with the point $P = 0$. The $\dot{m}_1'$ stream is formed in the new layer (K_1, D_1, L_1).

The problem of optimizing the shape of the A_1 rectangle is the same as the A_0 problem that we just solved. First, we note that when the number of A_0 elements assembled into A_1 is large, the composite material of Fig. 4.18 is analogous to the

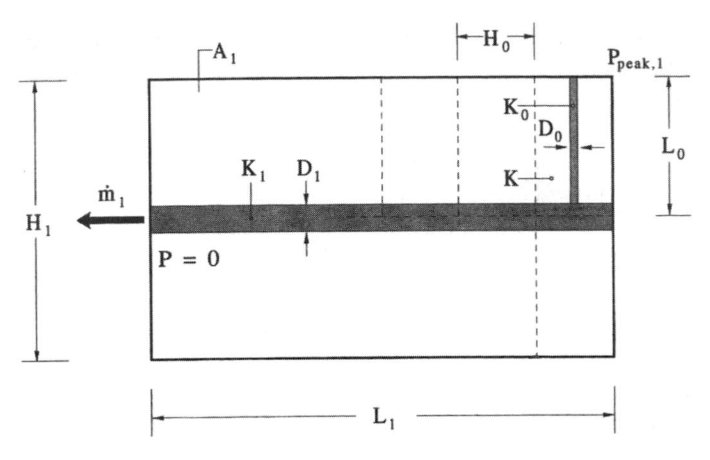

Figure 4.18. The first assembly (A_1) of elements of size A_0, and the new high-permeability layer K_1.

composite of Fig. 4.17, provided that the permeability K of Fig. 4.17 is replaced by an equivalent (volume averaged) permeability (K_{e1}) in Fig. 4.18. The K_{e1} value is obtained by writing that the pressure drop across an A_0 element [Eq. (4.168)] is equal to the pressure drop over the distance $H_1/2$ in the K_{e1} medium [this second pressure drop can be read off Eq. (4.162), after replacing H_0 with H_1, y with $H_1/2$, and K with K_{e1}]. The result is $K_{e1} = K_0\phi_0$: this value is then used in place of K_0, in an analysis that repeats the steps executed in Eqs. (4.166)–(4.168) for the A_0 optimization problem.

A clearer alternative to this analysis begins with the observation that the peak pressure ($P_{peak.1}$) in Fig. 4.18 is due to two contributions: the flow through the upper-right corner element ($P_{peak.0}$) and the flow along the (K_1, D_1) layer:

$$P_{peak,1} = \dot{m}''' A_0 \frac{\nu}{K}\frac{1}{2}(\widetilde{K}_0\phi_0)^{-1/2} + \dot{m}_1' \nu \frac{L_i}{2K_1 D_1}. \tag{4.171}$$

This expression can be rearranged by using the first of Eqs. (4.168) and $H_1 = 2L_0$:

$$\frac{P_{peak,1}}{\dot{m}''' A_1 \nu / K} = \frac{1}{4\widetilde{K}_0\phi_0}\frac{H_1}{L_1} + \frac{1}{2\widetilde{K}_1\phi_1}\frac{L_1}{H_1}. \tag{4.172}$$

The corner pressure $P_{peak,1}$ can be minimized by selecting the H_1 / L_1 shape of the A_1 rectangle. The resulting expressions for the optimized geometry (H_1/L_1, $\widetilde{H}_1$, $\widetilde{L}_1$) are listed in Table 4.1. The minimized peak pressure ($\Delta\widetilde{P}_1$) is divided equally between the flow through the corner A_0 element and the flow along the collecting (K_1, D_1) layer. In other words, as in the case of the A_0 element, the geometric optimization of the A_1 assembly is ruled by a principle of *equipartition* of pressure drop between the two main paths of the assembly (Lewins, 2003).

Table 4.1. The optimized geometry of the elemental area A_0 and the subsequent assemblies when the channel permeabilities are unrestricted (Note: $C_i = K_i \phi_i$).

i	H_i/L_i	$\widetilde{H}_i$	$\widetilde{L}_i$	$n_i = A_i/A_{i-1}$	$\Delta \widetilde{P}_i$
0	$2C_0^{-1/2}$	$2^{1/2}C_0^{-1/4}$	$2^{-1/2}C_0^{1/4}$	—	$\frac{1}{2}C_0^{-1/2}$
1	$(2C_0/C_1)^{1/2}$	$2^{1/2}C_0^{1/4}$	$C_0^{-1/4}C_1^{1/2}$	$(2C_1)^{1/2}$	$(2C_0C_1)^{-1/2}$
2	$(2C_1/C_2)^{1/2}$	$2C_0^{-1/4}C_1^{1/2}$	$2^{1/2}C_0^{-1/4}C_2^{1/2}$	$2(C_2/C_0)^{1/2}$	$(2C_1C_2)^{-1/2}$
$i \geq 2$	$(2C_{i-1}/C_i)^{1/2}$	$2^{i/2}C_0^{-1/4}C_{i-1}^{1/2}$	$2^{(i-1)}/2C_0^{-1/4}C_i^{1/2}$	$2(C_1/C_{i-2})^{1/2}$	$(2C_{i-1}C_i)^{-1/2}$

4.18.4. Higher-Order Constructs

The assembly and area shape optimization procedure can be repeated for larger assemblies ($A_2, A_3, \ldots$). Each new assembly (A_i) contains a number (n_i) of assemblies of the immediately smaller size (A_{i-1}), the flow of which is collected by a new high-permeability layer (K_i, D_i, L_i). As in the drawing shown in Fig. 4.17 for A_1, it is assumed that the number of constituents n_i is sensibly larger than 2. The analysis begins with the statement that the maximum pressure difference sustained by A_i is equal to the pressure difference across the optimized constituent (A_{i-1}) that occupies the farthest corner of A_i, and the pressure drop along the K_i central layer:

$$P_{peak,i} = P_{peak,i-1} + \dot{m}_i' \nu \frac{L_i}{2K_i D_i}. \tag{4.173}$$

The geometric optimization results are summarized in Table 4.1, in which we used $C_i = \widetilde{K}_i \phi_i$ for the dimensionless flow conductance of each layer. The optimal shape of each rectangle $H_i \times L_i$ is ruled by the pressure-drop equipartition principle noted in the optimization of the A_0 and A_1 shapes.

Beginning with the second assembly, the results fall into the pattern represented by the recurrence formulas listed for $i \geq 2$. If these formulas were to be repeated *ad infinitum* in both directions—toward large A_i and small A_i—then the pattern formed by the high-permeability paths (K_i, D_i) would be a fractal. Natural tree-shaped flows and those predicted by constructal theory are not fractal. In the present solution to the volume-to-point flow problem, the construction begins with an element of finite size, A_0 and ends when the given volume (A) is covered. Access to the infinity of points contained by the given volume is not made by making A_0 infinitely small. Instead, all the points of the given volume are reached by a diffusive flow that bathes A_0 *volumetrically*, because the permeability K of the material that fills A_0 is the lowest of all the permeabilities of the composite porous medium. Constructal theory is the clearest statement that the geometry of nature is not fractal (Bejan, 1997c) and the first theory that predicts the multitude of natural flow structures that could be described as "fractal-like" structures (Poirier, 2003; Rosa *et al.*, 2004).

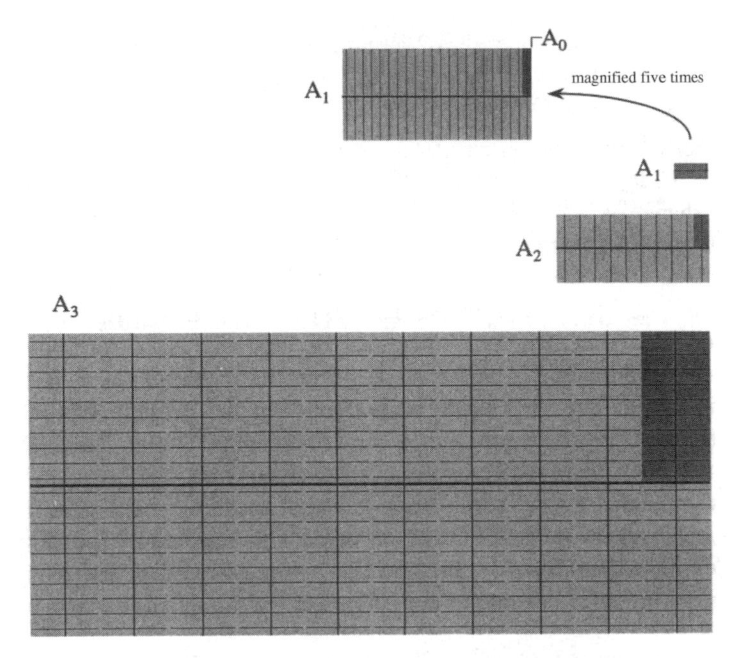

Figure 4.19. Composite medium tree architecture for minimal volume-to-point flow resistance when $C_0 = 100$ and $C_i/C_{i-1} = 10$ for $i = 1, 2$, and 3.

Figure 4.19 illustrates the minimal-resistance architecture recommended by the results of Table 4.1. At each level of assembly, the calculated number of constituents n_i was rounded off to the closest even number. The optimal design of the composite porous medium contains a tree network of high-permeability layers ($K_0, K_1, K_2, \ldots$), where the interstitial spaces are filled with low-permeability material (K). The actual shape of the tree depends on the relative size of the flow conductance parameters C_i. The conductance increase ratio C_i/C_{i-1} is essentially equal to the permeability ratio K_i/K_{i-1}, because the volume fraction ($\phi_i \ll 1$) is expected to vary little from one assembly to the next, cf. the comment made above Eq. (4.167). In other words, the conductance parameters C_i can be specified independently because the porous-medium characteristics of the materials that fill the high-permeability channels have not been specified.

Several trends are revealed by constructions such as Fig. 4.19. When the conductance ratio C_i/C_{i-1} is large, the number n_i is large, the optimal shape of each assembly is slender ($H_i/L_i < 1$), and the given volume is covered "fast," i.e., in a few large steps of assembly and optimization. When the ratio C_i/C_{i-1} is large but decreases from one assembly to the next, the number of constituents decreases and the shape of each new assembly becomes closer to square.

Combining the limit $C_i/C_{i-1} \to 1$ with the n_i formula of Table 4.1, we see that the number *two* (i.e., dichotomy, bifurcation, pairing) emerges as a result of geometric optimization of volume-to-point flow. Note that the actual value $n_i = 2$

is not in agreement with the $n_i > 2$ assumption that was made in Fig. 4.18 and the analysis that followed. This means that when $C_i/C_{i-1} \sim 1$ is of order 1, the analysis must be refined by using, for example, a Fig. 4.18 in which the length of the (K_1, D_1) layer is not L_1 but $(n_1/2 - 1)H_0 + H_0/2$. In this new configuration the right-end tip of the (K_1, D_1) layer is absent because the flow rate through it would be zero. To illustrate this feature of the tree network, in Fig. 4.19 the zero-flow ends of the central layers of all the assemblies have been deleted.

4.18.5. The Constructal Law of Geometry Generation in Nature

The point-to-volume resistance can be minimized further by varying the angle between tributaries (D_{i-1}) and the main channel (D_i) of each new volume assembly. This optimization principle is well known in physiology where the work always begins with the assumption that a tree network of tubes *exists*. It can be shown numerically that the reductions in flow resistance obtained by optimizing the angles between channels are small relative to the reductions due to optimizing the shape of each volume element and assembly of elements. In this section we fixed the angles at 90° and focused on the optimization of volume shape. It is the optimization of shape subject to volume constraint—the consistent use of this principle at every volume scale—that is responsible for the emergence of a tree network between the volume and the point. We focused on the optimal shapes of building blocks because our objective was to discover a single optimization principle that can be used to explain the origin of tree-shaped networks in natural flow systems. The objective was to find the physics principle that was missing in the treelike images generated by assumed fractal algorithms.

In summary, we solved in general terms the fundamental fluid mechanics problem of minimizing the flow resistance between one point and a finite-size volume. A single optimization principle—the optimization of the shape of each volume element such that its flow resistance is minimized—is responsible for all the geometric features of the point-to-volume flow path. One of these features is the geometric structure—the tree network—formed by the portions with higher permeabilities $(K_0, K_1, \ldots)$. The interstices of the network, i.e., the infinity of points of the given volume, are filled with material of the lowest permeability (K) and are touched by a flow that diffuses through the K material.

The most important conclusion is that the larger picture, the optimal overall performance, structure, and working mechanism can be described in a purely deterministic fashion; that is, if the resistance-minimization principle is recognized as law. This law can be stated as follows (Bejan, 1996b, 1997a):

> For a finite-size system to persist in time (to live), it must evolve in such a way that it provides easier access (less resistance) to the imposed currents that flow through it.

This statement has two parts. First, it recognizes the natural tendency of imposed global currents to construct paths (shapes, structures) for better access through constrained open systems. The second part accounts for the evolution of the structure, which occurs in an identifiable direction that can be aligned with time itself. Small size and shapeless flow (diffusion) are followed in time by larger sizes and organized flows (streams). The optimized complexity continues to increase in time. Optimized complexity must not be confused with maximized complexity.

How important is the constructal approach to the minimal-resistance design, i.e., this single geometric optimization principle that allows us to anticipate the tree architecture seen in so many natural systems? In contemporary physics a significant research volume is being devoted to the search for universal design principles that may explain organization in animate and inanimate systems. In this search, the tree network is recognized as the symbol of the challenge that physicists and biologists face (Kauffman, 1993, pp. 13 and 14): "Imagine a set of identical round-topped hills, each subjected to rain. Each hill will develop a particular pattern of rivulets which branch and converge to drain the hill. Thus the particular branching pattern will be unique to each hill, a consequence of particular contingencies in rock placement, wind direction, and other factors. The particular history of the evolving patterns of rivulets will be unique to each hill. But viewed from above, the statistical features of the branching patterns may be very similar. Therefore, we might hope to develop a theory of the statistical features of such branching patterns, if not of the particular pattern on one hill."

The constructal approach outlined in this section is an answer to the challenge articulated so well by Kauffman. It introduces an engineering flavor in the current debate on natural organization, which until now has been carried out in physics and biology. By training, engineers begin the design of a device by first understanding its purposes. The size of the device is always finite, never infinitesimal. The device must function (i.e., fulfill its purpose) subject to certain constraints. Finally, to analyze (describe) the device is not sufficient: to optimize it, to construct it, and to make it work are the ultimate objectives. All these features—purpose, finite size, constraints, optimization, and construction—can be seen in the network constructions reported in this section. The resulting tree networks are entirely deterministic, and consequently they represent an alternative worthy of consideration in fields outside engineering. The progress in this direction is summarized in a new book (Bejan, 1997c).

The short discussion here is confined to hydrodynamic aspects. For conduction, convection, turbulence, and other flows with structure, the reader is referred to the new books that review the growing interest in constructal theory (Bejan, 2000; Rosa *et al.*, 2004; Bejan *et al.*, 2004). For example, constructal trees were designed for chemically reactive porous media by Azoumah *et al.* (2004), and constructal theory was used to predict the basic features and dimensions of Bénard convection and nucleate boiling (Nelson and Bejan, 1998).

The place of the constructal law as a self-standing law in thermodynamics is firmly established (Bejan and Lorente, 2004). The constructal law is distinct from the second law. For example, with respect to the time evolution of an isolated thermodynamic system, the second law states that the system will proceed toward a state of equilibrium ("nothing moves," maximum entropy at constant energy). In this second-law description, the system is a black box, without configuration.

With regard to the same isolated system, the constructal law states that the currents that flow in order to bring the system to equilibrium will seek and develop paths of maximum access. In this way, the system develops its flow configuration, which endows the system with the ability to approach its equilibrium the fastest.

The constructal law is the law of geometry generation, whereas the second law is the law of entropy generation. The constructal law can be stated in several equivalent ways: a principle of flow access maximization (or efficiency increase), as in the original statement quoted above, a principle of flow compactness maximization (miniaturization), and a principle of flow territory maximization, as in the spreading of river deltas, living species, and empires (Bejan and Lorente, 2004).

In sum, constructal theory originates from the engineering of porous and complex flow structures and now unites engineering, physics, biology, and social organization (Poirier, 2003; Rosa *et al.*, 2004).

4.19. Constructal Multiscale Flow Structures

The tree-shaped flow structures of Section 4.18 are examples of "designed" porous structures with multiple length scales, which are organized hierarchically and distributed nonuniformly. Another class of designed porous media stems from an early result of constructal theory: the prediction of optimal spacings for the internal flow structure of volumes that must transfer heat and mass to the maximum (Bejan, 2000; Section 4.15). Optimal spacings have been determined for several configurations, for example, arrays of parallel plates (e.g., Fig. 4.20). In each configuration, the reported optimal spacing is a single value, that is, a *single length scale* that is distributed uniformly through the available volume.

Is the stack of Fig. 4.20 the best way to pack heat transfer into a fixed volume? It is, but only when a single length scale is to be used, that is, if the structure is to be *uniform*. The structure of Fig. 4.20 is uniform, because it does not change from $x = 0$ to $x = L_0$. At the most, the geometries of single-spacing structures vary periodically, as in the case of arrays of cylinders and staggered plates.

Bejan and Fautrelle (2003) showed that the structure of Fig. 4.20 can be improved if more length scales (D_0, D_1, $D_2, \ldots$) are available. The technique consists of placing more heat transfer in regions of the volume HL_0 where the boundary layers are thinner. Those regions are situated immediately downstream of the entrance plane $x = 0$. Regions that do not work in a heat transfer sense either must be put to work or eliminated. In Fig. 4.20, the wedges of fluid contained between the tips of opposing boundary layers are not involved in transferring heat. They can be involved if heat-generating blades of shorter lengths (L_1) are installed on their planes of symmetry. This new design is shown in Fig. 4.21.

Each new L_1 blade is coated by Blasius boundary layers with the thickness $\delta(x) \cong 5x(Ux/\nu)^{-1/2}$. Because δ increases as $x^{1/2}$, the boundary layers of the L_1

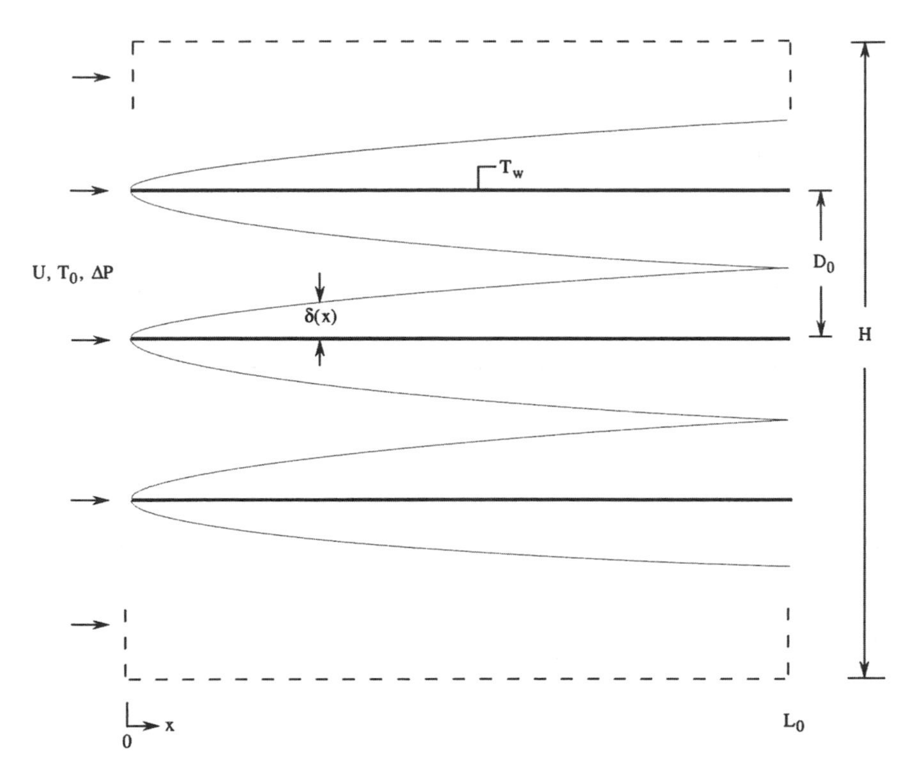

Figure 4.20. Optimal package of parallel plates with one spacing (Bejan and Fautrelle, 2003).

blade merge with the boundary layers of the L_0 blades at a downstream position that is approximately equal to $L_0/4$. The approximation is due to the assumption that the presence of the L_1 boundary layers does not significantly affect the downstream development ($x > L_0/4$) of the L_0 boundary layers. This assumption is made for the sake of simplicity. The order-of-magnitude correctness of this assumption comes from geometry: the edges of the L_1 and L_0 boundary layers must intersect at a distance of order

$$L_1 \cong \frac{1}{4} L_0. \tag{4.174}$$

Note that by choosing L_1 such that the boundary layers that coat the L_1 blade merge with surrounding boundary layers at the downstream end of the L_1 blade, we once more invoke the maximum packing principle of constructal theory. We are being consistent as constructal designers, and because of this every structure with merging boundary layers will be optimal, no matter how complicated.

The wedges of isothermal fluid (T_0) remaining between adjacent L_0 and L_1 blades can be populated with a new generation of even shorter blades, $L_2 \cong L_1/4$. Two such blades are shown in the upper-left corner of Fig. 4.21. The length scales become smaller (L_0, L_1, L_2), but the shape of the boundary layer region is the

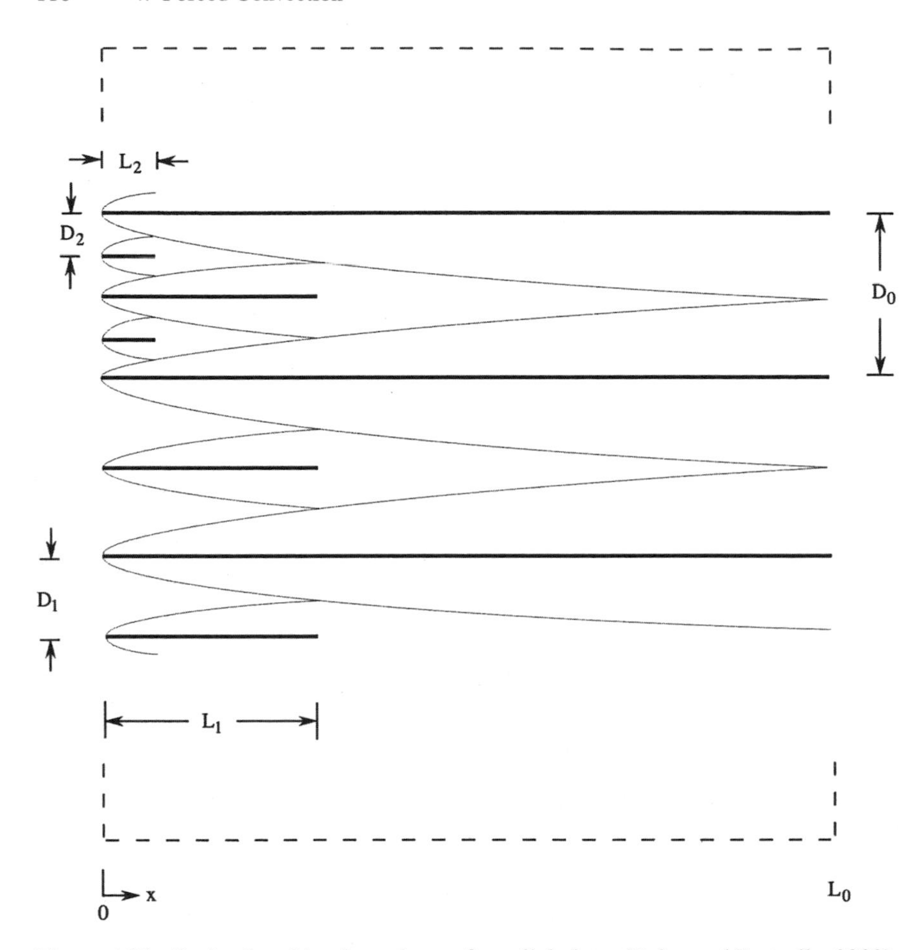

Figure 4.21. Optimal multiscale package of parallel plates (Bejan and Fautrelle, 2003).

same for all the blades, because the blades are all swept by the same flow (U). The merging and expiring boundary layers are arranged according to the algorithm

$$L_i \cong \frac{1}{4} L_{i-1}, \quad D_i \cong \frac{1}{2} D_{i-1} \quad (i = 1, 2, \ldots, m), \tag{4.175}$$

where we show that m is finite, not infinite. In other words, as in all the constructal tree structures, the image generated by the algorithm is not a fractal [cf. Bejan (1997c, p. 765)]. The sequence of decreasing length scales is finite, and the smallest size (D_m, L_m) is known, as shown in Bejan and Fautrelle (2003) and Bejan *et al.* (2004). The global thermal conductance of the multiscale package is

$$\frac{q'}{k\,\Delta T} \cong 0.36 \frac{H}{L_0}\,\mathrm{Be}^{1/2}\left(1 + \frac{m}{2}\right)^{1/2} \tag{4.176}$$

where q' is the total heat transfer rate installed in the package (W/m, per unit length in the direction perpendicular to Fig. 4.21), k is the fluid thermal conductivity, and ΔT is the temperature difference between the plates (assumed isothermal) and the fluid inlet. The dimensionless pressure and difference is

$$\mathrm{Be} = \frac{\Delta P L_0^2}{\mu\alpha}, \tag{4.177}$$

where μ and α are the fluid viscosity and thermal diffusivity.

Bejan and Fautrelle (2003) also showed that the optimized complexity increases with the imposed pressure difference (Be),

$$2^m \left(1 + \frac{m}{2}\right)^{1/4} \cong 0.17\ \mathrm{Be}^{1/4}. \tag{4.178}$$

As Be increases, the multiscale structure becomes more complex *and* finer. The monotonic effect of m is accompanied by diminishing returns: each smaller length scale (m) contributes to global performance less than the preceding length scale (m − 1). The validity of the novel design concept sketched in Fig. 4.21 was demonstrated through direct numerical simulations and optimization for multiscale parallel plates (Bello-Ochende and Bejan, 2004) and multiscale parallel cylinders in cross-flow (Bello-Ochende and Bejan, 2005a). A related natural convection situation was treated by Bello-Ochende and Bejan (2005b).

Forced convection was used in Bejan and Fautrelle (2003) only for illustration, that is, as a language in which to describe the new concept. A completely analogous multiscale structure can be deduced for laminar natural convection. The complete analogy that exists between optimal spacings in forced and natural convection was described by Petrescu (1994). In brief, if the structure of Fig. 4.20 is rotated by 90° counterclockwise and if the flow is driven upward by the buoyancy effect, then the role of the overall pressure difference ΔP is played by the difference between two hydrostatic pressure heads, one for the fluid column of height L_0 and temperature T_0 and the other for the L_0 fluid column of temperature T_w. If the Boussinesq approximation applies, the effective ΔP due to buoyancy is

$$\Delta P = \rho g \beta \Delta T\ L_0, \tag{4.179}$$

where β is the coefficient of volumetric thermal expansion and g is the gravitational acceleration aligned vertically downward (against x in Fig. 4.20). By substituting the ΔP expression (4.179) into the Be definition (4.177) we find that the dimensionless group that replaces Be in natural convection is the Rayleigh number Ra $= g\beta\Delta T L_0^3/(\alpha\nu)$. Other than the Be → Ra transformation, all the features that are due to the generation of multiscale blade structure for natural convection should mirror, at least qualitatively, the features described for forced convection in this section. The validity of the constructal multiscale concept for volumes packed with natural convection is demonstrated numerically in da Silva and Bejan (2005).

Finally, the hierarchical multiscale flow architecture constructed in this section is a theoretical comment on fractal geometry. Fractal structures are generated by assuming (postulating) certain algorithms. In much of the current fractal literature, the algorithms are selected such that the resulting structures resemble flow structures observed in nature. For this reason, fractal geometry is descriptive, not predictive (Bejan, 1997c; Bradshaw, 2001). Fractal geometry is not a theory.

4.20. Optimal Spacings for Plates Separated by Porous Structures

Taking the concept of Fig. 4.20 even closer to traditional porous media, consider the optimization of spacings between plates that sandwich a porous medium (Bejan, 2004a). For example, the channels may be occupied by a metallic foam such that the saturated porous medium has a thermal conductivity (k_m) and a thermal diffusivity (α_m) that are much higher than their pure fluid properties (k_f, α_f). We consider both natural convection and forced convection with Boussinesq incompressible fluid and assume that the structures are fine enough that Darcy flow prevails in all cases. The analysis is another application of the intersection of asymptotes method (Lewins, 2003).

The natural convection configuration is shown in Fig. 4.22. This time each D-thin space is filled with the assumed fluid-saturated porous structure. The width in the direction perpendicular to Fig. 4.22 is W. The effective pressure difference

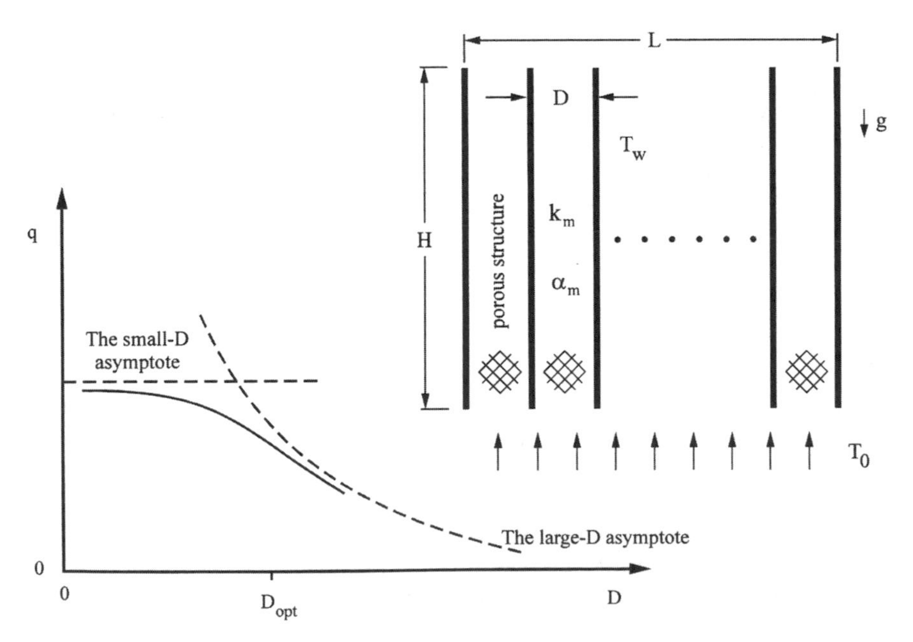

Figure 4.22. Volume filled with vertical heat-generating plates separated by a fluid-saturated porous medium, and the effect of the channel spacing on the global thermal conductance (Bejan, 2004a).

that drives the flow is due to buoyancy:

$$\Delta P = \rho H g \beta (T_w - T_0). \tag{4.180}$$

This ΔP estimate is valid in the limit where the spacing D is sufficiently small so that the temperature in the channel porous medium is essentially the same as the plate temperature T_w. In this limit, the heat current extracted by the flow from the $H \times L$ volume is $q = \dot{m} c_P (T_w - T_0)$, with $\dot{m} = \rho\ ULW$ and Darcy's law, $U = K \Delta P / \mu H$, where K is the permeability of the structure. In conclusion, the total heat transfer rate in the small-D limit is independent of the spacing D,

$$q = \rho c_P (T_w - T_0) LW (K \Delta P)/\mu H. \tag{4.181}$$

In the opposite limit, D is large so that the natural convection boundary layers that line the H-tall plates are distinct. The heat transfer rate from one boundary layer is $\overline{h}\, HW(T_w - T_0)$, where $\overline{h}\, H/k = 0.888\ \mathrm{Ra}_H^{-1/2}$, and Ra_H is the Rayleigh number for Darcy flow, $\mathrm{Ra}_H = K g \beta H (T_w - T_0)/\alpha_m \nu$. The number of boundary layers in the $H \times L$ volume is $2L/D$. In conclusion, the total heat transfer rate decreases as D increases,

$$q = 1.78 (L/D) W k (T_w - T_0) \mathrm{Ra}_H^{1/2}. \tag{4.182}$$

For maximal thermal conductance $q/(T_w - T_0)$, the spacing D must be smaller than the estimate obtained by intersecting asymptotes (4.181) and (4.182)

$$D_{opt}/H \lesssim 1.78\ \mathrm{Ra}_H^{-1/2}. \tag{4.183}$$

The simplest design that has the highest possible conductance is the design with the fewest plates (i.e., the one with the largest D_{opt}); hence $D_{opt}/H \cong 1.78\,\mathrm{Ra}_H^{-1/2}$ for the recommended design. Contrary to Fig. 4.22, however, q does not remain constant if D decreases indefinitely. There exists a small enough D below which the passages are so tight (tighter than the pores) that the flow is snuffed out. An estimate for how large D should be so that Eq. (4.183) is valid is obtained by requiring that the D_{opt} value for natural convection when the channels are filled only with fluid, $D_{opt}/H \cong 2.3[g\beta H^3 (T_w - T_0)/\alpha_f \nu]^{-1/4}$ must be smaller than the D_{opt} value of Eq. (4.171). We find that this is true when

$$\frac{H^2}{K} \frac{\alpha}{\alpha_f} > \mathrm{Ra}_H, \tag{4.184}$$

in which, normally, $\alpha/\alpha_f \gg 1$ and $H^2/K \gg 1$.

The forced convection configuration can be optimized similarly (Bejan, 2004a). The flow is driven by the imposed ΔP through parallel-plate channels of length L and width W. It is found that the forced-convection asymptotes have the same behavior as in Fig. 4.22. The highest conductance occurs to the left of the intersection of the two asymptotes, when

$$D_{opt}/L \lesssim 2.26\,\mathrm{Be}_p^{-1/2} \tag{4.185}$$

and where Be_p is the porous medium Bejan number, $\mathrm{Be}_p = (\Delta P\ K)/\mu \alpha_{\mathrm{m}}$. This forced-convection optimization is valid when the D_{opt} estimate for the channel with pure fluid is smaller than the D_{opt} value provided by Eq. (4.185) when

$$\frac{L^2}{K}\frac{\alpha}{\alpha_f} > \mathrm{Be}_p. \tag{4.186}$$

In summary, Eqs. (4.183) and (4.185) provide estimates for the optimal spacings when the channels between heat-generating plates are filled with a fluid-saturated porous structure. The relevant dimensionless groups are Ra_H, Be_p, K/H^2, K/L^2, and α_m/α_f. The symmetry between Eqs. (4.183) and (4.185) and between Eqs. (4.184) and (4.186) reinforces Petrescu's (1994) argument that the role of the Bejan number in forced convection is analogous to that of the Rayleigh number in natural convection.

These results are most fundamental and are based on a simple model and a simple analysis: Darcy flow and the intersection of asymptotes method. The same idea of geometry optimization deserves to be pursued in future studies of "designed porous media," based on more refined models and more accurate methods of flow simulation.

5
External Natural Convection

Numerical calculation from the full differential equations for convection in an unbounded region is expensive, and hence approximate solutions are important. For small values of the Rayleigh number Ra, perturbation methods are appropriate. At large values of Ra thermal boundary layers are formed, and boundary layer theory is the obvious method of investigation. This approach forms the subject of much of this chapter. We follow to a large extent the discussion by Cheng (1985a), supplemented by recent surveys by Pop and Ingham (2000, 2001) and Pop (2004).

5.1. Vertical Plate

We concentrate our attention on convection in a porous medium adjacent to a heated vertical flat plate, on which a thin thermal boundary layer is formed when Ra takes large values. Using the standard order-of-magnitude estimation, the two-dimensional boundary layer equations take the form

$$\frac{\partial u}{\partial x} + \frac{\partial v}{\partial y} = 0, \tag{5.1}$$

$$u = -\frac{K}{\mu}\left[\frac{\partial P'}{\partial x} - \rho g \beta (T - T_\infty)\right], \tag{5.2}$$

$$\frac{\partial P'}{\partial y} = 0, \tag{5.3}$$

$$\sigma \frac{\partial T}{\partial t} + u\frac{\partial T}{\partial x} + v\frac{\partial T}{\partial y} = \alpha_m \frac{\partial^2 T}{\partial y^2}. \tag{5.4}$$

Here the subscript ∞ denotes the reference value at a large distance from the heated boundary and P' denotes the difference between the actual static pressure and the local hydrostatic pressure. It has been assumed that the Oberbeck-Boussinesq approximation and Darcy's law are valid. For later convenience of comparison, the x axis has been taken in the direction of the main flow (in this case vertically upwards, Fig. 5.1, left) and the y axis normal to the boundary surface and into the porous medium. Near the boundary, the normal component of seepage velocity (v) is small compared with the other velocity component (u), and derivatives with respect to y of a quantity are large compared with derivatives of that quantity with respect to x. Accordingly no term in v appears in Eq. (5.3) and the term in $\partial^2 T/\partial x^2$ has been omitted from Eq. (5.4).

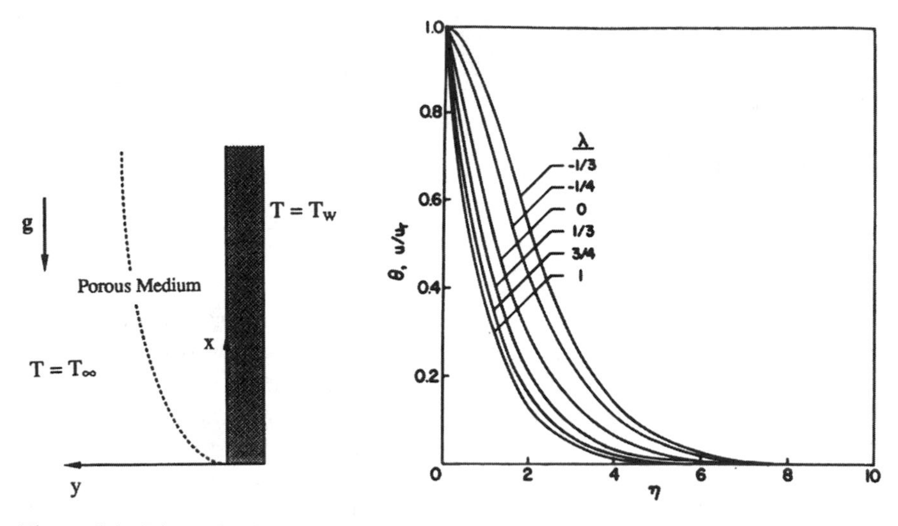

Figure 5.1. Dimensionless temperature and vertical velocity versus the similarity variable for natural convection adjacent to a vertical heated surface (Cheng and Minkowycz, 1977).

Eliminating P' between Eqs. (5.2) and (5.3) and introducing the stream-function ψ defined by

$$u = \frac{\partial \psi}{\partial y}, \qquad v = -\frac{\partial \psi}{\partial x}, \tag{5.5}$$

we reduce Eqs. (5.1)–(5.4) to the pair

$$\frac{\partial^2 \psi}{\partial y^2} = \frac{g\beta K}{\nu}\frac{\partial T}{\partial y}, \tag{5.6}$$

$$\frac{\partial^2 T}{\partial y^2} = \frac{1}{\alpha_m}\left(\sigma\frac{\partial T}{\partial t} + \frac{\partial \psi}{\partial y}\frac{\partial T}{\partial x} - \frac{\partial \psi}{\partial x}\frac{\partial T}{\partial y}\right). \tag{5.7}$$

This pair of equations must be solved subject to the appropriate boundary conditions.

5.1.1. Power Law Wall Temperature: Similarity Solution

We now concentrate our attention on the situation when the wall temperature T_w is a power function of distance along the plate, because in this case a similarity solution can be obtained. Accordingly, we take

$$T_w = T_\infty + Ax^\lambda, \quad x \geq 0. \tag{5.8}$$

For $x < 0$ we suppose that either there is no plate or that $T_w = T_\infty$ on the plate. The set of boundary conditions then is

$$y = 0: \quad v = 0, \quad T = T_\infty + Ax^\lambda, \quad x \geq 0, \tag{5.9}$$

$$y \to \infty: u = 0, T = T_\infty. \tag{5.10}$$

One can easily check that a steady-state solution of Eqs. (5.6)–(5.10) is given by

$$\psi = \alpha_m (\mathrm{Ra}_x)^{1/2} f(\eta), \tag{5.11}$$

$$\frac{T - T_\infty}{T_w - T_\infty} = \theta(\eta), \tag{5.12}$$

where

$$\eta = \frac{y}{x} \mathrm{Ra}_x^{1/2}, \tag{5.13}$$

$$\mathrm{Ra}_x = \frac{g\beta K (T_w - T_\infty) x}{\nu \alpha_m}, \tag{5.14}$$

provided that the functions $f(\eta)$, and $\theta(\eta)$ satisfy the ordinary differential equations

$$f'' - \theta' = 0, \tag{5.15}$$

$$\theta'' + \frac{(1+\lambda)}{2} f\theta' - \lambda f'\theta = 0, \tag{5.16}$$

and the boundary conditions

$$f(0) = 0, \quad \theta(0) = 1, \tag{5.17}$$

$$f'(\infty) = 0, \quad \theta(0) = 0. \tag{5.18}$$

In terms of the similarity variable η, the seepage velocity components are

$$u = u_r f'(\eta), \tag{5.19}$$

$$v = \frac{1}{2} \left[\frac{\alpha_m g\beta K (T_w - T_\infty}{\nu x} \right]^{1/2} [(1-\lambda)\eta f' - (1+\lambda) f], \tag{5.20}$$

where the characteristic velocity u_r is defined by

$$u_r = \frac{g\beta K (T_w - T_\infty)}{\nu}. \tag{5.21}$$

Integrating Eq. (5.15) and using Eq. (5.18) we get

$$f' = \theta. \tag{5.22}$$

This implies that the normalized vertical velocity u/u_r and the normalized temperature θ are the same function of η. Their common graph is shown in Fig. 5.1. Another implication is that in this context, Eqs. (5.2) and (5.3) formally may be replaced by

$$u = \frac{g\beta K}{\nu}(T - T_\infty). \tag{5.23}$$

From Eq. (5.13) we see that the boundary layer thickness δ is given by

$$\frac{\delta}{x} = \frac{\eta_T}{\mathrm{Ra}_x^{1/2}}, \tag{5.24}$$

where η_T is the value of η at the edge of the boundary layer, conventionally defined as that place where θ has a value 0.01. Values of η_T, for various values of

Table 5.1. Values of η_T and $-\theta'(0)$ for various values of λ for the heated vertical plate problem (after Cheng and Minkowycz, 1977)

λ	η_T	$-\theta'(0)$	$\overline{\mathrm{Nu}}/\overline{\mathrm{Ra}}^{1/2}$	
– 1/3	7.2	0		
– 1/4	6.9	0.162	0.842	
0	6.3	0.444	0.888	isothermal
1/4	5.7	0.630	1.006	
1/3	5.5	0.678	1.044	uniform flux
1/2	5.3	0.761	1.118	
3/4	4.9	0.892	1.271	
1	4.6	1.001	1.416	

λ, are given in Table 5.1. For the case of constant wall temperature ($\lambda = 0$), δ is proportional to $x^{1/2}$.

The local surface heat flux at the heated plate is

$$q'' = -k_m \left(\frac{\partial T}{\partial y}\right)_{y=0} = k_m A^{3/2} \left(\frac{g\beta K}{\nu\alpha_m}\right)^{1/2} x^{(3\lambda-1)/2}[-\theta'(0)]. \tag{5.25}$$

Clearly $\lambda = 1/3$ corresponds to uniform heat flux. In dimensionless form, Eq. (5.25) is

$$\frac{\mathrm{Nu}_x}{\mathrm{Ra}_x^{1/2}} = -\theta'(0), \tag{5.26}$$

where the local Nusselt number is defined by $\mathrm{Nu}_x = hx/k$ and where h is the local heat transfer coefficient $q''/(T_w - T_\infty)$. The values of $[-\theta'(0)]$ also are listed in Table 5.1. In particular, we note that $[-\theta'(0)] = 0.444$ when $\lambda = 0$.

The total heat transfer rate through a plate of height L, expressed per unit length in the direction perpendicular to the plane (x, y), is

$$L\overline{q}'' = q' = \int_0^L q''(x)dx = k_m A^{3/2} \left(\frac{g\beta K}{\nu\alpha_m}\right)^{1/2} \left(\frac{2}{1+3\lambda}\right) L^{(1+3\lambda)/2}[-\theta'(0)]. \tag{5.27}$$

This result can be rewritten as

$$\frac{\overline{\mathrm{Nu}}}{\overline{\mathrm{Ra}}^{1/2}} = \frac{2(1+\lambda)^{3/2}}{1+3\lambda}[-\theta'(0)], \tag{5.28}$$

where Nu and Ra are based on the L-averaged temperature difference

$$\overline{\mathrm{Nu}} = \frac{q'}{k_m(\overline{T_w - T_\infty})}, \quad \overline{\mathrm{Ra}} = \frac{g\beta K L(\overline{T_w - T_\infty})}{\nu\alpha_m},$$

$$(\overline{T_w - T_\infty}) = \frac{1}{L}\int_0^L (T_w - T_\infty)dx. \tag{5.29}$$

Xu (2004) has treated the same problem by means of homotopy analysis.

5.1.2. Vertical Plate with Lateral Mass Flux

If the power law variation of wall temperature persists but now we have an imposed lateral mass flux at the wall given by $v = ax^n$, $(x = 0)$, then a similarity solution exists for the case $n = (\lambda - 1)/2$. Equations (5.11)–(5.18) apply, with the exception that Eq. (5.17a) is replaced by

$$f(0) = f_w = 2a(\alpha_m g\beta KA/\nu)^{-1/2}(1+\lambda)^{-1}. \tag{5.30}$$

The thermal boundary layer thickness is still given by Eq. (5.24) but now η_T is an increasing function of the injection parameter f_w (Cheng, 1977b). This problem has applications to injection of hot water in a geothermal reservoir. The practical case of constant discharge velocity at uniform temperature has been treated by different methods by Merkin (1978) and Minkowycz and Cheng (1982).

The solution for the related problems where the heat flux (rather than the temperature) is prescribed at the wall can be deduced from the present solution via a certain change of variables (Cheng, 1977a) or obtained directly. Of course we already have the solution for constant prescribed heat flux, with the wall temperature related to the heat flux, via the parameter A with $\lambda = 1/3$, through Eqs. (5.8) and (5.25). From Eq. (5.24) we see that the boundary layer thickness is proportional to $x^{1/3}$ in this case.

Similarity solutions for a vertical permeable surface were developed by Chaudhary *et al.* (1995a,b) for the class with heat flux proportional to x^{μ} and mass flux proportional to $x^{(\mu-1)/3}$, where μ is a constant.

5.1.3. Transient Case: Integral Method

For transient natural convection in a porous medium, similarity solutions exist for only a few unrealistic wall temperature distributions. For more realistic boundary conditions, approximate solutions can be obtained using an integral method. Integrating Eq. (5.4) across the thermal boundary layer and using Eqs. (5.1) and (5.2), we obtain

$$\sigma\frac{\partial}{\partial t}\int_0^\infty \Phi(x,y,t)dy + \frac{g\beta K}{\nu}\frac{\partial}{\partial x}\int_0^\infty \Phi^2(x,y,t)dy = -\alpha_m\left(\frac{\partial\Phi}{\partial y}\right)_{y=0}. \tag{5.31}$$

where $\Phi = T - T_\infty$. The Karman-Pohlhausen integral method involves assuming an explicit form of Φ that satisfies the temperature boundary conditions, namely $\Phi = T_w - T_\infty$ at $y = 0$ and $\Phi \to 0$ as $y \to \infty$. The integrals in Eq. (5.31) are then determined and the resulting equation for the thermal boundary layer thickness δ becomes a first-order partial differential equation of the hyperbolic type which can be solved by the method of characteristics.

For the case of a step increase in wall temperature, Cheng and Pop (1984) assume that the temperature distribution is of the form

$$\Phi = (T_w - T_\infty)\mathrm{erfc}(\zeta) \tag{5.32}$$

where $\zeta = y/\delta(x, t)$. The results of the method of characteristics show that during the interval before the steady state is reached one has

$$\delta = 2\left(\frac{\alpha_m t}{\sigma}\right)^{1/2}, \tag{5.33}$$

$$\frac{T - T_\infty}{T_w - T_\infty} = \operatorname{erfc}\left[\frac{y}{2}\left(\frac{\sigma}{\alpha_m t}\right)^{1/2}\right] = \frac{\nu u}{g\beta K(T_w - T_\infty)}, \tag{5.34}$$

$$q''_w = k\left(\frac{\sigma}{\pi\alpha_m t}\right)^{1/2}(T_w - T_\infty), \tag{5.35}$$

for $t < T_{ss}$, with $t_{ss} = \sigma x^2/\alpha_m K_1 \mathrm{Ra}_x (K_1 = 2 - 2^{1/2} = 0.5857)$, denoting the time at which steady state is reached. This time interval is related to the propagation of the leading edge effect, which is assumed to travel with the local velocity. In Eq. (5.34), u is the x component of the seepage velocity.

Equations (5.33)–(5.34) are independent of x and are similar in form to the solution for the transient heat conduction problem. During the initial stage when the leading edge effect is not being felt, heat is transferred by transient one-dimensional heat conduction. After the steady state is reached, we have

$$\frac{\delta}{x} = \frac{2.61}{\mathrm{Ra}_x^{1/2}}, \tag{5.36}$$

$$\frac{T - T_\infty}{T_w - T_\infty} = \operatorname{erfc}\left(\frac{K_1^{1/2} y\, \mathrm{Ra}_x^{1/2}}{2x}\right) = \frac{\nu u}{g\beta K(T_w - T_\infty)}, \tag{5.37}$$

$$q''_w = \frac{k(T_w - T_\infty)}{x}(K_1/\pi \mathrm{Ra}_x)^{1/2}, \tag{5.38}$$

Equation (5.38) can be written in dimensionless form as

$$\frac{\mathrm{Nu}_x}{\mathrm{Ra}_x^{1/2}} = 0.431, \tag{5.39}$$

which compares favorably with the exact similarity solution where the constant is equal to 0.444 (see Table 5.1). Comparison of Eq. (5.36) with Eq. (5.24) for $\lambda = 0$ shows that the integral method considerably underestimates the steady-state thermal boundary layer thickness. This is due to the error in the assumed temperature profile in the integral-method formulation.

For flow past a suddenly cooled wall, similarity solutions were obtained by Ingham and Merkin (1982) for the case of small and large dimensionless times, and these were joined by a numerical solution. A detailed study of the transient problem for the case where the wall temperature varies as x^λ was made by Ingham and Brown (1986). They found that for $\lambda < -1/2$ no solution of the unsteady boundary layer equations was possible, and that for $-1/2 < \lambda < 1$ the parabolic partial differential equation governing the flow is singular. For $-1/2 < \lambda < -1/3$ the velocity achieves its maximum value within the boundary layer (instead of on the boundary).

For the case $\lambda = 0$, Haq and Mulligan (1990a) have integrated the unsteady boundary layer equations numerically. Their results confirm that during the initial stage, before the effects of the leading edge are influential at a location, heat transfer is governed by conduction. They show that in a Darcian fluid the local Nusselt number decreases with time monotonically to its steady-state value. The effect of inertia was considered by Chen *et al.* (1987). They found that the effect of quadratic drag increases the momentum and thermal boundary layer thicknesses and reduces the heat transfer rate at all times (cf. Section 5.1.7.2).

The situation where the permeability varies linearly along the plate was treated by Mehta and Sood (1992a). As one would expect, they found that increase in permeability results in higher rate of heat transfer at the wall and in decreased time to reach the steady state at any location on the plate.

The case of wall heating at a rate proportional to x^λ was examined by Merkin and Zhang (1992). The similarity equations that hold in the limit of large t were shown to have a solution only for $\lambda > -1$. Numerical solutions were obtained for a range of possible values of λ.

Harris *et al.* (1996, 1997a,b) have treated the case of a jump to a uniform flux situation and the case where the surface temperature or the surface heat flux suddenly jumps from one uniform value to another. Pop *et al.* (1998) reviewed work on transient convection heat transfer in external flow. Techniques for solving the boundary layer equations that arise in such circumstances were discussed by Harris and Ingham (2004). Khadrawi and Al-Nimr (2005) have examined the effect of the local inertial term for a domain partly filled with porous material. The Brinkman model was employed in the numerical study by K. H. Kim *et al.* (2004).

5.1.4. Effects of Ambient Thermal Stratification

When the porous medium is finite in the x and y directions, the discharge of the boundary layer into the rest of the medium leads in time to thermal stratification. If the temperature profile at "infinity" is as in Fig. 5.2, and if $T_0 - T_{\infty,0}$ remains

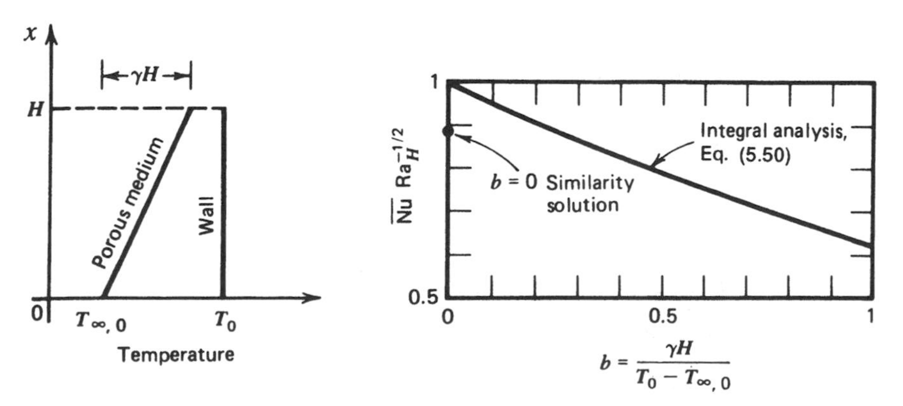

Figure 5.2. Heat transfer from a vertical isothermal wall to a linearly stratified porous medium (Bejan, 1984).

fixed, then as the positive temperature gradient $\gamma = dT_\infty/dx$ increases, the average temperature difference between the wall and the porous medium decreases. Thus we should expect a steady decrease in the total heat transfer rate as γ increases. We apply the integral method to the solution of this problem (Bejan, 1984).

The Darcy law relation (5.6) integrates to give

$$T = \frac{\nu}{g\beta K}u + \text{function(x)}. \tag{5.40}$$

We assume a vertical velocity profile of the form

$$u = u_0(x)\exp\left[-\frac{y}{\delta_T(x)}\right]. \tag{5.41}$$

Then, using Eq. (5.40) and the temperature boundary conditions

$$T(x,0) = T_0, \quad T(x,\infty) = T_{\infty,0} + \gamma x, \tag{5.42}$$

we see that the corresponding temperature profile is

$$T(x,y) = (T_0 - T_{\infty,0} - \gamma x)\exp(-y/\delta_T) + T_{\infty,0} + \gamma x, \tag{5.43}$$

and the maximum (wall) vertical velocity is

$$u_0 = \frac{g\beta K}{\nu}(T_0 - T_{\infty,0} - \gamma x). \tag{5.44}$$

The integral form of the boundary layer energy equation, obtained by integrating Eq. (5.4) from $y = 0$ to $y = \infty$, is

$$v(x,\infty)T(x,\infty) + \frac{d}{dx}\int_0^\infty uT\,dy = -\alpha_m\left(\frac{\partial T}{\partial y}\right)_{y=0}, \tag{5.45}$$

where $T(x,\infty) = T_{\infty,0} + \gamma x$, and from the mass conservation equation,

$$v(x,\infty) = -\frac{d}{dx}\int_0^\infty u\,dy. \tag{5.46}$$

Substituting the assumed u and T profile into the energy integral equation (5.46) yields

$$\frac{d\delta_*}{dx_*} = \frac{2}{\delta_*(1 - bx_*)}, \tag{5.47}$$

in terms of the dimensionless quantities

$$b = \frac{\gamma H}{T_0 - T_{\infty,0}}, \quad x_* = \frac{x}{H}, \quad \delta_* = \frac{\delta_T}{H}\left[\frac{g\beta H^3(T_0 - T_{\infty,0})}{\nu\alpha_m}\right]^{1/2}. \tag{5.48}$$

Integrating Eq. (5.47), with $\delta_*(0) = 0$, we obtain

$$\delta_*(x_*) = \left[-\frac{4}{b}\ln(1 - bx_*)\right]^{1/2}. \tag{5.49}$$

As $b \to 0$ this gives the expected result $\delta_* \sim x_*^{1/2}$. The average Nusselt number (over the wall height H) is given by

$$\frac{\overline{\mathrm{Nu}}}{\mathrm{Ra}_H^{1/2}} = \int_0^1 \frac{(1 - bx_*)dx_*}{[-(4/b)\ln(1 - bx_*)]^{1/2}}, \tag{5.50}$$

where $\overline{\mathrm{Nu}}$ and Ra_H are based on the maximum (i.e., *starting*) temperature difference

$$\overline{\mathrm{Nu}} = \frac{q''H}{k(T_0 - T_{\infty,0})}, \quad \mathrm{Ra}_H = \frac{g\beta KH}{\nu\alpha_m}(T_0 - T_{\infty,0}). \tag{5.51}$$

Equation (5.50) is plotted in Fig. 5.2. As expected, $\mathrm{Nu}/\mathrm{Ra}_H^{1/2}$ decreases monotonically as b increases. The above approximate integral solution gives $\mathrm{Nu}/\mathrm{Ra}_H^{1/2} = 1$ at $b = 0$, whereas the similarity solution value for this quantity is 0.888, a discrepancy of 12.5 percent.

The same phenomenon was studied numerically, without the boundary layer approximation, by Angirasa and Peterson (1997b) and Ratish Kumar and Singh (1998). The case of a power law variation of wall temperature was discussed by Nakayama and Koyama (1987c) and Lai *et al.* (1991b). The stratification problem has also been treated by Tewari and Singh (1992) and (with quadratic drag effects included) by Singh and Tewari (1993). In their study of an isothermal surface with stratification on the Brinkman-Forchheimer model, Chen and Lin (1995) found that a flow reversal is possible in certain circumstances. The same model, with the effect of variable porosity and thermal dispersion included, was employed by Hung *et al.* (1999). The case of variable wall heat flux was analyzed by Hung and Cheng (1997). An MHD problem was analyzed by Chamkha (1997g).

5.1.5. Conjugate Boundary Layers

When one has a vertical wall between two porous media (or between a porous medium and a fluid reservoir) and a temperature difference exists between the two systems, we may have a pair of conjugate boundary layers, one on each side of the wall, with neither the temperature nor the heat flux specified on the wall but rather to be found as part of the solution of the problem. Bejan and Anderson (1981) used the Oseen linearization method to analytically solve the problem of a solid wall inserted in a porous medium. They found that the coefficient in the $\mathrm{Nu}/\mathrm{Ra}_H^{1/2}$ proportionality decreases steadily as the wall thickness parameter ω increases, where ω is defined as

$$\omega = \frac{W}{H}\frac{k_m}{k_w}\mathrm{Ra}_H^{1/2}. \tag{5.52}$$

In this dimensionless group W and H are the width and height of the wall cross section, k_m and k_w are the conductivities of the porous medium and wall material, respectively, and Ra_H is the Rayleigh number based on H and the temperature difference between the two systems, $\Delta T = T_{\infty 2} - T_{\infty 1}$. The overall Nusselt number

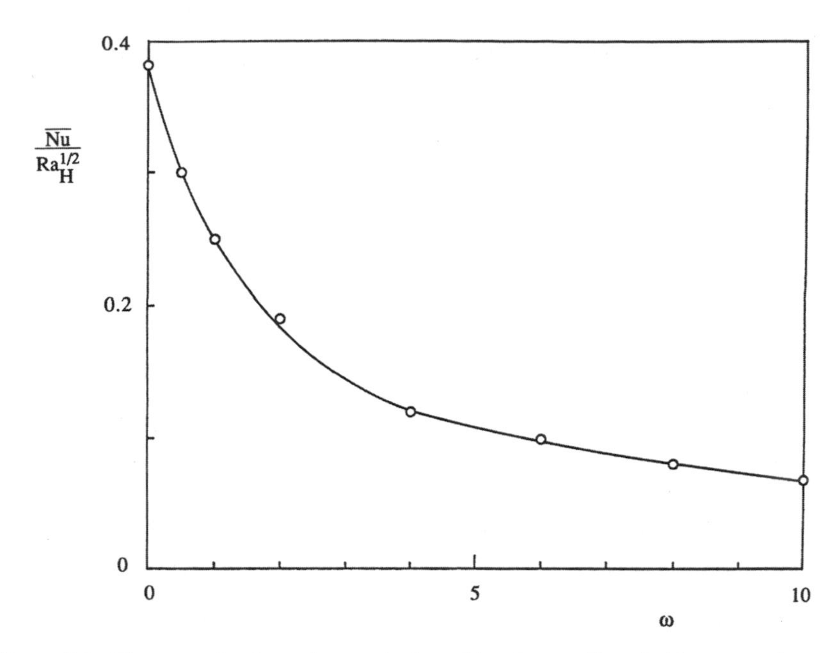

Figure 5.3. Heat transfer through a vertical partition inserted in a fluid-saturated porous medium (Bejan and Anderson, 1981; Bejan, 1984).

Nu is based on the wall-averaged heat flux $\bar{q}''$ and the overall temperature difference ΔT,

$$\overline{\text{Nu}} = \frac{\bar{q}'' H}{k_m \Delta T}. \tag{5.53}$$

The variation of $\overline{\text{Nu}}/\text{Ra}_H^{1/2}$ with ω is shown in Fig. 5.3. In the limit of negligible wall thermal resistance ($\omega \to 0$) the overall Nusselt number reduces to

$$\overline{\text{Nu}} = 0.383\,\text{Ra}_H^{1/2}. \tag{5.54}$$

The case of wall between a porous medium and a fluid reservoir was solved by Bejan and Anderson (1983). Their heat transfer results are reproduced in Fig. 5.4. The value of dimensionless group

$$B = \frac{k_m \text{Ra}_H^{1/2}}{k_a \text{Ra}_{Ha}^{1/4}} \tag{5.55}$$

determines whether the conjugate problem is dominated by porous medium convection (small B) or pure fluid convection (large B). Here k_a and Ra_{Ha} represent the fluid conductivity and Rayleigh number on the side of the pure fluid (which typically is air).

Pop and Merkin (1995) showed that the boundary-layer equations can be made dimensionless so that the thermal conductivity ratio is scaled out of the problem,

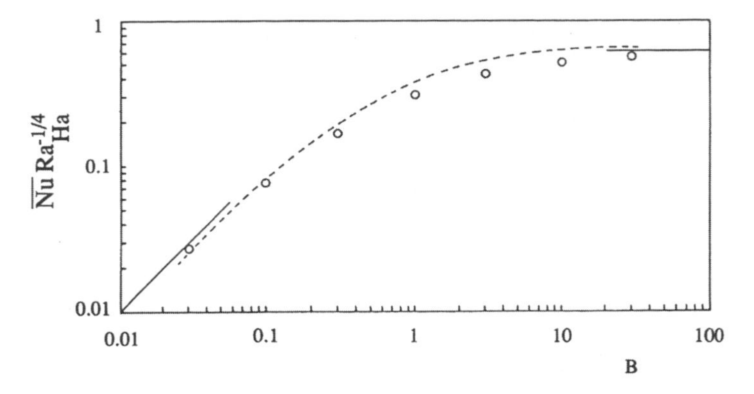

Figure 5.4. Heat transfer through the interface between a porous medium and a fluid reservoir (Bejan and Anderson, 1983; Bejan, 1984).

and thus just one solution of the transformed nonsimilar boundary layer equations need be computed. This they did by a finite-difference scheme.

The above analysis of Bejan and Anderson is limited to the case of a thin plate. The thin plate assumption was dropped by Vynnycky and Kimura (1994). They considered a wall of thickness a and with a segment of height b conducting and the remainder insulating; the aspect ratio is $\lambda = a/b$. They constructed an approximate one-dimensional solution based on the assumption of a boundary layer of thickness δ. The average boundary heat flux is given by

$$q'' = k_w \frac{(T_c - \bar{T}_b)}{a} = k_m \frac{\bar{T}_h - T_\infty}{\delta}, \tag{5.56}$$

where $\bar{T}_b$ is the average interface temperature and T_c is the constant temperature at the far side of the conducting wall. If Ra denotes the Rayleigh number based on $T_c - T_\infty$ and Ra* that is based on $\bar{T}_b - T_\infty$, and $\bar{\theta}_b = (T - T_\infty)/(T_c - T_\infty)$, so that $\text{Ra}^* = \text{Ra}\bar{\theta}_b$, then

$$\frac{\delta}{b} = 1.126(\text{Ra}^*)^{-1/2}, \tag{5.57}$$

from the isothermal entry in Table 5.1. Combining Eqs. (5.56) and (5.57) one has

$$\sigma_c X^3 + X^2 - 1 = 0 \tag{5.58}$$

where $X = \bar{\theta}_b^{1/2}$ and $\sigma_c = \lambda \text{Ra}^{1/2}/1.126k$ where $k = k_w/k_m$. The quantity σ_c may be regarded as a conjugate Biot number. Conjugate effects are small if $\sigma_c \ll 1$. For a given σ_c, Eq. (5.58) is readily solved to give $\bar{\theta}_b$ and then the average Nusselt number can be obtained from

$$\overline{\text{Nu}} = \frac{q''a}{k_m(T_c - T_\infty)} = 0.888 \frac{\bar{T}_b - T_\infty}{T_c - T_\infty} (\text{Ra}^*)^{1/2} = 0.888 \bar{\theta}_b^{3/2} \text{Ra}^{1/2}. \tag{5.59}$$

Vynnycky and Kimura (1994) showed that this formula agrees well with numerical computations in typical cases.

Kimura *et al.* (1997) show how the same ideas can be applied to the problem of a wall between two reservoirs, the extension (to a thick partition) of the work of Bejan and Anderson (1983). Kimura and Pop (1992b, 1994) treated convection around a cylinder or a sphere in a similar fashion. A transient one-dimensional model for conjugate convection from a vertical conducting slab was developed by Vynnycky and Kimura (1995). They obtained analytical solutions for two parameter regimes, (i) $Ra \gg 1$, $\Gamma \ll Ra$, and (ii) $\Gamma \gg 1$, $Ra \ll \Gamma$, where $\Gamma = [(\rho c)_m/(\rho c)_f](\alpha_w/\alpha_m)$. Regime (i) implies that the temperature and velocity within the boundary layer adjust themselves instantaneously to conditions in the conducting plate and time dependency arises through variation of the conjugate boundary temperature. The value of Γ affects the development but not the steady state. Regime (ii) corresponds to the case where conduction dominates convection in the early stages of flow development in the porous medium. Vynnycky and Kimura (1995) also checked their analytical solutions against numerical solutions.

The case of conjugate natural convection heat transfer between two porous media at different temperatures separated by a vertical wall was treated by Higuera and Pop (1997). They obtained asymptotic and numerical solutions. The corresponding case for a horizontal wall was examined by Higuera (1997). Conjugate convection from vertical fins was studied numerically by Vaszi *et al.* (2003). A transient problem involving a vertical plate subjected to a sudden change in surface heat flux was analyzed by Shu and Pop (1998). Another transient problem involving the cooling of a thin vertical plate was analyzed by Méndez *et al.* (2004). The topic of conjugate natural convection in porous media was reviewed by Kimura *et al.* (1997).

5.1.6. Higher-Order Boundary Layer Theory

The above boundary layer theory arises as a first approximation for large values of Rayleigh number, when expansions are made in terms of the inverse one-half power of the Rayleigh number. At this order, the effects of entrainment from the edge of the boundary layer, the axial heat conduction, and the normal pressure gradient are all neglected.

The magnitudes of these effects have been investigated using higher-order asymptotic analysis by Cheng and Chang (1979), Chang and Cheng (1983), Cheng and Hsu (1984), and Joshi and Gebhart (1984). They found that the ordering of the eigenfunction terms in the perturbation series was dependent on the wall temperature parameter λ. They also found that the coefficients of the eigenfunctions cannot be determined without a detailed analysis of the leading edge effect. Therefore, they truncated the perturbation series at the term where the leading edge effect first appeared. They found that the effect of entrainments from the edge of the thermal boundary layer was of second order while those of axial heat conduction and normal pressure gradient were of third order. For the case of the isothermal vertical plate with $\lambda = 0$, the second-order corrections for both the Nusselt number and the vertical velocity are zero and the leading edge effect appears in the third-order terms. For other values of λ, both the second- and third-order corrections in the

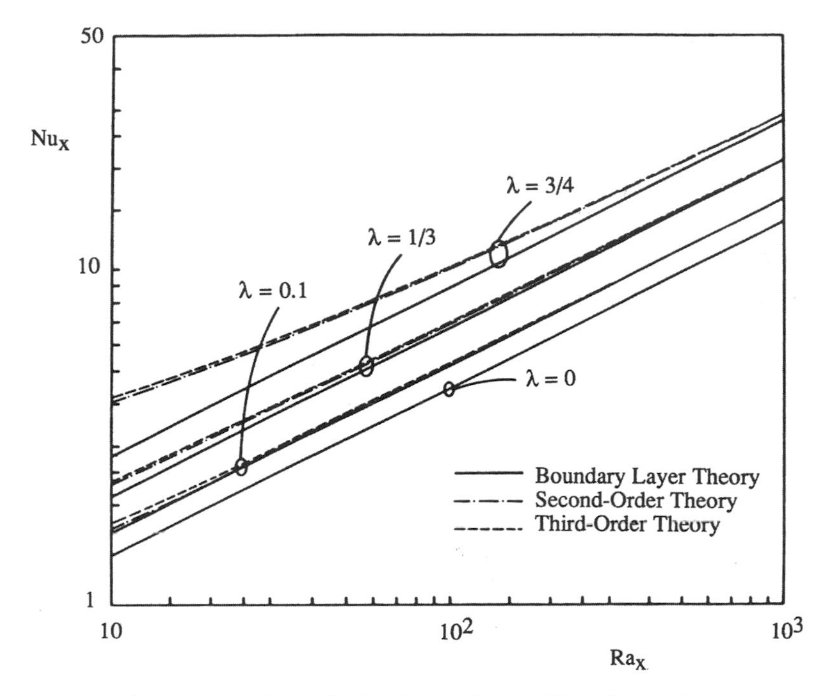

Figure 5.5. Higher-order theoretical values of local Nusselt number versus local Rayleigh number for natural convection about a vertical flat plate in a porous medium (Cheng and Hsu, 1984).

Nusselt number are positive and the leading edge effect appears in the fourth-order terms.

The slight increase in the surface heat flux in the higher-order theories is mainly due to the fact that entrainments from the outer flow induce a flow parallel to the heated surface. The higher-order theory has a profound effect on the velocity profiles but has a relatively small effect on the temperature distribution, and hence on the surface heat flux. Figure 5.5 illustrates the higher-order effects on the local Nusselt number Nu_x. It is evident that for small wall temperature variations ($\lambda = 1/3$) the boundary layer theory is quite accurate even at small Rayleigh numbers.

Pop *et al.* (1989) have shown that for the case of uniform wall heat flux the leading edge effects enter the second and subsequent order problems. They cause an increase of the streamwise vertical velocity near the outer edge of the boundary layer and a consequent increase in heat transfer rate by an amount comparable with entrainment effects, the combination producing a 10 percent increase at $\mathrm{Ra}_x = 100$ and a greater amount at smaller Ra_x.

5.1.7. Effects of Boundary Friction, Inertia, and Thermal Dispersion

So far in this chapter it has been assumed that Darcy's law is applicable and the effects of the no-slip boundary condition, inertial terms, and thermal dispersion are negligible. We now show that all of these effects are important only at high Rayleigh numbers. The effects of boundary friction and inertia tend to decrease the heat transfer rate while that of thermal dispersion tends to increase the heat transfer rate.

5.1.7.1. *Boundary Friction Effects*

To investigate the boundary friction effect Evans and Plumb (1978) made some numerical calculations using the Brinkman equation. They found that the boundary effect is negligible if the Darcy number Da ($\mathrm{Da} = K/L^2$, where L is the length of the plate) is less than 10^{-7}. For higher values of Da their numerical results yield a local Nusselt number slightly smaller than those given by the theory based on Darcy's law.

Hsu and Cheng (1985b) and Kim and Vafai (1989) have used the Brinkman model and the method of matched asymptotic expansions to reexamine the problem. Two small parameters that are related to the thermal and viscous effects govern the problem. For the case of constant wall temperature these are $\varepsilon_T = \mathrm{Ra}^{-1/2}$ and $\varepsilon_v = \mathrm{Da}^{1/2}$, where Ra is the Rayleigh number based on plate length L and temperature difference $T_w - T_\infty$, and Da is the Darcy number $K/L^2\varphi$. For the case of constant wall heat flux, $\varepsilon_T = \mathrm{Ra}^{-1/3}$, where Ra is now the Rayleigh number based on L and the heat flux q''_w; here we concentrate on the case of constant T_w. Cases (a) $\varepsilon_v \ll \varepsilon_T$ and (b) $\varepsilon_v \gg \varepsilon_T$ must be treated separately.

In geophysical and engineering applications it is usually case (a) that applies. Dimensional analysis shows that three layers are involved: the inner momentum boundary layer with a constant thickness of $O(\varepsilon_v)$, the middle thermal layer with a thickness of $O(\varepsilon_T)$, which is inversely proportional to the imposed temperature difference, and the outer potential region of $O(1)$. The asymptotic analysis of Hsu and Cheng (1985b) gives the local Nusselt number in the form

$$\mathrm{Nu}_x = C_1 \mathrm{Ra}_x^{1/2} - C_2 \mathrm{Ra}_x \mathrm{Da}_x^{1/2}, \tag{5.60}$$

where $\mathrm{Da}_x = K/x^2$ is the local Darcy number, and the constants C_1 and C_2 are related to the dimensionless temperature gradients at the wall appearing in the first-order and second-order problems. The values of these constants depend on the wall temperature. Equation (5.60) can be rewritten as

$$\mathrm{Nu}_x/\mathrm{Ra}_x^{1/2} = C_1 - C_3 P_{nx}, \tag{5.61}$$

where $C_3 = C_2/C_1$ and P_{nx} is the local no-slip parameter given by

$$P_{nx} = \mathrm{Ra}_x^{1/2} \mathrm{Da}_x^{1/2} = \left[\frac{g\beta K^2 (T_w - T_\infty)^{1/2}}{\nu \alpha_m x} \right]. \tag{5.62}$$

Equation (5.61) is plotted in Fig. 5.6. It is clear that the deviation from Darcy's law becomes appreciable at high local Rayleigh numbers only for high local Darcy numbers (i.e., near the leading edge) and for large wall temperature

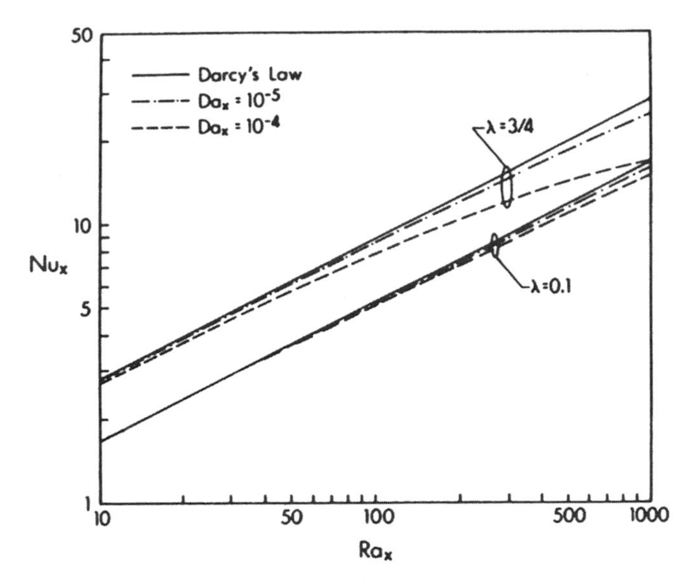

Figure 5.6. Boundary effects on the local Nusselt number for natural convection about a vertical surface in a porous medium (Hsu and Cheng, 1985b, with permission from Pergamon Press).

variations. This conclusion is in accordance with the numerical results of Evans and Plumb (1978) and is confirmed by further calculations by Hong *et al.* (1987).

For case (b) where $\varepsilon_v \gg \varepsilon_T$, Kim and Vafai (1989) find that the local Nusselt number Nu_x is given by

$$\mathrm{Nu}_x = 0.5027\,\mathrm{Da}_x^{-1/4}\mathrm{Ra}_x^{1/4} = 0.5027(\mathrm{Ra}_f\varphi)^{1/4}, \tag{5.63}$$

where Ra_f is the standard Rayleigh number for a viscous fluid (independent of permeability), as expected for a very sparse medium. Numerical studies using the Brinkman model were conducted by Beg *et al.* (1998) and Gorla *et al.* (1999b). The last study included the effect of temperature-dependent viscosity applied to the plume above a horizontal line source (either isolated or on an adiabatic vertical wall) as well as to a vertical wall with uniform heat flux.

5.1.7.2. Inertial Effects

Forchheimer's equation with a quadratic drag term was introduced into the boundary layer theory by Plumb and Huenefeld (1981). Equation (5.23) is replaced by

$$u + \frac{\chi}{\nu}u^2 = \frac{g\beta K}{\nu}(T - T_\infty), \tag{5.64}$$

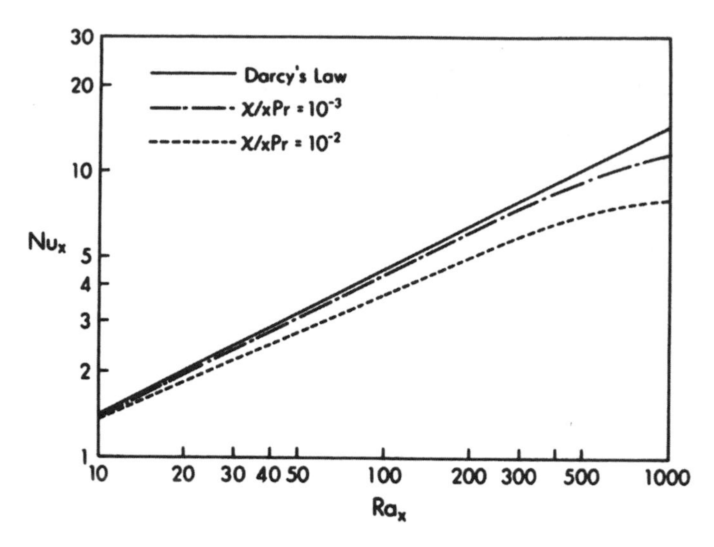

Figure 5.7. Non-Darcy inertial effects on the local Nusselt number for natural convection about a vertical surface in a porous medium (Cheng, 1985a, with permission from Hemisphere Publishing Corporation).

in which the coefficient χ has the units [m] and represents the group $c_F K^{1/2}$ seen earlier on the right-hand side of Eq. (1.12). In place of Eq. (5.15) one now has

$$f'' + \mathrm{Gr}^*(f'^2)' - \theta' = 0, \tag{5.65}$$

where

$$\mathrm{Gr}^* = \frac{g\beta\chi K(T_w - T_\infty)}{\nu^2}. \tag{5.66}$$

It is clear that a similarity solution exists if and only if the Grashof number Gr* is a constant, which requires that T_w is constant. Plumb and Huenefeld's results are displayed in Fig. 5.7, which as expected shows that the effect of quadratic drag is to slow down the buoyancy-induced flow and so retard the heat transfer rate.

The alternative analysis of Bejan and Poulikakos (1984) is based on the observation that at sufficiently large Rayleigh numbers, and hence large velocities, the quadratic term on the left-hand side of Eq. (5.60) will dominate the linear term. Scale analysis then indicates that the boundary layer thickness δ is of the order

$$\delta \sim H\mathrm{Ra}_\infty^{-1/4}, \tag{5.67}$$

where H is a characteristic length scale and the "large Reynolds number limit" Rayleigh number Ra_∞ is defined as

$$\mathrm{Ra}_\infty = \frac{g\beta K^2(T_w - T_\infty)}{\chi\alpha_m^2}. \tag{5.68}$$

The introduction of the nondimensional variables

$$x_* = \frac{x}{H}, \qquad y_* = \frac{y}{H}\mathrm{Ra}_\infty^{1/4},$$

$$u_* = \frac{uH}{\alpha_m \mathrm{Ra}_\infty^{1/2}}, \qquad v_* = \frac{vH}{\alpha_m \mathrm{Ra}_\infty^{1/4}}, \tag{5.69}$$

$$\theta = \frac{T - T_\infty}{T_w - T_\infty},$$

yields

$$u_* \frac{\partial \theta}{\partial x_*} + v_* \frac{\partial \theta}{\partial y_*} = \frac{\partial^2 \theta}{\partial y_*^2}, \tag{5.70}$$

$$G \frac{\partial u_*}{\partial y_*} + \frac{\partial (u_*^2)}{\partial y_*} = \frac{\partial \theta}{\partial y_*}, \tag{5.71}$$

where G is the new dimensionless group

$$\mathrm{G} = \nu[\chi g \beta K (T_w - T_\infty)]^{-1/2} = (\mathrm{Gr}^*)^{-1/2}. \tag{5.72}$$

The Forchheimer regime corresponds to $G \to 0$. Then Eq. (5.71) and the outer condition $\theta \to 0$ as $y \to \infty$ yields

$$u_* = \theta^{1/2}. \tag{5.73}$$

The appropriate similarity variable is

$$\eta = \frac{y_*}{x_*^{1/2}}. \tag{5.74}$$

The dimensionless streamfunction ψ defined by $u_* = \partial\psi/\partial y_*$, $v_* = -\partial\psi/\partial x_*$ is now given by

$$\psi = x_*^{1/2} F(\eta), \tag{5.75}$$

where

$$F(\eta) = \int_0^\eta \theta^{1/2} d\eta. \tag{5.76}$$

The boundary layer equations reduce to the system

$$\theta^{1/2} = F', \qquad -\frac{1}{2} F\theta' = \theta'', \tag{5.77}$$

with the conditions

$$\theta(0) = 1,\ F(0) = 0, \quad \text{and} \quad \theta \to 0 \text{ as } \eta \to \infty. \tag{5.78}$$

This system is readily integrated using a shooting technique. One finds that $\theta'(0) = -0.494$, and so the local Nusselt number becomes

$$\mathrm{Nu}_x = \frac{q'' x}{(T_w - T_\infty)k_m} = 0.494 \mathrm{Ra}_{\infty x}^{1/4}. \tag{5.79}$$

On the right-hand side, $\mathrm{Ra}_{\infty x}$ is obtained from expression (5.68) for Ra_∞ by replacing H by x. This formula for Nu_x differs radically from its Darcy counterpart, listed in Table 5.1, $\mathrm{Nu}_x = 0.444\mathrm{Ra}_x^{1/2}$.

The case of uniform heat flux can be treated similarly. One now finds that the boundary layer thickness is

$$\delta \sim H\mathrm{Ra}_{\infty x}^{-1/5}, \tag{5.80}$$

where Ra_∞^* is the flux-based Rayleigh number for the large Reynolds number limit,

$$\mathrm{Ra}_\infty^* = \frac{g\beta K H^3 q''}{\chi \alpha_m^2 k_m}. \tag{5.81}$$

The corresponding local Nusselt number is

$$\mathrm{Nu}_x = \frac{q'' x}{(T_w - T_\infty)k_m} = 0.804\,\mathrm{Ra}_{\infty^* x}^{1/5} \tag{5.82}$$

For intermediate values of the Forchheimer parameter, similarity solutions do not exist, but nonsimilarity results have been obtained by Bejan and Poulikakos (1984), Kumari *et al.* (1985) (including the effect of wall mass flux), Hong *et al.* (1985), Chen and Ho (1986), Hong *et al.* (1987) (including the effects of nonuniform porosity and dispersion), and Kaviany and Mittal (1987) (for the case of high permeability media). The combination of effects of inertia and suction on the wall was analyzed by Banu and Rees (2000) and by Al-Odat (2004a) for an unsteady situation. The combined effect of inertia and spanwise pressure gradient was examined by Rees and Hossain (1999). In this case the resulting flow is three-dimensional but self-similar, and the boundary layer equations are supplemented by an algebraic equation governing the magnitude of the spanwise velocity field. It was found that the inertial effects serve to inhibit the spanwise flow near a heated surface.

5.1.7.3. Thermal Dispersion Effects

Following Cheng (1985a) one can introduce the effects of thermal dispersion by expressing the heat transfer per unit volume, by conduction and dispersion, in the form

$$\frac{\partial}{\partial x}\left((k_m + k_x')\frac{\partial T}{\partial x}\right) + \frac{\partial}{\partial y}\left((k_m + k_y')\right)\frac{\partial T}{\partial y}.$$

With x denoting the streamwise direction, $k_x{}'$ and $k_y{}'$ are the longitudinal and transverse thermal dispersion coefficients, respectively. Cheng (1981a) assumed that the dispersion coefficients were proportional to the velocity components and to the Forchheimer coefficient χ, so

$$k_x' = a_L \frac{\chi}{\nu}\,|u|\,, \quad k_y' = a_T \frac{\chi}{\nu}\,|v|\,, \tag{5.83a}$$

where a_L and a_T are constants found by matching with experimental data. With this formulation, Cheng found that the effect of thermal dispersion was to decrease the surface heat flux.

On the other hand, Plumb (1983) assumed that the longitudinal coefficient was negligible and the transverse coefficient was proportional to the streamwise velocity component,

$$k'_x = 0, \quad k'_y = C\rho c_P u d. \tag{5.83b}$$

In the k'_y expression, d is the grain diameter and C is a constant found by matching experimental heat transfer data. In this formulation the surface heat flux is given by

$$q''_w = -\left[(k + k'_y)\frac{\partial T}{\partial y}\right]_{y=0} = -[k + C\rho c_P u(x,0)d]\frac{\partial T}{\partial y}(x,0). \tag{5.84a}$$

The second term inside the square brackets of the last term is always positive since $u(x, 0)$ is positive. In dimensionless form this equation is

$$\frac{\mathrm{Nu}_x}{\mathrm{Ra}_x^{1/2}} = -[1 + C\mathrm{Ra}_d f'(0)]\theta'(0) \tag{5.84b}$$

where

$$\theta = \frac{T - T_\infty}{T_w - T_\infty}, \quad f'(\eta)\frac{u}{u_r}, \quad \mathrm{Ra}_d = \frac{g\beta K(T_w - T_\infty)d}{\nu\alpha_m}. \tag{5.85}$$

The dimensionless velocity slip on the wall $f'(0)$ and the dimensionless temperature gradient at the wall $\theta'(0)$ are functions of Gr* and $C\mathrm{Ra}_d$. Plumb's numerical results are shown in Figs. 5.8 and 5.9. They show that both inertial and dispersion effects tend to decrease the temperature gradient at the wall but the combined effects either may increase or decrease the Nusselt number.

Hong and Tien (1987) included the effect of a Brinkman term to account for the no-slip boundary condition. As expected, this substantially reduces the dispersion effect near the wall.

5.1.8. Experimental Investigations

Evans and Plumb (1978) investigated natural convection about a plate embedded in a medium composed of glass beads with diameters ranging from 0.85 to 1.68 mm. Their experimental data, which is shown in Figs. 5.10 and 5.11, is in good agreement with the theory for $\mathrm{Ra}_x < 400$. When $\mathrm{Ra}_x > 400$, temperature fluctuations were observed and the Nusselt number values became scattered.

Similar experiments were undertaken by Cheng *et al.* (1981) with glass beads of 3 mm diameter in water. They observed that temperature fluctuations began in the flow field when the non-Darcy Grashof number Gr* attained a value of about 0.017. They attributed the fluctuations to the onset of non-Darcian flow. After the onset of temperature fluctuations the experimentally determined Nusselt number began to level off and deviate from that predicted by the similarity solution based on Darcy's law. The decrease in Nu_x was found to be substantially larger

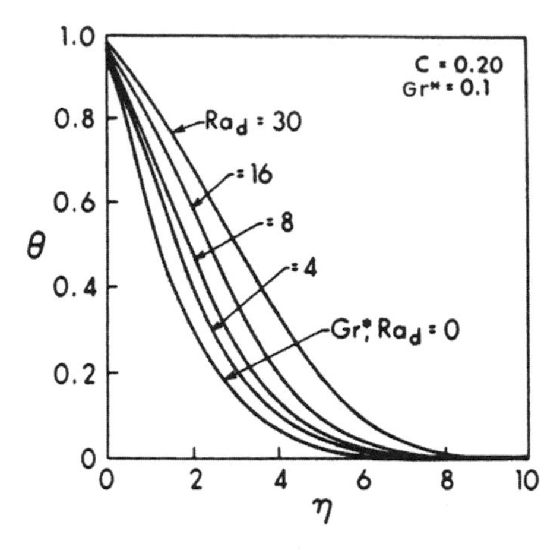

Figure 5.8. Combined effects of inertia and thermal dispersion on dimensionless temperature profiles for natural convection adjacent to a vertical heated surface (Plumb, 1983).

than that predicted by Plumb and Huenefeld's (1981) theory. Cheng (1981a) originally attributed the decrease in Nu_x to the effect of thermal dispersion, but in Cheng (1985a) he announced that this attribution was erroneous. The discrepancy remains ill understood.

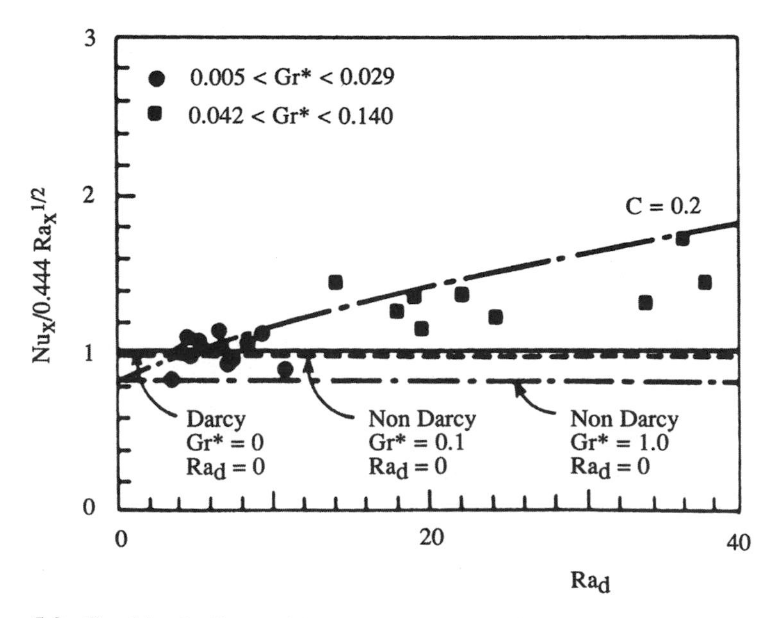

Figure 5.9. Combined effects of inertia and thermal dispersion on the local Nusselt number for natural convection adjacent to a vertical heated surface (Plumb, 1983).

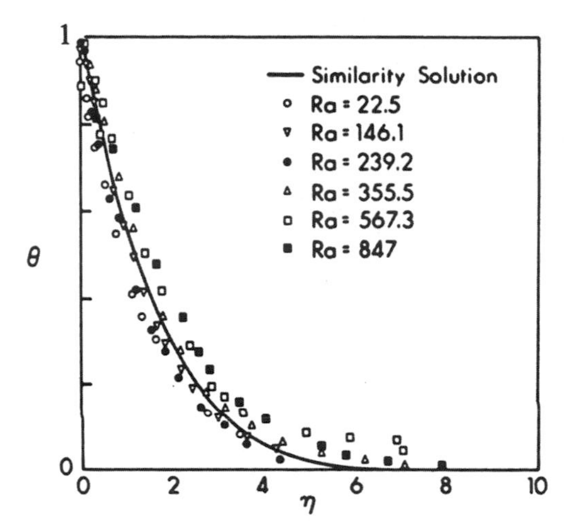

Figure 5.10. Dimensionless temperature profiles for natural convection about an isothermal vertical heated surface (Evans and Plumb, 1978).

Huenefeld and Plumb (1981) performed experiments on convection about a vertical surface with uniform heat flux, the medium being glass beads saturated with water. They observed that temperature fluctuations occurred when the non-Darcy Grashof number Gr* attained a value of about 0.03. Their results are illustrated in

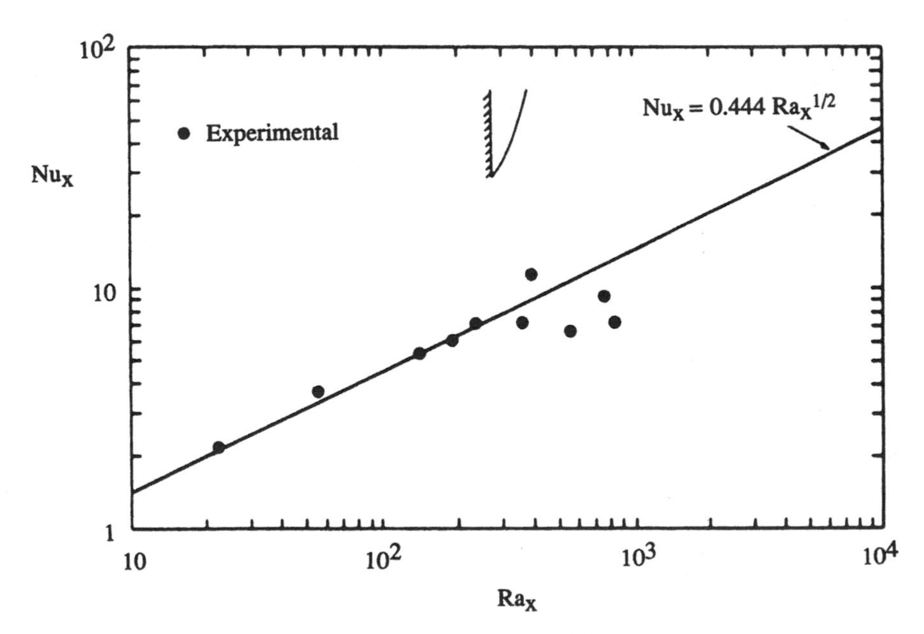

Figure 5.11. Local Nusselt number versus local Rayleigh number for natural convection about an isothermal vertical heated surface (Evans and Plumb, 1978).

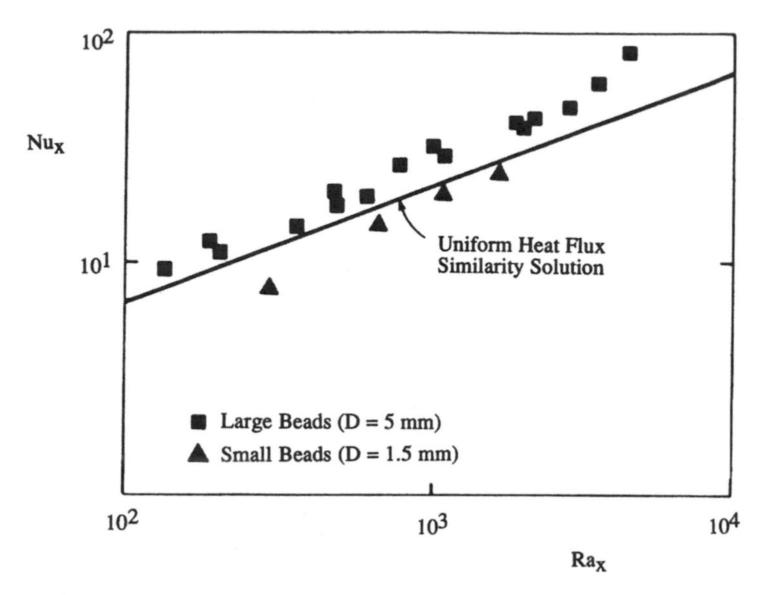

Figure 5.12. Local Nusselt number versus local Rayleigh number for a vertical surface with uniform heat flux (Huenefeld and Plumb, 1981).

Fig. 5.12. The experimental data for the larger beads (diameter 5 mm) are above the predicted values from the Darcy theory, while those of the smaller beads (diameter 1.5 mm) are below the predicted values.

Kaviany and Mittal (1987) performed experiments with high permeability polyurethane foams saturated with air. Except when the permeability was relatively low, they found good agreement between their results and calculations made using a Brinkman-Forchheimer formulation. In their experiments inertial effects were not significant because the Rayleigh numbers were not very high.

Imadojemi and Johnson (1991) reported results of experiments with water-saturated glass beads or irregular shaped gravel. They found that they were unable to obtain an effective correlation of the form $\mathrm{Nu} = A\ \mathrm{Ra}^n$. Rather, they found that A and n varied with the medium and with the heat flux. The mass transfer coefficients found experiementally by Rahman *et al.* (2000) were found to agree closely with those predicted using a Brinkman model. Rahman and Badr (2002) repeated this experimental work for the case of a vertical wavy surface.

5.1.9. Further Extensions of the Theory

5.1.9.1. Particular Analytical Solutions

The homotopy analysis method has been used by Liao and Pop (2004) to obtain explicit analytical solutions of similarity boundary-layer equations. The case of a vertical plate with wall temperature (relative to ambient) varying as $x^{-1/3}$ (i.e., the case $m = -1/3$) yields a hyperbolic tangent solution that was shown by Magyari *et al.* (2003c) to belong to a one-parameter family of multiple solutions that can be expressed in terms of Airy functions. Magyari and Keller (2004b) obtained exact analytical solutions for the cases $m = 1$ and $m = -1/3$ for the backward boundary layer that arises over a cooled (but upward projecting) vertical plate. Some existence and uniqueness results pertaining to the classic boundary equations were reported by Belhachmi *et al.* (2000, 2001). Belhachmi *et al.* (2003) suggested two complementary numerical methods to compute similarity solutions. Magyari and Keller (2000) obtained some special exact analytical solutions for the extended problem where there is variable lateral mass flux. Further special analytical solutions, for unsteady convection for the cases of exponential and power-law time-dependence of the surface temperature, were obtained by Magyari *et al.* (2004).

5.1.9.2. Non-Newtonian Fluids

Non-Newtonian fluid flow has been treated by H. T. Chen and Chen (1988a), Haq and Mulligan (1990b), Pascal (1990), Shenoy (1992, 1993a), Hossain and Nakayama (1994), Beithou *et al.* (2001), El-Hakiem and El-Amin (2001b), El-Amin (2003a), El-Amin *et al.* (2003), Kim (2001a,b), and Hassanien *et al.* (2004, 2005). Of these papers, those by Kim (2001a) and El-Amin (2003a) included the effect of a magnetic field.

5.1.9.3. Local Thermal Nonequilibrium

The classic Cheng-Minkowycz theory was extended to a two-temperature model by Rees and Pop (2000c), using a model introduced by Rees and Pop (1999). The effect of local thermal nonequilibrium (LTNE) was found to modify substantially the behavior of the flow relative to the leading edge, where the boundary layer is composed of two distinct asymptotic regions. At increasing distances from the leading edge the difference between the temperatures of the solid and fluid phases decreases to zero, i.e., thermal equilibrium is attained. Mohamad (2001) independently treated the same problem. In commenting on this paper Rees and Pop (2002) emphasized the importance of undertaking a detailed asymptotic analysis of the leading edge region in order to obtain boundary conditions for the solid-phase temperature field that are capable of describing accurately its behaviour outside the computational domain.

Rees (2003) solved numerically the full equations of motion, and thus investigated in detail how the elliptical terms in the governing equations are manifested. In general it is found that at any point in the flow the temperature of the solid phase is higher than that of the fluid phase, and thus the thermal field of the solid phase is of greater extent than that of the fluid phase. The extension to the Brinkman model was made by Haddad *et al.* (2004), while Haddad *et al.* (2005) reconsidered flow with the Darcy model. Rees *et al.* (2003a) considered forced convection past a heated horizontal circular cylinder. Rees and Pop (2005) reviewed work on LTNE in porous media convection.

5.1.9.4. Volumetric Heating due to Viscous Dissipation, Radiation or Otherwise

Volumetric heating due to the effect of viscous dissipation was analyzed by Magyari and Keller (2003a–c). In their first two papers they observed that the opposing effect of viscous dissipation allows for a parallel boundary-layer flow along a cold vertical plate. In their third paper they considered a quasiparallel flow involving a constant transverse velocity directed perpendicularly toward the wall. They observed that even in the case where the wall temperature equals the ambient temperature thermal convection is induced by the heat released by the viscous dissipation. They examined in detail the resulting self-sustaining wall jets. The development of the asymptotic viscous profile that results was studied by Rees, Magyari, and Keller (2003b). The vortex instability of the asymptotic dissipation profile was analyzed by Rees *et al.* (2005a). The case of an exponential wall temperature was studied by Magyari and Rees (2005). The general effect of viscous dissipation, which reduces heat transfer, was investigated by Murthy and Singh (1997a), who also took thermal dispersion effects into account. The effect of variable permeability was added by Hassanien *et al.* (2005). A survey of work on the effect of viscous dissipation was made by Magyari *et al.* (2005b).

Volumetric heating due to the absorption of radiation was studied by Chamkha (1997a), Takhar *et al.* (1998), Mohammadien *et al.* (1998), Mohammadien and El-Amin (2000), Raptis (1998), Raptis and Perdikis (2004), Hossain and Pop (2001), El-Hakiem (2001a), El-Hakiem and El-Amin (2001a), Chamkha *et al.* (2001), Mansour and El-Shaer (2001), Mansour and Gorla (2000a,c), and Israel-Cookey *et al.* (2003). Some more general aspects of volumetric heating were considered by Chamkha (1997d), Bakier *et al.* (1997), Postelnicu and Pop (1999), and Postelnicu *et al.* (2000).

5.1.9.5. Anisotropy and Heterogeneity

Anisotropic permeability effects have been analyzed by Ene (1991) and Rees and Storesletten (1995). The latter found that the boundary-layer thickness was altered and a spanwise fluid drift induced by the anisotropy. As Storesletten and Rees (1998) demonstrated, anisotropic thermal diffusivity produces no such drift. An analytical and numerical study of the effect of anisotropic permeability was reported by Vasseur and Degan (1998).

The effect of variable permeability, enhanced within a region of constant thickness, was treated analytically and numerically by Rees and Pop (2000a). They found that near the leading edge the flow is enhanced and the rate of heat transfer is much higher than in the uniform permeability case. Further downstream the region of varying permeability is well within the boundary layer, and in this case the flow and heat transfer is only slightly different from that in the uniform case. Convection over a wall covered with a porous substrate was analyzed by Chen and Chen (1996). Convection from an isothermal plate in a porous medium layered in a parallel fashion, with discrete changes in either the permeability or the diffusivity of the medium, was studied by Rees (1999). He supplemented his numerical work with an asymptotic analysis of the flow in the far-downstream limit.

5.1.9.6. *Wavy Surface*

The case of a wavy surface has been analyzed by Rees and Pop (1994a, 1995a,b, 1997). In the last paper they considered the full governing equations and derived the boundary layer equations in a systematic way. They found that, for a wide range of values of the distance from the leading edge, the boundary layer equations for the three-dimensional flow field are satisfied by a two-dimensional similarity solution.

5.1.9.7. *Time-dependent Gravity or Time-dependent Heating*

The effect of g-jitter was analyzed by Rees and Pop (2000b, 2003) for the cases of small and large amplitudes. Their numerical and asymptotic solutions show that the g-jitter effect is eventually confined to a thin layer embedded within the main boundary layer, but it becomes weak at increasing distances from the leading edge. The case of time-periodic surface temperature oscillating about a constant mean was studied by Jaiswal and Soundalgekar (2001). The more general case of oscillation about a mean that varies as the nth power of the distance from the leading edge was analyzed by Hossain *et al.* (2000). They considered low- and high-frequency limits separately and compared these with a full numerical solution, for $n \leq 1$. They noted that when $n = 1$ the flow is self-similar for any prescribed frequency of modulation. Temperature oscillations also were studied using a Forchheimer model by El-Amin (2004b). A vertical wall with suction varying in the horizontal direction and with a pulsating wall temperature was studied by Chaudhary and Sharma (2003). A nonequilibrium model was used by Saeid and Mohamad (2005a) in their numerical study of the effect of a sinusoidal plate temperature oscillation with respect to time about a nonzero mean.

5.1.9.8. *Newtonian Thermal Boundary Condition*

The case of surface heating with a boundary condition of the third kind was studied by Lesnic *et al.* (1999) and Pop *et al.* (2000). They obtained fully numerical, asymptotic, and matching solutions.

5.1.9.9. *Other Aspects*

A comprehensive listing of similarity solutions, including some for special transient situations, was presented by Johnson and Cheng (1978). The cases of arbitrary wall temperature and arbitrary heat flux have been treated using a Merk series technique by Gorla and Zinalabedini (1987) and Gorla and Tornabene (1988).

Merkin and Needham (1987) have discussed the situation where the wall is of finite height and the boundary layers on each side of the wall merge to form a buoyant wake. Singh *et al.* (1988) have studied the problem when the prescribed wall temperature is oscillating with time about a nonzero mean. Zaturska and Banks (1987) have shown that the boundary layer flow is stable spatially.

The asymptotic linear stability analysis of Lewis *et al.* (1995) complements the direct numerical simulation of Rees (1993) in showing that the flow is stable at locations sufficiently close to the leading edge. In the asymptotic regime also the wave disturbances decay, but the rate of decay decreases as the distance downstream of the leading edge increases.

The effect of temperature-dependent viscosity has been examined theoretically by Jang and Leu (1992), for a steady flow, and by Mehta and Sood (1992b) and Rao and Pop (1994), for a transient flow.

For the case of prescribed heat flux, Kou and Huang (1996a,b) have shown how three cases are related by a certain transformation, and Wright *et al.* (1996) have treated another special case. Ramanaiah and Malarvizhi (1994) have shown how three situations are related. Nakayama and Hossain (1994) have shown that both local similarity and integral methods perform excellently for a nonisothermal plate. A perturbation approach to the nonuniform heat flux situation was used by Seetharamu and Dutta (1990) and Dutta and Seetharamu (1993). Bradean *et al.* (1996, 1997a) have given an analytical and numerical treatment for a periodically heated and cooled vertical or horizontal plate. For a vertical plate, a row of counter-rotating cells forms close to the surface, but when the Rayleigh number increases above about 40 the cellular flow separates from the plate. For a horizontal plate the separation does not occur.

Merkin and Needham (1987) have discussed the situation where the wall is of finite height and the boundary layers on each side of the wall merge to form a buoyant wake. Singh, Misra, and Narayan (1988) have studied the problem when the prescribed wall temperature is oscillating with time about a nonzero mean.

Seetharamu and Dutta (1990) used a perturbation approach to treat the case of arbitrary wall temperature. Herwig and Koch (1990) examined the asymptotic situation when the porosity tends to unity. Ramaniah and Malarvizhi (1991) presented some exact solutions for certain cases. Chandrasekhara *et al.* (1992) and Chandrasekhara and Nagaraju (1993) have treated a medium with variable porosity, with surface mass transfer or radiation. Pop and Herwig (1992) presented an asymptotic approach to the case where fluid properties vary, while Na and Pop (1996) presented a new accurate numerical solution of the Cheng-Minkowycz equation equivalent to (5.15)–(5.18). Rees (1997b) discussed the case of parallel layering, with respect to either permeability or thermal diffusivity, of the medium. The numerical solution of the nonsimilar boundary layer equations was supplemented by an asymptotic analysis of the flow in the far downstream limit. Rees (1997b) examined the three-dimensional boundary layer on a vertical plate where the surface temperature varies sinusoidally in the horizontal direction. The effect of an exothermic reaction was studied by Minto *et al.* (1998).

The study of the influence of higher-order effects on convection in a wedge bounded by a uniformly heated plane and one cold or insulated by Storesletten and Rees (1998) revealed that generally instability occurs too close to the leading edge for the basic flow to be represented adequately either by the leading order boundary layer theory used in previous papers or even by the most accurate higher-order theory obtained using matched expansions. This is a chastening result.

The effect of lateral mass flux was studied analytically and numerically by Dessaux (1998). Unsteady convection was studied by Al-Nimr and Massoud (1998), and also with the effect of a magnetic field by Helmy (1998). Convection along a vertical porous surface consisting of a bank of parallel plates with constant gaps was studied experimentally by Takatsu *et al.* (1997).

5.2. Horizontal Plate

For high Rayleigh number natural convection flow near the edge of an upward-facing heated plate a similarity solution was obtained by Cheng and Chang (1976), for the case of a power law wall temperature distribution given by Eq. (5.8). This leads to the formulas

$$\frac{\delta}{x} = \frac{\eta_T}{\mathrm{Ra}_x^{1/3}}, \tag{5.86}$$

$$\frac{\mathrm{Nu}_x}{\mathrm{Ra}_x^{1/3}} = -\theta'(0), \tag{5.87}$$

$$\frac{\overline{\mathrm{Nu}}}{\overline{\mathrm{Ra}}^{1/3}} = \frac{3(1+\lambda)^{4/3}}{(1+4\lambda)}[-\theta'(0)], \tag{5.88}$$

for the thermal boundary layer thickness δ, the local Nusselt number Nu_x, and the overall Nusselt number $\overline{\mathrm{Nu}}$. Table 5.2 lists values of η_T and $[-\theta'(0)]$ for selected values of λ. It should be noted that in practice the assumption of quiescent flow outside the boundary layer on an upward-facing heated plate is unlikely to be justified, and such a boundary layer is better modeled as a mixed convection problem.

One would expect a natural convection boundary layer to form on a cooled plate facing upward, or on a warm plate facing downward. This situation was analyzed by Kimura *et al.* (1985). Relative to a frame with x axis horizontal and y axis vertically upward, we suppose that the plate is at $-l \leq x \leq l$, $y = 0$ and is at constant temperature T_w, ($T_w < T_\infty$). The plate length is $2l$.

Table 5.2. Values of η_T and $-\theta'(0)$ for various values of λ for an upward-facing heated horizontal plate (Cheng and Chang, 1976)

λ	η_T	$-\theta'(0)$
0	5.5	0.420
1/2	5.0	0.816
1	4.5	1.099
3/2	4.0	1.351
2	3.7	1.571

The mass and energy equations (5.1) and (5.4) still stand, but now the Darcy equations are

$$u = -\frac{K}{\mu}\frac{\partial P'}{\partial x}, \tag{5.89}$$

$$v = -\frac{K}{\mu}\left[\frac{\partial P'}{\partial y} + \rho g\beta(T_\infty - T)\right]. \tag{5.90}$$

Eliminating P' we get

$$\frac{\partial u}{\partial y} = \frac{g\beta K}{\nu}\frac{\partial}{\partial x}(T_\infty - T). \tag{5.91}$$

The boundary conditions are

$$\begin{aligned} y = 0 :\quad & v = 0 \text{ and } T = T_w, \\ y \to \infty : u = 0,\ & T = T_\infty \text{ and } \frac{\partial T}{\partial y} = 0. \end{aligned} \tag{5.92}$$

The appropriate Rayleigh number is based on the plate half-length l,

$$\text{Ra} = \frac{g\beta Kl(T_\infty - T_w)}{\nu\alpha_m}. \tag{5.93}$$

Scaling analysis indicates that the boundary layer thickness must be of order

$$\delta \sim l\ \text{Ra}^{-1/3} \tag{5.94}$$

The Nusselt number defined by

$$\text{Nu} = \frac{q'}{k_m(T_\infty - T_w)} \tag{5.95}$$

in which q'[W/m] is the heat transfer rate into the whole plate, is of order

$$\text{Nu} \sim \text{Ra}^{1/3}. \tag{5.96}$$

This is in contrast to the $\text{Nu} \sim \text{Ra}^{1/2}$ relationships for a vertical plate. Kimura *et al.* (1985) solved the boundary layer equations approximately using an integral method. Their numerical results shown in Fig. 5.13 confirm the theoretical trend (5.92).

Ramaniah and Malarvizhi (1991) noted a case in which an exact solution could be obtained. Wang *et al.* (2003c) reported an explicit, totally analytic and uniformly valid solution of the Cheng-Chang equation that agreed well with numerical results. Modifications of the Cheng and Chang (1976) analysis include those made by Chen and Chen (1987) for a non-Newtonian power-law fluid, by Lin and Gebhart (1986) for a fluid whose density has a maximum as the temperature is varied, by Minkowycz *et al.* (1985b) for the effect of surface mass flux and by Vedha-Nayagam *et al.* (1987) for the effects of surface mass transfer and variation of porosity. The combination of power-law fluid and thermal radiation was considered by Mohammadein and El-Amin (2001).

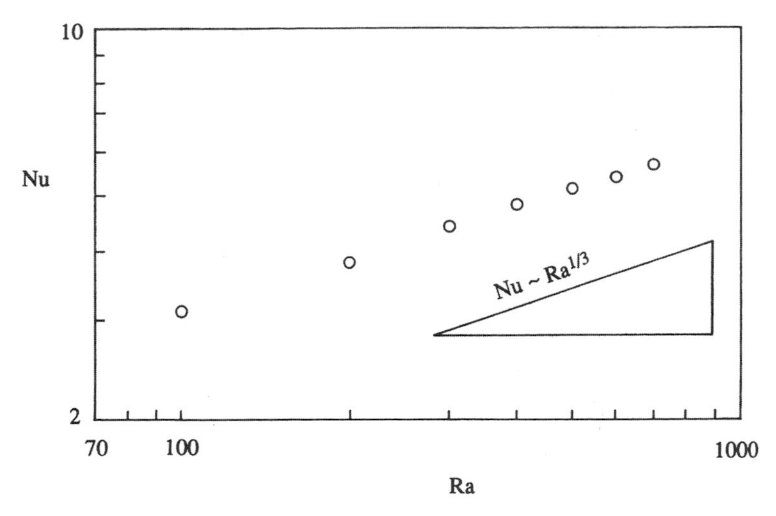

Figure 5.13. Nusselt number versus Rayleigh number for convection on a cooled horizontal plate of finite length facing upward (Kimura *et al.*, 1985).

Ingham *et al.* (1985a) have studied the transient problem of a suddenly cooled plate. Harris *et al.* (2000) studied analytically and numerically the transient convection induced by a sudden change in surface heat flux.

Merkin and Zhang (1990a) showed that for the case of wall temperature proportional to x^m a solution of the similarity equations is possible only for $m > -2/5$. For a non-Newtonian power law fluid, Mehta and Rao (1994) treated the case of a power law wall temperature and Chamkha (1997c) studied the case of uniform wall heat flux, while the effect of surface mass flux was added by Gorla and Kumari (2003). Pop and Gorla (1991) studied a heated horizontal surface, the fluid being a gas whose thermal conductivity and dynamic viscosity are proportional to temperature. They obtained a similarity solution for the case of constant wall temperature. The effect of temperature-dependent viscosity also was studied by Kumari (2001a,b), and by Postelnicu *et al.* (2001) for the case of internal heating already treated by Postelnicu and Pop (1999). Similarity solutions for convection adjacent to a horizontal surface with an axisymmetric temperature distribution were given by Cheng and Chau (1977), and El-Amin *et al.* (2004) added the effects of a magnetic field and lateral mass flux. Lesnic *et al.* (2000, 2004) studied analytically and numerically the case of a thermal boundary condition of mixed type (Newtonian heat transfer). The case of wall temperature varying as a quadratic function of position was studied, as a steady or unsteady problem, by Lesnic and Pop (1998a).

The singularity at the edge of a downward-facing heated plate was analyzed by Higuera and Weidman (1995) and the appropriate boundary condition deduced. They considered both constant temperature and constant flux boundary conditions and they treated a circular disk as well as an infinite strip. They also gave solutions for a slightly inclined plate maintained at constant temperature.

Convection below a downward facing heated horizontal surface also was treated numerically by Angirasa and Peterson (1998b). Convection from a heated upward-facing finite horizontal surface was studied numerically by Angirasa and Peterson (1998a). Two-dimensional flows were found for $40 \leq \mathrm{Ra} \leq 600$, and the correlation $\mathrm{Nu} = 3.092\ \mathrm{Ra}^{0.272}$ was obtained. At higher Rayleigh numbers the flow becomes three-dimensional with multiple plume formation and growth.

Rees and Bassom (1994) found that waves grow beyond a nondimensional distance 28.90 from the leading edge, whereas vortices grow only beyond 33.47. This stability analysis was based on a parallel flow approximation. Because of the inadequacy of this approximation, Rees and Bassom (1993) performed numerical simulations of the full time-dependent nonlinear equations of motion. They found that small-amplitude disturbances placed in the steady boundary layer propagated upstream much faster than they were advected downstream. With the local growth rate depending on the distance downstream, there is a smooth spatial transition to convection. For the problem where the temperature of the horizontal surface is instantaneously raised above the ambient, they found a particularly violent fluid motion near the leading edge. A strong thermal plume is generated, which is eventually advected downstream. The flow does not settle down to a steady or time-periodic state. The evolving flow field exhibits a wide range of dynamic behavior including cell merging, the ejection of hot fluid from the boundary layer, and short periods of relatively intense fluid motion accompanied by boundary-layer thinning and short wavelength waves.

Rees (1996a) showed that when the effects of inertia are sufficiently large, the leading order boundary layer theory is modified, and he solved numerically the resulting nonsimilar boundary layer equations. He showed that near the leading edge inertia effects then dominate, but Darcy flow is reestablished further downstream. The effects of inertia in the case of a power-law distribution of temperature were analyzed by Hossain and Rees (1997).

The Brinkman model was employed by Rees and Vafai (1999). They showed that for a constant temperature surface, both the Darcy and Rayleigh numbers can be scaled out of the boundary layer equations leaving no parameters to vary. They studied these equations using both numeric and asymptotic methods. They found that near the leading edge the boundary layer has a double-layer structure: a near-wall layer where the temperature adjusts from the wall temperature to the ambient and where Brinkman effects dominate and an outer layer of uniform thickness that is a momentum adjustment layer. Further downstream, these layers merge, but the boundary layer eventually regains a two-layer structure; in this case a growing outer layer exists, which is identical to the Darcy flow case for the leading order term and an inner layer of constant thickness resides near the surface where the Brinkman term is important.

Convection induced by a horizontal wavy surface was analyzed by Rees and Pop (1994b). They focused their attention on the case where the waves have an $O(\mathrm{Ra}^{-1/3})$ amplitude, where Ra is based on the wavelength and is assumed large. They found that a thin near-wall boundary layer develops within the basic boundary layer as the downstream distance is increased and they gave an asymptotic analysis that determines the structure of this layer. They found that when the wave amplitude is greater than approximately 0.95 $\mathrm{Ra}^{-1/3}$, localized regions of reversed flow occur at the heated surface.

The case of a sinusoidally (lengthwise) heated and cooled horizontal surface was studied by Bradean *et al.* (1995a), when at large distances from the plate there is either constant temperature or zero heat flux. Bradean *et al.* (1996, 1997a) examined cases of unsteady convection from a horizontal (or vertical) surface that is suddenly heated and cooled sinusoidally along its length. They obtained an analytical solution valid for small times and any value of Ra, and a numerical solution matching this to the steady-state solution (when this exists). The flow pattern is that of a row of counterrotating cells situated close to the surface. When the surface is vertical and for $\mathrm{Ra} > 40$ (approximately), two recirculating regions develop at small times at the point of collision of two boundary layers that flow along the surface. However, for $40 < \mathrm{Ra} < 150$ the steady-state solution is unstable and at very large time the solution is periodic in time. When the surface is horizontal, the collision of convective boundary layers occurs without separation. As time increases, the height of the cellular flow pattern increases and then decreases to its steady-state value. The heat penetrates infinitely into the porous medium and the steady state is approached later in time as the distance from the surface increases.

Numerical and similarity solutions for the boundary layer near a horizontal surface with nonuniform temperature and mass injection or withdrawal were reported by Chaudhary *et al.* (1996). In their study the temperature and mass flux varied as x^{μ} and $x^{(\mu-2)/3}$, respectively, where μ is a constant. The conjugate problem of boundary layer natural convection and conduction inside a horizontal plate of finite thickness was solved numerically by Lesnic *et al.* (1995). The conjugate problem for convection above a cooled or heated finite plate was studied numerically by Vaszi *et al.* (2001a, 2002a).

5.3. Inclined Plate

Again we take the x axis along the plate and the y axis normal to the plate. In the boundary layer regime $\partial T/\partial x \ll \partial T/\partial y$, and the equation obtained by eliminating the pressure between the two components of the Darcy equation, reduces to

$$\frac{\partial^2 \psi}{\partial y^2} = \frac{g_x \beta K}{\nu} \frac{\partial T}{\partial y}, \tag{5.97}$$

where g_x is the component of $\boldsymbol{g}$ parallel to the plate. This is just Eq. (5.6) with g replaced by g_x. With this modification, the analysis of Section 5.1 applies to the inclined plate problem unless the plate is almost horizontal, in which case g_x is small compared with the normal component g_y.

The case of small inclination to the horizontal was analyzed by Ingham *et al.* (1985b) and Rees and Riley (1985). Higher-order boundary layer effects, for the case of uniform wall heat flux, were incorporated by Ingham and Pop (1988). Jang and Chang (1988d) performed numerical calculations for the case of a power

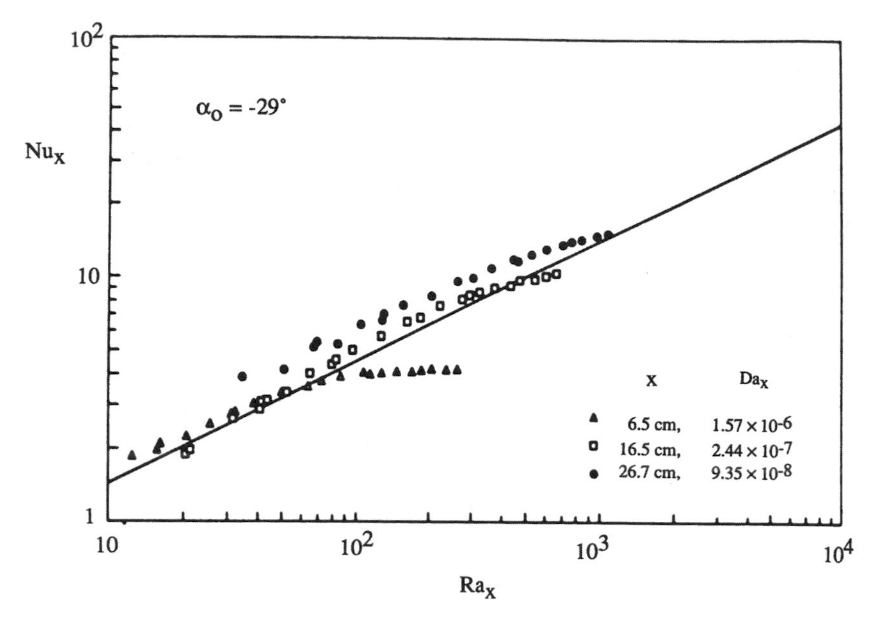

Figure 5.14. Local Nusselt number versus local Rayleigh number for a downward-facing heated inclined plate (Lee, 1983; Cheng, 1985a, with permission from Hemisphere Publishing Corporation).

function distribution of wall temperature. They found that, as the inclination to the horizontal increases, both the velocity and temperature boundary layer thicknesses decrease, and the rate of surface heat transfer increases. Jang and Chang (1989) have analyzed the case of double diffusion and density maximum.

In their experiments on natural convection from an upward-facing inclined isothermal plate to surrounding water-filled glass beads, Cheng and Ali (1981) found that large amplitude temperature fluctuations exist in the flow field at high Rayleigh numbers, presumably because of the onset of vortex instability. Cheng (1985a) also reported on experiments by himself, R. M. Fand, and H. M. Lee for a downward-facing isothermal plate with inclinations of 29° and 45°. Their results are presented in Fig. 5.14, which shows a leveling off of the local Nusselt number Nu_x from the $Ra_x^{1/2}$ dependence at high values of the local Rayleigh number Ra_x. It is not certain whether this deviation is due to the boundary friction effect, the inertial effect, or inaccuracy in the experimental determination of Nu_x.

The effect of lateral surface mass flux, with a power law variation of lateral surface velocity and wall temperature, was studied by Dwiek *et al.* (1994). The use of a novel inclination parameter enabled Pop and Na (1997) to describe all cases of horizontal, inclined, and vertical plates by a single set of transformed boundary layer equations. Hossain and Pop (1997) studied the effect of radiation. Shu and Pop (1997) obtained a numerical solution for a wall plume arising from a line source embedded in a tilted adiabatic plane. MHD convection with thermal stratification was studied by Chamkha (1997e) and Takhar *et al.* (2003). The effects of variable porosity and solar radiation were discussed by Chamkha *et al.* (2002). The effects of lateral mass flux and variable permeability were analyzed by Rabadi and Hamdan (2000). Conjugate convection from a slightly inclined plate was studied analytically and numerically by Vaszi *et al.* (2001b). Lesnic *et al.* (2004) studied analytically and numerically the case of a thermal boundary condition of mixed type (Newtonian heat transfer) on a nearly horizontal surface.

The linear stability of a thermal boundary layer with suction in an anisotropic porous medium was discussed by Rees and Storesletten (2002). The effects of inertia and nonparallel flow were incorporated in the analysis of Zhao and Chen (2002). These effects stabilize the flow.

5.4. Vortex Instability

For an inclined or a horizontal upward-facing heated surface embedded in a porous medium, instability leading to the formation of vortices (with axes aligned with the flow direction) may occur downstream as the result of the top-heavy situation. Hsu *et al.* (1978) and Hsu and Cheng (1979) have applied linear stability analysis for the case of a power law variation of wall temperature, on the assumption that the basic state is the steady two-dimensional boundary layer flow discussed above. They showed that the length scale of vortex disturbances is less than that for the undisturbed thermal boundary layer, and as a result certain terms in the three-dimensional disturbance equations are negligible.

The simplified equations for the perturbation amplitudes were solved on the basis of local similarity assumptions (the disturbances being allowed to have a weak dependence in the streamwise direction). It was found that the critical value for the onset of vortex instability in natural convection about an inclined isothermal surface with inclination α_0 to the vertical is given by

$$\mathrm{Ra}_{x,a} \tan^2 \alpha_0 = 120.7, \tag{5.98}$$

where

$$\mathrm{Ra}_{x,a} = \frac{g\beta K(T_w - T_\infty)(\cos \alpha_0)x}{\nu\alpha_m}. \tag{5.99}$$

It follows that the larger the inclination angle with respect to the vertical, the more susceptible the flow to vortex instability, and in the limit of zero inclination angle (vertical heated surface) the flow is stable to this type of disturbance.

For the case of a horizontal heated plate, a similar analysis shows that the critical value is $\mathrm{Ra}_x = 33.4$, where Ra_x is defined as in Eq. (5.14). More precise calculations, including pressure and salinity effects, and including the effect of the normal component of the buoyancy force in the main flow, were made by Jang and Chang (1987, 1988a). Chang and Jang (1989a,b) have examined the non-Darcy effects. The effect of inertia is to destabilize the flow to the vortex mode of disturbance, while the other non-Darcy terms lead to a stabilizing effect. The effect of inertia was also considered by Lee *et al.* (2000) in their study involving an inclined plate. Jang and Chen (1993a,b, 1994) studied the effect of dispersion (which stabilizes the vortex mode) and the channeling effect of variable porosity (which destabilizes it). The effect of variable viscosity was studied by Jang and Leu (1993) and Leu and Jang (1993). Nield (1994c) pointed out that their implication that this property variation produced a destabilizing effect was invalid. Jang and Lie (1992) and Lie and Jang (1993) treated a mixed convection flow. For a horizontal plate, Hassanien *et al.* (2004b,c) considered the effect of variable permeability for the case of variable wall temperature, and the effect of inertia in the case of surface mass flux.

The above studies have been made on the assumption of parallel flow. Bassom and Rees (1995) pointed out the inadequacy of this approach and reexamined the problem using asymptotic techniques that use the distance downstream as a large parameter. The parallel-flow theories predict that at each downstream location there are two possible wavenumbers for neutral stability, and one of these is crucially dependent on nonparallelism within the flow. The nonparallel situation and inertial effects have been treated analytically and numerically by Zhao and Chen (2002, 2003) for the case of horizontal and inclined pates. They found that the nonparallel flow model predicts a more stable flow than the parallel flow model. They also noted that as the inclination relative to the horizontal increases, or the inertia effect as measured by a Forchheimer number increases, the surface heat transfer rate decreases and the flow becomes more stable.

Comprehensive and critical reviews of thermal boundary layer instabilities were made by Rees (1998, 2002c). In the first study he pointed out an inconsistency in the analysis of Jang and Chang (1988a, 1989) and Jang and Lie (1992) that negates their claim that their analysis is valid for a wide range of inclinations; rather, it applies for a near-horizontal surface only. Rees (1998) also noted that the analysis of Jang and Chen (1993a,b) involves a nongeneric formula for permeability variation. The basic difficulty is that a contradiction is entertained by asserting simultaneously that x, the nondimensional streamwise distance, is asymptotically large (so that the boundary layer approximation is valid) and that finite values of x are to be computed as a result of approximating the stability equations, and in general this critical value of x is far too small for the boundary layer approximation to be valid.

One way out of the impass is to carry out fully elliptic simulations. This was the avenue taken by Rees and Bassom (1993) in their description of wave instabilities in a horizontal layer. The second way out is to consider heated surfaces that are very close to the vertical but which remain upward facing. In such cases the critical distance recedes to large distances from the leading edge, and therefore instability arises naturally in a regime where the boundary layer approximation is valid. This was the avenue taken by Rees (2001, 2002a) in his study of the linear and nonlinear evolution of vortex instabilities in near-vertical surfaces. Rees found that even under these favorable circumstances the concept of neutral stability is difficult to define. The reason is that the evolution of vortices is governed by a parabolic partial differential equation system rather than an ordinary differential equation system. As a result the point at which instability is "neutral" depends on whether instability is defined as the value of x at which the thermal energy has a local minimum a x increase, or where the surface rate of heat transfer or the maximum disturbance temperature have minima. Whenever vortices grow, they attain a maximum strength and then decay again, and there is an optimum disturbance amplitude that yields the largest possible response downstream. When applied to developing flows such as boundary layers these three criteria yield different results. In addition of the wavelength of the vortex, the location of the initiating disturbance and its shape also alter the critical value of x.

The linear and nonlinear evolution of vortex instabilities in near-vertical surfaces was studied by Rees (2001, 2002a). He found that the strength of the resulting convection depends not only on the wavelength of the vortex disturbance but also on the amplitude of the disturbance and its point of introduction into the boundary layer. Whenever vortices grow, they attain a maximum strength and then decay again. There is an optimum disturbance amplitude that yields the largest possible response downstream. The later study by Rees (2004b) involved the destabilizing of an evolving vortex using subharmonic disturbances. He found that the onset of the destabilization is fairly sudden, but its location depends on the size of the disturbance. Rees also looked at the evolution of isolated thermal vortices. He found then that developing vortices induce a succession of vortices outboard of the current local pattern until the whole spanwise domain is filled with a distinctive wedge shaped pattern.

5.5. Horizontal Cylinder

5.5.1. Flow at High Rayleigh Number

We now consider steady natural convection about an isothermal cylinder, at temperature T_w and with radius r_0, embedded in a porous medium at temperature T_∞. We choose a curvilinear orthogonal system of coordinates, with x measured along the cylinder from the lower stagnation point (in a plane of cross section), y measured radially (normal to the cylinder), and φ the angle that the y axis makes with the downward vertical. This system is presented in Fig. 5.15.

If curvature effects and the normal component of the gravitational force are neglected, the governing boundary layer equations are

$$\frac{\partial^2 \psi}{\partial y^2} = \frac{g\beta K}{\nu} \sin\varphi \frac{\partial}{\partial y}(T - T_\infty) \tag{5.100}$$

$$\alpha_m \frac{\partial^2 T}{\partial y^2} = \frac{\partial \psi}{\partial y}\frac{\partial T}{\partial x} - \frac{\partial \psi}{\partial x}\frac{\partial T}{\partial y}. \tag{5.101}$$

It is easily checked that the solution of Eqs. (5.100) and (5.101), subject to the

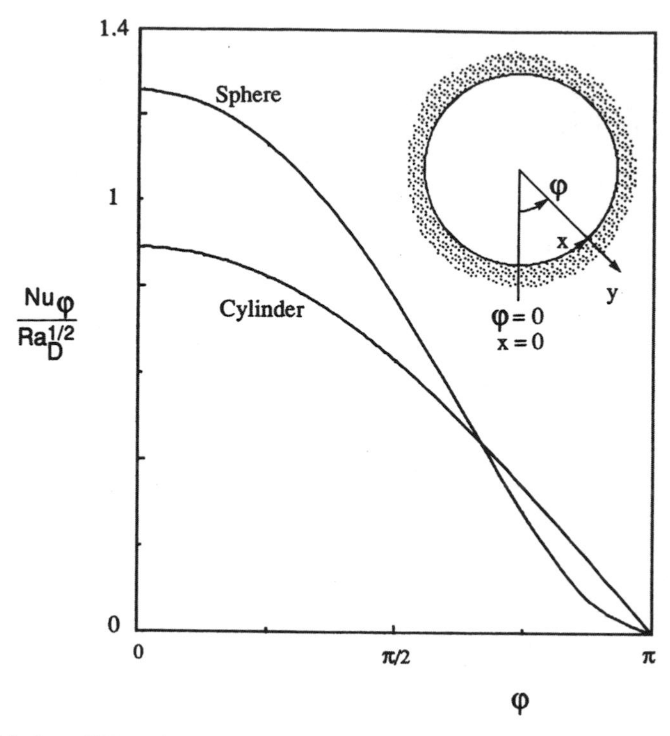

Figure 5.15. Local Nusselt number variation for high Rayleigh number natural convection over a horizontal cylinder and a sphere.

boundary conditions (5.9) and (5.10) with $\lambda = 0$, are given by

$$\psi = \left[\frac{g\beta K}{\nu}(T_w - T_\infty)\alpha_m r_0\right]^{1/2}(1-\cos\varphi)^{1/2} f(\eta), \tag{5.102}$$

$$T - T_\infty = (T_w - T_\infty)\theta(\eta), \tag{5.103}$$

$$\eta = \left[\frac{g\beta K(T_w - T_\infty)}{\nu\alpha_m r_0}\right]^{1/2}\frac{y\sin\varphi}{(1-\cos\varphi)^{1/2}}, \tag{5.104}$$

where f and θ satisfy Eqs. (5.15)–(5.18) with $\lambda = 0$. Accordingly, the local surface heat flux is

$$q''_w = -k_m\left(\frac{\partial T}{\partial y}\right)_{=0} = 0.444 k_m (T_w - T_\infty)^{3/2}\left(\frac{g\beta K}{\nu\alpha_m r_0}\right)^{1/2}\frac{\sin\varphi}{(1-\cos\varphi)^{1/2}}, \tag{5.105}$$

which can be expressed in dimensionless form as

$$\frac{\text{Nu}_\varphi}{\text{Ra}_D^{1/2}} = 0.628\frac{\sin\varphi}{(1-\cos\varphi)^{1/2}}, \tag{5.106}$$

where

$$\mathrm{Nu}_{\varphi} = \frac{q''_w D}{k_m(T_w - T_\infty)} \tag{5.107}$$

and

$$\mathrm{Ra}_D = \frac{g\beta K(T_w - T_\infty)D}{\nu\alpha_m}, \tag{5.108}$$

with D denoting the diameter of the cylinder. This result is plotted in Fig. 5.15. The average surface heat flux is

$$\bar{q}'' = \frac{1}{\pi}\int_0^{\pi} q''_w(\varphi)d\varphi = 0.565\, k_m(T_w - T_\infty)^{3/2}\left(\frac{g\beta K}{\nu\alpha_m D}\right)^{1/2}, \tag{5.109}$$

which in dimensionless form is

$$\frac{\mathrm{Nu}}{\mathrm{Ra}_D^{1/2}} = 0.565, \tag{5.110}$$

where

$$\overline{\mathrm{Nu}} = \frac{\bar{q}'' D}{k}(T_w - T_\infty). \tag{5.111}$$

The present problem is a special case of convection about a general two-dimensional heated body analyzed by Merkin (1978). The generalization to a non-Newtonian power law fluid was made by Chen and Chen (1988b) and for the Forchheimer model by Kumari and Jayanthi (2004).

The conjugate steady convection from a horizontal circular cylinder with a heated core was investigated by Kimura and Pop (1992b). The method of matched asymptotic expansions was applied by Pop *et al.* (1993a) to the transient problem with uniform temperature. They found that vortices then form at both sides of the cylinder. An extension of this work to a cylinder of arbitrary cross section was reported by Tyvand (1995). For the circular cylinder, a numerical treatment was reported by Bradean *et al.* (1997b), and further work on transient convection was discussed by Bradean *et al.* (1998a). They found that as convection becomes more dominant, a single hot cell forms vertically above the cylinder and then rapidly moves away. Free convection about a cylinder of elliptic cross section was treated by Pop *et al.* (1992b). Transient convection about a cylinder with constant surface flux heating was dealt with by Pop *et al.* (1996). A problem involving unsteady convection driven by an nth-order irreversible reaction was examined by Nguyen *et al.* (1996). Natural and forced convection around line sources of heat and heated cylinders was analyzed by Kurdyumov and Liñán (2001). Convection near the stagnation point of a two-dimensional cylinder, with the surface temperature oscillating about a mean above ambient, was analyzed by Merkin and Pop (2000).

Empirical heat transfer correlation equations, with viscous dissipation taken into account, were reported by Fand *et al.* (1994).

5.5.2. Flow at Low and Intermediate Rayleigh Number

The experimental results obtained by Fand *et al.* (1986) on heat transfer in a porous medium consisting of randomly packed glass spheres saturated by either water or silicone oil suggested the division of the Rayleigh number range into a low Ra (and hence low Reynolds number Re) Darcy range and a high Ra Forchheimer range. Fand *et al.* (1986) proposed the following correlation formulas:

For $(0.001 < \mathrm{Re}_{max} = 3)$,

$$\mathrm{Nu}\,\mathrm{Pr}^{0.0877} = 0.618\mathrm{Ra}^{0.698} + 8.54 \times 10^6 \mathrm{Ge}\,\mathrm{sech}\,\mathrm{Ra}, \tag{5.112}$$

while for $(3 < \mathrm{Re}_{max} = 100)$,

$$\mathrm{Nu}\,\mathrm{Pr}^{0.0877} = 0.766\mathrm{Ra}^{0.374}\left(\frac{C_1 D}{C_2}\right)^{0.173}. \tag{5.113}$$

In these correlations,

$$\mathrm{Re}_{max} = \frac{Dv_{max}}{V}, \quad \mathrm{Nu} = \frac{hD}{k_m}, \quad \mathrm{Pr} = \frac{\mu c_P}{k_m}, \quad \mathrm{Ge} = \frac{g\beta D}{c_P},$$

$$\mathrm{Ra} = \frac{g\beta KD(T_w - T_\infty)}{\nu\alpha_m} \tag{5.114}$$

where D is the diameter of the cylinder, v_{max} is the maximum velocity, h is the heat transfer coefficient, and C_1 and C_2 are the dimensional constants appearing in Forchheimer's equation expressed in the form

$$-\frac{dP}{dx} = C_1\mu u + C_2\rho u^2 \tag{5.115}$$

The correlation formulas (5.112) and (5.113) may be compared with the Darcy model boundary layer formula

$$\mathrm{Nu} = 0.565\mathrm{Ra}^{1/2} \tag{5.116}$$

and the Forchheimer model boundary layer formula found by Ingham and quoted by Ingham and Pop (1987c),

$$\mathrm{Nu} \propto \mathrm{Ra}^{1/4}\left(\frac{\nu D\chi}{\alpha_m K}\right)^{1/2}. \tag{5.117}$$

The effect of d/D, the ratio of particle diameter to cylinder diameter, was investigated experimentally by Fand and Yamamoto (1990). They noted that the reduction in the heat transfer coefficient due to wall porosity variation increases with d/D.

Ingham and Pop (1987c) have performed finite-difference calculations for streamlines, isotherms, and Nusselt numbers for Ra up to 400. Their results for an average Nusselt number $\overline{\mathrm{Nu}}$ defined by

$$\overline{\mathrm{Nu}} = -\frac{1}{2\pi}\int_0^{2\pi} \left.\frac{\partial\Theta}{\partial r}\right|_{r=1} d\theta \tag{5.118}$$

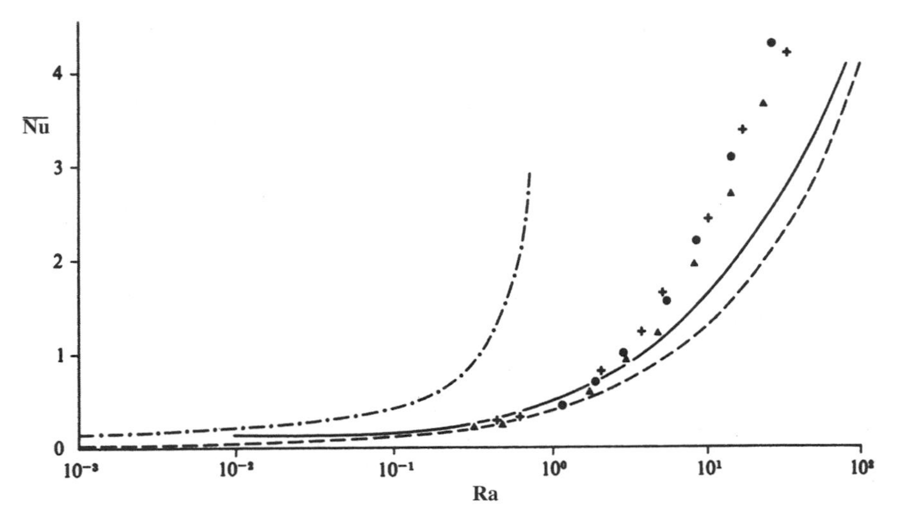

Figure 5.16. The variation of the mean Nusselt number with Rayleigh number. ——, numerical solution; – – – –, boundary layer solution; – · – · – · –, small Rayleigh number solution; •, Δ, +, experimental results using spheres of diameter 2, 3, and 4 mm, respectively (Ingham and Pop, 1987c, with permission from Cambridge University Press).

are given in Fig. 5.16. The dimensionless temperature difference is defined as $\Theta = (T - T_\infty)/(T_w - T_\infty)$.

5.6. Sphere

5.6.1. Flow at High Rayleigh Number

With the x and y axes chosen in a vertical diametral plane of the sphere, and with x measured along the sphere from the lower stagnation point and y measured radially outward from the surface (Fig. 5.15), the governing boundary layer equations are

$$\frac{1}{r}\frac{\partial^2 \psi}{\partial y^2} = \frac{g\beta K \sin\varphi}{\nu}\frac{\partial}{\partial y}(T - T_\infty), \tag{5.119}$$

$$\alpha_m \frac{\partial^2 T}{\partial y^2} = \frac{1}{r}\left(\frac{\partial \psi}{\partial y}\frac{\partial T}{\partial x} - \frac{\partial \psi}{\partial x}\frac{\partial T}{\partial y}\right). \tag{5.120}$$

The streamfunction ψ is defined by

$$ru = \frac{\partial \psi}{\partial y}, \qquad rv = -\frac{\partial \psi}{\partial x}, \tag{5.121}$$

where $r = r_0 \sin\varphi$ and r_0 is the radius of the sphere. Again, the boundary conditions are given by Eqs. (5.9) and (5.10), with $\lambda = 0$. The problem admits the similar solution (Cheng, 1985a)

$$\psi = \alpha_m \left[\frac{g\beta K(T_w - T_\infty) r_0^3}{\nu \alpha_m} \left(\frac{\cos^3 \varphi}{3} - \cos\varphi + \frac{2}{3} \right) \right]^{1/2} f(\eta), \quad (5.122)$$

$$\frac{T - T_\infty}{T_w - T_\infty} = \theta(\eta), \quad (5.123)$$

$$\eta = \frac{y}{r_0} \left[\frac{g\beta K(T_w - T_\infty) r_0}{\nu \alpha_m} \right]^{1/2} \frac{\sin^2 \varphi}{[(\cos^3 \varphi)/3 - \cos\varphi + 2/3]^{1/2}}, \quad (5.124)$$

where f and θ atisfy Eqs. (5.15)–(5.18) with $\lambda = 0$. Accordingly the local surface heat flux is given by

$$q_w'' = 0.444 k_m (T_w - T_\infty)^{3/2} \left(\frac{g\beta K}{\nu \alpha_m r_0} \right)^{1/2} \frac{\sin^2 \varphi}{[(\cos^3 \varphi)/3 - \cos\varphi + 2/3]^{1/2}} \quad (5.125)$$

which in dimensionless form is

$$\frac{\mathrm{Nu}_\varphi}{\mathrm{Ra}_D^{1/2}} = 0.628 \frac{\sin^2 \varphi}{[(\cos^3 \varphi)/3 - \cos\varphi + 2/3]^{1/2}}. \quad (5.126)$$

This result is plotted in Fig. 5.15, which shows that the local heat transfer rate for a sphere is higher than that for a horizontal cylinder except near the upper stagnation point. The average surface heat flux is

$$\begin{aligned} \bar{q}'' &= \frac{1}{4\pi r_0^2} \int_0^\pi 2\pi r_0^2 \, q_w''(\varphi) \sin\varphi d\varphi \quad (5.127) \\ &= \frac{0.888}{3^{1/2}} k_m (T_w - T_\infty)^{3/2} \left(\frac{g\beta K}{\nu \alpha_m r_0} \right)^{1/2}, \end{aligned}$$

which in dimensionless form reduces to

$$\frac{\mathrm{Nu}}{\mathrm{Ra}_D^{1/2}} = 0.724. \quad (5.128)$$

This problem is a special case of the natural convection about a general axisymmetric heated body embedded in a porous medium, analyzed by Merkin (1979). The extension to include the effect of normal pressure gradients on convection in a Darcian fluid about a horizontal cylinder and a sphere has been provided by Nilson (1981). The extension to a non-Newtonian power law theory was made by Chen and Chen (1988b).

Conjugate steady convection from a solid sphere with a heated core of uniform temperature was investigated by Kimura and Pop (1994). The transient problem, where either the temperature or the heat flux of the sphere is suddenly raised and subsequently maintained at a constant value, was treated numerically for both small and large values of the Rayleigh number by Yan *et al.* (1997).

The analogous problem of convective mass transfer from a sphere was studied experimentally by Rahman (1999). MHD convection over a permeable sphere with internal heat generation was analyzed by Yih (2000a).

5.6.2. Flow at Low Rayleigh Number

This topic was first studied by Yamamoto (1974). When Ra is small, we can use a series expansion in powers of Ra. Using a spherical polar coordinate system (r, θ, φ, and a Stokes streamfunction ψ), we can write the governing equations in nondimensional form for the case of constant surface temperature T_w,

$$\frac{1}{\sin\theta}\frac{\partial^2\psi}{\partial r^2}+\frac{1}{r^2}\frac{\partial}{\partial\theta}\left(\frac{1}{\sin\theta}\frac{\partial\psi}{\partial\theta}\right)=\mathrm{Ra}\left(\cos\theta\frac{\partial\Theta}{\partial r}+r\sin\theta\frac{\partial\Theta}{\partial r}\right). \quad (5.129)$$

$$\frac{\partial\psi}{\partial\theta}\frac{\partial\Theta}{\partial r}-\frac{\partial\psi}{\partial r}\frac{\partial\Theta}{\partial\theta}=\sin\theta\frac{\partial}{\partial r}\left(r^2\frac{\partial\Theta}{\partial r}\right)+\frac{\partial}{\partial\theta}\left(\sin\theta\frac{\partial\Theta}{\partial\theta}\right). \quad (5.130)$$

In these equations we have used the definitions

$$\Theta=\frac{T-T_\infty}{T_w-T_\infty},\qquad \mathrm{Ra}=\frac{g\beta Ka(T_w-T_\infty)}{\nu\alpha_m}, \quad (5.131)$$

and r is the nondimensional radial coordinate scaled with a, the radius of the sphere. The boundary and symmetry conditions are

$$\begin{aligned}
&r=1: && \Theta=1, && \frac{\partial\psi}{\partial\theta}=0,\\
&r\to\infty: && \Theta=0, && \frac{\partial\psi}{\partial\theta}=0, && \frac{\partial\psi}{\partial r}=0,\\
&\theta=0,\pi: && \frac{\partial\Theta}{\partial\theta}=0, && \frac{\partial\psi}{\partial r}=0, && \frac{\partial}{\partial\theta}\left(\frac{1}{\sin\theta}\frac{\partial\psi}{\partial\theta}\right)=0.
\end{aligned} \quad (5.132)$$

The solution is obtained by writing

$$(\psi,\Theta)=(\psi_0,\Theta_0)+\mathrm{Ra}(\psi_1,\Theta_1)+\mathrm{Ra}^2(\psi_2,\Theta_2)+\ldots, \quad (5.133)$$

substituting and solving in turn the problems of order 0, 1, 2, ... in Ra. One finds that

$$\psi_0=0,\qquad \Theta_0=\frac{1}{r}, \quad (5.134)$$

$$\psi_1=\frac{1}{2}(r-r^{-1})\sin^2\theta,\qquad \Theta_1=\frac{1}{4}(2r^{-1}-3r^{-2}+r^{-3})\cos\theta, \quad (5.135)$$

$$\psi_2=\frac{1}{24}(4r-9+6r^{-1}-r^2)\sin^2\theta\cos\theta, \quad (5.136)$$

$$\begin{aligned}
\Theta_2=&-\frac{13}{180}r^{-1}+\frac{11}{240}r^{-3}\ln r+\frac{31}{224}r^{-3}-\frac{13}{144}r^{-4}\\
&+\frac{27}{1120}r^{-5}+\left(\frac{5}{48}r^{-1}-\frac{3}{8}r^{-2}+\frac{11}{80}r^{-3}\ln r\right.\\
&\left.+\frac{223}{672}r^{-3}-\frac{1}{12}r^{-4}+\frac{5}{224}r^{-5}\right)\cos 2\theta.
\end{aligned} \quad (5.137)$$

Table 5.3. The overall Nusselt number for an isothermal sphere embedded in a porous medium (Pop and Ingham, 1990)

$\frac{1}{2}\mathrm{Ra}_D$	Boundary layer solution	Numerical solution
1	0.5124	2.1095
10	1.6024	2.8483
20	2.2915	3.2734
40	3.2407	3.9241
70	4.2870	5.0030
100	5.1240	5.8511
150	6.2756	7.0304
200	7.2464	8.2454

Working from the second-order approximation $\psi = \mathrm{Ra}\,\psi_1 + \mathrm{Ra}^2 \quad \psi_2$, Ene and Poliševski (1987) found that, whereas for Ra < 3 the streamline pattern was unicellular, for Ra > 3 a second cell appears below the sphere. This is apparently an artifact of their solution, resulting from the nonconvergence of the series for Ra > 3. No second cell was found by Pop and Ingham (1990).

For convection around a sphere that is suddenly heated and subsequently maintained at a constant heat flux or constant temperature, asymptotic solutions were obtained by Sano and Okihara (1994), Sano (1996), and Ganapathy (1997).

5.6.3. Flow at Intermediate Rayleigh Number

In addition to obtaining a second-order boundary layer theory for large Ra, Pop and Ingham (1990) used a finite-difference scheme to obtain numerical results for finite values of Ra. Their results are shown in Table 5.3 and Fig. 5.17. They expressed their heat transfer results in terms of a mean Nusselt surface $\overline{Nu}$ defined by

$$\overline{\mathrm{Nu}} = -\frac{1}{2}\int_0^{\pi} \left(\frac{\partial \Theta}{\partial r}\right)_{r=1} \sin\theta d\theta, \tag{5.138}$$

5.7. Vertical Cylinder

For the problem of natural convection about a vertical cylinder with radius r_0, power law wall temperature, and embedded in a porous medium, similarity solutions do not exist. An approximate solution was obtained by Minkowycz and Cheng (1976). For a given value of the power law exponent λ, Eq. (5.8), they found that the ratio of local surface heat flux of a cylinder (q''_c) to that of a flat plate (q'') is a nearly linear function of a curvature parameter ξ,

$$\frac{q''_c}{q''} = 1 + C'\xi, \tag{5.139}$$

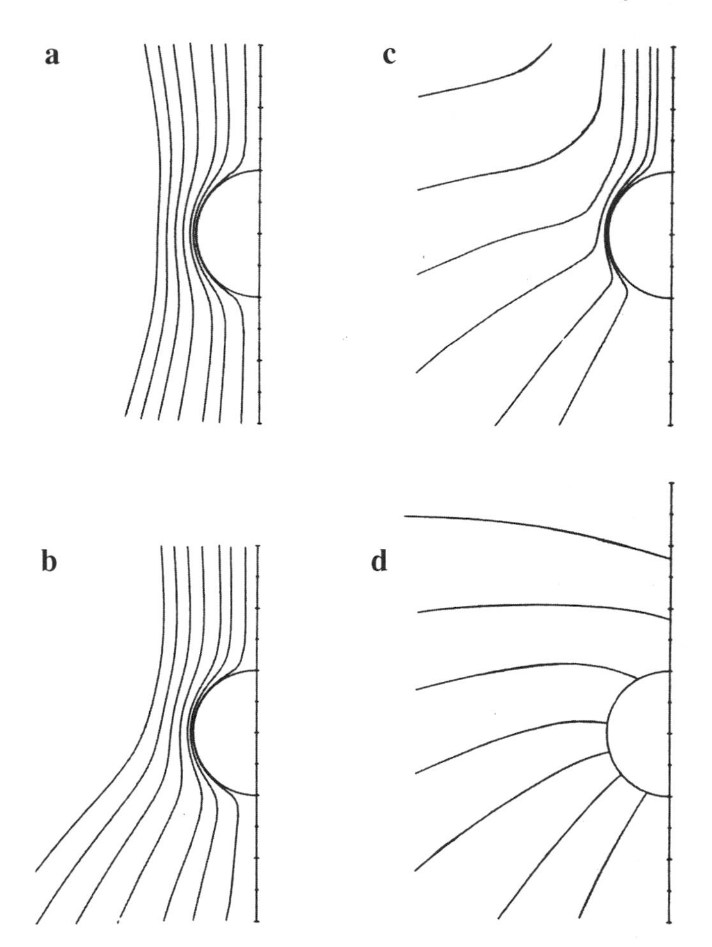

Figure 5.17. The streamlines in the vicinity of a sphere: (a) Ra = 1, (b) Ra = 10, (c) Ra = 100, and (d) asymptotic solution (Pop and Ingham, 1990, with permission from Hemisphere Publishing Corporation).

where

$$\xi = \frac{2x}{r_0 \mathrm{Ra}_x^{1/2}}, \tag{5.140}$$

where x denotes the vertical coordinate along the axis of the cylinder and q'' is given by Eq. (5.25). The values of the positive constant C' are given in Table 5.4.

The ratio of average heat fluxes $\bar{q}_c''/\bar{q}''$ turns out to be independent of λ, and is given approximately by

$$\frac{\bar{q}_c''}{\bar{q}''} = 1 + 0.26\xi_L, \tag{5.141}$$

where $\xi_L = 2L/r_0\mathrm{Ra}_L^{1/2}$, and L is the height of the cylinder. The average heat flux for the vertical plate $(\bar{q}'')$ is given by Eq. (5.27).

Table 5.4. Values of the constant C' in Eq. (5.139) for various values of the power law exponent λ (Cheng, 1984a)

λ	C'
0	0.30
1/4	0.23
1/3	0.21
1/2	0.20
3/4	0.17
1	0.15

A detailed solution was obtained by Merkin (1986). Magyari and Keller (2004a) showed that the flow induced by a nonisothermal vertical cylinder approaches the shape of Schlichting's round jet as the porous radius tends to zero. The effects of surface suction or blowing were examined by Huang and Chen (1985); suction increases the rate of heat transfer. The transient problem has been analyzed by Kimura (1989b).

Asymptotic analyses and numerical calculations for this problem were reported by Bassom and Rees (1996). They showed that when $\lambda < 1$, the asymptotic flowfield for the leading edge of the cylinder takes on a multiple layer structure. However, for $\lambda > 1$, only a simple single layer is present far downstream, but a multiple layer structure exists close to the leading edge of the cylinder.

The effects of surface suction or blowing were examined by Huang and Chen (1985); suction increases the rate of heat transfer. Inertial effects and those of suction were analyzed by Hossain and Nakayama (1993). The case of suction with a non-Newtonian fluid (for a vertical plate or a vertical cylinder) was investigated by Pascal and Pascal (1997). The transient problem has been analyzed by Kimura (1989b), while Libera and Poulikakos (1990) and Pop and Na (2000) have treated a conjugate problem. The effect of thermal stratification was added by Chen and Horng (1999) and Takhar *et al.* (2002). An analogous mass transfer problem was studied experimentally by Rahman *et al.* (2000). The effect of local thermal nonequilibrium was investigated by Rees *et al.* (2003a). The effect of radiation was studied numerically by Yih (1999e).

5.8. Cone

We consider an inverted cone with semiangle γ and take axes in the manner indicated in Fig. 5.18. The boundary layer develops over the heated frustum $x = x_0$. In terms of the streamfunction ψ defined by

$$u = \frac{1}{r}\frac{\partial \psi}{\partial y}, \quad v = -\frac{1}{r}\frac{\partial \psi}{\partial x}, \tag{5.142}$$

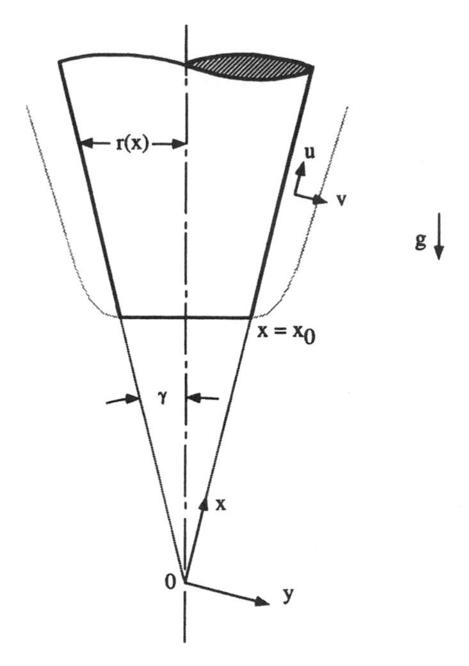

Figure 5.18. Coordinate system for the boundary layer on a heated frustum of a cone (Cheng *et al.*, 1985).

the boundary layer equations are

$$\frac{1}{r}\frac{\partial^2\psi}{\partial y^2} = \frac{g\beta K}{\nu}\frac{\partial T}{\partial y}, \tag{5.143}$$

$$\frac{1}{r}\left(\frac{\partial\psi}{\partial y}\frac{\partial T}{\partial x} - \frac{\partial\psi}{\partial x}\frac{\partial T}{\partial y}\right) = \alpha_m\frac{\partial^2 T}{\partial y^2}. \tag{5.144}$$

For a thin boundary layer we have approximately $r = x\sin\gamma$. We suppose that either a power law of temperature or a power law of heat flux is prescribed on the frustum. Accordingly, the boundary conditions are

$$\begin{aligned} y \to \infty: &\quad u = 0,\ T = T_\infty, \\ y = 0,\ x_0 \le x < \infty: &\ u = 0,\ \text{and either } T = T_w = T_\infty + (x - x_0)^\lambda \\ &\text{or} - k_m \left.\frac{\partial T}{\partial y}\right|_{y=0} = q''_w = A(x - x_0)^\lambda \end{aligned} \tag{5.145}$$

For the case of a full cone ($x_0 = 0$) a similarity solution exists. In the case of prescribed T_w, we let

$$\psi = \alpha_m r \mathrm{Ra}_x^{1/2} f(\eta), \tag{5.146}$$

$$T - T_\infty = (T_w - T_\infty)\theta(\eta), \tag{5.147}$$

$$\eta = \frac{y}{x}\mathrm{Ra}_x^{1/2}, \tag{5.148}$$

where

$$\text{Ra}_x = \frac{g\beta K \cos\gamma (T_w - T_\infty)x}{\nu\alpha_m}. \tag{5.149}$$

The dimensionless momentum and energy equations are

$$f' = \theta, \tag{5.150}$$

$$\theta'' + \left(\frac{\lambda + 3}{2}\right) f\theta' - \lambda f'\theta = 0, \tag{5.151}$$

with boundary conditions

$$f(0) = 0, \theta(0) = 1, \ \theta(\infty) = 0. \tag{5.152}$$

The local Nusselt number is given by

$$\text{Nu}_x = \text{Ra}_x^{1/2}[-\theta'(0)], \tag{5.153}$$

for which computed values of $\theta'(0)$ are given in Table 5.5.

The case of a cone with prescribed uniform heat flux q''_w is handled similarly. We begin with the dimensionless variables

$$\psi = \alpha_m r \hat{\text{Ra}}^{1/3} f(\hat{\eta}), \tag{5.154}$$

$$T - T_\infty = \frac{q''_w x}{k_m} \hat{\text{Ra}}_x^{-1/3} \hat{\theta}(\hat{\eta}). \tag{5.155}$$

$$\hat{\eta} = \frac{y}{x} \hat{\text{Ra}}_x^{1/3}, \tag{5.156}$$

where the Rayleigh number is based on heat flux,

$$\hat{\text{Ra}}_x = \frac{g\beta K \cos\gamma q''_w x^2}{\nu\alpha_m k_m}. \tag{5.157}$$

The governing equations become

$$\hat{f} = \hat{\theta}, \tag{5.158}$$

$$\hat{\theta}'' + \left(\frac{\lambda + 5}{2}\right) \hat{f}'\hat{\theta}' - \left(\frac{2\lambda + 1}{3}\right) \hat{f}'\hat{\theta} = 0, \tag{5.159}$$

Table 5.5. Values of $\theta'(0)$ and $\hat{\theta}(0)$ for calculating the local Nusselt number on a vertical cone embedded in a porous medium (Cheng *et al.*, 1985)

λ	$\theta'(0)$	$\hat{\theta}(0)$
0	−0.769	1.056
1/3	−0.921	0.992
1/2	−0.992	0.965

subject to

$$\hat{f}'(0) = 0, \ \hat{\theta}'(0) = -1, \ \hat{\theta}(\infty) = 0. \tag{5.160}$$

The local Nusselt number solution to this problem is

$$\text{Nu}_x = \hat{\text{Ra}}^{1/3}[\hat{\theta}(0)]^{-1}. \tag{5.161}$$

with the computed values of $\hat{\theta}(0)$ given in Table 5.5. The local Nusselt number is defined in the usual way: $\text{Nu}_x = q''x/k_m(T_w - T_\infty)$. Note that in the present (constant q'') configuration the cone temperature T_w is a function of x.

No similarity solution exists for the truncated cone, but Cheng *et al.* (1985) obtained results using the local nonsimilarity method. Pop and Cheng (1986) included the curvature effects that become important when the cone is slender. Vasantha *et al.* (1986) treated non-Darcy effects for a slender frustum and Nakayama *et al.* (1988a) also have considered inertial effects.

A cone with a point heat source at the apex was considered by Afzal and Salam (1990). Pop and Na (1994, 1995) studied convection on an isothermal wavy cone or frustum of a wavy cone, for large Ra, under the assumption that the wavy surface has amplitude and wavelength of order one. They presented results for the effect of the sinusoidal surface on the wall heat flux.

Rees and Bassom (1991) examined convection in a wedge-shaped region bounded by two semi-infinite surfaces, one heated isothermally and the other insulated. For the particular cases (i) a vertical heated surface with a wedge angle of π, and (ii) a horizontal upward-facing surface with a wedge angle of $3\pi/2$, the equations on the Darcy model reduce to the classic ordinary differential equations.

Non-Darcy hydromagnetic convection over a cone or wedge was studied by Chamkha (1996). Variable viscosity and thermal conductivity effects on convection from a cone or wedge were studied numerically by Hassanien *et al.* (2003b). For convection over a cone, the effect of uniform lateral mass flux was studied by Yih (1997, 1998b) for the case of Newtonian or non-Newtonian fluids, and with a Forchheimer effect by Kumari and Jayanthi (2005) for a non-Newtonian fluid.

5.9. General Two-Dimensional or Axisymmetric Surface

Nakayama and Koyama (1987a) have shown how it is possible to obtain similarity solutions to the boundary layer equations for flow about heated two-dimensional or axisymmetric bodies of arbitrary shape provided that the wall temperature is a power function of a variable ξ, which is a certain function of the streamwise coordinate x. Then the governing equations reduce to those for a vertical flat plate. They thus generalized Merkin's (1979) results for the isothermal case.

A simple analysis of convection about a slender body of revolution with its axis vertical was given by Lai *et al.* (1990c). In terms of cylindrical polar coordinates with x in the axial direction and r in the radial direction, the governing boundary layer equations are

$$\frac{\partial}{\partial r}\left(\frac{1}{r}\frac{\partial \psi}{\partial r}\right) = \frac{g\beta K}{\nu}\frac{\partial T}{\partial r}, \tag{5.162}$$

$$\frac{\partial \psi}{\partial r}\frac{\partial T}{\partial x} - \frac{\partial \psi}{\partial x}\frac{\partial T}{\partial r} = \alpha_m \frac{\partial}{\partial r}\left(r\frac{\partial T}{\partial r}\right). \tag{5.163}$$

The boundary conditions at the body surface $[r = R(x)]$ and far from the surface $(r \to \infty)$ are, respectively,

$$T = T_w(x) = T_\infty + Ax^\lambda,\ v = 0, \tag{5.164}$$

$$T = T_\infty, \quad u = 0. \tag{5.165}$$

Suitable similarity variables are defined by

$$\eta = \mathrm{Ra}_x\left(\frac{r}{x}\right)^2, \tag{5.166}$$

$$\psi = \alpha_m x f(\eta), \tag{5.167}$$

$$T - T_\infty = (T_w - T_\infty)\theta(\eta), \tag{5.168}$$

where

$$\mathrm{Ra}_x = \frac{g\beta K(T_w - T_\infty)x}{\nu\alpha_m}. \tag{5.169}$$

If we set $\eta = a_{nc}$, where a_{nc} is a numerically small constant, we have prescribed the surface of a slender body, given by

$$r = \left(\frac{\nu\alpha_m a_{nc}}{g\beta K A}\right)^{1/2} x^{(1-\lambda)/2}. \tag{5.170}$$

This represents a cylinder when $\lambda = 1$, a paraboloid when $\lambda = 0$ and a cone when $\lambda = -1$. The resulting equations are

$$2f' = \theta, \tag{5.171}$$

$$2\eta\theta'' + (2 + f)\theta' - \lambda f'\theta = 0 \tag{5.172}$$

with boundary conditions

$$\eta = a_{nc}: \quad \theta = 1,\ f + (\lambda - 1)\eta f' = 0, \tag{5.173}$$

$$\eta \to \infty: \quad \theta = 0,\ f' = 0. \tag{5.174}$$

These equations can be easily solved numerically and the local Nusselt number is then given by

$$\frac{\mathrm{Nu}}{\mathrm{Ra}^{1/2}} = -2a_{nc}^{1/2}\theta'(a_{nc}). \tag{5.175}$$

Inertial effects were examined by Ingham (1986) and Nakayama *et al.* (1989, 1990b). The effects of a stratified medium were discussed by Nakayama and Koyama (1989) and those of viscous dissipation by Nakayama and Pop (1989). Convection from a nonisothermal axisymmetric surface was analyzed by Mehta and Sood (1994). Flow of non-Newtonian power law fluids over nonisothermal bodies of arbitrary shape was studied by Nakayama and Koyama (1991) and by Wang *et al.* (2002) for the case of permeable bodies. Certain wall temperature distributions lead to similarity solutions. A general transformation procedure, for the transient problem and the Forchheimer model, was presented by Nakayama *et al.* (1991). With this the local similarity assumption was adapted to produce solutions for a range of geometries. Power law fluid flow, with or without yield stress, also was discussed by Yang and Wang (1996). Similarity solutions for convection due to internal heating were obtained by Bagai (2003, 2004) for the cases of constant or variable viscosity.

5.10. Horizontal Line Heat Source

5.10.1. Flow at High Rayleigh Number

5.10.1.1. Darcy Model

At high Rayleigh number the flow about a horizontal line source of heat takes the form of a vertical plume. For steady flow the governing boundary layer equations are again Eqs. (5.6) and (5.7). The boundary conditions (5.10) still apply, but Eq. (5.9) is replaced by the symmetry conditions

$$y = 0: \quad \frac{\partial^2 \psi}{\partial y^2} = \frac{\partial T}{\partial y} = 0. \tag{5.176}$$

We now have a homogeneous system of equations, and a nontrivial solution exists only if a certain constraint holds. In the present problem this arises from the global conservation of energy and takes the form

$$q' = \rho_\infty c_P \int_{-\infty}^{\infty} \frac{\partial \psi}{\partial y}(T - T_\infty)dy, \tag{5.177}$$

where q' is the prescribed heat flux per unit length and c_P is the specific heat of the convected fluid at constant pressure. Consistent with the boundary layer approximation, the axial heat conduction term is omitted from Eq. (5.173).

It is easily checked that the solution of the present problem is (Wooding, 1963)

$$\psi = \alpha_m \hat{\mathrm{Ra}}_x^{1/3} f(\eta), \tag{5.178}$$

$$T - T_\infty = \frac{q'}{\rho_\infty c_P \alpha_m} \hat{\mathrm{Ra}}_x^{-1/3} \theta(\eta), \tag{5.179}$$

$$\eta = \frac{y}{x} \hat{\mathrm{Ra}}_x^{1/3}, \tag{5.180}$$

where

$$\hat{\mathrm{Ra}}_x = g\beta K q' x / \mu \alpha_m^2 c_P. \tag{5.181}$$

The functions f and θ satisfy the differential equations

$$f' - \theta = 0, \tag{5.182}$$

$$\theta'' + \frac{1}{3}(f\theta)' = 0, \tag{5.183}$$

the boundary conditions

$$f(0) = \theta'(0) = 0, \tag{5.184}$$

$$f'(\pm\infty) = \theta(\pm\infty) = 0, \tag{5.185}$$

and the constraint

$$\int_{-\infty}^{\infty} f'(\eta)\theta(\eta)d\eta = 1. \tag{5.186}$$

The nontrivial solution of Eqs. (5.182)–(5.186) is

$$\psi = \alpha_m \hat{\mathrm{Ra}}_x^{1/3} B \tanh\left(\frac{B\eta}{6}\right), \tag{5.187}$$

$$T - T_\infty = \frac{q'}{\rho_\infty c_P \alpha_m} \hat{\mathrm{Ra}}_x^{-1/3} \frac{B^2}{6} \operatorname{sech}^2\left(\frac{B\eta}{6}\right), \tag{5.188}$$

where $B = (9/2)^{1/3} = 1.651$. The dimensionless temperature profile $\theta(\eta)$ is illustrated in Fig. 5.19.

The problem of a line source situated at the vertex of a solid wedge, together with higher-order boundary layer effects, was analyzed by Afzal (1985). The effect of material anisotropy on convection induced by point or line sources was studied by Rees *et al.* (2002). They showed that the path of the plume centerline is strongly affected by the anisotropy and the presence of impermeable bounding surfaces. A line source situated in an anisotropic medium also was studied by Degan and Vasseur (2003). They noted that the minimum (maximum) intensity of the plume is attained if the medium is oriented with its principal axis with high permeability parallel (perpendicular) to the vertical.

5.10.1.2. Forchheimer Model

When quadratic drag is taken into account, Eq. (5.23) is replaced by

$$u + \frac{\chi}{\upsilon}u^2 = \frac{\beta K}{\nu}(T - T_\infty), \tag{5.189}$$

where $u = \partial\psi/\partial y$. Following Ingham (1988), we introduce nondimensional quantities defined by

$$x = Xl, \quad y = Yl\left(\frac{\mathrm{Fo}}{\hat{\mathrm{Ra}}}\right)^{1/5}, \tag{5.190}$$

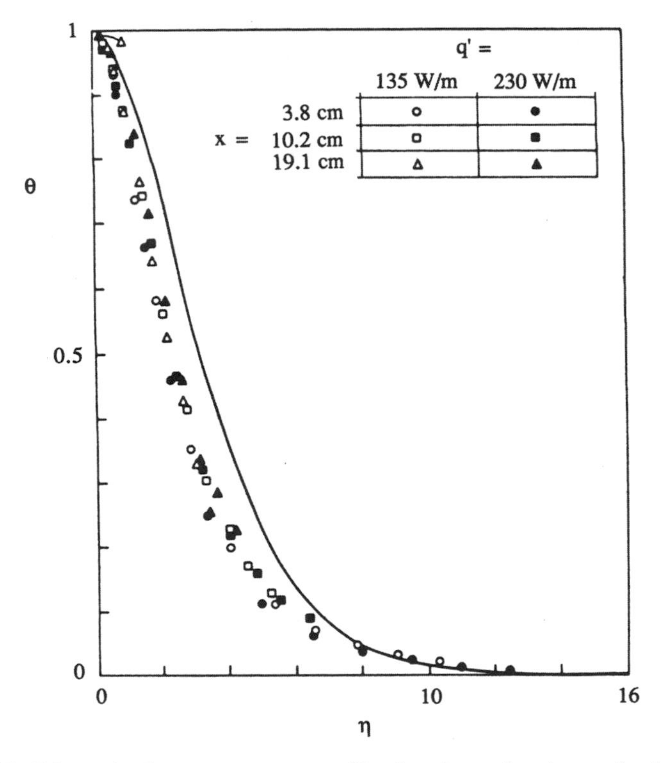

Figure 5.19. Dimensionless temperature profiles for plume rise above a horizontal line source of heat in a porous medium (Lee, 1983; Cheng, 1985a, with permission from Hemisphere Publishing Corporation).

$$\psi = \alpha_m \left(\frac{\hat{\mathrm{Ra}}}{\mathrm{Fo}}\right)^{1/5} \Psi, \quad T - T_\infty = \frac{\chi a_m^2}{g\beta K l^2}\left(\frac{\hat{\mathrm{Ra}}}{\mathrm{Fo}}\right)^{4/5} \Theta, \tag{5.191}$$

$$\hat{\mathrm{Ra}} = \frac{g\beta K l q'}{\nu \alpha_m k_m}, \quad \mathrm{Fo} = \frac{\chi \alpha_m}{\nu l}, \tag{5.192}$$

where l is a characteristic length scale. Substitution into Eq. (5.189) and the steady-state form of Eq. (5.7) gives

$$\left(\frac{\partial \psi}{\partial Y}\right)^2 = \Theta, \tag{5.193}$$

$$\frac{\partial \psi}{\partial Y}\frac{\partial \Theta}{\partial X} - \frac{\partial \psi}{\partial X}\frac{\partial \Theta}{\partial Y} = \frac{\partial^2 \Theta}{\partial Y^2}, \tag{5.194}$$

when a term $\hat{\mathrm{Ra}}^{-2/5}\mathrm{Fo}^{-3/5}\partial\Psi/\partial Y$ in Eq. (5.193) has been neglected. Since the boundary layer thickness is of order $l(\mathrm{Fo}/\hat{\mathrm{Ra}})^{1/5}$, we are requiring that $\hat{\mathrm{Ra}}\,\mathrm{Fo}^{-1} \gg 1$ and $\hat{\mathrm{Ra}}^{2/5}\mathrm{Fo}^{3/5} \gg 1$. The boundary conditions and the source energy constraint are

$$Y = 0: \qquad \frac{\partial^2 \Psi}{\partial Y^2} = 0, \qquad \frac{\partial \Theta}{\partial Y} = 0,$$
$$Y \to \infty: \qquad \frac{\partial \Psi}{\partial Y} \to 0, \qquad \Theta \to 0, \tag{5.195}$$
$$\int_{-\infty}^{\infty} \frac{\partial \Psi}{\partial Y} \Theta dY = 1.$$

We now introduce the similarity transformation

$$\Psi = X^{2/5} f(\eta), \quad \Theta = X^{-2/5} g(\eta), \quad \eta = Y/X^{3/5}, \tag{5.196}$$

and then the system (5.193)–(5.195) becomes

$$(f')^2 = g, \tag{5.197}$$

$$g'' = -\frac{2}{5}(f'g + fg'), \tag{5.198}$$

$$g'(0) = 0, \quad f'(\infty) = 0, \quad \int_{-\infty}^{\infty} f' g d\eta = 1. \tag{5.199}$$

These equations have the analytical solution

$$f = C \tanh \frac{C}{10}\eta, \quad g = \frac{C^4}{100} \operatorname{sech}^4 \frac{C}{10}\eta, \tag{5.200}$$

where C $= (8 \times 10^{3/2}/3)^{1/4} = 3.03$. Comparison of Eq. (5.200) with Eq. (5.188) shows that a sech^2 function is replaced by a sech^4 function and this means that the Forchheimer model leads to a more sharply peaked temperature profile than does the Darcy model.

This conclusion is in accordance with the experiments reported by Cheng (1985a), carried out by himself, R. M. Fand, and D. K. Chui, on the plume rise from a horizontal line source of heat embedded in 3-mm diameter glass beads saturated with silicone oil. Their results are presented in Figures 5.19 and 5.20. The work of Ingham (1988) was extended by Rees and Hossain (2001) to intermediate distances from the source by computing the smooth transition between the inertia-dominated and the inertia-free regimes.

An experimental and analytic study of the buoyant plume above a concentrated heat source in a stratified porous medium was made by Masuoka *et al.* (1986). In experiments with a two-layer system two kinds of glass spheres of different diameter were employed, with water as the saturating fluid. They found that their similarity solution broke down near the interface.

The effect of dispersion was added by Lai (1991b). The wall plume was studied by Leu and Jang (1994) using a Brinkmann-Forchheimer model. The wall plume has a lower peak velocity and a higher maximum temperature than the corresponding free plume. The case of a non-Newtonian power law fluid was examined by Nakayama (1993b).

Masuoka *et al.* (1995b) reported an experimental and analytical study of the effects of a horizontal porous layer on the development of the buoyant plume

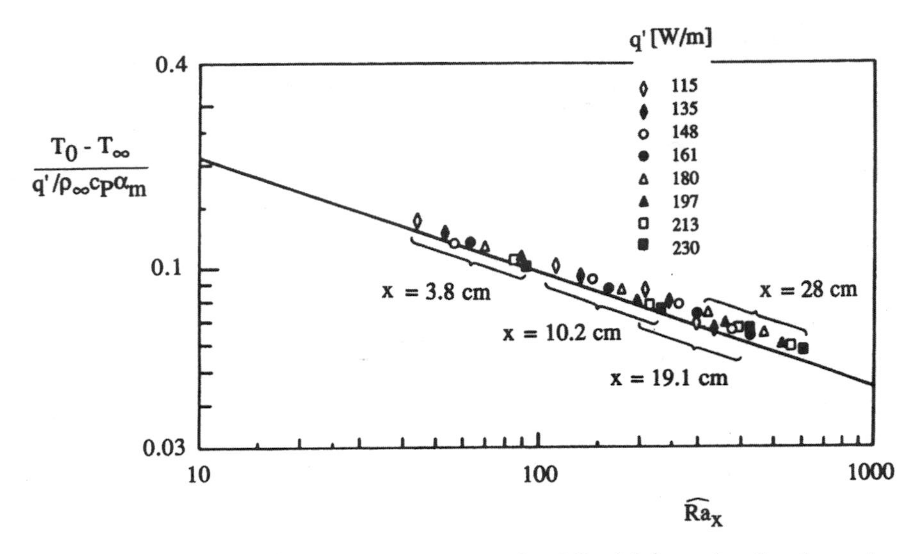

Figure 5.20. Dimensionless temperature versus local Rayleigh number for plume rise above a horizontal line source of heat in a porous medium (Lee, 1983; Cheng, 1985a, with permission from Hemisphere Publishing Corporation).

arising from a line heat source in an infinite fluid space. They observed an expansion of the plume at the lower interface and a contraction at the upper interface of the permeable layer. Their theoretical model incorporated the Beavers-Joseph slip boundary condition and they interpreted the fairly good agreement between their experimental and numerical results as confirming the validity of that condition.

5.10.2. Flow at Low Rayleigh Number

Following Nield and White (1982) we introduce polar coordinates (r, θ) with origin at the source and the plane $\theta = 0$ horizontal. The seepage velocity is (v_r, v_θ). The equations for mass conservation, Darcy flow, and transient energy conservation are

$$\frac{\partial}{\partial r}(r v_r) + \frac{\partial v_\theta}{\partial \theta} = 0, \tag{5.201}$$

$$v_r = -\frac{K}{\mu}\left(\frac{\partial P}{\partial r} + \rho g \sin\theta\right), \tag{5.202}$$

$$v_\theta = -\frac{K}{\mu}\left(\frac{1}{r}\frac{\partial P}{\partial \theta} + \rho g \cos\theta\right), \tag{5.203}$$

$$\frac{1}{\alpha_m}\left(\sigma\frac{\partial T}{\partial t} + v_r\frac{\partial T}{\partial r} + \frac{v_\theta}{r}\frac{\partial T}{\partial \theta}\right) = \frac{\partial^2 T}{\partial r^2} + \frac{1}{r}\frac{\partial T}{\partial r} + \frac{1}{r^2}\frac{\partial^2 T}{\partial \theta^2}. \tag{5.204}$$

Introducing the streamfunction $\psi(r, \theta)$ by

$$v_r = \frac{1}{r}\frac{\partial \psi}{\partial \theta}, \quad v_\theta = -\frac{\partial \psi}{\partial r} \tag{5.205}$$

and eliminating the pressure between the two Darcy equations, we obtain, in nondimensional form,

$$r_* \frac{\partial^2 \psi_*}{\partial r_*^2} + \frac{\partial \psi_*}{\partial r_*} + \frac{1}{r_*}\frac{\partial^2 \psi_*}{\partial \theta^2} = \hat{\text{Ra}}\left(\sin\theta \frac{\partial T_*}{\partial \theta} - r_* \cos\theta \frac{\partial T_*}{\partial r_*}\right). \tag{5.206}$$

$$\frac{\partial T_*}{\partial t_*} + \frac{1}{r_*}\left(\frac{\partial \psi_*}{\partial \theta}\frac{\partial T_*}{\partial r_*} - \frac{\partial \psi_*}{\partial r_*}\frac{\partial T}{\partial \theta}\right) = \frac{\partial^2 T_*}{\partial r_*^2} + \frac{1}{r_*}\frac{\partial T_*}{\partial r_*} + \frac{1}{r_*^2}\frac{\partial^2 T_*}{\partial \theta^2}, \tag{5.207}$$

where

$$t_* = \frac{t\alpha_m}{K\sigma}, \quad r_* = \frac{r}{K^{1/2}}, \quad T_* = \frac{(T - T_\infty)k_m}{q'}, \quad \psi_* = \frac{\psi}{\alpha_m}, \tag{5.208}$$

$$\hat{\text{Ra}} = \frac{g\beta K^{3/2} q'}{\nu \alpha_m k_m}. \tag{5.209}$$

The initial conditions, boundary conditions, and energy balance constraint are

$$t = 0: v_r = v_\theta = 0, T = T_\infty, \tag{5.210}$$

$$r \to \infty: v_r = v_\theta = 0, T = T_\infty, \tag{5.211}$$

$$\theta = \pm\frac{\pi}{2}: v_\theta = \frac{\partial v_r}{\partial \theta} = \frac{\partial T}{\partial \theta} = 0, \tag{5.212}$$

$$\lim_{r \to 0}\left[-k_m(2\pi r)\frac{\partial T}{\partial r}\right] = q'. \tag{5.213}$$

The last equation implies that T is of order $\ln r$ as $r \to 0$, and Eq. (5.204) then implies that v_r is of order $r^{-1} \ln r$ and v_θ is of order r^{-1}. The above conditions are readily put in nondimensional form.

For sufficiently small values of Ra we can expand ψ_* and T_* as power series in Ra,

$$(\psi_*, T_*) = (\psi_{*0}, T_{*0}) + \text{Ra}\,(\psi_{*1}, T_{*1}) + \text{Ra}^2(\psi_{*2}, T_{*2}) + \cdots. \tag{5.214}$$

When we substitute the above equations, collect the terms of the same power of Ra, and solve in terms the problems of order 0, 1, 2, ... in Ra, we find the zero-order conduction solution

$$\psi_{*0} = 0, \quad T_{*0} = -\frac{1}{4\pi}\text{Ei}(-\eta^2), \tag{5.215}$$

with

$$\eta = \frac{r_*}{2t_*^{1/2}}, \tag{5.216}$$

and then the first-order solution

$$\psi_{*_1} = \frac{t_*^{1/2}}{4\pi} \cos\theta \left[\frac{\exp(-\eta^2) - 1}{\eta} + \eta \operatorname{Ei}(-\eta^2) \right], \tag{5.217}$$

$$T_{*_1} = t_*^{1/2} \frac{\sin\theta}{16\pi^2} \left\{ (\ln\eta)[(\gamma - 2)\eta - \eta^3] + \eta(\ln\eta)^2 + \eta\frac{2-\gamma}{2} + \eta^3\frac{3-\gamma}{2} + \ldots \right\}, \tag{5.218}$$

where $\gamma = 0.5772\ldots$ is Euler's constant. In Fig. 5.21 a set of streamlines $\psi_*/t_*^{1/2}$ have been plotted. We see that the flow pattern for small Rayleigh numbers consists of an expanding vortex whose radius increases with time as $t_*^{1/2}$ and whose core is situated at $\eta = 0.567$ in the horizontal plane containing the source.

Since the momentum equation is linear, we can superpose solutions for sources and use the method of images to deduce the flow field due to the presence of a line source near an insulated vertical wall. We assume that the insulated vertical wall is given by the y axis of a Cartesian system and the line source is located at $x = d$, $y = 0$. The flow field is equivalent to that produced by a pair of line sources, of equal strength, positioned at $x = \pm d$, $y = 0$. The expression for ψ_{*1} is

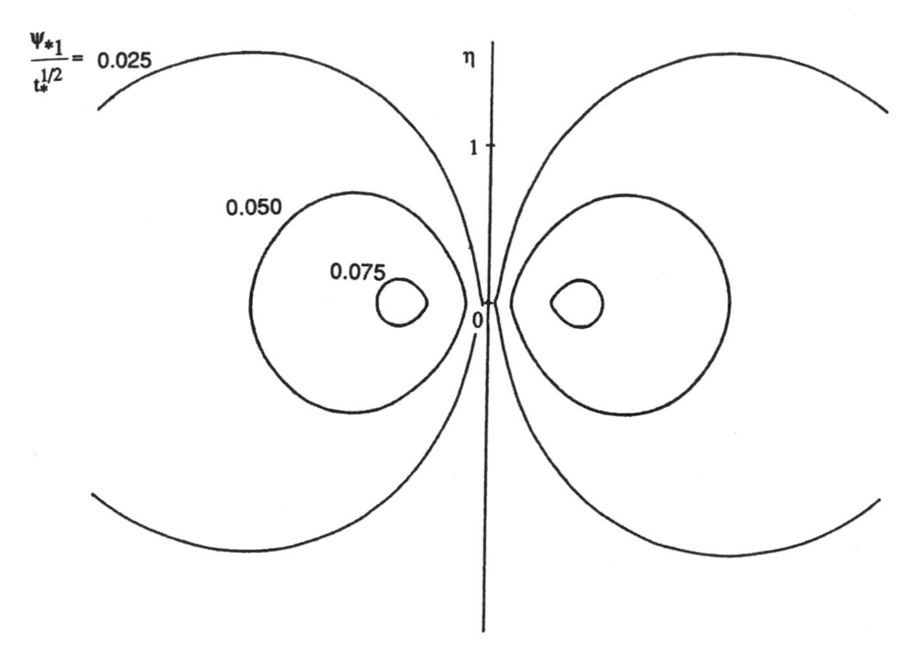

Figure 5.21. Streamlines drawn at constant increments of $\psi_{*1}/t_*^{1/2}$, for transient natural convection around a horizontal line heat source (Nield and White, 1982).

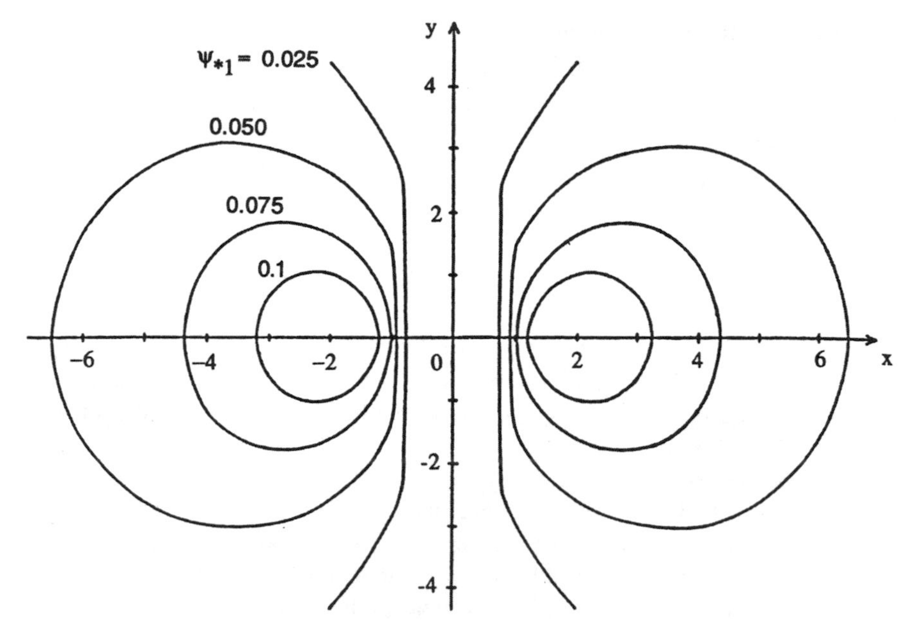

Figure 5.22. Streamlines drawn at constant increments of ψ_{*1}, for transient natural convection around a pair of line heat sources of equal strength, at (1,0) and (–1,0) at time $\tau_* = 1$ (Nield and White, 1982).

now

$$\psi_{*_1} = \frac{\tau^{1/2}}{4\pi}(S_+ + S_-), \tag{5.219}$$

where

$$S_\pm = \frac{2\tau^{1/2}(X \pm 1)}{(X \pm 1)^2 + Y^2}\left\{\exp\left[-\frac{(X \pm 1)^2 + Y^2}{4\tau}\right] - 1\right\} - \frac{X \pm 1}{2\tau^{1/2}}\int_{[(X\pm 1)^2+Y^2]/4\tau}^{\infty} \frac{\exp(-\xi)}{\xi}d\xi, \tag{5.220}$$

where $\tau = t\alpha_m/d$, $X = x/d$, $Y = y/d$. From this expression the streamlines were plotted in Fig. 5.22. Since the energy equation is nonlinear, it is not possible to superpose the solutions for T_{*1}.

5.11. Point Heat Source

5.11.1. Flow at High Rayleigh Number

Following Wooding (1963) and Bejan (1984), we consider the slender plume above a point source of constant strength, placed at an impermeable horizontal boundary. We take cylindrical polar coordinates (r, θ, z) with the origin at the source and the z axis vertically upward. The problem has axial symmetry, and the seepage velocity $(v_r, 0, v_z)$ is given in terms of a Stokes streamfunction Ψ by

$$v_r = \frac{1}{r}\frac{\partial \psi}{\partial z}, \qquad v_z = -\frac{1}{r}\frac{\partial \psi}{\partial r}. \tag{5.221}$$

The boundary layer equations for momentum and energy and the boundary conditions are

$$\frac{1}{r}\frac{\partial^2 \psi}{\partial r^2} = -\frac{g\beta K}{\nu}\frac{\partial T}{\partial r}, \tag{5.222}$$

$$\frac{\partial \psi}{\partial z}\frac{\partial T}{\partial r} - \frac{\partial \psi}{\partial r}\frac{\partial T}{\partial z} = \alpha_m \frac{\partial}{\partial r}\left(r\frac{\partial T}{\partial r}\right), \tag{5.223}$$

$$r = 0: \quad \frac{\partial \psi}{\partial z} = \frac{\partial T}{\partial r} = 0, \tag{5.224}$$

$$r \to \infty: \quad \frac{\partial \psi}{\partial r} = 0, T = T_\infty, \tag{5.225}$$

$$z = 0: \quad \frac{\partial \psi}{\partial r} = 0. \tag{5.226}$$

If q [W] is the strength of the source, energy conservation requires that

$$q = \int_0^\infty \rho_\infty c_P v_z (T - T_\infty) 2\pi r \, dr. \tag{5.227}$$

These equations admit the similarity solution

$$\Psi = -\alpha_m z f(\eta), \tag{5.228}$$

$$T - T_\infty = \frac{q}{k_m r}\widetilde{\mathrm{Ra}}^{-1/2}\theta(\eta), \tag{5.229}$$

$$\eta = \widetilde{\mathrm{Ra}}^{1/2}\frac{r}{z}, \tag{5.230}$$

where $\widetilde{\mathrm{Ra}}$ is the Rayleigh number based on source strength,

$$\widetilde{\mathrm{Ra}} = \frac{g\beta K q}{\nu \alpha_m k_m}. \tag{5.231}$$

The functions f and θ satisfy the differential equations

$$f'' - \theta' = 0, \tag{5.232}$$

$$\eta^2\theta'' + \eta(f - 1)\theta' + (1 - f + \eta f')\theta = 0, \tag{5.233}$$

the boundary conditions

$$f(0) = \theta(0) = 0, \tag{5.234}$$

$$f''(\infty) = f'(\infty) = \theta(\infty) = 0, \tag{5.235}$$

and the constraint

$$\int_0^\infty \frac{f'\theta}{\eta} d\eta = \frac{1}{2\pi}. \tag{5.236}$$

When the boundary conditions (5.231) are utilized, Eq. (5.228) integrates to give $f' = \theta$, and so Eq. (5.229) becomes

$$\frac{d}{d\eta}\left(f'' - \frac{f'}{\eta} + \frac{ff'}{\eta}\right) = 0. \tag{5.237}$$

Integrating this equation and invoking Eq. (5.230), we have

$$ff' = f' - \eta f''. \tag{5.238}$$

The solution satisfying the boundary conditions is

$$f = \frac{(C\eta)^2}{1 + (C\eta/2)^2}, \tag{5.239}$$

and the constraint (5.236) requires that $C = \pi^{-1/2}/4 = 0.141$. Wooding (1985) has extended the boundary layer equations to account for large density differences, dispersion, and convection in the presence of tidal oscillations.

Lai (1990b) showed that a similarity solution could be found for the case of a power law variation of centerline temperature. The problem was treated using the Forchheimer model by Degan and Vasseur (1995). As one would expect, inertial effects tend to reduce the buoyancy-induced flow. Inertial effects, together with those of thermal dispersion, also were discussed by Leu and Jang (1995). The case of a non-Newtonian power law fluid was examined by Nakayama (1993a). Higuera and Weidman (1998) noted that the case of natural convection far downstream of a heat source on a solid wall led to a parameter free differential equation problem.

5.11.2. Flow at Low Rayleigh Number

We now consider a point heat source of strength q [W] in an unbounded domain. We introduce spherical polar coordinates (r, θ, φ), with θ the "colatitude" and φ the "longitude," and with the line $\theta = 0$ vertically upward. We have an axisymmetric problem with no dependence on φ. The equations for mass conservation, Darcy flow, and transient energy conservation are

$$\frac{\partial}{\partial r}(r^2 v_r \sin\theta) + \frac{\partial}{\partial \theta}(r v_\theta \sin\theta) = 0, \tag{5.240}$$

$$v_r = -\frac{K}{\mu}\left(\frac{\partial P}{\partial r} + \rho g \cos\theta\right), \tag{5.241}$$

$$v_\theta = -\frac{K}{\mu}\left(\frac{1}{r}\frac{\partial P}{\partial \theta} - \rho g \sin\theta\right), \tag{5.242}$$

$$\frac{1}{\alpha_m}\left(\sigma\frac{\partial T}{\partial t} + v_r\frac{\partial T}{\partial r} + \frac{v_\theta}{r}\frac{\partial T}{\partial \theta}\right) = \frac{1}{r^2}\frac{\partial}{\partial r}\left(r^2\frac{\partial T}{\partial r}\right) + \frac{1}{r^2 \sin\theta}\frac{\partial}{\partial \theta}\left(\sin\theta\frac{\partial T}{\partial t}\right). \tag{5.243}$$

Introducing the Stokes streamfunction $\Psi(r, \theta)$ by

$$v_r = \frac{1}{r^2 \sin\theta}\frac{\partial\Psi}{\partial\theta}, \; v_\theta = -\frac{1}{r\sin\theta}\frac{\partial\Psi}{\partial r}, \tag{5.244}$$

and eliminating the pressure between the two Darcy equations, we get, in nondimensional variables,

$$\frac{1}{r_*^2}\frac{\partial}{\partial\theta}\left(\frac{1}{\sin\theta}\frac{\partial\Psi_*}{\partial\theta}\right) + \frac{1}{\sin\theta}\frac{\partial^2\Psi_*}{\partial r_*^2} = \widetilde{\mathrm{Ra}}\left(\cos\theta\frac{\partial T_*}{\partial\theta} + r_*\sin\theta\frac{\partial T_*}{\partial r_*}\right), \tag{5.245}$$

$$\begin{aligned}\frac{\partial T_*}{\partial t_*} &+ \frac{1}{r_*^2\sin\theta}\left(\frac{\partial\Psi_*}{\partial\theta}\frac{\partial T_*}{\partial r_*} - \frac{\partial\Psi_*}{\partial r_*}\frac{\partial T_*}{\partial\theta}\right)\\ &= \frac{1}{r_*^2}\frac{\partial}{\partial r_*}\left(r_*^2\frac{\partial T_*}{\partial r_*}\right) + \frac{1}{r_*^2\sin\theta}\frac{\partial}{\partial\theta}\left(\sin\theta\frac{\partial T_*}{\partial\theta}\right),\end{aligned} \tag{5.246}$$

where

$$\begin{gathered} t_* = \frac{t\alpha_m}{K\sigma}, \quad r_* = \frac{r}{K^{1/2}}, \quad T_* = \frac{(T - T_\infty)k_m K^{1/2}}{q} \\ \Psi_* = \frac{\Psi}{\alpha_m K^{1/2}}, \quad \widetilde{\mathrm{Ra}} = \frac{g\beta K q}{\nu\alpha_m k_m}. \end{gathered} \tag{5.247}$$

The initial conditions for this transient problem are

$$t = 0: \quad v_r = v_\theta = 0, \quad T = T_\infty.$$

The appropriate boundary conditions are

$$\begin{gathered} r \to \infty: \; v_r = v_\theta = 0, T = T_\infty, \\ \theta = 0, \pi: \; v_\theta = \frac{\partial v_r}{\partial\theta} = \frac{\partial T}{\partial\theta} = 0, \end{gathered} \tag{5.248}$$

together with the fact that v_r, v_θ, and T are of order $1/r$ as $r \to 0$. This is required by the balance of terms in the above differential equations, together with the energy balance constraint

$$\lim_{r\to 0}\left[-k_m(4\pi r^2)\frac{\partial T}{\partial r}\right] = q. \tag{5.249}$$

The above conditions are readily put in nondimensional form. For sufficiently small values of $\widetilde{\mathrm{Ra}}$ we can expand Ψ_* and T_* as power series in $\widetilde{\mathrm{Ra}}$,

$$(\Psi_*, T_*) = (\Psi_{*_0}, T_{*_0}) + \tilde{\mathrm{R}}\mathrm{a}(\Psi_{*_1}, T_{*_1}) + \cdots. \tag{5.250}$$

We can then substitute into the above equations and equate terms in like powers of $\widetilde{\mathrm{Ra}}$, thus obtaining subproblems at order $\widetilde{\mathrm{Ra}}_0, \widetilde{\mathrm{Ra}}_1, \widetilde{\mathrm{Ra}}_2, \ldots$. The zero-order problem yields the conduction solution

$$T_{*_0} = \frac{1}{4\pi r}\operatorname{erfc}\eta, \tag{5.251}$$

$$\Psi_{*0} = 0, \tag{5.252}$$

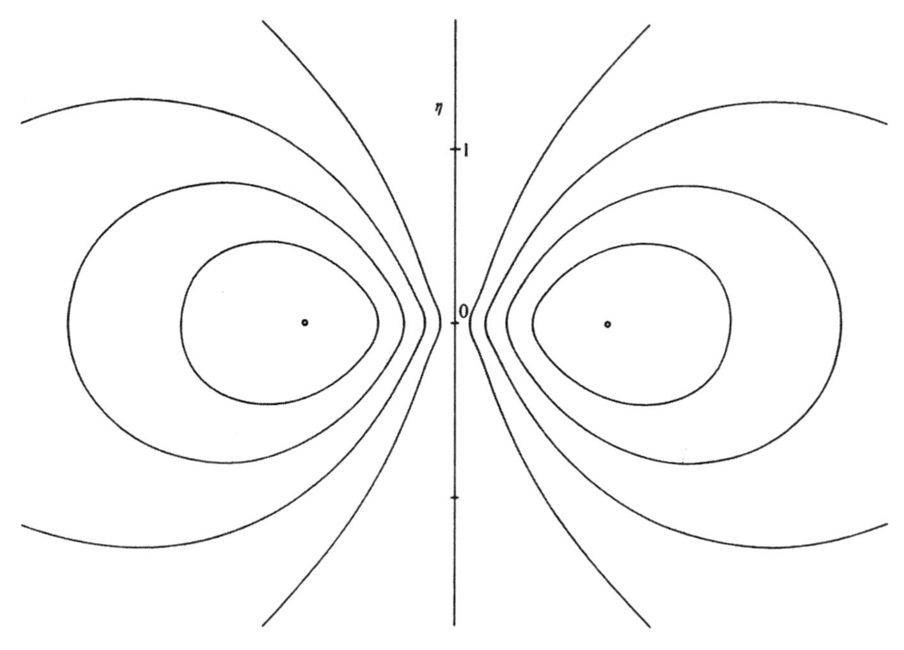

Figure 5.23. Transient natural convection flow pattern about a point heat source. The lines correspond to equal increments of $\psi_{*_1}/t_*^{1/2}$ (Bejan, 1978, 1984).

where $\eta = r_*/2t_*^{1/2}$. The first-order problem yields (Bejan, 1978)

$$\Psi_{1*} = \frac{1}{8\pi} t_*^{1/2} \sin^2\theta \left(2\eta \operatorname{erfc}\eta + \frac{1}{\eta}\operatorname{erf}\eta - \frac{2}{\pi^{1/2}} e^{-\eta^2} \right), \tag{5.253}$$

$$T_{*_1} = \frac{\cos\theta}{64\pi^2 t_*^{1/2}} \left(\frac{1}{\eta} - \frac{4}{3\pi^{1/2}} + \frac{6}{5\pi^{1/2}}\eta^2 - \frac{16}{45\pi}\eta^3 - \frac{152}{315\pi^{1/2}}\eta^4 + \ldots \right). \tag{5.254}$$

Figure 5.23, based on Eq. (5.253), shows that as soon as the heat source is turned on a vortex ring forms about the source. The radius of the core of the vortex is given by $\eta = 0.881$, i.e., the physical radius grows with time as the group $1.762(\alpha_m t/\sigma)^{1/2}$.

Unlike the line-source problem of Section 5.9.2, the present point source problem has a steady-state small $\widetilde{\text{Ra}}$ solution with

$$\psi_* = \frac{r_*}{8\pi}\left[\sin^2\theta\widetilde{\text{Ra}} + \frac{1}{24\pi}\sin\theta\sin 2\theta\widetilde{\text{Ra}}^2 - \frac{5}{18432\pi^3}(8\cos^4\theta - 3)\widetilde{\text{Ra}}^3 + \ldots\right], \tag{5.255}$$

$$T_* = \frac{1}{4\pi r_*}\left[1 + \frac{1}{8\pi}\cos\theta\widetilde{\text{Ra}} + \frac{5}{768\pi^2}\cos 2\theta\widetilde{\text{Ra}}^2 + \frac{1}{55296\pi^3}\cos\theta(47\cos^2\theta - 30)\widetilde{\text{Ra}}^3 + \ldots\right]. \tag{5.256}$$

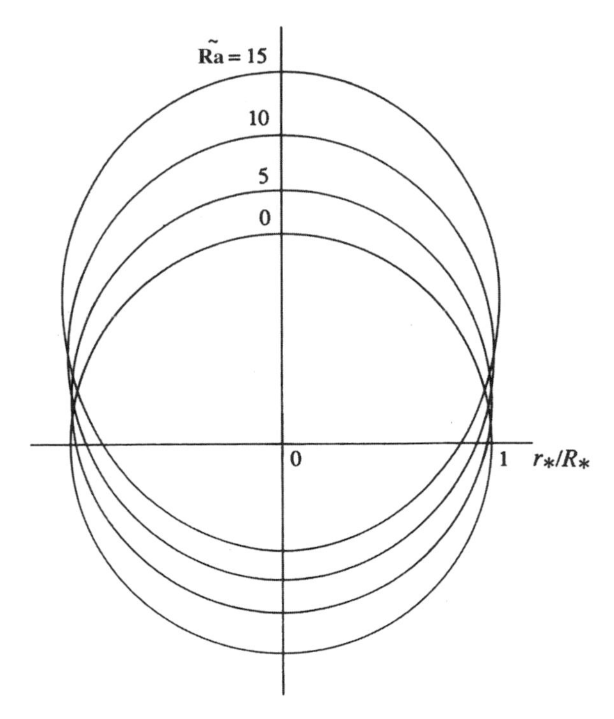

Figure 5.24. Steady temperature distribution around a point heat source; the lines represent the $(4\pi R_*)\theta = 1$ isotherm, for increasing values of Ra (Bejan, 1978, 1984).

This solution gives valid results for source strength Rayleigh numbers $\widetilde{\text{Ra}}$ up to about 20. The temperature field is illustrated in Fig. 5.24 in which a curve represents the isothermal surface $T_* = 1/4\pi R_*$, where R_* is a fixed nondimensional distance from the origin. The figure shows that the warm region, originally spherical about the point source, shifts upward and becomes elongated like the flame of a candle as $\widetilde{Ra}$ increases.

Whereas Bejan (1978) used the source condition (5.249), which requires the heat flux to be uniformly distributed over an isothermal source, Ene and Poliševski (1987) took

$$\lim_{r\to 0} \int_{S_r} \left(-k_m \frac{\partial T}{\partial r}\right) d\sigma = q, \tag{5.257}$$

where S_r is the sphere of radius r. Equation (5.257) implies that $\partial T/\partial r$ varies with θ in a special way (determined by the overall problem) as $r \to 0$. It appears to be the more appropriate condition. Both Eqs. (5.249) and (5.257) are based on the assumption that the convective heat transport at the source is negligible [compare Eq. (5.264)].

Hickox (1981) has utilized the fact that the momentum equation is linear in Ra to investigate certain other geometries by superposing sources. Ganapathy and Purushothaman (1990) discussed the case of an instantaneous point source.

The Brinkman term affects the solution at radial distances up to $O(K^{1/2})$ from the source, where at small times it slows the rate of momentum transfer. Puroshothaman *et al.* (1990) dealt with a pulsating point heat source. Ganapathy (1992) treated an instantaneous point source that is enveloped by a solid sphere, which is itself surrounded by a porous medium.

5.11.3. Flow at Intermediate Rayleigh Number

For the steady situation Hickox and Watts (1980) obtained results for arbitrary values of $\tilde{\text{Ra}}$, for both the semi-infinite region considered by Wooding and Bejan and the infinite region. For the infinite region, with spherical polar coordinates and the streamfunction defined as in Eq. (5.244), one can put

$$\eta = \cos\theta, \quad \psi = \alpha_m r f(\eta), \quad T - T_\infty = \frac{\alpha_m}{gK\beta}\frac{g(\eta)}{r}. \tag{5.258}$$

The problem reduces to the solution of the differential equations

$$f'' = -(\eta g)', \tag{5.259}$$

$$(fg)' = g'' - (\eta^2 g')', \tag{5.260}$$

subject to the symmetry and boundary conditions

$$f(1) = 0, \quad f(-1) = 0, \tag{5.261}$$

$$g, g' \text{ bounded as } \eta \to \pm 1, \tag{5.262}$$

and the constraint

$$\int_{-1}^{1} (1 - f')g\,d\eta = 2\pi\widetilde{\text{Ra}}. \tag{5.263}$$

The last equation arises from the requirement that the energy flux, integrated over a sphere centered at the origin, should equal q, so

$$\int_0^\pi \left[(\rho c_P)_f v_r (T - T_\infty) - k_m \frac{\partial T}{\partial r}\right] 2\pi r^2 \sin\theta\, d\theta = q. \tag{5.264}$$

Hickox and Watts (1980) integrated Eqs. (5.259)–(5.263) numerically. They treated the semi-infinite region in a similar fashion, but using a different similarity transformation. Some representative plots of isotherms and streamlines are presented in Figs. 5.25 and 5.26.

5.12. Other Configurations

5.12.1. Fins Projecting from a Heated Base

The problem of high Rayleigh number convection about a long vertical thin fin with a heated base can be treated as a conjugate conduction-convection problem. Various geometries have been considered. Pop *et al.* (1985) obtained a similarity

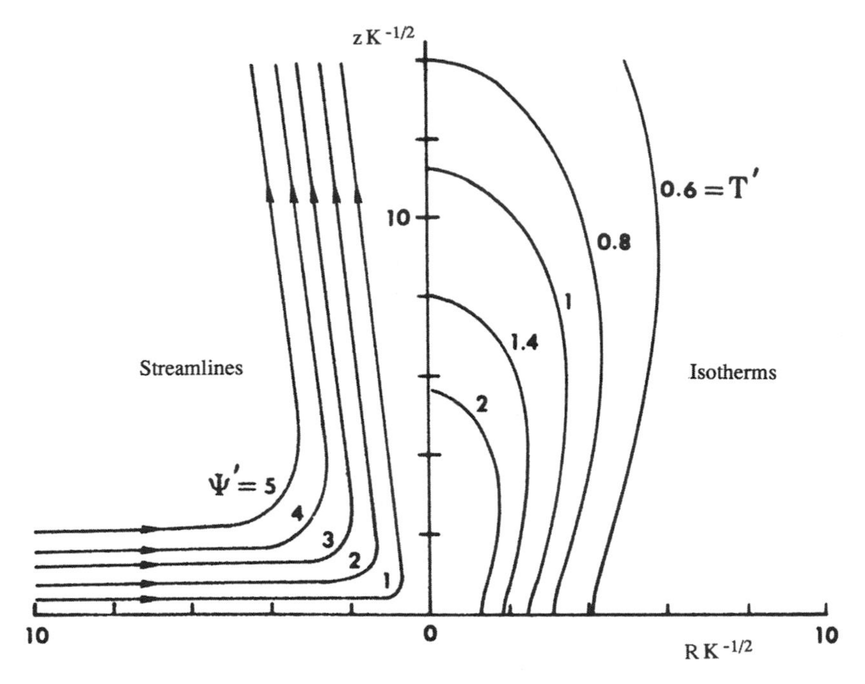

Figure 5.25. Streamlines and isotherms for a point source at the base of a semi-infinite region, Ra = 10, $\psi' = \psi/\alpha_m K^{1/2}$, $T' = (T - T_\infty)g\beta K^{3/2}/\nu\alpha_m$ (Hickox and Watts, 1980).

solution for a vertical plate fin projecting downward from a heated horizontal plane base at constant temperature for the case of the conductivity-fin thickness product varying as a power function of distance from a certain specified origin. They also dealt with the similar problem of a vertical plate extending from a heated horizontal cylindrical base at constant temperature.

Pop *et al.* (1986) used a finite-difference numerical method for the former geometry but with constant conductivity and fin thickness, and Liu and Minkowycz (1986) investigated the influence of lateral mass flux in this situation. Gill and Minkowycz (1988) examined the effects of boundary friction and quadratic drag. Hung *et al.* (1989) have incorporated non-Darcy effects in their study of a transient problem. The above studies all have been of a vertical plate fin. The case of a vertical cylindrical fin was analyzed by Liu *et al.* (1987b); again the effect of lateral mass flux was included. Convection from a slender needle, for the case where the axial wall thickness varies as a power function of distance from the leading edge, was analyzed by Peng *et al.* (1992).

Conjugate convection about a vertical plate fin was studied by Hung (1991) using the Brinkman-Forchheimer model. Chen and Chiou (1994) added the effects of thermal dispersion and nonuniform porosity. Conjugate convection of a non-Newtonian fluid about a vertical plate was studied by Pop and Nakayama

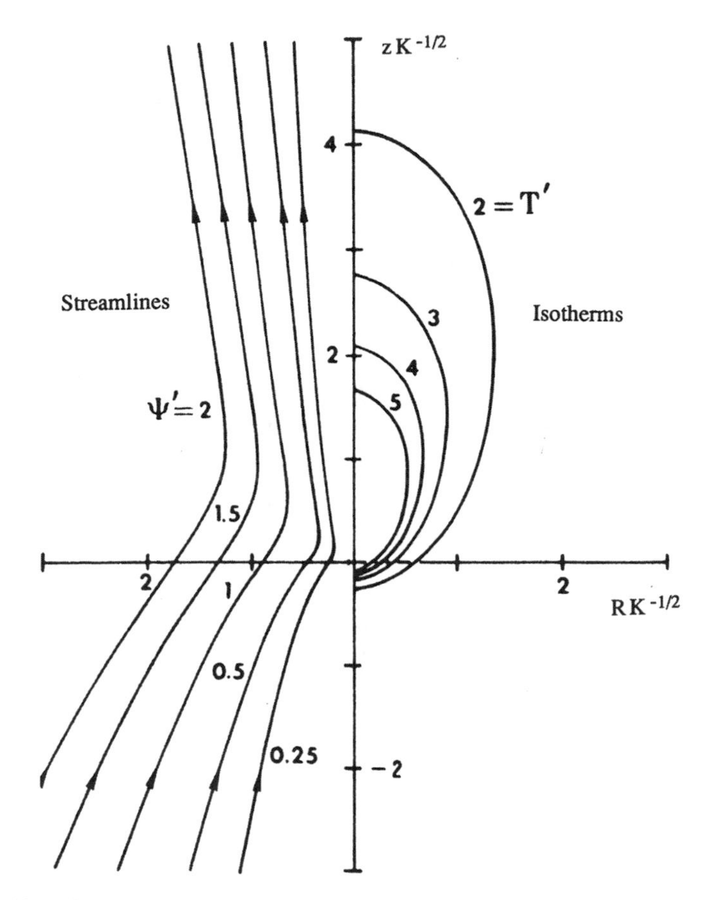

Figure 5.26. Streamlines and isotherms for a point source in an infinite region, Ra = 10, ψ' and T' defined as in Fig. 5.25 (Hickox and Watts, 1980).

(1994), while the corresponding problem for a vertical cylindrical fin was treated by Hossain *et al.* (1995). Further work on conjugate convection from vertical plate fins was reported by Vaszi *et al.* (2002b, 2004) and Pop and Nakayama (1999).

5.12.2. Flows in Regions Bounded by Two Planes

The Darcy flow in a corner region bounded by a heated vertical wall and an insulated inclined wall was analyzed by Daniels and Simpkins (1984), while Riley and Rees (1985) analyzed the non-Darcy flow in the exterior region bounded by a heated inclined wall and an inclined wall that was either insulated or cooled. In each of these two papers the heated wall was at constant temperature. Hsu and Cheng (1985a) analyzed the Darcy flow about an inclined heated wall with a power law of variation of temperature and an inclined unheated isothermal wall.

The particular case of the Darcy flow in the "stably heated" corner between a cold horizontal wall and a hot vertical wall situated above the horizontal wall (or between a hot horizontal wall and a cold vertical wall situated below the horizontal wall) was studied by Kimura and Bejan (1985). Their scale analysis and numerical solutions showed that the single-cell corner flow becomes increasingly more localized as the Rayleigh number increases. At the same time the mass flow rate engaged in natural convection and the conduction-referenced Nusselt number increase.

Liu *et al.* (1987a) found a local similarity solution for flow in the corner formed by two mutually perpendicular vertical plates for the case when both plates are at the same constant wall temperature. Earlier solutions by Liu and Ismail (1980) and Liu and Guerra (1985) (the latter with an arbitrary angle between the vertical plates) had been obtained under an asymptotic suction assumption. Two other problems involving perpendicular planes were studied by Ingham and Pop (1987a,b). Pop *et al.* (1997) performed calculations for convection in a Darcian fluid in a horizontal L-shaped corner, with a heated isothermal vertical plate joined to a horizontal surface that is either adiabatic or held at ambient temperature.

5.12.3. Other Situations

The problem of the cooling of a circular plate situated in the bottom plane boundary of a semi-infinite region was analyzed as a boundary layer problem by McNabb (1965). The boundary layer flow near the edge of a horizontal circular dish in an unbounded region was studied by Merkin and Pop (1989). A numerical study on various models of convection in open-ended cavities was reported by Ettefagh *et al.* (1991).

The subject of conjugate natural convection in porous media has been reviewed by Kimura *et al.* (1997). They discussed various configurations including slender bodies, rectangular slabs, horizontal cylinders, and spheres. Three-dimensional stagnation point convection on a surface on which heat is released by an exothermic reaction was analyzed by Pop *et al.* (2003). The topic of chemically driven convection in porous media was reviewed by Pop *et al.* (2002); other relevant papers include those by Mahmood and Merkin (1998) and Merkin and Mahmood (1998). The effect of local thermal nonequilibrium or g-jitter on convective stagnation point flow was analyzed by Rees and Pop (1999, 2001). Convection from a cylinder covered with an orthotropic porous layer in cross-flow was investigated numerically by Abu-Hijleh (2001a)

5.13. Surfaces Covered with Hair

The two-temperatures porous medium model described in Section 4.10 was also used in the theoretical study of natural convection heat transfer from surfaces covered with hair (Bejan, 1990b). With reference to a vertical surface (Fig. 5.27) the boundary layer equations for energy conservation and Darcy flow are

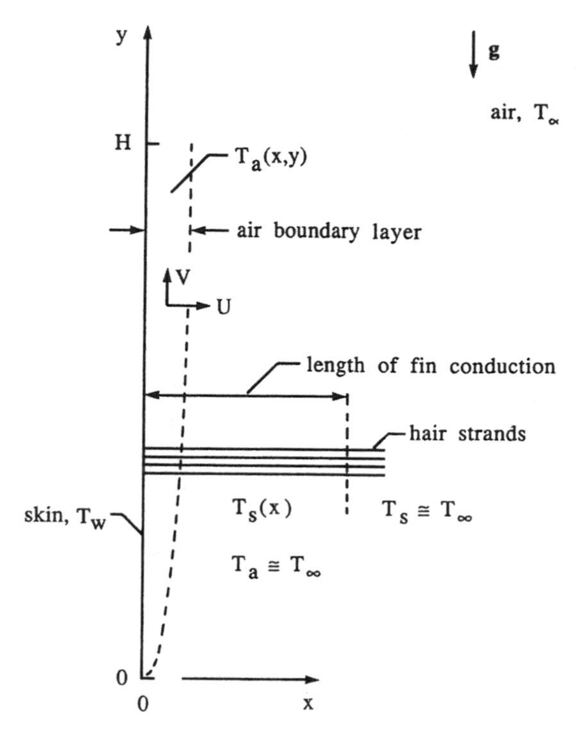

Figure 5.27. Vertical skin area, air boundary layer, and hair strands that act as fins (Bejan, 1990b).

written as

$$\rho c_P \left(U \frac{\partial T_a}{\partial x} + V \frac{\partial T_a}{\partial y} \right) = k_a \frac{\partial^2 T_a}{\partial x^2} + nhp_s(T_s - T_a), \tag{5.265}$$

$$\frac{\partial V}{\partial x} = \frac{g\beta K}{\nu\varphi} \frac{\partial T_a}{\partial x}. \tag{5.266}$$

The porosity φ appears in the denominator in Eq. (5.266), because in this model V is the air velocity averaged only over the space occupied by air. The rest of the notation is defined in Fig. 5.27 and Section 4.10. For example, n is the hair density (strands/m^2).

The boundary layer heat transfer analysis built on this model showed that the total heat transfer rate through a skin area of height H is minimized when the hair strand diameter reaches the optimal value

$$\frac{D_{opt}}{H} = \left(\frac{1 - \varphi}{0.444} \right)^{1/2} \left(\frac{k_s}{k_a} \frac{f_2}{\varphi f_1 \mathrm{Ra}_f} \right)^{1/4}. \tag{5.267}$$

The Ra_f factor in the denominator is the Rayleigh number for natural convection in open air, $\mathrm{Ra}_f = g\beta H^3(T_w - T_\infty)/\nu\alpha_a$. The minimum heat transfer rate that corresponds to D_{opt} is

$$\frac{q'_{min}}{k_a(T_w - T_\infty)} = 1.776(1-\varphi)^{1/2}\left(\varphi f_1 f_2 \frac{k_s}{k_a}\mathrm{Ra}_f\right)^{1/4}. \tag{5.268}$$

The factors f_1 and f_2 are both functions of porosity, and result from having modeled the permeability and strand-air heat transfer coefficient by

$$K = D^2 f_1(\varphi), \quad h = \frac{k_a}{D} f_2(\varphi). \tag{5.269}$$

It is important to note that since Ra_f is proportional to H^3, Eq. (5.267) states that the optimal strand diameter is proportional to $H^{1/4}$. The theoretical results for a vertical surface covered with hair were tested in an extensive series of numerical experiments (Lage and Bejan, 1991).

Analogous conclusions are reached in the case where instead of the vertical plane of Fig. 5.27, the skin surface has the shape of a long horizontal cylinder of diameter D_0. The optimal hair strand diameter is

$$\frac{D_{opt}}{D_0} = 1.881(1-\varphi)^{1/2}\left(\frac{k_s}{k_a}\frac{f_2}{\varphi f_1 \mathrm{Ra}_{fo}}\right)^{1/4}, \tag{5.270}$$

where $\mathrm{Ra}_{fo} = g\beta D_0^3(T_w - T_\infty)/\nu\alpha_a$. In the case where the body shape approaches a sphere of diameter D_0, the optimal hair strand diameter has a similar form,

$$\frac{D_{opt}}{D_0} = 2.351(1-\varphi)^{1/2}\left(\frac{k_s}{k_a}\frac{f_2}{\varphi f_1 \mathrm{Ra}_{fo}}\right)^{1/4}. \tag{5.271}$$

Equations (5.270) and (5.271) show that D_{opt} increases as $D_o^{1/4}$. Combined with Eq. (5.267), they lead to the conclusion that when the heat transfer mechanism is boundary layer natural convection, the optimal hair strand diameter increases as the vertical dimension of the body (H, or D_0) raised to the power 1/4.

6
Internal Natural Convection: Heating from Below

6.1. Horton-Rogers-Lapwood Problem

We start with the simplest case, that of zero flow through the fluid-saturated porous medium. For an equilibrium state the momentum equation is satisfied if

$$-\nabla P + \rho_f \mathbf{g} = 0. \tag{6.1}$$

Taking the curl of each term yields

$$\nabla \rho_f \times \mathbf{g} = 0. \tag{6.2}$$

If the fluid density ρ_f depends only on the temperature T, then this equation implies that $\nabla T \times \mathbf{g} = 0$. We conclude that a necessary condition for equilibrium is that the temperature gradient is vertical (or zero). Intrapore convection may increase effective conductivity of the medium. We thus have a special interest in the problem of a horizontal layer of a porous medium uniformly heated from below. This problem, the porous-medium analog of the Rayleigh-Bénard problem, was first treated by Horton and Rogers (1945) and independently by Lapwood (1948).

With reference to Fig. 6.1, we take a Cartesian frame with the z axis vertically upward. We suppose that the layer is confined by boundaries at $z = 0$ and $z = H$, the lower boundary being at uniform temperature $T_0 + \Delta T$ and the upper boundary at temperature T_0. We thus have a layer of thickness H and an imposed adverse temperature gradient $\Delta T/H$. We suppose that the medium is homogeneous and isotropic, that Darcy's law is valid, and that the Oberbeck-Boussinesq approximation is applicable, and we also make the other standard assumptions (local thermal equilibrium, negligible heating from viscous dissipation, negligible radiative effects, etc.). The appropriate equations are, cf. Eqs. (1.1), (1.10), (2.3), and (2.20),

$$\nabla \cdot \mathbf{v} = 0, \tag{6.3}$$

$$c_a \rho_0 \frac{\partial \mathbf{v}}{\partial t} = -\nabla P - \frac{\mu}{K}\mathbf{v} + \rho_f \mathbf{g}, \tag{6.4}$$

$$(\rho c)_m \frac{\partial T}{\partial t} + (\rho c_P)_f \mathbf{v} \cdot \nabla T = k_m \nabla^2 T, \tag{6.5}$$

$$\rho_f = \rho_0[1 - \beta(T - T_0)]. \tag{6.6}$$

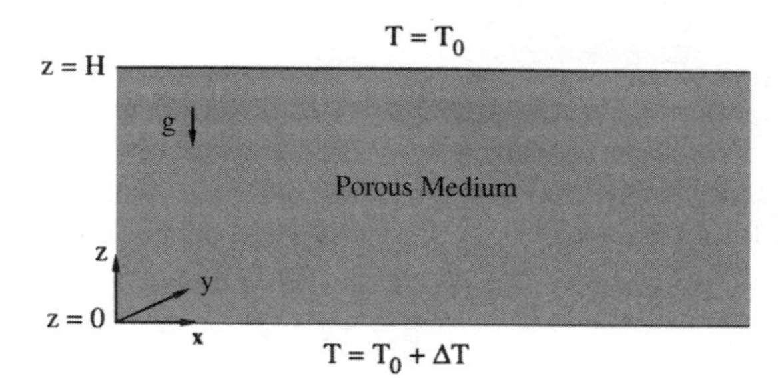

Figure 6.1. The Horton-Rogers-Lapwood problem: infinite horizontal porous layer heated from below.

The reader is reminded that $\mathbf{v}$ is the seepage velocity, P is the pressure, μ the dynamic viscosity, K the permeability, c the specific heat, k_m the overall thermal conductivity, and β the thermal volume expansion coefficient.

We observe that Eqs. (6.3)–(6.6) have a basic steady-state solution, which satisfies the boundary conditions $T = T_0 + \Delta T$ at $z = 0$ and $T = T_0$ at $z = H$. That solution is

$$\mathbf{v}_b = 0, \tag{6.7}$$

$$T_b = T_0 + \Delta T\left(1 - \frac{z}{H}\right), \tag{6.8}$$

$$P_b = P_0 - \rho_0 g\left[z + \frac{1}{2}\beta\Delta T\left(\frac{z^2}{H} - 2z\right)\right]. \tag{6.9}$$

It describes the "conduction state," one in which the heat transfer is solely by thermal conduction.

6.2. Linear Stability Analysis

We now examine the stability of this solution and assume that the perturbation quantities (those with primes) are small. We write

$$\mathbf{v} = \mathbf{v}_b + \mathbf{v}', \quad T = T_b + T', \quad P = P_b + P'. \tag{6.10}$$

When we substitute into Eqs. (6.3)–(6.5) and neglect second-order small quantities we obtain the linearized equations [note $\mathbf{v}' = (u', v', w')$]

$$\nabla \cdot \mathbf{v}' = 0, \tag{6.11}$$

$$c_a\rho_0\frac{\partial \mathbf{v}'}{\partial t} = -\nabla P' - \frac{\mu}{K}\mathbf{v}' - \beta\rho_0 T'\mathbf{g}, \tag{6.12}$$

$$(\rho c)_m\frac{\partial T'}{\partial t} - (\rho c_P)_f\frac{\Delta T}{H}w' = k_m\nabla^2 T'. \tag{6.13}$$

Nondimensional variables are introduced by choosing H, $\sigma H^2/\alpha_m$, α_m/H, ΔT, and $\mu\alpha_m/K$ as scales for length, time, velocity, temperature, and pressure, respectively. Here α_m is a thermal diffusivity defined by

$$\alpha_m = \frac{k_m}{(\rho c_P)_f} = \frac{k_m}{k_f}\alpha_f, \tag{6.14a}$$

where $\alpha_f = k_f/(\rho c_P)_f$ is the thermal diffusivity of the fluid phase. It is convenient to define the *heat capacity ratio*

$$\sigma = \frac{(\rho c)_m}{(\rho c_P)_f} \tag{6.14b}$$

and put

$$\begin{aligned} \hat{\mathbf{x}} &= \frac{\mathbf{x}}{H}, \quad \hat{t} = \frac{\alpha_m t}{\sigma H^2}, \quad \hat{\mathbf{v}} = \frac{H'_{\mathbf{v}}}{\alpha_m}, \\ \hat{T} &= \frac{T}{\Delta T}, \quad \hat{P} = \frac{KP'}{\mu\alpha_m}, \end{aligned} \tag{6.15}$$

with $\hat{\mathbf{x}} = (x, y, z)$. Substituting Eqs. (6.11)–(6.13) we get

$$\nabla \cdot \hat{\mathbf{v}} = 0, \tag{6.16}$$

$$\gamma_a \frac{\partial \hat{\mathbf{v}}}{\partial \hat{t}} = -\nabla \hat{P} - \hat{\mathbf{v}} + \mathrm{Ra}\hat{T}\mathbf{k}, \tag{6.17}$$

$$\frac{\partial \hat{T}}{\partial \hat{t}} - \hat{w} = \nabla^2 \hat{T}, \tag{6.18}$$

where $\mathbf{k}$ is the unit vector in the z direction and

$$\mathrm{Ra} = \frac{\rho_0 g \beta K H \Delta T}{\mu\alpha_m}, \quad \mathrm{Pr}_m = \frac{\mu}{\rho_0 \alpha_m}, \quad \gamma_a = \frac{c_a K}{\sigma \mathrm{Pr}_m H^2}. \tag{6.19}$$

In Eq. (6.19) Ra is the Rayleigh-Darcy number (or Rayleigh number, for short), Pr_m is an overall Prandtl number, and γ_a is a nondimensional acceleration coefficient. In most practical situations the Darcy number K/H^2 will be small and as a consequence γ_a also will be small. Accordingly, we take $\gamma_a = 0$ unless otherwise specified. Note that the Rayleigh-Darcy number is the product of the Darcy number and the usual Rayleigh number for a clear viscous fluid.

Operating on Eq. (6.17) twice with curl, using Eq. (6.16) and taking only the z component of the resulting equation, we obtain

$$\nabla^2 \hat{w} = \mathrm{Ra}\nabla_H^2 \hat{T}, \tag{6.20}$$

where $\nabla_H^2 = \partial^2/\partial x^2 + \partial^2/\partial y^2$. Equations (6.18) and (6.20) contain just two dependent variables, $\hat{w}$ and $\hat{t}$. Since the equations are linear, we can separate the variables. Writing

$$\left(\hat{w}, \hat{T}\right) = [W(\hat{z}), \theta(\hat{z})] \exp\left(s\hat{t} + il\hat{x} + im\hat{y}\right) \tag{6.21}$$

and substituting into Eqs. (6.18) and (6.20), we obtain

$$(D^2 - \alpha^2 - s)\theta = -W, \tag{6.22}$$

$$(1 + \gamma_a s)(D^2 - \alpha^2)W = -\alpha^2 \mathrm{Ra}\,\theta \tag{6.23}$$

where

$$D \equiv \frac{d}{d\hat{z}} \quad \text{and} \quad \alpha = (l^2 + m^2)^{1/2}. \tag{6.23$'$}$$

In these equations α is an overall horizontal wavenumber. This pair of ordinary differential equations forms a fourth-order system, which must be solved subject to four appropriate boundary conditions.

Various types of boundaries can be considered. If both boundaries are impermeable and are perfect thermal conductors, then we must have $w' = 0$ and $T' = 0$ at $z = 0$ and $z = H$, and so

$$W = \theta = 0 \quad \text{at} \quad \hat{z} = 0 \quad \text{and} \quad \hat{z} = 1. \tag{6.24}$$

The homogeneous equations (6.22) and (6.24) form an eigenvalue system in which Ra may be regarded as the eigenvalue. In order for the solution to remain bounded as $x, y \pm 8$, the wavenumbers l and m must be real, and hence the overall wavenumber α must be real. In general s can be complex, $s = s_r + i\omega$. If $s_r > 0$, then perturbations of the form (6.21) grow with time, i.e., we have instability. The case $s_r = 0$ corresponds to marginal stability. In general ω gives the frequency of oscillations, but in the present case it is easily proven that $\omega = 0$ when $s_r > 0$, so when the disturbances grow with time they do so monotonically. In other words, the so-called principle of exchange of stabilities is valid.

For the case of marginal stability we can put $s = 0$ in Eqs. (6.22) and (6.23), which become

$$(D^2 - \alpha^2)\theta = -W, \tag{6.25}$$

$$(D^2 - \alpha^2)W = -\alpha^2 \text{Ra}\theta. \tag{6.26}$$

Eliminating θ we have

$$(D^2 - \alpha^2)^2 W = \alpha^2 \text{Ra} W, \tag{6.27}$$

with

$$W = D^2 W = 0 \text{ at } \hat{z} = 0 \text{ and } \hat{z} = 1. \tag{6.28}$$

We see immediately that $W = \sin(j\pi\hat{z})$ is a solution, for $j = 1, 2, 3, \ldots,$ if

$$\text{Ra} = \frac{(j^2\pi^2 + \alpha^2)^2}{\alpha^2}. \tag{6.29}$$

Clearly Ra is a minimum when $j = 1$ and $\alpha = \pi$, i.e., the critical Rayleigh number is $\text{Ra}_c = 4\pi^2 = 39.48$ and the associated critical wavenumber is $\alpha_c = \pi$. For the higher-order modes ($j = 2, 3, \ldots$), $\text{Ra}_j = 4\pi^2 j^2$ and $\alpha_{cj} = j\pi$. An alternative to the derivation of critical Rayleigh number is constructal theory (Nelson and Bejan, 1998; Bejan, 2000), which yields $\text{Ra}_c = 12\pi = 37.70$ (see Section 6.26).

In conclusion, for $\text{Ra} < 4\pi^2$ the conduction state remains stable. When Ra is raised above $4\pi^2$, instability appears as convection in the form of a cellular motion with horizontal wavenumber π.

In this way linear stability theory predicts the size of the convection cells but it says nothing about their horizontal plan-form, because the eigenvalue problem is degenerate. The (x, y) dependence can be given by any linear combination of terms of the form $\exp\,(ilx + imy)$ where $l^2 + m^2 = \alpha^2$. In particular, dependence on $\sin\alpha x$ corresponds to convection rolls whose axes are parallel to the y axis; dependence on $\sin(\alpha x/\sqrt{2})\sin(\alpha y/\sqrt{2})$ corresponds to cells of square plan form, and dependence on $\cos\alpha x + 2\cos(\alpha x/2)\cdot\cos(\sqrt{3}\,\alpha y/2)$ corresponds to cells of hexagonal plan form. In each case the nondimensional horizontal wavelength is $2\pi/\alpha_c = 2$. Since the height of the layer is 1, this wavelength is the width of a pair of counterrotating rolls of square vertical cross section. Further, linear theory does not predict whether, in a hexagonal cell, fluid rises in the center and descends near the sides or vice versa; nonlinear theory is needed to predict which situation will occur.

Equation (6.29) has been obtained for the case of impermeable conducting boundaries. For other boundary conditions the eigenvalue problem must in general be solved numerically, but there is one other case when a numerical calculation is not necessary. It is made possible by the fact that the critical wavenumber is zero, and so an expansion in powers of α^2 works.

That special case is when both boundaries are perfectly insulating, i.e., the heat flux is constant on the boundaries. When the boundaries are also impermeable, we have

$$W = D\theta = 0 \quad \text{at} \quad \hat{z} = 0 \quad \text{and} \quad \hat{z} = 1. \tag{6.30}$$

Writing

$$(W, \theta, \mathrm{Ra}) = (W_0, \theta_0, \mathrm{Ra}_0) + \alpha^2(W_1, \theta_1, \mathrm{Ra}_1) + \cdots \tag{6.31}$$

substituting Eqs. (6.25), (6.26), and (6.30) and equating powers of α^2, we obtain in turn systems of various orders. For the zero-order system we find that

$$\begin{gathered} D^2 W_0 = 0, \qquad D^2\theta_0 + W_0 = 0, \\ W_0 = D\theta_0 = 0 \quad \text{at} \quad \hat{z} = 0, 1 \end{gathered}$$

This system has the solution $W_0 = 0$, $\theta_0 = $ constant, and without loss of generality we can take $\theta_0 = 1$. The order α^2 system is

$$\begin{gathered} D^2 W_1 = W_0 - \mathrm{Ra}_0\theta_0 = -\mathrm{Ra}_0, \qquad D^2\theta_1 + W_1 = \theta_0 = 1, \\ W_1 = D\theta_1 = 0 \text{ at } \hat{z} = 0, 1. \end{gathered}$$

With the arbitrary factor suitably chosen, these equations yield in succession

$$\begin{gathered} W_1 = -\tfrac{1}{2}\,\mathrm{Ra}_0\,(\hat{z}^2 - \hat{z}) \\ \left\langle 1 + \tfrac{1}{2}\,\mathrm{Ra}_0\,(\hat{z}^2 - \hat{z})\right\rangle = 0. \end{gathered}$$

This implies that $\mathrm{Ra}_0 = 12$. From the order α^4 system Ra_1 can be calculated. It turns out to be positive, so it follows that $\mathrm{Ra}_c = 12$, $\alpha_c = 0$.

Table 6.1. Values of the critical Rayleigh number Ra_c and the corresponding critical wavenumber α_c for various boundary conditions (after Nield, 1968). The terms free, conducting, and insulating are equivalent to constant pressure, constant temperature, and constant heat flux, respectively.

IMP: impermeable ($K = 8$) FRE: free ($K = 0$)
CON: conducting ($L = 8$) CHF: constant heat flux ($L = 0$)

K_l	K_u	L_l	L_u	Ra_c	α_c
IMP	IMP	CON	CON	$39.48 = 4\pi^2$	$3.14 = \pi$
IMP	IMP	CON	CHF	27.10	2.33
IMP	IMP	CHF	CHF	12	0
IMP	FRE	CON	CON	27.10	2.33
IMP	FRE	CHF	CON	17.65	1.75
IMP	FRE	CON	CHF	$9.87 = \pi^2$	$1.57 = \pi/2$
IMP	FRE	CHF	CHF	3	0
FRE	FRE	CON	CON	12	0
FRE	FRE	CON	CHF	3	0
FRE	FRE	CHF	CHF	0	0

More generally, one can impose boundary conditions

$$DW - K_l W = 0,\ D\theta - L_l\theta \quad \text{at} \quad \hat{z} = 0, \qquad (6.32)$$
$$DW + K_u W = 0,\ D\theta + L_u\theta = 0 \quad \text{at} \quad \hat{z} = 1.$$

The subscripts l and u refer to lower and upper boundaries, respectively. Here L_l and L_u are Biot numbers, taking the limit values 0 for an insulating boundary and 8 for a conducting boundary. The coefficients K_l and K_u take discrete values, 0 for a boundary at constant pressure (as for the porous medium bounded by fluid), and 8 for an impermeable boundary. Critical values for various combinations are given in Table 6.1 after Nield (1968), with a correction. [The traditional term "insulating" refers to perturbations. This is somewhat confusing terminology, so following Rees (2000) we now refer to this as the constant heat flux condition. Also, strictly speaking, the constant pressure condition refers to a hydrostatic situation in the exterior region.] As one would expect, Ra_c and α_c both decrease as the boundary conditions are relaxed. Calculations for intermediate values of the Biot numbers L_l and L_u were reported by Wilkes (1995). The onset of gas convection in a moist porous layer with the top open to the atmosphere was analyzed by Lu *et al.* (1999). They found that the critical Rayleigh number was then less than the classical value of π^2. The open-top problem for a vertical fault was analyzed by Malkovsky and Pek (2004).

Tyvand (2002) demonstrated that the open boundary condition, traditionally known as the constant temperature boundary condition, corresponds to requiring that the surrounding fluid is hydrostatic. Just as the kinematic condition on an impermeable boundary is $\mathbf{n} \cdot \mathbf{v} = 0$, the condition on an open boundary is $\mathbf{v} \times \mathbf{n} = 0$.

6.3. Weak Nonlinear Theory: Energy and Heat Transfer Results

The nonlinear nondimensional perturbation equations are

$$\gamma_a \frac{\partial \mathbf{v}}{\partial t} = -\nabla P - \mathbf{v} + \mathrm{Ra}\, T\, \mathbf{k}, \tag{6.33}$$

$$\frac{\partial T}{\partial t} - w + \mathbf{v} \cdot \nabla T = \nabla^2 T \tag{6.34}$$

in which, for convenience, we have dropped the carets. These equations can be compared with the linear set (6.17) and (6.18).

We can obtain equations involving energy balances by multiplying Eqs. (6.33) and (6.34) by $\mathbf{v}$ and θ, respectively, and averaging over the fluid layer. We use the notation

$$\langle f \rangle = \int_0^1 \bar{f} dz,$$

where the bar denotes an average over (x, y) values at a given value of z. Using the fact that all expressions that can be written as a divergence vanish because of the boundary conditions and because contributions from the sidewalls become negligible in the limit of an infinitely extended layer, we obtain

$$\frac{1}{2}\gamma_a \frac{\partial}{\partial t}\langle \mathbf{v} \cdot \mathbf{v} \rangle = \langle \mathrm{Ra}\, wT \rangle - \langle \mathbf{v} \cdot \mathbf{v} \rangle, \tag{6.35}$$

$$\frac{1}{2}\frac{\partial}{\partial t}\langle T^2 \rangle = \langle wT \rangle - \langle |\nabla T|^2 \rangle. \tag{6.36}$$

For steady or statistically stationary convection the left-hand sides of these two equations are zero. Then Eq. (6.35) expresses the balance between the work done by the buoyancy force and the viscous dissipation, while Eq. (6.36) represents a similar relationship between the convective heat transfer and the entropy production by convection.

That $\langle wT \rangle$ represents the convective part of the heat transport can be demonstrated as follows. The horizontal mean of Eq. (6.34) is

$$\frac{\partial \overline{T}}{\partial t} + \frac{\partial}{\partial z}(\overline{wT}) = \frac{\partial^2 \overline{T}}{\partial z^2}. \tag{6.37}$$

For a steady temperature field, integration with respect to z and use of the boundary conditions gives

$$\frac{\partial \overline{T}}{\partial z} = \overline{wT} - \langle wT \rangle. \tag{6.38}$$

Since the normal component of the velocity (w) vanishes at the boundary, the entire heat flux is transported by conduction at the boundary. Thus the expression

$$-\left.\frac{\partial \overline{T}}{\partial z}\right|_{z=1} = \langle wT \rangle$$

represents the convective contribution to the heat transport. The Nusselt number Nu is defined as the ratio of the heat transports with and without convection. Therefore we conclude that

$$\mathrm{Nu} = 1 + \langle wT \rangle. \tag{6.39}$$

From Eq. (6.35) it follows that under stationary conditions $\langle wT \rangle = 0$ and so Nu = 1. Also, under the same conditions, we see from Eqs. (6.35) and (6.36) that

$$\mathrm{Ra} = \frac{\langle |\mathbf{v}|^2 \rangle \langle |\nabla T|^2 \rangle}{\langle wT \rangle^2}. \tag{6.40}$$

The right-hand side has a positive minimum value, and it follows that steady or statistically stationary convection can exist only above a certain positive value of Ra. The right-hand side can be interpreted as a functional of the trial fields $\mathbf{v}$ and T. When this functional is minimized subject to the constraints of the continuity equation (6.16) and the boundary conditions, the energy stability limit Ra_E is obtained. No steady or statistically stationary form of convection is possible for $\mathrm{Ra} < \mathrm{Ra}_E$; further details on this are given by Joseph (1976). The Euler equations corresponding to the variational problem that determine Ra_E turn out to be mathematically identical to the linearized steady version of Eqs. (6.16)–(6.18). Thus finite amplitude "subcritical instability" is not possible, and the criterion $\mathrm{Ra}=\mathrm{Ra}_c$ provides not only a sufficient condition for instability but also a necessary one.

We also note that the total nondimensional mean temperature gradient $\partial T_{total}/\partial z$ is given by

$$\frac{\partial \bar{T}_{total}}{\partial z} = -1 + \overline{wT} - \langle wT \rangle \tag{6.41}$$

and that it is related to the conduction-referenced Nusselt number,

$$\mathrm{Nu} = -\left| \frac{\partial \bar{T}_{total}}{\partial z} \right|_{z=0}. \tag{6.42}$$

We also note that the effect of convection is to increase the temperature gradient near each boundary and decrease it in the remainder of the layer.

From Eq. (6.34) in the steady case we have

$$\int_0^1 \overline{wT} \frac{\partial \overline{T}_{total}}{\partial z} dz = \langle T \nabla^2 T \rangle$$

and after using Eq. (6.41),

$$\langle wT \rangle + \langle T \nabla^2 T \rangle = \int_0^1 (\overline{wT})^2 dz - \langle wT \rangle^2. \tag{6.43}$$

If we now substitute for w and T the solutions of the linearized equations, we obtain an expression for the amplitude A of the disturbances corresponding to the jth mode,

$$A = (\mathrm{Ra} - \mathrm{Ra}_{cj})^{1/2}. \tag{6.44}$$

At the same time we can compute the Nusselt number from Eq. (6.39).

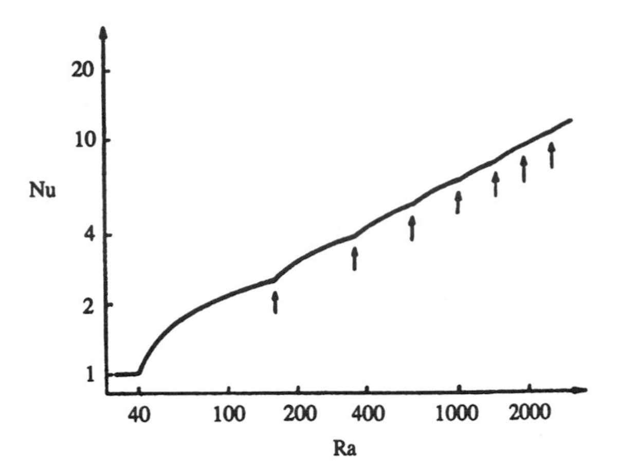

Figure 6.2. The theoretical relationship Nu(Ra) given by Eq. (6.45) (Bories, 1987, with permission from Kluwer Academic Publishers).

If we assume that the various modes contribute independently to the Nusselt number, we then obtain

$$\mathrm{Nu} = 1 + \sum_{j=1}^{\infty} k_j \left(1 - \frac{\mathrm{Ra}_{cj}}{\mathrm{Ra}}\right), \tag{6.45}$$

where $\mathrm{Ra}_{cj} = 4j^2\pi^2$, $k_j = 2$ for $\mathrm{Ra} > \mathrm{Ra}_{cj}$, and $k_j = 0$ for $\mathrm{Ra} < \mathrm{Ra}_{cj}$, for the case of two-dimensional rolls. As Rudraiah and Srimani (1980) showed, other plan forms lead to smaller values of k_1, and hence may be expected to be less favored at slightly supercritical Rayleigh numbers. The Nusselt number relationship (6.45) is plotted in Fig. 6.2. It predicts values of Nu that generally are lower than those observed. It leads to the asymptotic relationship $\mathrm{Nu} \to (2/3\pi)\,\mathrm{Ra}^{1/2}$ as $\mathrm{Ra} \to \infty$ (Nield, 1987b). Similar results for the case of constant flux boundaries rather than isothermal boundaries were obtained by Salt (1988). As expected, this change leads to an increase in Nu, the change becoming smaller as Ra increases (because more and more modes then contribute).

Expression (6.45) may be compared with the result of Palm *et al.* (1972), who performed a perturbation expansion in powers of a perturbation parameter ζ defined by

$$\zeta = \left(1 - \frac{\mathrm{Ra}_c}{\mathrm{Ra}}\right)^{1/2}.$$

Their sixth-order result is

$$\mathrm{Nu} = 1 + 2\lambda\left[\zeta^2 + \left(1 - \frac{17}{24}\lambda\right)\zeta^4 + \left(1 - \frac{17}{24}\lambda + \frac{191}{288}\lambda^2\right)\zeta^6\right], \tag{6.46}$$

where $\lambda = (1 - \zeta^6)^{-1}$. Equation (6.46) predicts well the observed heat transfer for $\mathrm{Ra}/\mathrm{Ra}_c < 5$.

Using a variational formulation based on the Malkus hypothesis that the physical realizable solution is the one that maximizes the heat transport (see also Section 6.24), Busse and Joseph (1972) and Gupta and Joseph (1973) obtained upper bounds on Nu. These were found to be in good agreement with the experimental data of Combarnous and Le Fur (1969) and Buretta and Berman (1976) for Ra values up to 500 (see Section 6.9). Further work on bounds on heat transport was reported by Doering and Constantin (1998) and Vitanov (2000).

An expansion in powers of $(\text{Ra} - \text{Ra}_c)^{1/2}$ to order 34 was carried out by Grundmann and Mojtabi (1995) and Grundmann *et al.* (1996). They thus computed with great precision the values of Nu at a few values of Ra.

6.4. Weak Nonlinear Theory: Further Results

We briefly outline the perturbation approach that is applicable to convection in both clear fluids and in porous media. It has been presented in detail by Busse (1985). The analysis starts with the series expansions

$$\mathbf{v} = \varepsilon[\mathbf{v}^{(0)} + \varepsilon\mathbf{v}^{(1)} + \varepsilon^2\mathbf{v}^{(2)} + \ldots], \tag{6.47}$$

$$\text{Ra} = \text{Ra}_c + \varepsilon\text{Ra}^{(1)} + \varepsilon^2\text{Ra}^{(2)} + \ldots, \tag{6.48}$$

and analogous expressions for T and P, and involves the successive solutions of linear equations corresponding to each power of ε. These expressions are substituted into Eqs. (6.16)–(6.18). Since only steady solutions are examined, the $\partial/\partial t$ terms vanish and in the order ε^1 problem we have the same equations as for the linear problem treated in Section 6.2. The general solution to that problem is expressed as

$$w^{(0)} = f(z, \alpha)\sum_n c_n \exp{(i\mathbf{k}_n \cdot \mathbf{r})}, \tag{6.49}$$

where $\mathbf{r}$ is the position vector and the horizontal wavenumber vectors $\mathbf{k}_n$ satisfy $|\mathbf{k}_n| = \alpha$ for all n.

In the order ε^2 and higher-order problems, inhomogeneous linear equations arise, and the solvability condition determines the coefficients $\text{Ra}^{(n)}$ and provides constraints on the choice of coefficients c_n. In this fashion possible solutions, representing two-dimensional rolls and hexagons, are determined. There still exist many such solutions. The stability of each of these is examined by superposing arbitrary infinitesimal disturbances $\tilde{\mathbf{v}}$, $\tilde{T}$ on the steady solution $\mathbf{v}$, T. By subtracting the steady equations from the equations for $\mathbf{v} + \tilde{\mathbf{v}}$, $T + \tilde{T}$, the following stability problem is obtained:

$$\sigma\gamma_a\mathbf{v} = -\nabla\tilde{P} + \text{Ra}\tilde{T}\mathbf{k} - \tilde{\mathbf{v}}, \tag{6.50a}$$

$$\tilde{\sigma}\tilde{T} + \tilde{\mathbf{v}}\cdot\nabla T + \mathbf{v}\cdot\tilde{T} = \tilde{w} + \nabla^2\tilde{T} \tag{6.50b}$$

$$\nabla\cdot\tilde{\mathbf{v}} = 0, \tag{6.50c}$$

$$\tilde{w} = \tilde{T} = 0 \quad \text{at} \quad \hat{z} = 0.1. \tag{6.50d}$$

These equations are based on the observation that since the stability problem is linear, the time dependence can be assumed to be of the form $\exp(\tilde{\sigma} t)$. The steady solution is unstable when an eigenvalue $\tilde{\sigma}$ with a positive real part exists.

The eigenvalue problem (6.50) can be solved by expanding $\tilde{v}$, $\tilde{T}$, and $\tilde{\sigma}$ as power series in ε analogous to Eq. (6.47). By considering coefficients up to $\tilde{\sigma}^{(2)}$ in the series for $\tilde{\sigma}$, one can demonstrate that all steady solutions are unstable with the exception of two-dimensional rolls. Moreover, it is found that at small but finite values of $\mathrm{Ra} - \mathrm{Ra}_c$ rolls corresponding to a finite range of wave-numbers are stable.

The main conclusion to be drawn from such results is that a spectrum of different steady convection modes is physically realizable and the asymptotic state of a convection layer in general will depend on the initial conditions.

Although two-dimensional rolls are favored when the physical problem has vertical symmetry about the midplane, it is found that hexagons are favored when there is a significant amount of asymmetry, whether it is due to different boundary conditions at top and bottom or due to property variations with temperature or other heterogeneities. Hexagons also are favored when the basic temperature profile is not linear, as when convection is produced by a volume distribution of heat sources rather than by heating from below. Two-dimensional rolls rarely have been observed in experiments on Rogers-Horton-Lapwood convection, even in circumstances when they might have been expected [as in one experiment reported by Lister (1990)].

The direction of motion in a hexagonal cell is influenced by property variations. Other things being equal, motion at the center of a cell is in the direction of increasing kinematic viscosity. In liquids the kinematic viscosity decreases as the temperature increases, so the liquid rises in the center of a cell. In gases the reverse is the case, so gas sinks in the center of a cell. Further reading on this is provided by Joseph (1976, p. 112).

We conclude this section with the results of a study of the stability of convection rolls to three-dimensional disturbances made by Joseph and Nield and reported in Joseph (1976, Chapter XI). The various types of possible disturbances are graphically labeled as parallel rolls, cross-rolls, sinuous (or zig-zag) rolls, and varicose rolls. Joseph and Nield found that the sinuous rolls and the cross-rolls are the ones that effectively restrict the range $\alpha_1(\varepsilon) < \alpha(\varepsilon) < \alpha_2(\varepsilon)$ for which the convection rolls of wavenumber $\alpha(\varepsilon)$ are stable. For the case of impermeable conducting boundaries, the stability boundary for cross-rolls in the neighborhood of the critical point $(\alpha_c, \mathrm{Ra}_c)$, where $\alpha_c = \pi$, $\mathrm{Ra}_c = 4\pi^2$, is given by

$$\frac{\mathrm{Ra}}{\mathrm{Ra}_c} - 1 = \frac{10}{3}\left(\frac{\alpha}{\alpha_c} - 1\right)^2 \tag{6.51}$$

and that for the sinuous rolls is given by

$$\frac{\mathrm{Ra}}{\mathrm{Ra}_c} - 1 = \frac{12}{19^{1/2}}\left(1 - \frac{\alpha}{\alpha_c}\right)^{1/2}, \quad \alpha < \alpha_c\,. \tag{6.52}$$

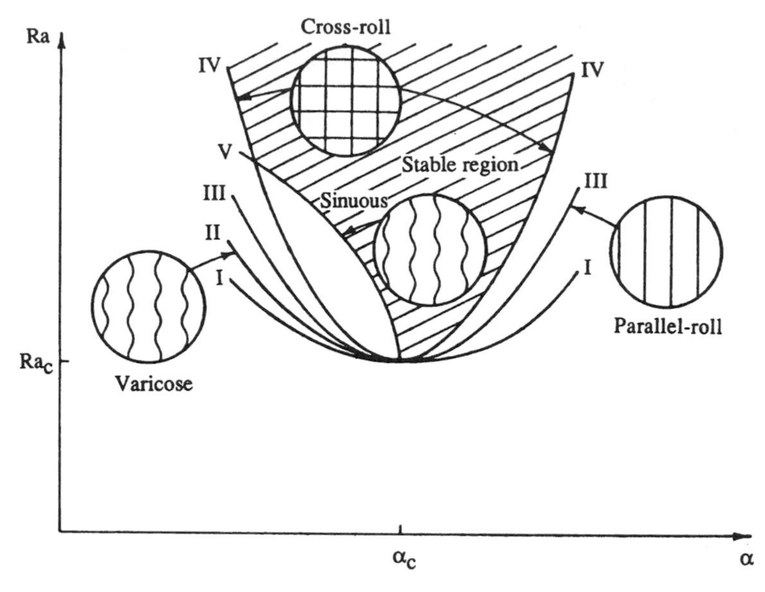

Figure 6.3. Sketch of parabolic approximations of (α, Ra) projections of nonlinear neutral curves for roll convection, valid in a neighborhood of (α_c, Ra_c). The neutral curve for the rest state with a constant temperature gradient is shown as I. Curves II, III, IV, and V are nonlinear neutral curves for different convection disturbances. Sinuous and varicose instabilities occur only when $\alpha < \alpha_c$ (Joseph, 1976, with permission from Springer Verlag).

For comparison, the neutral curve for the basic conduction solution is

$$\frac{Ra}{Ra_c} - 1 = \left(\frac{\alpha}{\alpha_c} - 1\right)^2. \tag{6.53}$$

Equation (6.52) determines the lower limit of the range of wavenumbers for stable rolls and Eq. (6.51) the upper limit for Ra values near Ra_c (Fig. 6.3).

For larger values of Ra numerical calculations are necessary to determine the range of wavenumbers for stability. In this way Straus (1974) calculated a balloon-shaped curve in the (α, Ra) plane. The points situated inside the balloon correspond to stable rolls (Fig. 6.4).

The stability of two-dimensional convection has been analyzed further by De la Torre Juárez and Busse (1995) for Ra values up to 20 times the critical. Some of their results are displayed in Figures 6.5–6.7. In Fig. 6.5, the Nusselt number is plotted against Ra for fixed $\alpha = \alpha_c$. At Ra $= 391 \pm 1$ the steady solution becomes unstable and is replaced by an oscillatory solution with a higher Nusselt number; the frequency also is given in the figure. At Ra $= 545$ this even solution becomes unstable. For a given Rayleigh number, the Nusselt number varies with the wavenumber as shown in Fig. 6.6. The results of stability analysis are shown in Fig. 6.7. This figure shows that there is an oscillatory instability predicted for small

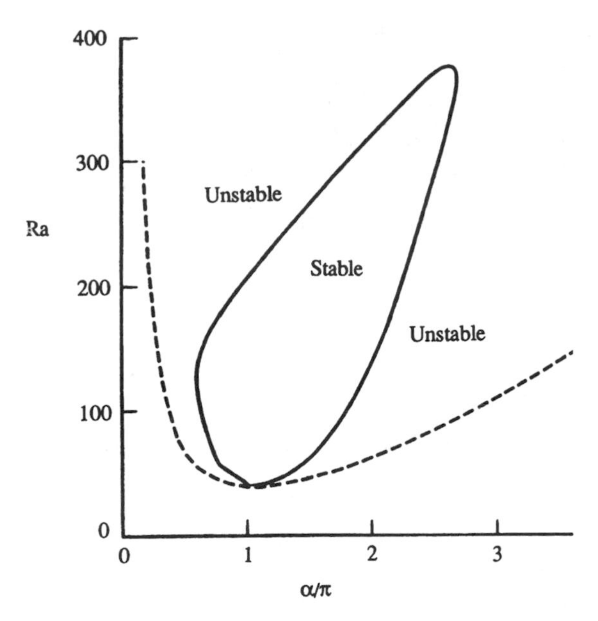

Figure 6.4. Regions of stable and unstable two-dimensional rolls. The dashed line is the neutral stability curve obtained from the linear stability analysis (Straus, 1974, with permission from Cambridge University Press).

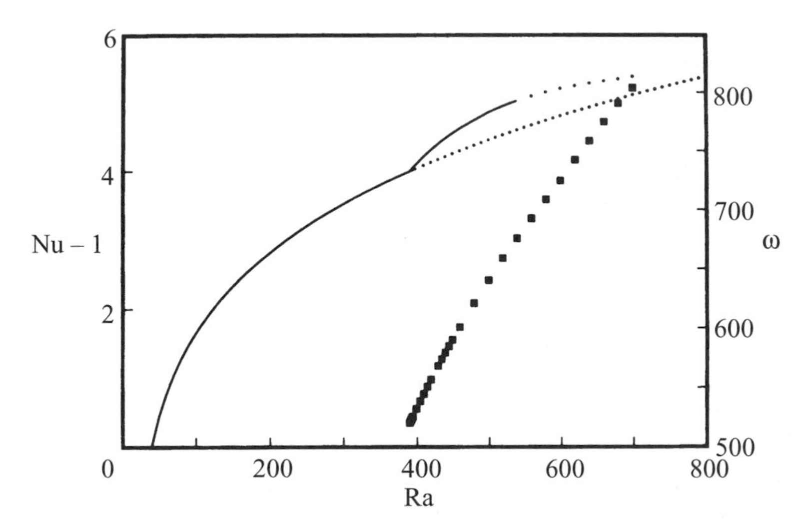

Figure 6.5. Average value of the Nusselt number of the steady and oscillatory solutions as a function of the Rayleigh number for a fixed wavenumber $\alpha = \alpha_c$. The unstable stationary solutions are represented by dots. The frequency of the oscillatory solutions is denoted by squares (De la Torre Juárez and Busse, 1995, with permission from Cambridge University Press).

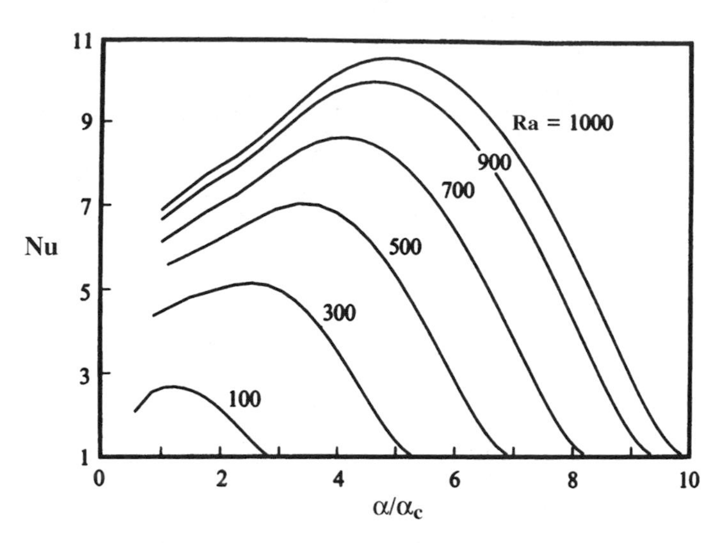

Figure 6.6. Value of the Nusselt number of the steady solutions as a function of the wavenumber α for different values of the Rayleigh number (De la Torre Juárez and Busse, 1995, with permission from Cambridge University Press).

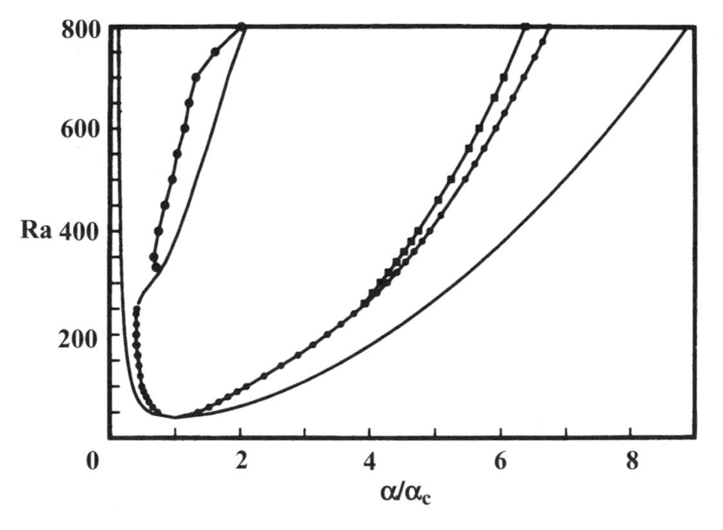

Figure 6.7. Regions of stability of the steady solutions as a function of the Rayleigh number and the wavenumber. The neutral curve is the outer solid line; the Eckhaus instability is plotted as a line with solid circles; the transitions to the different oscillatory instabilities are plotted as a solid line with squares at high wavenumbers and as a solid line at low wavenumbers; the stability limits of the stationary oscillatory solutions are plotted as a solid line with open circles (De la Torre Juárez and Busse, 1995, with permission from Cambridge University Press).

wavenumbers $\alpha \sim \alpha_c$. This oscillatory state has been observed in experiments with Hele-Shaw cells.

De la Torre Juàrez and Busse also carried out direct numerical integrations in time of the solutions in the unstable regions. They found that the Eckhaus instability limiting the band of stable wavenumbers at low supercritical Rayleigh numbers is replaced by a sideband instability corresponding to odd-parity perturbations as the Rayleigh number increases. This instability leads to a 3:1 jump in the wavelength. A third instability of oscillatory character occurs at high wavenumbers, which is also related to a 3:1 resonance mechanism and tends to change the wavelength by a finite amount. The fourth instability yields an oscillatory state of even parity for low wavenumbers and for Rayleigh numbers above Ra = 218. In the region where even oscillatory solutions exist, they lose stability through the growth of odd oscillatory modes. In one case the odd modes grow while the existing even oscillatory solution persists, yielding a noncentrosymmetric state with several temporal frequencies. In a second case, occurring at Ra above 790, steady convection bifurcates into a regular oscillatory state where the odd modes dominate the even modes; this is related to an asymmetry between the rising hot and the falling cold plumes.

Nisse and Néel (2005) have investigated the stability of rolls with intermediate wavelength (those not unstable to the cross-roll, Eckhaus and zig-zag instabilities). They proved that such rolls are spectrally stable.

The effect of quadratic drag was studied by Rees (1996b). He found that rolls with a wavenumber less than the critical value are no longer unconditionally unstable. Also the Eckhaus (parallel-roll) and zigzag (sinuous) stability bounds are less restrictive than in the absence of quadratic drag, but the opposite is true for the cross-roll instability.

The results discussed so far in this section have been based on the assumption that the porous medium is bounded by impermeable isothermal (perfectly conducting) planes. Riahi (1983) has shown that when the boundaries have finite thermal conductivity, the convection phenomenon is different. He found that cells of square plan form are preferred in a bounded region Γ of the (λ_b, λ_t) space, where λ_b and λ_t are the ratios of the thermal conductivities of the lower and upper boundaries to that of the fluid and two-dimensional rolls are favored only outside Γ.

For the case of uniform heat-flux on the boundaries, Néel and Lyubimov (1995) have proved the existence of periodic solutions for a class of nonlinear regular vector fields.

The results in this section bear on the choice of wavenumber to use in numerical simulations. Since the theory does lead to a unique value and since the Malkus hypothesis (that the selected wavenumber is that which maximizes heat transfer) is now known to be unsatisfactory, Nield (1997b) has suggested that in most cases it is probably satisfactory to take $\alpha = \alpha_c$ in the simulations..

Adomian's decomposition method and weak nonlinear theory were compared by Vadasz (1999a), who explained the experimental observation of hysteresis from steady convection to chaos to steady state (see also Auriault, 1999; Vadasz, 1999b). The Adomian method was further used by Vadasz and Olek (1999a, 2000a) to discuss convection for low and moderate Prandtl number, and its application to the solution of the Lorenz equations was investigated by Vadasz and Olek (2000b).

Weak turbulence in small and moderate Prandtl number convection was reviewed and elucidated by Vadasz (2003). The computational recovery of the homoclinic orbit was discussed by Vadasz and Olek (1999b), while the compatibility of analytical and computational solutions was discussed by Vadasz (2001b). The question of whether the transitions involved in porous media natural convection could be smooth was examined by Vadasz *et al.* (2005). The results of their examination suggest that the transitions inevitably are sudden. A comprehensive review of the subject of weak turbulence and transitions to chaos was made by Vadasz (2000b).

An unconditional stability result for the case of a cubic dependence of density on temperature, with the Forchheimer equation, was obtained by Carr (2003). Further work on oscillatory convection regimes was reported by Holzbecker (2001), while Holzbecher (2004b) treated a mixed boundary condition appropriate for open-top enclosures. He noted that at 16.51 the critical Rayleigh number is then much lower than the classical value. Holzbecher (2005a) studied both free and forced convection for open-top enclosures. Cosymmetric families of steady states and their collision were investigated by Karasozen and Tsybulin (2004).

A review of some aspects of nonlinear convection was made by Rudraiah *et al.* (2003). A comprehensive review of other matters, including methods for calculating eigenvalues, is contained in the book by Straughan (2004b).

6.5. Effects of Solid-Fluid Heat Transfer

At sufficiently large Rayleigh numbers, and hence sufficiently large velocities, one can expect that local thermal equilibrium will break down, so that the temperatures T_s and T_f in the solid and fluid phases are no longer identical. Instead of a single energy equation (2.3) or (6.5) one must revert to the pair of equations (2.1) and (2.2). Following Bories (1987), we consider the case of constant conductivities k_s and k_f and no heat sources, but we modify Eqs. (2.1) and (2.2) by allowing for heat transfer between the two phases. Accordingly we have

$$(1-\varphi)(\rho c)_s \frac{\partial T_s^*}{\partial t^*} = k_{es} \nabla^{*^2} T_s^* - h(T_s^* - T_f^*), \tag{6.54}$$

$$\varphi(\rho c_P)_f \frac{\partial T_f^*}{\partial t^*} + (\rho c_P)_f \mathbf{v}^* \cdot \nabla^* T_f^* = k_{ef} \nabla^{*^2} T_f^* - h(T_f^* - T_s^*). \tag{6.55}$$

In these equations asterisks denote dimensional quantities and h is a heat transfer coefficient, while k_{es} and k_{ef} are effective conductivities. In the purely thermal conduction limit $k_{es} = (1-\varphi)k_s$ and $k_{ef} = \varphi k_f$. Equations (6.3), (6.4), and (6.6) still stand. We choose H for length scale, $(\rho c)_m H^2/k_m$ for time scale, $k_m\ /(\rho c_P)_f H$ for velocity scale, ΔT for temperature scale, and $\mu k_m / K(\rho c_P)_m$ for pressure scale. Then Eqs. (6.54) and (6.55) take nondimensional forms:

$$(1-\varphi M)(1+\Lambda)\frac{\partial T_s}{\partial t} = \nabla^2 T_s - \Lambda\chi(T_s - T_f), \tag{6.56}$$

$$\varphi M(1+\Lambda^{-1})\frac{\partial T_f}{\partial t} + (1+\Lambda^{-1})\mathbf{v}\cdot\nabla T_f = \nabla^2 T_f - \chi(T_f - T_s), \tag{6.57a}$$

where

$$M = \frac{(\rho c_P)_f}{(\rho c)_m}, \quad \Lambda = \frac{k_{ef}}{k_{es}}, \quad \chi = \frac{hH^2}{\varphi k_f}. \tag{6.57b}$$

Combarnous (1972) calculated the Nusselt number Nu as a function of Ra, Λ, and χ. He found that for a given value of Λ, Nu is an increasing value of χ which tends, when $h \to \infty$, toward the value given in the local equilibrium model. This trend is expected, because the limit corresponds to perfect transfer between solid and fluid phase.

When the parameter χ defined in Eq. (6.58) is maintained constant, Nu tends toward the local equilibrium value as Λ increases, i.e., as the contribution of heat conduction by the solid phase becomes negligible. When heat conduction through the solid phase becomes very large, the Nusselt number decreases; in fact, Nu $\to$ 1 as $\Lambda \to 0$.

The computed temperature distributions show that $|T_s - T_f|$ takes relatively large values in the upper part of the upward current and the lower part of the downward current. This illustrates the role of the solid phase as a heat exchanger. Another point follows from the fact that χ is the product of a local heat transfer factor $hd_p^2/\varphi k_f$ and $(H/d_p)^2$, where d_p is the pore scale. When the scale factor H/d_p is large, the porous medium behaves as a thorough blend of solid and fluid phases. When it is small, the porous medium is effectively more heterogeneous.

Banu and Rees (2002) demonstrated that both the critical Rayleigh number and the wavenumber are modified by thermal nonequilibrium. For intermediate values of the interphase heat transfer coefficient, the critical wavenumber is always greater than π, the classic value. Postelnicu and Rees (2003) incorporated from drag and boundary effects. For the case of stress-free boundaries they obtained the expression

$$\mathrm{Ra} = \frac{(\pi^2+\alpha^2)^2}{\alpha^2}\left[1+\mathrm{Da}(\pi^2+\alpha^2)\right]\left[\frac{(\pi^2+\alpha^2)+\chi(1+\gamma)}{(\pi^2+\alpha^2+\gamma\chi)}\right], \tag{6.58}$$

where $\gamma = \phi\Lambda(1-\phi)$. The critical Rayleigh number is obtained on minimization with respect to variation of α. Clearly $\mathrm{Ra_c}$ is an increasing function of Da, and for $\mathrm{Da} = 0$ it is an increasing function of χ from the base value $4\pi^2$ with the amount of increase decreasing as γ increases.

Boundary effects were also considered by Malashetty *et al.* (2005b). The situation where there is heat generation in the solid phase in a square enclosure was studied numerically by Baytas (2003, 2004). An anisotropic layer was considered by Malashetty *et al.* (2005a).

6.6. Non-Darcy, Dispersion, and Viscous Dissipation Effects

Corresponding to the Darcy equation (6.17), the linear Brinkman equation is

$$\gamma_a \frac{\partial \mathbf{v}}{\partial t} = -\nabla P - \mathbf{v} + \tilde{\mathrm{D}}\mathrm{a}\nabla^2\mathbf{v} + \mathrm{Ra}\,\theta\mathbf{k}. \tag{6.59}$$

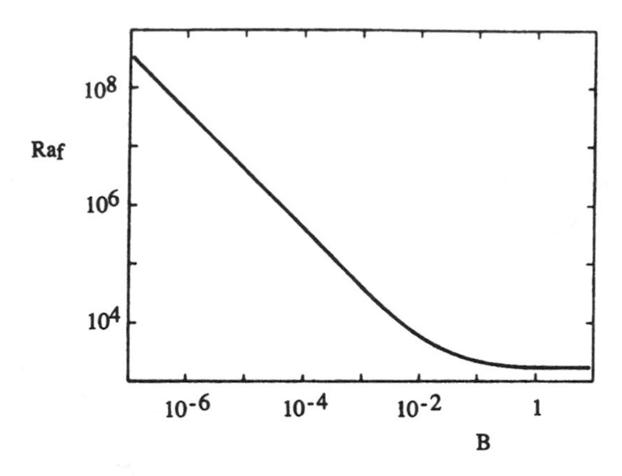

Figure 6.8. Critical fluid Rayleigh number $\mathrm{Ra}_f = \mathrm{Ra}H^2/K$ versus the Brinkman coefficient $B = (\tilde{\mu}/\mu)K/H^2 = (\tilde{\mu}/\mu)\,\mathrm{Da} = \tilde{\mathrm{D}}\mathrm{a}$. The figure illustrates the transition from the Darcy limit to the clear fluid limit (Walker and Homsy, 1977).

Here $\tilde{\mathrm{D}}\mathrm{a}$ is a Darcy number defined by

$$\tilde{\mathrm{D}}\mathrm{a} = \frac{\tilde{\mu}}{\mu}\frac{K}{H^2} = \frac{\tilde{\mu}}{\mu}\frac{K}{d_p^2}\left(\frac{d_p}{H}\right)^2 \tag{6.60}$$

and d_p is a characteristic length on the pore scale. From the Carman-Kozeny equation (1.5) we see that K/d_p^2 is of the order of unity unless φ is close to 1. Also $\tilde{\mu}/\mu$ is of the order of unity, while d_p/H is small if the porous medium is properly represented by a continuum. It follows that B is normally very small, and thus the Brinkman term is important only in boundary layers where $\nabla^2\mathbf{v}$ is large. In conclusion, in naturally occurring media the net effect of the Brinkman term is to alter the critical Rayleigh number by a small amount. An apparent exception to this statement was reported by Lebon and Cloot (1986); they failed to distinguish between a constant-pressure boundary and a stress-free boundary. Detailed calculations are given by Walker and Homsy (1977), and Fig. 6.8. The Darcy result holds if the Darcy number $\mathrm{Da} = K/H^2 < 10^{-3}$. For $\mathrm{Da} > 10$, $\mathrm{Ra} \sim 1708\,\mathrm{Da}$, the clear fluid limit. Rees (2002b) performed a perturbation analysis for small Darcy number (defined to include the viscosity ratio) and obtained the approximation

$$\mathrm{Ra}_c = 4\pi^2 + 8\pi^2\mathrm{Da}^{1/2} + [8\pi^4 + 12\pi^2 + 4\pi^3 3^{1/2}\tanh(3^{1/2}\pi/2)]\mathrm{Da}, \tag{6.60a}$$

$$\mathrm{a}_c = \pi + \pi\mathrm{Da}^{1/2} \tag{6.60b}$$

The Forchheimer equation that replaces Eq. (6.17) is

$$\gamma_a \frac{\partial \mathbf{v}}{\partial t} = -\nabla P - \mathbf{v} - F\,|\mathbf{v}|\,\mathbf{v} + \mathrm{Ra}\,\theta\,\mathbf{k}, \tag{6.61}$$

where F is a Forchheimer coefficient defined by

$$F = \frac{c_F \rho_f K^{1/2} \alpha_m}{\mu H} Q = \frac{c_F}{\mathrm{Pr}_f}\frac{k_m}{k_f}\left(\frac{K}{H^2}\right)^{1/2} Q. \tag{6.62}$$

Table 6.2. Approximate values showing the dependence of Nusselt number Nu and nondimensional r.m.s. velocity $\overline{Q}$ on the Rayleigh number Ra (Nield and Joseph, 1985).

Ra	Nu	$\overline{Q}$
10^2	3	15
3×10^2	6	40
10^3	10	100
3×10^3	14	200
10^4	20	340

In these equations $\mathrm{Pr}_f = \mu/\rho_f \alpha_f$ is the Prandtl number of the fluid and Q is a Péclet number expressing the ratio of a characteristic velocity of the convective motion to the velocity scale α_m/H (with which we are working). In particular, if we take Q to be the r.m.s. average $\bar{Q}$, then we can use information given by Palm *et al.* (1972) to deduce that (Table 6.2):

$$\overline{Q} = [\mathrm{Ra}\,(\mathrm{Nu} - 1)]^{1/2}. \tag{6.63}$$

We can conclude that the Forchheimer term can be significant, even for modest Rayleigh numbers, for thin layers of media for which $\mathrm{Pr}_f\ (k_f/k_m)$ is small. For example, if we take the values $c_F = 0.1$, $\mathrm{K} = 10^{-3}\mathrm{cm}^2$, $H = 1$ cm, which are appropriate for a 1-cm-thick layer of a medium of metallic fibers, and the value Ra = 300 that is typical for a transition to oscillatory convection (Section 6.8), then quadratic drag is significant if $\mathrm{Pr}_f k_f/k_m$ is of order 0.1 or smaller. In other situations rather large Rayleigh numbers are needed before quadratic drag becomes important.

The effect of quadratic drag was shown by Nield and Joseph (1985) to cause the nose of the bifurcation curve in the (Ra, ε) plane to be sharpened; the standard pitchfork bifurcation is modified to straight lines intercepting the zero amplitude axis. Here ε is a measure of the amplitude of the disturbance. He and Georgiadis (1990) confirmed the sharpening. Rees (1996b) undertook a third-order analysis that showed that at higher Rayleigh numbers the usual square root behavior is restored. He also developed a full weakly nonlinear stability analysis and found that inertia causes some wavenumbers less than the critical value to regain stability, but the cross-roll instability is more effective and reduces the stable wavenumber range. The effect of quadratic drag on higher-order transitions was studied numerically by Strange and Rees (1996). They expressed their results in terms of a parameter $G = F/Q$. They found that at Rayleigh numbers below a second critical value a steady cellular pattern exists, but the amplitude of the motion and the corresponding rate of heat transfer decrease sharply as G increases. At the second critical Rayleigh number, whose value increases almost linearly with G, the preferred mode of convection is time periodic. The mechanism of Kimura *et al.* (1986), where waves orbit each cell, also applies when quadratic drag is present.

Néel (1998) considered how a horizontal pressure gradient affects convection in the presence of inertia and boundary friction effects. Her formulation leads to a cubic (rather than quadratic) drag term, and she found that this inertial effect leads to an increase in the critical Rayleigh number.

We saw in Section 2.2.3 that the effect of thermal dispersion was to increase the effective conductivity of the porous medium. Instead of Eq. (6.18) we now have

$$\frac{\partial T}{\partial t} - \hat{w} = \nabla \cdot [(1 + D^*)\nabla \hat{T}]. \tag{6.64}$$

where D^* is the ratio of dispersive to stagnant conductivity. According to the model for a packed bed of beads adopted by Georgiadis and Catton (1988), $D^* = \mathrm{Di}\,|\mathbf{v}|$, where

$$\mathrm{Di} = \frac{C d_b}{(1 - \varphi)H}. \tag{6.65}$$

Here d_b is the mean bead diameter and C is a dispersion coefficient whose value depends on the type of packing. Georgiadis and Catton performed calculations with the value $C = 0.36$, which was chosen to give the best fit to experimental data.

Since the term $D^*\nabla T\ Di\,|\mathbf{v}|\,\nabla T$ is of second order, it is clear that dispersion does not affect the critical Rayleigh number, but it does have nonlinear effects that decrease the overall Nusselt number significantly for coarse materials (Neichloss and Degan, 1975). Kvernvold and Tyvand (1980) showed that dispersion expands the stability balloon of Straus (1974) (Fig. 6.4), i.e., it causes two-dimensional rolls to remain stable to cross-roll instabilities for Rayleigh numbers larger than those in the absence of dispersion.

The effect of viscous dissipation and inertia on hexagonal cell formation was studied by Magyari *et al.* (2005b). They show that when viscous dissipation is present, the temperature profile loses its up/down symmetry when convection occurs, and this causes hexagonal cells rather than parallel rolls to occur in the case of a layer of infinite horizontal extent. This is because the lack of symmetry allows two rolls, whose axes are at 60° to one another, to interact and reinforce a roll at 60° to each of them, thus providing the hexagonal pattern. Hexagonal is subcritical, i.e., it appears at Rayleigh numbers below $4\pi^2$. However, when Ra is sufficiently above $4\pi^2$ the rolls are reestablished as the preferred pattern of convection. When the Forchheimer terms are included, the range of Rayleigh numbers over which hexagons exist and are stable decreases and the hexagons are eventually extinguished. This result is qualitatively similar to that resulting when the layer is tilted at increasing angles from the horizontal, although there are two main orientations of hexagonal solutions in this case. The rolls that form when hexagons are destabilized are longitudinal rolls that may be regarded as streamwise vortices like those considered by Rees *et al.* (2005a).

6.7. Non-Boussinesq Effects

So far we have neglected the work done by pressure changes. When we allow for this, we replace Eq. (6.5) by

$$(\rho c)_m \frac{\partial T}{\partial t} + (\rho c_P)_f \mathbf{v} \cdot \nabla T + \beta T \left(\frac{\partial P}{\partial t} + \mathbf{v} \cdot \nabla P \right) = k_m \nabla^2 T, \tag{6.66}$$

where the coefficient of thermal expansion β and isothermal compressibility β_P are given by

$$\beta = -\frac{1}{\rho}\left(\frac{\partial \rho}{\partial T}\right)_P, \tag{6.67}$$

$$\beta_P = \frac{1}{\rho}\left(\frac{\partial \rho}{\partial P}\right)_T. \tag{6.68}$$

The basic steady-state solution is given by the hydrostatic equations

$$\mathbf{v}_b = 0, \quad T_b = T_0 + \Delta T\left(1 - \frac{z}{H}\right), \quad \frac{dP_b}{dz} = -\rho_b g, \tag{6.69a,b,c}$$

$$\frac{d\rho_b}{dz} = \beta_{Pb}\rho_b\frac{dP_b}{dz} - \beta_b\rho_b\frac{dT_b}{dz} = -\beta_{Pb}\rho_b^2 g + \beta_b\rho_b\frac{\Delta T}{H}. \tag{6.69d}$$

The two-dimensional linearized time-independent perturbation equations are

$$\frac{\partial u'}{\partial x} + \frac{\partial w'}{\partial z} + w'\left(\beta_b\frac{\Delta T}{H} - \beta_{Pb}\rho_b g\right) = 0, \tag{6.70}$$

$$\rho_b w'\left(-c_{Pb}\frac{\Delta T}{H} + \beta_b T_b g\right) = k_m\left(\frac{\partial^2 T'}{\partial x^2} + \frac{\partial^2 T'}{\partial z^2}\right). \tag{6.71}$$

$$\frac{\partial P'}{\partial x} + \mu_b\frac{u'}{K} = 0, \tag{6.72}$$

$$\frac{\partial P'}{\partial z} + \mu_b\frac{w'}{K} = -\rho' g, \tag{6.73}$$

$$\rho' = \beta_{Pb}\rho_b P' - \beta_b\rho_b T'. \tag{6.74}$$

In the Boussinesq approximation the term $-\beta_b\rho_b T'$ in the equation for ρ' is retained, but otherwise β_b and β_{Pb} are set equal to zero, while ρ_b, c_{Pb}, T_b, and μ_b are regarded as constants. As a second approximation, one also can retain the term $\beta_b T_b g$ in Eq. (6.71), the left-hand side of which can be written as $\rho_b c_{Pb}(\beta_b T_b g/c_{Pb} - \Delta T/H)w'$.

The end result is that the critical Rayleigh number value is the same as before, provided that in the definition of Rayleigh number one replaces the applied temperature gradient $\Delta T/H$ by the difference between that and the adiabatic gradient $\beta_b T_b g/c_{Pb}$. Thus the prime effect of compressibility is stabilizing, and the other non-Boussinesq effects have only a comparatively minor effect on the critical Rayleigh number. Details for the case when the fluid is water are discussed by Straus and Schubert (1977) and for the case of an ideal gas by Nield (1982). For a moist ideal gas of 100% humidity, flow and heat transfer are strongly coupled. Zhang *et al.* (1994), using a perturbation analysis, showed that Ra_c then depends heavily on the vapor pressure; the moist gas is much less stable than a dry gas, because of the large latent heat carried by the former. A rarefied gas was considered by Parthiban and Patil (1996), but their analysis is flawed (see Nield, 2001c). A finite amplitude analysis was reported by Stauffer *et al.* (1997). The impact of thermal expansion on transient convection was studied by Vadasz (2001c,d).

It is usually a straightforward adjustment to allow for the variations of fluid properties with temperature. This is exemplified by the numerical investigations of Gartling and Hickox (1985). The effect of viscosity variation was explicitly examined by Blythe and Simpkins (1981) and Patil and Vaidyanathan (1981). Morland *et al.* (1977) examined variable property effects in an elastic porous matrix. Nonlinear stability analysis for the case of temperature-dependent viscosity was reported by Richardson and Straughan (1993) and Qin and Chadam (1996) incorporating Brinkman and inertial terms, respectively. Payne and Straughan (2000b) addressed the Forchheimer equation and obtained unconditional nonlinear stability bounds close to the linear stability ones using a viscosity linear in the temperature. They also extended the analysis to a viscosity quadratic in temperature and to a penetrative convection situation. For the Forchheimer model nonlinear stability was analyzed using Lyapunov's direct method by Capone (2001).

Nield (1996) showed that the effect of temperature-dependent viscosity on the onset of convection was well taken into account provided the Rayleigh number was defined in terms of the viscosity at the average temperature. This result is in accord with concept of effective Rayleigh number Ra_{eff} introduced by Nield (1994c); for this parameter the quantities appearing in the numerator of Ra are replaced by their arithmetic mean values and those that appear in the denominator are replaced by their harmonic mean values. A detailed theoretical and numerical study of the effect of temperature-dependent viscosity was reported by Lin *et al.* (2003).

6.8. Finite-Amplitude Convection: Numerical Computation and Higher-Order Transitions

Starting with Holst and Aziz (1972), the governing equations for natural convection have been solved using a range of numerical techniques (finite differences, finite element, spectral method). Out of necessity, these calculations must be made in a finite domain, so a preliminary decision must be made about conditions on lateral boundaries. It is presumed that these vertical boundaries are placed to coincide at the cell boundaries, where the normal (i.e., horizontal) component of velocity and the normal component of heat flux are both zero.

Caltagirone *et al.* (1981) performed calculations using the spectral method and obtained the following results:

(a) For $\mathrm{Ra} < 4\pi^2$, the perturbation induced by initial conditions decreased and the system tended to the pure conduction solution, as expected.
(b) For $4\pi^2 < \mathrm{Ra} < 240$ to 300 the initial perturbation developed to give a stable convergent solution that does not depend on the intensity or nature of this perturbation. Various stable convective rolls were observed: counterrotating rolls (two-dimension), superposition of counterrotating rolls (three-dimension), and polyhedral cells (three-dimension).
(c) For $\mathrm{Ra} > 240$ to 300 a stable regime was not reached.

Transition to the fluctuating convection regime is characterized by an increase of heat transfer relative to the stable solutions. The oscillations appear to be caused by the instability of the thermal boundary layers at the horizontal boundaries. The existence of the oscillating state had been deduced from a stability analysis of finite-amplitude two-dimensional solutions by Straus (1974), whose results are illustrated in Fig. 6.3. It was also demonstrated through numerical calculations by Horne and O'Sullivan (1974a). The oscillations have been shown by Caltagirone (1975) and Horne and Caltagirone (1980) to be associated with the continuous creation and disappearance of cells.

Or (1989) has extended the computations to the situation where the viscosity is allowed to be temperature dependent. The vertical asymmetry thereby introduced makes mixed modes significant. Or (1989) also examined stability with respect to a class of disturbances that have a $\pi/2$ phase shift relative to the basic state. He found little difference in transition parameters for the in-phase and phase-shifted oscillatory instabilities. It is noteworthy that the temperature dependence of viscosity provides a mechanism for generating a mean flow.

Further studies of higher-order transitions have been made by Aidun (1987), Aidun and Steen (1987), Kimura *et al.* (1986, 1987), Caltagirone *et al.* (1987), Steen and Aidun (1988), and Caltagirone and Fabrie (1989). The last study, based on a pseudospectral method, concluded that in a two-dimensional square cavity the following sequence occurs: From the second bifurcation, occurring at $\mathrm{Ra} = 390$, the flow becomes periodic. Between 390 and 600 the phenomenon is single-periodic and only the frequency f_2 incommensurable with f_1 introduces a quasiperiodic regime QP_1. When Ra increases further, the flow again becomes periodic (state P_2) up to $\mathrm{Ra} = 1000$, where the appearance of frequencies f_2 and f_3 give a second quasiperiodic regime QP_2.

The second regime QP_2 can be maintained up to $\mathrm{Ra} = 1500$, after which the single convecting roll splits up into two unsteady convecting rolls by entering a chaotic restructuring regime. This sequence is subject to hysteresis as Ra is lowered. The frequency f_1 varies as Ra^2, f_2 as $\mathrm{Ra}^{5/2}$, and f_3 as $\mathrm{Ra}^{3/2}$.

The periodic window between $\mathrm{Ra} = 600$ and 1000 corresponds to third-order locking of the oscillators corresponding to f_1 and f_2. The oscillators spring up and develop within the thermal boundary layer near the horizontal walls, and the evolution with Ra^2 of f_1 corroborates the fact that the observed instabilities are due to the loss of stability in the boundary layer. The earlier study by Kimura *et al.* (1987) revealed a rather different picture; for example, the second quasiperiodic regime was not found, and f_1 varied as $\mathrm{Ra}^{7/8}$. The work of Kladias and Prasad (1990) suggests that when non-Darcy effects are taken into account the second quasiperiodic regime does not exist.

Kladias and Prasad (1989b, 1990) have made numerical studies of oscillatory convection using a Brinkman-Forchheimer equation. They found that whereas the channeling effect (due to porosity variation) substantially reduced the critical Rayleigh number for the onset of steady convection, the opposite occurred with the critical Rayleigh number for the transition to oscillatory convection. This is primarily due to the fact that the core of the cavity becomes more or less stagnant, whereas the thermal activity and fluid motion is concentrated within thin boundary layers along the walls. While the effects of mean porosity and specific heat ratio are insignificant for steady convection, they are quite significant in the random fluctuating regime. In a square cavity steady convection is characterized by a single cell, but the flow pattern for fluctuating convection is complex and dependent on the fluid Prandtl number Pr_f. For example, four cells can exist with pairs on the diagonals alternately attaching and detaching with time. This results in a large variation in Nusselt number with time. Generally an increase in $\mathrm{Pr}_f(> 10)$ increases the amplitude of fluctuation, whereas a decrease in $\mathrm{Pr}_f(< 0.1)$ results in a more stable flow. Otero *et al.* (2004) have studied numerically the case of infinite Darcy-Prandtl number and high Rayleigh number. Their results include a derivation of an upper bound on the heat transport: $\mathrm{Nu} \leq 0.0297\times \mathrm{Ra}$.

For the special case of constant flux imposed on the horizontal boundaries the situation is markedly different. The analytical and numerical study of Kimura *et al.* (1995) revealed that the unicellular set up when Ra exceeds 12 remains a stable mode as the aspect ratio A increases, in contrast to the constant-temperature case where multicellular convection is the preferred mode for $A > 2^{1/2}$. Further, the unicellular flow remains as Ra increases to 311.53, above which nonoscillatory longitudinal disturbances can grow. At sufficiently large Ra (above about 640 for $A = 8$, with a critical frequency $f = 22.7$) there is a transition to oscillatory flow, according to the numerical calculations; linear stability theory predicts a Hopf bifurcation with transverse disturbances at $\mathrm{Ra} = 506.07$ with frequency $f = 22.1$.

Vadasz and Olek (1998) have shown that when a Darcy equation with timewise inertia term is taken, and with suitable scaling, the system of partial differential equations can be approximated by the same famous system of ordinary equations treated by Lorenz but with different values of the parameters. Their work described for centrifugally driven convection extends to the gravitational situation.

Further numerical studies using a unified finite approach exponential-type scheme have been reported by Llagostera and Figueiredo (1998) and Figueiredo and Llagostera (1999). Bilgen and Mbaye (2001) have treated a cavity with warm bottom and warm top and with additional lateral cooling.

This discussion of finite-amplitude convection is continued in Section 6.15.1.

6.9. Experimental Observations

6.9.1. Observations of Flow Patterns and Heat Transfer

Qualitative results for two-dimensional free convection were obtained using the Hele-Shaw cell analogy by Elder (1967a) and Bories (1970a,b). In a Hele-Shaw cell the isothermal lines can be observed by interferometry by using the fact that the refractive index of a liquid is a function of density and so of temperature. The streamlines can be visualized by strioscopy, i.e., by using light diffracted from aluminum particles suspended in the liquid. These experiments confirmed the theoretical value for the critical wavenumber and the fact that the wavenumber increases with Ra in accordance with calculations based on the Malkus hypothesis.

Direct visualization of three-dimensional flow in a porous medium was made by Bories and Thirriot (1969). They observed the accumulation of aluminum scattered on a thin liquid layer overlying the medium. The cells appeared to have approximately hexagonal cross section (away from the lateral boundaries) with the fluid rising in the center of each cell. The observations were checked by *in situ* temperature recordings. For slightly supercritical Rayleigh number values the dimensions of the cells were about the same as those predicted by linear theory. Howle *et al.* (1997) reported further visualization studies.

Many authors have performed experimental work in layers bounded by impermeable isothermal planes using conventional experimental cells (Schneider, 1963; Elder 1967a,b; Katto and Masuoka, 1967; Combarnous and Le Fur, 1969; Bories, 1970a; Combarnous, 1970; Yen, 1974; Kaneko *et al.*, 1974; Buretta and Berman, 1976). These have been concerned largely with heat transfer, but some experimenters have measured temperatures in the median plane of the layer in order to observe the boundaries of convective cells. In experiments reported by Combarnous and Bories (1975) it was found that the cells were not as regular as those obtained with a fluid clear of solid material. Again, polygons were observed away from the lateral boundaries; the cell sizes were consistent with linear theory and the wavenumber increased slightly with Rayleigh number. This change of wavenumber is consistent with the observations in a Hele-Shaw cell (see two paragraphs above) but it is in the opposite direction to that found in experiments with a clear fluid. Nield (1997b) tentatively ascribed the difference as an effect of dispersion.

The experimental heat transfer results of several of these workers, together with curves showing results from the upper bound analysis of Gupta and Joseph (1973) and the numerical calculations of Straus (1974) and Combarnous and Bia

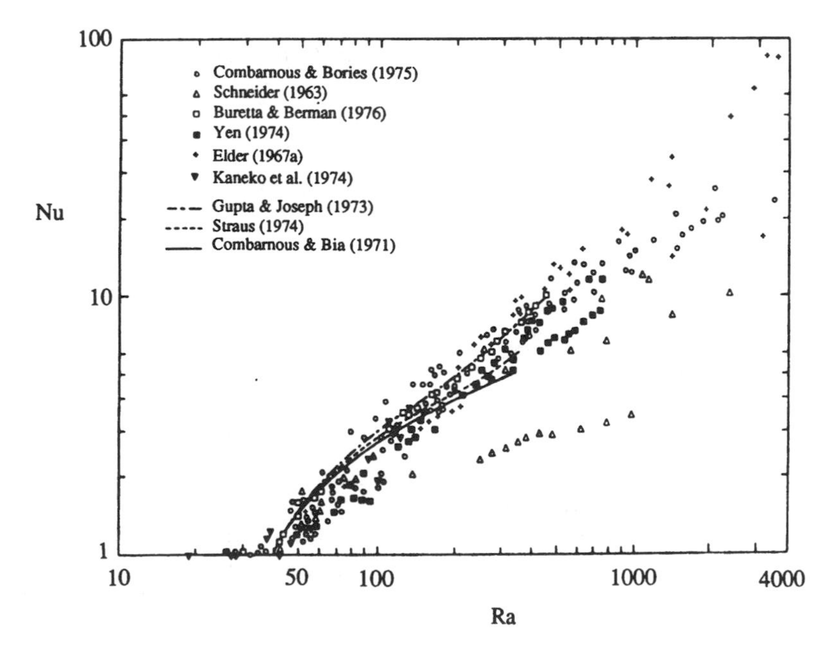

Figure 6.9. Compilation of experimental, analytical, and numerical results of Nusselt number versus Rayleigh number for convective heat transfer in a horizontal layer heated from below (Cheng, 1978, with permission from Academic Press).

(1971), are displayed in Fig. 6.9. The theoretical results are in agreement with experimental results for glass-water, glass-oil, and heptane-sand systems, but considerably overestimate the heat transfer for steel-oil, lead-water, and ethanol-sand systems. Possible reasons for this discrepancy are discussed below.

We note that the theoretical critical Rayleigh number $Ra_c \approx 40$ (defined as the Ra value for which Nu departs from the value 1) is confirmed by numerous experiments. A precise test for Ra_c was made by Katto and Masuoka (1967), who used nitrogen as the saturating fluid in order to reduce the temperature difference required for a large variation in Rayleigh number, and thus reduce the effect of property variation with temperature. Both the kinematic viscosity and thermal diffusivity of a gas are almost inversely proportional to the pressure, and so Ra can be varied through a large range by varying the pressure. Katto and Masuoka found satisfactory agreement between theory and experiment. Kaneko *et al.* (1974) observed $Ra_c \approx 28$ for ethanol-sand systems, but it is likely that the reduction in Ra_c was due to a nonlinear basic temperature profile (see Section 6.11). Close *et al.* (1985) found that Ra_c remains near 40 even when the layer depth is as small as two particle diameters.

When Ra is slightly supercritical, Nu increases linearly with Ra. For some systems (e.g., glass-water) the range of linearity is quite extensive, and for these Elder (1967a) proposed the correlation

$$\mathrm{Nu} = \frac{\mathrm{Ra}}{40}. \tag{6.75}$$

An extensive investigation, using glass beads, lead spheres, and sand as solids and silicone oil and water as fluids, was carried out by Combarnous and Le Fur (1969). This study showed that when the Rayleigh number reaches 240–280, there was a noticeable increase in the slope of the *Nu* versus *Ra* curve. Caltagirone *et al.* (1971) noted that it was apt to call the new regime the "fluctuating convective state," since the temperature field was continually oscillating. This fluctuating state also was observed in Hele-Shaw cell experiments by Horne and O'Sullivan (1974a). The transition is in accord with the numerical results discussed in Section 6.8. We recall that the transition is caused by instability of boundary layers at the horizontal boundaries, and that the fluctuating state is one in which convection cells continually appear and disappear, the number of cells doubling and halving.

Lein and Tankin (1992a) used the Christiansen filter concept to visualize the convection in test sections with different aspect ratios. They found that the width-to-height ratio of the convection cells did not vary with Ra for an impermeable upper boundary, but it did increase significantly for a permeable upper boundary.

Further experiments were conducted by Kazmierczak and Muley (1994). They found an increased heat transfer for a "clear top layer" compared with that for a completely packed layer, the increase being due to channeling and which Nield (1994a) showed was consistent with predictions based on the model of a clear fluid layer on top of the porous medium layer (Sec. 6.19.1). They also did experiments with the bottom wall temperature changed cyclically and found that the modulation could either increase or decrease the heat transfer.

Using magnetic resonance imaging, a noninvasive technique that yields quantitative velocity information, Shattuck *et al.* (1997) examined the onset of convection in a bed packed with monodisperse spheres in circular rectangular and hexagonal planforms. Disordered media, prepared by pouring spheres into a container, are characterized by regions of close packing separated by grain barriers and isolated defects that lead to locally larger porosity and permeability, and so to spatial variations in Ra. The authors found that stable localized convective regions exist for $\mathrm{Ra} < \mathrm{Ra_c}$, and these remain as pinning sites for convection patterns in the ordered regions as Ra increases above $\mathrm{Ra_c}$ up to $5\mathrm{Ra_c}$, the highest value studied in such media. In ordered media, with deviations from close packing only near the vertical walls, stable localized convection appears at $0.5\mathrm{Ra_c}$ in the wall regions. Different stable patterns are observed in the bulk for the same Ra after each recycling below $\mathrm{Ra_c}$, even for similar patterns of small rolls in the wall regions. As expected, roll-like structures are observed that relax rapidly to stable patterns between $\mathrm{Ra_c}$ and $5\mathrm{Ra_c}$, but the observed wavenumber was found to be 0.7π instead of the π predicted from linear stability theory. As Ra grows above $\mathrm{Ra_c}$ it was found that the volume of upflowing to the volume of downflowing regions decreases and leads to a novel time-dependent state, rather than the expected cross-rolls; this state begins at $6\mathrm{Ra_c}$ and is observed up to $8\mathrm{Ra_c}$, the largest Ra studied, and is probably linked to departures from the Boussinesq approximation. Further, it was found that the slope (S) of the Nusselt number curve is 0.7 rather than the predicted value of 2. [For comparison, Elder (1967a) found $S = 1$. Howle *et al.* (1997) found a slope between 0.53 and 1.35, depending on the medium, while Close *et al.* (1985) found that S decreases as d/H increases.] Further experiments involving nuclear magnetic resonance plus numerical simulations were reported by Weber *et al.* (2001), Kimmich *et al.* (2001), and Weber and Kimmich (2002).

In related work, Howle *et al.* (1997) used a modified shadow graphic technique to observe pattern formation at the onset of convection. They found that for ordered porous media, constructed from grids of overlapping bars, convective onset is characterized by a sharp bifurcation to straight parallel rolls whose orientation is determined by the number of bar layers, N_b; for odd N_b the roll arc are perpendicular to the direction of the top and bottom bars, but for even N_b they are at 45° to the bars. In a disordered system, produced by stacking randomly drilled disks separated by spaces, a rounded bifurcation to convection, with localized convection near onset, is observed, and the flow patterns take on one of several different cellular structures after each recycling through onset. The observations suggest that the mechanism of Zimmerman *et al.* (1993) (involving spatial fluctuations in Ra) and of Braester and Vadasz (1993) (involving continuous spatial variations of permeability and thermal diffusivity) may both be operating. Howle (2002) has reviewed work on convection in ordered and disordered porous layers.

6.9.2. Correlations of the Heat Transfer Data

The outstanding question posed by the experimental results is how one can best explain the spread of points in the Nu versus Ra plot, Fig. 6.6. There are two theoretical approaches to the matter. The first explanation, put forward by Combarnous (1972), elaborated by Combarnous and Bories (1974) and modified by Chan and Banerjee (1981), is based on the effect of solid-fluid heat transfer (see Section 6.5). A drawback to using this approach is that it is difficult to make an independent assessment of the heat transfer coefficient h. It turns out that this theory predicts some but not all of the observed reduction in Nu values (below those predicted from the simple Darcy local-thermal equilibrium model).

For this reason Prasad *et al.* (1985) decided that the solid-fluid heat transfer model was of limited use. They proposed the use of an effective conductivity:

$$k_e = \omega k_f + (1 - \omega)k_m, \tag{6.76}$$

where $(1 - \omega)$ is the ratio obtained by dividing the overall pure-conduction heat transfer estimate by the total heat transfer rate. This procedure, which is based on the argument that somehow or other the influence of the porous medium conductivity k_m decreases and that of the fluid-phase conductivity k_f increases, is quite successful in correlating the data, but it is *ad hoc*.

The second explanation is that put forward by Somerton (1983), Catton (1985), and Georgiadis and Catton (1986). These authors showed that the data spread can be substantially reduced by taking into account the effect of fluid inertia (the quadratic drag) which inevitably becomes increasingly important as Ra increases. Jonsson and Catton (1987) presented a power law correlation of Nu in terms of Ra and Pr_e, where Pr_e is an effective Prandtl number that can be defined, in terms of the quantities that appear in Eq. (6.62), by

$$\mathrm{Pr}_e = \frac{\mathrm{Pr}_f}{c_F}\frac{k_f}{k_m}\left(\frac{K}{H^2}\right)^{1/2}. \tag{6.77}$$

Close (1986) suggested that the data be brought in line with theory by means of the formula

$$\frac{\mathrm{Nu}}{\mathrm{Nu}_i} = 1.572 \times 10^{-2} \times \mathrm{Ra}_f^{0.344}\left(\frac{k_f}{k_s}\right)^{0.227}\left(\frac{H}{d_p}\right)^{0.446}\left(\frac{\varphi}{1-\varphi}\right)^{0.496}\mathrm{Pr}_f^{0.279}, \tag{6.78}$$

where Nu_i is given by expression (6.45) and Ra_f is a standard (non-Darcy) Rayleigh number based on the properties of the fluid and a layer thickness d_p (the pore diameter). Formula (6.78) is successful for $\mathrm{Nu} < 10$, but there are discrepancies for $\mathrm{Nu} > 10$. Close noted that the near equality of the exponents of k_f/k_s and Pr_f in Eq. (6.78) meant that Somerton's claim that it is neglect of inertial terms rather than solid-fluid heat transfer that causes the spread of data is not necessarily correct, and it is likely that both are involved.

Wang and Bejan (1987) strengthened the case for the inertial explanation by introducing the dimensionless group

$$\mathrm{Pr}_p = \mathrm{Pr}_e\,\frac{H^2}{K} \tag{6.79}$$

which arises naturally from the following scale analysis. At large Ra the quadratic drag term dominates over the linear term in the Forchheimer equation

$$\mathbf{v} + \frac{\chi}{\nu} |\mathbf{v}| \, \mathbf{v} = \frac{K}{\mu}(-\nabla P + \rho_f \mathbf{g}), \tag{6.80}$$

where $\chi = c_F K^{1/2}$ and $\nu = \mu/\rho_f$. The flow consists of a core counterflow plus boundary layers as shown in Fig. 6.10. In the core the vertical inertia scales as $\chi v^2/v$ and the boundary term scales as $(K/\mu)\rho g\beta\Delta T$, so the momentum balance requires

$$\frac{\chi}{\nu} v^2 \sim \frac{K}{\mu} \rho g \beta \Delta T. \tag{6.81}$$

The energy equation (Bejan, 1984) is a balance between upward enthalpy flow gradient $(v\Delta T/H)$ and lateral thermal diffusion between the two branches of the counterflow $\alpha_m \Delta T/L^2$, so

$$v\frac{\Delta T}{H} \sim \alpha_m \frac{\Delta T}{L^2}. \tag{6.82}$$

The balance between vertical enthalpy flow through the core $(\rho v L c_P \Delta T)$ and vertical thermal diffusion through the end region of height δ_H and width L requires

$$\rho v L c_P T \sim k_e L \Delta T/\delta_H. \tag{6.83}$$

The scales that emerge as solutions to the system (6.81)–(6.83) are

$$L \sim (\alpha_m H)^{1/2} \left(\frac{\chi}{g\beta K \Delta T} \right)^{1/4}, \tag{6.84}$$

$$v \sim \left(\frac{g\beta K \Delta T}{\chi} \right)^{1/2}, \tag{6.85}$$

$$\delta_H \sim \alpha_m \left(\frac{\chi}{g\beta K \Delta T} \right)^{1/2}. \tag{6.86}$$

We note in passing that these equations imply that L/H varies as $\mathrm{Ra}^{-1/4}$ and v varies as $\mathrm{Ra}^{1/2}$. The heat transfer rate in the Forchheimer flow limit therefore must scale as

$$\mathrm{Nu} \sim \frac{H}{\delta_H} \sim (\mathrm{Ra}\,\mathrm{Pr}_p)^{1/2}. \tag{6.87}$$

In contrast, heat flow in the Darcy flow limit scales as

$$\mathrm{Nu} \sim \frac{1}{40} \mathrm{Ra}. \tag{6.88}$$

Thus the transition from Darcy to Forchheimer flow occurs at the intersection of Eqs. (6.87) and (6.88),

$$\mathrm{Ra} \sim \mathrm{Pr}_p \tag{6.89}$$

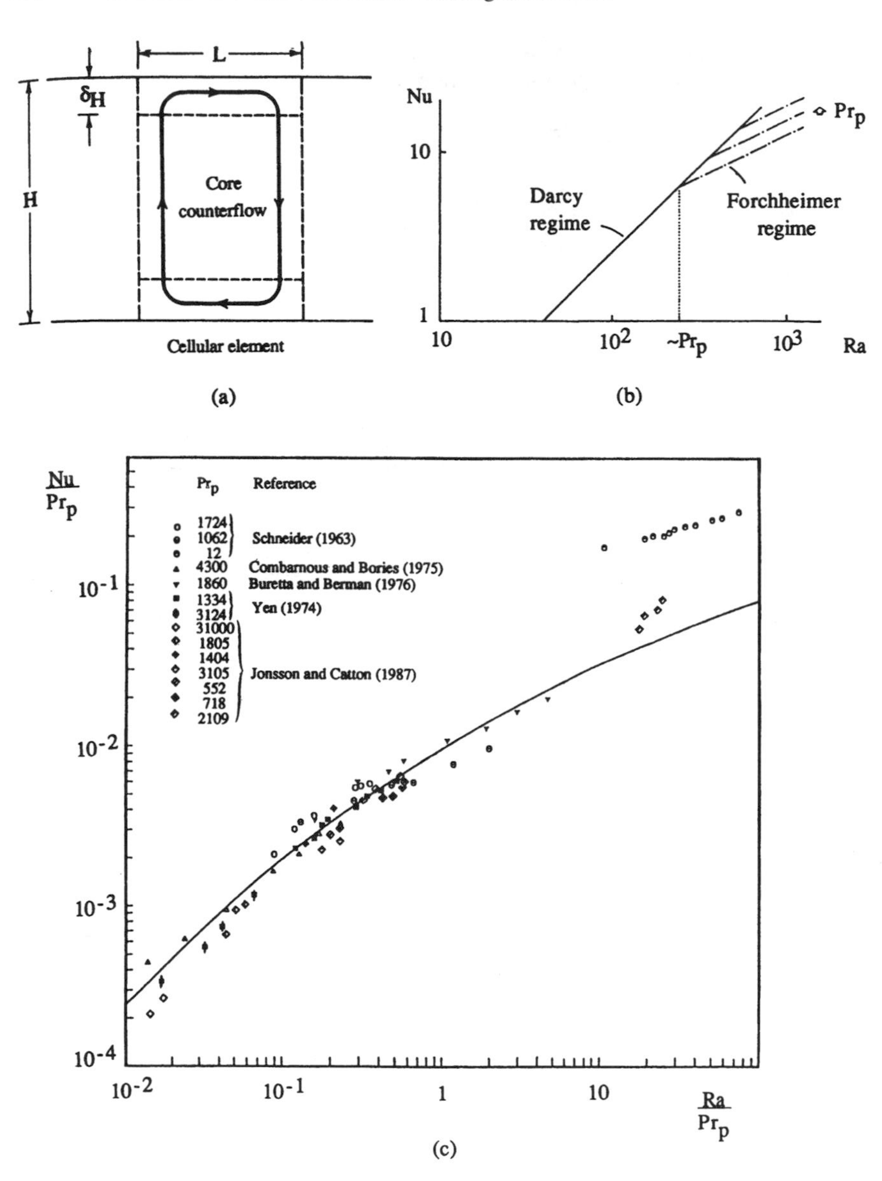

Figure 6.10. (a) Convective roll dimensions. (b) The asymptotes of the function Nu(Ra, Pr_p) suggested by scale analysis. (c) Heat transfer data, for convective heat transfer in a horizontal layer heated from below (Wang and Bejan, 1987).

from which we deduce

$$\frac{\mathrm{Nu}}{\mathrm{Pr}_p} \sim \frac{1}{40}\frac{\mathrm{Ra}}{\mathrm{Pr}_p}, \quad 40 < \mathrm{Ra} < \mathrm{Pr}_p, \tag{6.90}$$

$$\frac{\mathrm{Nu}}{\mathrm{Pr}_p} \sim \left(\frac{\mathrm{Ra}}{\mathrm{Pr}_p}\right)^{1/2}, \quad \mathrm{Ra} > \mathrm{Pr}_p. \tag{6.91}$$

An important feature of Eqs. (6.90) and (6.91) is that they are both of the form $\mathrm{Nu/Pr}_p = f(\mathrm{Ra/Pr}_p)$. This motivates the plotting of $\mathrm{Nu/Pr}_p$ against $\mathrm{Ra/Pr}_p$ to produce the graph shown in the lower part of Fig. 6.10. The agreement is good, with the notable exception of Schneider's (1963) data for $\mathrm{Pr}_p = 12$ in the top right corner of the figure. A line through this subset of data has the correct slope but is clearly too high and possibly the deduced Pr_p value of 12 is not correct. With this subset ignored, Wang and Bejan obtained the correlation

$$\mathrm{Nu} = \left\{ \left(\frac{\mathrm{Ra}}{40} \right)^n + \left[c(\mathrm{Ra}\,\mathrm{Pr}_p)^{1/2} \right]^n \right\}^{1/n}. \tag{6.92a}$$

where n and c are two empirical constants,

$$n = -1.65 \quad \text{and} \quad c = 1896.4. \tag{6.92b}$$

The simplicity of Eq. (6.92) in comparison with Eq. (6.78) is obvious.

We note that Kladias and Prasad (1989a,b, 1990) published the results of numerical calculations of the Nusselt number in which they have investigated the effects of Darcy number, Prandtl number, and conductivity ratio. They presented their results in terms of a fluid Rayleigh number and a fluid Prandtl number. We find this unhelpful for our present purpose, which is to summarize how the various effects act in concert rather than in isolation. Kladias and Prasad have made an important advance by showing that allowance for porosity variations brings the computed Nusselt numbers in better agreement with experimental observations. However, their (1989b) claim that Ra_c increases as Pr decreases was refuted by Lage *et al.* (1992), who showed numerically that Ra_c is independent of Pr, as the linear stability analysis indicates. [As Rees (2000) pointed out, this result is obvious when the momentum equation is scaled so that Pr appears only in the nonlinear terms, but it is not so obvious with other scalings.] Lage *et al.* proposed the correlation (accurate to within 2 percent)

$$\frac{\mathrm{Nu} - 1}{\mathrm{Ra/Ra}_c - 1} = \left[(C_1\,\mathrm{Pr}^2)^{-m} + C_2^{-m} \right]^{-m}, \tag{6.93a}$$

where

$$\varphi = 0.4,\ C_1 = 172\,\mathrm{Da}^{-0.516},\ C_2 = 0.295\,\mathrm{Da}^{-0.121},\ m = 0.4, \tag{6.93b}$$

$$\varphi = 0.7,\ C_1 = 30\,\mathrm{Da}^{-0.501},\ C_2 = 1.21\,\mathrm{Da}^{-0.013},\ m = 0.7. \tag{6.93c}$$

On the basis of scale analysis, Lage (1993a) obtained the following general scale for the Nusselt number:

$$\mathrm{Nu} \sim \frac{(L/H)}{2} \left\{ \frac{\sigma}{\tau} + \frac{-\Pi + [\Pi^2 + 2\varphi^2 \mathrm{Ra}\,\mathrm{Pr}\,E]^{1/2}}{2E} \right\}^{1/2}, \tag{6.94a}$$

where

$$E = 1 + \varphi J A(\mathrm{Pr})\,\mathrm{Pr} + \frac{0.143\varphi^{1/2}}{\mathrm{Da}^{1/2}} \tag{6.94b}$$

$$\Pi = \frac{\varphi}{\tau} + \frac{\varphi^2\,\mathrm{Pr}}{\mathrm{Da}} + \frac{\varphi J A(\mathrm{Pr})\,\mathrm{Pr}\,\sigma}{\tau}, \tag{6.94c}$$

and the function $A(\mathrm{Pr})$ takes the value 1 for $\mathrm{Pr} \geq 1$ and Pr^{-1} for $\mathrm{Pr} < 1$, and L and H are horizontal and vertical length scales, respectively, while τ is the characteristic time and J denote the viscosity ratio μ/μ_{eff}. The coefficient 0.143 arises from the assumption that c_F takes a form proposed by Ergun (1952). As Rees (2000) pointed out, the criterion for the onset of convection depends on Da/φ rather than just Da, and this dependence also may be observed in Eq. (6.94a,b,c). Additional experimental work has been reported by Ozaki and Inaba (1997).

6.9.3. Further Experimental Observations

Experiments by Lister (1990) in a large porous slab (3 m in diameter, 30 cm thick), using two quite different media (a matrix of rubberized curled coconut fiber and clear polymethylmethacrylate beads), have revealed several new phenomena. With the clear beads it was possible to visually observe the flows at the upper boundary. The boundary conditions were symmetrical (both impermeable and conducting) and so rolls were to be expected. Lister found that convection began in a hexagonal pattern and there was only a slight tendency to form rolls at slightly supercritical Rayleigh numbers.

Lister suggested that the asymmetry of the onset (one boundary maintained at a constant temperature, the other slightly heated) and the shape of the apparatus (hexagonal) could both be involved in the appearance of hexagons rather than rolls. At higher Rayleigh numbers the pattern of convection became very complex, irregular, and three-dimensional, without developing any obvious temporal instabilities. The visualization provided direct confirmation that the horizontal wavenumber of the convection cells increased with the Rayleigh number, approximately as $(\mathrm{Ra} + C)^{0.5}$, where C is a constant.

The Nu versus Ra curves obtained with the two media were substantially different. This conclusion was unexpected. The only feature that they had in common was a central section where the slope on a log/log graph was slightly over 0.5. On the graph for the fiber experiment this section was preceded by a slope close to 1 and followed by a slope close to 0.33. This last value is about the same as other experimenters have observed for convection in a clear liquid, so the result is expected because the fiber-filled medium had a porosity close to 100 percent. The temperature measured at a point in the fill 25 mm below the top boundary was unsteady at conditions representative of the upper two segments of the graph.

On the other hand, the Nusselt number for the bead fill jumps upward just above onset (where $\mathrm{Ra} = 4\pi^2$), rapidly settles to a slope of 0.52, and then gradually breaks upward again to a slope greater than 1 at the highest values (about 2000) for Ra reached in the experiment. Lister reported that increases in conductivity and permeability close to the boundary were not large enough to cause this increase in slope. He concluded that a new phenomenon, lateral thermal dispersion, appears to be responsible.

The phenomenon becomes important when the boundary layers become comparable in size with the diffusion length of the lateral dispersion, namely the bead size. The pores between beads are interlacing channels, i.e., they continually join and separate again, occasionally juxtaposing flows that would otherwise be separated by a substantial thermal-diffusion distance. This greatly enhances interchannel thermal contact, and the use of beads with an irregular shape (they were slightly rounded short cylinders of 3 mm diameter and length in Lister's experiment) means that there will be some actual flow exchange between channels. In this way the effective thermal diffusivity can be raised, but only if the flow velocity is sufficient to juxtapose channel streamlines more frequently than they would diffuse into equilibrium with each other by conduction. This means that lateral thermal dispersion has no effect on heat transfer at the onset of convection nor when the pores are sufficiently fine.

6.10. Effect of Net Mass Flow

On the Darcy model, if the basic flow is changed from zero velocity to a uniform flow in the x direction with speed U, then the eigenvalue problem of linear stability analysis is not altered if dispersion is negligible, since all the equations involved are invariant to a change to coordinate axes moving with speed U, a result noted by Prats (1966). Now some degeneracy is removed in that now longitudinal rolls (i.e., rolls with axes parallel to the x axis) are favored over other patterns of convection, in other words, such disturbances grow faster than other disturbances for the same Rayleigh number and overall horizontal wavenumber.

On the Forchheimer model the situation is different, as Rees (1998) pointed out. Now, for the usual boundary conditions

$$\mathrm{Ra}_c = \pi^2[(1+F)^{1/2} + (1+2F)^{1/2}]^2. \tag{6.95a}$$

where F is given by Eq. (6.62) with the Péclet number Q based on the throughflow. The critical wave number is given by

$$\alpha_c = \pi \left(\frac{1+2F}{1+F} \right)^{1/4}. \tag{6.95b}$$

Rees noted that this result provides a means of testing the validity of Eq. (1.12) compared with, for example, Eq. (2.57) of Kaviany (1995). Kubitscheck and Weidman (2003) have analyzed a problem where the bottom wall is heated by forced convection. Delache *et al.* (2002) have studied the effect of inertia and transverse aspect ratio on the pattern of flow. Time-periodic convective patterns have been studied numerically and analytically by Néel (1998) and Dufour and Néel (1998, 2000). Here various end-wall boundary conditions are imposed and the resulting flow patterns investigated. They found an entry effect whereby increasing flow rates yield increasing distances before strong travel-wave convection is obtained. A nonlinear instability study using the Brinkman model was performed by Lombardo and Mulone (2003). An experimental study related to aquifer thermal energy storage was performed by Nakagano *et al.* (2002). Numerical simulations related to diagenesis in layers of sedimentary rock were reported by Raffensperger and Vlassopoulos (1999), but it appears that they ignored the possibility of longitudinal rolls.

The effect of net mass flow with mean speed U in the z direction was studied by Sutton (1970) and Homsy and Sherwood (1976). This effect is more significant, because this alters the dimensionless temperature gradient from -1 to $F(z)$ where

$$F(z) = -\frac{\text{Pe}\exp(\text{Pe})}{\exp(\text{Pe}) - 1}, \tag{6.96}$$

where Pe is the Péclet number for the flow,

$$\text{Pe} = \frac{UH}{\alpha_m}. \tag{6.97}$$

Equation (6.26) is unchanged, but Eq. (6.25) is replaced by

$$(D^2 - \alpha^2 - PeD)\hat{\theta} = F(z)\hat{W}. \tag{6.98}$$

Before discussing quantitative results, we consider some qualitative ones. When Pe is large, the effect of the throughflow is to confine significant thermal gradients to a thermal boundary layer at the boundary toward which the throughflow is directed. The effective vertical length scale L is then the small boundary layer thickness rather than the thickness H of the porous medium, and so the effective Rayleigh number, which is proportional to L, is much less than the actual Rayleigh number Ra. Larger values of Ra thus are needed before convection begins. Thus the effect of large throughflow is stabilizing.

Within the bulk of the medium a large part of the heat transport can be effected by the throughflow alone, and the value of the temperature gradient at which convection cells are required is increased. The effective Rayleigh number is largely independent of the boundary conditions at the boundary from which the throughflow comes.

The situation for small values of Pe is more complex. The case of insulating boundaries is readily amenable to approximate analysis. On the assumption that the effect of Pe does not appreciably alter the shape of the eigenfunctions, one can obtain analytical formulas for the critical Rayleigh number for various combinations of boundary conditions.

For example, for the case in which both boundaries are impermeable and insulating, Nield (1987a) obtained the formula

$$\text{Ra}_c = \frac{2\,\text{Pe}^2}{\text{Pe}\coth(\text{Pe}/2) - 2}. \tag{6.99}$$

Clearly Ra_c is an even function of Pe and for positive Pe is an increasing function of Pe. Hence throughflow is stabilizing for all values of Pe, and the direction of flow does not matter. For small values of Pe we have

$$\text{Ra}_c = 12 + \frac{1}{5}\text{Pe}^2. \tag{6.100}$$

On the other hand, when the lower boundary is impermeable and insulating and the upper boundary is insulating and free (at constant pressure),

$$\text{Ra}_c = \frac{2\text{Pe}[\exp(\text{Pe}) - 1]}{2\text{Pe} + 2 + (\text{Pe})^2\exp(\text{Pe})}. \tag{6.101}$$

For small values of Pe:

$$\mathrm{Ra}_c = 3\left(1 - \frac{1}{8}\mathrm{Pe}\right) \tag{6.102}$$

showing that the case of downflow (Pe $<$ 0) is stabilizing and that upflow of small magnitude is destabilizing. A similar picture is painted by the numerical results for conducting boundaries by Jones and Persichetti (1986).

For symmetrical situations, where the lower and upper boundaries are of the same type, Ra_c is an even function of Pe and throughflow is stabilizing by a degree that is independent of the flow direction. When the boundaries are of different types, throughflow in one direction is clearly destabilizing for small values of Pe since $d\mathrm{Ra}_c/d\mathrm{Pe}$ at $\mathrm{Pe} = 0$ is not zero. The destabilization occurs when the throughflow is away from the more restrictive boundary. The throughflow then decreases the temperature gradient near the restrictive boundary and increases it in the rest of the medium. Effectively the applied temperature drop acts across a layer of smaller thickness, but the stabilizing effect of this change is more than made up by the destabilization produced by changing the effective boundary condition to a less restrictive one. A similar phenomenon, arising when the vertical symmetry is removed by the temperature dependence of viscosity or by some nonuniformity of the permeability, was found by Artem'eva and Stroganova (1987). Khalili and Shivakumara (1998, 2003), Shivakumara (1999) and Khalili *et al.* (2002) have extended the linear stability theory to consider the effects of internal heat generation and anisotropy and also boundary and inertial effects. A study of the stability of the solutions given by linear stability theory, together with a numerical study to confirm the findings, was conducted by Zhao *et al.* (1999b).

Wu *et al.*(1979) have used numerical methods to study the case of maximum density effects with vertical throughflow, while Quintard and Prouvost (1982) studied throughflow with viscosity variations that lead to Rayleigh-Taylor instability. The nonlinear stability analysis of Riahi (1989) for the case of large Pe shows that subcritical instability exists and this is associated with up-hexagons, which are stable for amplitude ε satisfying $|\varepsilon| = 0.35$. For $|\varepsilon| = 0.4$, squares too are stable, and the realized flow pattern depends on initial conditions. A general nonlinear analysis was reported by van Duijn *et al.* (2002). Their predictions were in good agreement with the results of laboratory experiment with Hele-Shaw cells of Wooding *et al.* (1997a,b).

6.11. Effect of Nonlinear Basic Temperature Profiles

6.11.1. General Theory

Nonlinear basic temperature profiles can arise in various ways, notably by rapid heating or cooling at a boundary or by a volumetric distribution of heat sources. When the former is the case, the profile is time-dependent, but one can investigate instability on the assumption that the profile is quasistatic, i.e., it does not change significantly on the timescale of the growth of small disturbances. It is found that with a curved temperature profile it is possible for the critical Rayleigh number to be less than that for a linear profile. Indeed, in the case of the parabolic profile arising from a uniform volume distribution of sources, the critical value Ra_c can be arbitrarily small. But when the profiles are restricted to ones in which the gradient does not change sign, the question of which profile leads to the least Ra_c is not trivial. The question can be answered readily for the case of insulating (constant heat flux) boundaries because then an analytic expression for Ra_c can be found.

The problem is to minimize Ra_c with respect to the class of nondimensional adverse temperature gradients $f(\hat{z})$ satisfying

$$f(\hat{z}) \geq 0, \quad \langle f(\hat{z}) \rangle = 1. \tag{6.103}$$

where $\langle f(\hat{z}) \rangle$ denotes the integral of $f(\hat{z})$ with respect to $\hat{z}$, from $\hat{z} = 0$ to $\hat{z} = 1$. Nield (1975) shows that the problem reduces to maximizing $\langle W_0 \theta_0 f(\hat{z}) \rangle$ where W_0 and θ_0 are normalized eigenfunctions. For example, in the case of impermeable insulating boundaries it is found that $W_0 = \hat{z} - \hat{z}^2$, $\theta_0 = 1$, and

$$Ra_c = \frac{2}{\langle (\hat{z} - \hat{z}^2) f(\hat{z}) \rangle}. \tag{6.104}$$

The expression $(\hat{z} - \hat{z}^2)$ has its maximum when $\hat{z} = 1/2$, and consequently the function $f(\hat{z})$, which minimizes Ra_c subject to the constraints (6.103), is the Dirac delta function

$$f(\hat{z}) = \delta\left(\hat{z} - \frac{1}{2}\right).$$

The corresponding minimum value is $Ra_c = 8$. This may be compared with the value $Ra_c = 12$ for the linear temperature profile. More generally, the step-function temperature profile whose gradient is $f(\hat{z}) = \delta(\hat{z} - \varepsilon)$ gives

$$Ra_c = \frac{2}{\varepsilon - \varepsilon^2}. \tag{6.105}$$

For piecewise linear temperature profiles whose gradient is of the form

$$f(\hat{z}) = \begin{cases} \varepsilon^{-1}, & 0 \leq \hat{z} < \varepsilon \\ 0, & \varepsilon < \hat{z} \leq 1 \end{cases} \tag{6.106}$$

one finds that

$$Ra_c = \frac{12}{3\varepsilon - 2\varepsilon^2}, \tag{6.107}$$

The case of the linear temperature profile is given by $\varepsilon = 1$, $Ra_c = 12$, as expected. As ε varies the minimum of expression (6.107) is attained at $\varepsilon = 3/4$, and then $Ra_c = 32/3$. Nield (1975) showed that this provides the minimum for Ra_c subject to

$$f(z) \geq 0, \quad df/d\hat{z} \leq 0 \text{ (almost everywhere)}, \quad \langle f(\hat{z}) \rangle = 1. \tag{6.108}$$

An extension of the above theory, incorporating the Brinkman term, was made by Vasseur and Robillard (1993). An extension to the case of permeable boundaries was reported by Thangaraj (2000).

6.11.2. Internal Heating

When a volumetric heat source q''' is present, Eq. (6.5) is replaced by

$$(\rho c)_m \frac{\partial T}{\partial t} + (\rho c_P)_f \mathbf{v} \cdot \nabla T = k_m \nabla^2 T + q'''. \tag{6.109}$$

The steady state is given by

$$\mathbf{v}_b = 0 \quad \text{and} \quad k_m \nabla^2 T_b = -q'''. \tag{6.110}$$

If q''' is constant, then the basic steady-state temperature distribution is parabolic,

$$T_b = -\frac{q''' z^2}{2k_m} + \left(\frac{q''' H}{2k_m} - \frac{\Delta T}{H} \right) z + T_0 + \Delta T. \tag{6.111}$$

In place of Eq. (6.13), one has

$$(\rho c)_m \frac{\partial T'}{\partial t} + (\rho c)_f \left| \frac{q'''}{2k_m} (H - 2z) - \frac{\Delta T}{H} \right| w' = k_m \nabla^2 T'. \tag{6.112}$$

Equations (6.11) and (6.12) still stand. If instead of ΔT we now choose $\text{Ra}\Delta T$ as temperature scale, then in terms of the new nondimensional variables one has, for monotonic instability,

$$\nabla^2 \hat{w} = \nabla_H^2 \hat{T}, \tag{6.113}$$

$$\nabla_H^2 \hat{\hat{T}} = [\text{Ra}_I (1 - 2\hat{z}) - \text{Ra}] \hat{w}, \tag{6.114}$$

where $\hat{\hat{T}} = \text{Ra}\,\hat{T}$. The new nondimensional parameter is the internal Rayleigh number Ra_I defined by

$$\text{Ra}_I = \frac{H^2 q'''}{2k_m \Delta T}, \quad \text{Ra} = \frac{g\beta K H^2 q'''}{2\nu \alpha_m k_m}. \tag{6.115}$$

We can refer to the original Ra as the external Rayleigh number, to distinguish it from the internal Rayleigh number Ra_I.

Equations (6.113) and (6.114), which now contain a nonconstant coefficient, may be solved numerically by using, for example, the Galerkin method. The stability boundary in the (Ra, Ra_I) plane, Fig. 6.11, was calculated by Gasser and Kazimi (1976) for the case of impermeable conducting boundaries. When $\text{Ra}_I = 0$, the critical value of Ra is $4\pi^2$. When $\text{Ra} = 0$, the critical value of Ra_I is 470. Changing the thermal boundary condition at the lower boundary has a marked effect on the critical value of Ra_I; Buretta and Berman (1976) gave the estimate 32.8 for the case of an insulating lower boundary. Within experimental error, this was in agreement with their experiments, which involved a copper sulfate solution saturating a bed of spherical glass beads.

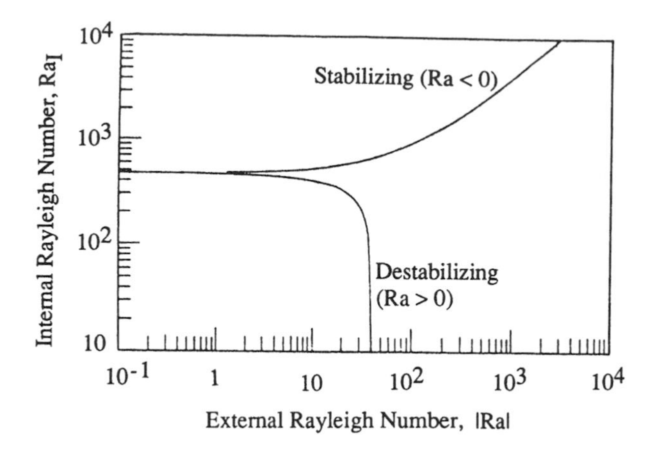

Figure 6.11. Critical internal Rayleigh number versus external Rayleigh number for stabilizing and destabilizing temperature differences (Gasser and Kazimi, 1976).

These experiments by Buretta and Berman revealed an interesting effect. Their Nu versus Ra diagram showed a bifurcation into two branches with different slopes. There was also a jump from the lower branch to the upper at some Ra value that increased with bead size. Subsequent experiments by Hardee and Nilson (1977), Rhee *et al.* (1978), and Kulacki and Freeman (1979) failed to reproduce the jump. The data obtained by Kulacki and Freeman tended to correlate with the lower branch of Buretta and Berman's curve, but those of the other experimenters tended to correlate with the upper branch (Fig. 6.12). It appears that the discrepancy is still unresolved, but it may be related to the unusual bifurcation structure found by He

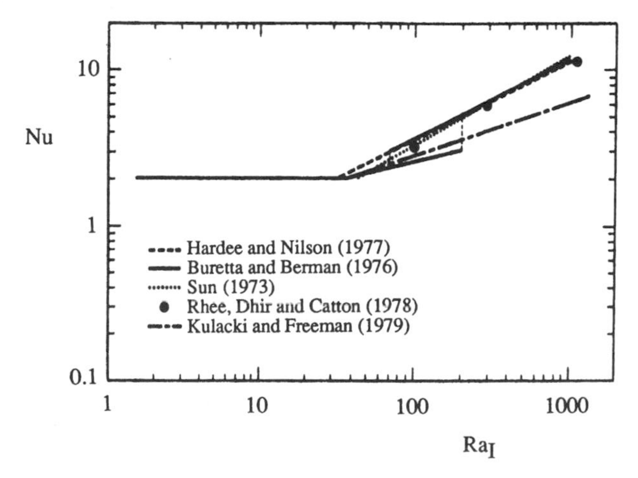

Figure 6.12. Comparison of heat transfer results for internal heating (Kulacki and Freeman, 1979).

and Georgiadis (1990), which arises from the effect of hydrodynamic dispersion in the case of uniform volumetric heating.

Various authors have made analytical or numerical extensions of the problem. Kulacki and Ramchandani (1975) varied the thermal boundary conditions. Tveitereid (1977) carried out a nonlinear stability analysis. He found that down-hexagons (downward flow in the centers of the cells) were stable for Ra up to 8 Ra_c, up-hexagons were stable for all values of Ra, and two-dimensional rolls were stable for $3\mathrm{Ra}_c < \mathrm{Ra} < 7\mathrm{Ra}_c$. His computed Nu versus Ra curves correlated quite well with the upper branch of Buretta and Berman's curve. Rudraiah *et al.* (1980, 1982) carried out calculations of Ra_{Ic} for various boundary conditions using the Brinkman equation. A nonlinear (energy) stability analysis was carried out by Ames and Cobb (1994), who thereby estimated the Ra band for possible subcritical instabilities.

Somerton *et al.* (1984) performed calculations that indicated that the wave number for convection decreases with increasing internal Rayleigh number. Kaviany (1984a) discussed a transient case when the upper surface temperature is decreasing linearly with time. Hadim and Burmeister (1988, 1992) have modeled a solar pond by allowing q''' to vary exponentially with depth, including the effect of vertical throughflow. Rionera and Straughan (1990) added the effect of gravity varying in the vertical direction. Their analysis, based on the energy method, revealed the possibility of subcritical convection Stubos and Buchlin (1993) numerically simulated the transient behavior of a liquid-saturated core debris bed with internal dissipation. Parthiban and Patil (1995) have extended the theory to the case of inclined gradients (see Section 7.9). A bifurcation study employing the Brinkman model was carried out by Choi *et al.* (1998).

The problem with a fluid undergoing a zero-order exothermic reaction was analyzed by Malashetty *et al.* (1994): the chemical reaction leads to increased instability. With determination of the conditions for the spontaneous combustion of a coal stockpile in mind, Bradshaw *et al.* (1991) used an approximate analysis to obtain convection patterns. They found that down-hexagons and two-dimensional rolls are the stable plan forms, and using a continuation procedure they obtained a simple criterion for the point of ignition in the layer, one given by a Frank-Kamenetskii parameter exceeding 5.17.

Lu and Zhang (1997) studied the onset of convection in a mine waste dump, in which there is active oxidation of pyritic materials, the rock being filled with moist gas. They took into account the effects of compressibility, latent heat, and a volumetric heat source varying exponentially with depth. Royer and Flores (1994) presented a novel way of dealing with Darcy flow in an anisotropic and heterogeneous medium. The combination of internal heat sources and vertical throughflow was treated by Yoon *et al.* (1998). A study involving external radiative incidence and imposed downward convection was reported by Liu (2003). A general study of radiative heat transfer was reported by Park *et al.* (1996).

The case where the volumetric heating is due to the selective absorption of radiation was studied by Hill (2003, 2004a,b), employing both linear and nonlinear stability analysis and also numerically, for each of the Darcy, Forchheimer, and

Brinkman models. Convection with a non-Newtonian (power law) fluid at a large internal Rayleigh number was treated numerically by Kim and Hyun (2004).

Transient effects and heat transfer correlations for turbulent heat transfer were reported by Kim and Kim (2002) and Kim *et al.* (2002a,b). Jimenez-Islas *et al.* (2004) conducted a numerical study of natural convection with grain in cylindrical silos.

6.11.3. Time-Dependent Heating

The case where the temperature imposed on the lower boundary is timewise periodic was analyzed by Chhuon and Caltagirone (1979). The thermal boundary conditions are now $T = T_0$ at $z = H$ and

$$T = T_0 + \Delta T(1 + \beta \sin \omega^* t) \quad \text{at} \quad z = 0. \tag{6.116}$$

For the basic state the nondimensional equations, expressed in terms of the same scales as in Section 6.2, are $\mathbf{v}_b = 0$ and

$$\frac{\partial \hat{T}_b}{\partial \hat{t}} = \frac{\partial^2 \hat{T}_b}{\partial \hat{z}^2}, \tag{6.117}$$

$$\hat{T}_b = 1 \quad \text{at} \quad \hat{z} = 1, \tag{6.118}$$

$$\hat{T}_b = 1 + \beta \sin \omega \hat{t} \quad \text{at} \quad \hat{z} = 0, \tag{6.119}$$

$$\omega = \frac{\sigma H^2}{\alpha_m} \omega^*. \tag{6.120}$$

The solution of the system of equations (6.117)–(6.120) is

$$\hat{T}_b = (1 - \hat{z}) + \beta \alpha(\hat{z}) \sin(\omega t + \varphi(\hat{z})) \tag{6.121}$$

where

$$\alpha(\hat{z}) = |q|, \quad \varphi(\hat{z}) = \operatorname{Arg} q,$$
$$q(\hat{z}) = \frac{\sinh[k(1+i)(1-\hat{z})]}{\sinh[k(1+i)]}, k = \left(\frac{\omega}{2}\right)^{1/2}. \tag{6.122}$$

If we take perturbations on this steady state and, instead of Eq. (6.23), take

$$(\hat{w}, \hat{T}) = [W(\hat{z}, \hat{t}), \theta(\hat{z}, \hat{t})] \exp(il\hat{x} + im\hat{y}), \tag{6.123}$$

we obtain

$$\frac{\partial \theta}{\partial \hat{t}} = (D^2 - \alpha^2)\theta - W \frac{\partial \hat{T}_b}{\partial \hat{z}}, \tag{6.124}$$

$$\gamma_a \frac{\partial}{\partial \hat{t}} (D^2 - \alpha^2) W = -\text{Ra}\alpha^2 \theta - (D^2 - \alpha^2) W. \tag{6.125}$$

In the case of impermeable isothermal boundaries the boundary conditions are

$$W = \theta = 0 \quad \text{at} \quad \hat{z} = 0 \quad \text{and} \quad \hat{z} = 1. \tag{6.126}$$

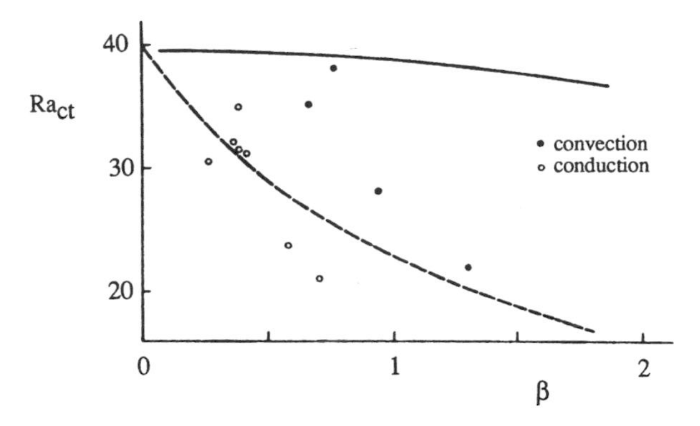

Figure 6.13. Critical Rayleigh numbers versus amplitude for frequency $f = 4.23$; •, ∘ experiment, – – – linear theory, —— Floquet theory (Chhuon and Caltagirone, 1979).

Chhuon and Caltagirone then set $\gamma_a = 0$, solved the system (6.124–6.126) using the Galerkin method, and examined the stability of solutions of the resulting ordinary differential equations using the Floquet theory. In this way they obtained the value of a critical Rayleigh number Ra_{ct} as a function of wavenumber α, amplitude β, and frequency $f = 2\pi/\omega$. They also performed experiments and compared their observations with their Floquet theory calculations and with calculations based on linear theory by Caltagirone (1976a), shown in Fig. 6.13.

In comparing the theories there is the problem that for the linear theory the stability criterion taken was $a(t) < 0$, $a(t)$ being the amplitude of the temperature perturbation. Both theories give the frequency range 1 to 100 as that over which Ra_{ct} varies significantly, but whereas in the Floquet theory Ra_{ct} varies only slightly with f, in the linear theory Ra_{ct} varies from 40 as $f \to \infty$ to $40/(\beta + 1)$ at $f = 0$. Both theories predict destabilization from the stationary case. The Floquet theory breaks down when $f \to 0$, since the critical Rayleigh numbers must necessarily approach $4\pi^2/(\beta + 1)$.

As Fig. 6.13 shows, convective phenomena are observed for Rayleigh numbers between those given by the two theories. Additional numerical calculations by Chhuon and Caltagirone showed that during part of the period of oscillation, the effect of convection is that the initial perturbation is attenuated considerably and then it increases. At high frequencies both theories agree that the temperature oscillation has no effect on the stability of the layer.

The Brinkman model was employed by Rudraiah and Malashetty (1990). They concluded that modulation could advance or delay the onset of convection according to whether the variation of top and bottom temperatures were in phase or out of phase. An extension to the Forchheimer model was made by Malashetty and Wadi (1999), while Malashetty and Basavaraja (2002) combined an oscillatory wall temperature with an oscillatory gravitational field. Néel and Nemrouch (2001) examined the stability of a layer with an open top and a pulsating temperature imposed at the upper boundary, using the Darcy model. The case of an oscillatory thermal condition at the top was also studied numerically by Holzbecher (2004c).

Other authors have been concerned with situations where the imposed surface temperature varies monotonically with time. Now amplification of disturbances inevitably occurs at some stage, and the interest is in determining an onset time by which the growth factor has reached some specified criterion, say 1000. Caltagirone (1980) investigated the case when the lower surface is subjected to a sudden rise in temperature. He used linear theory, energy-based theory, and a two-dimensional numerical model. Kaviany (1984a) made a theoretical and experimental investigation of a layer with a lower surface temperature increasing linearly with time. His second paper (Kaviany, 1984b) involved both time-dependent cooling of the upper surface and uniform internal heating. An alternative treatment of this problem was reported by Yoon, Choi and Yoo (1992). They predicted an onset time τ_c given by

$$\tau_c = 6.55(\mathrm{DaRa})^{-2/3}, \tag{6.127}$$

where $\mathrm{Da} = K/\varphi H^2$, and found that the experimental data of Kaviany (1984b) indicated that the convection is detectable at time $4\tau_c$. A study of the most unstable disturbance corresponding to momentary instability, based on an optimization over the range of possible initial perturbations, was made by Green (1990). A prediction of the time required for the onset of convection in a porous medium saturated with oil with a layer of gas underlying the oil was made by Rashidi *et al.* (2000). The study of the onset of transient convection by Tan *et al.* (2003) is flawed [see Nield (2004a)]. Propagation theory was employed by Kim *et al.* (2004) to study the onset of convection in a transient situation with a suddenly applied constant heat flux at the bottom of the layer. A further theoretical study was reported by Kim and Kim (2005). Linear and global stability analyses of the extension of the Caltagirone (1980) problem to the case of an anisotropic medium were made by Ennis-King *et al.* (2005), for both thin and thick slabs. For a thick slab they found that the increase of τ_c as γ (the ratio of vertical permeability to horizontal permeability) decreases is given approximately by $(1 + \sqrt{\gamma})^4/16\gamma^2$. Their study is applicable to the geological storage of carbon dioxide, for which the timescale can vary from less than a year (for high-permeability formations) to decades or centuries (for low-permeability ones). Geological details are provided by Ennis-King and Paterson (2005).

The possibility of feedback control of the conduction state was demonstrated theoretically by Tang and Bau (1993). The temperature perturbation θ at some horizontal cross-section is monitored. The controller momentarily modifies the perturbation temperature distribution of the heated base in proportion to a linear combination of θ_0 and its time-derivative. Thus the controller slightly reduces/increases the bottom temperature at locations where the fluid tends to ascend/descend. Once the disturbance has disappeared, the bottom temperature is restored to its nominal value. This simple procedure suppresses the first even mode and so delays the onset of convection until the first odd mode is unstable, giving a fourfold increase in the critical Rayleigh number. More general issues were discussed by Bau (1993).

The effects of a sinusoidal temperature distribution, as a wave with wavelength that of the incipient Bénard cells superimposed on the hot temperature of the lower plate, were studied numerically by Mamou *et al.* (1996). For a given value of Ra, the cells move with the imposed wave if the velocity of the latter remains below a critical value, but at higher velocity the cell motion is irregular and fluctuates. Ganapathy and Purushothaman (1992) had analyzed previously a similar problem with a moving thermal boundary condition at the upper surface. The effects of adding small-amplitude traveling thermal waves, of the same amplitude and phase at the top and bottom boundaries, were examined by Banu and Rees (2001). At sufficiently low Rayleigh numbers Ra the induced flow follows the motion of the thermal wave, but at higher Ra this form of convection breaks down and there follows a regime where the flow travels more slowly on the average and does not retain the forcing periodicity. At much higher Ra (or for large wave speeds at moderate Ra) two very different timescales appear in the numerical simulations. Hossain and Rees (2003) treated the variant problem where the sidewalls have the same cold temperature as the upper surface. Now the flow becomes weaker as the Darcy number decreases from the pure fluid limit toward the Darcy flow limit, and the number of cells that form in the cavity varies primarily with the aspect ratio and is always even due to the symmetry imposed by the cold sidewalls.

The onset of convection induced by volumetric heating, with the source strength varying exponentially with depth and also varying with time, was analyzed by Nield (1995); he added a term

$$q''' e^{\beta(z/H-1)}[1 + \varepsilon e^{i\omega(t-t'')}] \tag{6.128}$$

to the right hand side of Eq.(6.5), and showed that, for the case of conducting boundaries, instability occurs when

$$\mathrm{Ra} + \mathrm{Ra}_I f(\beta)[4\pi^2 \varepsilon (16\pi^2 + \omega^2)^{-1/2} - 1] > 4\pi^2, \tag{6.129}$$

where $f(\beta) = 2(1 - e^{-\beta})/(4\pi^2 + \beta^2)$, Ra_I is given by Eq.(6.115), and ω is given by Eq. (6.120), the most unstable conduction-state temperature profile occurring at the end of the cooling phase of a cycle if β is positive. Nield also gave results for other thermal boundary conditions. He also investigated the case of square-wave periodic heating, both for a steady state and the transient situation after the heating is suddenly switched on. He showed that the square-wave time-periodic source leads to a more unstable situation than a sinusoidal time-periodic source of the same amplitude, and that transient on-off heating leads to greater instability than the corresponding steady state.

6.11.4. Penetrative Convection or Icy Water

Mamou *et al.* (1999) used linear stability analysis with the Brinkman model to study the onset of convection in a rectangular porous cavity saturated by icy water. They also obtained numerical results for finite-amplitude convection. These results indicate that subcritical convection is possible when the upper stable layer extends over more than one half of the cavity depth and demonstrate the existence of multiple solutions for a certain parameter range. Penetrative convection in a horizontally isotropic porous layer was investigated by Carr and de Putter (2003) using alternatively an internal heat sink model or a quadratic temperature law. They performed linear and nonlinear stability analyses and showed that their two models led to the same predicted instability boundaries. Carr and Straughan (2003) numerically calculated the onset of convection in two-layer system with icy water underlying a porous medium with patterned ground in mind. Straughan (2004a) studied an interesting resonant situation. He showed that there is a parametric range in which the convection may switch from the lower part of the layer to being prominent in the upper part of the layer. Mahidjiba *et al.* (2000b,c, 2002, 2003) applied linear stability analysis to an anisotropic porous medium saturated with icy water. They introduced an inversion parameter γ and an orientation θ of the principal axes. They found that the presence of a stable layer near the upper boundary for $\gamma < 2$ changes drastically the critical Rayleigh number, and an asymptotic situation is reached when $\gamma \leq 1$. For that asymptotic solution, and with $\theta = 0°$ or $90°$, the incipient flow field consists of primary convective cells near the lower boundary with superposed layers of secondary cells. For $0° < \theta < 90°$, primary and secondary cells coalesce to form obliquely elongated cells.

6.12. Effects of Anisotropy

The material in this section and the next is based on the review by McKibbin (1985). The criterion for the onset of convection in a layer with anisotropic permeability and which has impermeable upper and lower boundaries was obtained by Castinel and Combarnous (1975). They also reported results from experiments using glass fiber materials saturated with water. The experimental values of Ra_c agreed reasonably well with the predictions.

Epherre (1975) allowed both permeability and thermal conductivity to be anisotropic. If one defines Ra in terms of the vertical permeability K_V and the vertical thermal conductivity k_V of the medium, so that

$$Ra = \frac{g\beta K_V H \Delta T}{\nu \alpha_V}, \tag{6.130}$$

where $\alpha_V = k_V/(\rho c_P)_f$, then the critical value of Ra for the onset of two-dimensional convection (rolls) of cell width/depth ratio L is

$$Ra_c(L) = \frac{\pi^2(\xi + L^2)(\eta + L^2)}{\xi L^2}, \tag{6.131}$$

where $\xi = K_H/K_V$ and $\eta = k_H/k_V$. The subscript H refers to quantities measured in the horizontal direction. As L varies, the minimum value of Ra is attained when $L = L_c = (\xi\eta)^{1/4}$,

$$Ra_{c,min} = \pi^2 \left[1 + \left(\frac{\eta}{\xi}\right)^{1/2}\right]^2. \tag{6.132}$$

These analyses were extended by Kvernvold and Tyvand (1979) to steady finite-amplitude convection. They found that for two-dimensional flow the Nusselt number Nu depends on ξ and η only in the ratio ξ/η. They also found that if Nu is graphed as a function of Ra/Ra_c, the various curves start out from the point (1,1) at the same slope, which is equal to 2.0 (Fig. 6.14). Nield (1997b) pointed out that Eq. (6.132) is equivalent to $Ra_{Ec} = 4\pi^2$, where Ra_E is an equivalent Rayleigh

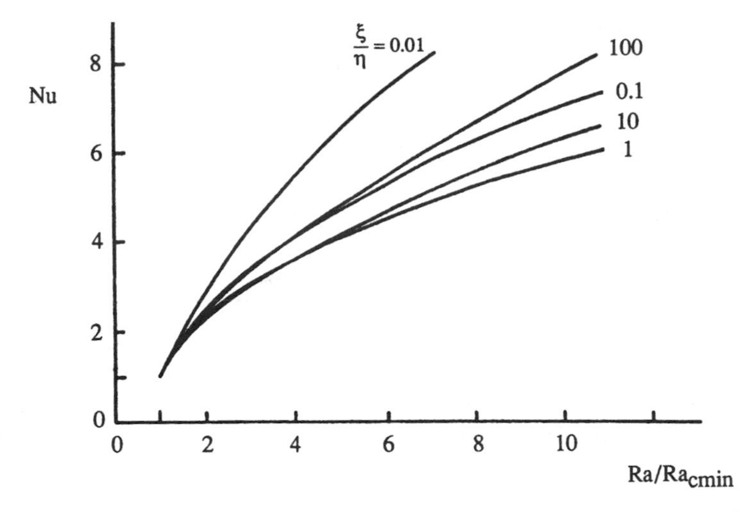

Figure 6.14. Nusselt number versus Ra/Ra$_{c,min}$ for various anisotropy ratios ξ, η (McKibbin, 1985; after Kvernvold and Tyvand, 1979).

number defined as in Eq. (6.121) but with K/α_m replaced by the square harmonic-mean square-root of K_v/α_v and K_H/α_H, and in Fig. 6.14 the quantity Ra/Ra$_{c,min}$ is equivalent to $\mathrm{Ra}_E/4\pi^2$.

Wooding (1978) noted that in a geothermal system with a ground structure composed of many successively laid down strata of different permeabilities, the overall horizontal permeability may be up to ten times as large as the vertical component. He extended the linear analysis to three-dimensional convection in a layer in which the permeability is anisotropic and also may vary with depth. He treated both impermeable and free (constant-pressure) upper boundaries. As expected, the free boundaries yield a smaller Ra$_c$ than the impermeable boundaries, but the difference becomes small when $\xi = K_H/K_V$ becomes large because then vertical flow is more difficult than horizontal flow.

A study of the fraction r of the total flow that recirculates within an anisotropic layer at the onset of convection was conducted by McKibbin *et al.* (1984). It was extended by McKibbin (1986a) to include a condition of the form $P + \lambda \partial P/\partial n = 0$ at the upper boundary, where λ is a parameter taking the limiting values 0 for a constant-pressure boundary and ∞ for an impermeable boundary. He found that there is always some recirculation of the fluid within the porous layer provided that λ is finite (Fig. 6.15a). In the case $\lambda = 0$ there is a stagnation point on the surface as well as in the interior of the layer (Fig. 6.15b). McKibbin calculated Ra$_{c,min}$, L_c, σ_*, and r for various values of λ, as functions of ξ/η. The results show that the recirculation diminishes as $\xi/\eta \to 0$ and there is full recirculation as $\xi/\eta \to \infty$. Here σ_* is the slope coefficient which appears in the heat transfer relationship (for slightly supercritical conditions)

$$\mathrm{Nu} = 1 + \sigma_* \left(\frac{\mathrm{Ra}}{\mathrm{Ra_c}} - 1 \right). \tag{6.133}$$

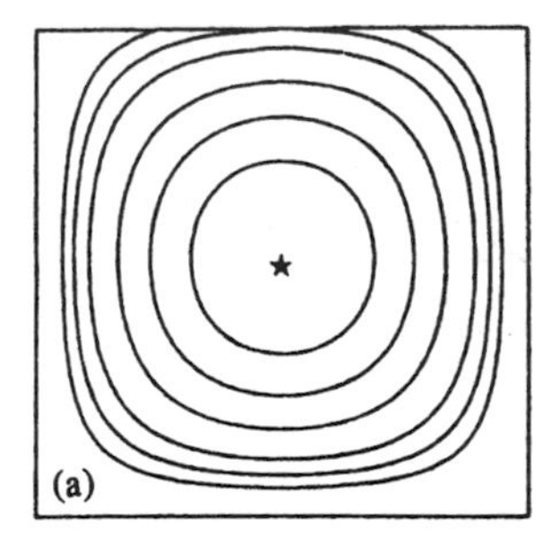

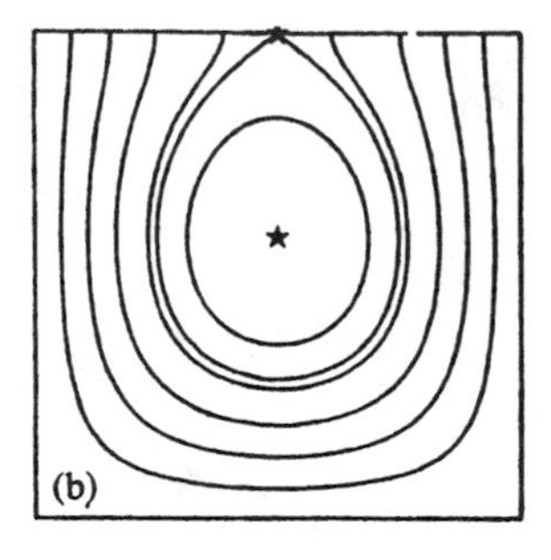

Figure 6.15. Streamline patterns at the onset of convection in an anisotropic layer with a kinematic boundary condition of the form $P + \lambda\, \partial P/\partial n = 0$ at the upper surface, for the case $\xi = 2$, $\eta = 1$, and $L = 1$. (a) $\lambda = 1$, recirculating fraction of flow $= 0.869$; (b) $\lambda = 0$, recirculating fraction of flow $= 0.290$. The stagnation point is marked with an asterisk (McKibbin, 1985; after McKibbin, 1986a).

The effects of dispersion, in addition to anisotropic permeability, were studied by Tyvand (1977, 1981). He found that the combined effects of anisotropy and dispersion may be much stronger than the separate effects.

Tyvand and Storesletten (1991) have analyzed the situation when the anisotropic permeability is transversely isotropic but the orientation of the longitudinal principal axes is arbitrary. The flow patterns now have either a tilted plane of motion or tilted cell walls if the transverse permeability is larger or smaller than the longitudinal permeability. Storesletten (1993) treated a corresponding problem where there is anisotropic thermal diffusivity. Zhang *et al.* (1993) studied numerically convection in a rectangular cavity with inclined principal axes of permeability.

A nonlinear stability analysis of the situation of Tyvand and Storesletten (1991), but with a quadratic density law, was conducted by Straughan and Walker (1996a). They obtained the dramatic result that, in contrast to the Boussinesq situation, the effect of anisotropy is to make the bifurcation into convection occur via an oscillatory instability.

The effect of anisotropy of the dispersive part of the effective thermal conductivity tensor, with a Forchheimer term included in the momentum equation, was investigated using numerical simulation by Howle and Georgiadis (1994), for two-dimensional steady cellular convection. They used the formula of Lage *et al.* (1992), Eq. (6.93), to determine experimental values of Ra_c and then plotted Nu versus Ra/Ra_c, thereby greatly reducing the divergence of experimental results found for the usual Nu versus Ra plot. They found that dispersion increased the net heat transfer after a Rayleigh number $\sim$ 100–200, and as the degree of anisotropy is increased, the wall averaged Nusselt number is decreased.

Joly and Bernard (1995) have computed values of Ra_c for an anisotropic porous medium bounded by anisotropic impermeable domains. Qin and Kaloni (1994) computed Ra_c values for the case of anisotropic permeability on the Brinkman model. A numerical study of the effects of anisotropic permeability and layering in seafloor hydrothermal systems was made by Rosenberg *et al.* (1993).

Linear stability analysis was applied to a conjugate problem with solid boundary plates by Gustafson and Howle (1999), and the results compared favorably with experiment. Mahidjiba *et al.* (2000c) applied linear and weak nonlinear analysis to a layer of finite lateral extent, and Mamou *et al.* (1998a) treated a layer for the case of constant heat flux on the boundaries. The effects of anisotropy on convection in both horizontal and inclined layers were studied by Storesletten (2004). The effects of nonuniform thermal gradient and transient effects were studied by Degan and Vasseur (2003), who studied a layer heated from the bottom with a constant heat flux and with the other surfaces insulated. The effect of radiative transfer was studied by Devi *et al.* (2002).

Anisotropy effects in general have been reviewed by Storesletten (1998). The later survey by Storesletten (2004) discussed various models for the anisotropy. It was noted that for horizontal layers, anisotropy affects the critical Rayleigh number and the critical wavenumber, but even the inclusion of three-dimensional anisotropy does not lead to any essentially new flow patterns at the onset of convection, provided that one of the principal axes of anisotropy is normal to the layer. When none of the principal axes are vertical, then new flow patterns, either with tilted plane of motion or with tilted as well a curved lateral walls, appear. For inclined layers, anisotropy has a strong influence on the preferred flow structure at the onset of convection. When the permeability is transversely isotropic, there are two cases. A permeability minimum in the longitudinal direction leads to longitudinal rolls for all inclinations. A permeability maximum in the longitudinal direction leads to transverse rolls when the inclination is less than a critical value and longitudinal rolls when the inclination is greater than that critical value. In the general case with anisotropy both in permeability and thermal diffusivity, either longitudinal rolls are favored for all inclinations or there is a transition from transverse rolls at lower inclinations to longitudinal rolls at higher inclinations via oblique rolls.

6.13. Effects of Heterogeneity

6.13.1. General Considerations

Extending previous work by Donaldson (1962), McKibbin (1983) calculated the criterion for the onset of convection and estimates of preferred cell width and heat transfer for two-dimensional convection in a system consisting of a permeable layer overlying an impermeable layer, the base of the impermeable layer being isothermal. McKibbin's results showed that, compared with a homogeneous permeable system of the same total depth, the presence of the impermeable layer increases the overall temperature difference required for instability, as well as reducing the subsequent heat flux when convection occurs. The critical value of a Rayleigh number based on the parameters of the permeable stratum is decreased by the presence of the impermeable layer, because of the relaxation of the thermal boundary condition at the base of the permeable stratum.

The marginal stability for a layer in which the thermal conductivity and the reciprocal of the permeability both vary linearly with depth (to an arbitrary extent) was studied by Green and Freehill (1969). Ribando and Torrance (1976) carried out numerical calculations of finite-amplitude convection for an exponential variation with depth of the ratio μ/K of viscosity to permeability. As expected, the strongest convection takes place in regions of small μ/K. A more general formulation of the onset problem where both the group μ/K and the thermal diffusivity vary with depth was made by Rubin (1981). A further study is that by Malkovski and Pek (1999).

6.13.2. Layered Porous Media

Studies of convection in general layered systems have been made by several investigators. The most comprehensive are those by McKibbin and O'Sullivan (1980, 1981), who studied both the onset of convection and subsequent heat transfer for a multilayered system bounded below by an isothermal impermeable surface and above by an isothermal surface that was either impermeable or at constant pressure. Two-dimensional flow patterns and associated values of Ra_c, cell width, and initial slope σ_* of the Nusselt number graph were calculated for two- and three-layer systems over a range of layer thickness and permeability ratios. The results show that significant permeability differences are required to force the layered system into an onset mode different from that for a homogeneous system. They also show that increasing contrasts ultimately lead to transition from "large-scale" convection (occurring through the entire system) to "local" convection confined mainly to fewer layers. Another conclusion is that σ_* depends strongly on the cell width (Fig. 6.16). An experimental study, using a Hele-Shaw cell modeling a three-layered system, by Ekholm (1983) yielded results in qualitative agreement with the theory of McKibbin and O'Sullivan (1980). The assumption of two-dimensionality used by many authors was examined by Rees and Riley (1990), who found criteria governing when the preferred flow patterns are three-dimensional and presented detailed results of the ranges of stable wavenumbers.

Gjerde and Tyvand (1984) studied a layer with permeability $K(z)$ of the form $K(z) = K_V/(1 + a \sin N\pi z)$, where K_V, a, and N are constants. They found that local convection never occurs in this smoothly stratified model.

Masuoka *et al.* (1988) made a numerical and theoretical examination of convection in layers with peripheral gaps. Hickox and Chu (1990) numerically simulated a geothermal system using a model involving three horizontal layers of finite horizontal extent. Delmas and Arquis (1995) reported an experimental and numerical investigation of convection in a layer with solid conductive inclusions.

For permeability fields that are anisotropic, layered, or both, Rosenberg and Spera (1990) performed time-dependent numerical simulations in a two-dimensional square box. They found that the time to steady state was proportional to the square-root of the kinetic energy. Their heat transfer results were consistent with previous results.

Masuoka *et al.* (1991, 1994, 1995a) made experimental and theoretical studies of the use of a thermal screen, consisting of a row of heat pipes with a very

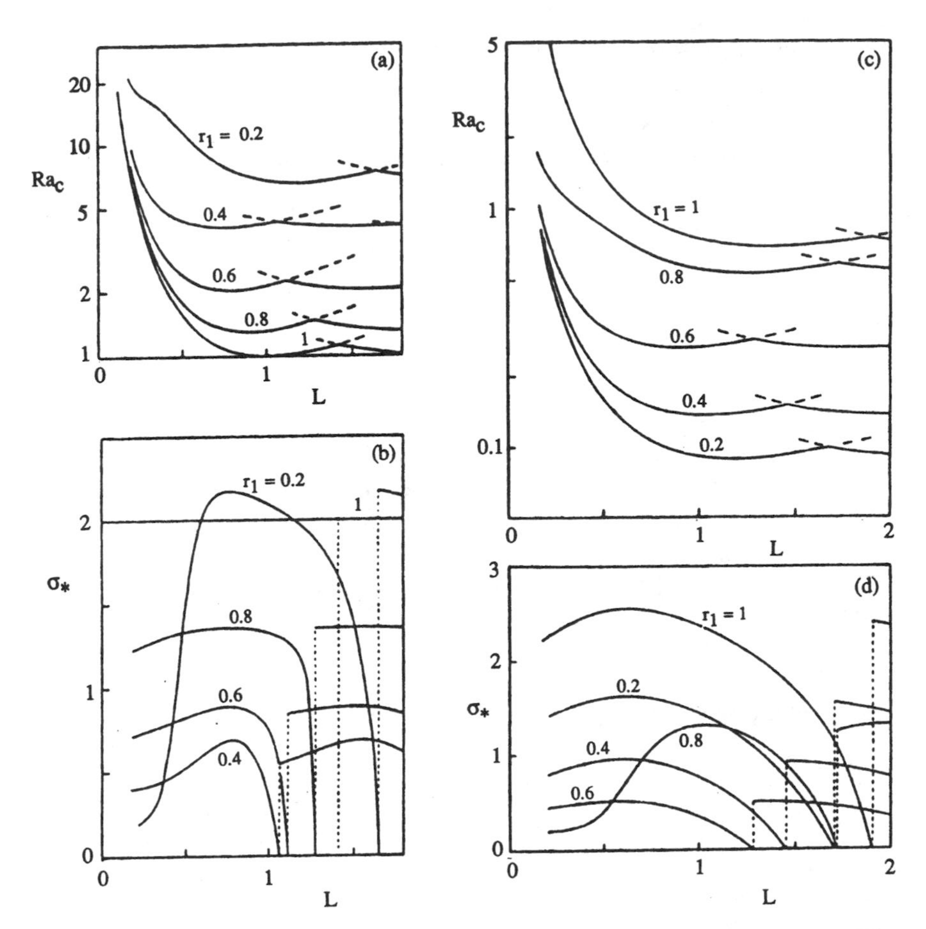

Figure 6.16. Variation of Ra_c and $\sigma*$ with system width L for two-layer systems. The lower layer occupies a fraction r_1 of the total depth, and the permeability contrast between upper and lower layers is K_2/K_1. (a), (b) Closed top, $K_2/K_1 = 0.1$, (c), (d) open top, $K_2/K_1 = 10$ (McKibbin, 1985; after McKibbin and O'Sullivan, 1981).

high effective thermal conductivity placed part way up the layer, in improving the insulation effect of a porous layer. The screen suppresses the onset of convection by making the temperature field more uniform.

Leong and Lai (2001) investigated the feasibility of using a lumped system approach using an effective Rayleigh number, on the basis of numerical calculations with two layers. Their results were generally anticipated by Nield (1994c). A similar study for a layered vertical porous annulus was made by Ngo and Lai (2000). Leong and Lai (2004) studied two or four layers in a rectangular cavity whose aspect ratio was either 0.2 or 5.0. They found that the convection is always initiated in the more permeable sublayer, and this convection penetrates to the less permeable sublayer as the Rayleigh number is further increased.

6.13.3. Analogy between Layering and Anisotropy

Wooding (1978) noted that there is a correspondence between layering and anisotropy in porous media. In a system in which the permeability K varies with the vertical coordinate z, the average horizontal and vertical permeabilities, in a layer of thickness H, are given by

$$\overline{K}_H = \frac{1}{H}\int_0^H K(z)dz, \quad \overline{K}_V = H/\int_0^H \frac{dz}{K(z)} \tag{6.134}$$

and so, since the arithmetic mean exceeds the harmonic mean,

$$\xi = \frac{\overline{K}_H}{\overline{K}_V} > 1. \tag{6.135}$$

A similar result applies for thermal conductivity. It implies that layering implies anisotropy with $\xi > 1, \eta > 1$. The result holds whether the layering is continuous or not, but the question is whether or not a transition to local convection will cause the analogy to break down. McKibbin and Tyvand (1982) explored this question. They concluded that the analogy is likely to be reliable in a continuously layered system, and also in a discretely layered system provided that the contrast between the layers is not too great.

McKibbin and Tyvand (1983, 1984) studied systems in which every second layer is very thin. If these thin layers have very small permeability (i.e., the layers are "sheets"), convection is large scale except when the sheets are almost impermeable. If the thin layers have very high permeability (i.e., the layers are "cracks"), then local convection is almost absent, so the analogy is more likely to be reliable for modeling. However, there is one feature of the crack problem that has no counterpart in the anisotropic model: there is a strong horizontal flow in the cracks (Fig. 6.17) and this affects the analogy.

6.13.4. Heterogeneity in the Horizontal Direction

The configuration where the porous medium consists of a number of homogeneous vertical slabs or columns of different materials is more difficult to study in comparison with the horizontally layered problem, and so far few studies have been published. McKibbin (1986b) calculated critical Rayleigh numbers, streamlines, and the variation of heat flux across the surface for a few examples involving inhomogeneity of permeability and thermal conductivity. His results are shown in Figs. 6.18–6.20. Here Ra_i denotes the Rayleigh number for material i,

$$\mathrm{Ra}_i = \frac{g\beta K_i H \Delta T}{\nu \alpha_{mi}}. \tag{6.136}$$

L is the horizontal to vertical aspect ratio of the entire system and r_i is the fraction of the total width occupied by material i.

Figure 6.18 shows that as the permeability contrast increases, so does Ra_{1c}, indicating, as expected, that a larger overall temperature gradient is required to

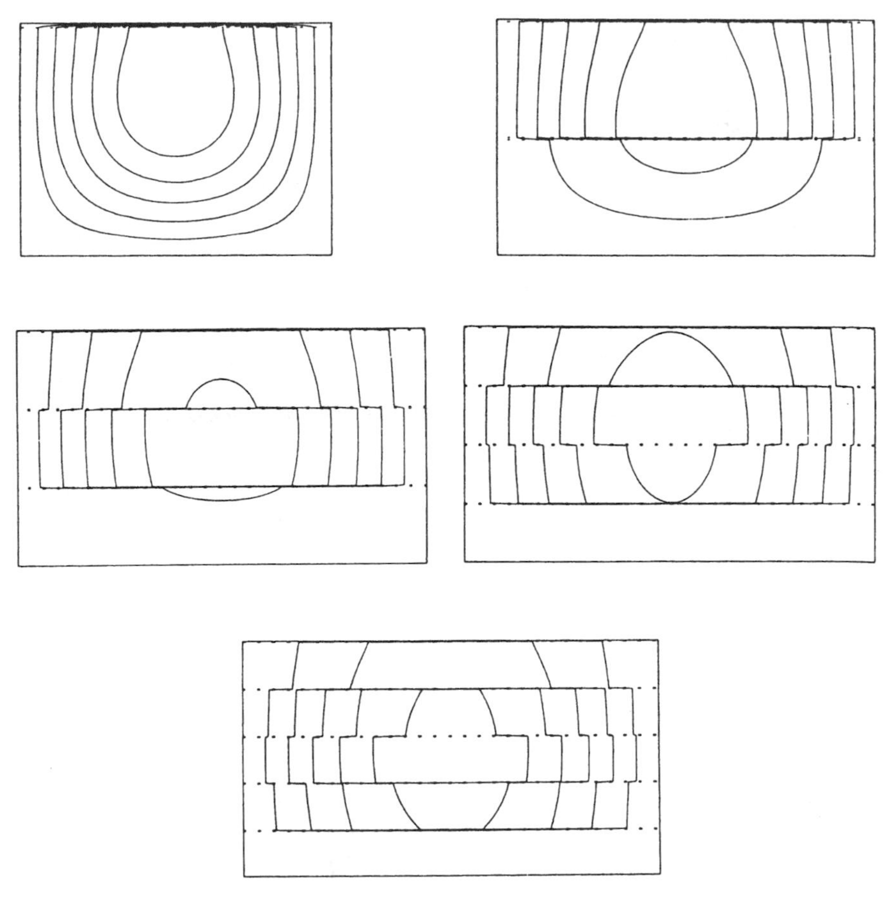

Figure 6.17. Streamlines at onset of convection in a system with thin, very permeable layers (cracks). The ratio of the thickness of each crack to that of the intervening material layers is 0.02 and the equivalent induced anisotropy in each case is $\xi = 10$ (McKibbin and Tyvand, 1984, with permission from Pergamon Press).

destabilize the conductive state of the system. One example of the streamline flow pattern is illustrated in Fig. 6.19a. Here the small amount of flow in the less permeable layer is reflected in the small and almost even increase in heat transfer at the surface due to convection. At the same time, the stronger flow in the more permeable section has a marked effect on the surface heat flux. Figure 6.19b illustrates a small thermal conductivity contrast. The strength of flow is slightly greater in the less conductive region. The jump in heat flux is due to the greater conductivity of material 2.

In the case of a thin, more permeable stratum cutting an otherwise homogeneous medium, as the thin stratum becomes more permeable there is a sudden transition from approximately square flow cells to a flow pattern where a very strong flow

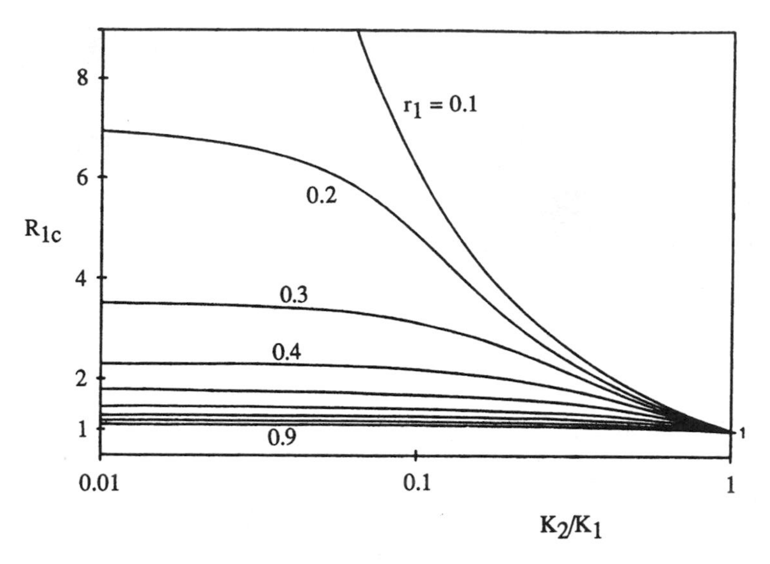

Figure 6.18. The critical Rayleigh number $R_{1c} = \mathrm{Ra}_{1c}/4\pi^2$ for a two-layer system with $L = 1, r_1 = 0.1$ (0.1) 0.9, and $0.01 = K_2/K_1 = 1.0$ (McKibbin, 1986b, with permission from Kluwer Academic Publishers).

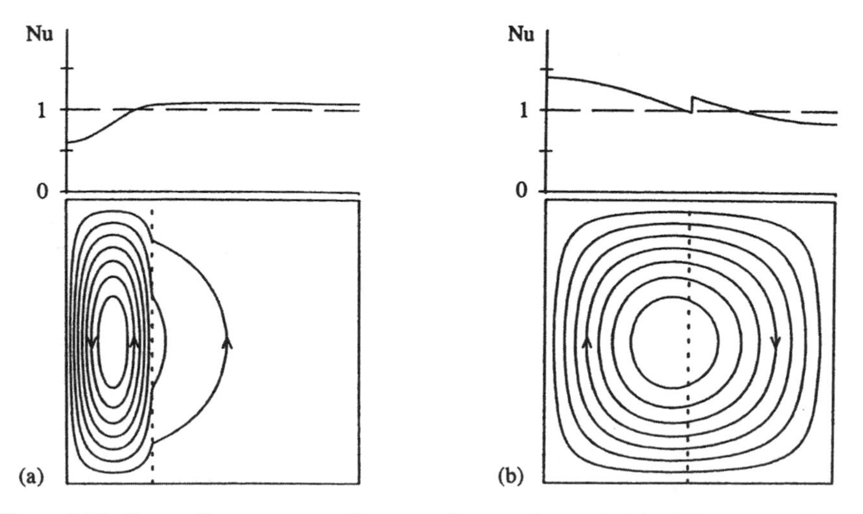

Figure 6.19. Streamline patterns at the onset of convection and typical Nusselt number at the surface, for slightly supercritical flow. In each case the overall aspect ratio $L = 1$. The subscripts 1 and 2 indicate regions numbered from left to right. $R = \mathrm{Ra}/4\pi^2$, where Ra is the Rayleigh number. (a) Permeability ratio $K_2/K_1 = 0.1$, $R_{1c} = 3.132$, $R_{2c} = 0.313$; (b) Thermal conductivity ratio $k_2/k_1 = 1.2$, $R_{1c} = 1.084$, $R_{2c} = 0.903$ (McKibbin, 1986b, with permission from Kluwer Academic Publishers).

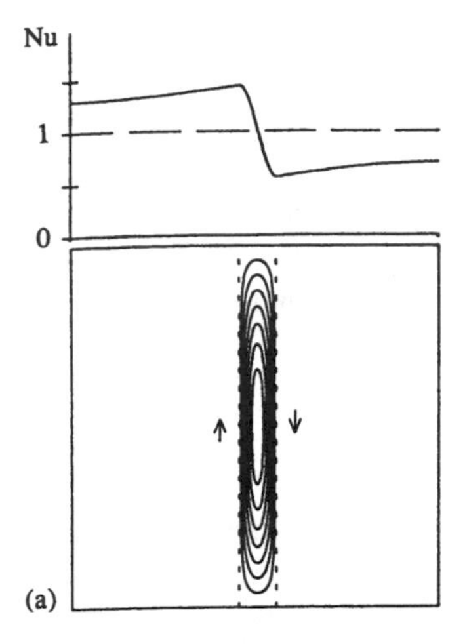

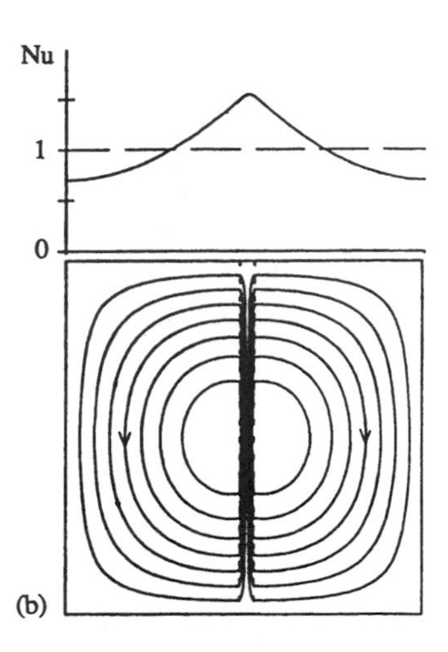

Figure 6.20. Streamline patterns and Nusselt number at the surface for the case of a centrally placed narrow stratum, permeability ratio $K_2/K_1 = 100$, and $K_3 = K_1$. (a) $r_2 = 0.1$, $R_{1c} = 0.267$, $R_{2c} = 26.7$; (b) $r_2 = 0.04$, $R_{1c} = 0.665$, $R_{2c} = 66.5$, where $R = \mathrm{Ra}/4\pi^2$ (McKibbin, 1986b, with permission from Kluwer Academic Publishers).

takes place up (or down) the permeable fault. For a narrower fault the permeability contrast needed for transition is greater. An example is shown in Fig. 6.20. The contrast between the flow patterns and the surface heat flux patterns is remarkable. This is different from the case of horizontal layering, where the spatial distribution of surface heat flux remains basically the same for all configurations, even though permeability and/or conductivity contrasts are great (McKibbin and Tyvand, 1984).

An approximate analysis of convective heat transport in vertical slabs or columns of different permeabilities was made by Nield (1987b). He took advantage of the fact that, when Darcy's law is applicable, one can superpose solutions of the eigenvalue problem for a single slab to obtain a feasible solution of the equations for the overall problem with the slabs placed side by side. This is, of course, an artificial flow since extra constraints have been imposed on the eigenvalue problem. In general, the actual flow will be one in which the convection induced in one slab will penetrate into adjacent slabs; one would expect that the actual flow would be more efficient at transporting heat than the artificial flow. This procedure leads to a lower bound for the true overall heat flux and an upper bound on a critical Rayleigh number.

Nield discussed some sample situations and also established a general result. If the heat transfer is given by $\mathrm{Nu} = g(\mathrm{Ra})$, then for sufficiently large values of Ra the second derivative $g''(\mathrm{Ra})$ is usually negative. It then follows that if Ra is supercritical everywhere, then for small and gradual variations in Ra with horizontal position the effect of inhomogeneity is to decrease the heat flux by a factor

$$1 + \left[\frac{g''(\overline{\text{Ra}})}{2g(\overline{\text{Ra}})} \right] \sigma_{\text{Ra}}^2 \tag{6.137}$$

relative to that for an equivalent homogeneous layer with the same Rayleigh number average $\overline{\text{Ra}}$. In the above expression, σ_{Ra}^2 is the variance of the Rayleigh number distribution. In particular, if we take $g(\text{Ra}) = \alpha \text{Ra}^{\beta}$, where $0 < \beta < 1$, then the reduction factor is

$$1 - \frac{1}{2}\beta(1-\beta)\frac{\sigma_{\text{Ra}}^2}{\overline{\text{Ra}}^2}. \tag{6.138}$$

Gounot and Caltagirone (1989) analyzed the effect of periodic variations in permeability. They showed that short-scale fluctuations had the same effect on stability as anisotropy. As expected, the variability causes the critical Rayleigh number based on the mean permeability to be raised and the Nusselt numbers to be lowered relative to the homogeneous values.

Vadasz (1990) used weakly nonlinear theory to obtain an analytic solution of the bifurcation problem for a heterogeneous medium for the case of heat leakage through the sidewalls. He showed that if the effective conductivity function $k_m(x, y, z)$ is not of the form $f(z)h(x, y)$, then horizontal temperature gradients (and hence natural convection) always must be present. A comprehensive study of convection in a layer with small spatial variations of permeability and effective conductivity was made by Braester and Vadasz (1993). For certain conductivity functions a motionless state is possible, and the stability of this was examined using weak nonlinear theory. A smooth transition through the critical Rayleigh number was found. Heterogeneity of permeability plays a relatively passive role compared with heterogeneity of thermal conductivity. For a certain range of supercritical Ra, symmetry of conductivity function produces symmetry of flow.

Convective stability for a horizontal layer containing a vertical porous segment having different properties was studied by Wang (1994). Convection in a rectangular box with a fissure protruding part way down from the top was treated numerically by Debeda *et al.* (1995).

6.14. Effects of Nonuniform Heating

O'Sullivan and McKibbin (1986) have performed a perturbation analysis and numerical calculations to investigate the effect of small nonuniformities in heating on convection in a horizontal layer. They found that $O(\varepsilon^3)$ variations in heating of the bottom generally produce variations of the same order in convection amplitude. However, if the distribution of the heating nonuniformity happens to have a wavelength equal to the wavelength of the preferred convection mode, then $O(\varepsilon^3)$ variations in heating produce an $O(\varepsilon)$ effect on the amplitude of convection at Rayleigh numbers within $O(\varepsilon^2)$ of the critical Rayleigh number Ra_c. This

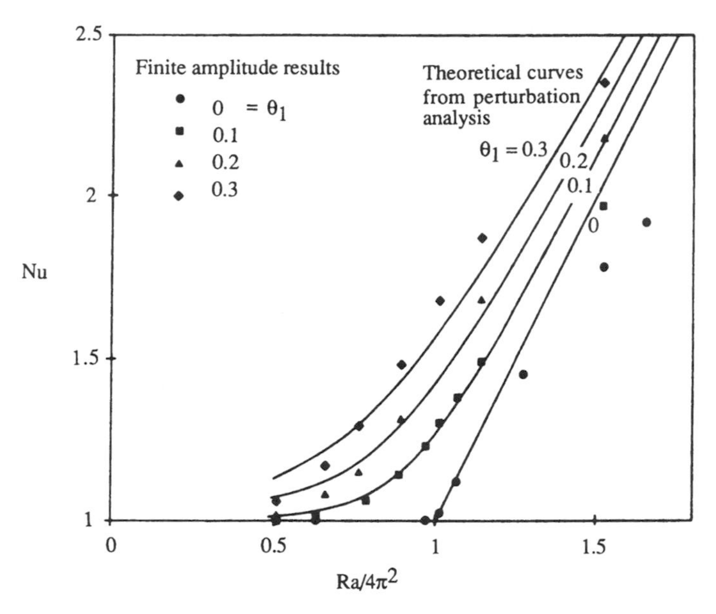

Figure 6.21. Effect of nonuniform heating on heat transfer. The calculations refer to a two-dimensional square container (length-to-height ratio = 1) with impermeable top and bottom and with insulated sidewalls. The dimensionless temperature distribution on the bottom is assumed to be $T(x, 0) = 1 + \theta_1 \cos \pi x$, $(0 \leq x \leq 1)$. (O'Sullivan and McKibbin, 1986).

produces a smoothing of the Nu versus Ra curve in the vicinity of the critical Rayleigh number, as shown in Fig. 6.21.

Rees and Riley (1989a,b) and Rees (1990), using weakly nonlinear theory, have considered in turn the consequences of excitation of near-resonant wavelength, nonresonant wavelength, and long-wavelength forms. When the modulations on the upper and lower boundaries are in phase, at the near-resonant wavelength, steady rolls with spatially deformed axes or spatially varying wavenumbers evolve. Rolls with a spatially varying wavenumber also evolve when the modulations are π out of phase. For a wide range of nonresonant wavelengths, a three-dimensional motion with a rectangular planform results from a resonant interaction between a pair of oblique rolls and the boundary forcing. Symmetric modulations of large wavelength can result in patterns of transverse and longitudinal rolls that do not necessarily have the same periodicity as the thermal forcing, but the most unstable transverse roll does have the same periodicity. For certain ranges of values of modulation wavelength the first mode to appear as Ra is increased is a rectangular cell of large-aspect-ratio plan form. This mode is a linear superposition of two rolls equally aligned at a small angle away from the direction of the longitudinal roll.

The effect of slightly nonuniform heating at the bottom of a parallelopipedic box on the onset of convection was analyzed by Néel (1992). Depending on the symmetry or otherwise of the heating, the nonuniformity can change the predicted pattern of steady convection for a particular choice of aspect ratios or even result in oscillatory convection.

A perturbation method was employed by Riahi (1993a) to study three-dimensional convection resulting when a nonuniform temperature with amplitude L^* is prescribed at the lower wall. When the wavelength γ_{bn} of the nth mode of modulation is equal to the critical wavelength γ_c for all n, regular or nonregular solutions in the form of multimodal pattern convection can become preferred in some range of L^*, provided the wave-vectors of such pattern are contained in the set of wave vectors representing the boundary temperature. There can be critical values L^*_c of L^* below which the preferred pattern is different from the one for $L^* > L^*_c$. For γ_{bn} equal to a constant different from γ_c, some three-dimensional solution in the form of multimodal convection can be preferred, even if the boundary modulation is one-dimensional, provided that the wavelength of the modulation is not too small. There are qualitatively similar results when the location (rather than the temperature) of the bottom boundary (and hence the depth) is modulated.

Riahi (1996) then extended his analysis to the case of a continuous finite bandwidth of modes. He found that the results were qualitatively similar to those for the discrete case. He also noted that large-scale flow structures are quite different from the small-scale flow structures in a number of cases and in particular they can exhibit kinks and can be nonmodal in nature. The resulting flow patterns can be affected accordingly, and they can provide quite unusual and nonregular three-dimensional preferred patterns. In particular, they are multiples of irregular rectangular patterns and they can be nonperiodic.

Rees and Riley (1986) conducted a two-dimensional simulation of convection in a symmetric layer with wavy boundaries. In this case the onset of convection is abrupt and is delayed by the presence of the nonuniformity. However, the onset of time-periodic flow takes place at much smaller Rayleigh numbers than those corresponding to the uniform layer. The mechanism generating unsteady flow is no longer a thermal boundary layer instability but rather a cyclical interchange between two distinct modes that support each other via the imperfection, and its onset is not a Hopf bifurcation. At relatively high amplitudes of the wavy surface, the basic flow may bifurcate directly to unsteady flow. Also, Riahi (1999), Ratish Kumar *et al.* (1997, 1998), Ratish Kumar (2000), Ratish Kumar and Shalini (2003, 2004c) have studied convection in a cavity with a wavy surface. The undulations generally lead to a reduced heat transfer.

Yoo and Schultz (2003) analyzed the small Rayleigh number convection in a layer whose lower and upper walls have sinusoidal temperature distributions with a phase difference. They found that for a given wavenumber, an out-of-phase configuration yields minimum heat transfer on the walls, and that maxiumum heat transfer occurs at the wavenumber value 2.286 with an in-phase configuration. Capone and Rionero (2003) considered the nonlinear stability of a vertical steady flow driven by a horizontal periodic temperature gradient.

6.15. Rectangular Box or Channel

6.15.1. Linear Stability Analysis, Bifurcation Theory, and Numerical Studies

In a horizontal layer, with vertical heating, the lateral boundaries of the convection cells are vertical and there is no heat transfer across them. This means that, assuming that slip is allowed on a rigid wall, an impermeable insulating barrier can be placed at a cell boundary without altering the flow. Consequently, Ra_c remains at $4\pi^2$ (for the case of impermeable conducting horizontal boundaries) if the nondimensional width and breadth ($L_x/H = h_1$ and $L_y/H = h_2$, for the box $0 = x = L_x, 0 = y = L_y, 0 = z = H$) of a rectangular box are integral multiples of $2\mathrm{p}/\pi_\mathrm{c}$. For other values of width and breadth the value of Ra_c is raised above $4\pi^2$. This is because the minimization of $(\pi^2 + \alpha^2)^2/\alpha^2$, where $\alpha^2 = l^2 + m^2$, is now over discrete values of the wavenumbers l and m rather than over continuous values. Eigenmodes are represented by the stream functions

$$\Psi_{pqr} = \sin p\pi \frac{\hat{x}}{h_1} \sin q\pi \frac{\hat{y}}{h_2} \sin r\pi\hat{z} \tag{6.139}$$

for integers p, q, and r. The corresponding Rayleigh numbers are

$$\mathrm{Ra}_{pqr} = \pi^2 \left(b + \frac{r^2}{b}\right)^2, \tag{6.140}$$

where

$$b = \left[\left(\frac{p}{h_1}\right)^2 + \left(\frac{q}{h_2}\right)^2\right]^{1/2}. \tag{6.141}$$

Thus the critical Rayleigh number is given by

$$\mathrm{Ra}_c = \pi^2 \min_{(p,q,r)} \left(b + \frac{r^2}{b}\right)^2 = \pi^2 \min_{(p,q)} \left(b + \frac{1}{b}\right)^2. \tag{6.142}$$

The minimization problem over the sets of nonnegative integers p and q for a set of values h_1 and h_2 has been solved by Beck (1972) and the results are displayed in Figs. 6.22 and 6.23. Figure 6.22 shows that the value of Ra_c rapidly approaches $4\pi^2$ as either h_1 or h_2 becomes large, so that the lateral walls have little effect on the critical Rayleigh number except in tall boxes with narrow bases, for which $h_1 \ll 1$ and $h_2 \ll 1$. The preferred cellular mode (p, q) as a function of (h_1, h_2) is shown in Fig. 6.23. Note the symmetry with respect to the line $h_1 = h_2$. The modal exchange between the rolls $(p, 0)$ and $(p + 1, 0)$ occurs at $h_1 = [p(p + 1)]^{1/2}$. Two-dimensional rolls are preferred when the height is not the smallest dimension (i.e., when $h_1 < 1$ or $h_2 < 1$) and that a roll that has the closest approximation to a square cross section is preferred.

Using techniques of bifurcation theory, Riley and Winters (1989) have investigated the mechanics of modal exchanges as Ra and $h_1 \equiv h$ vary, for a two-dimensional cavity. They use a synthesis of degree theory, symmetry arguments,

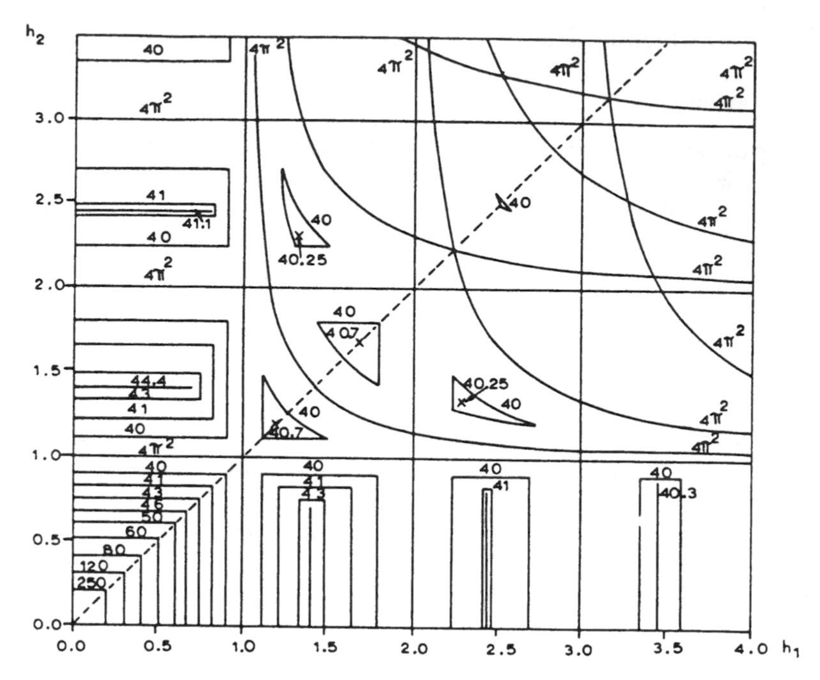

Figure 6.22. Variation of the critical Rayleigh number Ra_c in an enclosed three-dimensional porous medium as a function of h_1 and h_2 (Beck, 1972).

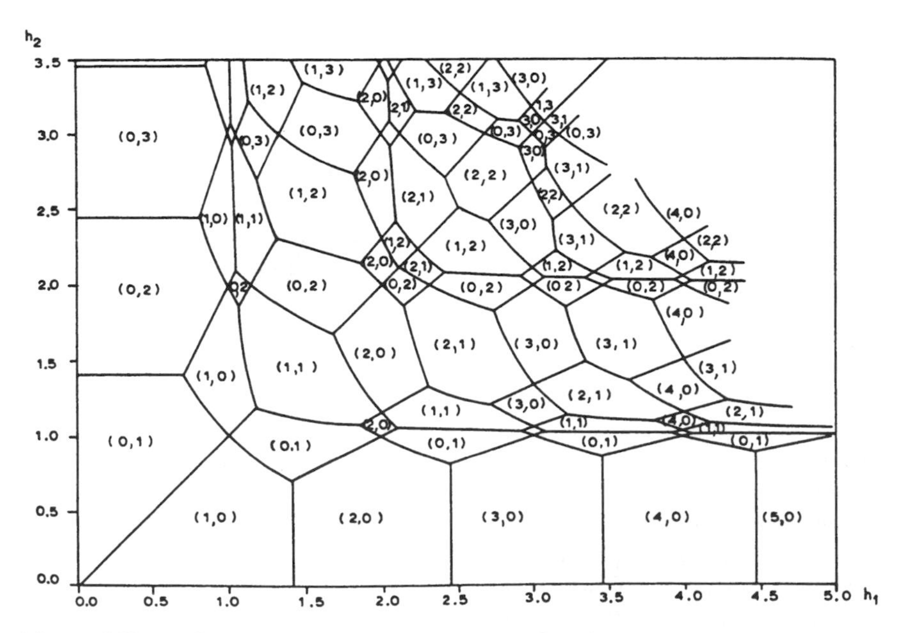

Figure 6.23. Preferred cellular mode (p, q) as a function of h_1 and h_2 in a three-dimensional box filled with a porous medium (Beck, 1972).

and continuation methods. They show that as h increases for fixed Ra, primary bifurcations (from conduction states) occur and then secondary bifurcations. At a secondary bifurcation, a previously unstable mode can regain stability. Thus the behavior of physical bifurcations is intimately connected with that of unphysical ones, and the stability boundary for one-cell flows turns out to be quite complicated. Impey *et al.* (1990) extended the work of Riley and Winters (1989) to include the effects of small tilt and small side-wall heat flux.

Beck's work has been extended to other types of boundary conditions. Tewari and Torrance (1981) considered the case of a permeable upper boundary; their results are as expected. A general feature is that when both breadth and width become small, α_c becomes large and $\mathrm{Ra}_c \sim \alpha_c^2$. In this case the perturbation quantities w and θ vary only slowly with the vertical coordinate z, and the horizontal components are negligible. When the lateral boundaries are not insulating, new features appear. Chelghoum *et al.* (1987) found that if lateral boundaries are conducting rather than insulating, Ra_c is raised and two-dimensional rolls [modes of type $(p, 0)$ and $(0, q)$] are eliminated in favor of modes of type $(p, 1)$ and $(1, q)$, and when h_1 and h_2 are not small, the modal picture is complicated.

Convection in rectangular boxes has been the topic of many numerical studies. Some of this work has been referred to in Section 6.8. For the two-dimensional case, Horne and O'Sullivan (1978a) and Horne and Caltagirone (1980) reported studies of oscillatory convection, while Schubert and Straus (1982), for a square cavity, found a succession of transitions as Ra is increased.

The three-dimensional case has been studied by Horne (1979), Straus and Schubert (1979, 1981), and Caltagirone *et al.* (1981). A noteworthy discovery was that different steady structures develop with time, the final form depending on the initial conditions. An analytical study by Steen (1983) has complemented and corrected results of the numerical studies. The top and bottom of the box are taken to be isothermal and the sides insulated. Steen showed that for a cubic box convection first occurs at $\mathrm{Ra} = 4\pi^2$, and then a two-dimensional roll cell grows to a finite-amplitude pattern with Ra increasing. Immediately above criticality it is the only stable pattern; the three-dimensional state found by Zebib and Kassoy (1978) is unstable. Another three-dimensional pattern comes into existence as a linear mode grows at $\mathrm{Ra} = 4.5\pi^2$. It remains unstable from birth until $\mathrm{Ra} = 4.87\pi^2$, when it gains stability and begins to compete with the two-dimensional pattern. These two- and three-dimensional patterns remain the only stable states up to a value of Ra (about 1.5 Ra_c) when other modes become important. Steen calculated that, provided all disturbances of unit norm are equally likely, there is a 21 percent chance that the three-dimensional pattern will be selected at $\mathrm{Ra} = 50$.

Other work on pattern selection and bifurcation in rectangular boxes has been reported by Steen (1986), Kordylewski and Borkowska-Pawlak (1983), Borkowska-Pawlak and Kordylewski (1982,1985), Kordylewski *et al.* (1987), Vincourt (1989a,b), and Néel (1990a,b). The study by Riley and Winters (1991) focused on the destabilization, through Hopf bifurcations (leading to time-periodic convection), of the various stable convective flow patterns. There is a complex evolution of the Hopf bifurcation along the unicellular branch as the aspect ratio h increases. Steady unicellular flow is stable for a range of Ra values that is $(4\pi^2, 390.7)$ at $h = 1$ and becomes increasingly narrow and finally disappears when h exceeds 2.691. Riley and Winters also obtained an upper stability bound for steady multicellular flows. They found that stable m cells exist only for $h < 2.691\mathrm{m}$.

An argument of Howard type, based on the Bénard-Rayleigh instability in boundary layers at the top and bottom surfaces, leads to the asymptotic scaling laws Nu $\sim$ Ra and $f \sim \mathrm{Ra}^2$ for the mean Nusselt number and the characteristic frequency f. For convection in a square, Graham and Steen (1994) computationally studied the regime from Ra = 600 to 1250. They found that as Ra increases a series of traveling waves with spatial wavenumber n appear, each born at a Hopf bifurcation. Modal interactions of these lead to quasiperiodic mixed modes (whose complicated behavior was studied by Graham and Steen (1992)). The Ra range studied is characterized by thermal plumes and overall follows the asymptotic scaling behavior, but the plumes drive resonant instabilities that lead to windows of quasiperiodic, subharmonic, or weakly chaotic behavior. The plume formation is disrupted in these windows, causing deviations from the simple scaling behavior. The instability is essentially a phase modulation of the plume formation process. Graham and Steen argue that each instability corresponds to a parametric resonance between the timescale for plume formation and the characteristic convection timescale of the flow. A computational comparison between classic Galerkin and approximate inertial manifold methods was made by Graham *et al.* (1993). Extensions of this work involving Gevrey regularity were conducted by Ly and Titi (1999) and Oliver and Titi (2000). A stability analysis based on a generalized integral transform technique involving transitions in the number of cells was carried out by Alves, Cotta, and Pontes (2002).

Rees and Tyvand (2004a,b) considered convection in cavities with conducting boundaries. In this case linear stability analysis leads to a Helmholtz equation that governs the critical Rayleigh number and makes it independent of the orientation of the porous cavity. They numerically solved the eigenvalue equation for cavities of various shapes. Rees and Tyvand (2004c) found that for a two-dimensional cavity with one lateral wall thermally conducting and the other thermally insulating and open, the mode of onset of convection is oscillatory in time, corresponding to a disturbance traveling as a wave through the box from the impermeable wall to the open wall.

General surveys of this subject have been done by Rees (2000), Tyvand (2002), and Straughan (2004b). Straughan (2001a) has discussed the calculations of eigenvalues associated with porous convection. In particular, Tyvand (2002) considered a two-dimensional rectangular container with closed and conducting top and bottom and with various combinations of kinematic and thermal boundary conditions on the left- and right-hand walls. His results for the values of the critical Rayleigh number are presented in Table 6.3. The corresponding streamline patterns may be found in Tyvand (2002). For a three-dimensional box with general lateral boundary conditions no simple analytical solution is possible.

Table 6.3. Values of the critical Rayleigh number Ra_c for various lateral boundary conditions, for the onset of convection in a rectangle of height H and width L.The top and bottom are assumed impermeable and conducting (after Tyvand, 2002).

IMP: impermeable (closed) FRE: free (open)
CON: conducting INS: insulating

Left-hand wall	Right-hand wall	Ra_c
IMP/INS	IMP/INS	$\pi^2 \min [(nH/L) + (L/nH)]^2$
FRE/CON	FRE/CON	$\pi^2 \min [(nH/L) + (L/nH)]^2$
IMP/CON	IMP/CON	$4\pi^2 [1 + H^2/L^2]$
FRE/INS	FRE/INS	$4\pi^2 [1 + H^2/L^2]$
IMP/INS	FRE/CON	$\pi^2 \min [(nH/2L) + (2L/nH)]^2$
IMP/INS	IMP/CON	$4\pi^2 [1 + H^2/4L^2]$
FRE/CON	IMP/CON	$4\pi^2 [1 + H^2/4L^2]$
FRE/CON	FRE/INS	$4\pi^2 [1 + H^2/4L^2]$
IMP/INS	FRE/INS	$4\pi^2 [1 + H^2/4L^2]$
FRE/INS	IMP/CON	unknown

Following earlier work by Schubert and Straus (1979) and Horne and O'Sullivan (1978a), three-dimensional convection in a cube was treated by Kimura *et al.* (1989). They found a transition from a symmetric steady state (S) to a partially nonsymmetric steady state (S$'$, vertical symmetry only) at Ra about 550. At Ra of 575 the flow became oscillatory ($P^{(1)}$) with a single frequency that increased with Ra. It became quasiperiodic at a value of Ra between 650 and 680, returned to a simple periodic state in a narrow range about Ra = 725, and then became quasiperiodic again. Thus the three-dimensional situation was similar to the two-dimensional one, except for the higher critical Ra at the onset of oscillations (575 vs. 390) and a corresponding higher frequency (175 vs 82.5) and except for the transition $S \rightarrow S'$; however, this was dependent on step size and it was possible that it might not occur prior to $S \rightarrow P^{(1)}$ for sufficiently small steps in Ra.They also noted that the (time-averaged) Nusselt number for the three-dimensional flows was generally greater than that for the two-dimensional flows.

The transition from steady to oscillatory convection in a cube was found by Graham and Steen (1991) to occur at Ra = 584 and to involve a traveling wave instability in which seven pairs of thermal blobs circulate around the cube. They also observed a correspondence between the three-dimensional convection and two-dimensional flow in a box of square planform but with aspect ratio $2^{-1/2}$.

Further numerical calculations for convection in a cube were performed by Stamps *et al.* (1990), For the case of insulated vertical sides, they found simply periodic oscillations with frequency $f \propto Ra^{3.6}$ appearing for Ra between 550 and 560 and irregular fluctuations once Ra exceeded a value between 625 and 640. When heat is transferred through the vertical sides of the cube, three different flow patterns could occur, depending on Ra and the rate of heat transfer. Sezai

(2005) used the Brinkman-Forchheimer model in his treatment of a cube with impermeable adiabatic walls. He carried out computations for Ra up to 1000. He observed a total of ten steady-flow patterns, of which five show oscillatory behavior for some Rayleigh-number range.

The general topic of oscillatory convection in a porous medium has been reviewed by Kimura (1998). Analysis of the onset of convection in a sector-shaped box [analogous to that of Beck (1972) for a rectangular box] was reported by Wang (1997). The case of a box with a rigid top or a constant pressure top and with constant flux bottom heating was analyzed by Wang (1999b, 2002).

6.15.2. Thin Box or Slot

Geological faults can be modeled by boxes that are short in one horizontal dimension but long in the other two dimensions. Convection in such boxes has been studied by Lowell and Shyu (1978), Lowell and Hernandez (1982), Kassoy and Zebib (1978), and Murphy (1979). Lowell and Shyu (1978) were concerned with the effect of a pair of conducting lateral boundaries (the other pair being insulated). Lowell and Hernandez (1982) used finite-difference techniques to investigate finite-amplitude convection. They found that in containers with prescribed wall temperatures the flow was weakly three-dimensional but with the general appearance of two-dimensional transverse rolls. In containers bounded by impermeable blocks of finite thermal conductivity, a flow pattern similar to that for containers with prescribed wall temperatures tended to be set up, but asymmetrical initial perturbations tended to give rise to slowly evolving flows. Kassoy and Zebib (1978) studied the development of an isothermal slug flow entering the fault at large depth. An entry solution and the subsequent approach to the fully developed slows were obtained for the case of large Rayleigh number.

Convection in a thin vertical slot has been analyzed by Kassoy and Cotte (1985), Weidmann and Kassoy (1986), and Wang *et al.* (1987). They found that the appearance of slender fingerlike convection cells is characteristic of motion in this configuration, and the streamline pattern is extremely sensitive to the value of Ra. For the case of large wavenumber and insulated sidewalls, Lewis *et al.* (1997) present asymptotic analyses for weakly nonlinear and highly nonlinear convection. They found that three separate nonlinear regimes appear as the Rayleigh number increases but convection remains unicellular. On the other hand, for the case of perfectly conducting boundaries and with a linearly decreasing temperature profile imposed at the sidewalls, Rees and Lage (1996) found that for all cell ratios the onset problem is degenerate in the sense that any combinations of an odd mode and an even mode is destabilized simultaneously at the critical Rayleigh number This degeneracy persists even into the nonlinear regime. For the case of particular linear distributions of temperature on the vertical walls, Storesletten and Pop (1996) obtained an analytical solution. Some implications for hydrothermal circulation for hydrothermal circulation along mid-ocean ridges or for the thermal regime in crystalline basements and for heat recovery experiments were discussed by Rabinowicz *et al.* (1999) and Tournier *et al.* (2000).

6.15.3. Additional Effects

The effect of large-scale dependence of fluid density on heat transfer has been numerically investigated by Marpu and Satyamurty (1989). Nilsen and Storesletten (1990) have analyzed two-dimensional convection in horizontal rectangular channels with the lateral walls (as well as the horizontal boundaries) permeable and conducting. They have treated both isotropic and anisotropic media. They showed that Ra_c depends on the anisotropy-aspect ratios ξ and η defined by

$$\xi = \frac{K_H}{K_V}\left(\frac{H}{L}\right)^2, \quad \eta = \frac{\alpha_{mH}}{\alpha_{mV}}\left(\frac{H}{L}\right)^2, \tag{6.143}$$

where K_H, and K_V are the horizontal and vertical permeabilities, α_{mH} and α_{mV} are the horizontal and vertical thermal diffusivities, and L and H are the horizontal and vertical dimensions of the channel.

For the case $\xi = \eta$, which includes the isotropic situation,

$$\mathrm{Ra}_c = 4\pi^2(1+\xi). \tag{6.144}$$

This may be compared with the result $\mathrm{Ra}_c = 4\pi^2$ for insulating walls; as expected, the effect of conductivity of the walls is stabilizing. There are two possible cell patterns, each with symmetrical streamlines. For $n = 2, 3, 4, \ldots$, they consist of n and $n+1$ cells, respectively, if

$$(n-1)^2 - 1 < \xi^{-1} \leq n^2 - 1. \tag{6.145}$$

The conclusion of weakly nonlinear stability analysis is that both structures are stable against two-dimensional perturbations. Compositions of this pair of flow patterns are possible, so the flow is not uniquely determined by the boundary conditions.

The situation is similar for the case $\xi \neq \eta$, but now there is only a single steady flow pattern (stable against two-dimensional disturbances) which consists of n cells if $\xi < \eta$ and $n+1$ cells if $\xi > \eta$, where

$$(n-1)^2 - 1 < (\xi\eta)^{-1/2} < n^2 - 1. \tag{6.146}$$

The problem of convection induced by internal heat generation in a box was given a theoretical and experimental treatment by Beukema and Bruin (1983).

The theory in this section has been based on the assumption that the sidewalls are perfectly insulating. Vadasz *et al.* (1993) showed that for perfectly conducting sidewalls convection occurs regardless of the Rayleigh number and regardless of whether the fluid is heated from below, except for a particular sidewall temperature variation. When there is no temperature difference between the sidewalls, and with heating from below, a subcritical flow results mainly near the sidewalls and this amplifies and extends over the entire domain under supercritical conditions. The authors treated cases with heating from above as well as heating from below.

Weak nonlinear theory was applied by Vadasz and Braester (1992) to the case of imperfectly insulated sidewalls. There is now a smooth transition of the amplitude of convection with increase of Ra from subcritical values, but a three branch bifurcation develops at higher Ra values, with two branches stable. For slightly supercritical Ra, the amplitude and direction of the convection currents are uniquely determined by the heat leakage through the lateral walls. In this situation there is weak convection at relatively low Rayleigh numbers and this grows sharply in strength near the classical critical Rayleigh number; a second stable flow exists within the weakly nonlinear regime if the Rayleigh number is sufficiently large.

Convection in a square box with a wavy bottom was studied numerically by Murthy *et al.* (1997). They found flow separation and attachment on the walls of the box for Ra around 50 and above, with the manifestation of cycles of unicellular and bicellular clockwise and counterclockwise flows. The counterflow on the wavy wall hinders heat transfer into the system by an amount that increases with wave amplitude or wave number.

The effect of harmonic oscillation of the gravitational acceleration was studied numerically by Khallouf *et al.* (1996). Numerical studies involving a transient situation or an oscillating boundary were reported by Jue (2001a,b).

6.16. Cylinder or Annulus

6.16.1. Vertical Cylinder or Annulus

Following Wooding (1959), for the case of a thin circular cylinder one can assume that θ and w are independent of z, and then Eq. (6.17) gives

$$\frac{dP}{d\hat{z}} = \mathrm{Ra}\,\hat{T} - \hat{w} = C, \tag{6.147}$$

where C is a "separation of variables" constant that can be taken as zero. For marginal stability, Eq. (6.18) reduces to

$$\nabla_H^2 \hat{T} = -\hat{w}. \tag{6.148}$$

Eliminating $\hat{w}$ from Eqs. (6.145) and (6.146) gives

$$\nabla_H^2 \hat{T} + \mathrm{Ra}\,\hat{T} = 0. \tag{6.149}$$

The solutions of this equation, which are periodic functions of φ and are finite at $\hat{r} = 0$, where $(\hat{r}, \varphi)$ are polar coordinates, have the form

$$\hat{T}_n = C_n J_n(\lambda \hat{r}) \cos n\varphi \quad (n = 0, 1, 2, \ldots), \tag{6.150}$$

where $\lambda = \mathrm{Ra}^{1/2}$ and J_n is the Bessel function of order n. The eigenvalues for this problem are determined by the temperature boundary conditions. For example, if we have an insulated surface at $\hat{r} = r_0/H$, so that $\partial\hat{T}/\partial\hat{r} = 0$, then

$$J_n'\left(\lambda \frac{r_o}{H}\right) = 0. \tag{6.151}$$

The smallest possible value of λ is attained when $n = 1$ (corresponding to flow antisymmetric with respect to a diameter) and the critical Rayleigh number is

$$\mathrm{Ra}_c = \lambda_1^2 \frac{H^2}{r_o^2} = 3.390 \frac{H^2}{r_o^2}. \tag{6.152}$$

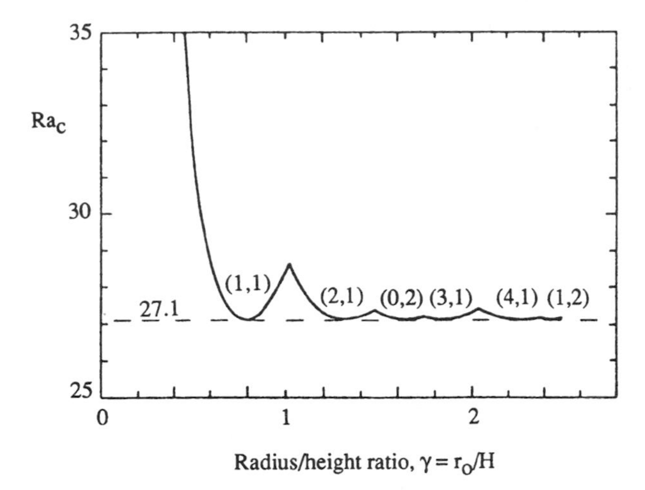

Figure 6.24. Critical Rayleigh number and the preferred convective modes (m, p) at the onset of convection in a vertical cylinder. The temperature perturbation is of the form $\theta = \Theta(z J_m(Z_{mp} r/\gamma) \cos m\varphi$, where Z_{mp} is the pth zero of $J_{m'}(x)$. The dashed line indicates the value for an infinite horizontal layer with isothermal boundaries, the lower impermeable and the upper permeable (Bau and Torrance, 1982b).

This can be written as $\tilde{\text{Ra}}_c = 3.390$, where

$$\tilde{\text{Ra}} = \text{Ra}\frac{r_o^2}{H^2} = \frac{g\beta K r_o^2 \Delta T}{H\nu\alpha_m}. \tag{6.153}$$

Further analysis of convection in a vertical cylinder was reported by Zebib (1978) and Bories and Deltour (1980) (who considered the effects of finite conduction in the surrounding medium) for the case of impermeable boundaries and by Bau and Torrance (1982b) for the case of a permeable upper boundary. The variation of Ra_c versus aspect (radius to height) ratio γ for the latter case is shown in Fig. 6.24. The preferred mode is asymmetric except for a limited range of γ. Experiments by Bau and Torrance (1982b) for situations with γ in the range 0.2–0.3 confirmed the prediction that the mode of onset of convection was asymmetric. Their heat transfer data for moderately supercritical convection was in accord with their calculations. When Ra reached a value 5.5 Ra_c, there was a transition to oscillatory convection (like that occurring in a horizontal layer).

Convection in the annulus between vertical coaxial cylinders was analyzed by Bau and Torrance (1981). Again the preferred mode of convection is asymmetric. Experiments with this geometry, with constant heat flux on the inner cylinder and constant temperature on the outer and with a permeable, constant pressure upper surface, were reported by Reda (1983). The measured distribution of temperature was in accord with numerical predictions. These results are pertinent to the design of nuclear waste repositories.

A numerical and experimental study of two-dimensional convection was reported by Charrier-Mojtabi *et al.* (1991). The experiments, in which the

Christiansen effect was employed for visualization, were in good agreement with the numerical results.

The onset of convection in a cylindrical enclosure with constant flux bottom heating and either an impermeable or permeable top was analyzed by Wang (1998b, 1999c). Convection in a cylindrical enclosure filled by a heat generating porous medium was studied numerically by Das *et al.* (2003). Tyvand (2002) noted that the lateral boundary conditions employed in the papers by Zebib (1978), Bau and Torrance (1982b), and Wang (1998b) are identical and that the transformation $(\text{Ra}/4\pi^2, \pi x, \pi y) \rightarrow (\text{Ra}/27.21, 2.33x, 2.33y)$ allows one to deduce the results of the second and third papers from the results of the first. The same transformation allows the results of Tewari and Torrance (1981) to be deduced from those of Beck (1972). Tyvand (2002) also studied convection in a vertical hexagonal cylinder with impermeable boundaries, a conducting top and bottom, and insulating lateral walls.

6.16.2. Horizontal Cylinder or Annulus

Lyubimov (1975) considered Rayleigh-Benard convection in a circular horizontal porous cylinder but he did continue the analysis to identify preferred modes and the critical Rayleigh number. Storesletten and Tveitereid (1987) analyzed two-dimensional convective motion in a circular horizontal cylinder. They calculated Ra_c to be 46.265, where Ra is now defined as

$$\text{Ra} = \frac{g\beta \Delta T K r_o}{\nu \alpha_m}, \tag{6.154}$$

where r_0 is the radius of the cylinder and ΔT is the temperature difference across the vertical diameter. At moderately supercritical Rayleigh numbers they found two steady flow patterns consisting of two or three cells, respectively, both structures being stable. The first mode involves two counterrotating cells with strictly vertical motions (upward or downward) in the middle. The second mode consists of three cells: one dominating central roll occupying most of the area flanked by two smaller rolls. In their numerical study, Robillard *et al.* (1993) obtained 46.6 as the critical value. The situation for a cylinder of length L with insulated ends was studied by Storesletten and Tveitereid (1991). For $L > 0.86$, a unique three-dimensional flow appears at the onset of convection, while for $L < 0.86$ the flow is two-dimensional with two or three rolls, each flow being stable, but with thermal forcing the flow is uniquely determined. The effect of weak rotation was studied by Zhao *et al.* (1996).

A bifurcation study of two-dimensional convection was made by Bratsun and Lyubimov (1995). The degeneracy (infinite number of solutions) is removed when fluid seeps through the boundaries either vertically or horizontally. At large Ra a quasiperiodic solution branches from a limit cycle for both types of seepage. The reduction of heat transfer in horizontal eccentric annuli, involving a transition from tetracellular to bicellular flow patterns, was studied numerically by Barbosa Mota and Saatdjian (1997). A numerical treatment of a horizontal annulus filled with an anisotropic porous medium was reported by Aboubi *et al.* (1998). Convection in a thin horizontal shell of finite length with impermeable walls was examined by Tyvand (2002), who also considered a similar problem with a thin spherical shell.

6.17. Internal Heating in Other Geometries

In Section 6.11.2 we discussed internal heating in an infinite horizontal layer. We now discuss internal heating in other geometrical configurations.

Blythe *et al.* (1985a) analyzed two-dimensional convection driven by uniformly distributed heat sources within a rectangular cavity whose vertical side walls are isothermal and whose horizontal boundaries are adiabatic. In the limit of large internal Rayleigh number $\mathrm{Ra_I}$ [defined in Eq. (6.114)] they found that boundary layers of thickness of order $\mathrm{Ra}_I^{-1/3}$ formed on the side walls, the internal core being stratified in the vertical direction. Further work on this geometry is the numerical studies by Haajizadeh *et al.* (1984) and Prasad (1987). The latter obtained heat transfer results for Ra_I up to 10^4 and for aspect ratios A in the range 0.5 to 20. These authors reported unicellular flow for the entire range of Ra_I and A and stratification in the upper layers of the cavity. Prasad (1987) also examined the effect of changing the boundary conditions on the horizontal walls from adiabatic to isothermally cooled.

Banu *et al.* (1998) noted that in the situation described by Blythe *et al.* (1985a) the upper part of the cavity is unstably stratified and so the flow described by Blythe *et al.* is unlikely to be realized in practice. The numerical study of Banu *et al.* (1998) showed that incipient unsteady flow occurs at values of $\mathrm{Ra_I}$ that are highly dependent on the aspect ratio of the cavity. The convective instabilities of the time-dependent motion are confined to the top of the cavity and for tall thin cavities the critical $\mathrm{Ra_I}$ is proportional to the inverse third power of the aspect ratio. For a shallow cavity the flow may become chaotic and it loses left/right symmetry. In this situation downward-pointing plumes are generated whenever there is sufficient room near the top of the cavity and subsequently travel toward the nearer side wall.

Vasseur *et al.* (1984) discussed convection in the annular space between horizontal concentric cylinders. Their calculations showed that at small Ra_I values a more or less parabolic temperature profile is established across the annulus, resulting in two counterrotating vortices (both with axes centered on the horizontal mid-plane) in each half-cavity. Under the effect of weak and moderate convection, the maximum temperature within the porous medium can be considerably higher than that induced by pure conduction. At large Ra_I values, the flow structure consists of a thermally stratified core and two boundary layers, with a thickness and heat transfer rate of the order of $\mathrm{Ra}_I^{-1/3}$ and $\mathrm{Ra}_I^{1/3}$, respectively. Now the inner radius replaces H in the definition of Ra_I.

Numerical studies of two-dimensional convection in a horizontal annulus with flow across a permeable outer or inner boundary were reported by Burns and Stewart (1992) and Stewart and Burns (1992), while the case where both boundaries are permeable was treated by Stewart *et al.* (1994).

Convection in a vertical cylinder of finite height was studied by Stewart and Dona (1988). They took the bottom to be adiabatic and the remaining boundaries isothermal. Their numerical results for height (H) to radius (R) ratio 2 showed compression of isotherms near the top and side of the cylinder as Ra_I increased. They defined Ra_I with R^2H replacing H^3 in Eq. (6.115). They found that single-cell flow occurred until Ra_I was about 7000. At higher Ra_I a smaller reverse flow region formed near the top and axis, and the transition was accompanied by the position of maximum temperature moving off the axis. Dona and Stewart (1989) treated the same problem, but including the effects of quadratic drag and the variation of density and viscosity with temperature for Ra_I values up to 7000. For such values the property variations have a significant effect, but the effect of quadratic drag is small.

Prasad and Chui (1989) made a numerical study of convection in a vertical cylinder with the vertical wall isothermal and the horizontal boundaries either adiabatic or isothermally cooled. When the horizontal walls are insulated, the flow in the cavity is unicellular and the temperature field in the upper region is highly stratified. However, if the top boundary is cooled, there may exist a multicellular flow and an unstable thermal stratification in the upper region of the cylinder. Under the influence of weak convection, the maximum temperature in the cavity can be considerably higher than that induced by pure conduction (as in the horizontal annulus problem mentioned above). The local heat flux on the wall is generally a strong function of Ra_I, the aspect ratio, and the wall boundary conditions.

The effect of water density maximum on heat transfer in a vertical cylinder, with adiabatic bottom and isothermal sides and top, was modeled numerically by Weiss *et al.* (1991). A linear stability analysis of convection in a vertical annulus was presented by Saravanan and Kandaswamy (2003a).

Weinitschke *et al.* (1990) and Islam and Nandakumar (1990) have conducted studies of two-dimensional bifurcation phenomena in rectangular ducts with uniform heat generation. Multiple steady states appear as the internal Rayleigh number is increased up to several thousand. In the second paper the evaluation with time of these multiple states is examined. The solution structure is complicated. The effect of tilt was treated by Ryland and Nandakumar (1992). A bifurcation study of convection generated by an exothermic chemical reaction was made by Islam (1993). Heat and mass transfer in a semi-infinite cylindrical enclosure, with permeable or impermeable boundaries, were treated by Van Dyne and Stewart (1994). A numerical study using the Brinkman model for eccentric or oval enclosures was reported by Das *et al.* (2003).

A problem related to astrophysics was studied by Zhang *et al.* (2005). This problem is concerned with pore water convection within carbonaceous chondrite parent bodies. These are modeled as spherical bodies within which the gravitational field is radial and varies with radial distance and the viscosity is allowed to vary with temperature. The linear stability analysis leads to the determining of a critical Rayleigh number as a function of the central temperature. Zhang *et al.* (2005) found that the nonlinearity from the viscosity-temperature dependence removed a degeneracy in the azimuthal variation of the mode of convection.

6.18. Localized Heating

Numerical calculations are called for in more complex situations, as when only part of the bottom boundary of a container is heated. The prototypical problem is convection in a rectangular cavity of height H and width $2L$, of which the central section (of the bottom) of width $2D$ is heated. One can define the aspect ratio of the half cavity A and the heated length fraction s by

$$A = \frac{L}{H}, \quad s = \frac{D}{L}. \tag{6.155}$$

The boundaries are assumed to be impermeable. Various thermal boundary conditions can be considered in turn (see Table 6.4). If, for example, one considers the boundary conditions of Prasad and Kulacki (1986), and the nondimensional variables are taken to be

$$X = \frac{x}{H}, Y = \frac{y}{H}, \theta = \frac{T - T_c}{T_h - T_c}, \Psi = \frac{\psi}{\alpha_m}, \tau = \frac{\alpha_m t}{\sigma H^2}. \tag{6.156}$$

then one has to solve

$$\frac{1}{A^2}\frac{\partial^2 \Psi}{\partial X^2} + \frac{\partial^2 \Psi}{\partial Y^2} = \mathrm{Ra}\,\frac{\partial \theta}{\partial X}, \tag{6.157}$$

$$\frac{\partial \theta}{\partial \tau} + \frac{\partial \Psi}{\partial X}\frac{\partial \theta}{\partial Y} - \frac{\partial \Psi}{\partial Y}\frac{\partial \theta}{\partial X} = \frac{\partial^2 \theta}{\partial X^2} + A^2\frac{\partial^2 \theta}{\partial Y^2} \tag{6.158}$$

subject to appropriate initial conditions (for the nonsteady problem) and the boundary conditions

$$\begin{aligned} \theta &= 1 \quad \text{for} \quad 0 \le |X| < s, Y = 0, \\ \frac{\partial \theta}{\partial Y} &= 0 \quad \text{for} \quad s < |X| \le 1, Y = 0, \\ \theta &= 0 \quad \text{for} \quad Y = 1, \\ \frac{\partial \theta}{\partial X} &= 0 \quad \text{for} \quad X = -1 \text{ or } 1. \end{aligned} \tag{6.159}$$

Table 6.4. Thermal boundary conditions for localized heating in a rectangular cavity.

	Central bottom	Outer bottom	Sides	Top
Elder (1967a,b)	$T = T_h$	$T = T_c$	$T = T_c$	$T = T_c$
Horne and O'Sullivan (1974a, 1978b)	$T = T_h$	$T = T_c$	$\frac{\partial T}{\partial n} = 0$	$T = T_c$
Prasad and Kulacki (1986, 1987); Robillard *et al.* (1988)	$T = T_h$	$\frac{\partial T}{\partial n} = 0$	$\frac{\partial T}{\partial n} = 0$	$T = T_c$
Rajen and Kulacki (1987)	$\frac{\partial T}{\partial n} = -\frac{q''}{k_m}$	$\frac{\partial T}{\partial n} = 0$	$\frac{\partial T}{\partial n} = 0$	$T = T_c$
El-Khatib and Prasad (1987)	$T = T_h$	$T = T_c$	$T = T_c + \frac{y}{H}(T_t - T_c)$	$T = T_t$

This system is readily solved using finite differences. Because of the symmetry of the problem, computations need be made for only the right half of the domain.

The pioneering numerical and experimental study by Elder (1967a) for steady convection demonstrated that more than one cell exists in the half cavity for $s = 1.5$ and the Nusselt number is a function of s and the number of cells. Elder (1967b) also studied the transient problem. He noted (see Fig. 6.25) an alternation between periods of slow gradual adjustment and periods of rapid change of flow patterns.

The numerical results of Horne and O'Sullivan (1974a) for time-dependent boundary conditions indicate that when the lower boundary is partially heated, the system is self-restricting and it settles down into a steady multicellular flow or a periodic oscillatory flow, depending on Ra and the amount of boundary that is heated. At high Ra oscillatory flow is the norm. Typical flow patterns are shown in Fig. 6.26. Approximately mushroom-shaped isotherms predominate. The effects of temperature-dependent viscosity and thermal expansion coefficient on the temperature and flow fields were studied by Horne and O'Sullivan (1978b). They found that in some cases the acceleration of the flow in certain areas, due to a decrease in viscosity, causes localized thermal instabilities.

Further numerical calculations were reported by Prasad and Kulacki (1986, 1987). For the case $D/H > 1$, they noted the appearance at small Ra of a circulation near the heated segment and the development of further cells as Ra increases. Further increase of Ra does not increase the number of cells, but it strengthens existing cells and leads to the formation of boundary layers. The outermost cell extends to the side wall. Within the inner cells plumes are formed at large Ra, the isotherms taking the characteristic mushroom shape (as with uniform heating). Because Prasad and Kulacki considered only the steady problem, they did not observe any oscillatory behavior.

Prasad and Kulacki (1986, 1987) also made calculations of heat transfer rates. As expected, the local Nusselt number has peaks where hot fluid rises. The peak value increases with the size of heat source until a new cell is formed. The overall Nusselt number based on the heated segment (Nu_s) decreases with s (for fixed $\mathrm{Ra} > 1000$) until $s = 0.4$ and then remains steady, the steadiness indicating that the heat transfer rate is then proportional to the area of the heat source. The overall Nusselt number based on the entire cavity width (Nu_L) increases monotonically with s. Both overall Nusselt numbers increase with Ra, the rate of increase being approximately uniform (on a log-log scale) when $\mathrm{Ra} > 100$, the boundary layer regime. In this regime the slope of the $\ln(\mathrm{Nu}_L)$ versus $\ln(\mathrm{Ra})$ curve increases gradually with s. When s is close to 1, the overall Nusselt numbers increase rapidly with Ra in the vicinity of $\mathrm{Ra} = 40$, as expected.

El-Khatib and Prasad (1987) extended the calculations to include the effects of linear thermal stratification, expressed by the parameter

$$S = \frac{T_t - T_c}{T_h - T_c}. \tag{6.160}$$

See the last line of Table 6.3 for definitions of T_t, T_c, and T_h. El-Khatib and Prasad did calculations for $A = 1$, $s = 0.5$, $0 = S = 10$, and Ra up to 1000. They found that an increase in S for a fixed Ra reduces the convective velocities, and hence the

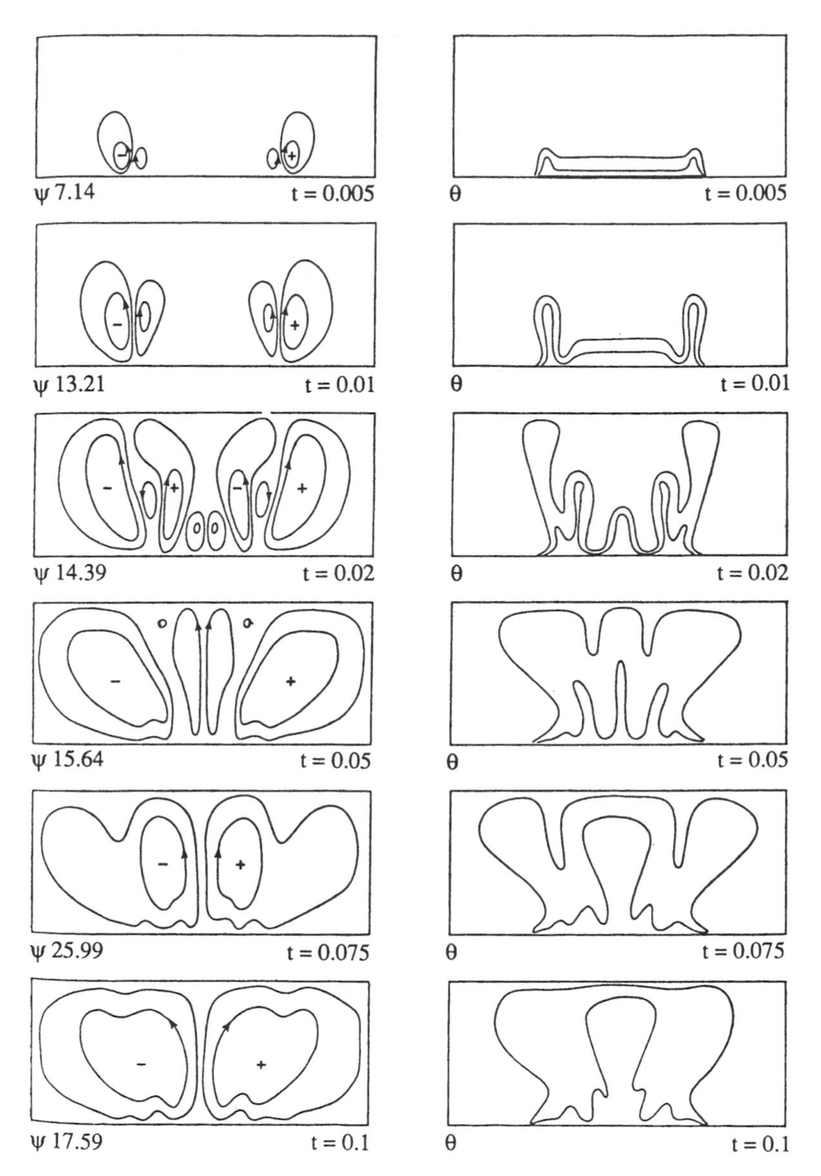

Figure 6.25. Streamlines and isotherms for a localized heater problem at various times; Ra $= 400$, $A = 2$, $s = 0.5$ (Elder, 1967b, with permission from Cambridge University Press).

energy lost by the heat source. In fact, for sufficiently large S at least part of the heated segment may gain energy. A similar situation pertains to the top surface. For $S > 1$ the energy gained by the upper surface is almost independent of Ra.

Rajen and Kulacki (1987) reported experimental and numerical results for $A =$ 16 or 4.8, and $s = 1, 1/2$, or $1/12$, with the boundary conditions given in Table 6.3.

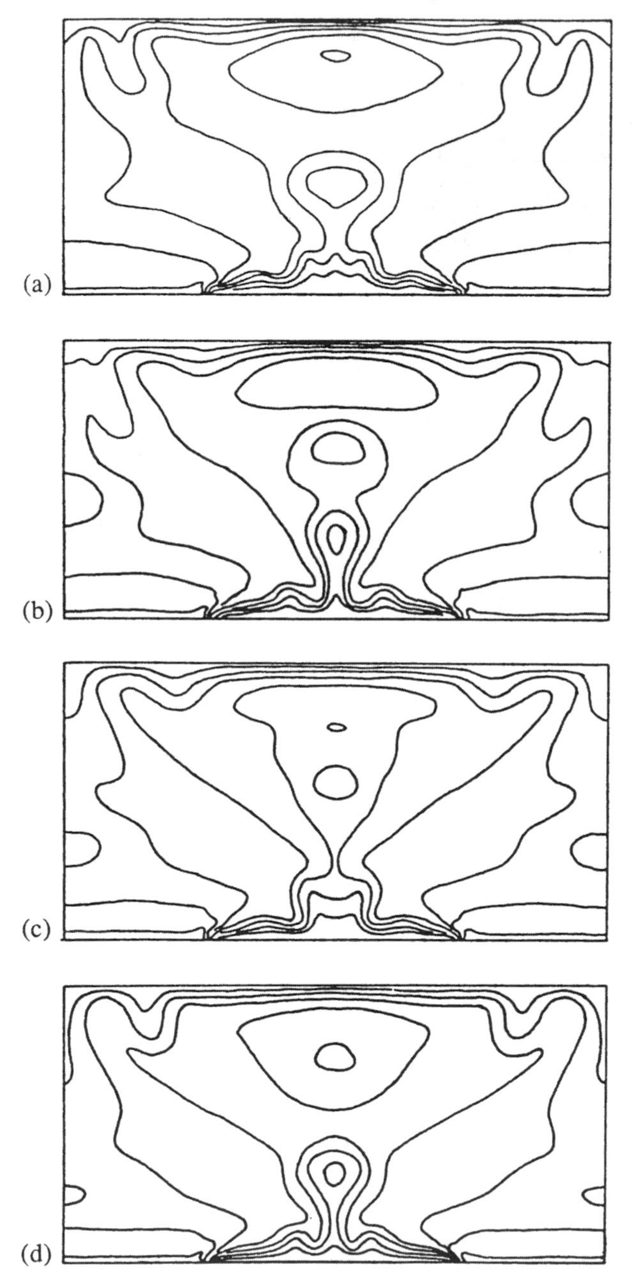

Figure 6.26. Plots of computed isotherms during a single oscillation for a localized heating problem; $\mathrm{Ra} = 750$, $A = 1$, $s = 0.5$ (Horne and O'Sullivan, 1974a, with permission from Cambridge University Press).

Their observations of Nusselt number values were in very good agreement with the predicted values.

Robillard *et al.* (1988) performed calculations for the case when the heat source is not symmetrically positioned. Merkin and Zhang (1990b) treated numerically a similar situation. A variant of the Elder short heater problem with a spatially sinusoidal distribution along the hot plate was studied numerically by Saeid (2005).

6.19. Superposed Fluid and Porous Layers

Convection in a system consisting of a horizontal layer of porous medium and a superposed clear fluid layer has been modeled in two alternative ways. In the two-domain approach the porous medium and clear fluid are considered separately and the pair of solutions is coupled using matching conditions at the interface. In the single-domain approach the fluid is considered as a special case of a porous medium, the latter being modeled using a Brinkman-Forchheimer equation. The second approach is subject to the caveat about use of the Brinkman equation mentioned in Section 1.6, but in most situations discussed in this section the two approaches are expected to yield qualitatively equivalent results for the global temperature and velocity fields. An exception is when the depth of the porous layer is not large in comparison with the particle/pore diameter.

6.19.1. Onset of Convection

6.19.1.1. Formulation

We start by considering a porous layer of depth d_m superposed by clear fluid of depth d_f, the base of the porous medium being at temperature T_l and the top of

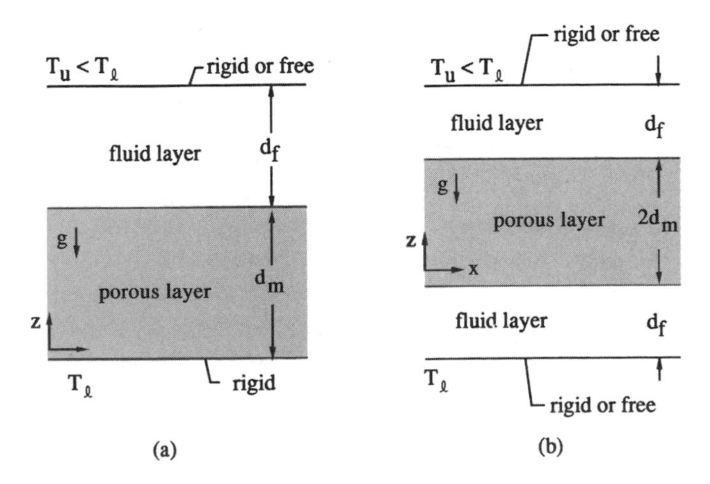

Figure 6.27. Composite fluid-layer porous-layer systems.

the clear fluid region at temperature T_u (Fig. 6.27a). We suppose that flow in the porous medium is governed by Darcy's equation and that in the clear fluid by the Navier-Stokes equation. The combined system has a basic steady-state conduction solution given by

$$\mathbf{V} = 0,\ T = T_b \equiv T_u - \beta_f(z - d_m - d_f),\ P = P_b \text{ for the fluid,} \tag{6.161}$$

$$\mathrm{v}_m = 0,\ T_m = T_{bm} \equiv T_l - \beta_m z,\ P_m = P_{bm} \text{ for the porous layer.} \tag{6.162}$$

Here β_f and β_m are the temperature gradients. Continuity of temperature and heat flux at the interface requires that

$$T_u + \beta_f d_f = T_l - \beta_m d_m = T_i \text{ and } k_f \beta_f = k_m \beta_m, \tag{6.163a,b}$$

where T_i is the interface temperature, and hence

$$\beta_f = \frac{k_m(T_l - T_u)}{k_m d_f + k_f d_m},\quad \beta_m = \frac{k_f(T_l - T_u)}{k_m d_f + k_f d_m}. \tag{6.164a,b}$$

In terms of perturbations from the conduction state, $T' = T - T_b$, $P' = P - P_b$, etc., the linearized perturbation equations in time-independent form are

$$\nabla \cdot \mathbf{V}' = 0, \tag{6.165a}$$

$$\frac{1}{\rho_0}\nabla P' = \nu \nabla^2 \mathbf{V}' + g\beta T'\mathbf{k}, \tag{6.165b}$$

$$\beta_f w' + \alpha_f \nabla^2 T' = 0, \tag{6.165c}$$

$$\nabla \cdot \mathbf{v}'_m = 0, \tag{6.166a}$$

$$\frac{1}{\rho_0}\nabla P'_m = -\frac{\upsilon}{K}\mathbf{v}'_m + g\beta T'_m\mathbf{k}, \tag{6.166b}$$

$$\beta_m w'_m + \alpha_m \nabla^2 T'_m = 0. \tag{6.166c}$$

In the fluid layer appropriate nondimensional variables are

$$\hat{\mathbf{x}} = \frac{(\mathbf{x} - d_m\mathbf{k})}{d_f},\ \hat{\mathbf{V}} = \frac{d_f \mathbf{v}'}{\alpha_f},\ \hat{P} = \frac{d_f^2 P'}{\mu \alpha_f},\ \hat{T} = \frac{T'}{\beta_f d_f}. \tag{6.167}$$

Substituting Eq. (6.163) and dropping the carets we have for the fluid layer

$$\nabla \cdot \mathbf{V} = 0, \tag{6.168a}$$

$$\nabla P = \nabla^2 \mathbf{V} + \mathrm{Ra}_f T\mathbf{k}, \tag{6.168b}$$

$$w + \nabla^2 T = 0, \tag{6.168c}$$

where

$$\mathrm{Ra}_f = \frac{g\beta\beta_f d_f^4}{\nu \alpha_f}. \tag{6.169}$$

Eliminating P, we reduce the equations for the fluid layer to

$$\nabla^4 w + \mathrm{Ra}\, \nabla_H^2 T = 0, \tag{6.170a}$$

$$w + \nabla^2 T = 0. \tag{6.170b}$$

Here ∇_H^2 denotes the horizontal Laplacian as in Eq. (6.20). Similarly, for the porous medium, we put

$$\hat{\mathbf{x}}_m = \frac{\mathrm{x}}{d_m}, \hat{\mathbf{v}}_m = \frac{d_m \mathbf{v}'}{\alpha_m}, \hat{P}_m = \frac{K P_m'}{\mu \alpha_m}, \hat{T}_m = \frac{T_m'}{\beta_m d_m}. \tag{6.171}$$

Substituting Eq. (6.162), dropping the carets, and eliminating P_m, we have for the porous medium

$$\nabla_m^2 w_m - \mathrm{Ra}_m \nabla_{Hm}^2 T_m = 0, \tag{6.172a}$$

$$w_m + \nabla_m^2 T_m = 0, \tag{6.172b}$$

where

$$\mathrm{Ra}_m = \frac{g\beta\beta_m K d_m^2}{\nu \alpha_m}. \tag{6.173}$$

We now separate the variables by letting

$$\begin{Bmatrix} w \\ T \end{Bmatrix} = \begin{Bmatrix} W(z) \\ \theta(z) \end{Bmatrix} f(x, y), \begin{Bmatrix} w_m \\ T_m \end{Bmatrix} = \begin{Bmatrix} w_m(z_m) \\ \theta_m(z_m) \end{Bmatrix} f_m(x_m, y_m), \tag{6.174}$$

where

$$\nabla_H^2 f + \alpha^2 f = 0, \quad \nabla_{Hm}^2 f_m + \alpha_m^2 f_m = 0. \tag{6.175}$$

Since the dimensional horizontal wavenumber must be the same for the fluid layer and the porous medium if matching is to be achieved, the nondimensional horizontal wavenumbers α and α_m are related by $\alpha/d_f = \alpha_m/d_m$, and so

$$\alpha_m = \hat{d}\alpha, \quad \text{where} \quad \hat{d} = d_m/d_f. \tag{6.176}$$

The reader should not be confused by the use (in this section only) of the symbol α_m for both thermal diffusivity and horizontal wavenumber. He or she should note that Chen and Chen (1988, 1989) have used $\hat{d}$ to denote d_f/d_m. Equations (6.164) and (6.166) yield

$$(D^2 - \alpha^2)^2 W - \mathrm{Ra}_f \alpha^2 \theta = 0, \tag{6.177a}$$

$$(D^2 - \alpha^2)\theta + W = 0 \tag{6.177b}$$

and

$$(D_m^2 - \alpha_m^2) W_m + \mathrm{Ra}_m \alpha_m^2 \theta_m = 0, \tag{6.178a}$$

$$(D_m^2 - \alpha_m^2)\theta_m + W_m = 0, \tag{6.178b}$$

where $D = d/dz$ and $D_m = d/dz_m$. We match the solutions of Eqs. (6.177) and (6.178) at the fluid/porous-medium interface by invoking the continuity of temperature, heat flux, normal velocity (note that it is the Darcy velocity and not the intrinsic velocity which is involved), and normal stress. The Beavers-Joseph condition supplies the fifth matching condition. Thus we have at $z_m = 1$

(or $z = 0$),

$$T = \varepsilon_T T_m, \quad \frac{\partial T}{\partial z} = \frac{\partial T_m}{\partial z_m}, \quad \varepsilon_T w = w_m, \tag{6.179a,b,c}$$

$$\varepsilon_T \hat{d}^3 \mathrm{Da}\left(3\nabla_H^2 \frac{\partial w}{\partial z} + \frac{\partial^3 w}{\partial z^3}\right) = -\frac{\partial w_m}{\partial z_m}, \tag{6.179d}$$

$$\varepsilon_T \hat{d}\left(\frac{\partial w}{\partial z} - \Delta \frac{\partial^2 w}{\partial z^2}\right) = -\frac{\partial w_m}{\partial z_m}, \tag{6.179e}$$

where $\varepsilon_T = \beta_m d_m / \beta_f d_f = k_f d_m / k_m d_f = \hat{d}/\hat{k}$,

$$\mathrm{Da} = \frac{K}{d_m^2}, \quad \Delta = \hat{d}\frac{\mathrm{Da}^{1/2}}{\alpha_{BJ}}, \quad \hat{d} = \frac{d_m}{d_f}, \quad \hat{k} = \frac{k_m}{k_f}. \tag{6.180}$$

Equation (6.179d) is derived from the condition

$$-P + 2\mu \frac{\partial w}{\partial z} = -P_m \tag{6.181}$$

and Eq. (6.179e) is derived from the Beavers-Joseph condition

$$\frac{\partial u}{\partial z} = \frac{\alpha_{BJ}}{K^{1/2}}(u - u_m). \tag{6.182}$$

The remaining boundary conditions come from the external conditions. For example, if the fluid-layer/porous-medium system is bounded above and below by rigid conducting boundaries, then one has

$$\begin{aligned} w = \frac{\partial w}{\partial z} = T = 0 \quad &\text{at} \quad z = 1, \\ w_m = T_m = 0 \quad &\text{at} \quad z_m = 0. \end{aligned} \tag{6.183a,b,c}$$

The tenth-order system (6.176) and (6.178) now can be solved subject to the ten constraints (6.179) and (6.180). Note that the fluid Rayleigh number Ra_f and the Rayleigh-Darcy number Ra_m are related by

$$\mathrm{Ra}_m = \hat{d}\varepsilon_T^2 \,\mathrm{Da}\,\mathrm{Ra}_f = \hat{d}^4 \hat{k}^{-2}\,\mathrm{Da}\,\mathrm{Ra}_f. \tag{6.184}$$

Hence the critical Rayleigh-Darcy number Ra_m can be found as a function of four parameters, $\hat{d}$, $\hat{k}$, Da, and α_{BJ}, or, alternatively, $\hat{d}$, ε_T, Da, and Δ.

6.19.1.2. Results

As in Section 6.2, the case of constant heat flux boundaries yields a closed form for the stability criterion. The critical wavenumber is zero and the stability criterion for the case of a free top and an impermeable bottom is given by (Nield, 1977)

$$\begin{aligned} &\varepsilon_T \left\{3 + 24\Delta + \mathrm{Da}\,\hat{d}^2\left[84 + 384\hat{d} + 300\varepsilon_T\hat{d} + 720\Delta\hat{d}(1+\varepsilon_T)\right]\right\}\mathrm{Ra}_{fc} \\ &+ \hat{d}^2\left[320 + 960\Delta + \mathrm{Da}\,\hat{d}^2(960 + 240\hat{d})\right. \\ &\left.+ \varepsilon_T^{-1}(300 + 720\Delta + 720\,\mathrm{Da}\,\hat{d}^2)\right]\mathrm{Ra}_{mc} \\ &= \left[960 + 2880\Delta + 2880\,\mathrm{Da}\,\hat{d}^2(1+\hat{d})\right](\varepsilon_T + \hat{d}^2). \end{aligned} \tag{6.185}$$

If we let $\hat{d} \to \infty$ with ε_T, Da, and Δ finite, Eq. (6.185) gives $\mathrm{Ra}_{mc} \to 12$, the expected value for a porous medium between two impermeable boundaries.

A similar analysis has been performed for a system consisting of a porous medium layer of thickness $2d_m$ sandwiched between two fluid layers, each of thickness d_f, Fig. 6.24b. The following stability criteria have been obtained.

Rigid top and rigid bottom (Nield, 1983):

$$\varepsilon_T\,[8 + 18\Delta + (15 + 45\Delta)\,\varepsilon_T]\,\mathrm{Ra}_{fc} + \left\{120(1+\Delta)\,\hat{d}^2 + 180\hat{d} + 60\frac{\hat{d}}{\varepsilon_T} + \frac{1}{\mathrm{Da}}\left[\left(\frac{30 + 120\Delta}{\hat{d}}\right) + \left(\frac{15 + 45\Delta}{\hat{d}\varepsilon_T}\right)\right]\right\} \mathrm{Ra}_{mc} = 360(1+\Delta)(\varepsilon_T + \hat{d}^2). \tag{6.186}$$

Free top and free bottom (Pillatsis *et al.* 1987):

$$\varepsilon_T\left[192 + 360\Delta(1 + 2\hat{d}) + 720\,\mathrm{Da}\,\hat{d}^3 + 300\hat{d}\right]\mathrm{Ra}_{fc} + \hat{d}^2\left\{480 + \left(\frac{60}{\mathrm{Da}\,\hat{d}^4}\right)[5 + 8\hat{d} + 12\Delta(1 + 2\hat{d}) + 24\hat{d}^3\,\mathrm{Da}]\right\}\mathrm{Ra}_{mc} = 1440(\varepsilon_T + \hat{d}^2). \tag{6.187}$$

As $\hat{d} \to 0$ (with $\Delta \to 0$), Eqs. (6.186) and (6.187) yield $\mathrm{Ra}_{fc} = 45$ and 7.5, the critical Rayleigh numbers for a fluid layer of depth $2d_f$ between rigid-rigid and free-free boundaries, respectively. As $\hat{d} \to \infty$ they yield $\mathrm{Ra}_{mc} = 3$, the critical Rayleigh-Darcy number for a porous layer of depth $2d_m$ between impermeable boundaries. These results are as expected. Figure 6.28 shows the results of calculations based on Eqs. (6.186) and (6.187).

For isothermal boundaries the critical wavenumber is no longer zero, and numerical calculations are needed. Pillatsis *et al.* (1987) and Taslim and Narusawa (1989) have employed power series in z to obtain the stability criterion. They treated the fluid/porous-medium, the fluid/porous-medium/fluid, and the porous-medium/fluid/porous-medium situations. The results are in accord with expectations. A rigid boundary at the solid-fluid interface suppresses the onset of convection compared with a free boundary. The presence of a fluid layer increases instability in the porous medium and α_{mc} decreases as the effect of the fluid layer becomes more significant, as it does when the fluid layer thickens. The parameter α_{BJ} has a significant effect only when the Darcy number is large. The effect of the Jones modification to the Beavers-Joseph condition is minimal.

Nield (1994a) has shown that the above theory is consistent with observations of increased heat transfer due to channeling reported by Kazmierzak and Muley (1994). Further calculations by Chen and Chen (1988) show that the marginal stability curves are bimodal for a fluid/porous-medium system with d_f/d_m small (Fig. 6.29). The critical wavenumber jumps from a small to a large value as d_f/d_m increases from 0.12 to 0.13. Chen and Chen noted that the change correlated with a switch from porous-layer-dominated convection to fluid-layer-dominated

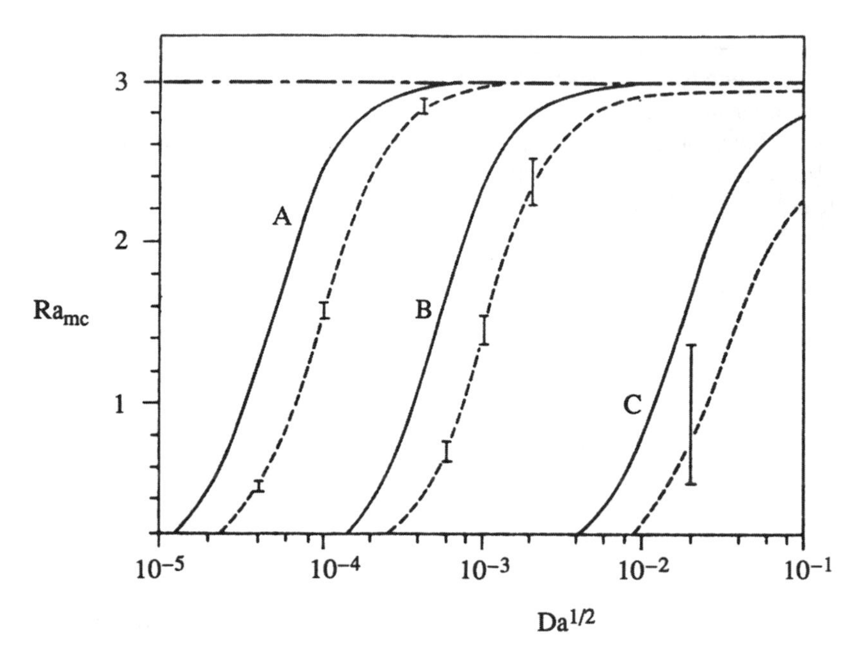

Figure 6.28. Critical Rayleigh number for a porous layer sandwiched between two fluid layers (Fig. 6.24b) for the constant flux case. (A) $\hat{d} = d_m/d_f = 500$, (B) $\hat{d} = 100$, (C) $\hat{d} = 10$; $k_f/k_m = 1$ and $\Delta = 0.05$ for all curves; —— rigid boundaries, – – – free boundaries, – · – $\mathrm{Ra}_{mc} = 3$ corresponding to the case of a porous layer alone. The vertical bars denote the range of Ra_{mc} when the conductivity ratio k_f/k_m is varied from 10 to 0.1 (Pillatsis *et al.*, 1987).

convection. Numerical calculations for supercritical convection by Kim and Choi (1996) are in good agreement with the predictions from linear stability theory.

The experiments of Chen and Chen (1989) generally confirmed the theoretical predictions. They employed a rectangular enclosure with 3-mm glass beads and a glycerin-water solution of varying concentrations to produce a system with $0 \leq d_f/d_m \leq 1$. They observed that Ra_{mc} does decrease significantly as d_f/d_m increases. They also estimated the size of convective cells from temperature measurements. They found that the cells were three-dimensional and that the critical wavenumber increased eightfold when d_f/d_m was increased from 0.1 to 0.2.

Somerton and Catton (1982) used the Brinkman equation. Their results are confined to high Darcy numbers, $K/(d_f + d_m)^2 > 37 \times 10^{-4}$, and thick fluid layers, $d_f/d_m = 0.43$. Vasseur *et al.* (1989) also employed the Brinkman equation in their study of a shallow cavity with constant heat flux on the external boundaries. An extra isothermal condition at the interface mentioned in their paper was in fact not used in the calculations.

A nonlinear computational study, using a Brinkman-Forchheimer equation, was made by Chen and Chen (1992). The effect of rotation on the onset of convection was analyzed by Jou *et al.* (1996). The effect of vertical throughflow was treated

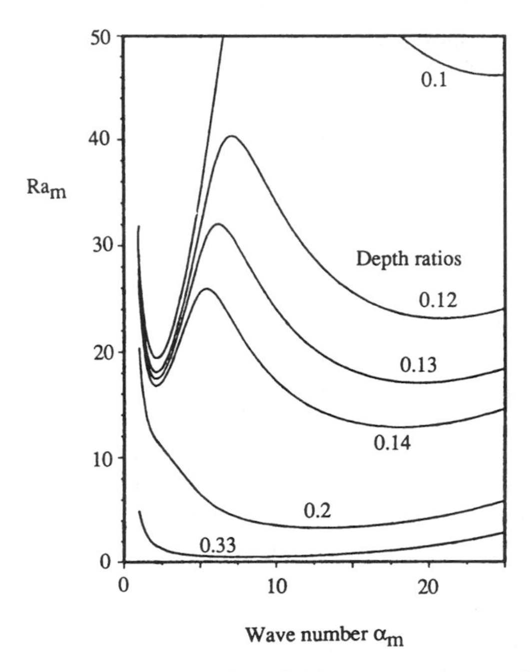

Figure 6.29. Marginal stability curves for a fluid-superposed porous layer heated from below for isothermal rigid boundaries, with the thermal conductivity ratio $k_f/k_m = 0.7$, $\mathrm{Da} = 4 \times 10^{-6}$, and $\alpha_{BJ} = 0.1$, for various values of the depth ratio d_f/d_m (Chen and Chen, 1988).

by Chen (1991). The effect of anisotropy was studied by Chen *et al.* (1991) and Chen and Hsu (1991) and that of viscosity variation by Chen and Lu (1992b). A fluid layer sandwiched between two porous layers of different permeabilities was analyzed by Balasubramanian and Thangaraj (1998). The case where the bottom boundary is heated by a constant flux was analyzed by Wang (1999a), who found that the critical Rayleigh number for the porous layer increases with the thickness of the solid layer, a result opposite to that when the heating is at constant temperature. Reacting fluid and porous layers were analyzed by McKay (1998a). The effect of property variation was incorporated by Straughan (2002). A comparison of the one and two domains approaches to handling the interface was made by Valencia-López and Ochoa-Tapia (2001). Significant differences between the predicted overall average Nusselt numbers were found when the Rayleigh and Darcy numbers were large enough. The characteristic based split algorithm was used in the numerical study of interface problems by Massarotti *et al.* (2001).

6.19.2. Flow Patterns and Heat Transfer

Heat transfer rates for a fluid-layer/porous-layer system were calculated by Somerton and Catton (see Catton, 1985) using the power integral method. Both streamlines and heat transfer rates were calculated by numerical integration of the time-dependent equations by Poulikakos *et al.* (1986) and Poulikakos (1987a). Laboratory experiments, in a cylindrical cavity heated from below, have been reported by Catton (1985), Prasad *et al.* (1989a), and Prasad and Tian (1990). Prasad and his colleagues performed both heat transfer and flow visualization experiments, the latter with transparent acrylic beads and a liquid matched for index of refraction. There is qualitative agreement between calculations and observations. For example, in a cell of aspect ratio and $d_f/d_m = 1$ there is a transition from a two-cell pattern to a four-cell pattern with an increase in Rayleigh number or Darcy number. In the two-cell pattern the flow extends well into the porous layer, while in the four-cell pattern the flow is concentrated in the fluid layer.

Once convection starts, the Nusselt number Nu always increases with Rayleigh number for fixed η, where η denotes the fraction of the depth occupied by the porous medium. For small particle size γ and/or small Rayleigh number, Nu decreases monotonically with η; otherwise the dependence of Nu on η is complex. The complexity is related to the variation in the number of convection cells that occur. In general, Nu depends on at least six parameters: Ra, Pr, γ, k_m/k_f, η, and A.

Further experiments, involving visualization as well as heat transfer studies, were made by Prasad *et al.* (1991). They found that flow channels through large voids produce highly asymmetric and complicated flow structures. Also, Nu first decreases from the fluid heat transfer rates with an increase in η and reaches a minimum at η_{min}. Any further increase in porous layer height beyond η_{min} augments the heat transfer rate and the Nu curves show peaks. Prasad (1993) observed the effects of varying thermal conductivity and Prandtl number.

Even more complicated is the situation when one has volumetric heating of the porous medium as well as an applied vertical temperature gradient. This situation was studied numerically, using the Brinkman equation, by Somerton and Catton (1982) and Poulikakos (1987a,b); the latter also included the Forchheimer term. Poulikakos studied convection in a rectangular cavity whose bottom was either isothermal or adiabatic. For the aspect ratio $H/L = 0.5$, he noted a transition from two to four cells as Ra_I and Da increase. Related experiments were reported by Catton (1985) and others. Further experimental and numerical work was conducted by Schulenberg and Müller (1984). Serkitjis (1995) conducted experimental (and numerical) work on convection in a layer of polystyrene pellets, of spherical or cylindrical shape, below a layer of air. He found that the occurrence of natural convection in the air space has only a marginal effect on heat transfer in the porous medium. A numerical study of transient convection in a rectangular cavity was reported by Chang and Yang (1995).

The subject of this section has been extensively reviewed by Prasad (1991). Further complexity arises if chemical reactions are involved. Examples are found in the papers by Hunt and Tien (1990) and Viljoen *et al.* (1990).

6.19.3. Other Configurations and Effects

A hydrothermal crystal growth system was modeled by Chen *et al.* (1999), on the assumption that the growth process is quasisteady. The flow through a fluid-sediment interface in a benthic chamber was computed by Basu and Khalili (1999). The addition of vertical through flow was studied by Khalili *et al.* (2003). Convective instability in a layer saturated with oil and a layer of gas underlying it was analyzed by Kim *et al.* (2003a).

The effect of gravity modulation was added by Malashetty and Padmavathi (1997). Convection in a square cavity partly filled with a heat-generating porous medium was studied analytically and numerically by G. B. Kim *et al.* (2001). Convection induced by the selective absorption of radiation was analyzed by Chang (2004). Penetrative convection resulting from internal heating was studied by Carr (2004). It was found that a heat source in the fluid layer has a destabilizing effect on the porous medium but one in the fluid has a stabilizing effect on the fluid, while the effects on their respective layers depends strongly on the overall temperature difference and the strength and type of heating in the opposite layer. It also was found that the initiating cell pattern is not necessarily the strongest one. A horizontal plane Couette flow problem was analyzed by Chang (2005).

A surface tension (Marangoni) effect on the onset of convection was analyzed by Nield (1998c). A similar situation was treated by Hennenberg *et al.* (1997), Rudraiah and Prasad (1998), Straughan (2001b), Desaive *et al.* (2001), and Saghir *et al.* (2002, 2005b) using the Brinkman model. The last authors reported numerical studies for the combined buoyancy and surface tension situation. Kozak *et al.* (2004) included the effect of evaporation at the free surface. Straughan (2004b) pointed out that the results of Chen and Chen (1988) lend much support for the two layer model of Nield (1998c).

6.20. Layer Saturated with Water Near 4°C

Poulikakos (1985b) reported a theoretical investigation of a horizontal porous layer saturated with water near 4°C, when the temperature of the top surface is suddenly lowered. The onset of convection has been studied using linear stability analysis (Sun *et al.*, 1970) and time-dependent numerical solutions of the complete governing equations (Blake *et al.*, 1984). In both studies, the condition for the onset of convection is reported graphically or numerically for a series of discrete cases. The numerical results of Blake *et al.* (1984) for layers with $T_c = 0$°C on the top and 5°C $= T_h =$ 8°C on the bottom can be used to derive (Bejan, 1987)

$$\frac{gKH}{\nu\alpha_m} > 1.25 \times 10^5 \exp[\exp(3.8 - 0.446T_h)] \tag{6.188}$$

as an empirical dimensionless criterion for the onset of convection. In this criterion the bottom temperature T_h is expressed in degrees Celsius.

Finite-amplitude heat and fluid flow results for Rayleigh numbers $g\gamma K(T_h - T_c)^2 H/\nu\alpha_m$ of up to 10^4 (i.e., about 50 times greater than critical) also have been reported by Blake *et al.* (1984). In the construction of this Rayleigh number γ is the coefficient in the parabolic model for the density of cold water, $\rho = \rho_{ref}[1 - \gamma(T - 3.98°\mathrm{C})^2]$, namely, $\gamma = 8 \times 10^{-6}(°\mathrm{C})^{-2}$.

Nonlinear changes in viscosity (as well as density) were treated numerically by Holzbecher (1997). He found that a variety of flow patterns (e.g., two or four cells in a two-dimensional square domain) are possible, depending on the choice of maximum and minimum temperatures.

6.21. Effects of a Magnetic Field

Despite the absence of experimental work and a lack of practical applications, several theoretical papers, including those by Patil and Rudraiah (1973), Rudraiah and Vortmeyer (1978), and Rudraiah (1984) have been published on magnetohydrodynamic convection in a horizontal layer. The simplest case is that of an applied vertical magnetic field and electrically conducting boundaries. Oscillatory convection is a possibility under certain circumstances, but this is ruled out if the thermal diffusivity is smaller than the magnetic resistivity, and this condition is met by a large margin under most terrestrial conditions. On the Darcy model, for the case of thermally conducting impermeable boundaries, the Rayleigh number at the onset of nonoscillatory instability for disturbances of dimensionless wavenumber a is given by

$$\mathrm{Ra} = \frac{(\pi^2 + \alpha^2)(\pi^2 + \alpha^2 + Q\pi^2)}{\alpha^2}, \tag{6.189}$$

where

$$Q = \frac{\sigma B^2 K}{\mu}. \tag{6.190}$$

Here B is the magnetic induction and σ is the electrical conductivity. The parameter Q has been called the Chandrasekhar-Darcy number; it is the Darcy number K/H^2 times the usual Chandrasekhar number, which in turn is the square of the Hartmann number. Some workers use a Hartmann-Darcy number equal to $Q^{1/2}$. It is clear that the effect of the magnetic field is stabilizing. The critical Rayleigh number again is found by taking a minimum as a varies. Because of the practical difficulties of achieving a large magnetic field, Q is almost always much less than unity, and so the effect of the magnetic field is negligible. Bergman and Fearn (1994) discussed an exceptional situation, namely convection in a mushy zone at the Earth's inner-outer core boundary. They concluded that the magnetic field may be strong enough to act against the tendency for convection to be in the form of chimneys.

The problem for the case of the Brinkman model and isoflux boundaries was treated by Alchaar *et al.* (1995a). In this paper and in Bian *et al.* (1996a) the effect of a horizontal magnetic field was studied, but these treatments are incomplete because only two-dimensional disturbances were considered and so the most unstable disturbance may have been overlooked.

Further studies of MHD convection have been reported by Goel and Agrawal (1998) for a viscoelastic dusty fluid, by Sunil and Singh (2000) for a

Rivlin-Ericksen fluid, and by Sunil *et al.* (2003a) for throughflow and rotation effects. Sekar and Vaidyanathan (1993), Vaidyanathan *et al.* (2002a–c), Ramanathan and Surenda (2003), Ramanathan and Suresh (2004), and Sunil *et al.* (2004a) have treated ferroconvection with a magnetic-field-dependent viscosity for an isotropic or anisotropic medium with and without rotation. A nonlinear stability problem for a ferromagnetic fluid was treated by Qin and Chadam (1995). The effect of dust particles on ferroconvection was added by Sunil *et al.* (2004c, 2005b).

6.22. Effects of Rotation

The subject of flow in rotating porous media has been reviewed in detail by Vadasz (1997a, 1998b, 2000a, 2002), whose treatment is followed here. On the Darcy model, constant density flow in a homogeneous porous medium is irrotational, and so the effect of rotation on forced convection is normally unimportant. For natural convection the situation is different. For a homogeneous medium the momentum equation (with Forchheimer and Brinkman terms omitted) can be written in the dimensionless form

$$\frac{\mathrm{Da}}{\varphi\,\mathrm{Pr}}\frac{\partial v}{\partial t} = -\nabla p - \mathrm{Ra}T\nabla(e_g.\mathbf{X}) + \mathrm{Ra}_\omega T\mathbf{e}_\omega \times (\mathbf{e}_\omega \times \mathbf{X}) + \frac{1}{\mathrm{Ek}}\mathbf{e}_\omega \times \mathbf{v}. \quad (6.191)$$

Here $\mathbf{e}_g$ and $\mathbf{e}_\omega$ are unit vectors in the direction of gravity and rotation, respectively, and $\mathbf{X}$ is the position vector. The new parameters are the rotational Rayleigh number Ra_ω and the Ekman-Darcy number Ek defined by

$$\mathrm{Ra}_\omega = \left(\frac{\omega^2 H}{g}\right)\mathrm{Ra}, \quad \mathrm{Ek} = \frac{\varphi\mu}{2\omega\rho K}, \quad (6.192)$$

where ω is the dimensional angular velocity of the coordinate frame with respect to which motion is measured. Normally $\mathrm{Ek} \gg 1$ and then the Coriolis term is negligible, but it can cause secondary flow in an inhomogeneous medium. Generally the Coriolis effect is analogous to that of anisotropy (Palm and Tyvand, 1984). The appearance of the porosity in the expression for Ek should be noted, because some authors have overlooked this factor. This error was pointed out by Nield (1999), who also discussed the analogy between (i) Darcy flow in an isotropic porous medium with a magnetic or rotation effect present, and (ii) flow in a medium with anisotropic permeability.

For the case $\mathrm{Ra}/\mathrm{Ra}_\omega \ll 1$, i.e. when the centrifugal force dominates over gravity, Vadasz (1992, 1994b) considered a two-dimensional problem for a rectangular domain with heating from below and rotation about a vertical boundary. He first showed that for small height-to-breadth aspect ratio H/L, the Nusselt number is given approximately by

$$\mathrm{Nu} = \frac{1}{24}(H/L)\,\mathrm{Ra}_\omega. \quad (6.193)$$

He then relaxed this condition, reduced the problem to that of solving an ordinary differential equation, and found that Nu increases faster with Ra_ω than Eq.(6.191) would indicate.

The Coriolis effect on the Horton-Rogers-Lapwood problem has been investigated by several authors. On the Darcy model, one finds that the critical Rayleigh number is given by

$$Ra_c = \pi^2[(1 + Ek^{-2})^{1/2} + 1]^2. \tag{6.194}$$

Using the Brinkman model, Friedrich (1983) performed a linear stability analysis and a nonlinear numerical study. On this model, convection sets in as an oscillatory instability for a certain range of parameter values. Patil and Vaidyanathan (1983) dealt with the influence of variable viscosity on linear stability. A nonlinear energy stability analysis was performed by Qin and Kaloni (1995). A study of the heat transfer produced in nonlinear convection was made by Riahi (1994), following the procedure of Gupta and Joseph (1973). In terms of a Taylor-Darcy number Ta defined by $Ta = 4/Ek^2$, he found the following results. For $Ta \ll O(1)$, the rotational effect is not significant. For $O(1) \ll Ta \ll O(Ra^{1/2} \log Ra)$, the Nusselt number Nu decreases with increasing Ta for a given Ra. For $O(Ra^{1/2} \log Ra) \ll Ta \ll O(Ra)$, Nu is proportional to $(Ra/Ta) \log (Ra/Ta)$. For $Ta = O(Ra)$, Nu becomes O(1) and the convection is inhibited entirely by rotation for $Ta > Ra/\pi^2$.

The weak nonlinear analysis of Vadasz (1998b) showed that, in contrast to the clear fluid case, overstable convection is possible for all values of Pr (not just $Pr < 1$) and that the critical wavenumber in the plane containing the streamlines for stationary convection is dependent on rotation. It also showed that the effect of viscosity is destabilizing for high rotation rates. As expected, there is a pitchfork bifurcation for stationary convection and a Hopf bifurcation for overstable convection and rotation retards heat transfer (except for a narrow range of small values of $\varphi Pr/Da$, where rotation enhances the heat transfer associated with overstable convection). Bounds on convective heat transfer in a rotating porous layer were obtained by Wei (2004). A sharp nonlinear threshold for instability was obtained by Straughan (2001c). Bresch and Sy (2003) presented some general mathematical results for convection in rotating porous media.

A study of Coriolis effects on the filtration law in rotating porous media was made by Auriault *et al.* (2002a). Alex and Patil (2000a) analyzed an anisotropic medium. Desaive *et al.* (2002) included a study of Kuppers-Lortz instability for the case of Coriolis effects. Vadasz and Govender (2001) and Govender (2003a,c) treated in turn the Coriolis effect for monotonic convection and oscillating convection induced by gravity and centrifugal forces, each in a rotating porous layer distant from the axis of rotation.

Various non-Newtonian fluid have been considered. A Rivlin-Ericksen fluid was analyzed by Krishna (2001). A micropolar fluid was treated by Sharma and Kumar (1998). An electrically conducting couple-stress fluid was studied by Sunil *et al.* (2002).

6.23. Other Types of Fluids

The onset of convection in a horizontal layer of a medium saturated with a micropolar fluid was studied by Sharma and Gupta (1995). Coupling between thermal and micropolar effects may introduce oscillatory motions. A nonlinear analysis with a micropolar medium was reported by Siddheshwar and Krishna (2003). A similar problem for a ferromagnetic fluid was studied by Vaidyanathan *et al.* (1991). The corresponding problem with a non-Newtonian power-law fluid, with constant-flux boundary conditions, was treated analytically and numerically by Amari *et al.* (1994). The effect of suspended particles was treated by Mackie (2000). Viscoelastic fluids were studied by Prakash and Kumar (1999a), Sri Krishna (2001), and Yoon *et al.* (2003, 2004), and also by Prakash and Kumar (1999a,b) for the case of variable gravity, Sharma and Kango (1999) for the MHD case, by Kumar (1999) with the addition of suspended particles, while Kim *et al.* (2003b) conducted a nonlinear analysis. Other papers involving non-Newtonian fluids have been mentioned in the previous section.

6.24. Effects of Vertical Vibration and Variable Gravity

The subject of thermovibrational convection is of current interest in connection with the study of the behavior of materials in a microgravity environment as on a spacecraft, where residual accelerations (g-jitter) may have undesirable effects. The term thermovibrational convection refers to the appearance of a mean flow in a fluid-filled cavity having temperature heterogeneities. In this case, by proper selection of frequency and amplitude of vibration one may observe significant modifications in the stability threshold of convective motions. Historically, there have been two schools of thought on treating this type of problem. The first group apply linear stability analysis to the system of hydrodynamic equations in its original form, and thus obtain a set of coupled linear differential equations with periodic coefficients. The second group apply the time-averaging method and now a periodic coefficient does not appear explicitly in the governing equations. In this approach, which is valid for the case of high frequency and small amplitude, the temperature, pressure, and velocity fields may be decomposed into two parts, the first of which varies slowly with time, while the second part varies rapidly with time and has a zero mean over a vibrational period. This method leads to substantial simplifications in the mathematical formulation and even in some cases provides us with analytical relationships for the onset of convection. It enables a more in-depth analysis of the control parameters and consequently a better understanding of vibrational effect. The validity of a time-averaged method has been proved mathematically as well as experimentally, c.f. Gershuni and Lyubimov (1998). Several theoretical papers, including those of Zenkovskaya (1992), Zenkovskaya and Rogovenko (1999), and Bardan and Mojtabi (2000), have been published on thermovibrational convection in porous media by applying this method. The simplest case is that of an infinite horizontal porous layer with height H that undergoes a vertical vibration of sinusoidal form, which is characterized by amplitude (b) and frequency (ω). As a first step, the simultaneous effects of vibration and gravitational acceleration may be considered; the vibration vector is parallel to the gravitational acceleration. The boundaries of the layer are kept at constant but different temperatures. Adopting the Darcy model, the Rayleigh number at the onset of stationary convection, can be expressed as:

$$\mathrm{Ra} = \frac{(\pi^2 + \alpha^2)^2}{\alpha^2} + \mathrm{Ra_v}\frac{\alpha^2}{\alpha^2 + \pi^2}, \tag{6.195}$$

where

$$\mathrm{Ra_v} = \frac{(R\mathrm{Ra})^2}{2B\omega^{*2}}. \tag{6.196}$$

and α is the dimensionless wave number.

Here R is an acceleration ratio $(b\omega^2/g)$, B may be considered as a sort of inverse Darcy-Prandtl number ($B = a_* K/\varepsilon\nu\sigma H^2$), and ω^* is the dimensionless frequency. The stability diagram in the $\mathrm{Ra_c}$-R plane reveals that vibration increases the stability threshold and reduces the critical wave number. Another interesting result obtained from Eq.(6.195) is that under microgravity conditions the layer is linearly/infinitely stable. It was shown mathematically by Zenkovskaya (1992) that the transition toward an oscillatory convection in this case is not possible. This problem was also treated by Razi *et al.* (2002) by using the direct method. In these papers, the authors showed that the stability analysis led to a Mathieu equation. An analogy between the stability behaviors of the thermofluid problem with that of an inverted pendulum under the effect of vertical vibration was made, cf. Razi *et al.* (2005). It may be recalled that vertical vibration may stabilize an inverted pendulum, which is in an unstable position. Based on a scale analysis reasoning, the domain of validity of time-averaged method was found. Razi *et al.* explained why the transient term should be kept in the momentum equation at high frequency. In addition, they argued that the time-averaged method only gives the harmonic response and they predicted the existence of a subharmonic response. Thus these studies bridged the gap between the two schools of thought on thermovibrational problems. The outcome of these analyses can be interpreted in the context of constructal theory (Bejan, 2000) as follows: among the many combinations between frequency and amplitude of vibration it is the high-frequency and small amplitude that provide the stabilizing effect.

The finite amplitude case was studied by Bardan and Mojtabi (2000), Mojtabi (2002), Bardan *et al.* (2004), and Razi *et al.* (2005). Their weakly nonlinear analysis shows that the bifurcation at the transition point is of the supercritical pitchfork type. Mojtabi *et al.* (2004) examined the case of variable directions of vibration in the limiting case of high-frequency and small amplitude. They concluded that when the direction of vibration is perpendicular to the temperature gradient, the vibration has a destabilizing effect. They also predicted the onset of convection in microgravity conditions.

The alternative school of thought is represented by the papers by Malashetty and Padmavathi (1998) (who included non-Darcy effects) and Govender (2004b, 2005c–f). The latter presented the results of both linear and weak nonlinear analysis with emphasis on the transition from synchronous to subharmonic motions, and he treated the cases of low frequency and a layer heated from above.

Herron (2001) analyzed the onset of convection in a porous medium heated internally and with the gravitational field varying with distance through the layer. He proved that oscillatory is not possible as long as the gravity field and the integral of the heat sources have the same sign. Kim *et al.* (2005) studied the transient convection resulting from a sudden imposition of gravity.

6.25. Bioconvection

Bioconvection is concerned with pattern formation in suspensions of microorganisms, such as bacteria and algae, due to up-swimming of the microorganisms. The microorganisms are denser than water and on the average they swim upward. When they congregate the system becomes top-heavy and instability as convection may result. Microorganisms respond to various stimuli. Gravitaxis refers to swimming in the opposite sense as gravity. Gyrotaxis is swimming directed by the balance between the torque due to gravity acting on a bottom-heavy cell and the torque due to viscous forces arising from local shear flows. Oxytaxis corresponds to swimming up an oxygen concentration gradient.

Kuznetsov and co-workers have analyzed various aspects of bioconvection in a porous medium, sufficiently sparse so that the microorganisms can swim freely. Gravitaxis was considered by Kuznetsov and Jiang (2001, 2003) and Kuznetsov and Avramenko (2003a) with and without cell deposition and declogging. Further studies of gravitaxis were conducted by Nguyen *et al.* (2004) and Nguyen-Quang *et al.* (2005). A falling plume involving the bioconvection of oxytactic bacteria was treated by Kuznetsov *et al.* (2003a, 2004). The oxytactic situation with superposed fluid and porous layers was studied by Avramenko and Kuznetsov (2005). A falling plume was also studied numerically by Becker *et al.* (2004). Gyrotaxis was studied by Kuznetsov and Avramenko (2002, 2003b, 2005), Nield *et al.* (2004c), and Avramenko and Kuznetsov (2004). Work on bioconvection in porous media was reviewed by Kuznetsov (2005).

6.26. Constructal Theory of Bénard Convection

In this section we take a closer look at the phenomenon of convection in a porous layer heated from below. Our objective is to show that most of the features of the flow can be determined based on a simple method: the intersection of asymptotes (Nelson and Bejan, 1998). This method was originally used for the optimization of spacings for compact cooling channels for electronics (Bejan, 1984); see also Lewins (2003) and Bejan *et al.* (2004).

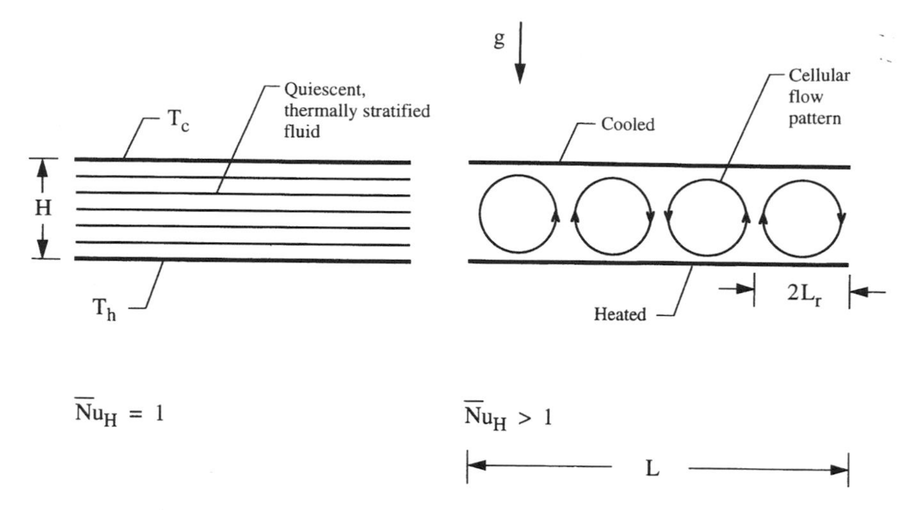

Figure 6.30. Horizontal porous layer saturated with fluid and heated from below (Nelson and Bejan, 1998).

Assume that the system of Fig. 6.30 is a porous layer saturated with fluid and that if present the flow is two-dimensional and in the Darcy regime. The height H is fixed and the horizontal dimensions of the layer are infinite in both directions. The fluid has nearly constant properties such that its density-temperature relation is described well by the Boussinesq linearization. The volume averaged equations that govern the conservation of mass, momentum and energy are

$$\frac{\partial u}{\partial x} + \frac{\partial v}{\partial y} = 0 \tag{6.195}$$

$$\frac{\partial u}{\partial y} - \frac{\partial v}{\partial x} = -\frac{Kg\beta}{\nu}\frac{\partial T}{\partial x} \tag{6.196}$$

$$u\frac{\partial T}{\partial x} + v\frac{\partial T}{\partial y} = \alpha_m\left(\frac{\partial^2 T}{\partial x^2} + \frac{\partial^2 T}{\partial y^2}\right) \tag{6.197}$$

The horizontal length scale of the flow pattern ($2L_r$), or the geometric aspect ratio of one roll, is unknown. The method consists of analyzing two extreme flow configurations—many counterflows vs. few plumes—and intersecting these asymptotes for the purpose of maximizing the global thermal conductance of the flow system, i.e., by invoking the constructal law, Bejan (1997c, 2000).

6.26.1. The Many Counterflows Regime

In the limit $L_r \to 0$ each roll is a very slender vertical counterflow, Fig. 6.31. Because of symmetry, the outer planes of this structure ($x = \pm L_r$) are adiabatic: they represent the center planes of the streams that travel over the distance H. The scale analysis of the $H \times (2L_r)$ region indicates that in the $L_r/H \to 0$ limit the

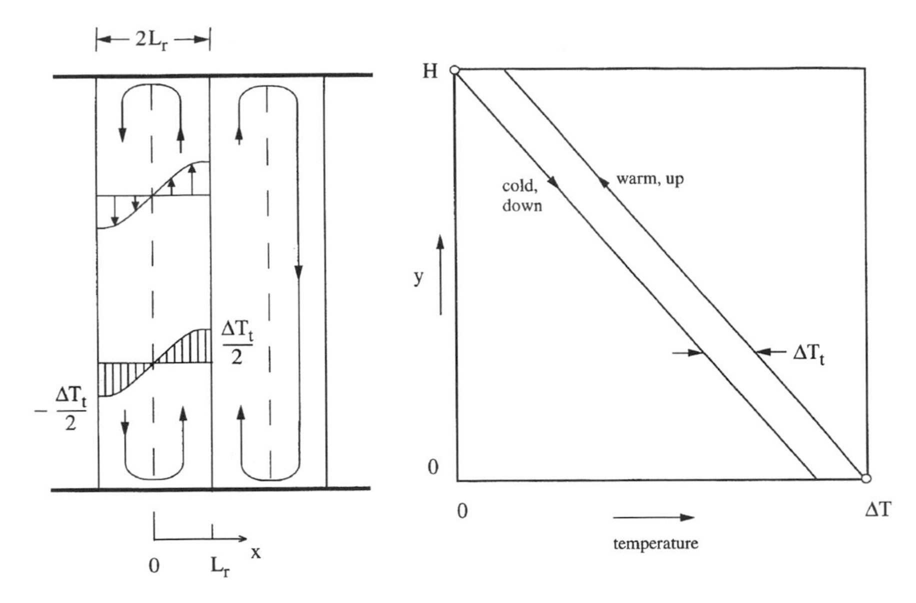

Figure 6.31. The extreme in which the flow consists of many vertical and slender counterflows (Nelson and Bejan, 1998).

horizontal velocity component u vanishes. This scale analysis is not shown because it is well known as the defining statement of fully developed flow. Equations (6.195)–(6.197) reduce to

$$\frac{\partial v}{\partial x} = \frac{kg\beta}{\nu}\frac{\partial t}{\partial x}, \tag{6.198}$$

$$v\frac{\partial t}{\partial y} = \alpha_m \frac{\partial^2 t}{\partial x^2}, \tag{6.199}$$

which can be solved exactly for v and T. The boundary conditions are $\partial T/\partial x = 0$ at $x = \pm L_r$, and the requirement that the extreme (corner) temperatures of the counterflow region are dictated by the top and bottom walls, $T(-L_r, H) = T_c$ and $T(L_r, 0) = T_h$. The solution is given by

$$v(x) = \frac{\alpha_m}{2H}\left[\mathrm{Ra}_H - \left(\frac{\pi H}{2L_r}\right)^2\right]\sin\left(\frac{\pi x}{2L_r}\right) \tag{6.200}$$

$$T(x, y) = \frac{\nu}{Kg\beta}v(x) + \frac{\nu}{Kg\beta}\left(2\frac{y}{H} - 1\right)\frac{\alpha_m}{2H}\left[\mathrm{Ra}_H - \left(\frac{\pi H}{2L_r}\right)^2\right], \\ + (T_h - T_c)\left(1 - \frac{y}{H}\right), \tag{6.201}$$

where the porous-medium Rayleigh number $\mathrm{Ra}_H = Kg\beta H(T_h - T_c)/(\alpha_m \nu)$ is a specified constant. The right side of Fig. 6.31 shows the temperature distribution along the vertical boundaries of the flow region ($x = \pm L_r$): the vertical temperature

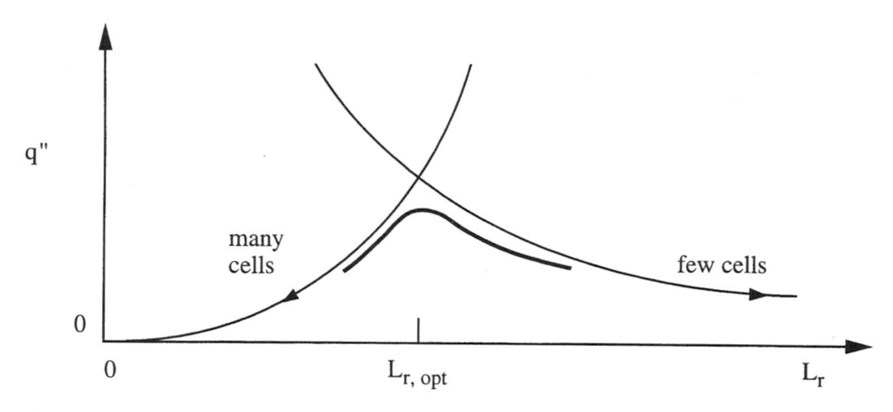

Figure 6.32. The intersection of asymptotes method: the geometric maximization of the thermal conductance of a fluid-saturated porous layer heated from below (Nelson and Bejan, 1998).

gradient $\partial T/\partial y$ is independent of altitude. The transversal (horizontal) temperature difference (ΔT_t) is also a constant,

$$\Delta T_t = T\,(x = L_r) - T\,(x = -L_r) = \frac{\nu}{Kg\beta}\frac{\alpha_m}{H}\left[\mathrm{Ra}_H - \left(\frac{\pi H}{2L_r}\right)^2\right]. \quad (6.202)$$

The counterflow convects heat upward at the rate q′, which can be calculated using Eqs. (6.200) and (6.201):

$$q' = \int_{-L}^{L} (\rho c_P)_f\, vT\, dx \quad (6.203)$$

The average heat flux convected in the vertical direction, $q'' = q'/(2L_r)$, can be expressed as an overall thermal conductance

$$\frac{q''}{\Delta T} = \frac{k_m}{8H\mathrm{Ra}_H}\left[\mathrm{Ra}_H - \left(\frac{\pi H}{2L_r}\right)^2\right]^2. \quad (6.204)$$

This result is valid provided the vertical temperature gradient does not exceed the externally imposed gradient, $(-\partial T/\partial y) < \Delta T/H$. This condition translates into

$$\frac{L_r}{H} > \frac{\pi}{2}\mathrm{Ra}_H^{-1/2}, \quad (6.205)$$

which in combination with the assumed limit $L_r/H \to 0$ means that the domain of validity of Eq. (6.204) widens when Ra_H increases. In this domain the thermal conductance $q''/\Delta T$ decreases monotonically as L_r decreases, cf. Fig. 6.32.

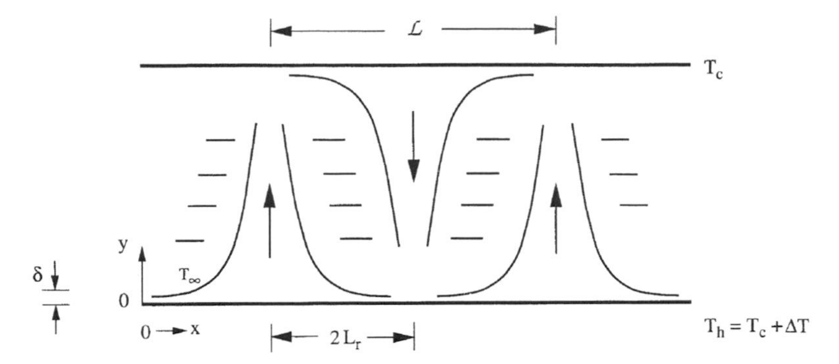

Figure 6.33. The extreme in which the flow consists of a few isolated plumes (Nelson and Bejan, 1998).

6.26.2. The Few Plumes Regime

As L_r increases, the number of rolls decreases and the vertical counterflow is replaced by a horizontal counterflow in which the thermal resistance between T_h and T_c is dominated by two horizontal boundary layers, as in Fig. 6.33. Let δ be the scale of the thickness of the horizontal boundary layer. The thermal conductance $q''/\Delta T$ can be deduced from the heat transfer solution for natural convection boundary layer flow over a hot isothermal horizontal surface facing upward or a cold surface facing downward. The similarity solution for the horizontal surface with power-law temperature variation (Cheng and Chang, 1976) can be used to develop an analytical result, as we show at the end of this section.

A simpler analytical solution can be developed in a few steps using the integral method. Consider the slender flow region $\delta \times (2L_r)$, where $\delta \ll 2L_r$, and integrate Eqs. (6.195) to (6.197) from y = 0 to $y \to \infty$, that is, into the region just above the boundary layer. The surface temperature is T_h, and the temperature outside the boundary layer is T_∞ (constant). The origin $x = 0$ is set at the tip of the wall section of length $2L_r$. The integrals of Eqs. (6.195) and (6.197) yield

$$\frac{d}{dx}\int_0^\infty u\,(T - T_\infty)\,dy = -\alpha_m \left(\frac{\partial T}{\partial y}\right)_{y=0} \tag{6.206}$$

The integral of Eq. (6.196), in which we neglect $\partial v/\partial x$ in accordance with boundary layer theory, leads to

$$u_0(x) = \frac{Kg\beta}{\nu}\frac{d}{dx}\int_0^\infty T\,dy, \tag{6.207}$$

where u_0 is the velocity along the surface, $u_0 = u(x,0)$. Reasonable shapes for the u and T profiles are the exponentials

$$\frac{u(x, y)}{u_0(x)} = \exp\left[-\frac{y}{\delta(x)}\right] = \frac{T(x, y) - T_\infty}{T_h - T_\infty} \tag{6.208}$$

which transform Eqs. (6.206) and (6.207) into

$$\frac{d}{dx}(u_0\delta) = \frac{2\alpha_m}{\delta} \tag{6.209}$$

$$u_0 = \frac{Kg\beta}{\nu}(T_h - T_\infty)\frac{d\delta}{dx} \tag{6.210}$$

These equations can be solved for $u_0(x)$ and $\delta(x)$,

$$\delta(x) = \left[\frac{9\alpha_m \nu}{Kg\beta(T_h - T_\infty)}\right]^{1/3} x^{2/3}. \tag{6.211}$$

The solution for $u_0(x)$ is of the type $u_0 \sim x^{-1/3}$, which means that the horizontal velocities are large at the start of the boundary layer and decrease as x increases. This is consistent with the geometry of the $H\times 2L_r$ roll sketched in Fig. 6.33, where the flow generated by one horizontal boundary layer turns the corner and flows vertically as a relatively narrow plume (narrow relative to $2L_r$), to start with high velocity (u_0) a new boundary layer along the opposite horizontal wall.

The thermal resistance of the geometry of Fig. 6.33 is determined by estimating the local heat flux $k(T_h - T_\infty)/\delta(x)$ and averaging it over the total length $2L_r$:

$$q'' = \left(\frac{3}{4}\right)^{1/3}\frac{k_m \Delta T}{H}\left(\frac{T_h - T_\infty}{\Delta T}\right)^{4/3} \mathrm{Ra}_H^{1/3}\left(\frac{H}{L_r}\right)^{2/3}. \tag{6.212}$$

The symmetry of the sandwich of boundary layers requires $T_h - T_\infty = (1/2)\Delta T$, such that

$$\frac{q''}{\Delta T} = \frac{3^{1/3}k}{4H}\mathrm{Ra}_p^{1/3}\left(\frac{H}{L_r}\right)^{2/3}. \tag{6.213}$$

The goodness of this result can be tested against the similarity solution for a hot horizontal surface that faces upward in a porous medium and has an excess temperature that increases as x^{λ}. The only difference is that the role that was played by $(T_h - T_\infty)$ in the preceding analysis is now played by the excess temperature averaged over the surface length $2L_r$. If we use $\lambda = 1/2$, which corresponds to uniform heat flux, then it can be shown that the solution of Cheng and Chang (1976) leads to the same formula as Eq. (6.213), except that the factor $3^{1/3} = 1.442$ is replaced by $0.816(3/2)^{4/3} = 1.401$. Equation (6.213) is valid when the specified Ra_H is such that the horizontal boundary layers do not touch. We write this geometric condition as $\delta(x = 2L_r) < H/2$ and, using Eq. (6.211), we obtain

$$\frac{L_r}{H} < \frac{1}{24}\mathrm{Ra}_H^{1/2}. \tag{6.214}$$

Since in this analysis L_r/H was assumed to be very large, we conclude that the L_r/H domain in which Eq. (6.213) is valid becomes wider as the specified Ra_H increases. The important feature of the "few rolls" limit is that the thermal conductance decreases as the horizontal dimension L_r increases. This second asymptotic trend has been added to Fig. 6.32.

6.26.3. The Intersection of Asymptotes

Figure 6.32 presents a bird's-eye view of the effect of flow shape on thermal conductance. Even though we did not draw completely $q''/\Delta T$ as a function of L_r, the two asymptotes tell us that the thermal conductance is maximum at an optimal L_r value that is close to their intersection. There is a family of such curves, one curve for each Ra_H. The $q''/\Delta T$ peak of the curve rises and the L_r domain of validity around the peak becomes wider as Ra_H increases. Looking in the direction of small Ra_H values we see that the domain vanishes (and the cellular flow disappears) when the following requirement is violated

$$\frac{1}{24} H\,\mathrm{Ra}_H^{1/2} - \frac{\pi}{2} H\,\mathrm{Ra}_H^{-1/2} \geq 0. \tag{6.215}$$

This inequality means that the flow exists when $Ra_H \geq 12\pi = 37.70$. This conclusion is extraordinary: it agrees with the stability criterion for the onset of two-dimensional convection, Eq. (6.29), namely $Ra_H > 4\pi^2 = 39.5$, which was derived based on a lengthier analysis and the assumption that a flow structure exists: the initial disturbances (Horton and Rogers, 1945; Lapwood, 1948).

We obtain the optimal shape of the flow, $2L_{r,opt}/H$, by intersecting the asymptotes (6.204) and (6.213):

$$\pi^2 \left(\frac{H}{2L_{r,opt}} \mathrm{Ra}_H^{-1/2} \right)^2 + 2^{5/6} 3^{1/6} \left(\frac{H}{2L_{r,opt}} \mathrm{Ra}_H^{-1} \right)^{1/3} = 1\,. \tag{6.216}$$

Over most of the Ra_H domain where Eq. (6.215) is valid, Eq. (6.216) is approximated well by its high Ra_H asymptote:

$$\frac{2L_{r,opt}}{H} \cong \pi\,\mathrm{Ra}_H^{-1/2}\,. \tag{6.217}$$

The maximum thermal conductance is obtained by substituting the $L_{r,opt}$ value in either Eq. (6.213) or Eq. (6.204). This estimate is an upper bound, because the intersection is above the peak of the curve. In the high-Ra_H limit (6.217) this upper bound assumes the analytical form

$$\left(\frac{q''}{\Delta T} \right)_{\max} \frac{H}{k_m} \leq \frac{3^{1/3}}{2^{4/3}\pi^{2/3}} \mathrm{Ra}_H^{2/3}. \tag{6.218}$$

Toward lower Ra_H values the slope of the $(q''/\Delta T)_{max}$ curve increases such that the exponent of Ra_H approaches 1. This behavior is in excellent agreement with the large volume of experimental data collected for Bénard convection in saturated porous media (Cheng, 1978). The less-than -1 exponent of Ra_H in the empirical $Nu(Ra_H)$ curve, and the fact that this exponent decreases as Ra_H increases, has attracted considerable attention from researchers during the last two decades, as we showed earlier in this chapter.

7
Internal Natural Convection: Heating from the Side

Enclosures heated from the side are most representative of porous systems that function while oriented vertically, as in the insulations for buildings, industrial cold-storage installations, and cryogenics. As in the earlier chapters, we begin with the most fundamental aspects of the convection heat transfer process when the flow is steady and in the Darcy regime. Later, we examine the special features of flows that deviate from the Darcy regime, flows that are time dependent, and flows that are confined in geometries more complicated than the two-dimensional rectangular space shown in Fig. 7.1. Some of the topics of this chapter have been reviewed by Oosthuizen (2000).

7.1. Darcy Flow between Isothermal Side Walls

7.1.1. Heat Transfer Regimes

Consider the basic scales of the clockwise convection pattern maintained by the side-to-side heating of the porous medium defined in Fig. 7.1. In accordance with the homogeneous porous medium model, we begin with the equations for the conservation of mass, Darcy flow, and the conservation of energy in the $H \times L$ space:

$$\frac{\partial u}{\partial x} + \frac{\partial v}{\partial y} = 0, \tag{7.1}$$

$$u = -\frac{K}{\mu}\frac{\partial P}{\partial x}, \tag{7.2}$$

$$v = -\frac{K}{\mu}\left(\frac{\partial P}{\partial y} + \rho g\right), \tag{7.3}$$

$$u\frac{\partial T}{\partial x} + v\frac{\partial T}{\partial y} = \alpha_m\left(\frac{\partial^2 T}{\partial x^2} + \frac{\partial^2 T}{\partial y^2}\right). \tag{7.4}$$

Note that in contrast to the system used in Section 5.1, the y axis is now vertically upward. By eliminating the pressure P between Eqs. (7.2) and (7.3) and by invoking the Boussinesq approximation $\rho \cong \rho_0[1 - \beta(T - T_0)]$ in the

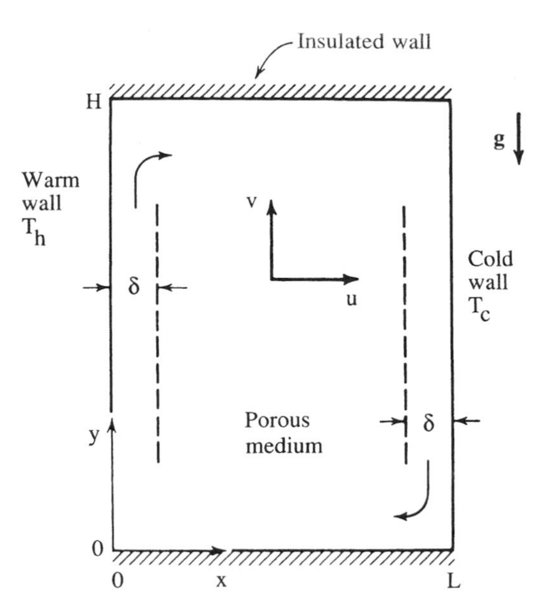

Figure 7.1. Two-dimensional rectangular porous layer held between differently heated side walls (Bejan, 1984).

body force term ρg of Eq. (7.3), we obtain a single equation for momentum conservation:

$$\frac{\partial u}{\partial y} - \frac{\partial v}{\partial x} = -\frac{Kg\beta}{\nu}\frac{\partial T}{\partial x}. \tag{7.5}$$

In this equation ν is the kinematic viscosity μ/ρ_0, which is assumed constant along with the other properties, the permeability K, the coefficient of volumetric thermal expansion β, and the porous-medium thermal diffusivity $\alpha_m = k_m/(\rho c_P)_f$.

The three equations (7.1), (7.4), and (7.5) hold in the entire domain $H \infty L$ subject to the boundary conditions indicated in the figure. The four walls are impermeable and the side-to-side temperature difference is $T_h - T_c = \Delta T$. Of special interest are the scales of the vertical boundary layers of thickness δ and height H. In each $\delta \times H$ region, the order-of-magnitude equivalents of Eqs. (7.1), (7.4), and (7.5) are

mass:
$$\frac{u}{\delta} \sim \frac{y}{H}, \tag{7.6}$$

energy:
$$\left(u\frac{\Delta T}{\delta}, v\frac{\Delta T}{H}\right) \sim \left(\alpha_m\frac{\Delta T}{\delta^2}, \alpha_m\frac{\Delta T}{H^2}\right), \tag{7.7}$$

momentum:
$$\left(\frac{u}{H}, \frac{v}{\delta}\right) \sim \frac{Kg\beta}{\nu}\frac{\Delta T}{\delta}. \tag{7.8}$$

To begin with, the mass balance (7.6) shows that the two scales on the left-hand side of Eq. (7.7) are of the same order, namely $v\Delta T/H$. On the right-hand side of Eq. (7.7), the second scale can be neglected in favor of the first, because the $\delta \times H$ region is a boundary layer (i.e., slender),

$$\delta \ll H. \tag{7.9}$$

In this way, the energy conservation statement (7.7) reduces to a balance between the two most important effects, the conduction heating from the side, and the convection in the vertical direction,

$$\underset{\text{longitudinal convection}}{v\frac{\Delta T}{H}} \sim \underset{\text{lateral conduction}}{\alpha_m \frac{\Delta T}{\delta^2}} \tag{7.10}$$

Turning our attention to the momentum scales (7.8), we see that the mass balance (7.6) implies that the ratio between (u/H) and (v/δ) is of the order $(\delta/H)^2 \ll 1$. We then neglect the first term on the left-hand side of Eq. (7.8) and find that the momentum balance reduces to

$$\frac{v}{\delta} \sim \frac{Kg\beta}{v}\frac{\Delta T}{\delta}, \tag{7.11}$$

Equations (7.10), (7.11), and (7.6) imply that the scales of the vertical boundary layer (Bejan, 1985) are

$$v \sim \frac{Kg\beta}{v}\Delta T \sim \frac{\alpha_m}{H}\mathrm{Ra}, \tag{7.12}$$

$$\delta \sim H\mathrm{Ra}^{-1/2}, \tag{7.13}$$

$$u \sim \frac{\alpha_m}{H}\mathrm{Ra}^{1/2}, \tag{7.14}$$

where Ra is the Rayleigh number based on height,

$$\mathrm{Ra} = \frac{g\beta K H \Delta T}{v\alpha_m}. \tag{7.15}$$

The total heat transfer rate from one side wall to the other is simply

$$q' \sim k_m H \frac{\Delta T}{\delta} \sim k_m \Delta T\, \mathrm{Ra}^{1/2}. \tag{7.16}$$

This heat transfer rate is expressed per unit length in the direction perpendicular to the plane $H \times L$. It can be nondimensionalized as the overall Nusselt number

$$\mathrm{Nu} = \frac{q'}{q'_c} \sim \frac{k_m \Delta T\, \mathrm{Ra}^{1/2}}{k_m H \Delta T/L} \sim \frac{L}{H}\mathrm{Ra}^{1/2}, \tag{7.17}$$

in which $q_c = k_m H \Delta T/L$ is the true heat transfer rate in the pure-conduction limit (i.e., in the absence of convection).

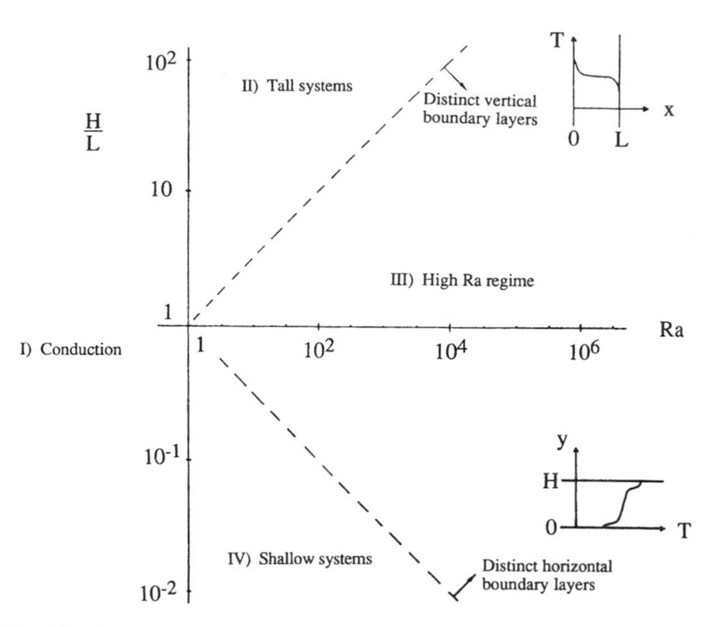

Figure 7.2. The four heat transfer regimes for natural convection in a two-dimensional porous layer heated from the side (Bejan, 1984).

Two requirements must be met if the results (7.12)–(7.17) are to be valid. First, the vertical boundary layers must be *slender*, which in view of Eqs. (7.9) and (7.13) means

$$\mathrm{Ra} \gg 1. \tag{7.18}$$

Second, the vertical boundary layers must be *distinct*, i.e., thinner than the layer itself, $\delta \ll L$. This second requirement can be rewritten [cf. Eq. (7.13)] as

$$\mathrm{Ra}^{1/2} \gg \frac{H}{L}. \tag{7.19}$$

The domain Ra, H/L in which the vertical boundary layers are distinct is indicated to the right of the rising dash line in Fig. 7.2.

The fluid completes its clockwise circulation in Fig. 7.1 by flowing along the horizontal boundaries. Whether or not these horizontal jets are distinct (thinner than H) can be determined using the scaling results (7.12)–(7.14). The volumetric flow rate of the horizontal jet is the same as that of the vertical boundary layer, namely $v\delta$. The two horizontal jets form a counterflow that carries energy by convection from left to right in Fig. 7.1, at the rate

$$q'_{(\rightarrow)} \sim v\delta(\rho c_P)_f \Delta T. \tag{7.20}$$

The heat transfer rate by thermal diffusion between these two jets, from top to bottom in Fig. 7.2, is

$$q'_{(\downarrow)} \sim k_m L \frac{\Delta T}{H}. \tag{7.21}$$

One horizontal jet travels the entire length of the porous layer (L) without experiencing a significant change in its temperature when the vertical conduction rate (7.21) is small relative to the horizontal convection rate (7.20). The inequality $q'_{(\downarrow)} \ll q'_{(\rightarrow)}$ yields

$$\frac{H}{L} \gg \mathrm{Ra}^{-1/2} \tag{7.22}$$

as the criterion for the existence of distinct horizontal layers. The parametric domain in which Eq. (7.22) is valid is indicated to the right of the descending dash line in Fig. 7.2. The structure of the horizontal layers contains additional features that have been analyzed systematically by Daniels *et al.* (1982).

Figure 7.2 summarizes the four regimes that characterize the heat transfer through a porous layer heated from the side. The results derived in this section recommend the adoption of the following heat transfer scales:

I. Pure conduction (no distinct boundary layers):

$$\mathrm{Nu} \cong 1, \quad q' \cong k_m H \frac{\Delta T}{L}. \tag{7.23}$$

II. Tall layers (distinct horizontal boundary layers only):

$$\mathrm{Nu} \gtrsim 1, \quad q' \gtrsim k_m H \frac{\Delta T}{L}. \tag{7.24}$$

III. High-Ra convection (distinct vertical and horizontal boundary layers):

$$\mathrm{Nu} \sim \frac{L}{H}\mathrm{Ra}^{1/2}, \quad q' \sim k_m H \frac{\Delta T}{H}. \tag{7.25}$$

IV. Shallow layers (distinct vertical boundary layers only):

$$\mathrm{Nu} \lesssim \frac{L}{H}\mathrm{Ra}^{1/2}, \quad q' \lesssim k_m H \frac{\Delta T}{\delta}. \tag{7.26}$$

In the remainder of this section we focus on regimes III and IV, in which the heat transfer rate can be significantly greater than the heat transfer rate associated with pure conduction. A more detailed classification of the natural convection regimes that can be present in a porous layer heated from the side was developed by Blythe *et al.* (1983).

7.1.2. Boundary Layer Regime

Weber (1975b) developed an analytical solution for the boundary layer regime by applying the Oseen linearization method. The focus of the analysis is the vertical boundary layer region along the left wall in Fig. 7.1, for which the momentum and energy equations are

$$\frac{\partial^2 \psi_*}{\partial x_*^2} = \frac{\partial T_*}{\partial x_*}, \tag{7.27}$$

$$\frac{\partial \psi_*}{\partial x_*}\frac{\partial T_*}{\partial y_*} - \frac{\partial \psi_*}{\partial y_*}\frac{\partial T_*}{\partial x_*} = \frac{\partial^2 T_*}{\partial x_*^2}. \tag{7.28}$$

These equations involve the streamfunction ψ now defined by $u = -\partial\psi/\partial y$ and $v = \partial\psi/\partial x$ and the dimensionless variables

$$x_* = \frac{x}{H}\mathrm{Ra}^{1/2}, \qquad y_* = \frac{y}{H}, \tag{7.29}$$

$$\psi_* = \frac{\psi}{\alpha_m \mathrm{Ra}^{1/2}}, \qquad T_* = \frac{T - (T_h + T_c)/2}{T_h - T_c}. \tag{7.30}$$

The solution begins with treating $\partial\psi_*/\partial y_*$ (the entrainment velocity) and $\partial T_*/\partial y_*$ as functions of y_* only. This leads to the exponential profiles

$$\psi_* = \psi_\infty(1 - e^{-\lambda x_*}), \tag{7.31}$$

$$T_* = T_\infty + \left(\frac{1}{2} - T\right)e^{-\lambda x_*}, \tag{7.32}$$

in which the core temperature T_8, the core streamfunction ψ_8, and the boundary layer thickness $1/\lambda$ are unknown functions of y_*. These unknowns are determined from three conditions, the equations obtained by integrating Eqs. (7.27) and (7.28) across the boundary layer,

$$\lambda\psi_\infty = \frac{1}{2} - T_\infty, \tag{7.33}$$

$$\frac{d}{dy_*}\left[\frac{1}{2\lambda}\left(\frac{1}{2} - T_\infty\right)^2\right] + \psi\frac{dT_\infty}{dy_*} = \lambda\left(\frac{1}{2} - T_\infty\right), \tag{7.34}$$

and the centrosymmetry of the entire flow pattern. The latter implies that ψ_8 must be an even function of $z = y_* - 1/2$ and that T_8 must be an odd function of altitude z. Note that z is measured away from the horizontal midplane of the rectangular space. The solution is expressed by

$$\psi^* = C(1 - q^2)\left\{1 - \exp\left[-\frac{x_*}{2C(1+q)}\right]\right\}, \tag{7.35}$$

$$T_* = \frac{1}{2}\left\{q + (1 - q)\exp\left[-\frac{x_*}{2C(1+q)}\right]\right\}, \tag{7.36}$$

where q is an implicit odd function of z:

$$z = C^2\left(q - \frac{1}{3}q^3\right). \tag{7.37}$$

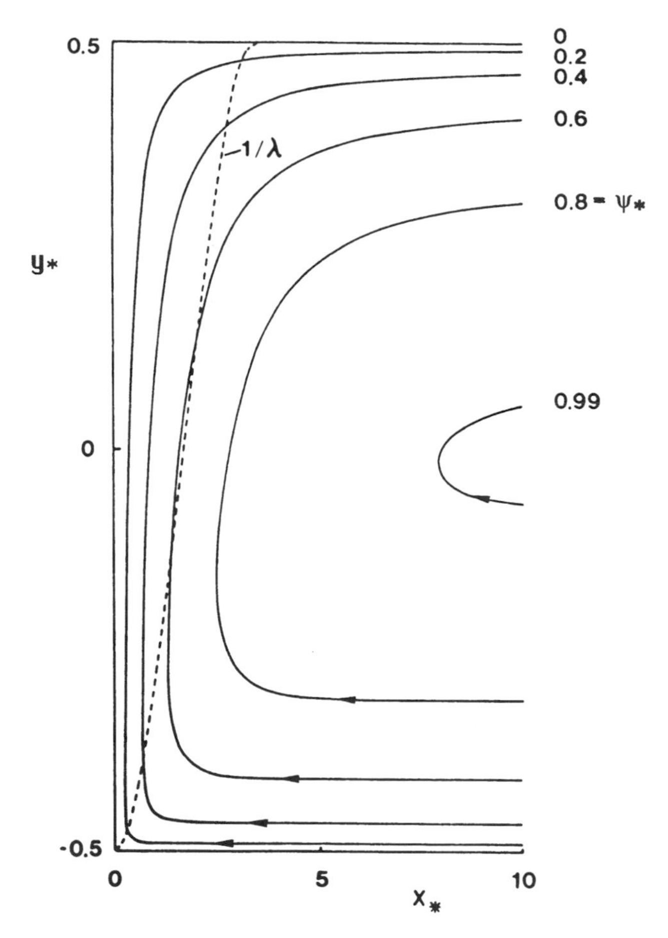

Figure 7.3. The streamlines near the heated wall in the boundary layer regime (Bejan, 1984).

Weber (1975b) determined the constant C by invoking the impermeable top and bottom conditions $\psi_* = 0$ at $z = \pm 1/2$ and obtained $C = 3^{1/2}/2 = 0.866$. The patterns of streamlines and isotherms that correspond to this solution were drawn later by Bejan (1984) and are reproduced in Figs. 7.3 and 7.4. These figures show a vertical boundary layer flow that discharges itself horizontally into a thermally stratified core region. The total heat transfer rate between the two side walls can be expressed as the conduction-referenced Nusselt number defined in Eq. (7.17), now given by

$$\mathrm{Nu} = 0.577 \frac{L}{H} \mathrm{Ra}^{1/2}. \tag{7.38}$$

The agreement between Weber's solution (7.38) and the order of magnitude prediction (7.17) is evident. Figure 7.5 shows a comparison between Eq. (7.38) and

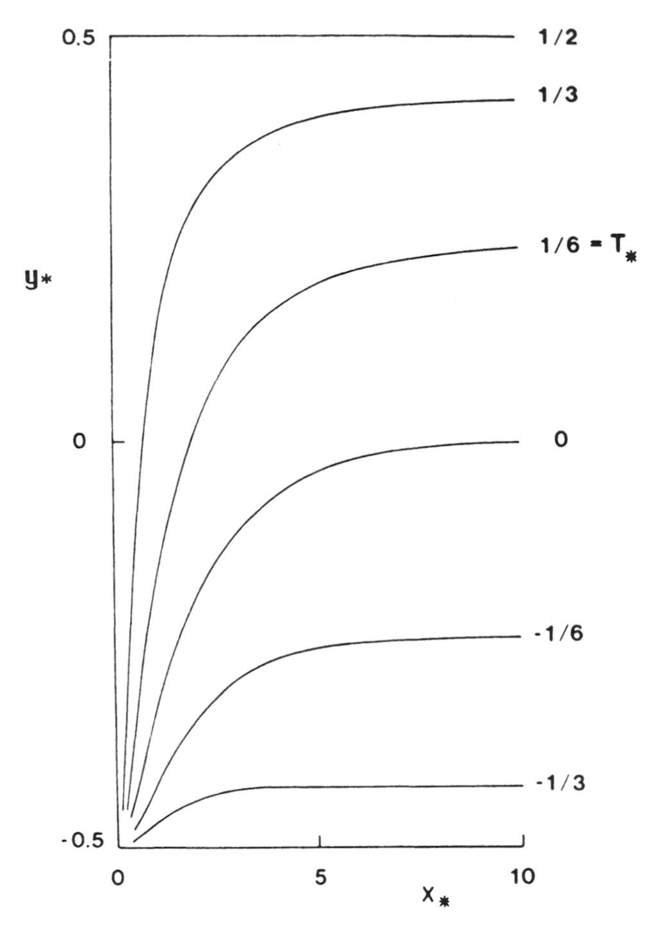

Figure 7.4. The isotherms near the heated wall in the boundary layer regime (Bejan, 1984).

experimental and numerical data collected from three sources (Schneider, 1963; Klarsfeld, 1970; Bankvall, 1974). The proportionality between Nu and $(L/H)\mathrm{Ra}^{1/2}$ anticipated from Eqs. (7.17) and (7.38) appears to be correct in the high Rayleigh number limit. It is important to also note that the boundary layer theory (7.38) consistently overpredicts the Nusselt number, especially at high Rayleigh numbers.

Bejan (1979) showed that the discrepancy between theory and empirical results can be attributed to the way in which the constant C was determined for the solutions (7.35)–(7.37). His alternative was to simultaneously invoke the impermeable and adiabatic wall conditions at $z = \pm 1/2$. This was approximately accomplished by setting the total vertical energy flow rate (convection + conduction) equal to zero at the top and the bottom of the porous layer. The C value that results from this

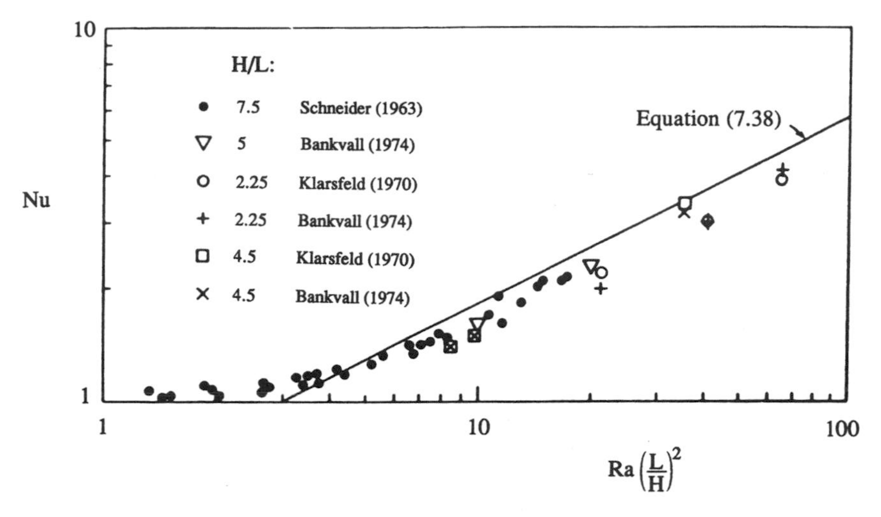

Figure 7.5. Theoretical, numerical, and experimental results for the heat transfer rate through a porous layer heated from the side (Bejan, 1984).

condition is given implicitly by

$$C = (1 - q_e^2)^{-2/3}\,\mathrm{Ra}^{-1/6}\left(\frac{H}{L}\right)^{-1/3} \tag{7.39}$$

in which q_e is itself a function of C,

$$\frac{1}{2} = C^2\left(q_e - \frac{1}{3}q_e^3\right). \tag{7.40}$$

Figure 7.6 shows the emergence of $\mathrm{Ra}(H/L)^2$ as a new dimensionless group that differentiates between various boundary layer regimes. The constant C approaches Weber's value $3^{1/2}/2$ as this new group approaches infinity. The same figure shows that the Nusselt number is generally below the value calculated with Eq. (7.38), where $0.577 = 3^{-1/2}$. An alternative presentation of this heat transfer information is given in Fig. 7.7, which shows that in the boundary layer regime Nu depends not only on $\mathrm{Ra}(L/H)^2$, cf. Eq. (7.17), but also on the aspect ratio H/L. This secondary effect is a reflection of the new group $\mathrm{Ra}(H/L)^2$ identified in Fig. 7.6.

An integral boundary layer solution that incorporates the same zero vertical energy flow condition was reported by Simpkins and Blythe (1980). The structure of the vertical boundary layer region near the top and bottom corners—neglected in the work reviewed here—was analyzed by Blythe *et al.* (1982). A numerical study of high Ra convection, yielding correlations for the heat transfer rate, was reported by Shiralkar *et al.* (1983). For tall cavities, Rao and Glakpe (1992) proposed a correlation of the form $\mathrm{Nu} = 1 + a(\mathrm{Ra})L/H$, for $H/L > H_m(\mathrm{Ra})$, where $a(\mathrm{Ra})$ and H_m (Ra) are quantities determined numerically. Ansari and Daniels (1993,

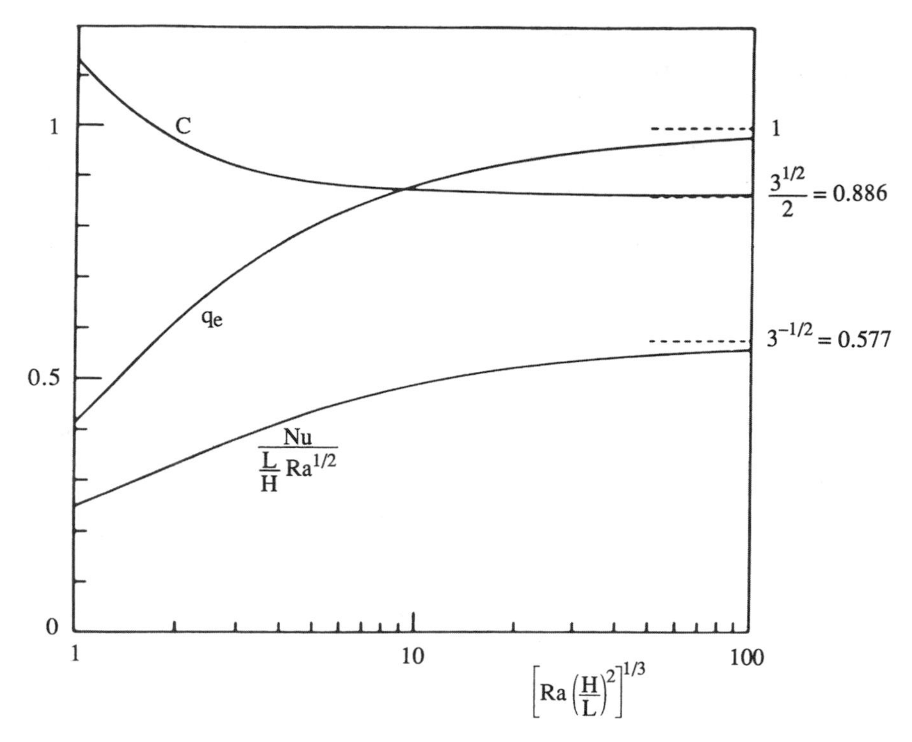

Figure 7.6. The effect of the group $Ra(H/L)^2$ on the solution for boundary layer natural convection in a porous layer heated from the side (Bejan, 1979).

1994) treated flow in tall cavities, taking into account the nonlinear flow that occurs near each end of the cavity. Their second study, which was concerned with the case of Ra and aspect ratio large and of the same order, led to the prediction of a position of minimum heat transfer across the cavity. A further study using a boundary domain integral method was reported by Jecl and Skerget (2000).

Masuoka *et al.* (1981) performed experiments with glass beads and water, the results of which were in agreement with a boundary layer analysis extended to take account of the vertical temperature gradient in the core and the apparent wall-film thermal resistance which is caused by a local increase in porosity near the wall.

7.1.3. Shallow Layer

Like the high-Ra regime III described in the preceding subsection, the natural convection in shallow layers (regime IV, Fig. 7.2) also can be characterized by heat transfer rates that are considerably greater than the heat transfer rate in the absence of a buoyancy effect. Regime IV differs from regime III in that the horizontal boundary layers are not distinct. The main characteristics of natural convection in a shallow layer are presented in Fig. 7.8: the vertical end layers are distinct and

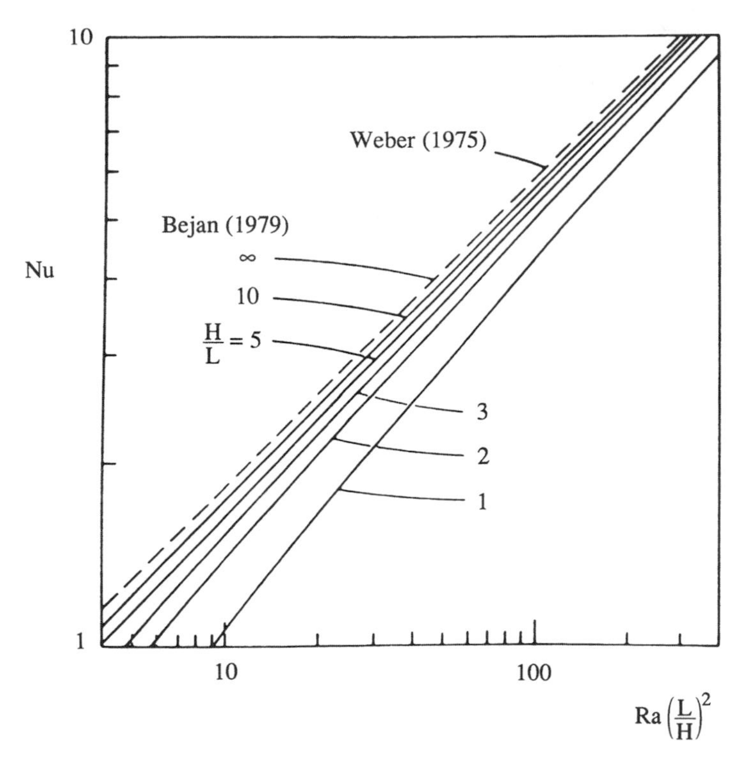

Figure 7.7. The heat transfer rate in the boundary layer regime (Nu > 1) in a porous layer heated from the side (Bejan, 1979).

a significant temperature drop is registered across the "core," that is, along the horizontal counterflow that occupies most of the length L.

The first studies of natural convection in shallow porous layers were published independently by Bejan and Tien (1978) and Walker and Homsy (1978). These studies showed that in the core region the circulation consists of a purely horizontal counterflow:

$$u = -\frac{\alpha_m}{H}\,\mathrm{Ra}\frac{H}{L}K_1\left(y* - \frac{1}{2}\right) \tag{7.41}$$

$$v = 0, \tag{7.42}$$

in which $y_* = y/H$. As shown in the lower part of Fig. 7.8, the core temperature varies linearly in the horizontal direction, while the degree of vertical thermal stratification is independent of x,

$$\frac{T - T_c}{T_h - T_c} = K_1\frac{x}{L} + K_2 + \mathrm{Ra}\left(\frac{H}{L}\right)^2 K_1^2\left(\frac{y_*^2}{4} - \frac{y_*^3}{6}\right). \tag{7.43}$$

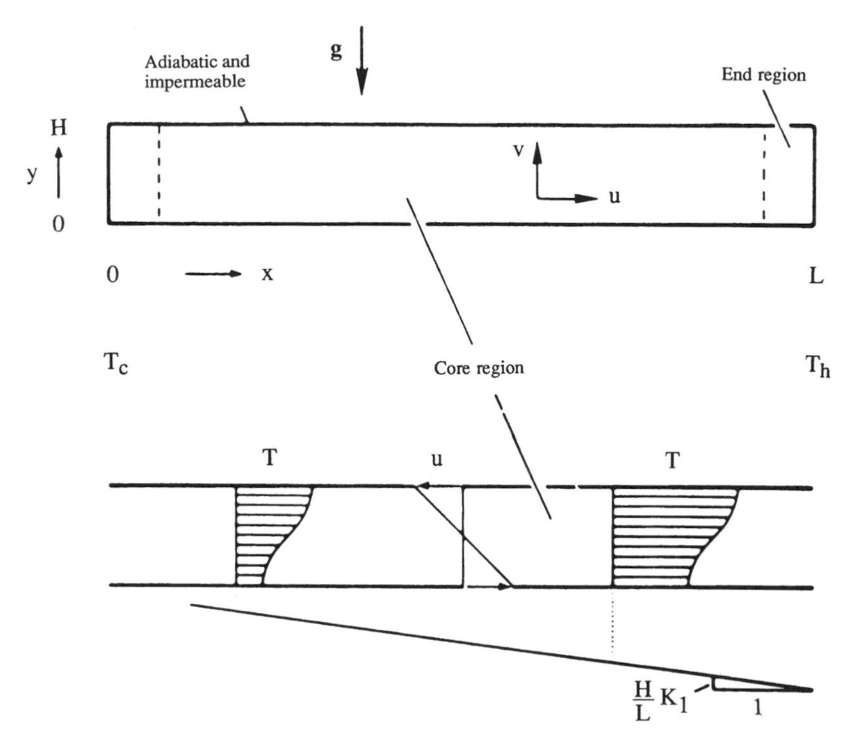

Figure 7.8. The structure of a horizontal porous layer subjected to an end-to-end temperature difference (Bejan and Tien, 1978).

The conduction-referenced Nusselt number for the total heat transfer rate from T_h to T_c is

$$\mathrm{Nu} = \frac{q'}{k_m \Delta T / L} = K_1 + \frac{1}{120} K_1^3 \left(\mathrm{Ra} \frac{H}{L} \right)^2 . \tag{7.44}$$

Parameters K_1 and K_2 follow from matching the core flows (7.41)–(7.43) to the vertical boundary layer flows in the two end regions. Bejan and Tien (1978) determined the function $K_1(H/L, \mathrm{Ra})$ parametrically by matching the core solution to integral solutions for the end regions. Their result is given implicitly by the system of equations

$$\frac{1}{120}\, \delta_e \mathrm{Ra}^2 K_1^3 \left(\frac{H}{L} \right)^3 = 1 - K_1, \tag{7.45}$$

$$\frac{1}{2} K_1 \frac{H}{L} \delta_e \left(\delta_e^{-2} - 1 \right) = 1 - K_1, \tag{7.46}$$

in which δ_e is the ratio end-region thickness/H. The Nusselt number based on this K_1 function and Eq. (7.44) has been plotted in Fig. 7.9, next to the numerical results published subsequently by Hickox and Gartling (1981), who also reported

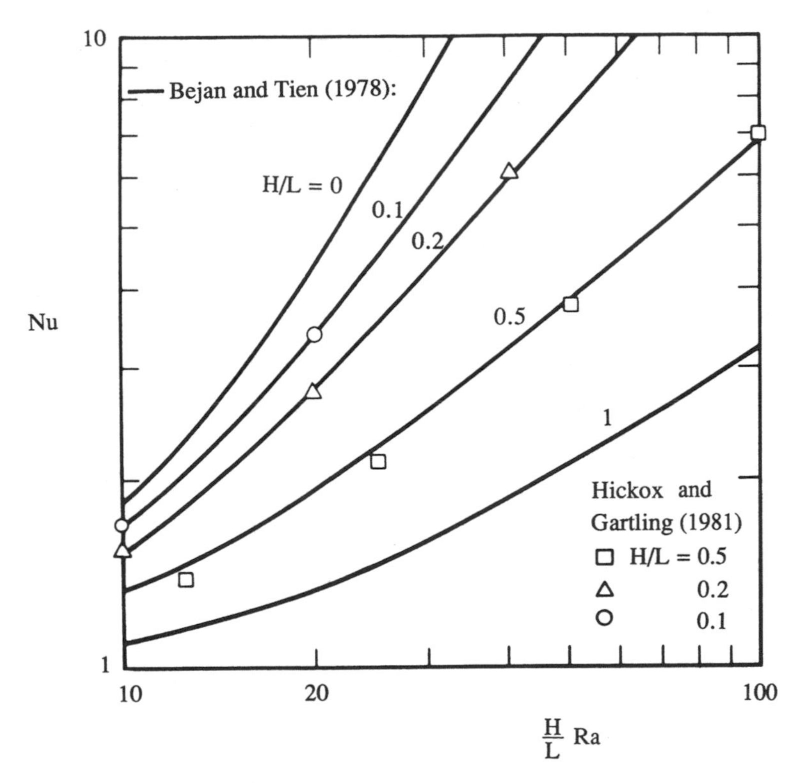

Figure 7.9. The heat transfer rate in a shallow porous layer with different and temperatures (Bejan, 1984).

representative patterns of streamlines and isotherms. Additional patterns can be seen in the paper by Daniels *et al.* (1986). In the infinitely shallow layer limit $H/L \to 0$, the horizontal counterflow accounts for the entire temperature drop from T_h to T_c and K_1 approaches 1. In the same limit Nu also approaches 1, cf. Eq. (7.44), with $K_1 = 1$:

$$\mathrm{Nu} = 1 + \frac{1}{120}\left(\mathrm{Ra}\frac{H}{L}\right)^2, \qquad \left(\frac{H}{L} \to 0\right). \tag{7.47}$$

It is important to note that the shallow-layer solution of Fig. 7.9 and Eqs. (7.44)–(7.47) approaches a proportionality of type $\mathrm{Nu} \sim (L/H)\mathrm{Ra}^{1/2}$ as Ra increases, which is in agreement with the scaling law (7.17). That proportionality (Bejan and Tien, 1978),

$$\mathrm{Nu} = 0.508\frac{L}{H}\mathrm{Ra}^{1/2} \qquad (\mathrm{Ra} \to \infty), \tag{7.48}$$

is nearly identical to Weber's (1975b) solution (7.38) for the high-Ra regime. In conclusion, the Nu(Ra, H/L) solution represented by Eqs. (7.44)–(7.48) and Fig. 7.9 is adequate for heat transfer calculations in both shallow and tall layers, at

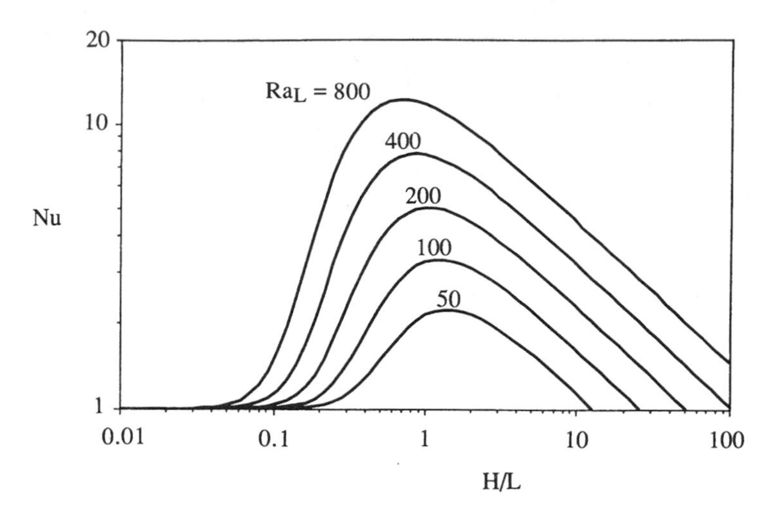

Figure 7.10. The effect of the height of the enclosure on the heat transfer rate through a porous layer heated from the side.

low and high Rayleigh numbers. This conclusion is stressed further in Fig. 7.10, which shows the full effect of the aspect ratio when the Rayleigh number based on the horizontal dimension $\text{Ra}_L = g\beta KL\Delta T/\nu\alpha_m$ is fixed (Bejan, 1980). The heat transfer rate reaches a maximum when the rectangular domain is nearly square.

This conclusion is relevant to the design of vertical double walls filled with fibrous or granular insulation held between internal horizontal partitions with the spacing H. In this design, the wall-to-wall spacing L is fixed while the number and positions of the horizontal partitions can change. The conclusion that the maximum heat transfer rate occurs when H is of order L also holds when the enclosure does not contain a porous matrix. In that case, the vertical spacing between partitions that corresponds to the maximum heat transfer rate is given approximately by $H/L \sim 0.1 - 1$ (Bejan, 1980).

Blythe *et al.* (1985b) and Daniels, Simpkins, and Blythe (1989) have analyzed the merged-layer regime which is defined by $L/H \to \infty$ at fixed $R_2 = \text{Ra}\ H^2/L^2$. In this limit the boundary layers on the horizontal walls merge and completely fill the cavity. The regime is characterized by a nonparallel core flow that provides the dominant structure over a wide range of R_2 values. The use of R_2 leads to the heat transfer correlation shown in Fig. 7.11.

7.1.4. Stability of Flow

Gill (1969) showed that linear stability analysis using the Darcy equation with no inertial terms leads to the prediction that the basic flow produced by differential heating of the walls of a vertical slab of infinite height is stable. Georgiadis and Catton (1985) claimed that instability was predicted when one included the time-wise acceleration term in the momentum equation, but Rees (1988) showed that

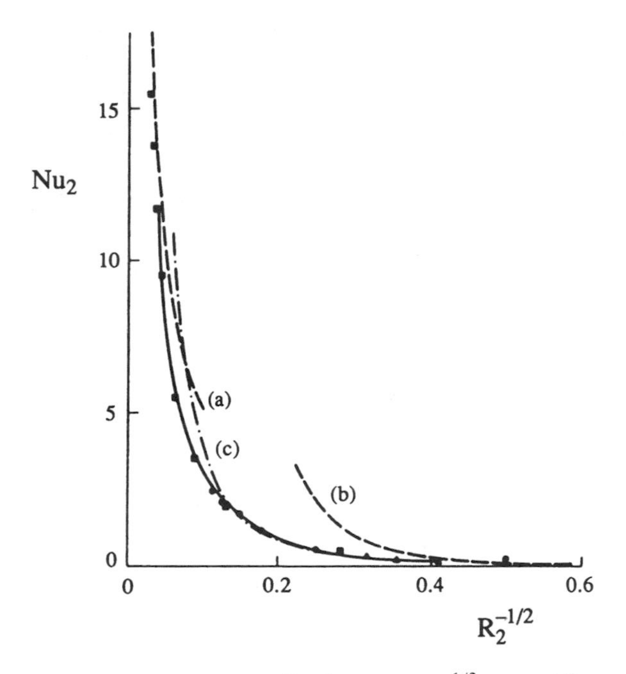

Figure 7.11. Variation of $\mathrm{Nu}_2 = \mathrm{Nu}H^2/L^2$ versus $R_2^{-1/2} = \mathrm{Ra}^{-1/2}L/H$. The solid line defines the merged-layer solution. Dashed lines show the asymptotic solutions (a) $\mathrm{Nu}_2 \sim 0.515R_2^{1/2}$, (b) $\mathrm{Nu}_2 \sim R_2^2/120$, and (c) $\mathrm{Nu}_2 \approx R_2^2/120\ (1 - 3\sigma_1 R_2)$ where $\sigma_1 \approx -0.07$. Results from numberical solutions by Hickox and Gartling (1981) and Prasad and Kulacki (1984b) are shown for various values of Ra and L/H : ■ $L/H = 2$; • $L/H = 5$; and ▲ $L/H = 10$ (Daniels *et al.*, 1989).

their analysis contained an error. The nonlinear analysis of Straughan (1988) predicts that the basic flow is stable provided that the initial disturbance is smaller than a certain threshold that is proportional to the inverse of the Rayleigh number.

The situation is dramatically changed when boundary friction is accounted for by means of the Brinkman equation. Kwok and Chen (1987) performed a linear stability analysis that led to predicted values $\mathrm{Ra}_c = 308.0$, $\alpha_c = 2.6$ if viscosity variations are ignored, and $\mathrm{Ra}_c = 98.3$, $\alpha_c = 1.6$ if viscosity variations are taken into account. In their experiment they observed a value 66.2 for the critical Rayleigh number Ra_c, which is based on the width L. They did not measure the critical vertical wavenumber α_c. The instability appears to be related to the fact that the basic vertical velocity profile is no longer linear. The disagreement between predicted and observed values of Ra_c presumably is due to the effect of porosity variation. A nonlinear analysis on the Brinkman model was performed by Qin and Kaloni (1993) for rigid or stress-free boundaries.

Riley (1988) has studied the effect of spatially periodic boundary imperfections. He found that out-of-phase imperfections enhance the heat transfer significantly.

The stability problem that arises for a rotating medium occupying a vertical slot, for which there is a horizontal body force due to the centrifugal acceleration and a positive temperature gradient in the same direction, was studied analytically by Vadasz (1994a). Convection in the form of superposed convection cells appears when a centrifugal Rayleigh number exceeds a certain value. Govender and Vadasz (1995) have shown that there is an analogy between this problem and natural convection in an inclined layer subject only to gravity. The results of experiments in a Hele-Shaw cell by Vadasz and Heerah (1996) showed qualitative agreement with the theory.

Rees and Lage (1996) considered a rectangular container where the impermeable bounding walls are held at a temperature that is a linearly decreasing function of height, the local temperature drop across the container being zero. They considered containers of finite aspect ratio and those of asymptotically large aspect ratio. For both cases, they found that modes bifurcate in pairs as the linear stability equations admit an infinite set of double eigenvalues. They analyzed the weakly nonlinear evolution of the primary pair of eigenmodes and found that the resulting steady-state flow is dependent on the form of the initial disturbance. For asymptotically tall boxes, their numerical and asymptotic analysis produced no evidence of persistently unsteady flow.

Kimura (1992) numerically studied convection in a square cavity with the upper half of a vertical wall cooled and the lower half heated, so that a cold current descends and fans out over a rising hot current. The unstable layer so formed appears to be associated with the onset of oscillations at Ra = 200. The effects of temperature-dependent thermal diffusivity and viscosity were included in a nonlinear stability analysis by Flavin and Rionero (1999).

7.1.5. Conjugate Convection

Conjugate convection in a rectangular cavity surrounded by walls of high relative thermal conductivity was examined by Chang and Lin (1994a). They reported that wall heat conduction effects decrease the heat transfer rate. The heat transfer through a vertical partition separating porous-porous or porous-fluid reservoirs at different temperatures was studied by Kimura (2003) on the basis of a simple one-dimensional vertically averaged model on the assumption that there is a linear increase in temperature in both of the reservoirs and the partition. He obtained results that are in general agreement with experiment. The steady-state heat transfer characteristics of a thin vertical strip with internal heat generation placed in a porous medium was studied by Méndez *et al.* (2002). A conjugate convection problem involving a thin vertical strip of finite length, placed in a porous medium, was studied by Martínez-Suástegui *et al.* (2003) using numerical and asymptotic techniques. A conjugate convection problem in a square cavity with horizontal conductive walls of finite thickness was studied numerically by Baytas *et al.* (2001). Mohamad and Rees (2004) have examined numerically conjugate convection in a porous medium attached to a wall held at a constant temperature.

7.1.6. Non-Newtonian Fluid

Convection in a rectangular cavity filled by a non-Newtonian power law fluid was studied theoretically and numerically by Getachew *et al.* (1996). They employed scaling arguments to delineate heat transfer regimes analogous to those discussed in Section 7.1.1 and verified their results using numerical calculations. A numerical study on the Brinkman-Forchheimer model was carried out by Hadim and Chen (1995). A further numerical study, using the boundary element method, was reported by Jecl and Skerget (2003). A numerical study of flow involving a couple-stress fluid was published by Umavarthi and Malashetty (1999), but the authors did not explain how the couple-stress is maintained on the scale of a representative elementary volume.

7.1.7. Other Situations

Convective heat transfer through porous insulation in a vertical slot with leakage of mass at the walls was analyzed by Burns *et al.* (1977). The effects of pressure stratification on multiphase transport across a vertical slot were studied by Tien and Vafai (1990b). The sidewall heating in shallow cavities with icy water was treated by Leppinen and Rees (2004). They considered a case in which the density maximum occurs somewhere between the sidewalls, and they treated the situation using asymptotic analysis valid in the limit of vanishing aspect ratio and Rayleigh number of O(1). In this case the flow is divided into two counterrotating cells whose size depends on the temperature giving the density maximum and the temperatures of the sidewalls. A study of entropy production for an MHD situation was made by Mahmud and Fraser (2004b). Thermal convection in a vertical slot with a spatially periodic thermal boundary condition was analyzed by Yoo (2003). Numerical studies of various problems involving lateral heating of square cavities were reported by Nithiarasu *et al.* (1999a,b, 2002).

7.2. Side Walls with Uniform Flux and Other Thermal Conditions

In the field of thermal insulation engineering, a more appropriate description for the side heating of the porous layer is the model where the heat flux q'' is distributed uniformly along the two side walls. In the high Rayleigh number regime (regime III, Fig. 7.2) the overall Nusselt number is given by (Bejan, 1983b)

$$\mathrm{Nu} = \frac{q''H}{k_m H\overline{\Delta T}/L} = \frac{1}{2}\left(\frac{L}{H}\right)^{4/5} \mathrm{Ra}_*^{2/5}, \tag{7.49}$$

In this Nu definition $\overline{\Delta T}$ is the height-averaged temperature difference that develops between the two side walls, $\overline{(T_h - T_c)}$, while Ra∗ is the Rayleigh number based on heat flux,

$$\mathrm{Ra}_* = \frac{g\beta K H^2 q''}{\nu \alpha_m k_m}. \tag{7.50}$$

Formula (7.49) is based on a matched boundary layer analysis that combines Weber's (1975b) approach with the zero energy flow condition for the top and bottom boundaries of the enclosure (Bejan, 1979). The solution obtained also showed that

(i) the vertical boundary layers have a constant thickness of order $H\,Ra_*^{-1/3}$;
(ii) the core region is motionless and linearly stratified, with a vertical temperature gradient equal to $(q''/k_m)\mathrm{Ra}_*^{-1/5}(H/L)^{2/5}$;
(iii) the temperature of each side wall increases linearly with altitude at the same rate as the core temperature, and so the local temperature difference between the side walls is independent of altitude; and
(iv) in any horizontal cut through the layer, there exists an exact balance between the net upflow of enthalpy and the net downward heat conduction.

The conditions that delineate the parametric domain in which Eq. (7.49) and regime III are valid are $\mathrm{Ra}_*^{-1/3} < H/L < \mathrm{Ra}_*^{1/3}$. This solution and the special flow features revealed by it are supported by numerical experiments performed in the range $100 = \mathrm{Ra} = 5000$ and $1 = H/L = 10$, which also are reported in Bejan (1983b).

The heat transfer by Darcy natural convection in a two-dimensional porous layer with uniform flux along one side and uniform temperature along the other side was investigated numerically by Prasad and Kulacki (1984a). Their set of thermal boundary conditions is a cross between those of Weber (1975b) and Bejan (1983b). The corresponding heat transfer process in a vertical cylindrical annulus with uniform heat flux on the inner wall and uniform temperature on the outer wall was studied experimentally by Prasad *et al.* (1986) and numerically by Prasad (1986). Dawood and Burns (1992) used a multigrid method to deal with three-dimensional convective heat transfer in a rectangular parallelepiped. Convection in a square cavity with one sidewall heated and the other cooled, with the heated wall assumed to have a spatial sinusoidal temperature variation about a constant mean value, was treated numerically by Saeid and Mohamad (2005b.)

An analytical and numerical study of the multiplicity of steady states that can arise in a shallow cavity was made by Kalla *et al.* (1999). The linear stability of the natural convection that arises in either a tall or shallow cavity was analyzed by Prud'homme and Bougherara (2001) and Prud'homme *et al.* (2003). Inverse problems, requiring the determination of an unknown sidewall flux, were treated by Prud'homme and Jasmin (2001) and Prud'homme and Nguyen (2001).

7.3. Other Configurations and Effects of Property Variation

7.3.1. Internal Partitions

The effect of horizontal and vertical internal partitions on natural convection in a porous layer with isothermal side walls was investigated numerically by Bejan

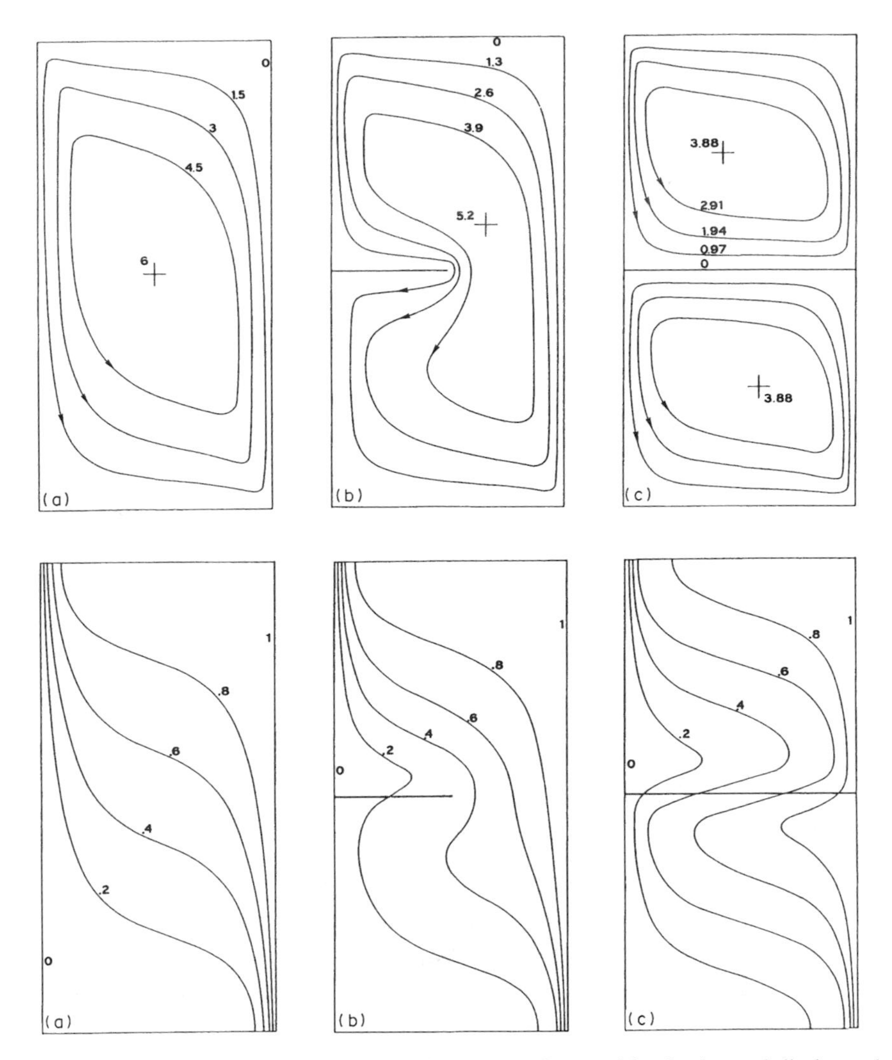

Figure 7.12. Streamlines and isotherms in a porous layer with a horizontal diathermal partition (Ra $= 400$, $H/L = 2$) (Bejan, 1983a).

(1983a). As an example, Fig. 7.12 shows the effect of a horizontal partition on the flow and temperature fields in regime III. In Fig. 7.12a the partition is absent and natural circulation is clearly in the boundary layer regime. When the horizontal midlevel partition is complete, the heat transfer rate decreases in predictable fashion as the height of each vertical boundary layer drops from H in Fig. 7.12a to $H/2$ in Fig. 7.12c. With the horizontal partition in place, the Nusselt number continues to scale as in Eq. (7.17); however, this time $H/2$ replaces H, and the Rayleigh number is based on $H/2$.

The insulation effect of a complete midplane vertical partition is illustrated in Fig. 7.13. The partition reduces the overall heat transfer rate by more than

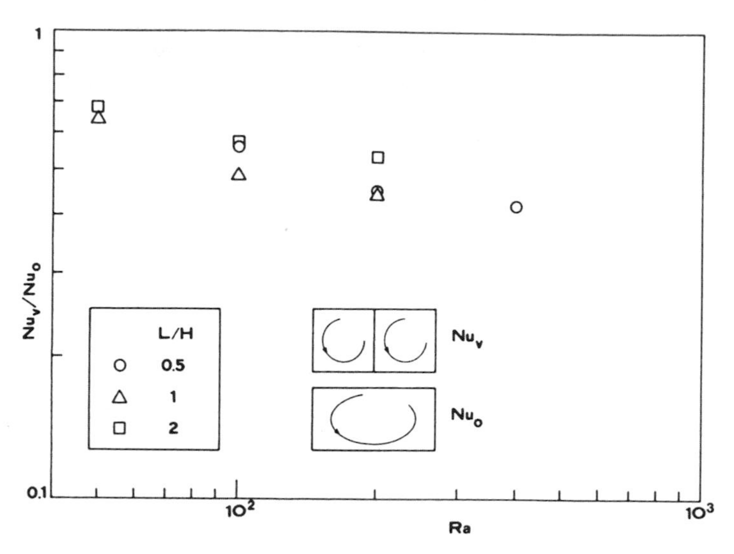

Figure 7.13. The reduction in overall heat transfer rate caused by a vertical diathermal partition (Bejan, 1983a).

50 percent as the Rayleigh number increases and vertical boundary layers form along all the vertical boundaries. This change can be expected in an order-of-magnitude sense: relative to the original system (without partitions), which has only two vertical boundary layers as thermal resistances between T_h and T_c, the partitioned system (Nu_ν in Fig. 7.13) has a total of four thermal resistances. The two additional resistances are associated with the conjugate boundary layers that form on the two sides of the partition. The thermal insulation effect associated solely with the conjugate boundary layers has been documented in Bejan and Anderson (1981) and in Section 5.1.5 of this book. Mbaye and Bilgen (1992, 1993) have studied numerically steady convection in a solar collector system that involves a porous wall.

7.3.2. Effects of Heterogeneity and Anisotropy

The preceding results apply to situations in which the saturated porous medium can be modeled as homogeneous. Poulikakos and Bejan (1983a) showed that the nonuniformity of permeability and thermal diffusivity can have a dominating effect on the overall heat transfer rate. For example, if the properties vary so much that the porous layer can be modeled as a vertical sandwich of vertical sublayers of different permeability and diffusivity (Fig. 7.14a), an important parameter is the ratio of the peripheral sublayer thickness (d_1) to the thermal boundary layer thickness (δ_1) based on the properties of the d_1 sublayer. Note that according to Eq. (7.14), δ_1 scales as $H\text{Ra}_1^{-1/2}$, where $\text{Ra}_1 = g\beta K_1 H(T_h - T_c)/\nu\alpha_{m,1}$ and the subscript 1 represents the properties of the d_1 sublayer.

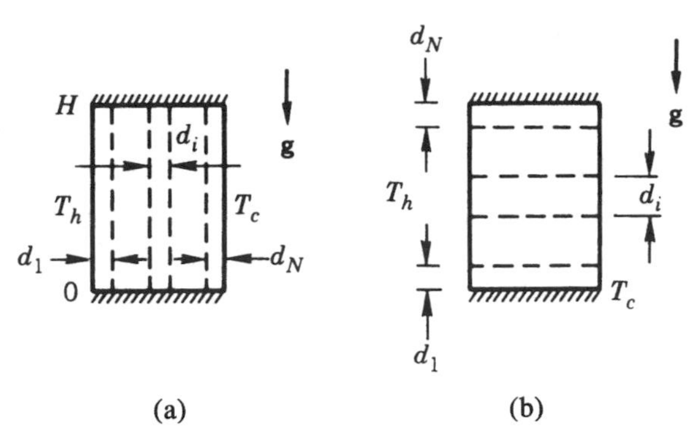

Figure 7.14. Layered porous media heated from the side: vertical sublayers (a), horizontal sublayers (b).

If the sublayer situated next to the right wall (d_N) has the same properties as the d_I sublayer, and if $\delta_1 < d_I$ and $\delta_N < d_N$, then the overall heat transfer rate can be estimated with the methods of Section 7.1 provided both Nu and Ra are based on the properties of the peripheral layers. An example of this kind is illustrated numerically in Fig. 7.15, where there are only three sublayers ($N = 3$), and the

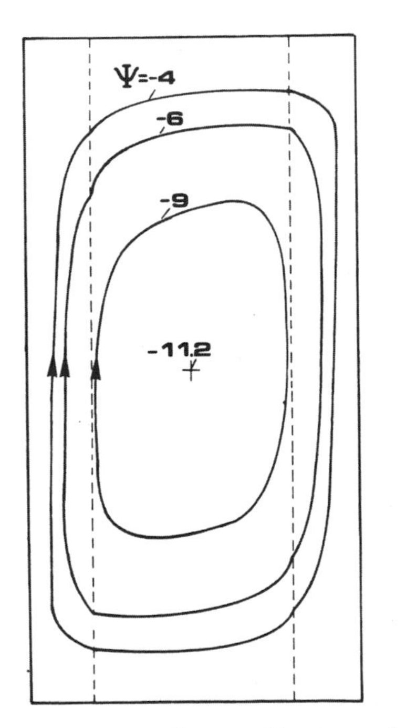

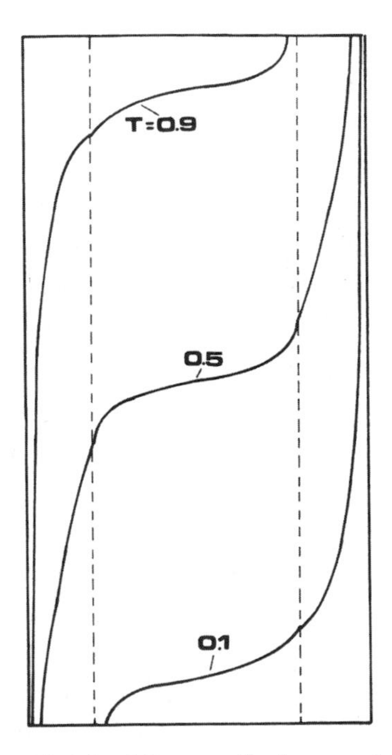

Figure 7.15. Streamlines and isotherms in a sandwich of three vertical porous layers heated from the side ($\mathrm{Ra}_1 = 200$, $H/L = 2$, $K_2/K_1 = 5$, $K_1 = K_3$, $N = 3$, and $\alpha_{m,1} = \alpha_{m,2} = \alpha_{m,3}$) (Poulikakos and Bejan, 1983a).

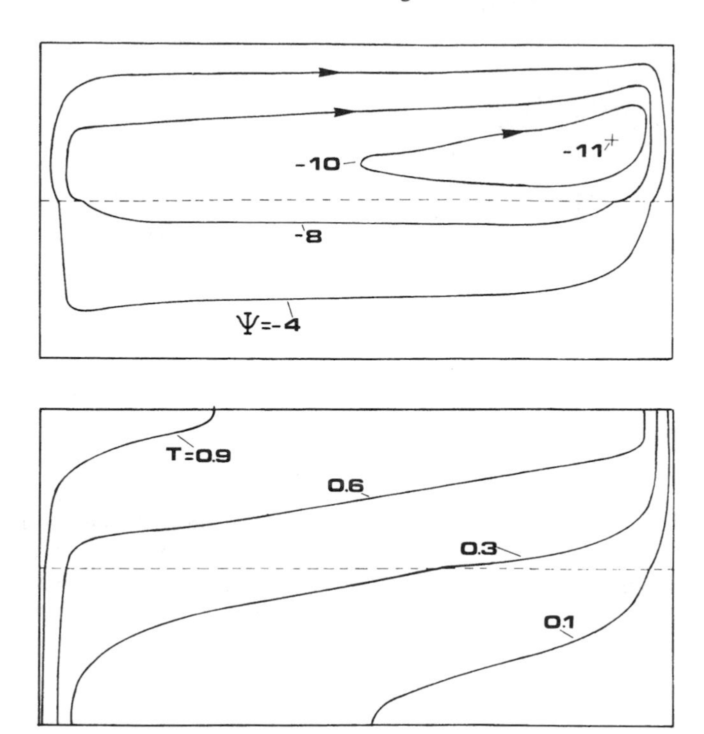

Figure 7.16. Streamlines and isotherms in a sandwich of two horizontal porous layers heated from the side ($Ra_1 = 150$, $H/L = 0.5$, $K_2/K_1 = 5$, $N = 2$, and $\alpha_{m,1} = \alpha_{m,2}$) (Poulikakos and Bejan, 1983a).

permeability of the core is five times greater than the permeability of the peripheral sublayers. The permeable core seems to "attract" the flow; this property renders the streamlines and isotherms almost horizontal and results in a vertically stratified core.

When the porous-medium inhomogeneity is such that the $H \times L$ system resembles a sandwich of N horizontal sublayers (Fig. 7.14b), the overall Nusselt number in the convection dominated regime is approximated by the correlation (Poulikakos and Bejan, 1983a)

$$\mathrm{Nu} \sim 2^{-3/2}\mathrm{Ra}_1^{1/2}\,\frac{L}{H}\sum_{i=1}^{N}\frac{k_i}{k_1}\left(\frac{K_i d_i \alpha_{m,1}}{K_1 d_1 \alpha_{m,i}}\right)^{1/2}, \tag{7.51}$$

where both Nu and Ra_1 are based on the properties of the bottom sublayer (d_1). This correlation was tested numerically in systems that contain two sublayers ($N = 2$). A sample of the computed streamlines and isotherms is presented in Fig. 7.16, for a case in which the upper half of the system is five times more permeable than the lower half. This is why the upper half contains most of the circulation. The discontinuity exhibited by the permeability K across the horizontal midplane causes cusps in the streamlines and the isotherms. The effect of nonuniformities in the thermal diffusivity of the porous medium in the two configurations of Fig. 7.14 also has been documented by Poulikakos and Bejan (1983a). A boundary layer analysis for a medium vertically layered in permeability was reported by Masuoka (1986).

In all the geometries discussed so far in this chapter, the walls that surrounded the saturated porous medium were modeled as impermeable. As a departure from the classic problem sketched in Fig. 7.8, the heat transfer through a shallow porous layer with both end surfaces permeable was predicted by Bejan and Tien (1978). Their theory was validated by subsequent laboratory measurements and numerical solutions conducted for Ra values up to 120 (Haajizadeh and Tien, 1983).

Lai and Kulacki (1988c) discussed convection in a rectangular cavity with a vertical permeable interface between two porous media of permeabilities K_1, K_2 and thermal conductivity k_1, k_2, respectively. The first medium was bounded by a heated face at constant heat flux and the second was bounded by a cooled isothermal face. The results of their calculations are generally in line with our expectations based on the material discussed in Section 6.13, but their finding of the existence of a second recirculating cell when $K_1/K_2 < 1$, $k_1/k_2 < 1$ is very surprising. A similar situation was treated numerically by Merrikh and Mohamad (2002).

Ni and Beckermann (1991a) have computed the flow in an anisotropic medium occupying a square enclosure. The horizontal permeability is denoted by K_x and the vertical permeability by K_y, and k_x, k_y are the corresponding thermal conductivities. Relative to the situation when the medium is isotropic with permeability K_x and thermal conductivity k_x, large K_y/K_x causes channeling along the vertical (isothermal) walls, a high flow intensity, and consequently a higher heat transfer rate Nu across the enclosure. Similarly, small K_y/k_x causes channeling along the horizontal (adiabatic) boundaries and a smaller Nu. Large k_y/k_x causes a higher flow intensity and a smaller Nu but small k_y/k_x has very little effect on the heat transfer pattern.

Non-Boussinesq variable-property effects were studied numerically by Peirotti *et al.* (1987) for the case of water or air. They found that these had a considerable impact on Nu. Kimura *et al.* (1993) presented an analysis, based on a perturbation method for small Ra, a rectangular cavity with anisotropy of permeability and thermal diffusivity. A numerical study for a rectangular cavity with a wall conduction effect and for anisotropic permeability and thermal diffusivity was performed by Chang and Lin (1994b). Degan *et al.* (1995) have treated analytically and numerically a rectangular cavity, heated and cooled with constant heat flux from the sides, with principal axes for permeability oblique to gravity and those for thermal conductivity aligned with gravity. They found that a maximum (minimum) heat transfer rate is obtained if the high permeability axis is parallel (perpendicular) to gravity, and that a large thermal conductivity ratio causes a higher flow intensity but a lower heat transfer. Degan and Vasseur (1996, 1997) and Degan *et al.* (1998a,b) presented a boundary layer analysis for the high Ra version of this problem and a numerical study on the Brinkman model. Egorov and Polezhaev

(1993) made a comprehensive theoretical (Darcy model) and experimental study for the anisotropic permeability problem. They found good agreement between their numerical results and experimental data for multilayer insulation. Vasseur and Robillard (1998) have reviewed the anisotropy aspects. The case of icy water was studied by Zheng *et al.* (2001). Further theoretical work, supplemented by experiments with a Hele-Shaw cell, was reported by Kimura and Okajima (2000) and Kimura *et al.* (2000).

7.3.3. Cylindrical or Annular Enclosure

Related to the two-dimensional convection phenomenon discussed so far in this chapter is the heat transfer through a porous medium confined by a horizontal cylindrical surface (Fig. 7.17a). The disk-shaped ends of the system are maintained at different temperatures. A parametric solution for heat transfer in this geometry was reported by Bejan and Tien (1978). The corresponding phenomenon in the porous medium between two horizontal concentric cylinders with different temperatures (Fig. 7.17b), was analyzed by Bejan and Tien (1979).

A basic configuration in the field of thermal insulation engineering is the horizontal annular space filled with fibrous or granular material (Fig. 7.18a). In this configuration the heat transfer occurs between the two concentric cylindrical surfaces of radii r_i and r_o, unlike in Fig. 7.17b where the cylindrical surfaces were insulated. Experimental measurements and numerical solutions for the overall heat transfer rate in the geometry of Fig. 7.18a have been reported by Caltagirone (1976b), Burns and Tien (1979), and Facas and Farouk (1983). The data of Caltagirone (1976b) in the range $1.19 \leq r_o/r_i \leq 4$ were correlated by Bejan (1987) on the basis of the scale-analysis procedure described in Bejan (1984, p. 194):

$$\mathrm{Nu} = \frac{q'}{q'_c} \cong 0.44\mathrm{Ra}_{r_i}^{1/2} \frac{\ln(r_o/r_i)}{1 + 0.916(r_i/r_o)^{1/2}}. \tag{7.52}$$

In the definition of the overall Nusselt number, the denominator is the conduction heat transfer rate $q'_c = 4\pi k_m(T_h - T_c)/\ln(r_o/r_i)$. The Rayleigh number is based

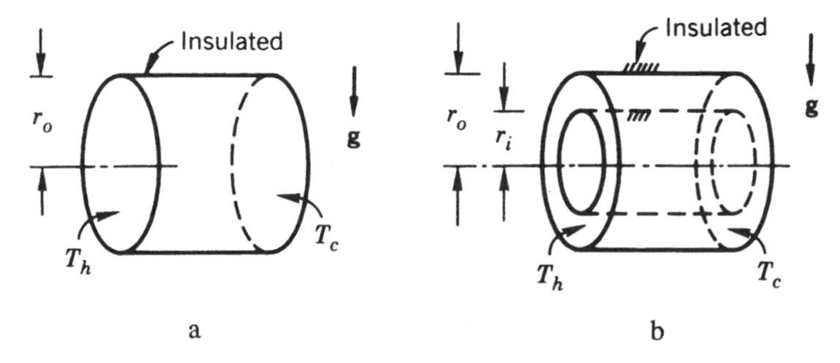

Figure 7.17. Confined porous medium with different end temperature: horizontal cylindrical enclosure (a) and horizontal cylindrical enclosure with annular cross section (b).

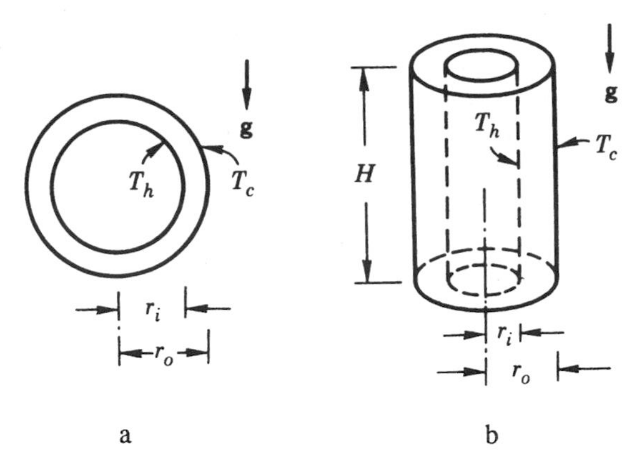

Figure 7.18. Radial heat transfer: horizontal cylindrical annulus or spherical annulus (a) and vertical cylindrical annulus (b).

on the inner radius, $\mathrm{Ra}_{ri} = g\beta K r_i(T_h - T_c)/\nu\alpha_m$. The correlation (7.52) is valid in the convection dominated regime, i.e., when $\mathrm{Nu} \gg 1$.

Transitions in the flow field in a horizontal annulus have been analyzed using a Galerkin method by Rao *et al.* (1987, 1988). As Ra is increased two-dimensional modes with one, two, and three cells on each side of the annulus appear in succession and the average Nusselt number increases with the number of cells. The extra cells appear near the top of the annulus. Three-dimensional modes also are possible with secondary flows in which the streamlines form a coaxial double helix, and these produce enhancement of the overall heat transfer resulting from a higher maximum local heat transfer rate in the upper part of the annulus.

Himasekhar and Bau (1988b) made a detailed bifurcation analysis for radii ratio values, 2, $2^{1/2}$, $2^{1/4}$, and $2^{1/8}$. Barbosa Mota and Saatdjian (1994, 1995) reported accurate numerical solutions for the Darcy model. For a radius ratio above 1.7 and for Rayleigh numbers above a critical value, they observed a closed hysteresis curve, indicating two possible solutions (two- or four-cell pattern) depending on initial conditions. For a radius ratio below 1.7 and as Ra is increased, the number of cells in the annulus increases without bifurcation and no hysteresis is observed. For very small radius ratios, steady-state regimes containing 2, 4, 6, and 8 cells are obtained in succession. For a radius ratio of 2, they found good agreement with experiment.

Charrier-Mojtabi and Mojtabi (1994, 1998) and Charrier-Mojtabi (1997) have numerically investigated both two- and three-dimensional flows for the Darcy model. They found that three-dimensional spiral flows are described in the vicinity of the transition from two-dimensional unicellular flows. They determined numerically the bifurcation points between two-dimensional unicellular flows and either two-dimensional multicellular flows or three-dimensional flows. Linear and nonlinear stability analyses were also performed by Charrier-Mojtabi and Mojtabi

(1998). These show that subcritical instability becomes increasingly likely as the radius ratio increases away from the value unity. For the cases of either isothermal or convective boundary conditions, Rajamani *et al.* (1995) studied the affects of both aspect ratio and radius ratio. They found that Nu always increases with radius ratio and Ra and it exhibits a maximum when the aspect ratio is about unity, the maximum shifting toward lesser aspect ratios as Ra increases.

For the case of the Darcy model and small dimensionless gap width $\varepsilon = (r_i - r_o)/r_i$, Mojtabi and Charrier-Mojtabi (1992) obtained an approximate analytical solution leading to the formula

$$\mathrm{Nu} = 1 + \frac{17}{40320}\mathrm{Ra}^2(\varepsilon^2 - \varepsilon^3), \tag{7.53}$$

where

$$\mathrm{Ra} - g\beta K(T_i - T_o)(r_i - r_o)/\nu\alpha_m. \tag{7.54}$$

A development up to order ε^{15} was given by Charrier-Mojtabi and Mojtabi (1998). Convection in a horizontal annulus with vertical eccentricity has been analyzed by Bau (1984a,c) for small Ra and by Himasekhar and Bau (1986) for large Ra for the case of steady two-dimensional flow. At low Ra there is an optimum eccentricity that minimizes the heat transfer, but generally the heat transfer decreases with eccentricity, independently of whether the heated inner cylinder is centered below or above the axis of the cooled outer cylinder. Highly accurate computations for this problem were reported by Barbosa Mota *et al.* (1994). A transient convection problem in an elliptical horizontal annulus was reported by Chen *et al.* (1990). A further numerical study of convection in such annuli was reported by Mota *et al.* (2000).

The heat transfer through an annular porous insulation oriented vertically (Fig. 7.18b) was investigated numerically by Havstad and Burns (1982), Hickox and Gartling (1985), and Prasad and Kulacki (1984c, 1985), and experimentally by Prasad *et al.* (1985). Havstad and Burns correlated their results with the five-constant empirical formula

$$\mathrm{Nu} \cong 1 + a_1\left[\frac{r_i}{r_o}\left(1 - \frac{r_i}{r_o}\right)\right]^{a_2} \mathrm{Ra}_{r_o}^{a_4}\left(\frac{H}{r_o}\right)^{a_5} \exp\left(-a_3\frac{r_i}{r_o}\right), \tag{7.55}$$

in which

$$\begin{aligned} &a_1 = 0.2196, \qquad a_4 = 0.9296,\\ &a_2 = 1.334, \qquad a_5 = 1.168\\ &a_3 = 3.702, \qquad \mathrm{Ra}_{r_o} = g\beta K r_o(T_h - T_c)/\nu\alpha_m. \end{aligned} \tag{7.56}$$

The overall Nusselt number is defined as in Eq. (7.52), $\mathrm{Nu} = q/q_c$, where $q_c = 2\pi k_m H(T_h - T_c)/\ln(r_o/r_i)$. The above correlation fits the numerical data in the range $1 = H/r_o = 20$, $0 = \mathrm{Ra}_{r0} < 150$, $0 < r_i/r_o = 1$, and $1 < \mathrm{Nu} < 3$.

For the convection-dominated regime (high Rayleigh numbers and $\mathrm{Nu} \gg 1$), the scale analysis of the boundary layers that form along the two cylindrical surfaces of Fig. 7.18b recommends the following correlation (Bejan, 1987):

$$\mathrm{Nu} = c_1 \frac{\ln(r_o/r_i)}{c_2 + r_o/r_i} \frac{r_o}{H} \mathrm{Ra}^{1/2}. \tag{7.57}$$

The Nusselt number is defined as in Eq. (7.52) and the Rayleigh number is based on height, $\mathrm{Ra} = g\beta K H(T_h - T_c)/\nu\alpha_m$. Experimental and numerical data are needed in the convection regime ($\mathrm{Nu} \gg 1$) in order to determine the constants c_1 and c_2. Havstad and Burns' (1982) data cannot be used because they belong to the intermediate regime $1 < \mathrm{Nu} < 3$ in which the effect of direct conduction from T_h to T_c is not negligible.

The experimental and numerical study of Reda (1986) treated a two-layered porous medium in a vertical annulus, with constant heat flux on the inner cylinder and constant temperature on the outer. Quasisteady convection in a vertical annulus, with the inner wall heated by a constant heat flux and the other walls adiabatic, was treated analytically and numerically by Hasnaoui *et al.* (1995). Also for a vertical annulus, Marpu (1995), Dharma Rao *et al.* (1996), and Satya Sai *et al.* (1997a) reported on numerical studies on the Brinkman-Forchheimer model. An asymptotic analysis for a shallow vertical annulus was presented by Pop *et al.* (1998) and Leppinen *et al.* (2004). Passive heat transfer augmentation in an annulus was studied by Iyer and Vafai (1999). The effect of local thermal nonequilibrium in convection in a vertical annulus was studied by Deibler and Bortolozzi (1998) and Bortolozzi and Deibler (2001). A numerical study of transient convection in a vertical annulus was reported by Shivakumar *et al.* (2002). Convection in a vertical annulus with an isothermal outer boundary and with a mixed inner boundary condition was treated by Jha (2005). Conjugate convection from a vertical cylindrical fin in a cylindrical enclosure was studied numerically by Naidu *et al.* (2004b). Convection in an elliptical vertical annulus was studied numerically by Saatdjian *et al.* (1999).

Rao and Wang (1991) studied both low and high Ra convection induced by internal heat generation in a vertical cylinder. Convection at large Ra is characterized by a homogeneous upward flow in the central part of the cylinder and a thin downward boundary layer at the cooled wall, with the effect of curvature of the boundary being negligible. This means that after introduction of a change of variable the results can be applied to enclosures with other than circular boundaries. Chang and Hsiao (1993) studied numerically convection in a vertical cylinder filled with an anisotropic medium with uniform high temperature on all boundaries except the cooler bottom. Lyubimov (1993) has summarized earlier Soviet work on the bifurcation analysis of two-dimensional convection in a cylinder of arbitrary shape, with the temperature specified on the boundary. The onset of convection in a vertical cylinder with a conducting wall was analyzed by Haugen and Tyvand (2003). Transient convection in a vertical cylinder with suddenly imposed or time-periodic wall heat flux was studied numerically by Slimi *et al.* (1998) and Amara *et al.* (2000). Transient convection in a vertical channel with the effect of radiation was studied numerically by Slimi *et al.* (2004).

Conjugate convection in a horizontal annulus was studied by Kimura and Pop (1991, 1992a). In their first paper they had isothermal boundaries but with a jump in heat flux at the fluid-solid interface, while in their second paper they used a Forchheimer model to study the case of the inner surface maintained at one temperature and the outer at a lower temperature.

Effects of rotation about the axis of a horizontal annulus were studied by Robillard and Torrance (1990) and Aboubi *et al.* (1995a). The former treated weak rotation, which generates a circulation relative to the solid matrix and thereby reduces the overall heat transfer. The latter examined the effect of a centrifugal force field for the case when the outer boundary is heated by a constant heat flux while the inner boundary is insulated. They performed a linear stability analysis and finite amplitude calculations which indicated the existence of multiple solutions differing by the number of cells involved.

Pan and Lai (1995, 1996) studied convection in a horizontal annulus with two subannuli for different permeabilities. They corrected (by satisfying the interface conditions more closely) the work by Muralidhar *et al.* (1986), thereby producing better agreement with experimental data. They noted that using a harmonic average permeability gives a better approximation to Nu than does an arithmetic average. Convection in a horizontal annulus with azimuthal partitions was studied numerically by Nishimura *et al.* (1996). Aboubi *et al.* (1995b) studied numerically convection in a horizontal annulus filled with an anisotropic medium, with principal axes of permeability inclined to the vertical. Three-dimensional anisotropy was incorporated into the model studied by Bessonov and Brailovskaya (2001). Convective flow driven by a constant vertical temperature gradient in a horizontal annulus was analyzed by Scurtu *et al.* (2001).

7.3.4. Spherical Enclosure

Another geometry that is relevant to the design of thermal insulations is the porous medium shaped as a spherical annulus (Fig. 7.18a). Heat is transferred radially between the two spherical walls that hold the porous material. Numerical heat transfer results for discrete values of the Rayleigh number and the geometric ratio r_i/r_o have been reported graphically by Burns and Tien (1979). From that set, the data that correspond to the convection dominated regime were correlated based on scale analysis by Bejan (1987),

$$\mathrm{Nu} = \frac{q}{q_c} \cong 0.756\,\mathrm{Ra}_{r_i}^{1/2}\frac{1 - r_i/r_o}{1 + 1.422(r_i/r_o)^{3/2}}. \tag{7.58}$$

The definitions used in Eq. (7.58) are $q_c = 4\pi k_m(T_h - T_c)/(r_i^{-1} - r_o^{-1})$ and $\mathrm{Ra}_{r_i} = g\beta K r_i(T_h - T_c)\nu\alpha_m$. The correlation (7.58) agrees within two percent with Burns and Tien's (1979) data for the convection regime represented by $\mathrm{Nu} \gtrsim 1.5$.

It is interesting to note that the scaling-correct correlation (7.58) can be restated in terms of the Rayleigh number based on insulation thickness,

$$\mathrm{Ra}_{r_o - r_i} = \frac{g\beta K(r_o - r_i)(T_h - T_c)}{\nu\alpha_m}. \tag{7.59}$$

The resulting expression that replaces Eq. (7.58) is

$$\mathrm{Nu} \cong 0.756\,\mathrm{Ra}_{r_o-r_i}^{1/2} \frac{[r_i/r_o - (r_i/r_o)^2]^{1/2}}{1 + 1.422(r_i/r_o)^{3/2}}. \tag{7.60}$$

This form can be differentiated to show that when Ra_{ro-ri} is fixed, the overall heat transfer rate (Nu) reaches a maximum value when $r_i/r_o = 0.301$. The existence of such a maximum was noted empirically by Burns and Tien (1979). An explanation for this maximum is provided by the boundary layer scale analysis on which the correlation (7.58) is based (Bejan, 1987). This maximum is the spherical-annulus analog of the maximum found in Fig. 7.10 for the heat transfer through a two-dimensional layer heated from the side. Future studies may show that similar Nu maxima occur in the cylindrical-annulus configurations of both Fig. 7.18a and Fig. 7.18b, when the Rayleigh number based on porous layer thickness Ra_{ro-ri} is constant. Convection in spherical annular sectors defined by an adiabatic radial wall was studied numerically by Baytas *et al.* (2002).

7.3.5. Porous Medium Saturated with Water Near 4°C

One class of materials that departs from the linear-density model used in the Boussinesq approximation (7.5) are the porous media saturated with cold water. The density of water at atmospheric pressure exhibits a maximum near 4°C. The natural convection in a cold-water saturated medium confined by the rectangular enclosure of Fig. 7.1 was described by Poulikakos (1984). As the equation of state in the Boussinesq approximation he used

$$\rho_m - \rho = \gamma \rho_m (T - T_m)^2 \tag{7.61}$$

with $T_m = 3.98$°C and $\gamma \cong 8 \times 10^{-6}\ \mathrm{K}^{-2}$ for pure water at atmospheric pressure. This parabolic density model is valid at temperatures ranging from 0°C to 10°C. Bejan (1987) showed that in the convection-dominated regime the Nusselt number correlation must have the form

$$\mathrm{Nu} = c_3 \frac{L/H}{\mathrm{Ra}_{\gamma h}^{-1/2} + c_4 Ra_{\gamma c}^{-1/2}}, \tag{7.62}$$

where the two Rayleigh numbers account for how T_h and T_c are positioned relative to the temperature of the density maximum T_m:

$$\mathrm{Ra}_{\gamma h} = \frac{g\gamma K H (T_h - T_m)^2}{\nu \alpha_m}, \quad \mathrm{Ra}_{\gamma c} = \frac{g\gamma K H (T_m - T_c)^2}{\nu \alpha_m}. \tag{7.63}$$

The overall Nusselt number Nu is referenced to the case of pure conduction, $\mathrm{Nu} = q'/q'_c$.

Poulikakos (1984) reported numerical Nu results in tabular form for the convection-dominated regime, primarily for the case $T_c = 0$°C, $T_h = 7.96$°C. By relying on these data, Bejan (1987) showed that when T_c and T_h are positioned symmetrically around T_m (i.e., when $\mathrm{Ra}_{\gamma h} = \mathrm{Ra}_{\gamma c}$) the correlation (7.62) reduces to

$$\mathrm{Nu} \cong 0.26 \frac{L}{H} \mathrm{Ra}_{\gamma h}^{1/2}. \tag{7.64}$$

In other words, this set of data indicates that in this case the two constants that appear in the general correlation (7.60) must satisfy the relationship $c_3 \cong 0.26(1 + c_4)$ in which, by symmetry, $c_4 = 1$. More experimental data for the high Rayleigh number range with asymmetric heating ($\mathrm{Ra}_{\gamma h} \neq \mathrm{Ra}_{\gamma c}$) are needed in order to determine c_3 uniquely. A numerical study of convection in a rectangular cavity saturated by icy water, with various boundary thermal boundary conditions on the sidewalls, was reported by Benhadji *et al.* (2003). The numerical study by Baytas *et al.* (2004) treated the case of a square cavity and a more complicated density state equation. The case where one vertical wall is heated differentially by an isothermal discrete heater and the other vertical wall is cooled to a constant temperature, with adiabatic horizontal walls, was studied numerically by Saeid and Pop (2004c).

7.3.6. Attic-Shaped Enclosure

In a saturated porous medium confined by a wedge-shaped impermeable enclosure cooled along the sloped wall (Fig. 7.19) the convective flow consists of a single cell. Like all the flows in porous media heated or cooled from the side, this particular flow exists even in the limit Ra → 0. The flow intensifies as the Rayleigh number based on height (Ra) increases. The bottom wall is heated, while the vertical wall is insulated.

The numerical solutions reported by Poulikakos and Bejan (1983b) show the development of a Bénard-type instability at sufficiently high Rayleigh numbers.

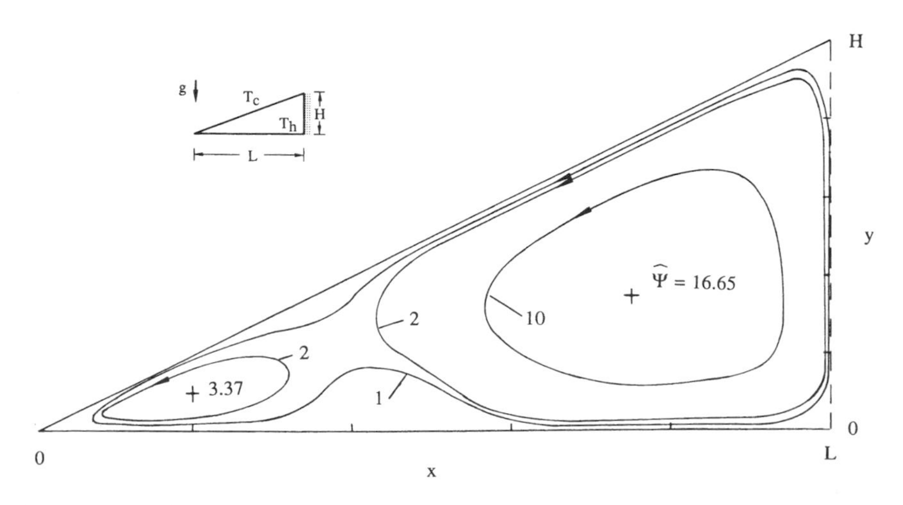

Figure 7.19. The flow pattern in an attic-shaped porous medium cooled the inclined wall ($H/L = 0.5$, Ra = 1000) (Poulikakos and Bejan, 1983b).

This instability is due to the heated bottom wall. In an enclosure with the aspect ratio $H/L = 0.2$, the instability occurs in the vicinity of Ra ~ 620. This critical Rayleigh number increases as H/L increases. Convection in trapezoidal enclosures was simulated using parallel computation by Kumar and Kumar (2004).

7.3.7. Other Enclosures

For the case of very small Rayleigh number, Philip (1982a,b, 1988) has obtained exact solutions for the flow pattern for a variety of two-dimensional (rectangular, elliptical, triangular, etc.) and axisymmetric (cylindrical, toroidal) cavities, for the case of uniform horizontal temperature gradient (which is radial for the axisymmetric situation). These have been obtained under the assumption of negligible convective heat transfer and so are of limited use on their own. They may be useful as the first stage in a perturbation analysis. Campos *et al.* (1990) studied numerically on the Brinkman model convection in a vertical annular enclosure partly filled with a vertical annular volume occupied by a porous medium. Asako *et al.* (1992) and Yamaguchi *et al.* (1993) reported numerical solutions with a Darcy model for three-dimensional convection in a vertical layer with a hexagonal honeycomb core that is either conducting or adiabatic. Chen and Wang (1993a,b) performed a convection instability analysis for a porous enclosure with either a horizontal or vertical baffle projecting part way into the enclosure. Lai (1993a,b, 1994) has performed calculations for the effects of inserting baffles of various sorts (radial and circumferential in horizontal annuli or pipes). Shin *et al.* (1994), with the aid of a transformation to bicylindrical coordinates, studied numerically two-dimensional convection in a segment of a circle, with the boundary inclined to the vertical.

Convection in a cavity with a dome (circular, elliptical, parabolic, etc.) on top was treated numerically by Das and Morsi (2003, 2005). Conjugate convection heat transfer from a vertical cylindrical fin in a cylindrical enclosure was treated numerically by Naidu *et al.* (2004a). A numerical solution procedure to study convection in a two-dimensional enclosure of arbitrary geometry was presented by Singh *et al.* (2000). Convection in an inclined trapezoidal enclosure with cylindrical top and bottom surfaces was studied numerically by Baytas and Pop (2001). Numerical investigations of convection in insulating layers in attics were carried out by Shankar and Hagentoft (2000). Convection in embankments built in permafrost has been modeled by Goering and Kumar (1996), Goering (2003), Jiang *et al.* (2004d), and Sun *et al.* (2005). Convection in a porous toroidal thermosyphon has been studied numerically by Jiang and Shoji (2002). Convection in a thin porous elliptical ring, located in an impermeable rock mass and subject to an inclined geothermal gradient, was treated by Ramazanov (2000). Fluid flow and heat transfer in partly divided cavities was studied numerically by Jue (2000). Convection in a reentrant rectopolygonal cavity was studied numerically and experimentally by Phanikumar and Mahajan (2002). Radiative effects on a MHD flow between infinite parallel plates with time-dependent suction were studied analytically by Alagoa *et al.* (1999). Convection from a wavy wall in a thermally stratified enclosure was treated numerically by Ratish Kumar and Shalini (2004a). Natural convection in a cavity with wavy vertical walls was studied by Misirlioglu *et al.* (2005). Convection driven by differential heating of the upper surface of a rectangular cavity was studied numerically and analytically by Daniels and Punpocha (2004). The case of a square cavity where one vertical wall is heated differentially by an isothermal discrete heater and the other vertical wall is cooled to a constant temperature, with adiabatic horizontal walls, was studied numerically by Saeid and Pop (2005b). A two temperature model was applied by Sanchez *et al.* (2005) to a problem with symmetrically connected fluid and porous layers.

7.3.8. Internal Heating

Steady natural convection in a two-dimensional cavity with uniform heat generation was simulated numerically by Du and Bilgen (1992) for the case of adiabatic horizontal walls and isothermal vertical walls at different temperatures. A further numerical treatment was reported by Das and Sahoo (1999). Steady convection in a rectangular enclosure with the top and one sidewall cold and the other nonisothermal and with the bottom heated at constant temperature was studied numerically by Hossain and Wilson (2002). Convection in a two-dimensional vertical cylinder with either (1) insulated top and bottom and cooled lateral walls or (2) all walls isothermally cooled was given a numerical treatment by Jiménez-Islas *et al.* (1999). A transient convection problem with sidewall heating was studied by Jue (2003). A dual reciprocity boundary element method was applied to a differentially and internally heated rectangular enclosure by Sarler (2000) and Sarler *et al.* (2000a,b, 2004a,b). A numerical and experimental study of three-dimensional convection in an anisotropic medium in a rectangular cavity was carried out by Suresh *et al.* (2005).

7.4. Penetrative Convection

In this section we turn our attention to buoyancy-driven flows that only partially penetrate the enclosed porous medium. One basic configuration in which this flow can occur is shown in Fig. 7.20a. The saturated porous medium is a two-dimensional layer of height H and length L, confined by a rectangular boundary. Three of the walls are impermeable and at the same temperature (for example, T_c), while one of the side walls is permeable and in communication with a fluid reservoir of a different temperature, T_h. In Fig. 7.20b the same layer is oriented vertically. In both cases, natural convection penetrates the porous medium over a length dictated by the Rayleigh number alone and not by the geometric ratio of the layer, H/L (Bejan, 1980, 1981). The remainder of the porous layer contains essentially stagnant and isothermal fluid.

7.4.1. Lateral Penetration

First consider the horizontal layer of Fig. 7.20a, in which the lateral penetration distance L_x is unknown. According to Eqs. (7.1), (7.4), and (7.5), the order-of-magnitude balances for mass, energy, and momentum are

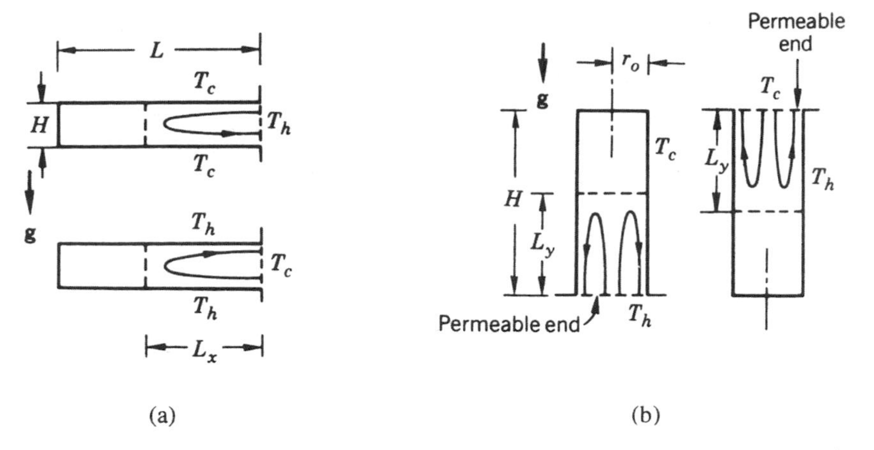

Figure 7.20. Lateral penetration (a) and vertical penetration (b) of natural convection into an isothermal porous space with one and permeable.

mass:
$$\frac{u}{L_x} \sim \frac{v}{H}, \tag{7.65}$$

energy:
$$u\frac{\Delta T}{L_x} \sim \alpha_m \frac{\Delta T}{H^2}, \tag{7.66}$$

momentum:
$$\frac{u}{H} \sim \frac{Kg\beta}{\nu}\frac{\Delta T}{L_x}. \tag{7.67}$$

In writing balances we have assumed that the penetration length L_x is greater than the vertical dimension H. The temperature difference ΔT is shorthand for $T_h - T_c$.

Equations (7.65)–(7.67) easily can be solved for the unknown scales u, ν, and L_x. For example, the penetration length is (Bejan, 1981)

$$L_x \sim H\,\mathrm{Ra}^{1/2}, \tag{7.68}$$

in which Ra is the Darcy modified Rayleigh number based on H and ΔT. The corresponding heat transfer rate q' [W/m] between the lateral fluid reservoir T_h and the T_c boundary of the porous medium scales as

$$q' \sim (\rho c_P)_f u H \Delta T \sim k_m \Delta T \mathrm{Ra}^{1/2}. \tag{7.69}$$

The heat transfer rate q' is expressed per unit length in the direction normal to the plane of Fig. 7.20a. All these results demonstrate that the actual length of the porous layer (L) has no effect on the flow and the heat transfer rate: L_x as well as q' are set by the Rayleigh number. The far region of length $L - L_x$ is isothermal and filled with stagnant fluid.

The actual flow and temperature fields associated with the lateral penetration phenomenon have been determined analytically as a similarity solution (Bejan, 1981). Figure 7.21 shows the dimensionless streamfunction and temperature for

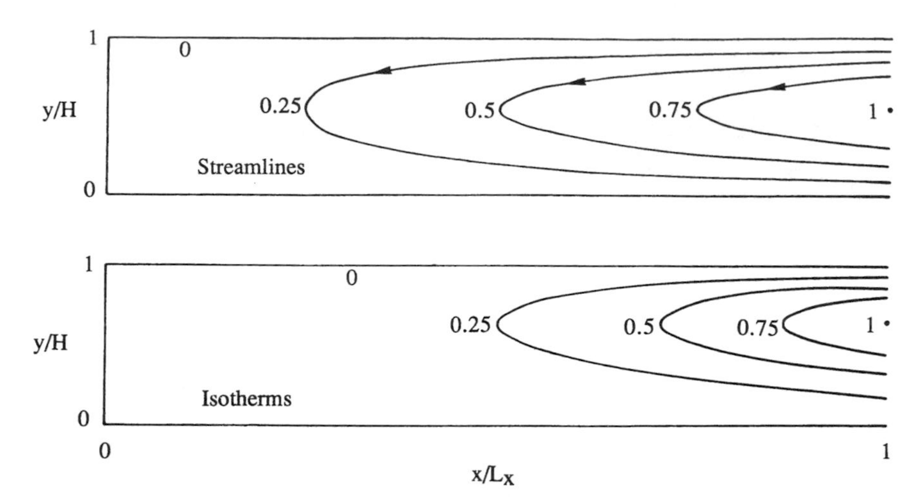

Figure 7.21. Streamlines and isotherms in the region of lateral penetration into a two-dimensional porous layer (Bejan, 1981).

only the region of length L_x. The penetration length and heat transfer rate predicted by this solution are

$$L_x = 0.158 H \mathrm{Ra}^{1/2}, \tag{7.70}$$

$$q' = 0.319 k_m \Delta T \mathrm{Ra}^{1/2}. \tag{7.71}$$

The results presented in this subsection are valid when $L_x < L$ and $L_x \gg H$, which translates into the following Ra range:

$$1 \ll Ra < \frac{L}{H}. \tag{7.72}$$

In the same paper, Bejan (1981) also documented the lateral penetration in an anisotropic porous medium in which the principal thermal conductivities are different and aligned with the x and y axes, $k_{m,x} \neq k_{m,y}$. He also showed that a similar partial penetration phenomenon occurs when the temperature of each of the two horizontal walls (Fig. 7.20a) varies linearly from T_h at one end to T_c at the other.

7.4.2. Vertical Penetration

In the vertical two-dimensional layer of Fig. 7.20b, it is the bottom or the top side that is permeable and in communication with a fluid reservoir of different temperature. In Chapter 6 we saw that in porous layers heated from below or cooled from above convection is possible only above a critical Rayleigh number. In the configuration of Fig. 7.20b; however, fluid motion sets in as soon as the smallest ΔT is imposed between the permeable horizontal boundary and the vertical walls. This motion is driven by the horizontal temperature gradient of order $\Delta T/L$.

If we write L_y for the unknown distance of vertical penetration and if we assume that $L_y \gg L$, we obtain the following order-of-magnitude balances

mass:
$$\frac{u}{L} \sim \frac{v}{L_y}, \tag{7.73}$$

energy:
$$u\frac{\Delta T}{L} \sim \alpha_m \frac{\Delta T}{L^2}, \tag{7.74}$$

momentum:
$$\frac{v}{L} \sim \frac{Kg\beta}{\nu}\frac{\Delta T}{L}, \tag{7.75}$$

The vertical penetration distance that results from this system of equations is (Bejan, 1984)

$$L_y \sim L\mathrm{Ra}_L, \tag{7.76}$$

in which Ra_L is the Rayleigh number based on the thickness $L : \mathrm{Ra}_L = g\beta KL\Delta T/\nu\alpha_m$. The scale of the overall heat transfer rate q' [W/m] through the permeable side of the porous layer is

$$q' \sim (\rho c_P)_f vL\Delta T \sim k_m \Delta T\, \mathrm{Ra}_L. \tag{7.77}$$

Once again, the physical extent of the porous layer (H) does not influence the penetrative flow, as long as H is greater than the penetration distance L_y. The latter is determined solely by the transversal dimension L and the imposed temperature difference ΔT. The vertical penetration distance and total heat transfer rate are proportional to the Rayleigh number based on the thickness L.

The vertical penetration of natural convection also was studied in the cylindrical geometry of Fig. 7.22, as a model of certain geothermal flows or the flow of air through the grain stored in a silo (Bejan, 1980). The vertical penetration distance and the total heat transfer rate q [W] are

$$\frac{L_y}{r_o} = 0.0847\, \mathrm{Ra}_{r_o}, \tag{7.78}$$

$$q = 0.255 r_o k_m \Delta T\, \mathrm{Ra}_{r_o}, \tag{7.79}$$

where r_o is the radius of the cylindrical cavity filled with saturated porous material and Ra_{ro} is the Rayleigh number based on radius, $\mathrm{Ra}_{ro} = g\beta K r_o \Delta T/\nu\alpha_m$. Figure 7.22 shows the streamlines in the region of height L_y, which is penetrated by natural convection. The region of height $H - L_y$, which is situated above this flow and is not shown in Fig. 7.22, is isothermal and saturated with motionless fluid.

The results presented in this subsection are valid when the penetrative flow is slender, $L_y \gg (L, r_o)$, and when L_y is shorter than the vertical dimension of the confined porous medium, $L_y < H$. These restrictions limit the Rayleigh number

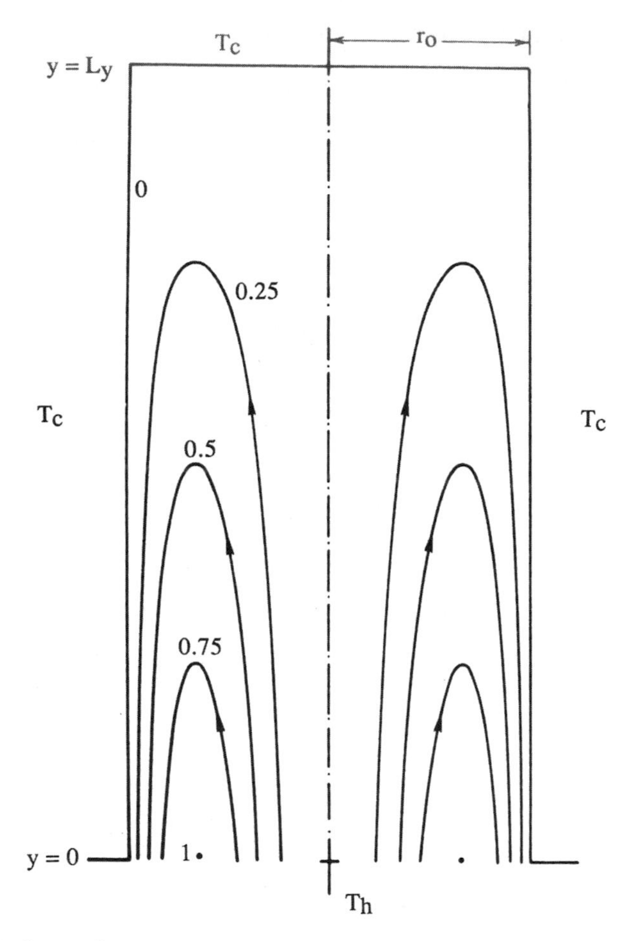

Figure 7.22. Streamlines in the region of vertical penetration into a cylindrical space filled with porous medium (Bejan, 1980).

domain that corresponds to these flows:

$$1 \ll \mathrm{Ra}_{(L,r_o)} < \frac{H}{(L, r_o)}. \tag{7.80}$$

7.4.3. Other Penetrative Flows

Two types of penetrative flows that are related to those of Figs. 7.20a and 7.20b are presented in Fig. 7.23. Poulikakos and Bejan (1984a) showed that in a porous medium that is heated and cooled along the same vertical wall the flow penetration can be either horizontal (Fig. 7.23a) or vertical (Fig. 7.23b). In the case of horizontal penetration, the penetration distance L_x and the total heat transfer rate q' are of the same order as in Eqs. (7.66) and (7.67). These scales are valid in the range

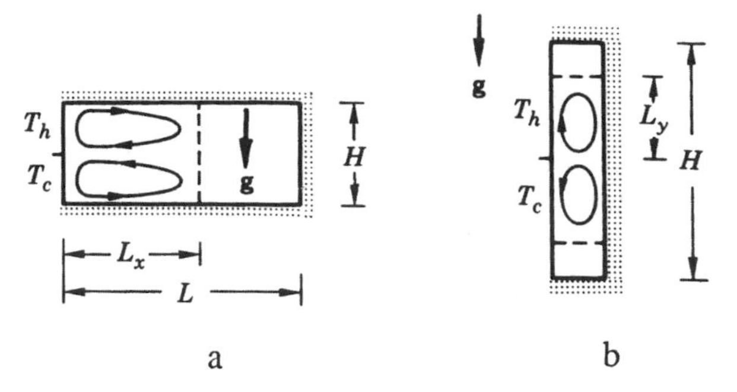

Figure 7.23. Incomplete horizontal penetration (a) and vertical penetration (b) in a porous layer heated and cooled along the same vertical side.

$1 \ll \mathrm{Ra} < L/H$, the Rayleigh number Ra being based on height. The scales of vertical penetration in Fig. 7.23b are different,

$$L_y \sim H \left(\frac{L}{H}\right)^{2/3} \mathrm{Ra}^{-1/3}, \tag{7.81}$$

$$q' \sim k_m (T_h - T_c) \left(\frac{L}{H} \mathrm{Ra}\right)^{1/3}. \tag{7.82}$$

in which Ra is again based on H. These scales are valid when $\mathrm{Ra} > H/L$. The two penetrative flows of Fig. 7.23 occur only when the heated section is situated above the cooled section of the vertical wall. When the positions of the T_h and T_c sections are reversed, the buoyancy-driven flow fills the entire $H \times L$ space (Poulikakos and Bejan, 1984a).

In a semi-infinite porous medium bounded from below or from above by a horizontal wall with alternating zones of heating and cooling (Fig. 7.24) the buoyancy-driven flow penetrates to a distance L_y into the medium (Poulikakos and Bejan, 1984b). This distance scales as $\lambda\, \mathrm{Ra}_\lambda^{1/2}$, where λ is the spacing between a heated zone and the adjacent cooled zone, and $\mathrm{Ra}_\lambda = g\beta K \lambda (T_h - T_c)/\nu\alpha_m$. Figure 7.24 shows a sample of the numerical results that have been developed for the range $1 \leq \mathrm{Ra}_\lambda \leq 100$.

7.5. Transient Effects

The work reviewed in the preceding sections dealt with steady-state conditions in which the flow is slow enough to conform to the Darcy model. In this section, we drop the steady-flow restriction and examine the time scales and evolution of the buoyancy-driven flow. The equations that govern the conservation of mass, momentum, and energy in Fig. 7.1 are, in order, Eqs. (7.1), (7.5), and, in

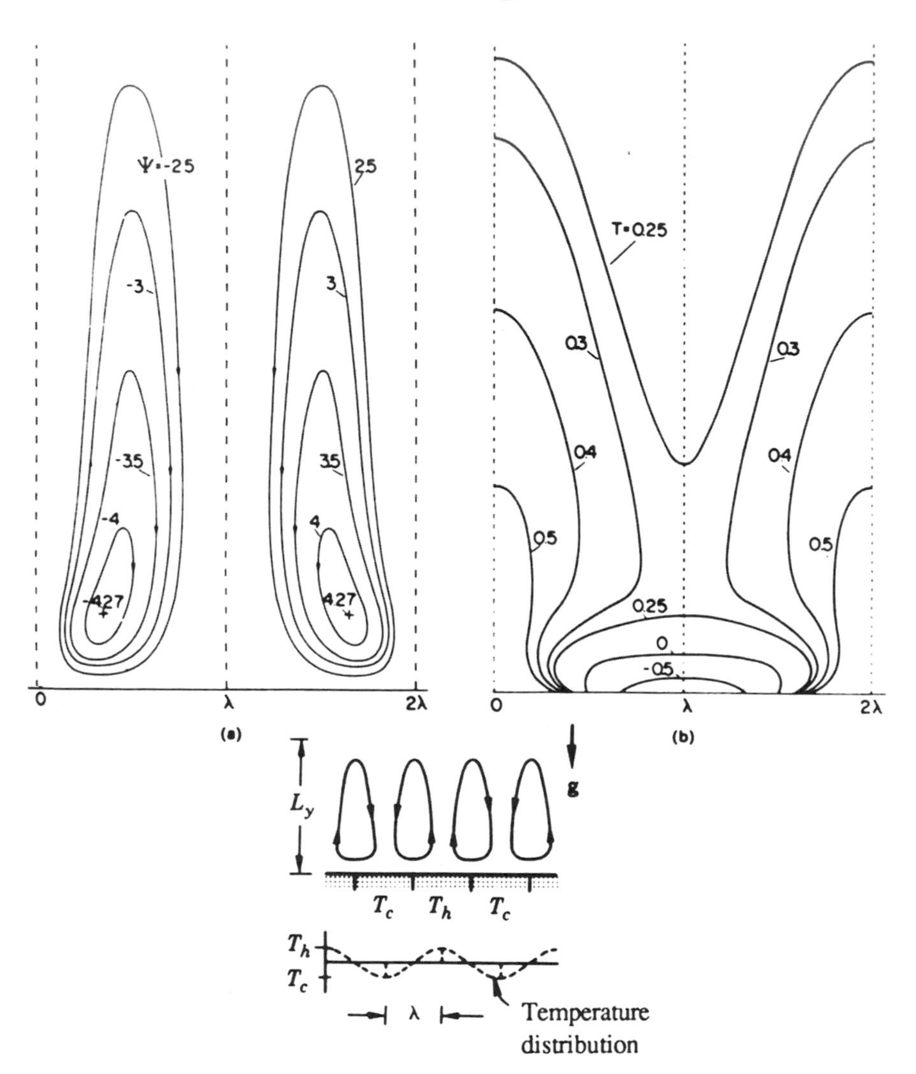

Figure 7.24. Streamlines and isotherms for the vertical penetration of natural convection in a semi-infinite porous medium bounded by a horizontal wall with alternating hot and cold spots ($Ra_\lambda = 100$) (Poulikakos and Bejan, 1984b).

place of (7.4),

$$\sigma \frac{\partial T}{\partial t} + u \frac{\partial T}{\partial x} + v \frac{\partial T}{\partial y} = \alpha_m \left(\frac{\partial^2 T}{\partial x^2} + \frac{\partial^2 T}{\partial y^2} \right). \tag{7.83}$$

Consider the two-dimensional saturated porous medium shown in Fig. 7.1, which is initially isothermal at $T_{avg} = (T_h + T_c)/2$ and saturated with motionless fluid. At the time $t = 0$, the temperatures of the two side walls are changed to $T_h = T_{avg} + \Delta T/2$ and $T_c = T_{avg} - \Delta T/2$, while the top and bottom walls remain insulated. All the walls are impermeable. Of special interest is the time needed by the flow and heat transfer to reach steady state, i.e., the time interval after which the flow regimes described in Section 7.1 become valid. This basic transient convection problem was studied by Poulikakos and Bejan (1983c).

By focusing on the vertical boundary layer that develops along the left-hand side of the rectangular system of Fig. 7.1, we note that initially the time-dependent thickness of this boundary layer $\delta(t)$ grows by pure conduction. With respect to the region of thickness δ and height H, the energy equation (7.83) dictates a balance between the side heating effect and the thermal inertia of the saturated porous medium,

$$\sigma \frac{\Delta T}{t} \sim \alpha_m \frac{\Delta T}{\delta^2}. \tag{7.84}$$

This balance yields the well-known penetration distance of pure conduction:

$$\delta \sim \left(\frac{\alpha_m t}{\sigma}\right)^{1/2}. \tag{7.85}$$

The growth of the conduction layer gives rise to a horizontal temperature gradient of order $\partial T/\partial x \sim \Delta T/\delta$. This development makes the buoyancy term in the momentum balance (7.5) finite. In fact, the scales of the three terms appearing in Eq. (7.5) are

$$\left(\frac{u}{H}, \frac{v}{\delta}\right) \sim \frac{Kg\beta}{\nu} \frac{\Delta T}{\delta}. \tag{7.86}$$

The mass conservation scaling (7.6) shows that the ratio of the two scales on the left-hand side of Eq. (7.86) is

$$\frac{u/H}{v/\delta} \sim \left(\frac{\delta}{H}\right)^2, \tag{7.87}$$

in other words, that u/H is negligible relative to v/δ. In conclusion, the momentum balance reduces to Eq. (7.11) and the vertical velocity scale turns out to be identical to the scale listed in Eq. (7.12) for the steady state. An interesting feature of the transient flow is that the vertical velocity scale is independent of time. The vertical flow rate however, $v\delta$, grows in time as $t^{1/2}$.

As soon as fluid motion is present, the energy equation (7.83) is ruled by the competition among three different scales:

$$\begin{array}{ccccc} \sigma \dfrac{\Delta T}{t}, & v\dfrac{\Delta T}{H} & \sim & \alpha_m \dfrac{\Delta T}{\delta^2}, \\ \text{Inertia} & \text{Convection} & & \text{Conduction} \\ (t^{-1}) & (t^0) & & (t^{-1}) \end{array} \tag{7.88}$$

The time dependence of each scale also is shown. Since the lateral conduction effect is always present, the convection scale eventually overtakes inertia on the left-hand side of Eq. (7.88). The time t when this changeover takes place, i.e., when the vertical boundary layer becomes convective, is given by

$$\sigma \frac{\Delta T}{t} \sim v \frac{\Delta T}{H}, \tag{7.89}$$

which in view of the v scale (7.12) yields

$$t \sim \frac{\sigma}{\alpha_m} H^2 \, \mathrm{Ra}^{-1}. \tag{7.90}$$

It is easy to verify that the boundary layer thickness (7.85), which corresponds to (and after) this time, is the steady-state scale determined earlier in Eq. (7.13). In conclusion, this transient-convection analysis reconfirms the criterion (7.19) for distinct vertical boundary layers.

By following the same approach Poulikakos and Bejan (1983c) traced the development of the horizontal boundary layers along the top and bottom walls of the enclosure. They found that the horizontal layers become "developed" earlier than the vertical layers when the enclosure is tall enough so that

$$\frac{H}{L} > \mathrm{Ra}^{1/6}. \tag{7.91}$$

The criterion for distinct horizontal boundary layers turns out to be the same as the inequality (7.22). In summary, the analysis of the time-dependent development of natural circulation in the two-dimensional system of Fig. 7.1 provides an alternative way to construct the four-regime map seen earlier in Fig. 7.2.

A comprehensive study of transient convection between parallel vertical plates on the Brinkman-Forchheimer model has been carried out by Nakayama *et al.* (1993). They obtained asymptotic solutions for small and large times and a bridging numerical solution for intermediate times. An MHD problem with suction or injection on one plate was treated by Chamkha (1997b). For convection in a rectangular enclosure, Lage (1993b) used scale analysis to obtain general heat transfer correlations. A further numerical study was reported by Merrikh and Mohamad (2000). An analytic study using Laplace transforms was conducted by Jha (1997). The effect of variable porosity was examined by Paul *et al.* (2001). Saeid and Pop (2004a,b) considered a transient problem arising from the sudden heating of one side wall and the sudden cooling of the other, with and without the effect of viscous dissipation. They found that the heat transfer was reduced as a result of the dissipation. Convection in a non-Newtonian power-law fluid was studied numerically by Al-Nimr *et al.* (2005).

A transient problem for convection between two concentric spheres was studied by Pop *et al.* (1993b). They obtained solutions, valid for short time, of the Darcy and energy equations using the method of matched asymptotic expansions. Nguyen *et al.* (1997b) treated a similar problem with a central fluid core surrounded by a porous shell. They performed numerical calculations on the Brinkman model. They found remarkable effects along the porous medium-fluid interface, but the overall heat flux was sensitive only to the ration of thermal conductivity of the solid matrix to that of the fluid.

Transient convection in a vertical annulus for various thermal boundary conditions was studied by Al-Nimr and Darabseh (1995) for the Brinkman model. Transient convection in a horizontal annulus, with the inner and outer cylinders maintained at uniform temperat ures, was examined by Pop *et al.* (1992a). They used the method of matched asymptotic expansions to obtain a solution valid for short times. Sundfor and Tyvand (1996) studied convection in a horizontal cylinder with a sudden change in wall temperature. An investigation of the effect of local thermal nonequilibrium was reported by Ben Nasrallah *et al.* (1997). Further numerical studies of the effect of local thermal nonequilibrium were carried out by Khadrawi and Al-Nimr (2003b) and Krishnan *et al.* (2004). A hybrid numerical-analytical solution for two-dimensional transient convection in a vertical cavity, based on a generalized transform technique, was presented by Alves and Cotta (2000). Similar three-dimensional studies were made by Neto *et al.* (2002, 2004) and Cotta *et al.* (2005).

7.6. Departure from Darcy Flow

7.6.1. Inertial Effects

The behavior of the flow and heat transfer process changes substantially as the flow regime departs from the Darcy limit. The effect of the quadratic drag on the heat transfer through the most basic configuration that opened this chapter (Fig. 7.1) was demonstrated by Poulikakos and Bejan (1985). In place of the momentum equation (7.5) they used the Forchheimer modification of Darcy's law,

$$\frac{\partial}{\partial y}(Bu) - \frac{\partial}{\partial x}(Bv) = -\frac{g\beta K}{\nu}\frac{\partial T}{\partial x}. \tag{7.92}$$

This follows from Eq. (1.12) by eliminating the pressure between the x and y momentum equations and by writing

$$B = 1 + \frac{\chi}{\upsilon}(u^2 + v^2)^{1/2}. \tag{7.93}$$

The Forchheimer term coefficient χ has the units [m] and is used as shorthand for the group $c_F K^{1/2}$, where c_F is defined by Eq. (1.12). The same notation was used in Eq. (5.60), in the analysis of the flow near a single vertical wall.

Poulikakos and Bejan (1985) analyzed the Darcy-Forchheimer convection phenomenon using three methods: scale analysis, a matched boundary layer analysis, and case-by-case numerical finite-difference simulations. The main results of the scale analysis for the convection regime III are summarized in Table 7.1, next to the scales derived for the Darcy limit in Section 7.1.1. The transition from Darcy flow to Forchheimer flow, i.e., to a flow in which the second term dominates on the right-hand side of Eq. (7.93), takes place when the dimensionless number G is smaller than O(1),

$$G = \nu[\chi g\beta K(T_h - T_c)]^{-1/2}. \tag{7.94}$$

Table 7.1. The scales of the vertical natural convection boundary layer in a porous layer heated from the side (Poulikakos and Bejan, 1985).

	Forchheimer regime $G \ll 1$	Darcy regime $G \gg 1$
Boundary layer thickness	$H\mathrm{Ra}_\infty{}^{-1/4}$	$H\mathrm{Ra}^{-1/2}$
Vertical velocity	$\frac{\alpha_m}{H}\mathrm{Ra}_\infty^{1/2}$	$\frac{\alpha_m}{H}\mathrm{Ra}$
Heat transfer rate	$k_m \Delta T \mathrm{Ra}_\infty^{1/2}$	$k_m \Delta T \mathrm{Ra}^{1/2}$

In the Forchheimer regime $G \ll 1$, the appropriate Rayleigh number is the large Reynolds number limit version encountered already in Eq. (5.64),

$$\mathrm{Ra}_\infty = \frac{g\beta K^2(T_h - T_c)}{\chi \alpha_m^2}. \tag{7.95}$$

The important heat transfer conclusion of the scale analysis is that the overall (conduction-referenced) Nusselt number defined in Eq. (7.17) scales as $(L/H)\mathrm{Ra}_8^{1/4}$ in the limit in which the effect of inertia dominates. A more accurate estimate was provided by an analytical solution in which Oseen linearized solutions for the two vertical boundary layers were matched to the same stratified core (Poulikakos and Bejan, 1985):

$$\mathrm{Nu} = 0.889\frac{L}{H}\,\mathrm{Ra}_\infty^{1/4}, \quad (G \ll 1). \tag{7.96}$$

This solution is the Forchheimer regime counterpart of the Oseen linearized solution derived by Weber (1975b) for the Darcy limit, namely Eq. (7.38). By intersecting Eq. (7.96) with Eq. (7.38) we learn that the transition from Darcy flow to Forchheimer flow occurs when $\mathrm{Ra}_\infty^{1/2} \sim \mathrm{Ra}$, which is another way of saying $G \sim \mathrm{O}(1)$. In fact the group G defined in Eq. (7.94) is the same as the ratio $\mathrm{Ra}_\infty^{1/2}/\mathrm{Ra}$.

Figure 7.25 shows Poulikakos and Bejan's (1985) finite-difference calculations for the overall heat transfer rate in the intermediate regime represented by $0.1 \leq G \leq 10$. In these calculations, the momentum equation contained the Darcy and Forchheimer terms shown in Eqs. (7.92)–(7.93). The numerical data agree well with Weber's formula in the Darcy limit $G \to \infty$. In the opposite limit, the numerical data fall slightly below the theoretical asymptote (7.96). This behavior has been attributed to the fact that the group $(H/L)\,\mathrm{Ra}_\infty{}^{-1/4}$, whose smallness describes the goodness of the boundary layer approximation built into the analysis that produced Eq. (7.96), increases steadily as G decreases at constant Ra (note that in Fig. 7.25 $\mathrm{Ra} = 4000$). In other words, constant-Ra numerical experiments deviate steadily from the boundary layer regime as G decreases. Indeed, Poulikakos and Bejan (1985) found better agreement between their $G < 1$ numerical data and Eq. (7.96) when the Rayleigh number was higher, $\mathrm{Ra} = 5000$.

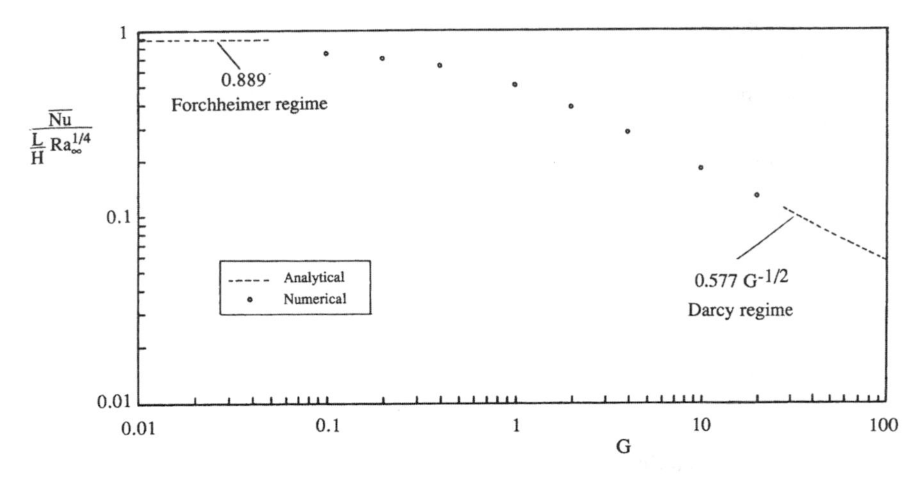

Figure 7.25. Numerical results for the total heat transfer rate through a porous layer heated from the side, in the intermediate Darcy-Forchheimer range ($H/L = 2$, $\mathrm{Ra} = 4000$, and $1.6 \times 10^5 \leq \mathrm{Ra}_\infty \leq 1.6 \times 10^9$) (Poulikakos and Bejan, 1985).

In a subsequent numerical study, Prasad and Tuntomo (1987) contributed additional numerical results for natural convection in the configuration treated by Poulikakos and Bejan (1985), which confirmed the reported theoretical scaling trends. Specifically, Prasad and Tuntomo included the Darcy and Forchheimer terms in the momentum equation and covered the range $1 \leq H/L \leq 20$, $10 \leq \mathrm{Ra} \leq 10^4$. They also pointed out that the progress toward the inertia-dominated regime ($G \to 0$ in Fig. 7.25) is accompanied by a proportional increase in the pore Reynolds number. This can be shown here by using the volume-averaged vertical velocity scale listed in Table 7.1, $v \sim (\alpha_m/H)\,\mathrm{Ra}_\infty{}^{1/2}$. The corresponding pore velocity scale is $v_p = v/\varphi \sim (\alpha_m/\varphi H)\,\mathrm{Ra}_\infty^{1/2}$. The pore Reynolds number is

$$\mathrm{Re}_p = \frac{v_p D_p}{\nu}, \tag{7.97}$$

in which D_p is the pore size. This Reynolds number can be rewritten in terms of G and the particle size d_p by invoking Eqs. (7.94) and (1.13):

$$\mathrm{Re}_p \sim \frac{D_p}{d_p}\,\frac{\beta^{1/2}}{c_F}\,\frac{(1-\varphi)}{\varphi^{5/2} G}. \tag{7.98}$$

Taking $\beta = 150$, $c_F \cong 0.55$, $D_p/d_p \sim \mathrm{O}(1)$, and $\varphi = 0.7$ as representative orders of magnitude in Eq. (7.96), the pore Reynolds number becomes approximately

$$\mathrm{Re}_P \sim \frac{C}{G}, \tag{7.99}$$

where C is a dimensionless coefficient of order 10.

This last Re_p expression reconfirms the notion that the effect of inertia becomes important when $\mathrm{Re}_p \sim \mathrm{O}(10)$, because $G < 1$ is the inertia-dominated domain revealed by Poulikakos and Bejan's (1985) theory. The pore Reynolds number domain $Re_p > 300$, in which the flow becomes turbulent (Dybbs and Edwards, 1984), corresponds to the range $G < 0.03$ at constant Ra. Prasad and Tuntomo (1987), however, went too far when they claimed that "the Forchheimer extended Darcy equation of motion will become invalid when G decreases below 0.1. The flow is then unsteady and chaotic (p. 311)." Their assertion is incorrect, because quadratic drag is a macroscopic phenomenon that does not change qualitatively when the flow in the pores becomes turbulent.

A numerical study of convection in a square cavity using the Forchheimer model was conducted by Saied and Pop (2005a). They confirmed the expectation that inertial effects slow down the convection currents and reduce the Nusselt number for a fixed value of the Rayleigh number.

7.6.2. Boundary Friction, Variable Porosity, Local Thermal Nonequilibrium, Viscous Dissipation, and Thermal Dispersion Effects

The effects of boundary friction incorporated in the Brinkman model has been studied by several authors, starting with Chan *et al.* (1970). For the shallow porous layer, with isothermal lateral walls and adiabatic top and bottom, the top being either rigid or free, Sen (1987) showed that the Brinkman term does not significantly affect the heat transfer rate until the Darcy number $Da = K/H^4$ exceeds $10^{-1/4}$, and then the Nusselt number Nu decreases as Da increases. As one would expect, the reduction is smaller for the case of a free upper surface than that of a rigid upper surface. Also, for a shallow cavity with various combinations of rigid or free upper and lower boundaries, Vasseur *et al.* (1989) studied the case of lateral heating with uniform heat flux, exploiting the fact that in this situation there is parallel flow in the core.

For cavities with aspect ratios of order unity, Tong and Subramanian (1985) performed a boundary layer analysis and Tong and Orangi (1986) carried out numerical calculations. Vasseur and Robillard (1987) studied the boundary layer regime for the case of uniform heat flux. The vertical cavity case was treated numerically by Lauriat and Prasad (1987). Again the chief result is that, because of the reduction in velocity near the wall, the Nusselt number Nu decreases as Da increases, the effect increasing as Ra increases. The variation porosity near the wall partly cancels the boundary friction effect. Numerical studies of this effect were conducted by Nithiarasu *et al.* (1997a, 1998) and Marcondes *et al.* (2001).

The combined effects of boundary friction and quadratic drag were studied numerically by Beckermann *et al.* (1986), David *et al.* (1988, 1991), Lauriat and Prasad (1989), and Prasad *et al.* (1992) for rectangular cavities; by Kaviany (1986) and Murty *et al.* (1989) for horizontal annuli (concentric and eccentric, respectively); and by David *et al.* (1989) for vertical annuli. The studies by David *et al.* (1988, 1989) included the effect of variable porosity, which increases the rate of heat transfer. The last paper reported excellent agreement between the numerical results and experimental data obtained by Prasad *et al.* (1985) for water-glass media at high Rayleigh numbers and large particle sizes. Extensive reviews of the topic of boundary friction, quadratic drag, and variable porosity were made by Prasad and Kladias (1991) and Lauriat and Prasad (1991). These authors noted that there remained a discrepancy between theory and experiment for the case of media with a highly conductive solid matrix, such as steel beads. The theoretical values were some 20 to 25 percent too high.

The comparative numerical studies of a heated square cavity on Darcy, Brinkman, and Brinkman-Forchheimer models by Misra and Sarkhar (1995) confirmed that boundary friction and quadratic drag lead to a reduction in heat transfer. Further numerical studies were conducted by Satya Sai *et al.* (1997b), Jecl *et al.* (2001), and Jecl and Škerget (2004).

The case of local thermal nonequilibrium (together with variable porosity and thermal dispersion) has been treated by Alazmi and Vafai (2000) Mohamad (2000), Al-Amiri (2002), and Baytas and Pop (2002). For a square enclosure heated at the left wall, the maximum difference between the fluid- and solid-phase temperatures occurs in the bottom left and upper right corners.

Rees (2004a) showed that the effect of viscous dissipation could result in single-cell convection being replaced by a two-cell flow as the dissipation parameter increases. At higher values of this parameter the maximum temperature within the cavity begins to exceed the highest boundary temperature and subsequently the flow becomes time-periodic.

Thermal dispersion effects were studied numerically by Beji and Gobin (1992) on the Brinkman-Forchheimer model. These cause a significant increase in the overall heat transfer, and when they are included a better agreement with the experimental data is obtained, particularly when the thermal conductivities of the fluid and the solid matrix are similar.

7.7. Fluid and Porous Regions

Several authors, all using the Brinkman equation, have calculated the flow in a laterally heated rectangular container partly filled by clear fluid and partly with a porous medium saturated by that fluid. In most of these studies the porous medium forms a vertical layer; the interface either can be impermeable to fluid or impermeable. Sathe *et al.* (1987) reported experimental results for a box divided in two with a vertical impermeable partition bounding the porous medium, which agreed with calculations made by Tong and Subramanian (1986). Sathe and Tong (1989) compared these results with calculations by Sathe *et al.* (1988) for the same problem with a permeable interface and with results for a cavity completely filled with porous medium and with a partitioned cavity containing solely clear fluid. Heat transfer is reduced by the presence of porous material having the same thermal conductivity as the fluid and by the presence of a partition. At low Da ($= 10^{-4}$) the first mechanism is more prominent while for high Da the second produces a greater insulating effect. The differences become accentuated at large Ra. Experiments by Sathe and Tong (1988) confirmed that partly filling an enclosure with porous medium may reduce the heat transfer more than totally filling it.

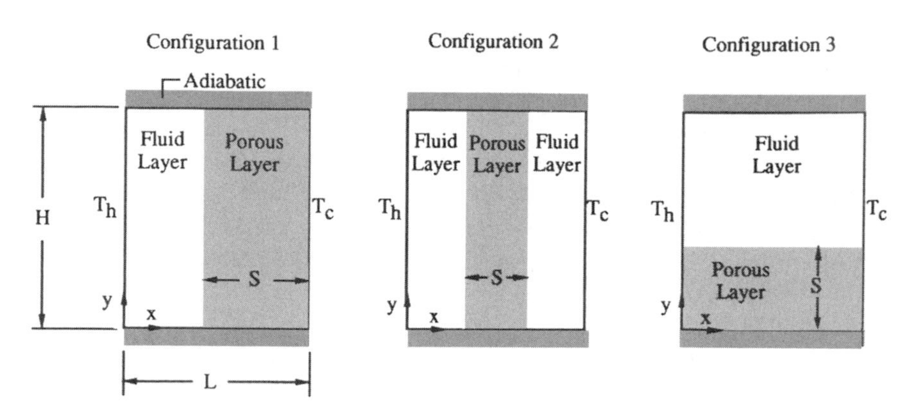

Figure 7.26. Definition sketch for fluid and porous regions in a vertical cavity.

The case of a rectangular cavity with a porous medium occupying the lower half, the interface being permeable, was studied numerically by Nishimura *et al.* (1986). The results agreed well with previous experiments by those authors. As one would expect, most of the flow and the heat transfer occurs in the fluid region.

The most comprehensive study available of flow and temperature fields is that by Beckermann *et al.* (1988). They performed calculations and experiments for the configurations shown in Fig. 7.26. In the experiments the beads were of glass or aluminum and the fluid was water or glycerin. A sample result is illustrated in Fig. 7.27. In all cases investigated, the temperature profiles indicated strong convection in the fluid layer but little in the porous layer. Figure 7.27 illustrates a situation with large beads of high thermal conductivity. For smaller aluminum beads (smaller Darcy number) there is less flow in the porous layer. For the case of glass beads (of small thermal conductivity) the situation is accentuated; for small beads there is almost no flow in the porous layer but for large beads there is a substantial amount of flow at the top and bottom of the porous layer, with the eddy centers in the fluid layer displaced toward the upper right and lower left corners.

A configuration similar to that of Fig. 7.27 is the vertical slot filled with air and divided along its vertical midplane by a permeable screen (Z. Zhang *et al.*, 1991). The screen is a venetian blind system made out of horizontal plane strips that can be rotated. In the nearly "closed" position, the strips almost touch and the air flow that leaks through it behaves as in Darcy or Forchheimer flow. On both sides of the partition the air circulation is driven by the temperature difference maintained between the two vertical walls of the slender enclosure. Zhang *et al.* showed numerically that there exists a ceiling value for the air flow conductance through the screen: above this value the screen pressure drop does not have a perceptible effect on the overall heat transfer rate. This ceiling value can be used for design purposes, e.g., in the calculation of the critical spacing that can be tolerated between two consecutive strips in the screen.

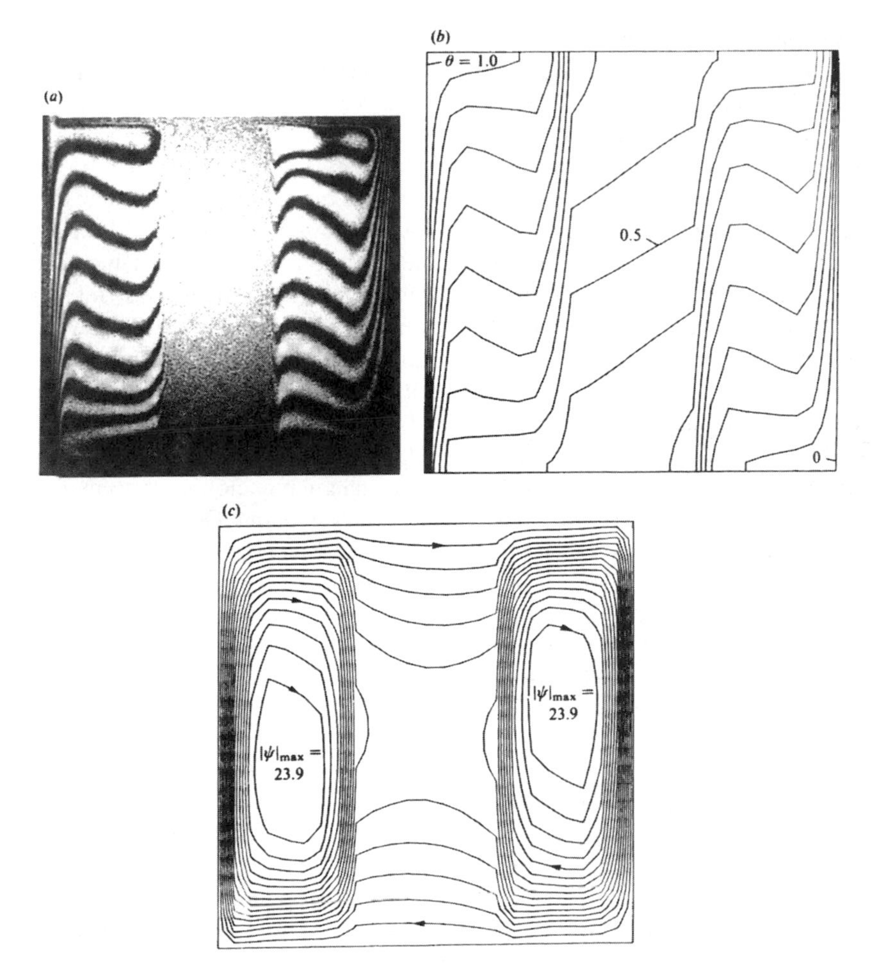

Figure 7.27. Experimental and predicted results for the configuration 2 shown in Fig. 7.26, with water and 6.35-mm aluminum breads: (a) photograph of interference fringe patterns; (b) predicted isotherms (equal increments); and (c) predicted stremalines (equal increments). $S/L = 0.33$, $\mathrm{Ra}_f = 3.70 \times 10^6$, $\mathrm{Da}_L = 1.534 \times 10^{-5}$, $\mathrm{Pr}_f = 6.44$, and $K_m/K_f = 37.47$ (Beckermann *et al.*, 1988, with permission from Cambridge University Press).

Du and Bilgen (1990) performed a numerical study of heat transfer in a vertical rectangular cavity partially filled with a vertical layer of uniform heat-generating porous medium and with lateral heating. They varied the aspect ratio of the cavity and the thickness and position of the porous layer.

Structures with solid walls separating cavities filled with porous materials and spaces filled with air are being contemplated in the advanced design of cavernous bricks and walls of buildings (Vasile *et al.*, 1998; Lorente *et al.*, 1996, 1998; Lorente and Bejan, 2002).

A numerical treatment of convection in a fluid-filled square cavity with differentially heated vertical walls covered by thin porous layers was studied numerically by Le Breton *et al.* (1991). They showed that porous layers having a thickness of the order of the boundary layer thickness were sufficient to reduce the overall Nusselt number significantly (by an amount that increased with increase of Ra) and thicker porous layers produced only a small additional decrease in heat transfer.

Three-dimensional convection in a rectangular enclosure containing a fluid layer overlying a porous layer was treated numerically on the Brinkman model by Singh *et al.* (1993). A comparison study of the Darcy, Brinkman, and Brinkman-Forchheimer models was carried out by Singh and Thorpe (1995). Convection in a rectangular cavity with a porous medium occupying half the lateral distance from heated to cooled wall was studied both theoretically (with an anisotropic medium incorporated) and experimentally (using perforated plates for the solid matrix which allowed flow visualization with the aid of dye) by Song and Viskanta (1994). Convection in a partly filled inclined rectangular enclosure, with uniform or localized heating of the bottom, was studied by Naylor and Oosthuizen (1995). They found that flow patterns were sensitive to small angles of inclination to the horizontal and that dual solutions were possible. Masuoka *et al.* (1994) investigated the channeling effect (due to porosity variation) with a model involving a thin fluid layer adjacent to a vertical porous medium layer. They found that convection was generally enhanced by the channeling effect, but for weak convection it is reduced by the thermal resistance near the wall.

A study that involves turbulence is that by L. Chen *et al.* (1998). They applied a κ-ε model to the fluid part of a partly filled enclosure. They found that when the flow is turbulent in the fluid region, the heat transfer in the porous region is dominated by convection and the penetration of the fluid into the porous region is more intensive than in the case of laminar flow.

A closed-form solution for natural convection in a rectangular cavity including a layer of porous medium adjacent to the heated side, with uniform heat flux from the sides, was obtained by Weisman *et al.* (1999). Mercier *et al.* (2002) obtained analytical expressions for a developing flow in similar circumstances. Fully developed convection in partly filled open-ended vertical channels was analyzed by Al-Nimr and Haddad (1999a); see also Nield (2001a). MHD convection in such channels was analyzed by Al-Nimr and Hader (1999b).

Paul and Singh (1998) studied convection in partly filled vertical annuli. An analytical study of convection in a partly filled vertical channel was performed by Paul *et al.* (1998). A numerical study of transient convection in a partly filled vertical channel was studied numerically by Paul *et al.* (1999). Transient convection in various domains partly filled with porous media was investigated analytically using Laplace transforms by Al-Nimr and Khadrawi (2003). A further study of convection in partly filled vertical channels was made by Khadrawi and Al-Nimr (2003a). Pseudosteady-state convection inside a spherical container partly filled with a porous medium was studied numerically by Zhang *et al.* (1999). Conjugate convection in a partly filled horizontal annulus was investigated by Aldoss *et al.* (2004). A two temperature model was applied by Sanchez *et al.* (2005) to a problem with symmetrically connected fluid and porous layers.

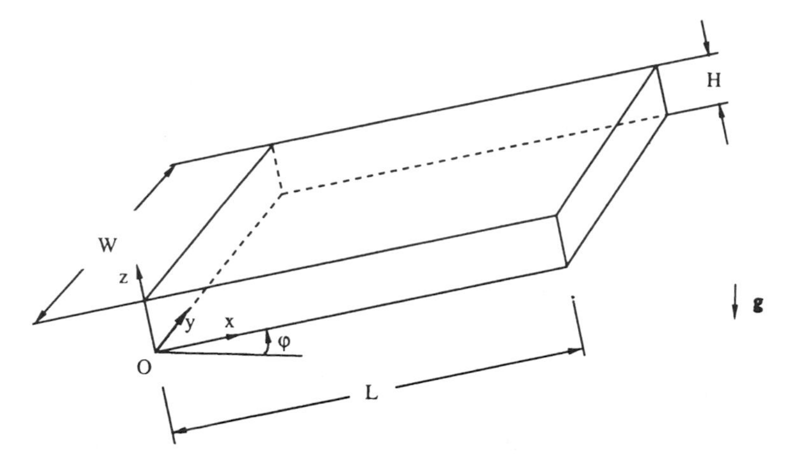

Figure 7.28. Definition sketch for a tilted box. Oy is horizontal and φ measures the inclination of Ox above the horizontal.

7.8. Sloping Porous Layer or Enclosure

The topic of this section has features discussed in Chapter 6 as well as those noted in the present chapter. We shall concentrate our attention on convection in the rectangular box shown in Fig. 7.28. Unless otherwise specified, the plane $z = 0$ is heated and the plane $z = H$ is cooled, and the other faces of the box are insulated. (Thus $\varphi = \pi$ corresponds to a box heated from above.)

We first consider the extension of the Horton-Rogers-Lapwood problem. The thermal boundary conditions are as in Section 6.1, namely $T = T_o + \Delta T$ at $z = 0$ and $T = T_o$ at $z = H$. The differential equations (6.3) – (6.6) have the basic steady-state solution given by Eqs. (6.8) and (6.9) and [in place of Eq. (6.7)]

$$\mathbf{v}_b = \frac{g\beta K \Delta T}{\nu}\left(\frac{1}{2} - \frac{z}{H}\right)\sin\varphi\,\mathbf{i}, \tag{7.100}$$

This describes a unicellular flow with an upward current near the hot plate and a downward current near the cold plate.

The perturbation equation (6.16) is unchanged, but Eqs. (6.17) and (6.18) are replaced by

$$\gamma_a\frac{\partial \mathbf{v}}{\partial \hat{t}} = -\nabla\hat{P} - \hat{\mathbf{v}} + \mathrm{Ra}\,\hat{T}(\sin\varphi\,\mathbf{i} + \cos\varphi\,\mathbf{k}), \tag{7.101}$$

$$\frac{\partial \hat{T}}{\partial \hat{t}} + (\mathrm{Ra}\sin\varphi)\left(\frac{1}{2} - \hat{z}\right)\frac{\partial \hat{T}}{\partial \hat{x}} - \hat{w} = \nabla^2\hat{T} \tag{7.102}$$

and instead of Eqs. (6.22) and (6.23) we now have

$$\left[D^2 - a^2 - s - ik(\mathrm{Ra}\sin\varphi)\left(\frac{1}{2} - \hat{z}\right)\right]\theta = -W, \tag{7.103}$$

$$(1 + \gamma_a s)(D^2 - a^2)W = -\mathrm{Ra}[a^2(\cos\varphi)\theta + ik(\sin\varphi)D\theta]. \tag{7.104}$$

Eliminating W, one gets

$$(1+\gamma_a s)(D^2-a^2)(D^2-a^2-s)\theta - \text{Ra}\,\alpha^2(\cos\varphi)\theta$$
$$-ik\,\text{Ra}\,\sin\varphi\left\{(1+\gamma_a s)\left[\left(\frac{1}{2}-\hat{z}\right)(D^2-\alpha^2)\theta - 2D\theta\right] + D\theta\right\} = 0. \tag{7.105}$$

For the case of conducting impermeable boundaries,

$$\theta = D^2\theta = 0 \quad \text{at} \quad \hat{z}=0, \quad \text{and} \quad \hat{z}=1. \tag{7.106}$$

The system (7.105) and (7.106) can be solved by the Galerkin method (Caltagirone and Bories, 1985) but an immediate result can be obtained for the case $k = 0$, because then the eigenvalue problem reduces to that for the horizontal layer but with Ra replaced by Ra cos φ. This case corresponds to longitudinal rolls (with axes up the slope) superposed on the basic flow, i.e., longitudinal helicoidal cells. A detailed examination shows that the basic unicellular flow is indeed stable for Ra cos $\varphi = 4\pi^2$. Caltagirone and Bories found that convection appears in the form of polyhedral cells for small inclinations φ and as longitudinal helicoidal cells for larger values of φ, for the range $4\pi^2 <$ Ra cos $\varphi <$ 240 to 280. When Ra cos φ exceeds 240 to 280 for small φ, one has a transition to a fluctuating regime characterized by the continuous creation and disappearance of cells (as for the horizontal layer), while for larger φ the transition is to oscillating rolls whose boundaries are no longer parallel planes. Experiments by Bories and Combarnous (1973), with a medium composed of glass beads and water, produced general agreement with the theory. The situation is summarized in Fig. 7.29.

Additional experimental results reported by Hollard *et al.* (1995) were in agreement with the prediction (based on scale analysis) of Bories (1993) that the inclination angle φ_T for the transition between polyhedral cells (or transverse rolls) and longitudinal rolls is given by solving the equation

$$\text{Ra}\sin\varphi = 2^{3/2}M(\text{Ra}\cos\varphi - 4\pi)^{1/2}, \tag{7.107}$$

where $M = 0.82$, $2^{3/2}$ or 2, for hexagonal cells, transverse roll or square cells, respectively. Hollard *et al.* (1995) also investigated the transition between the stationary and nonstationary flows by means of a spectral analysis of the temperature field.

In the above discussion we have assumed that the inclination φ is fixed prior to the experiment. When one changes φ with Ra held constant one observes hysteresis with respect to flow pattern transition (Kaneko *et al.* 1974) but the overall heat transfer appears to be almost independent of flow pattern. As predicted by the analysis of Weber (1975a), the Nusselt number correlates well with Ra cos φ. End effects modify the transition criteria, increasing the domain of stability of the basic flow (Jaffrenou *et al.*, 1974).

Inaba *et al.* (1988) performed experiments using media of several different materials for $0° \le \varphi \le 180°$, $5 \le L/H \le 32.7$, and $0.074 \le d_p/H \le 1.0$, where d_p is the particle diameter. These and previous experiments indicated the existence

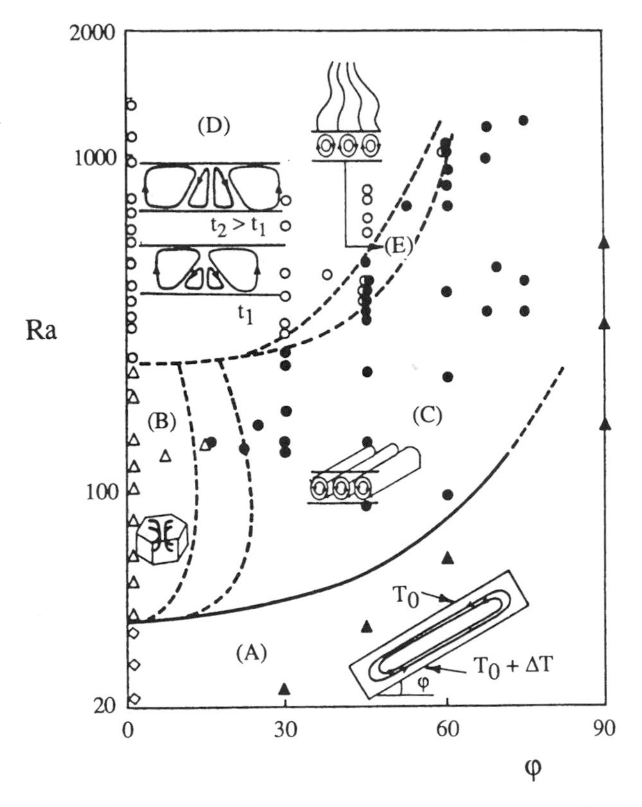

Figure 7.29. The different types of convective motion experimentally observed in a tilted porous layer: (A) unicellular flow; (B) polyhedral cells; (C) longitudinal stable coils; (D) fluctuating regime; and (E) oscillating longitudinal coils (Combarnous and Bories, 1975, with permission from Academic Press).

of a maximum heat transfer rate at $\varphi = 45°$ to $60°$ for Ra = 350. This motivates the following correlation formulas of Inaba *et al.* (1988), in which $\mathrm{Pr} = \nu/\alpha_m$.

For $60 \leq \mathrm{Ra}\cos(\varphi - 60°) \leq 4.5 \times 10^2$, $0° \leq \varphi \leq 15°$,

$$\mathrm{Nu} = 0.053\,\mathrm{Pr}^{0.13}\left(\frac{d_p}{H}\right)^{-0.20}[\mathrm{Ra}\cos(\varphi - 60°)]^{0.72}. \tag{7.108}$$

For $60 \leq \mathrm{Ra}\cos(\varphi - 60°) \leq 4.5 \times 10^2$, $15° \leq \varphi \leq 120°$,

$$\mathrm{Nu} = 0.024\,\mathrm{Pr}^{0.13}\left(\frac{L}{H}\right)^{-0.34}[\mathrm{Ra}\cos(\varphi - 60°)]^{0.52}. \tag{7.109}$$

For $4.5 \times 10^2 \leq \mathrm{Ra}\cos\varphi \leq 3 \times 10^4$, $0 \leq \varphi \leq 60°$,

$$\mathrm{Nu} = 0.067\,\mathrm{Pr}^{0.13}\left(\frac{d_p}{H}\right)^{-0.65}(\mathrm{Ra}\cos\varphi)^{0.52}. \tag{7.110}$$

For $4.5 \times 10^2 \leq \mathrm{Ra} \sin \varphi \leq 3 \times 10^4$, $60^\circ \leq \varphi \leq 120^\circ$,

$$\mathrm{Nu} = 0.062\,\mathrm{Pr}^{0.13} \left(\frac{L}{H}\right)^{-0.52} (\mathrm{Ra} \sin \varphi)^{0.46}. \tag{7.111}$$

The case of large L/H, W/H was examined numerically by Moya *et al.* (1987). They found that for small φ multiple solutions were possible. In addition to "natural" unicellular convection with flow up the heated wall and down the cooled wall, there also can exist an "antinatural" motion with circulation in the opposite direction. A bifurcation study by Riley and Winters (1990) shows that the appearance of the antinatural mode is associated with an isola. The various modal exchanges that occur as the aspect ratio of the tilted cavity varies were studied by Impey and Riley (1991).

There are further complications when L/H and W/H are of order unity. Pien and Sen (1989) showed by numerical calculation that there was hysteresis in the transition from an up-slope roll pattern to a cross-slope roll pattern as φ is varied, the Nusselt number being affected.

Detailed studies of the onset of convection in an inclined layer heated from below were reported by Rees and Bassom (1998, 2000). They included a full numerical solution of the linearized disturbance equations, and the results were used to motivate various asymptotic analyses. They found that at large Rayleigh numbers a two-dimensional instability only can arise when the angle that the layer makes with the horizontal is less than or equal to 31.30°, while the maximum inclination below which this instability may be possible is the slightly greater value 31.49°, which corresponds to a critical Rayleigh number of 104.30.

So far in our discussion the heated and cooled boundaries have been isothermal. Problems involving constant-flux heating have also been considered. For $90^\circ < \varphi < 180^\circ$ and the limit $L/H \to \infty$ an analytic parallel-flow solution was obtained by Vasseur *et al.* (1987). This solution is a good predictor of Nu for $L/H = 4$. Sen *et al.* (1987, 1988) have investigated the multiple steady states that occur when φ is small and all four faces of a rectangular enclosure are exposed to uniform heat fluxes, opposite faces being heated and cooled, respectively. Vasseur *et al.* (1988) showed that in the case $90^\circ < \varphi < 180^\circ$ the maximum temperature within the porous medium can be considerably higher than that induced by pure conduction. In this case the convection is considerably decreased when L/H is either very large or very small. A further study of constant-flux heating was made by Alex and Patil (2000b), using the Brinkman model.

The effect of the Brinkman boundary friction on heat transfer in an inclined box or layer was first calculated by Chan *et al.* (1970) and later by Vasseur *et al.* (1990). The additional effects of viscous dissipation were studied analytically by Malashetty *et al.* (2001). Flow in between concentric inclined cylinders was studied numerically and experimentally by Takata *et al.* (1982) for isothermal heating and by Wang and Zhang (1990) for constant flux on the inner cylinder.

The problem for a non-Newtonian (power-law) fluid was studied by Bian *et al.* (1994a,b). For a Newtonian fluid, the effect of a magnetic field was examined by Alchaar *et al.* (1995b) and Bian *et al.* (1996b). Because they considered two-dimensional disturbances only, their treatment may be incomplete.

The quasisteady convection produced by heating one side of a porous slab was studied by Robillard and Vasseur (1992). The case of a porous layer adjacent to a wall of finite thickness was investigated by Mbaye *et al.* (1993). A porous layer with an off-center diathermal partition was examined by Jang and Chen (1989). A numerical solution for convection in a cavity with a discrete heat source on one wall was obtained by Hsiao *et al.* (1994). An experimental investigation of a layer bounded by impervious domains of finite thermal conductivity in the presence of a vertical temperature gradient was conducted by Chevalier *et al.* (1996). The expected transition from two-dimensional to three-dimensional convection, as Ra increases, was found. A further numerical and experimental study of this configuration was reported by Chevalier *et al.* (1999).

Convection and dispersion in a reservoir with tilted fractures was studied theoretically and experimentally by Luna *et al.* (2004) under the assumption that the fluid thermal conductivity is very small compared with the rock conductivity.

A novel approach to convection in anisotropic inclined porous layers, which is able to deal with nonsymmetric multilayered systems, was presented by Trew and McKibbin (1994). The method involves the numerical summation of a series. A further study of the effect of anisotropic permeability on convection flow patterns was made by Storesletten and Tveitereid (1999). A layer anisotropic with respect to both permeability and diffusivity was analyzed by Rees and Postelnicu (2001) and Postelnicu and Rees (2001); the second paper was concerned with small angles to the horizontal. They found that often there is a smooth rather than an abrupt transition between longitudinal and transverse rolls as the governing parameters are varied. The effect of the Forchheimer drag was added by Rees *et al.* (2005b). The effect of anisotropy also was studied numerically by Cserepes and Lenkey (2004) for the case of an unconfined aquifer.

The effects of variable porosity and thermal dispersion were investigated numerically by Hsiao (1998). An analytical and experimental study of low-Rayleigh-number convection in long tilted fractures, embedded in an impermeable solid subjected to a vertical temperature gradient, was reported by Medina *et al.* (2002). Detailed numerical calculations for steady-state convection in an inclined porous cavity were made by Baytas and Pop (1999) and calculations of entropy generation were reported by Baytas (2000) and Baytas and Baytas (2005). MHD problems were studied numerically by Khanafer and Chamkha (1998) and Khanafer *et al.* (2000).

The case of volumetric heating in a porous bed adjacent to a fluid layer in an inclined enclose was investigated numerically by Chen and Lin (1997). In this case multiple steady-state solutions are possible. The combined effects of inclination, anisotropy, and internal heat generation on the linear stability of the basic parallel flow were analyzed by Storesletten and Rees (2004). They found that the preferred motion at the onset of convection depends strongly on the anisotropy ratio $\xi = K_L/K_T$. When $\xi < 1$ the preferred motion is longitudinal rolls for all inclinations. When $\xi > 1$ transverse rolls are preferred for small inclinations but at high inclinations longitudinal rolls are preferred, while at intermediate inclinations the preferred roll orientation varies smoothly between these two extremes. Convection in tilted cylindrical cavities embedded in rocks subject to a uniform temperature gradient was studied theoretically by Sanchez *et al.* (2005).

7.9. Inclined Temperature Gradient

We now discuss an extension to the Horton-Rogers-Lapwood problem. We suppose that a uniform horizontal temperature gradient β_H is imposed on the system, in addition to the vertical temperature gradient $\Delta T/H$. The boundary conditions used in Section 6.1 are now replaced by

$$T = T_o + \Delta T - \beta_H x \text{ at } z = 0, \quad T = T_o - \beta_H x \text{ at } z = H. \tag{7.112}$$

The basic steady-state solution, in nondimensional form, is now given by

$$u_b = \hat{\beta}_H \,\mathrm{Ra}\left(\hat{z} - \frac{1}{2}\right), \tag{7.113}$$

$$T_b = \frac{T_o}{\Delta T} + 1 - \hat{\beta}_H \hat{x} - \hat{z} - \frac{1}{12}\hat{\beta}_H^2 \,\mathrm{Ra}\,(\hat{z} - 3\hat{z}^2 + 2\hat{z}^3), \tag{7.114}$$

where

$$\hat{\beta}_H = \frac{\beta_H H}{\Delta T}. \tag{7.115}$$

Equation (6.23) is unchanged, but Eq. (6.22) is replaced by

$$(D^2 - \alpha^2 - s - ilu_b)\theta + i\hat{\beta}_H\left(\frac{l}{\alpha^2}\right)DW - WDT_b = 0. \tag{7.116}$$

The system (6.22), (6.24), and (7.116) can be solved using the Galerkin method. Some approximate results based on a low-order approximation and with γ_a assumed negligible were obtained by Nield (1991a). He found that longitudinal stationary modes ($l = 0, s = 0$) are the most unstable modes. For the first such mode the critical values are

$$\alpha_1 = \pi, \quad \mathrm{Ra}_1 = 4\pi^2 + \frac{\mathrm{Ra}_H^2}{4\pi^2}, \tag{7.117}$$

where the horizontal Rayleigh number Ra_H is defined by

$$\mathrm{Ra}_H = \hat{\beta}_H \mathrm{Ra} = \frac{g\beta K^2 \beta_H}{v\alpha_m}. \tag{7.118}$$

For small $\hat{\beta}_H$, Eq. (7.117) agrees with the approximation obtained by Weber (1974), namely

$$\mathrm{Ra} = 4\pi^2(1 + \hat{\beta}_H^2). \tag{7.119}$$

For the second mode, the critical values are

$$\alpha_2 = 2\pi, \quad \mathrm{Ra}_2 = 16\pi^2 + \frac{\mathrm{Ra}_H^2}{16\pi^2}. \tag{7.120}$$

We see that $\mathrm{Ra}_2 > \mathrm{Ra}_1$ for $\mathrm{Ra}_H < 8\pi^2$ but $\mathrm{Ra}_2 < \mathrm{Ra}_1$ when $\mathrm{Ra}_H > 8\pi^2$. Thus there is a transition from the first mode to the second as Ra_H increases.

The effect of increasing Ra_H is stabilizing because it distorts the basic temperature profile away from the linear one and ultimately changes the sign of its slope in the center of the channel. More accurate results, reported by Nield (1994d), showed that as Ra_H increases the critical value of Ra reaches a maximum and passes through zero. This means that the Hadley flow becomes unstable, even in the absence of an applied vertical gradient, when the circulation is sufficiently intense. The flow pattern changes from a single layer of cells to two or more superimposed layers of cells (superimposed on the Hadley circulation) as Ra_H increases. Yet more accurate results, together with the results of a nonlinear energy stability analysis, were reported by Kaloni and Qiao (1997). Two very accurate methods for determining the eigenvalues and eigenfunctions involved with such problems were discussed by Straughan and Walker (1996b).

Direct numerical simulations of supercritical Hadley circulation, restricted to transverse secondary flow, were performed by Manole and Lage (1995) and Manole *et al.* (1995). The results are in general accord with the linear stability analysis. Beyond a threshold value of Ra_H the Hadley circulation evolves to a time-periodic flow and the vertical heat transfer increases. The secondary flow emerges in the form of a traveling wave aligned with the Hadley flow direction. At low supercritical values of Ra, this traveling wave is characterized by the continuous drifting of two horizontal layers of cells that move in opposite directions. As Ra increases, the traveling wave becomes characterized by a single layer of cells drifting in the direction opposite to the applied horizontal temperature gradient. The extension to the anisotropic case, or to include the effect of internal heat sources, was made by Parthiban and Patil (1993,1995).

Nield (1990) also investigated the effect of adding a net horizontal mass flux Q in the x direction. This is destabilizing and at sufficiently large values of Q instability is possible in the absence of a vertical temperature gradient. Q also has the effect of smoothing out the transition from one mode to the next. More accurate results and a supplementary nonlinear analysis were reported by Qiao and Kaloni (1997). In this connection new computational methods described by Straughan and Walker (1996b) are useful. The effect of vertical throughflow was incorporated by Nield (1998b). The effect of a gravitational field varying with distance in the layer and with the additional effects of vertical through flow, or volumetric heating with or without anisotropy, was analyzed by Alex *et al.* (2001), Alex and Patil (2002a,b), and Parthiban and Patil (1997). Nonlinear instability studies, for the cases of vertical throughflow and variable gravity, were conducted by Qiao and Kaloni (1998) and Kaloni and Qiao (2001). Horizonal mass flux and variable gravity effects were considered by Saravan and Kandaswamy (2003b). The topic of this section has been reviewed by Lage and Nield (1998).

7.10. Periodic Heating

Lage and Bejan (1993) showed that when an enclosed saturated porous medium is heated periodically from the side, the buoyancy-induced circulation resonates to a well-defined frequency of the pulsating heat input. The resonance is characterized by maximum fluctuations in the total heat transfer rate through the vertical midplane of the enclosure. Lage and Bejan (1993) demonstrated this principle for an enclosure filled with a clear fluid and an enclosure filled with a fluid-saturated porous medium. They showed that the resonance frequency can be anticipated based on theoretical grounds by matching the period of the pulsating heat input to the period of the rotation (circulation) of the enclosed fluid. Below we outline Lage and Bejan's (1993) scale analysis of the resonance frequency in the Darcy and Forchheimer flow regimes.

Consider the two-dimensional configuration of Fig. 7.1 and assume that the flow is in the Darcy regime. The period of the fluid wheel that turns inside the enclosure is

$$w \sim \frac{4H}{v} \tag{7.121}$$

where ν is the scale of the peripheral velocity of the wheel and $4H$ is the wheel perimeter in a square enclosure. The velocity scale is given by Eq. (7.12),

$$v \sim \frac{\alpha_m}{H}\,\overline{\text{Ra}} \tag{7.122}$$

where $\overline{\text{Ra}} = g\beta K H(\overline{T}_h - T_c)/(v\alpha_m)$ is the Darcy modified Rayleigh number based on the average side-to-side temperature difference $(\overline{T}_h - T_c)$. The hot-side temperature (T_h) varies in time because the heat flux through that wall is administered in pulses that vary between q''_M (maximum) and zero. The cold-side temperature (T_c) is fixed.

The v scale can be restated in terms of the flux Rayleigh number $\text{Ra}_* = g\beta K H^2 q''_M/(v\alpha_m k_m)$ by noting that $\overline{\text{Ra}} = \text{Ra}_*/\overline{\text{Nu}}$, where in accordance with Eq. (7.49)

$$\overline{\text{Nu}} = \frac{q''_M H}{k(\overline{T}_h - T_c)} \sim \text{Ra}_*^{2/5} \tag{7.123}$$

Combining the relations listed between Eqs. (7.121) and (7.123) we obtain $v \sim (\alpha_m/H)\text{Ra}_*^{3/5}$ and the critical period for resonance (Lage and Bejan, 1993):

$$w \sim 4\frac{H^2}{\alpha_m}\,\text{Ra}_*^{-3/5} \quad \text{(Darcy)} \tag{7.124}$$

At higher Rayleigh numbers, when the Forchheimer term $(\chi/\nu)v^2$ is greater than the Darcy term (ν) on the left side of Eq. (7.90), the vertical velocity scale is (cf. Table 7.1):

$$v \sim \frac{\alpha_m}{H}\,\text{Ra}_\infty^{1/2}. \tag{7.125}$$

In this expression $\mathrm{Ra}_\infty = g\beta K H^2(\overline{T}_h - T_c)/(\chi\alpha_m^2)$ is the Forchheimer-regime Rayleigh number. Next, we introduce the flux Rayleigh number for the Forchheimer regime, $\mathrm{Ra}_{\infty *} = g\beta K H^3 q''_M/(\chi\alpha_m^2 k_m)$, and note that $\mathrm{Ra}_\infty = \mathrm{Ra}_{\infty *}/\overline{\mathrm{Nu}}$ and $\overline{\mathrm{Nu}} \sim Ra_{\infty *}^{1/5}$. These relations produce the following scaling law for the critical period (Lage and Bejan, 1993):

$$w \sim 4\frac{H^2}{\alpha_m}\mathrm{Ra}_{\infty *}^{-2/5} \quad \text{(Forchheimer)} \tag{7.126}$$

Three findings were extended and strengthened by subsequent numerical and theoretical studies of the resonance phenomenon. Antohe and Lage (1994) generalized the preceding scale analysis and produced a critical-frequency scaling law that unites the Darcy and Forchheimer limits [Eqs. (7.124) and (7.126)] with the clear fluid limit, which had been treated separately in Lage and Bejan (1993). The effect of the pulse amplitude was investigated more recently by Antohe and Lage (1996), who showed that the convection intensity within the enclosure increases linearly with the heating amplitude. The convection intensity decreases when the fluid Prandtl number increases or decreases away from a value of order one (Antohe and Lage, 1997a).

The corresponding phenomenon in forced convection was analyzed theoretically and numerically by Morega *et al.* (1995). Their study covered both the clear fluid (all Pr values) and saturated porous medium limits of the flow parallel to a plane surface with pulsating heating. The critical heat pulse period corresponds to the time scale of one sweep over the surface, i.e., the time of boundary layer renewal.

7.11. Sources in Confined or Partly Confined Regions

The problem of nuclear waste disposal has motivated a large number of studies of heat sources buried in the ground. An early review of the subject is that by Bau (1986).

The analyses of Bau (1984b) for small Ra and Farouk and Shayer (1988) for Ra up to 300 apply to a cylinder in the semi-infinite region bounded by a permeable plane. This geometry is applicable to the experiments conducted by Fernandez and Schrock (1982). The numerical work is aided by a preliminary transformation to bicylindrical coordinates.

Himasekhar and Bau (1987) obtained analytical and numerical solutions for convection induced by isothermal hot or cold pipes buried in a semi-infinite medium with a horizontal impermeable surface subject to a Robin thermal boundary condition. They (Himasekhar and Bau, 1988a) also made a theoretical and experimental study of convection around a uniform-flux cylinder embedded in a box. They found a transition from a two-dimensional steady flow to a three-dimensional oscillatory flow as the Rayleigh number increased. A similar problem with a sheath of different permeability surrounding the pipe was examined numerically by Ngo and Lai (2005). Hsiao *et al.* (1992) studied two-dimensional transient convection numerically on the Brinkman-Forchheimer model with thermal dispersion and nonuniform porosity allowed for. The effects of these two agencies increase the predicted heat flux, bringing it more in line with experimental data. Murty *et al.* (1994) used the Brinkman-Forchheimer model to study numerically convection around a buried cylinder using a penalty function method. Muralidhar (1992) summarized some analytical and numerical results for the temperature distribution around a cylinder (or an array of cylinders), for free or forced convection, with temperature or heat flux prescribed on the cylinder and Darcy's law assumed. Muralidhar (1993) made a numerical study of heat and mass transfer for buried cylinders with prescribed heat flux and leach rates. He obtained the temperature and concentration distributions on the surface of the containers under a variety of conditions. The case of a buried elliptic heat source with a permeable surface was studied numerically on the Darcy model by Facas (1995b). An ellipse with its minor axis horizontal yields much higher heat transfer rates than one with its major axis horizontal. The heat transfer depends little on the burial depth. A numerical study using the Brinkman-Forchheimer model of steady and transient convection from a corrugated plate of finite length placed in a square enclosure was performed by Hsiao and Chen (1994) and Hsiao (1995).

Anderson and Glasser (1990) fitted experimental steady-state temperature measurements in a porous medium containing a buried heater to a theoretical model vertical cylindrical source of finite height placed in a box with a pyramid lid with a constant heat transfer coefficient at the upper free surface. They derived a simple one-dimensional model relating power input to surface temperature irrespective of the values of permeability, source size, and depth and they showed that this was useful in monitoring the self-heating in stockpiles of cal, for example, and was consistent with the experiments.

Numerical modeling on the Brinkman-Forchheimer model of convection around a horizontal circular cylinder was carried out by Christopher and Wang (1993). They found that the presence of an impermeable surface above the cylinder significantly alters the flow field and reduces the heat transfer from the cylinder, while recirculating zones may develop above the cylinder, creating regions of low and high heat transfer rates. As expected, the Forchheimer term reduces the flow velocity and heat transfer, especially for the case of large Da.

Facas (1994, 1995a) has investigated numerically on the Darcy model convection around a buried pipe with two horizontal baffles attached and with a permeable bounding surface. They handled the complicated geometry using a body-fitted curvilinear coordinate system.

The case of a horizontal line heat source placed in an enclosure of rectangular cross section was studied numerically on the Darcy model by Desrayaud and Lauriat (1991). Their results indicated that the heat fluxes transferred to the walls and the source temperature vary strongly with the thermal conductivity of the side walls and the convective boundary condition at the ground. Further, for burial depths larger than the width of the cavity, the flow may be unstable to small disturbances and as a result the thermal plume may be deflected toward one of the side walls.

Oosthuizen and Naylor (1996a) studied numerically heat transfer from a cylinder placed on the vertical centerline of a square enclosure partially filled with a porous medium. Oosthuizen and Paul (1992), Oosthuizen (1995), and Oosthuizen and Naylor (1996b) used a finite element method to study heat transfer from a heated cylinder buried in a frozen porous medium in a square container, the flow being steady, two-dimensional, and with Darcy's law applicable and with either uniform temperature or heat flux specified on the cylinder and with one or more of the walls of the enclosure held at some subfreezing point temperature (or temperatures).

7.12. Effects of Rotation

The problem of stability of free convection in a rotating porous slab with lateral boundaries at different temperatures and rotation about a vertical axis so that the temperature gradient is collinear with the centrifugal body force was treated analytically by Vadasz (1994a, 1996a,b), first for a narrow slab adjacent to the center of rotation and then distant from the center of rotation. In the limit of infinite distance from the axis of rotation, the problem is analogous to that of gravitational buoyancy-induced convection with heating from below, the critical value of the centrifugal Rayleigh number $Ra_{\omega 0}$ [defined as in Eq. (6.190) being $4\pi^2$ for the case of isothermal boundaries]. At finite distance from the axis of rotation, a second centrifugal Rayleigh number $Ra_{\omega 1}$ (one proportional to that distance) enters the analysis. The stability boundary is given by the equation $(Ra_{\omega 1}/7.81\pi^2) + (Ra_{\omega 0}/4\pi^2) = 1$. The convection appears in the form of superimposed rolls.

The case where the axis of rotation is within the slab so that the centrifugal body force alternates in direction was treated by Vadasz (1996b). He found that the flow pattern was complex and that the critical centrifugal Rayleigh number and wavenumber increase significantly as the slab's cold wall moves significantly away from the rotation axis. This leads eventually to unconditional stability when the slab's hot wall coincides with the rotation axis. Unconditional stability is maintained when the axis of rotation moves away from the porous domain, so that the imposed temperature gradient opposes the centrifugal acceleration. Centrifugal convection with a magnetic fluid was analyzed by Saravanan and Yamaguchi (2005).

A further extension in which gravity as well as centrifugal forces are taken was made by Vadasz and Govender (1997). They considered a laterally heated vertical slab far away from the axis of rotation and calculated critical values of $Ra_{\omega 0}$ for various values of a gravitational Rayleigh number Ra_g.

A related problem involving a slowly rotating (large Ekman number) long box heated above and rotating about a vertical axis was analyzed by Vadasz (1993). Now the applied temperature gradient is orthogonal to the centrifugal body force and the interest is on the Coriolis effect. Vadasz employed an expansion in terms of small aspect ratio and small reciprocal Ekman number. He showed that secondary flow in a plane orthogonal to the leading free convection plane resulted. The controlling parameter is Ra_ω/Ek. The Coriolis effect in a long box subject to uniform heat generation was investigated analytically by Vadasz (1995). A nonlinear analysis using the Adomian decomposion method was employed by Olek (1998).

8
Mixed Convection

8.1. External Flow

8.1.1. Inclined or Vertical Plane Wall

We already have discussed one form of mixed convection in a horizontal layer, namely the onset of convection with throughflow when the heating is from below (see Section 6.10). In this chapter we discuss some more general aspects of mixed convection. Since we have dealt with natural convection and forced convection in some detail, our treatment of mixed convection in a porous medium [first treated by Wooding (1960)] can be brief. It is guided by the surveys by Lai *et al.* (1991a) and Lai (2000). We endorse the statement by Lai (2000) that despite the increased volume of research in this field, experimental results are still very few. In particular experimental data on thermal dispersion are very scarce and this is hindering the study of the functional relationship between effective thermal conductivity and thermal dispersion.

We start with a treatment of boundary layer flow on heated plane walls inclined at some nonzero angle to the horizontal. This configuration is illustrated in Fig. 8.1. The boundary layer equations [compare Eqs. (5.5) and (5.6)] for steady flow are

$$\frac{\partial^2 \psi}{\partial y^2} = \pm \frac{g_x \beta K}{\nu} \frac{\partial T}{\partial y} \tag{8.1}$$

$$\frac{\partial \psi}{\partial y}\frac{\partial T}{\partial x} - \frac{\partial \psi}{\partial x}\frac{\partial T}{\partial y} = \frac{\partial}{\partial y}\left(\alpha_m \frac{\partial T}{\partial y}\right). \tag{8.2}$$

Here $\pm g_x$ is the component of **g** in the positive x direction, i.e., the direction of the stream velocity $\mathbf{U}_\infty$ at infinity. The + sign corresponds to the case where the buoyancy force has a component "aiding" the general flow and the – sign to the "opposing" case. We seek a similarity solution and allow for suction/injection at the wall. Hence we take as boundary conditions the set

$$y = 0: T = T_\infty \pm Ax^\lambda,\ v = -\frac{\partial \psi}{\partial x} = ax^n, \tag{8.3}$$

$$y \to \infty: T = T_\infty,\ u = \frac{\partial \psi}{\partial y} = U_\infty = Bx^n, \tag{8.4}$$

where A, a, and B are constants. The exponent m is related to the angle of inclination $\gamma\pi/2$ (to the incident free stream velocity) by the relation $\gamma = 2m/(m+1)$.

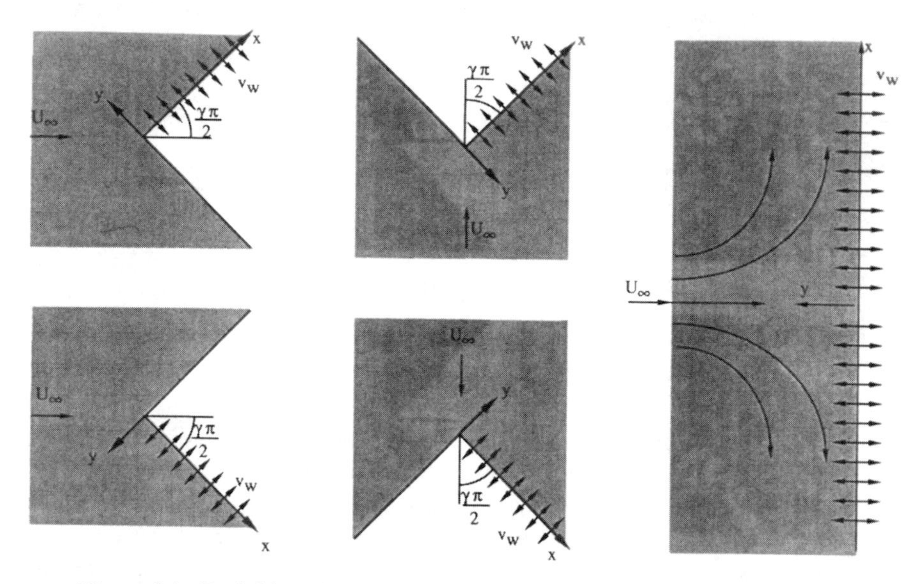

Figure 8.1. Definition sketch for mixed convection over an inclined surface.

We find that a similarity solution does exist if $\lambda = m$ and $n = (m-1)/2$. The range of possibilities includes the cases

$\lambda = m = 0,\quad n = -1/2$ (vertical isothermal wall injection $\propto x^{-1/2}$),
$\lambda = m = 1/3,\quad n = -1/3$ (wall sat 45° inclination, constant heat flux),
$\lambda = m = 1,\quad n = 0$ (stagnation flow normal to vertical wall (Fig .1.1e), linear temperature variation, uniform injection).

With the similarity variables

$$\eta = \left(\frac{U_\infty x}{\alpha_m}\right)^{1/2},\ f(\eta) = \frac{\psi}{(\alpha_m U_\infty x)^{1/2}},\ \theta(\eta) = \frac{T - T_\infty}{T_w - T_\infty} \tag{8.5}$$

and the wall suction parameter

$$f_w = -2a/(\alpha_m B)^{1/2}, \tag{8.6}$$

we obtain the system

$$f'' = \pm\frac{\mathrm{Ra}_x}{\mathrm{Pe}_x}\theta', \tag{8.7}$$

$$\theta'' = -\frac{\lambda+1}{2}f\theta' + \lambda f\theta, \tag{8.8}$$

$$\theta(0) = 1,\ f(0) = f_w,\ \theta(\infty) = 0,\ f'(\infty) = 1. \tag{8.9}$$

The numbers Ra_x and Pe_x are defined in Eq. (8.14). The quantity Ra/Pe has been called the mixed convection parameter by Holzbecher (2004a). For the case when this parameter is in the range [–3/2, 0] and the plate temperature varies

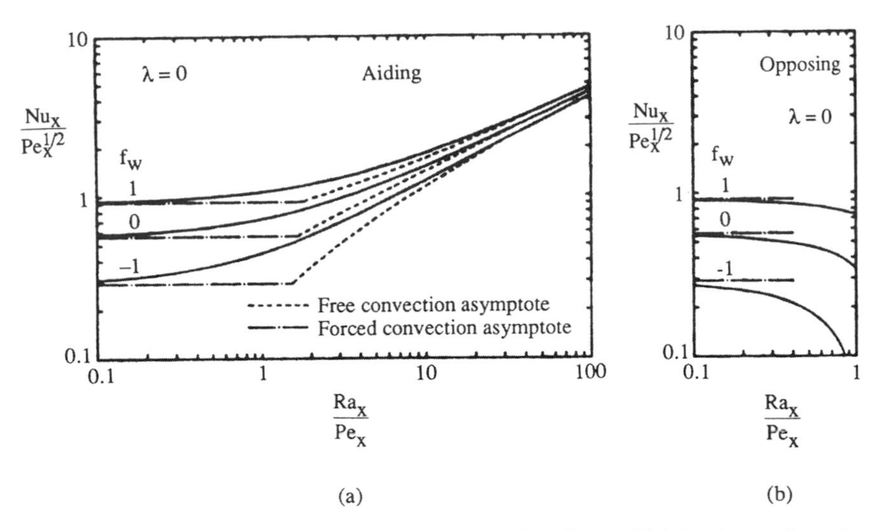

Figure 8.2. Nusselt numbers for aiding and opposing flow with injection and suction on a vertical flat plate (Lai and Kulacki, 1990d).

inverse-linearly with distance, exact dual solutions were obtained by Magyari *et al.* (2001b). Such solutions were first investigated by Merkin (1985). A special case that leads to a self-similar solution was studied by Magyari *et al.* (2002).

A positive f_w indicates withdrawal of fluid. The case of forced convection corresponds to letting $\mathrm{Ra}_x \to \infty$. The case of natural convection requires a different similarity variable. Lai and Kulacki (1990d) obtained and solved these equations. Their results for the Nusselt number are shown in Fig. 8.2 for the case $\lambda = 0$. Those for $\lambda = 1/3$ and $\lambda = 1$ are qualitatively similar; the effect of increasing λ is to raise the Nusselt number slightly. The case of adiabatic surfaces was analyzed by Kumari *et al.* (1988).

The effects of flow inertia and thermal dispersion were studied by Lai and Kulacki (1988a). Now Eq. (8.1) is replaced by

$$\frac{\partial^2 \psi}{\partial y^2} + \frac{\chi}{\nu}\frac{\partial}{\partial y}\left(\frac{\partial \psi}{\partial y}\right)^2 = \pm \frac{g_x \beta K}{\nu}\frac{\partial T}{\partial y}, \tag{8.10}$$

where $\chi = c_F K^{1/2}$, and in Eq. (8.2) α_m is replaced by α_e, the sum of a molecular diffusivity α_0 and a dispersive term $\alpha' = Cud_p$, where d_p is the mean pore diameter and C is a constant. We treat an isothermal vertical plate and we suppose that there is no suction. Equations (8.7)–(8.9) thus are replaced by

$$f'' + \mathrm{Fo}_x \mathrm{Re}_x[(f')^2]' = \pm \frac{\mathrm{Ra}_x}{\mathrm{Pe}_x}\theta', \tag{8.11}$$

$$\theta'' + \frac{1}{2} f\theta' + C\mathrm{Pe}_d(f''\theta' + f'\theta'') = 0, \tag{8.12}$$

$$\theta(0) = 1,\ f(0) = 0,\ \theta(\infty) = 0,\ f'(\infty) = 1. \tag{8.13}$$

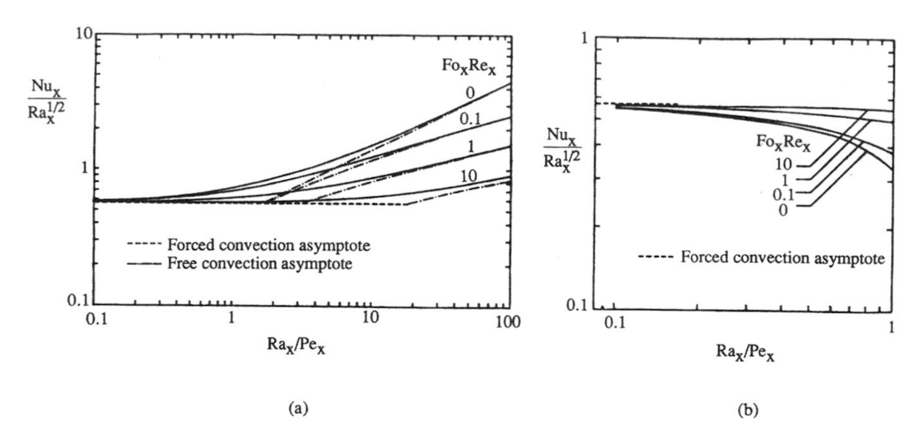

Figure 8.3. Nusselt numbers for aiding and opposing flow on a vertical plate with inertia effects (Lai and Kulacki, 1988a).

where

$$\mathrm{Fo}_x = \frac{c_F K^{1/2}}{x}, \quad \mathrm{Re}_x = \frac{U_\infty x}{\nu}, \quad \mathrm{Pe}_x = \frac{U_\infty x}{\alpha_m}, \quad \mathrm{Pe}_d = \frac{U_\infty d_p}{\alpha_m}, \tag{8.14}$$

$$\mathrm{Ra}_x = \frac{g_x \beta K x (T_w - T_\infty)}{\nu \alpha_m}, \quad \mathrm{Ra}_d = \frac{g_x \beta K d_p (T_w - T_\infty)}{\nu \alpha_m}. \tag{8.15}$$

The local Nusselt number Nu_x is given by

$$\frac{\mathrm{Nu}_x}{\mathrm{Pe}_x{}^{1/2}} = \left(\frac{\mathrm{Ra}_x}{\mathrm{Pe}_x{}^{1/2}}\right)^{1/2} \left\{-\left[1 + C\mathrm{Ra_d} f'(0)\right]\theta'(0)\right\}. \tag{8.16}$$

The results of the calculations of Lai and Kulacki (1988a) are shown in Figs. 8.3 and 8.4. The effect of quadratic drag is to reduce the aiding or opposing effect of buoyancy in increasing $\mathrm{Nu}_x/\mathrm{Pe}_x{}^{1/2}$, while that of thermal dispersion is (as expected) to increase the heat transfer. Non-Darcy effects also were treated by Gorin *et al.* (1988), Kodah and Al-Gasem (1998), Tashtoush and Kodah (1998), Elbashbeshy and Bazid (2000b) with variable surface heat flux, Elbashbeshy (2003) with suction or injection, and Murthy *et al.* (2004a) with suction or injection and the effect of radiation. The effect of variable permeability was studied by Mohammadein and El-Shaer (2004).

For a vertical surface, higher-order boundary layer theory (for Darcy flow) has been developed by Merkin (1980) and Joshi and Gebhart (1985). Merkin pointed out that in the case of opposing flow there is separation of the boundary layer downstream of the leading edge. Ranganathan and Viskanta (1984) included the effects of inertia, porosity variation, and blowing at the surface. They reported the rather unexpected result that porosity variation affected the Nusselt number by no more than 1 percent. Chandrasekhara and Namboodiri (1985) have studied the effect of variation of permeability and conductivity. Lai and Kulacki (1990c) have examined the effect of viscosity variation with temperature. They found that for liquids the Nusselt number values are greater than those for the constant viscosity

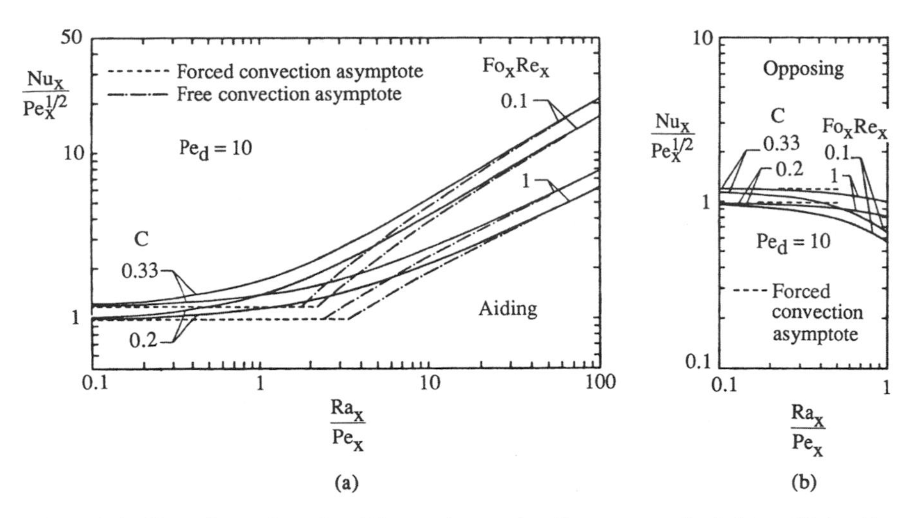

Figure 8.4. Nusselt numbers for aiding and opposing flow on a vertical plate with inertia and dispersion effects (Lai and Kulacki, 1988a).

case and for gases the reverse holds. Ramaniah and Malarvizhi (1990) have obtained a similarity solution for the combination of lateral mass flux and inertia when the linear Darcy drag term is negligible in comparison with the quadratic drag.

Chen and Chen (1990a) have studied the combined effects of quadratic drag, boundary friction, thermal dispersion, and nonuniform porosity and the consequent nonuniform conductivity for the case of aiding flow on a vertical surface. As expected, boundary friction reduces the velocity at the wall, inertia generally reduces the velocity, thermal dispersion has negligible effect on the velocity, and nonuniform porosity substantially increases the velocity just out from the wall. The temperature gradient at the wall is reduced by boundary friction and quadratic drag and increased by variable porosity; the overall effect is reduction. Consequently, the local Nusselt number is reduced by boundary friction and quadratic drag and increased by variable porosity; the overall effect is little change. The local Nusselt number Nu_x is increased about threefold by thermal dispersion. The effect of increase of Ra_x/Pe_x is to increase Nu_x and increase the amount of channeling. The effects of thermal dispersion and stratification were considered by Hassanien *et al.* (1998), while Hassanien and Omer (2002) considered the effect of variable permeability.

The case of a non-Newtonian power-law fluid has been treated by Wang *et al.* (1990), Nakayama and Shenoy (1992, 1993a), Gorla and Kumari (1996, 1998, 1999a–c), Kumari and Gorla (1996), Gorla *et al.* (1997), Mansour and Gorla (2000b), Ibrahim *et al.* (2000), and El-Hakiem (2001a,b), while Shenoy (1992) studied flow of an elastic fluid.

The magnetohydrodynamic case was examined by Aldoss *et al.* (1995) and Chamkha (1998). The effect of suction (which increases heat transfer) was treated by Hooper *et al.* (1994b) and Weidman and Amberg (1996). Conjugate convection was studied by Pop *et al.* (1995b) and Shu and Pop (1999). Stagnation point flow with suction or injection was treated by Yih (1999i).

Comprehensive nonsimilarity solutions were presented by Hsieh, Chen, and Armaly (1993a,b). Jang and Ni (1992) considered convection over an inclined plate. For a vertical plate, numerical work on non-Darcy models has been reported by Takhar *et al.* (1990), Lai and Kulaki (1991a), Yu *et al.* (1991), Shenoy (1993a), Chen *et al.* (1996), and Kodah and Duwairi (1996), and Takhar and Beg (1997). The numerical studies by Gorla *et al.* (1996) and Chen (1997a) have discussed the effect of such things as thermal dispersion, porosity variation, and variable conductivity. Thermal dispersion and viscous dissipation was discussed by Murthy and Singh (1997b) and Murthy (1998, 2001). The case of the plate temperature oscillating with time about a nonzero mean was studied by Vighnesan *et al.* (2001). Volumetric heating due to radiation was discussed by Bakier (2001a,b). The case of a piecewise heated wall was studied by Saeid and Pop (2005c).

Transient convection resulting from a sudden change in wall temperature was studied by Harris *et al.* (1998, 1999, 2002). The last paper allowed for a thermal capacity effect. They made a complete analysis of the steady-state solution (large times), obtained a series solution for small times, and then linked the two by a numerical solution for intermediate times. Transient convection near stagnation point flow was treated by Nazar *et al.* (2003a) and, using a homotopy analysis method that produces accurate uniformly valid series solutions, by Cheng *et al.* (2005). A transient problem involving suction or injection was studied by Al-Odat (2004b).

8.1.2. Horizontal Wall

For horizontal surfaces the situation is similar to that for vertical surfaces but now $\mathrm{Ra}_x/\mathrm{Pe}_x{}^{3/2}$ replaces $\mathrm{Ra}_x/\mathrm{Pe}_x$ as a measure of buoyancy to nonbuoyancy effects. Cheng (1977d) provided similarity solutions for the cases of (a) horizontal flat plate at zero angle of attack with constant heat flux and (b) stagnation point flow about a horizontal flat plate with wall temperature T_w varying as x^2.

Minkowycz *et al.* (1984) dealt with T_w varying as x^λ for arbitrary λ, using the local nonsimilarity method. Chandrasekhara (1985) extended Cheng's results to the case of variable permeability (which increases the heat transfer rate). Lai and Kulacki (1987, 1989a, 1990b) treated quadratic drag (for uniform U_x with T_w varying as $x^{1/2}$), thermal dispersion, and flow-injection/withdrawal, respectively. As in the case of the vertical wall, Nu_x is decreased by inertial effects and substantially increased by thermal dispersion effects; it is also enhanced by withdrawal of fluid across the surface. Chandrasekhara and Nagaraju (1988) and Bakier and Gorla (1996) included the effect of radiation. Kumari *et al.* (1990a) treated quadratic drag and extended the work of Lai and Kulacki (1987) to obtain some nonsimilarity solutions. The singularity associated with certain outer velocity profiles was investigated by Merkin and Pop (1997). Some new similarity solutions for specific outer velocity and wall temperature distributions were reported by Magyari *et al.* (2003a).

Ramaniah *et al.* (1991) and Elbashbeshy (2001) examined the effect of wall suction or injection. For the Forchheimer model, Yu *et al.* (1991) presented a universal similarity transformation. For the case of variable wall flux, calculations on the Brinkman model were performed by Aldoss *et al.* (1994b), while Chen (1996) used the Brinkman-Forchheimer model and also included the effects of porosity variation and thermal dispersion. On the Darcy model and for various thermal boundary conditions, Aldoss *et al.* (1993a,b, 1994a) presented nonsimilarity solutions for a comprehensive set of circumstances. A comprehensive analysis on the Brinkman-Forchheimer model was presented by Chen (1997b). The effect of velocity-dependent dispersion was studied by Thiele (1997). Nonsimilarity solutions were obtained for the case of variable surface heat flux by Duwairi *et al.* (1997) and Chen (1998a) and for the case of variable surface temperature by Chen (1998b). Non-Newtonian fluids were treated Kumari *et al.* (1997), Gorla *et al.* (1998), and Kumari and Nath (2004a). The effect of radiative flux was added by Kumari and Nath (2004b). The effect of temperature-dependent viscosity was discussed by Kumari (2001a). Convection above or below a horizontal plate was discussed by Lesnic and Pop (1998b).

Renken and Poulikakos (1990) presented experimental results of mixed convection about a horizontal isothermal surface embedded in a water-saturated bed of glass spheres. They measured the developing thermal boundary layer thickness and the local surface heat flux.

The onset of vortex instability for horizontal and inclined impermeable surfaces was studied by Hsu and Cheng (1980a,b). They found that the effect of the external flow is to suppress the growth of vortex disturbances in both aiding and opposing flows. For the inclined surfaces, aiding flows are more stable than opposing flows (for the same value of $\mathrm{Ra}_x/\mathrm{Pe}_x$). For the horizontal surfaces, stagnation-point aiding flows are more stable than parallel aiding flows. A case of unsteady convection near a stagnation point was analyzed by Nazar and Pop (2004). Jang *et al.* (1995) showed that the effect of blowing at the surface is to decrease Nu and make the flow more susceptible to vortex instability, while suction results in the opposite. The effect of variable permeability was treated by Hassanien *et al.* (2003c, 2004a). The effect of surface mass flux was studied by Murthy and Singh (1997c), together with thermal dispersion effects, and by Hassanien *et al.* (2004c) and Hassanien and Omer (2005).

The above theoretical papers have dealt with walls of infinite length. The case of a wall of finite length was studied analytically and numerically, on the Darcy model, by Vynnycky and Pop (1997). They observed flow separation for both heating and cooling.

8.1.3. Cylinder or Sphere

For an isothermal sphere or a horizontal cylinder in the presence of an otherwise uniform vertically flowing stream, Cheng (1982) obtained boundary layer equations in the form

$$\frac{1}{r^n}\frac{\partial^2\psi}{\partial y^2} = \frac{g\beta K}{\nu}\sin\left(\frac{x}{r_0}\right)\frac{\partial T}{\partial y}, \tag{8.17}$$

$$\frac{1}{r^n}\left(\frac{\partial\psi}{\partial y}\frac{\partial T}{\partial x} - \frac{\partial\psi}{\partial x}\frac{\partial T}{\partial y}\right) = \alpha_m\frac{\partial^2 T}{\partial y^2}, \tag{8.18}$$

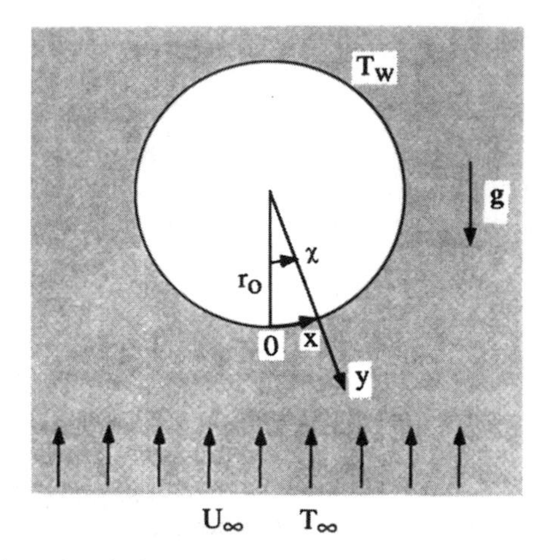

Figure 8.5. Definition sketch for mixed convection over a horizontal cylinder or a sphere.

where

$$u = \frac{1}{r^n}\frac{\partial \psi}{\partial y}, \quad v = -\frac{1}{r^n}\frac{\partial \psi}{\partial x}. \tag{8.19}$$

In these equations $n = 0$ for a horizontal cylinder, $n = 1$ for a sphere, and $r = r_0 \sin(x/r_0)$ where r_0 is the radius of the sphere or cylinder. These apply when x is measured from the lower stagnation point and y is in the normal (radial) direction. This configuration is sketched in Fig. 8.5.

The appropriate boundary conditions are

$$y = 0\colon T = T_w, \ \frac{\partial \psi}{\partial x} = 0, \tag{8.20}$$

$$y \to \infty\colon T = T_\infty, \ \frac{1}{r^n}\frac{\partial \psi}{\partial y} = U(x) = U_\infty A_n \sin\left(\frac{x}{r_0}\right), \tag{8.21}$$

where $U(x)$ is the tangential velocity on the surface (given by potential theory), so $A_0 = 2$ and $A_1 = 3/2$. The introduction of nondimensional variables defined by

$$\psi = \alpha r_0{}^n \left(A_n U_\infty r_0/\alpha_m\right) G_n(\chi) f(\eta), \tag{8.22}$$

$$\theta = \frac{T - T_\infty}{T_w - T_\infty}, \tag{8.23}$$

$$\eta = \left(A_n U_\infty r_0/\alpha_m\right)^{1/2} (y/r_0)\, H_n(\chi), \tag{8.24}$$

where

$$\chi = x/r_0, \ G_0(\chi) = (1 - \cos\chi)^{1/2}, \ G_1(\chi) = \left(\frac{\cos^3\chi}{3} - \cos\chi + \frac{2}{3}\right)^{1/2}, \tag{8.25}$$

$$H_0(\chi) = \sin\chi/G_0(\chi), \ H_1(\chi) = \sin^2\chi/G_1(\chi)$$

reduces the problem to finding the solution of

$$f'' = \frac{\text{Ra}}{\text{Pe}}\theta', \, \theta'' = -\frac{1}{2} f\theta', \tag{8.26}$$

$$f(0) = 0, \;\; \theta(0) = 1, \;\; f'(\infty) = 1, \;\; \theta(\infty) = 0, \tag{8.27}$$

which is the set [Eqs. (8.7)–(8.9)] for $\lambda = 0$, $f_w = 0$. Here Ra and Pe are based on r_0. Thus the solution for an isothermal sphere or horizontal cylinder can be deduced from that for a vertical plate.

Following the same approach, Huang *et al.* (1986) obtained the solution for the constant heat flux case. Minkowycz *et al.* (1985a) obtained approximate solutions for a nonisothermal cylinder or sphere using the local nonsimilarity method. Kumari *et al.* (1987) made more precise calculations for flow about a sphere. Badr and Pop (1988) considered aiding and opposing flows over a horizontal isothermal cylinder using a series expansion plus a finite-difference scheme. They found that for opposing flows there exists a recirculating flow zone just above the cylinder. For a similar situation, Badr and Pop (1992) studied the effect of varying the stream direction.

For horizontal cross flow over a horizontal cylinder below an impermeable horizontal surface, Oosthuizen (1987) performed a numerical study. He found that the presence of the surface has a negligible effect on heat transfer when the depth of the cylinder is greater than three times its diameter. The heat transfer is a maximum when the depth of the axis is about 0.6 times the diameter. The presence of the surface increases local heat transfer coefficients on the upper upstream quarter of the cylinder and decreases it on the upper downstream quarter, while buoyancy increases it on the upper upstream quarter and decreases it on the lower downstream quarter. The experiments by Fand and Phan (1987) were confined to finding correlations for overall Nusselt number data for horizontal cross flow over a horizontal cylinder.

The problem of longitudinal flow past a vertical cylinder was analyzed by Merkin and Pop (1987), who found that a solution of the boundary layer equations was possible only when $\text{Ra}/\text{Pe} = -1.354$, and that there is a region of reversed flow when $\text{Ra}/\text{Pe} < -1$. Here the minus sign indicates opposing flow. Reda (1988) performed experiments and a numerical analysis (without a boundary layer approximation) for opposing flow along a vertical cylinder of finite length. He found that buoyancy-induced upflow disappeared when $|\text{Ra}/\text{Pe}| = 0.5$. Ingham and Pop (1986a,b) analyzed the boundary layers for longitudinal flow past a vertical cylinder and horizontal flow past a vertical cylinder. For the case of a permeable vertical thin cylinder an exact solution was found by Magyari *et al.* (2005a). A three-dimensional problem involving the combined effects of wake formation and buoyancy on convection with cross-flow about a vertical cylinder was numerically simulated by Li and Kimura (2005).

Inertial effects on heat transfer along a vertical cylinder, with aiding or opposing flows, were analyzed by Kumari and Nath (1989a). As expected, their results show that inertial effects reduce heat transfer. Heat transfer still increases with buoyancy increase for aiding flows and decreases for opposing flows. Kumari and Nath

(1990) have studied inertial effects for aiding flow over a nonisothermal horizontal cylinder and a sphere. For a vertical cylinder, numerical studies on the Brinkman-Forchheimer model, with the effects of porosity variation and transverse thermal dispersion included, were reported by Chen *et al.* (1992) and for conjugate convection by Chen and Chen (1991), while nonsimilarity solutions were found by Hooper *et al.* (1994a) and Aldoss *et al.* (1996), the magnetohydrodynamic case was treated by Aldoss (1996), and Kumari *et al.* (1993) included the effect of thermal dispersion. A problem involving variable surface heat flux was analyzed by Pop and Na (1998). Further numerical studies by Zhou and Lai (2002) revealed that oscillatory flows occur for opposing flows at high Grashof number to Reynolds number ratios. The case of a non-Newtonian fluid was discussed by Mansour *et al.* (1997).

The double-diffusive and MHD problem for an unsteady (oscillatory or uniform acceleration) vertical flow over a horizontal cylinder and a sphere was analyzed by Kumari and Nath (1989b). MHD convection from a horizontal cylinder was also treated by Aldoss and Ali (1997). A substantial study of convection from a suddenly heated horizontal cylinder was reported by Bradean *et al.* (1998b). A correction to their results was pointed out by Diersch (2000). The Brinkman model was applied to the case of a horizontal cylinder by Nazar *et al.* (2003b).

For convection over a sphere, Tung and Dhir (1993) performed experiments and Nguyen and Paik (1994) carried out further numerical work. The latter considered variable surface temperature and variable surface heat flux conditions and they noted that recirculation was possible when the forcing flow opposed the flow induced by buoyancy, as in the case of cylinders. Unsteady convection around a sphere at low Péclet numbers for the case of sudden heating was analyzed by Sano and Makizono (1998). Unsteady mass transport from a sphere at finite Péclet numbers was studied by Feng and Michaelides (1999). Transient conjugate convection from a sphere with pure saline water was treated numerically by Paik *et al.* (1998).

8.1.4. Other Geometries

Introducing a general transformation, Nakayama and Koyama (1987b) showed that similarity solutions are possible for two-dimensional or axisymmetric bodies of arbitrary shape provided the external free-stream velocity varies as the product of the streamwise component of the gravitational force and the wall-ambient temperature difference. Examples are when $T_w - T_\infty$ varies as the same power function as U_∞ for a vertical wedge or a vertical cone. In these cases the problem can be reduced to the vertical plate problem solved by Cheng (1977c).

Invoking the slender body assumption, Lai *et al.* (1990c) have obtained similarity solutions for two other problems, namely accelerating flow past a vertical cylinder with a linear temperature variation along the axis and uniform flow over a paraboloid of revolution at constant temperature. They found that $\mathrm{Nu}_x/\mathrm{Pe}_x{}^{1/2}$ decreases with an increase in the dimensionless radius of a cylinder, but for paraboloids of revolution this is so only when $\mathrm{Ra}_x/\mathrm{Pe}_x$ is not too large.

Chen and Chen (1990b) have studied the flow past a downward projecting plate fin in the presence of a vertically upward free stream, incorporating the effects of quadratic drag, boundary friction, variable porosity, and thermal dispersion. A vertical cylindrical fin was investigated by Gill *et al.* (1992) and Aly *et al.* (2003).

For mixed convection in the thermal plume over a horizontal line heat source, Cheng and Zheng (1986) obtained a local similarity solution. They performed calculations for the thermal and flow fields and for heat transfer with the effects of transverse thermal dispersion and quadratic drag included. Further studies of this problem were reported by Lai (1991c) and Pop *et al.* (1995a). A line heat source embedded at the leading edge of a vertical adiabatic surface was examined by Jang and Shiang (1997). A heat source/sink effect on MHD convection in stagnation flow on a vertical plate was studied numerically by Yih (1998a).

Vargas *et al.* (1995) employed three different methods of solution for mixed convection on a wedge in a porous medium with Darcy flow. The methods were local nonsimilarity, finite elements in a boundary layer formulation, and finite elements in a formulation without boundary-layer approximations. For wedges with uniform wall temperature in the range $0.1 \leq \mathrm{Ra}_x/\mathrm{Pe}_x \leq 100$, the three methods produced results that are in very good agreement. New solutions were reported for wedges with half angles of 45°, 60°, and 90°. Convection over a wedge also has been treated by Kumari and Gorla (1997) for the case of a nonisothermal surface, by Mansour and Gorla (1998) and Mansour and El-Shaer (2004) for the case of a power law fluid with radiation, by Gorla and Kumari (2000) for a non-Newtonian fluid and with variable surface heat flux, and by Hassanien (2003) for variable permeability and with variable surface heat flux. Studies for the entire regime were carried out by Ibrahim and Hassanien (2000) for variable permeability and a nonisothermal surface and by Yih (2001a) with a radiation effect included. Transient convection resulting from impulsive motion from rest and a suddenly imposed wedge surface temperature was studied numerically by Bhattacharyya *et al.* (1998). Steady MHD convection with variable permeability, surface mass transfer, and viscous dissipation was investigated by Kumari *et al.* (2001).

Ingham and Pop (1991) treated a cylinder embedded to a wedge. Oosthuizen (1988b) studied a horizontal plate buried beneath an impermeable horizontal surface. Kimura *et al.* (1994) investigated heat transfer to ultralarge-scale heat pipes placed in a geothermal reservoir. Thermal dispersion effects on non-Darcy convection over a cone were studied by Murthy and Singh (2000). MHD convection from a rotating cone was studied by Chamkha (1999). The effect of radiation on convection from an isothermal cone was studied by Yih (2001b). The entire regime for convection about a cone was investigated by Yih (1999g).

A special geometry was considered in the early numerical and experimental study by Jannot *et al.* (1973). Heat transfer over a continuously moving plate was treated numerically by Elbashbeshy and Bazid (2000a).

8.1.5. Unified Theory

We now present the unified theory of Nakayama and Pop (1991) for mixed convection on the Forchheimer model about plane and axisymmetric bodies of arbitrary shape. The boundary layer equations are

$$\frac{1}{r^*}\frac{\partial r^* u}{\partial x} + \frac{\partial v}{\partial y} = 0, \tag{8.28}$$

$$\frac{\nu}{K}u + \frac{c_F}{K^{1/2}}u^2 = \frac{\nu}{K}u_\infty + \frac{c_F}{K^{1/2}}{u_\infty}^2 + g_x\beta(T - T_\infty), \tag{8.29}$$

$$u\frac{\partial T}{\partial x} + v\frac{\partial T}{\partial y} = \alpha_m\frac{\partial^2 T}{\partial y^2}, \tag{8.30}$$

with the boundary conditions

$$y = 0\colon v = 0,\ T = T_w(x), \tag{8.30a}$$

$$y = \infty\colon u = u_\infty(x) \text{ or } T = T_\infty, \tag{8.30b}$$

where

$$r^* = \begin{cases} 1, & \text{planebody,} \\ r(x), & \text{axisymmetric body} \end{cases} \tag{8.31}$$

and

$$g_x = g\left[1 - \left(\frac{dr}{dx}\right)^2\right]^{1/2}. \tag{8.32}$$

For the case of axisymmetric bodies it is assumed that the body radius $r(x)$ is large relative to the boundary layer thickness, so the transverse radial pressure gradient is negligible. Horizontal flat surfaces are excluded here; these require separate treatment.

The convective inertia term has been dropped from Eq. (8.28) because a scaling argument shows that the influence of this term is felt only very close to the leading edge, except for flow in highly permeable media. Nakayama (1995, 1998) also argued that for most porous materials the viscous boundary layer is confined for almost the entire surface to a thin layer close to the wall, so that the temperature distribution is essentially free from boundary viscous effects, and hence it is reasonable to drop the Brinkman term. However, Rees (private communication) noted that the analysis reported in Rees and Vafai (1999) for a uniformly heated horizontal plate indicates that the situation is more complicated, at least at intermediate values of x, than implied by Nakayama and Pop. Equation (8.28) gives

$$u = \frac{\nu}{2c_F K^{1/2}}\left\{\left[(1 + 2Re_K)^2 + 4\mathrm{Gr}_K\left(\frac{T - T_\infty}{T_w - T_\infty}\right)\right]^{1/2} - 1\right\}, \tag{8.33}$$

where

$$\mathrm{Re}_K(x) = c_F K^{1/2} u_\infty(x)/\nu, \tag{8.34}$$

and

$$\mathrm{Gr}_K(x) = c_F K^{3/2} g_x(x)\beta\left[T_w(x) - T_\infty\right]/\nu^2. \tag{8.35}$$

From Eqs. (8.30a) and (8.33), the wall velocity is

$$u_w = \frac{\nu}{2c_F K^{1/2}} \left\{ \left[(1 + 2\mathrm{Re}_K)^2 + 4Gr_K \right]^{1/2} - 1 \right\}, \tag{8.36}$$

Nakayama and Pop (1991) argued that it is this velocity, which depends on both external flow, that essentially determines convective heat transfer from the wall, and they introduced a modified Péclet number,

$$\mathrm{Pe}_x{}^* = \frac{u_w x}{\alpha_m} = \mathrm{Pe}_x \frac{\left[(1 + 2\mathrm{Re}_K)^2 + 4Gr_K \right]^{1/2} - 1}{2\mathrm{Re}_K}, \tag{8.37}$$

since the usual Péclet number is defined by

$$\mathrm{Pe}_x = \frac{u_\infty x}{\alpha_m}. \tag{8.38}$$

The energy equation (8.29) yields the scaling

$$u_w \frac{T_w - T_\infty}{x} \sim \alpha_m \frac{T_w - T_\infty}{\delta_T{}^2}, \tag{8.39}$$

where δ_T is the thermal boundary layer thickness. Hence one expects that for all convection modes,

$$\mathrm{Nu}_x \sim \frac{x}{\delta_T} \sim \mathrm{Pe}_x^{*1/2}, \tag{8.40}$$

where the local Nusselt number is defined as

$$\mathrm{Nu}_x = \frac{q''x}{k_m(T_w - T_\infty)}. \tag{8.41}$$

Nakayama and Pop (1991) also define

$$\mathrm{Ra}_x{}^* = \frac{K^{1/4} \left[g_x \beta (T_w - T_\infty) \right]^{1/2} x}{c_F{}^{1/2} \alpha_m} \tag{8.42}$$

and then identify the following regimes:

Regime I (Forced convection regime):

$$\mathrm{Nu}_x{}^2 \sim \mathrm{Pe}_x^* = \mathrm{Pe}_x \quad \text{for } \mathrm{Re}_K + \mathrm{Re}_K{}^2 \gg \mathrm{Gr}_K. \tag{8.43a}$$

Regime II (Darcy natural convection regime):

$$\mathrm{Nu}_x{}^2 \sim \mathrm{Pe}_x^* = \mathrm{Ra}_x \quad \text{for } \mathrm{Re}_K \ll \mathrm{Gr}_K \ll 1. \tag{8.43b}$$

Regime III (Forchheimer natural convection regime):

$$\mathrm{Nu}_x{}^2 \sim \mathrm{Pe}_x^* = \mathrm{Ra}_x \quad \text{for } \mathrm{Re}_K + \mathrm{Re}_K{}^2 \ll \mathrm{Gr}_K \text{ and } \mathrm{Gr}_K \gg 1. \tag{8.43c}$$

Regime IV (Darcy mixed convection regime):

$$\mathrm{Nu}_x{}^2 \sim \mathrm{Pe}_x^* = \mathrm{Pe}_x + \mathrm{Ra}_x \quad \text{for } \mathrm{Re}_K \sim \mathrm{Gr}_K \ll 1. \tag{8.43d}$$

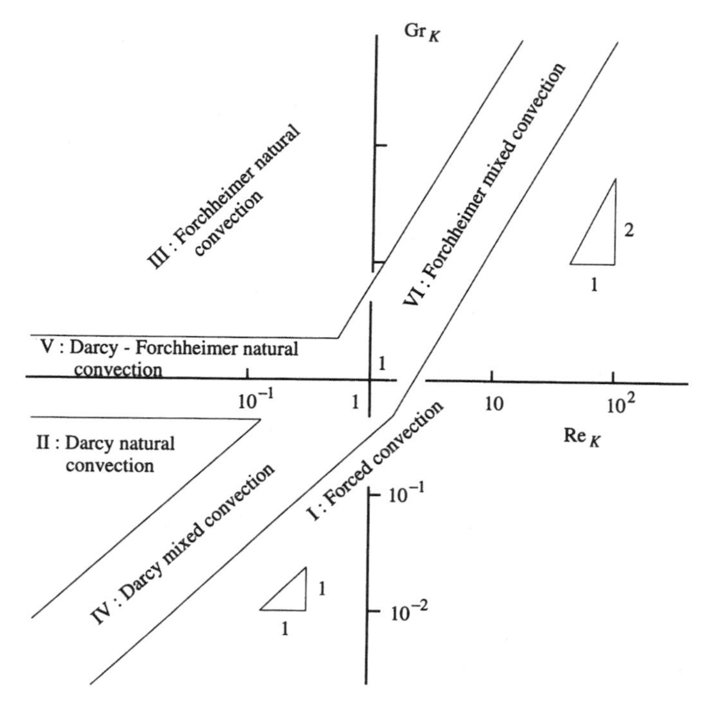

Figure 8.6. Convection flow regime map (Nakayama and Pop, 1991, with permission from Pergamon Press).

Regime V (Darcy-Forchheimer natural convection regime):

$$\mathrm{Nu}_x{}^2 \sim \mathrm{Pe}_x^* = \mathrm{Ra}_x \frac{(1+4\mathrm{Gr}_K)^{1/2}-1}{2\mathrm{Gr}_K} \quad \text{for } \mathrm{Gr}_K \sim 1 \text{ and } \mathrm{Re}_K \ll 1. \tag{8.43e}$$

Regime VI (Forchheimer mixed convection regime):

$$\mathrm{Nu}_x{}^2 \sim \mathrm{Pe}_x^* = \left(\mathrm{Pe}_x{}^2 + \mathrm{Ra}_x^{*2}\right)^{1/2} \quad \text{for } \mathrm{Gr}_K \sim \mathrm{Re}_K \gg 1. \tag{8.43f}$$

The situation is summarized in Fig. 8.6. The three macroscale parameters Pe_x, Ra_x, Ra_x^* and the two microscale parameters Re_K, Gr_K are related by

$$\frac{\mathrm{Ra}_x}{\mathrm{Pe}_x} = \frac{\mathrm{Gr}_K}{\mathrm{Re}_K}, \quad \frac{\mathrm{Ra}_x}{\mathrm{Ra}_x^*} = \mathrm{Gr}_K{}^{1/2}. \tag{8.44}$$

Nakayama and Pop (1991) then introduce the general transformations

$$f(x,\eta) = \frac{\psi}{\alpha_m r^* \left(\mathrm{Pe}_x^* I\right)^{1/2}}, \tag{8.45}$$

$$\theta(x,\eta) = \frac{T - T_\infty}{T_w - T_\infty}, \tag{8.46}$$

$$\eta = \frac{y}{x}\left(\frac{\mathrm{Pe}_x^*}{I}\right)^{1/2}, \tag{8.47}$$

where

$$I = \frac{\int_0^x (T_w - T_\infty)^2 u_w r^{*2} dx}{(T_w - T_\infty)^2 u_w r^{*2} x}. \tag{8.48}$$

The momentum and energy equations reduce to

$$f = \frac{\int_0^\eta \left[(1 + 2\mathrm{Re}_K)^2 + 4\mathrm{Re}_K \theta\right]^{1/2} d\eta - \eta}{\left[(1 + 2\mathrm{Re}_K)^2 + 4\mathrm{Re}_K \theta\right]^{1/2} - 1}, \tag{8.49}$$

and

$$\theta'' + \left(\frac{1}{2} - m_T I\right) f\theta' - m_T I \theta f' = Ix \left(f' \frac{\partial \theta}{\partial x} - \theta' \frac{\partial f}{\partial x}\right), \tag{8.50}$$

where

$$m_T(x) = \frac{d \ln(T_w - T_\infty)}{d \ln x} = \frac{x}{(T_w - T_\infty)} \frac{dT_w}{dx}. \tag{8.51}$$

The transformed boundary conditions are

$$\eta = 0: \quad \theta = 1, \tag{8.52a}$$

$$\eta \to \infty: \quad \theta = 0. \tag{8.52b}$$

Once the set of equations (8.49) and (8.50) subject to (8.52a) and (8.52b) have been solved, the local Nusselt number may be evaluated from

$$\mathrm{Nu}_x = -\theta'(x, 0) \left(\frac{\mathrm{Pe}_x{}^2}{I}\right)^{1/2}. \tag{8.53}$$

Nakayama and Pop (1991) then proceed to consider regimes I through VI in turn, seeking similarity solutions. In general these exist if and only if $T_w - T_\infty$ is a power function of the downstream distance variable ξ. They recover various results reported above in Chapter 4 (forced convection; regime I) and Chapter 5 (natural convection; regimes II, III, and V). For their other results, the reader is referred to the original paper and also the reviews by Nakayama (1995, 1998). These reviews include related material on the cases of convection over a horizontal plane, convection from line or point heat sources (Nakayama, 1993b, 1994), and also a study of forced convection over a plate on the Brinkman-Forchheimer model (Nakayama *et al*., 1990a).

8.2. Internal Flow: Horizontal Channel

8.2.1. Horizontal Layer: Uniform Heating

The problem of buoyancy-induced secondary flows in a rectangular duct filled with a saturated porous medium through which an axial flow is maintained was examined experimentally by Combarnous and Bia (1971) for the case of a large horizontal to vertical aspect ratio denoted by A. As predicted by linear stability theory (see Section 6.10), the axial flow did not affect the critical Rayleigh number for the onset of convective secondary flow nor the heat transfer. For Péclet numbers Pe less than about 0.7, cross rolls rather than longitudinal rolls were usually (but not always) the observed secondary motion. For larger values of Pe longitudinal rolls were always observed.

Islam and Nandakumar (1986) made a theoretical investigation of this problem. They used the Brinkman equation for steady fully developed flow and assumed negligible axial conduction, a constant rate of heat transfer per unit length, and an axially uniform heat flux, thus reducing the problem to a two-dimensional one that they solved numerically. Since axial conduction was neglected, their solutions are valid for large Pe values only. To save computational effort they assumed symmetry about the vertical midline of the duct, thus permitting only an even number of buoyancy-induced rolls. In our opinion this assumption is probably not justified; for the aspect ratios used ($0.6 < A < 3$) we would expect that the physically significant solution would sometimes be a single vortex roll. They treated two cases: bottom heating and heating all around the periphery. For each case they found a transition from a two-vortex pattern to a four-vortex pattern as the Grashof number Gr increased, with both two- and four-vortex solutions existing in a certain range of Gr. Further investigations by Nandakumar *et al.* (1987) indicated that the number of possible solutions depends sensitively on the aspect ratio. Islam and Nandakumar (1988) extended their analysis by including quadratic drag.

For a rectangular channel, Chou *et al.* (1992a) reported experimental and numerical work on the Brinkman-Forchheimer model and with variable porosity and thermal dispersion allowed for, while Chou and Chung (1995) allowed for the effect of variation of effective thermal conductivity. Hwang and Chao (1992) investigated numerically the case of finite wall conductivity. Chou *et al.* (1994) studied numerically the effect of thermal dispersion in a cylindrical tube. Islam (1992) investigated numerically the time evolution of the multicellular flows. His results show the presence of periodic, quasiperiodic, and chaotic behavior for increasingly high Grashof numbers (or Rayleigh numbers). An MHD problem was studied by Takhar and Ram (1994). Llagostera and Figueiredo (2000) numerically simulated mixed convection in a two-dimensional horizontal layer with a cavity of varying depth on the bottom surface and heated from below. Yokoyama *et al.* (1999) studied numerically and experimentally convection in a duct whose cross-section has a sudden expansion with heating on the lower downstream section. The onset of vortex instability in a layer, heated below with a stepwise change on the bottom boundary and with thermal dispersion, was studied using propagation theory by Chung *et al.* (2002). Unsteady convection involving internal heating and a moving lid was studied numerically by Khanafer and Chamkha (1999).

8.2.2. Horizontal Layer: Localized Heating

Prasad *et al.* (1988) and their colleagues have conducted a series of two-dimensional numerical studies to examine the effects of a horizontal stream on buoyancy-induced velocity and temperature fields in a horizontal porous layer discretely heated over a length D at the bottom and isothermally cooled at the top. The heated portion consisted of one or more sections of various sizes (nondimensional length $A = D/H$) and the heating was either isothermal or uniform-flux. Darcy's equation was used. The computations were carried out for the range $1 \leq \mathrm{Ra} \leq 500$, $0 \leq \mathrm{Pe} \leq 50$, the Rayleigh and Péclet numbers being based on the layer height. The domain was taken sufficiently long so that at the exit the flow could be assumed parallel and axial conduction could be neglected.

The results for the case of a single source of length $A = 1$ indicate that when the forced flow is weak (Pe small) a thermal plume rises above the heat source and a pair of counterrotating cells is generated above the source, the upstream cell being higher than the downstream one. The temperature field is approximately symmetric, fore and aft. As Pe is increased the isotherms lose their symmetry, the strength of the two recirculating cells becomes weaker, and the convective rolls and plume move downstream, the downstream roll being weaker than the upstream one. This is so for small values of Ra, but when $\mathrm{Ra} = 500$ there are two pairs of convective rolls along side each other.

The overall Nusselt number Nu increases monotonically with Pe as long as $\mathrm{Ra} = 10$, the increase being significant when $\mathrm{Pe} > 1$, but for $\mathrm{Ra} = 100$ the Nusselt number goes through a minimum before increasing when forced convection becomes dominant. The apparent reason for the decrease initially is because the enhancement in heat transfer by an increase in forced flow is not able to compensate for the reduction in buoyancy-induced circulation.

Further studies (Lai *et al.*, 1987a) indicated that Nu is increased significantly if the heat source is located on an otherwise isothermally cooled (rather than adiabatic) bottom surface, because this results in stronger buoyancy effects, but the effect is small if either buoyancy or forced convection dominates the other. Additional investigation (Lai *et al.*, 1987b) revealed that flow structure, temperature field, and heat transfer coefficients change significantly with the size of the heat source. If Ra is small, only two recirculating cells are produced, one near the leading edge and the other at the trailing edge of the heat source. At large Ra, the number of cells increases with the size of the source. All are destroyed by sufficient increase in forced flow. The transient problem has been discussed by Lai and Kulacki (1988b).

The extension to multiple heat sources was undertaken by Lai *et al.* (1990a). For the convective regime each source behaves more or less independently and the contributions to heat transfer are approximately additive. With the introduction of forced flow interaction occurs. Ultimately as Pe increases the buoyancy cells weaken and disappear, but at certain intermediate values of Ra and Pe the flow becomes oscillatory as cells are alternately generated and destroyed. A similar phenomenon was observed in the case of a long heat source. In general, the dependence of Nu on Ra and Pe for multiple sources is similar to that for a single source. The minimum in Nu that occurs at intermediate values of Pe is accentuated for large numbers of heat sources and tends to be associated with the oscillatory behavior; both effects involve an interaction between forcing flow and buoyancy.

Experiments performed by Lai and Kulacki (1991b) corroborated to a large extent the numerical results. In particular the observed overall Nusselt number data agreed quite well with the predicted values. When an effective thermal conductivity was introduced, the experimental data were correlated by

$$\frac{\mathrm{Nu}_D}{\mathrm{Pe}_D{}^{0.5}} = \left[1.895 + 0.200\left(\frac{\mathrm{Ra}_D}{\mathrm{Pe}_D{}^{1.5}}\right)\right]^{0.375}, \tag{8.54}$$

where the subscript D denotes numbers based on the heater length D. This is very close to the correlation obtained from the numerical solutions,

$$\frac{\mathrm{Nu}_D}{\mathrm{Pe}_D{}^{0.5}} = \left[1.917 + 0.210\left(\frac{\mathrm{Ra}_D}{\mathrm{Pe}_D{}^{1.5}}\right)\right]^{0.372}. \tag{8.55}$$

The experiments also verified the occurrence of oscillatory behavior. This was observed by recording the fluctuations in temperatures. A precise criterion for the appearance of oscillatory flow could not be determined, but the data available show that Ra_D has to exceed 10. A numerical study of oscillatory convection was reported by Lai and Kulacki (1991c). The experimental and numerical study by Yokoyama and Kulacki (1996) of convection in a duct with a sudden expansion just upstream of the heated region showed that the expansion had very little effect on the Nusselt number. A problem involving uniform axial heating and peripherally uniform wall temperature was studied numerically by Chang *et al.* (2004).

8.2.3. Horizontal Annulus

The problem of mixed convection in a horizontal annulus with isothermal walls, the inner heated and the outer cooled, was studied by Muralidhar (1989). His numerical results for radius ratio $r_0/r_i = 2$ and Ra = 500, Pe = 10 indicate that forced convection dominates in an entry length $x < (r_0 - r_i)$. Buoyancy increases the rate at which boundary layers grow and it determines the heat transfer rate once the annular gap is filled by the boundary layer on each wall.

Vanover and Kulacki (1987) conducted experiments in a porous annulus with $r_0/r_i = 2$, with the inner cylinder heated by constant heat flux and the outer cylinder isothermally cooled. The medium consisted of 1- and 3-mm glass beads saturated with water. In terms of Pe and Ra based on the gap width $(r_0 - r_i)$ and the temperature scale $q''(r_0 - r_i)/k_m$, their experimental data covered the range Pe < 520 and Ra < 830. They found that when Ra is large the values of Nu for mixed convection may be lower than the free convection values. They attributed this to restructuring of the flow as forced convection begins to play a dominant role. Muralidhar (1989) did not observe this phenomenon since he dealt only with Pe = 10. Vanover and Kulacki obtained the following correlations:

$$\text{Mixed convection } (6 < \mathrm{Pe} < 82)\text{: } \mathrm{Nu} = 0.619\,\mathrm{Pe}^{0.177}\mathrm{Ra}^{0.092}, \tag{8.56}$$

$$\text{Forced convection } (\mathrm{Pe} > 180)\text{: } \mathrm{Nu} = 0.117\,\mathrm{Pe}^{0.657}, \tag{8.57}$$

where the overall Nusselt number is normalized with its conduction value $\mathrm{Nu}_c = 1.44$ for an annulus with $r_0/r_i = 2$. Convection within a heat-generating horizontal annulus was studied numerically by Khanafer and Chamkha (2003).

8.2.4. Horizontal Layer: Lateral Heating

The flow produced by an end-to-end pressure difference and a horizontal temperature gradient in a horizontal channel was studied by Haajizadeh and Tien (1984) using perturbation analysis and numerical integration. The parameters are the Rayleigh number Ra, the channel aspect ratio L (length/height), and the dimensionless end-to-end pressure difference P which is equivalent to a Péclet number. Their results show that in the range $\mathrm{Ra}^2/L^3 \leq 50$ and $P \leq 1.5$, the heat transfer enhancement due to the natural convection and the forced flow can be simply added together. Even a small rate of throughflow has a significant effect on the temperature distribution and heat transfer across the channel. For $P/\mathrm{Ra} \geq 0.2$ the contribution of the natural convection to the Nusselt number is negligible.

8.3. Internal Flow: Vertical Channel

8.3.1. Vertical Layer: Uniform Heating

Hadim and Govindarajan (1988) calculated solutions of the Brinkman-Forchheimer equation for an isothermally heated vertical channel and examined the evolution of mixed convection in the entrance region. Viscous dissipation effects were analyzed by Ingham *et al.* (1990), for the cases of symmetric and asymmetrically heated walls. Further calculations on the Brinkman Forchheimer model were performed by Kou and Lu (1993a,b) for various cases of thermal boundary conditions, by Chang and Chang (1996) for the case of a partly filled channel, by Chen *et al.* (2000) for the case of uniform heat flux on the walls, and by Hadim (1994b) for the development of convection in a channel inlet. Umavathi *et al.* (2005) included the effect of viscous dissipation in their numerical and analytic study using the Brinkman-Forchheimer model and with various combinations of boundary conditions. They noted that viscous dissipation enhances the flow reversal in the case of downward flow and counters the flow in the case of upward flow. An MHD convection problem with heat generation or absoption was studied numerically by Chamkha (1997f). The effect of local thermal nonequilibrium was investigated by Saeid (2004).

An experimental study for the case of asymmetric heating of the opposing walls was conducted by Pu *et al.* (1999). The results indicated the existence of three convection regimes: natural convection, $105 < \mathrm{Ra/Pe}$; mixed convection, $1 < \mathrm{Ra/Pe} < 105$; and forced convection, $\mathrm{Ra/Pe} < 1$. Multiple solutions associated with the case of a linear axial temperature distribution were observed by Mishra *et al.* (2002).

A linear stability analysis of the mixed convection flow was reported by Chen and Chung (1998) and Chen (2004). It was found that the fully developed shear flow can become unstable under only mild heating conditions in the case of large Darcy number values (1 and 10^{-2}), and the critical Rayleigh number drops steeply when the Reynolds number reaches a threshold value that depends on the values of the Darcy and Forchheimer numbers. The critical Rayeigh number increases substantially for $Da = 10^{-4}$. For the case of an anisotropic channel, a linear stability analysis was conducted by Bera and Khalili (2002b). The convective cells may then be unicellular or bicellular.

For an anisotropic channel, aiding mixed convection was studied by Degan and Vasseur (2002). The effect of viscous dissipation was analyzed by Al-Hadhrami *et al.* (2002). For the case of wall temperature decreasing linearly with height, they found that at any value of the Rayleigh number there were two solutions mathematically, but only one of them is physically acceptable. The effects of a porous manifold on thermal stratification in a liquid storage tank, an unsteady problem, was treated numerically by Yee and Lai (2001). Problems involving multiple porous blocks were studied by Bae *et al.* (2004) and Huang *et al.* (2004a).

8.3.2. Vertical Layer: Localized Heating

Lai *et al.* (1988) performed a numerical study of the case when the heat source is a strip of height H (equal to the layer width) on an otherwise adiabatic vertical wall. The other wall was isothermally cooled; aiding or opposing Darcy flow was considered.

In the absence of a forced flow, a convection cell extends from near the bottom edge of the source to well above the top edge, and the higher the Rayleigh number Ra the larger is its extent and the stronger the circulation. When the forced flow is weak, buoyancy forces generally dominate the velocity field, but the acceleration caused by buoyancy forces deflects the main flow toward the heat source, so the circulation zone is pushed to the cold wall side. One consequence is that the vertical velocity in a thin layer on the heated segment increases. The aiding flow reattaches to the cold wall far downstream.

An increase in Pe moves the convective cell upward and this delays the separation of the main flow from the cold wall. When Pe becomes large, the strength of the circulation decreases substantially, the reattachment point moves upstream, and the center of the cell is pushed toward the cold wall. At a sufficiently high Péclet number ($Pe > 10$) the main flow does not separate from the cold wall and the effects of buoyancy forces become negligible.

The opposite trends are present when the forced flow is downward (opposing). When the main flow is weak, there is a circulation in the hot wall region and the main flow is directed toward the cold wall. As Pe increases both the separation and reattachment points move closer to the heat source, so that circulation is confined to the neighborhood of the source and the heat transfer is reduced from its free convection value. As Pe increases further, the circulation disappears and the heat transfer coefficient increases with Pe.

For both aiding and opposing flows, the average Nusselt number Nu increases with Ra, it being greater for aiding flows than for opposing flows. It increases monotonically with Pe for aiding flows, but for opposing flows it decreases with Pe until a certain value (which increases with Ra and increases from then on). The boundary layer formula for an isothermally heated vertical flat plate overpredicts the values of Nu for a channel if the flow is aiding and underpredicts them if the flow is opposing, the error being small in the forced convection regime. Further numerical work was reported by A. Hadim and Chen (1994) and H. A. Hadim and Chen (1994). A theoretical study of convection in a thin vertical duct with suddenly applied localized heating on one wall was reported by Pop *et al.* (2004).

8.3.3. Vertical Annulus: Uniform Heating

Muralidhar (1989) has performed calculations for aiding Darcy flow in a vertical annulus with height to gap ratio $= 10$ and $r_0/r_i = 2$, for $\mathrm{Ra} < 100$, $\mathrm{Pe} < 10$, with isothermal heating and cooling on the inner and outer walls, respectively. As expected, the average Nusselt number Nu increases with Ra and/or Pe. Muralidhar found a sharp change in Nu as Pe changed from 0 to 1. According to him, the circulation that exists at $\mathrm{Pe} = 0$ is completely destroyed when $\mathrm{Pe} > 0$, and is replaced by thin thermal boundary layers that give rise to large heat transfer rates. Hence, the jump in Nu from $\mathrm{Pe} = 0$ to $\mathrm{Pe} = 1$ is essentially a phenomenon related to inlet conditions of flow, and the jump can be expected to reduce as the length of the vertical annulus is reduced.

Parang and Keyhani (1987) solved the Brinkman equation for fully developed aiding flow in an annulus with prescribed constant heat flux q_i'' and q_0'' on the inner and outer walls, respectively. They found that the Brinkman term has a negligible effect if $\mathrm{Da}/\varphi = 10^{-5}$. For larger values of Da/φ it had a significant effect, which is more pronounced at the outer wall where the temperature is raised and the Nusselt number is reduced, the relative change increasing with Gr/Re.

In their experimental and numerical study, Clarksean *et al.* (1988) considered an adiabatic inner cylinder and an isothermally heated outer wall, with a radius ratio of about 12. Their numerical and experimental data showed the Nusselt number to be proportional to $(\mathrm{Ra}/\mathrm{Pe})^{-0.5}$ in the range $0.05 < \mathrm{Ra}/\mathrm{Pe} < 0.5$, wherein heat transfer is dominated by forced convection.

Choi and Kulacki (1992b) performed experimental and numerical work (on the Darcy model) that agreed in showing that Nu increases with either Ra or Pe when the imposed flow aids the buoyancy-induced flow, while when the imposed flow is opposing Nu goes through a minimum as Pe increases. They noted that under certain circumstances Nu for a lower Ra may exceed that for a higher Ra value. Good agreement was found between predicted and measured Nusselt numbers, which are correlated by expressing $\mathrm{Nu}/\mathrm{Pe}^{1/2}$ in terms of $\mathrm{Ra}/\mathrm{Pe}^{3/2}$.

Further numerical work, including non-Darcy effects, was reported by Kwendakwema and Boehm (1991), Choi and Kulacki (1993), Jiang *et al.* (1996) and Kou and Huang (1997) (for various thermal boundary conditions) and also by Du and Wang (1999). The experimental and numerical work of Jiang *et al.* (1994), for an inner wall at constant heat flux and the outer wall adiabatic, was specifically concerned with the effect of thermal dispersion and variable properties. Choi and Kulacki (1992a) reviewed work in this area. Density inversion with icy water was studied numerically by Char and Lee (1998) using the Brinkman-Forchheimer model.

8.3.4. Vertical Annulus: Localized Heating

Choi *et al.* (1989) have made calculations based on the Darcy model for convection in a vertical porous annulus, when a finite heat source (of height H equal to the annulus gap) is located on the inner wall. The rest of the inner wall is adiabatic and the outer wall is cooled at a constant temperature. They found that for both aiding and opposing flows the strength of the circulation decreases considerably as the radius ratio $\gamma = (r_0 - r_i)/r_i$ increases (with Ra and Pe fixed). Under the same circumstances the center of the cell moves toward the cold wall. The variations in Nu as Ra and Pe change are similar to those for the vertical layer channel. As γ increases, Nu increases toward the asymptotic value appropriate for a vertical cylinder. The following correlations were found.

Isothermal source, aiding flow:

$$\frac{\text{Nu}}{\text{Pe}^{0.5}} = \left(3.373 + \gamma^{0.566}\right)\left(0.0676 + 0.0320\frac{\text{Ra}}{\text{Pe}}\right)^{0.489}. \tag{8.58}$$

Isothermal source, opposing flow:

$$\frac{\text{Nu}}{\text{Pe}^{0.5}} = \left(2.269 + \gamma^{0.511}\right)\left(0.0474 + 0.0469\frac{\text{Ra}}{\text{Pe}}\right)^{0.509}. \tag{8.59}$$

Constant-flux source, aiding flow:

$$\frac{\text{Nu}}{\text{Pe}^{0.5}} = \left(7.652 + \gamma^{0.892}\right)\left(0.0004 + 0.0005\frac{\text{Ra}}{\text{Pe}^2}\right)^{0.243}. \tag{8.60}$$

Constant-flux source, opposing flow:

$$\frac{\text{Nu}}{\text{Pe}^{0.5}} = \left(4.541 + \gamma^{0.787}\right)\left(0.0017 + 0.0021\frac{\text{Ra}}{\text{Pe}^2}\right)^{0.253}, \tag{8.61}$$

where Nu, Ra, and Pe are defined in terms of the annular gap and either the temperature difference (for the isothermal source) or the temperature scale $q''H/k_m$ (for the constant-flux source). Nield (1993) noted that the final exponents in (8.31)–(8.34) are better replaced by 1/2, 1/2, 1/4, 1/4, since Nu should be independent of Pe as Ra tends to infinity. For the same reason, the final exponents in (8.27) and (8.28) should be 1/3.

The numerical and experimental study performed by Reda (1988) qualitatively supports the observations of Choi *et al.* (1989). In Reda's experiment the medium extended vertically from $z/\Delta r = 0$ to 4 and the heater from $z/\Delta r = 1.9$ to 3.1, where $\Delta r = r_0 - r_i$, the remainder of the inner wall being insulated, and the outer wall isothermally cooled. The forced flow was downward. Since the radius ratio was large (r_0/r_i approximately equal to 23) the effects of the outer wall on the temperature and flow fields were small. Reda found that buoyancy-induced circulation disappeared when Ra/Pe is approximately equal to 0.5, independent of the source length or power input.

The effects of quadratic drag and boundary friction were studied by Choi and Kulacki (1990). Their numerical results show that quadratic drag has a negligible effect on Nu, but boundary friction significantly changes the flow and temperature fields near the boundary and in highly porous media, as expected. For aiding flows the reduction of Nu becomes pronounced as either Ra or Pe increases. For opposing flows the interaction is complex.

9
Double-Diffusive Convection

In this chapter we turn our attention to processes of combined (simultaneous) heat and mass transfer that are driven by buoyancy. The density gradients that provide the driving buoyancy force are induced by the combined effects of temperature and species concentration nonuniformities present in the fluid-saturated medium. The present chapter is guided by the review of Trevisan and Bejan (1990), which began by showing that the conservation statements for mass, momentum, energy, and chemical species are the equations that have been presented here in Chapters 1–3. In particular the material in Section 3.3 is relevant. The new feature is that beginning with Eq. (3.26) the buoyancy effect in the momentum equation is represented by two terms, one due to temperature gradients and the other to concentration gradients. Useful review articles on double-diffusive convection include those by Mojtabi and Charrier-Mojtabi (2000, 2005), Mamou (2002b), and Diersch and Kolditz (2002).

9.1. Vertical Heat and Mass Transfer

9.1.1. Horton-Rogers-Lapwood Problem

The interesting effects in double-diffusive (or thermohaline, if heat and salt are involved) convection arise from the fact that heat diffuses more rapidly than a dissolved substance. Whereas a stratified layer involving a single-component fluid is stable if the density decreases upward, a similar layer involving a fluid consisting of two components, which can diffuse relative to each other, may be dynamically unstable. If a fluid packet of such a mixture is displaced vertically, it loses any excess heat more rapidly than any excess solute. The resulting buoyancy may act either to increase the displacement of the particle, and thus cause monotonic instability, or reverse the direction of the displacement and so cause oscillatory instability, depending on whether the solute gradient is destabilizing and the temperature gradient is stabilizing or vice versa.

The double-diffusive generalization of the Horton-Rogers-Lapwood problem was studied by Nield (1968). In terms of the temperature T and the concentration C, we suppose that the density of the mixture is given by Eq. (3.26),

$$\rho_f = \rho_0 \left[1 - \beta (T - T_0) - \beta_C (C - C_0) \right]. \tag{9.1}$$

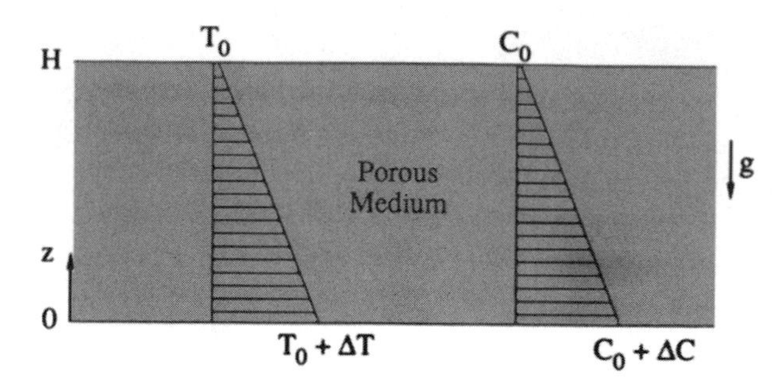

Figure 9.1. Infinite horizontal porous layer with linear distributions of temperature and concentration.

In this equation $\beta_C = -\rho_f{}^{-1}\partial\rho_f/\partial C$ is a concentration expansion coefficient analogous to the thermal expansion coefficient $\beta = -\rho_f{}^{-1}\partial\rho_f/\partial T$. We assume that β_C and β are constants. In most practical situations β_C will have a negative value.

As shown in Fig. 9.1, we suppose that the imposed conditions on C are

$$C = C_0 + \Delta C \text{ at } z = 0 \quad \text{and} \quad C = C_0 \text{ at } z = H. \tag{9.2}$$

The conservation equation for chemical species is

$$\varphi\frac{\partial C}{\partial t} + \mathrm{v}\cdot\nabla C = D_m\nabla^2 C \tag{9.3}$$

and the steady-state distribution is linear:

$$C_s = C_0 + \Delta C\left(1 - \frac{z}{H}\right). \tag{9.4}$$

Proceeding as in Section 6.2, choosing ΔC as concentration scale and putting $\hat{C} = C'/\Delta C$, and writing

$$\hat{C} = \gamma(z)\exp\left(st + il\hat{x} + im\hat{y}\right), \tag{9.5}$$

we obtain

$$\left[\mathrm{Le}^{-1}\left(D^2 - \alpha^2\right) - \frac{\varphi}{\sigma}s\right]\gamma = -W. \tag{9.6}$$

In place of Eq. (6.23) we now have, if γ_a is negligible,

$$\left(D^2 - \alpha^2\right)W = -\alpha^2\mathrm{Ra}(\theta + \mathrm{N}\gamma), \tag{9.7}$$

while Eq. (6.22) remains unchanged, namely

$$\left(D^2 - \alpha^2 - \mathrm{s}\right)\theta = -W. \tag{9.8}$$

The nondimensional parameters that have appeared are the Rayleigh and Lewis numbers

$$\mathrm{Ra} = \frac{g\beta K H\Delta T}{\nu\alpha_m}, \quad \mathrm{Le} = \frac{\alpha_m}{D_m}, \tag{9.9}$$

and the buoyancy ratio

$$N = \frac{\beta_C \Delta C}{\beta \Delta T}. \tag{9.10}$$

If both boundaries are impermeable, isothermal (conducting), and isosolutal (constant C), then Eqs. (9.6)–(9.8) must be solved subject to

$$W = \theta = \gamma = 0 \quad \text{at} \quad \hat{z} = 0 \quad \text{and} \quad \hat{z} = 1. \tag{9.11}$$

Solutions of the form

$$(W, \theta, \gamma) = (W_0, \theta_0, \gamma_0) \sin j\pi\hat{z} \tag{9.12}$$

are possible if

$$J(J + s)(J + \Phi s) = \mathrm{Ra}\, \alpha^2 (J + \Phi s) + \mathrm{Ra}_D \alpha^2 (J + s), \tag{9.13}$$

where

$$J = j^2\pi^2 + \alpha^2,\ \Phi = \frac{\varphi}{\sigma}\mathrm{Le}, \quad \mathrm{Ra}_D = N\mathrm{LeRa} = \frac{g\beta_C K H \Delta C}{\nu D_m}. \tag{9.14}$$

At marginal stability, $s = i\omega$ where ω is real, and the real and imaginary parts of Eq. (9.13) yield

$$J^2 - \Phi\omega^2 = (\mathrm{Ra} + \mathrm{Ra}_D)\,\alpha^2, \tag{9.15}$$

$$\omega[J^2(1 + \Phi) - (\Phi\mathrm{Ra} + \mathrm{Ra}_D)\alpha^2] = 0. \tag{9.16}$$

This system implies either $\omega = 0$ and

$$\mathrm{Ra} + \mathrm{Ra}_D = \frac{J^2}{\alpha^2}, \tag{9.17}$$

or

$$\Phi\mathrm{Ra} + \mathrm{Ra}_D = (1 + \Phi)\,\frac{J^2}{\alpha^2}, \tag{9.18}$$

and

$$\Phi\frac{\omega^2}{\alpha^2} = \frac{J^2}{\alpha^2} - (\mathrm{Ra} + \mathrm{Ra}_D)\,. \tag{9.19}$$

Since J^2/α^2 has the minimum value $4\pi^2$, attained when $j = 1$ and $\alpha = \pi$, we conclude that the region of stability in the (Ra, Ra_D) plane is bounded by the lines

$$\mathrm{Ra} + \mathrm{Ra}_D = 4\pi^2, \tag{9.20}$$

$$\Phi\mathrm{Ra} + \mathrm{Ra}_D = 4\pi^2\,(1 + \Phi)\,, \tag{9.21}$$

Equation (9.20) represents the boundary for monotonic or stationary instability, and Eq. (9.21) is the boundary for oscillatory instability with frequency ω given by

$$\Phi\frac{\omega^2}{\pi^2} = 4\pi^2 - (\mathrm{Ra} + \mathrm{Ra}_D)\,. \tag{9.22}$$

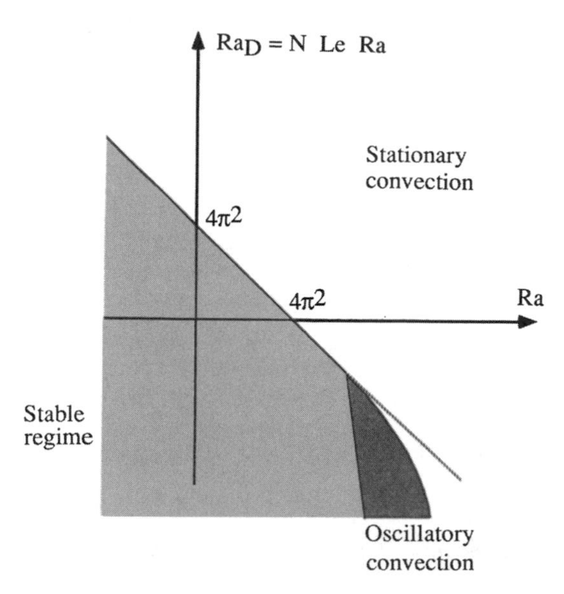

Figure 9.2. The stability and instability domains for double-diffusive convection in a horizontal porous layer.

Clearly the right-hand side of Eq. (9.22) must be nonnegative in order to yield a real value for ω.

If $\Phi = 1$, then the lines (9.20) and (9.21) are parallel, with the former being nearer the origin. Otherwise they intersect at

$$\mathrm{Ra} = \frac{4\pi^2\Phi}{\Phi - 1}, \quad \mathrm{Ra}_D = \frac{4\pi^2\Phi}{1 - \Phi}, \tag{9.23}$$

Illustrated in Fig. 9.2 is the case $\Phi > 1$, which corresponds to $\mathrm{Le} > \sigma/\varphi$.

The cases of other combinations of boundary conditions can be treated in a similar manner. If the boundary conditions on the temperature perturbation θ are formally identical with those of the solute concentration perturbation γ, then the monotonic instability boundary is a straight line:

$$\mathrm{Ra} + \mathrm{Ra}_D = \mathrm{Ra}_c. \tag{9.24}$$

One can interpret Ra as the ratio of the rate of release of thermal energy to the rate of viscous dissipation of energy and a similar interpretation applies to Ra_D. When the thermal and solutal boundary conditions are formally identical, the eigenfunctions of the purely thermal and purely solutal problems are identical, and consequently the thermal and solutal effects are additive. When the two sets of boundary conditions are different, the coupling between the thermal and solutal agencies is less than perfect and one can expect that the monotonic instability boundary will be concave toward the origin, since then $\mathrm{Ra} + \mathrm{Ra}_D = \mathrm{Ra}_c$ with equality occurring only when $\mathrm{Ra} = 0$ or $\mathrm{Ra}_D = 0$.

When Ra and Ra_D are both positive the double-diffusive situation is qualitatively similar to the single-diffusive one. When Ra and Ra_D have opposite signs there appear interesting new phenomena: multiple steady- and unsteady-state solutions, subcritical flows, periodic or chaotic oscillatory flows, traveling waves in relatively large aspect ratio enclosures, and axisymmetric flow structures. Such phenomena arise generally because the different diffusivities lead to different time scales for the heat and solute transfer. But similar phenomena can arise even when the thermal and solutal diffusivities are nearly equal because of the factor φ/σ (often called the normalized porosity). This is because heat is transferred through both the fluid and solid phases but the solute is necessarily transported through the fluid phase only since the porous matrix material is typically impermeable.

Experiments with a Hele-Shaw cell by Cooper *et al.* (1997, 2001) and Pringle *et al.* (2002) yielded results in agreement with the theory.

9.1.2. Nonlinear Initial Profiles

Since the diffusion time for a solute is relatively large, it is particularly appropriate to discuss the case when the concentration profile is nonlinear, the basic concentration distribution being given by

$$C_s = C_0 + \Delta C[1 - F_c(\hat{z})]. \tag{9.25}$$

The corresponding nondimensional concentration gradient is $f_c(\hat{z}) = F_c'(\hat{z})$, and satisfies $\langle f_c'(\hat{z})\rangle = 1$, where the angle brackets denote the vertical average. Then, in place of Eq. (9.6) one now has

$$\left[\mathrm{Le}^{-1}\left(D^2 - \alpha^2\right) - \frac{\varphi}{\sigma}s\right]\gamma = -f_c(\hat{z})\,W. \tag{9.26}$$

In the case of impermeable conducting boundaries, the Galerkin method of solution (trial functions of the form $\sin l\pi\hat{z}$ with $l = 1, 2, \ldots$) gives as the first approximation to the stability boundary for monotonic instability,

$$\mathrm{Ra} + 2\mathrm{Ra}_D \langle f_c(\hat{z}) \sin^2 \pi^2 \hat{z}\rangle = 4\pi^2. \tag{9.27}$$

For example, for the cosine profile with $F_c(\hat{z}) = (1 - \cos\pi\hat{z})/2$, and hence with $f_c = (\pi/2)\sin\pi\hat{z}$, we get

$$\mathrm{Ra} + \frac{4}{3}\mathrm{Ra}_D = 4\pi^2. \tag{9.28}$$

Similarly, for the step-function concentration, with $F_c(\hat{z}) = 0$ for $0 \le \hat{z} < {}^1/_2$ and $F_c(\hat{z}) = 1$ for ${}^1/_2 < \hat{z} \le 1$, so that $f_c(\hat{z}) = \delta(\hat{z} - {}^1/_2)$, we have

$$\mathrm{Ra} + \mathrm{Ra}_D = 4\pi^2. \tag{9.29}$$

The approximation leading to this result requires that $|\mathrm{Ra}_D|$ be small.

9.1.3. Finite-Amplitude Effects

Experiments in viscous fluids have shown that monotonic instability, associated with warm salty water above cool fresh water, appears in the form of "fingers" that grow downward from the upper part of the layer. More generally, fingering occurs when the faster diffusing component is stabilizing and the slower diffusing component is destabilizing. This situation is referred to as the fingering regime. On the other hand, oscillatory instability, associated with warm salty water below cool fresh water, gives rise to a series of convecting layers that form in turn, each on top of its predecessor. This situation is referred to as the diffusive regime.

In the case of a porous medium the questions are whether the fingers form fast enough before they are destroyed by dispersive effects and whether their width is large enough compared to the grain size for Darcy's law to be applicable. Following earlier work by Taunton *et al.* (1972), these questions were examined by Green (1984), who, on the basis of his detailed analysis, predicted that fluxes associated with double-diffusive fingering may well be important but horizontal dispersion may limit the vertical coherence of the fingers. In their visualization and flux experiments using a sand-tank model and a salt-sugar system Imhoff and Green (1988) found that fingering did indeed occur but it was quite unsteady, in contrast to the quasisteady fingering observed in a viscous fluid (Fig. 9.3). Despite the unsteadiness, good agreement was attained with the theoretical predictions. Imhoff and Green (1988) concluded that fingering could play a major role in the vertical transport of contaminants in groundwater.

That layered double-diffusive convection is possible in a porous medium was shown by Griffiths (1981). His experiments with a two-layer convecting system in a Hele-Shaw cell and a porous medium of glass spheres indicated that a thin "diffusive" interface is maintained against diffusive thickening, despite the lack of inertial forces. The solute and thermal buoyancy fluxes are approximately in the ratio $r = \varphi \, \mathrm{Le}^{-1/2}$. Griffiths explained the behavior of the heat flux in terms of a coupling between purely thermal convection within each convecting layer and diffusion through the density interface. Further experiments in a Hele-Shaw cell

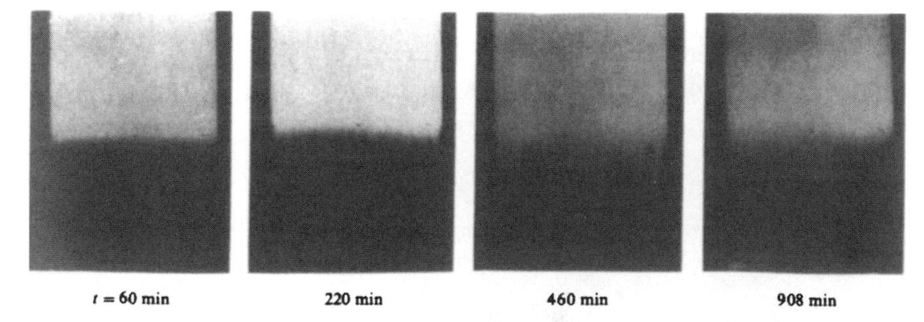

Figure 9.3. A series of pictures of finger growth. Dyed sugar solution (light color) overlies heavier salt solution (Imhoff and Green, 1988, with permission from Cambridge University Press).

by Pringle and Glass (2002) explored the influence of concentration at a fixed buoyancy ratio.

Rudraiah *et al.* (1982a) applied nonlinear stability analysis to the case of a porous layer with impermeable, isothermal, and isosolutal boundaries. They reported Nusselt and Sherwood numbers for Ra values up to 300 and Ra_D values up to 70. Their results show that finite-amplitude instability is possible at subcritical values of Ra.

Brand and Steinberg (1983a,b) and Brand *et al.* (1983) have obtained amplitude equations appropriate for the onset of monotonic instability and oscillatory instability and also for points in the vicinity of the lines of monotonic and oscillatory instability. Brand *et al.* (1983) found an experimentally feasible example of a codimensional-two bifurcation (an intersection of monotonic and oscillatory instability boundaries). Brand and Steinberg (1983b) predicted that the Nusselt number and also the "Froude" (Sherwood) number should oscillate with a frequency twice that of the temperature and concentration fields. Small-amplitude nonlinear solutions in the form of standing and traveling waves and the transition to finite amplitude overturning convection, as predicted by bifurcation theory, were studied by Knobloch (1986). Rehberg and Ahlers (1985) reported heat transfer measurements in a normal-fluid $He^3 - He^4$ mixture in a porous medium. They found a bifurcation to steady or oscillatory flow, depending on the mean temperature, in accordance with theoretical predictions.

Murray and Chen (1989) have extended the linear stability theory, taking into account effects of temperature-dependent viscosity and volumetric expansion coefficients and a nonlinear basic salinity profile. They also performed experiments with glass beads in a box with rigid isothermal lower and upper boundaries. These provide a linear basic-state temperature profile but only allow a nonlinear and time-dependent basic-state salinity profile. With distilled water as the fluid, the convection pattern consisted of two-dimensional rolls with axes parallel to the shorter side. In the presence of stabilizing salinity gradients, the onset of convection was marked by a dramatic increase in heat flux at a critical temperature difference ΔT. The convection pattern was three-dimensional, whereas two-dimensional rolls are observed for single-component convection in the same apparatus. When ΔT was then reduced from supercritical to subcritical values the heat flux curve completed a hysteresis loop.

For the case of uniform flux boundary conditions, Mamou *et al.* (1994) have obtained both analytical asymptotic and numerical solutions, the latter for various aspect ratios of a rectangular box. Both uniform flux and uniform temperature boundary conditions were considered by Mamou and Vasseur (1999) in their linear and nonlinear stability, analytical, and numerical studies. They identified four regimes dependent on the governing parameters: stable diffusive, subcritical convective, oscillatory, and augmenting direct regimes. Their results indicated that steady convection can arise at Rayleigh numbers below the supercritical value for linear stability, indicating the development of subcritical flows. They also demonstrated that in the overstable regime multiple solutions can exist. Also, their numerical results indicate the possible occurrence of traveling waves in an infinite horizontal enclosure.

A nonlinear stability analysis using the Lyapunov direct method was reported by Lombardo *et al.* (2001) and Lombardo and Mulone (2002). A numerical study of the governing and perturbation equations, with emphasis on the transition from steady to oscillatory flows and with an acceleration parameter taken into consideration, was conducted by Mamou (2003). The numerical and analytic study by Mbaye and Bilgen (2001) demonstrated the existence of subcritical oscillatory instabilities. The numerical study by Mohamad *et al.* (2004) for convection in a rectangular enclosure examined the effect of varying the lateral aspect ratio. Schoofs *et al.* (1999) discussed chaotic thermohaline convection in the context of low-porosity hydrothermal systems. Schoofs and Spera (2003) in their numerical study observed that increasing the ratio of chemical buoyancy to thermal buoyancy, with the latter kept fixed, led to a transition from steady to chaotic convection with a stable limit cycle appearing at the transition. The dynamics of the chaotic flow is characterized by transitions between layered and nonlayered patterns as a result of the spontaneous formation and disappearance of gravitationally stable interfaces. These interfaces temporally divide the domain in layers of distinct solute concentration and lead to a significant reduction of kinetic energy and vertical heat and solute fluxes. A scale analysis, supported by numerical calculations, was presented by Bourich *et al.* (2004c) for the case of bottom heating and a horizontal solutal gradient. The case of mixed boundary conditions (constant temperature and constant mass flux, or vice versa) was studied numerically by Mahidjiba *et al.* (2000a). They found that when the thermal and solute effects are opposing, the convection patterns differ markedly from the classic Bénard ones.

The linear stability for triply-diffusive convection was studied by Tracey (1996). For certain parameter values complicated neutral curves were found, including a heart-shaped disconnected oscillatory curve, and it was concluded that three critical Rayleigh numbers were involved. The energy method was used to obtain an unconditional nonlinear stability boundary and to identify possible regions of subcritical instability.

9.1.4. Soret Diffusion Effects

In the case of steep temperature gradients the cross coupling between thermal diffusion and solutal diffusion may no longer be negligible. The tendency of a solute to diffuse under the influence of a temperature gradient is known as the Soret effect.

In its simplest expression, the conservation equation for C now becomes

$$\varphi \frac{\partial C}{\partial t} + v \cdot \nabla C = D_m \nabla^2 C + D_{CT} \nabla^2 T, \tag{9.30}$$

where the Soret coefficient D_{CT} is treatable as a constant. If the Soret parameter S is defined as

$$S = -\frac{\beta_C D_{CT}}{\beta D_m}, \tag{9.31}$$

then the equation for the marginal state of monotonic instability in the absence of an imposed solutal gradient is

$$\mathrm{Ra} = \frac{4\pi^2}{1 + S(1 + \mathrm{Le})}. \tag{9.32}$$

The corresponding equation for marginal oscillatory instability is

$$\mathrm{Ra} = \frac{4\pi^2(\sigma + \varphi \mathrm{Le})}{\mathrm{Le}(\varphi + \sigma S)}. \tag{9.33}$$

The general situation, with both cross-diffusion and double-diffusion (thermal and solutal gradients imposed), was analyzed by Patil and Rudraiah (1980). Taslim and Narusawa (1986) showed that there is an analogy between cross-diffusion (Soret and Dufour effects) and double-diffusion in the sense that the equations can be put in mathematically identical form.

The linear analysis of Lawson *et al.* (1976), based on the kinetic theory of gases and leading to a Soret effect, was put forward to explain the lowering of the critical Rayleigh number in one gas due to the presence of another. This effect was observed in a binary mixture of helium and nitrogen by Lawson and Yang (1975). Lawson *et al.* (1976) observed that the critical Rayleigh number may be lower or greater than for a pure fluid layer depending upon whether thermal diffusion induces the heavier component of the mixture to move toward the cold or hot boundary, respectively. Brand and Steinberg (1983a) pointed out that with the Soret effect it is possible to have oscillatory convection induced by heating from above. Rudraiah and Siddheshwar (1998) presented a weak nonlinear stability analysis with cross-diffusion taken into account.

The experimental and numerical study of Benano-Melly *et al.* (2001) was concerned with Soret coefficient measurement in a medium subjected to a horizontal thermal gradient. The onset of convection in a vertical layer subject to uniform heat fluxes along the vertical walls was treated analytically and numerically by Joly *et al.* (2001). The Soret effect also was included in the numerical study by Nejad *et al.* (2001). Sovran *et al.* (2001) studied analytically and numerically the onset of Soret-driven convection in an infinite horizontal layer with an applied vertical temperature gradient. They found that for a layer heated from above, the motionless solution is infinitely linearly stable in $N > 0$, while a stationary bifurcation occurs in $N < 0$. For a layer heated from below, the onset of convection is steady or oscillatory depending on whether N is above or below a certain value that depends on Le and the normalized porosity. The numerical study of Faruque *et al.* (2004) of the situation where fluid properties vary with temperature, composition, and pressure showed that for lateral heating the Soret effect was weak, but with bottom heating the Soret effect was more pronounced.

Further studies of Soret convection, building on studies discussed in Section 1.9, were reported by Jiang *et al.* (2004a–c) and by Saghir *et al.* (2005a). Attention has been placed on thermogravitational convection, a topic treated by Estebe and Schott (1970). This refers to a coupling effect when a fluid mixture saturating a vertical porous cavity in a gravitational field is exposed to a uniform horizontal thermal gradient, and thermodiffusion produces a concentration gradient that leads to species separation. The porous media situation has been considered by Jamet *et al.* (1992) and Marcoux and Charrier-Mojtabi (1998). The numerical results of Marcoux and Mojtabi show the existence of a maximum separation corresponding to an optimal Rayleigh number as expected, but there remains a difference between the numerical results for that optimal value and experimental results of Jamet *et al.* (1992). The study by Jiang *et al.* (2004b) concentrated on the two-dimensional simulation of thermogravitation convection in a laterally heated vertical column with space-dependent thermal, molecular, and pressure diffusion coefficients taken as functions of temperature using the irreversible thermodynamics theory of Shukla and Firoozabadi. The numerical results reveal that the lighter fluid component migrates to the hot side of the cavity, and as the permeability increases the component separation in the thermal diffusion process first increases, reaches a peak, and then decreases. Jiang *et al.* (2004b) reported values of a separation ratio for a methane and n-butane mixture. Jiang *et al.* (2004c) explicitly investigated the effect of heterogenous permeability, something that strongly affects the Soret coefficient. Saghir *et al.* (2005a) have reviewed some aspects of thermodiffusion in porous media.

Soret-driven convection in a shallow enclosure and with uniform heat (or both heat and mass) fluxes was studied analytically and numerically by Bourich *et al.* (2002, 2004e–f, 2005) and Er-Raki *et al.* (2005). Depending on the values of Le and N, subcritical stationary convection may or may not be possible and parallel convective flow may or may not be possible.

An analytical and numerical study of convection in a horizontal layer with uniform heat flux applied at the horizontal walls, and with or without constant mass flux at those walls, was reported by Bahloul *et al.* (2003) and Boutana *et al.* (2004). A structural stability result was reported by Straughan and Hutter (1999).

9.1.5. Flow at High Rayleigh Number

The interaction between the heat transfer and mass transfer processes in the regime of strong convection was investigated on the basis of a two-dimensional model by Trevisan and Bejan (1987b). They used scale analysis to back up their numerical work. Figure 9.4 shows the main characteristics of the flow, temperature, and concentration fields in one of the rolls that form. This particular flow is heat transfer-driven in the sense that the dominant buoyancy effect is one due to temperature gradients ($N = 0$). The temperature field (Fig. 9.4b) shows the formation of thermal boundary layers in the top and bottom end-turn regions of the roll. The concentration field is illustrated in Figs. 9.4b–9.4d. The top and bottom concentration boundary layers become noticeably thinner as Le increases from 1 to 20.

The overall Nusselt numbers Nu and overall Sherwood number Sh are defined by

$$\mathrm{Nu} = \frac{\bar{q}''}{k_m \Delta T / H}, \quad \mathrm{Sh} = \frac{\bar{j}}{D_m \Delta C / H} \tag{9.34}$$

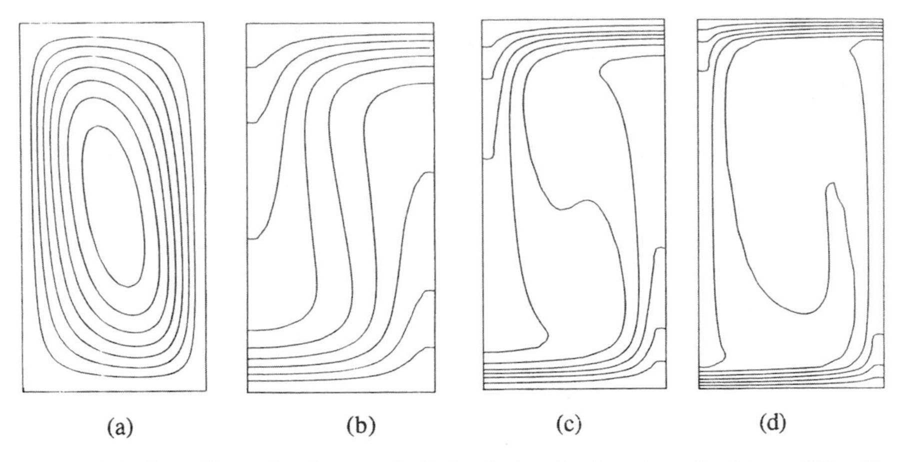

Figure 9.4. Two-dimensional numerical simulation for heat transfer-driven ($N = 0$) convection in a horizontal porous layer ($\mathrm{Ra} = 200$, $H/L = 1.89$). (a) Streamlines; (b) isotherms, also isosolutal lines for $\mathrm{Le} = 1$; (c) isosolutal lines for $\mathrm{Le} = 4$; and (d) isosolutal lines for $\mathrm{Le} = 20$ (Trevisan and Bejan, 1987b).

where $\bar{q}''$ and $\bar{j}$ are the heat and mass fluxes averaged over one of the horizontal boundaries. In heat transfer-driven convection, $|N| \ll 1$, it is found that the Nusselt number scales as

$$\mathrm{Nu} = \left(\mathrm{Ra}/4\pi^2\right)^{1/2}. \tag{9.35}$$

In the same regime the mass transfer scales are

$$\mathrm{Sh} \approx \mathrm{Le}^{1/2}\left(\mathrm{Ra}/4\pi^2\right)^{7/8} \quad \text{if} \quad \mathrm{Le} > \left(\mathrm{Ra}/4\pi^2\right)^{1/4}, \tag{9.36a}$$

$$\mathrm{Sh} \approx \mathrm{Le}^{2}\left(\mathrm{Ra}/4\pi^2\right)^{1/2} \quad \text{if} \quad \left(\mathrm{Ra}/4\pi^2\right)^{-1/4} < \mathrm{Le} < \left(\mathrm{Ra}/4\pi^2\right)^{1/4}, \tag{9.36b}$$

$$\mathrm{Sh} \approx 1 \quad \text{if} \quad \mathrm{Le} < \left(\mathrm{Ra}/4\pi^2\right)^{1/4}. \tag{9.36c}$$

The scales of mass transfer-driven flows, $|N| \gg 1$, can be deduced from these by applying the transformation $\mathrm{Ra} \rightarrow \mathrm{Ra}_D$, $\mathrm{Nu} \rightarrow \mathrm{Sh}$, $\mathrm{Sh} \rightarrow \mathrm{Nu}$, and $\mathrm{Le} \rightarrow \mathrm{Le}^{-1}$. The results are

$$\mathrm{Sh} \approx \left(\mathrm{Ra}_D/4\pi^2\right)^{1/2}, \tag{9.37}$$

and

$$\mathrm{Nu} \approx \mathrm{Le}^{-1/2}\left(\mathrm{Ra}_D/4\pi^2\right)^{7/8} \quad \text{if} \quad \mathrm{Le} < \left(\mathrm{Ra}_D/4\pi^2\right)^{-1/4}, \tag{9.38a}$$

$$\mathrm{Nu} \approx \mathrm{Le}^{-2}\left(\mathrm{Ra}_D/4\pi^2\right)^{1/2} \quad \text{if} \quad \left(\mathrm{Ra}_D/4\pi^2\right)^{-1/4} < \mathrm{Le} < \left(\mathrm{Ra}_D/4\pi^2\right)^{1/4}, \tag{9.38b}$$

$$\mathrm{Nu} \approx 1 \quad \text{if} \quad \mathrm{Le} > \left(\mathrm{Ra}_D/4\pi^2\right)^{1/4}. \tag{9.38c}$$

These estimates agree well with the results of direct numerical calculations.

Rosenberg and Spera (1992) performed numerical simulations for the case of a fluid heated and salted from below in a square cavity. As the buoyancy ratio N increases, the dynamics change from a system that evolves to a well-mixed steady state, to one that is chaotic with large amplitude fluctuations in composition, and finally to one that evolves to a conductive steady state. Their correlations for Nu and Sh were in good agreement with the results of Trevisan and Bejan (1987b).

Sheridan *et al.* (1992) found that their experimentally measured heat transfer data correlated well with $\mathrm{Nu} \sim (\mathrm{Ra\,Da}\,N)^{0.294}\mathrm{Ja}^{-0.45}$. Here Ja is the Jakob number, defined by $\mathrm{Ja} = c_p \Delta T / h_{fg} \Delta m$, where h_{fg} is the enthalpy of evaporation and m is the saturated mass ratio (vapor/gas).

9.1.6. Other Effects

9.1.6.1. Dispersion

If a net horizontal flow is present in the porous layer, it will influence not only the vertical solutal gradient but also the phenomenon of solute dispersion. Thermal dispersion also can be affected. In most applications α_m is greater than D_m, and as a consequence the solutal dispersion is more sensitive to the presence of through flow. The ultimate effect of dispersion is that the concentration distribution becomes homogeneous.

The stability implications of the anisotropic mass diffusion associated with an anisotropic dispersion tensor were examined by Rubin (1975) and Rubin and Roth (1978, 1983). The dispersion anisotropy reduces the solutal stabilizing effect on the inception of monotonic convection and at the same time enhances the stability of the flow field with respect to oscillatory disturbances. Monotonic convection appears as transverse rolls with axes perpendicular to the direction of the horizontal net flow, while oscillatory motions are associated with longitudinal rolls (axes aligned with the net flow), the rolls of course being superposed on that net flow.

Certain geological structures contain some pores and fissures of large sizes. In such cavernous media even very slow volume-averaged flows can deviate locally from the Darcy flow model. The larger pores bring about an intensification of the dispersion of solute and heat and because of the high pore Reynolds numbers, Re_p, the effect of turbulence within the pores. Rubin (1976) investigated the departure from the Darcy flow model and its effect on the onset of convection in a horizontal layer with horizontal through flow. This study showed that in the case of laminar flow through the pores ($\mathrm{Re}_p \ll 1$), the net horizontal flow destabilizes the flow field by enhancing the effect of solutal dispersion. A stabilizing effect is recorded in the intermediate regime ($\mathrm{Re}_p \approx 1$). In the inertial flow regime ($\mathrm{Re}_p \gg 1$) the stability characteristics become similar to those of monodiffusive convection, the net horizontal flow exhibiting a stabilizing effect.

9.1.6.2. Anisotropy and Heterogeneity

The onset of thermohaline convection in a porous layer with varying hydraulic resistivity ($r = \mu/K$) was investigated by Rubin (1981). If one assumes that the dimensionless hydraulic resistivity $\xi = r/r_0$ varies only in the vertical direction and only by a relatively small amount, the linear stability analysis yields the monotonic marginal stability condition

$$\mathrm{Ra} + \mathrm{Ra}_D = \pi^2 \left(\xi_H{}^{1/2} + \xi_V{}^{1/2}\right)^2. \tag{9.39}$$

In this equation ξ_H and ξ_V are the horizontal and vertical mean resistivities

$$\xi_H = \left(\int_0^1 \frac{d\hat{z}}{\xi}\right)^{-1}, \quad \xi_V = \int_0^1 \xi d\hat{z}, \tag{9.40}$$

and so $\xi_H \leq \xi_V$. The right-hand side of Eq. (9.39) can be larger or smaller than $4\pi^2$ depending on whether Ra is based on ξ_V or ξ_H. A similar conclusion is reached with respect to the onset of oscillatory motions.

The Galerkin method has been used by Rubin (1982a) in an analysis of the effects of nonhomogeneous hydraulic resistivity and thermal diffusivity on stability. The effect of simultaneous vertical anisotropy in permeability (hydraulic resistivity), thermal diffusivity, and solutal diffusivity was investigated by Tyvand (1980) and Rubin (1982b).

Chen (1992) and Chen and Lu (1992b) analyzed the effect of anisotropy and inhomogeneity on salt-finger convection. They concluded that the critical Rayleigh number for this is invariably higher than that corresponding to the formation of plumes in the mushy zone during the directional solidification of a binary solution (see Section 10.2.3). A numerical study of double-diffusive convection in layered anisotropic porous media was made by Nguyen *et al.* (1994).

Viscosity variations and their effects on the onset of convection were considered by Patil and Vaidyanathan (1982), who performed a nonlinear stability analysis using the Brinkman equation, assuming a cosine variation for the viscosity. The variation reduces the critical Rayleigh number based on the mean viscosity. Bennacer (2004) treated analytically and numerically a two-layer (one anisotropic) situation with vertical through mass flux and horizontal through heat flux.

9.1.6.3. Brinkman Model

The effect of porous-medium coarseness on the onset of convection was documented by Poulikakos (1986). With the Brinkman equation the critical Rayleigh number for the onset of monotonic instability is given by

$$\mathrm{Ra} + \mathrm{Ra}_D = \frac{\left(\alpha_c{}^2 + \pi^2\right)^2}{\alpha_c{}^2} \left[\left(\alpha_c^2 + \pi^2\right) \tilde{\mathrm{D}}\mathrm{a} + 1\right], \tag{9.41}$$

where the critical dimensionless horizontal wavenumber (α_c) is given by

$$\alpha_c^2 = \frac{\left(\pi^2 \tilde{\mathrm{D}}\mathrm{a} + 1\right)^{1/2} \left(9\pi^2 \tilde{\mathrm{D}}\mathrm{a} + 1\right)^{1/2} - \pi^2 \tilde{\mathrm{D}}\mathrm{a} - 1}{4\tilde{\mathrm{D}}\mathrm{a}}. \tag{9.42}$$

In terms of the effective viscosity $\tilde{\mu}$ introduced in Eq. (1.17), the Darcy number $\tilde{\text{Da}}$ is defined by

$$\tilde{\text{Da}} = \frac{\tilde{\mu}}{\mu}\frac{K}{H^2}. \tag{9.43}$$

Nonlinear energy stability theory was applied to this problem by Guo and Kaloni (1995b). Fingering convection, with the Forchheimer term as well as the Brinkman term taken into account, was treated numerically by Chen and Chen (1993). With Ra fixed, they found a transition from steady to time-periodic (and then to quasi-periodic) convection as Ra_D increases. An analytical solution based on a parallel-flow approximation and supported by numerical calculations was presented by Amahmid *et al.* (1999a). They showed that there is a region in the (N, Le) plane where a convective flow of this type is not possible for any Ra and Da values. A linear and nonlinear stability analysis leading to calculations of Nusselt numbers, streamlines, isotherms, and isohalines was presented by Shivakumara and Sumithra (1999).

9.1.6.4. Additional Effects

The situation in which one of the components undergoes a slow chemical reaction was analyzed by Patil (1982), while a convective instability that is driven by a fast chemical reaction was studied by Steinberg and Brand (1983). Further work involving chemical reactions was carried out by Subramanian (1994), Malashetty *et al.* (1994) and Malashetty and Gaikwad (2003). The effect of a third diffusing component was treated by Rudraiah and Vortmeyer (1982), Poulikakos (1985c), and Tracey (1998), who obtained some unusual neutral stability curves, including a closed approximately heart-shaped oscillatory curve disconnected from the stationary neutral curve, and thus requiring three critical values of Ra to describe the linear stability criteria. For certain values of parameters the minima on the oscillatory and stationary curves occur at the same Rayleigh number but different wavenumbers. Kalla *et al.* (2001a) studied a situation involving imposed vertical heat and mass fluxes and a horizontal heat flux that they treated as a perturbation leading to asymmetry of the bifurcation diagram. Multiple steady-state solutions, with different heat and mass transfer rates, were found to coexist. In their analytical studies Masuda *et al.* (1999, 2002) found that there is a range of buoyancy ratios N for which there is an oscillation between two types of solution: temperature dominated and concentration dominated.

The effect of rotation was included by Chakrabarti and Gupta (1981) and Rudraiah *et al.* (1986), for anisotropic media by Patil *et al.* (1989, 1990), and for a ferromagnetic fluid by Sekar *et al.* (1998). The effects of magnetic field and compressibility were studied by Sunil (1994, 1999, 2001), while Khare and Sahai (1993) combined the effects of a magnetic field and heterogeneity. Chamkha and Al-Naser (2002) studied numerically MHD convection in a binary gas. Papers on MHD convection with a non-Newtonian fluid are those by Sharma and Kumar (1996), and Sharma and Thakur (2000), Sharma and Sharma (2000), Sharma and

Kishor (2001), Sharma *et al.* (2001), and Sunil *et al.* (2001). Papers involving a rotating non-Newtonian fluid are those by Sharma *et al.* (1998, 1999a) and Sharma and Rana (2001, 2002). Non-Newtonian fluids permeated with suspended particles have been studied by Sharma *et al.* (1999b), Sunil *et al.* (2003b, 2004a,b), and Sharma and Sharma (2004). A ferromagnetic fluid was treated by Sunil *et al.* (2004b, 2005a,c,d).

The effect of vertical through flow was studied by Shivakumara and Khalili (2001) and that of horizontal through flow by Joulin and Ouarzazi (2000).

Subramanian and Patil (1991) combined anisotropy with cross-diffusion. The critical conditions for the onset of convection in a doubly diffusive porous layer with internal heat generation were documented by Selimos and Poulikakos (1985). The effect of heat generation or absorption was also studied by Chamkha (2002). Lin (1992) studied numerically a transient problem.

The effect of temporally fluctuating temperature on instability was analyzed by Ouarzazi and Bois (1994), Quarzazi *et al.* (1994), McKay (1998b, 2000), Ramazanov (2001), and Malashetty and Basvaraja (2004). The last study included the effect of anisotropy. The studies by McKay make use of Floquet theory. He demonstrated that the resulting instability may be synchronous, subharmonic, or at a frequency unrelated to the heating frequency.

The effect of vertical vibration was studied analytically and numerically by Sovran *et al.* (2000, 2002) and Jounet and Bardan (2001). Depending on the governing parameters, vibrations are found to delay or advance the onset of convection, and the resulting convection can be stationary or oscillatory. An intensification of the heat and mass transfers is observed at low frequency for sufficiently high vibration frequency. The onset of Soret-driven convection with a vertical variation of gravity was analyzed by Alex and Patil (2001) and Charrier-Mojtabi *et al.* (2004). The latter considered also horizontal vibration and reported that for both monotonic and oscillatory convection the vertical vibration has a stabilizing effect while the horizontal vibration has a destabilizing effect on the onset of convection.

The problem of convection in groundwater below an evaporating salt lake was studied in detail by Wooding *et al.* (1997a,b) and Wooding (2005). Now the convection is driven by the evaporative concentration of salts at the land surface, leading to an unstable distribution of density, but the evaporative groundwater discharge dynamically can stabilize this saline boundary layer. The authors investigated the nature, onset, and development (as fingers or plumes) of the convection. They reported the result of linear stability analysis, numerical simulation, and laboratory experimentation using a Hele-Shaw cell. The results indicate that in typical environments, convection will predominate in sediments whose permeability exceeds about 10^{-14} m^2, while below this threshold the boundary layer should be stabilized, resulting in the accumulation of salts at the land surface. A numerical model simulating this situation was presented by Simmons *et* al. (1999). A related problem involving the evaporation of groundwater was studied analytically and numerically by Gilman and Bear (1996). The groundwater flow pattern in the vicinity of a salt lake also has been studied numerically by Holzbecher (2005b).

9.2. Horizontal Heat and Mass Transfer

9.2.1. Boundary Layer Flow and External Natural Convection

The most basic geometry for simultaneous heat and mass transfer from the side is the vertical wall embedded in a saturated porous medium. Specified at the wall are the uniform temperature T_0 and the uniform concentration C_0. The temperature and concentration sufficiently far from the wall are T_∞ and C_∞.

The Darcy flow driven by buoyancy in the vicinity of the vertical surface can have one of the four two-layer structures shown in Fig. 9.5. The thicknesses δ, δ_T, and δ_C indicate the velocity, thermal, and concentration boundary layers. The relative size of these three thicknesses is determined by the combination (N, Le).

The heat and mass transfer from the vertical surface was determined first based on scale analysis (Bejan, 1984, pp. 335–338) and later based on the boundary layer similarity method (Bejan and Khair, 1985). The results of the scale analysis are summarized in Table 9.1. Each row in this table corresponds to one of the quadrants of the (N, Le) domain covered by Fig. 9.5. The v scale represents the largest vertical velocity, which in Darcy flow occurs right at the wall. By writing this time $\bar{q}''$ and $\bar{j}$ for the heat and mass fluxes averaged over the wall height H,

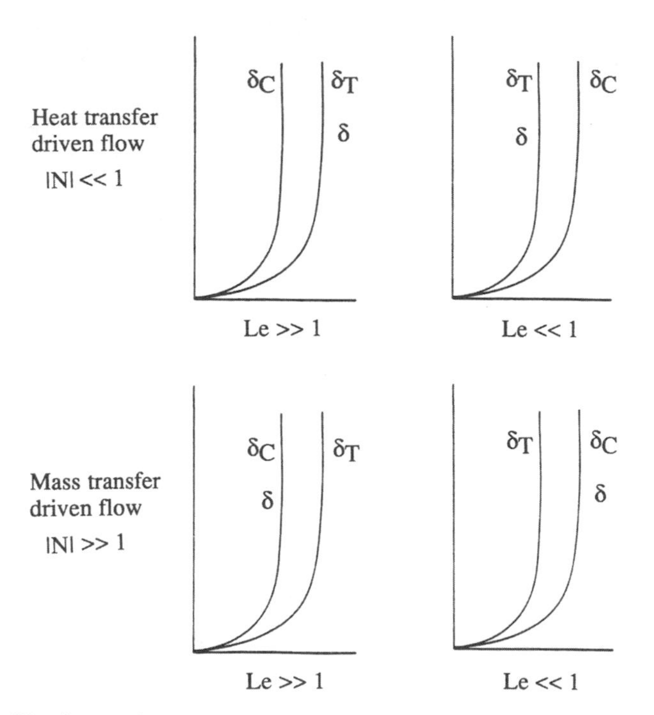

Figure 9.5. The four regimes of boundary layer heat and mass transfer near a vertical surface embedded in a porous medium (Bejan and Khair, 1985).

Table 9.1. The flow, heat, and mass transfer scales for the boundary layer near a vertical wall embedded in a porous medium (Bejan, 1984; Bejan and Khair, 1985).

Driving mechanism	v	Nu	Sh	Le domain
Heat transfer	$(\alpha_m/H)\,\mathrm{Ra}$	$\mathrm{Ra}^{1/2}$	$(\mathrm{Ra\,Le})^{1/2}$	$\mathrm{Le} \gg 1$
$(\vert N\vert \ll 1)$	$(\alpha_m/H)\,\mathrm{Ra}$	$\mathrm{Ra}^{1/2}$	$\mathrm{Ra}^{1/2}\,\mathrm{Le}$	$\mathrm{Le} \ll 1$
Mass transfer	$(\alpha_m/H)\,\mathrm{Ra}\vert N\vert$	$(\mathrm{Ra}\vert N\vert)^{1/2}$	$(\mathrm{Ra}\vert N\vert\,\mathrm{Le})^{1/2}$	$\mathrm{Le} \ll 1$
$(\vert N\vert \gg 1)$	$(\alpha_m/H)\,\mathrm{Ra}\vert N\vert$	$\mathrm{Le}^{-1/2}(\mathrm{Ra}\vert N\vert)^{1/2}$	$(\mathrm{Ra}\vert N\vert\,\mathrm{Le})^{1/2}$	$\mathrm{Le} \gg 1$

the overall Nusselt and Sherwood numbers are defined as

$$\mathrm{Nu} = \frac{\bar{q}''}{k_m\left(T_0 - T_\infty\right)/H}, \qquad \mathrm{Sh} = \frac{\bar{j}}{D_m\left(C_0 - C_\infty\right)/H}. \tag{9.44}$$

The similarity solution to the same problem was obtained by Bejan and Khair (1985) by selecting the nondimensional similarity profiles recommended by the scale analysis (Table 9.1):

$$u = -\frac{\alpha_m}{x}\mathrm{Ra}_x f'(\eta), \tag{9.45}$$

$$v = -\frac{\alpha_m}{2x}\mathrm{Ra}_x{}^{1/2}\left(f - \eta f'\right), \tag{9.46}$$

$$\theta(\eta) = \frac{T - T_\infty}{T_0 - T_\infty}, \qquad \eta = \frac{y}{x}\mathrm{Ra}_y{}^{1/2}, \tag{9.47}$$

$$c(\eta) = \frac{C - C_\infty}{C_0 - C_\infty}. \tag{9.48}$$

In this formulation, x is the distance measured along the wall and the Rayleigh number is defined by $\mathrm{Ra}_x = g\beta\, K x(T_0 - T_\propto)/\nu\alpha_m$. The equations for momentum, energy, and chemical species conservation reduce to

$$f'' = -\theta' - Nc', \tag{9.49}$$

$$\theta'' = \frac{1}{2}f\theta', \tag{9.50}$$

$$c'' = \frac{1}{2}fc'\mathrm{Le}, \tag{9.51}$$

with the boundary conditions $f = 0$, $\theta = 1$, and $c = 1$ at $\eta = 0$, and $(f, \theta, c) \to 0$ as $\eta \to \infty$. Equations (9.49)–(9.51) reinforce the conclusion that the boundary layer phenomenon depends on two parameters, N and Le.

Figure 9.6 shows a sample of vertical velocity and temperature (or concentration) profiles for the case $\mathrm{Le} = 1$. The vertical velocity increases and the thermal

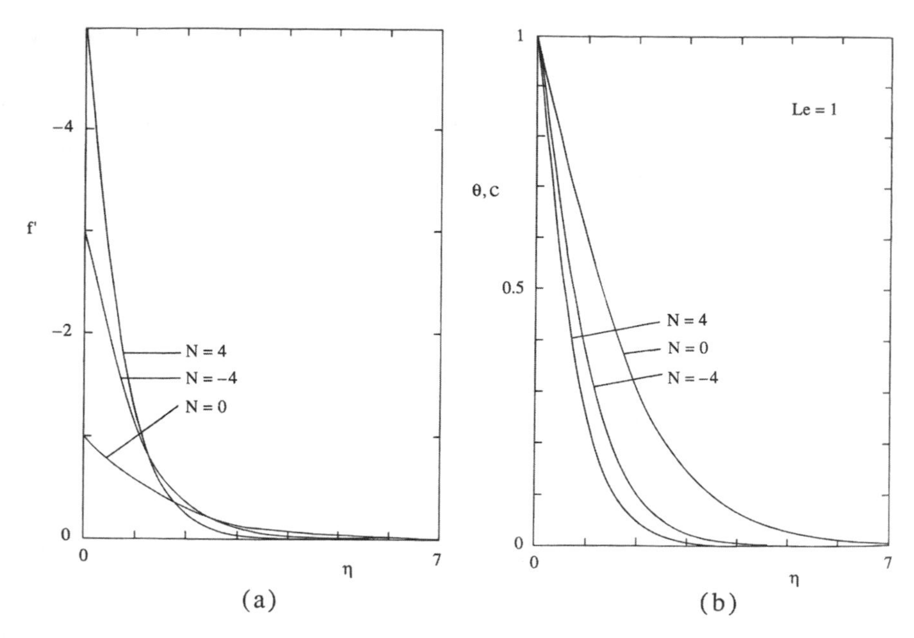

Figure 9.6. The buoyancy ratio effect on the Le = 1 similarity profiles for boundary layer heat and mass transfer near a vertical wall embedded in a porous medium. a) Velocity profiles and b) temperature and concentration profiles (Bejan and Khair, 1985).

boundary layer becomes thinner as $|N|$ increases. The same similarity solutions show that the concentration boundary layer in heat transfer-driven flows ($N = 0$) becomes thinner as Le increases, in good agreement with the trend anticipated by scale analysis.

The effect of wall inclination on the two-layer structure was described by Jang and Chang (1988b,c). Their study is a generalization of the similarity solution approach employed by Bejan and Khair (1985). The heat and mass transfer scales that prevail in the extreme case when the embedded H-long surface is horizontal are summarized in Table 9.2. A related study was reported by Jang and Ni (1989), who considered the transient development of velocity, temperature, and concentration boundary layers near a vertical surface.

The effect of flow injection on the heat and mass transfer from a vertical plate was investigated by Lai and Kulacki (1991d): see also the comments by Bejan (1992a). Raptis *et al.* (1981) showed that an analytical solution is possible in the case of an infinite vertical wall with uniform suction at the wall-porous-medium interface. The resulting analytical solution describes flow, temperature, and concentration fields that are independent of altitude (y). This approach was extended to the unsteady boundary layer flow problem by Raptis and Tzivanidis (1984). For the case of a non-Newtonian (power-law fluid), an analytical and numerical treatment was given by Rastogi and Poulikakos (1995). The case of a thermally stratified medium was studied numerically by Angirasa *et al.* (1997).

Table 9.2. The flow, heat, and mass transfer scales for the boundary layer near a horizontal wall embedded in a saturated porous medium (Jang and Chang, 1988b).

Driving mechanism	u	Nu	Sh	Le domain
Heat transfer	$(\alpha_m/H)\,\mathrm{Ra}^{2/3}$	$\mathrm{Ra}^{1/3}$	$\mathrm{Ra}^{1/3}\,\mathrm{Le}^{1/2}$	$\mathrm{Le} \gg 1$
$(\lvert N\rvert \ll 1)$	$(\alpha_m/H)\,\mathrm{Ra}^{2/3}$	$\mathrm{Ra}^{1/3}$	$\mathrm{Ra}^{1/3}\,\mathrm{Le}$	$\mathrm{Le} \ll 1$
Mass transfer	$(\alpha_m/H) \times (\mathrm{Ra}\lvert N\rvert)^{2/3}\,\mathrm{Le}^{-1/3}$	$(\mathrm{Ra}\,\lvert N\rvert)^{-1/3}\,\mathrm{Le}^{-1/6}$	$(\mathrm{Ra}\lvert N\rvert\,\mathrm{Le})^{1/3}$	$\mathrm{Le} \ll 1$
$(\lvert N\rvert \gg 1)$	$(\alpha_m/H) \times (\mathrm{Ra}\lvert N\rvert)^{2/3}\,\mathrm{Le}^{-1/3}$	$(\mathrm{Ra}\,\lvert N\rvert)^{-1/3}\,\mathrm{Le}^{-2/3}$	$(\mathrm{Ra}\lvert N\rvert\,\mathrm{Le})^{1/3}$	$\mathrm{Le} \gg 1$

The physical model treated by Bejan and Khair (1985) was extended to the case of a boundary of arbitrary shape by Nakayama and Hossain (1995). A further scale analysis of natural convection boundary layers driven by thermal and mass diffusion was made by Allain *et al.* (1992), who also made some corroborating numerical investigations. They noted the existence of flows that are heat driven even though the amplitude of the solutal convection is dominant.

An analytical-numerical study of hydrodynamic dispersion in natural convection heat and mass transfer near vertical surfaces was reported by Telles and Trevisan (1993). They considered flows due to a combination of temperature and concentration gradients and found that four classes of flows are possible according to the relative magnitude of the dispersion coefficients.

For convection over a vertical plate, the Forchheimer effect was analyzed by Murthy and Singh (1999); dispersion effects were studied by Khaled and Chamkha (2001), Chamkha and Quadri (2003), and El-Amin (2004a); and the effect of double stratification was discussed by Bansod *et al.* (2002) and Murthy *et al.* (2004b). Using homotopy analysis and the Forchheimer model, an analytic solution was obtained by Wang *et al.* (2003a). The effect of thermophoresis particle deposition was analyzed by Chamkha and Pop (2004). The case of power-law non-Newtonian fluids was treated numerically by Jumah and Majumdar (2000, 2001).

MHD convection was treated for a vertical plate by Cheng (1999, 2005), Chamkha and Khaled (2000c,d), Acharya *et al.* (2000), and Postelnicu (2004) with Soret and Dufour effects; for a cone or wedge by Chamkha *et al.* (2000); with heat generation or absorption effects for a cylinder or a cone by Chamkha and Quadri (2001, 2002); and for unsteady convection past a vertical plate by Kamel (2001) and Takhar *et al.* (2003). MHD convection of a micropolar fluid over a vertical moving plate was studied by Kim (2004). MHD convection for the case where the permeability oscillates with time about a nonzero mean was analyzed by Hassanien and Allah (2002).

Convection over a wavy vertical plate or cone was studied by Cheng (2000c,d) and Ratish Kumar and Shalini (2004b). Convection from a wavy wall in a thermally stratified enclosure with mass and thermal stratification was treated numerically by Ratish Kumar and Shalini (2005). A cone, truncated or otherwise, with variable wall temperature and concentration was analyzed by Yih (1999a,d) and Cheng (2000a). Convection above a near-horizontal surface and convection along a vertical permeable cylinder were analyzed by Hossain *et al.* (1999a,b). A horizontal permeable cylinder was considered by Yih (1999f). Li and Lai (1998), Bansod (2003), and Bansod *et al.* (2005) reexamined convection from horizontal plates. Also for a horizontal plate, Wang *et al.* (2003b) obtained an analytical solution for Forchheimer convection with surface mass flux and thermal dispersion effects. Abel *et al.* (2001) studied convection with a viscoelastic fluid flowing over a stretching sheet.

9.2.2. Enclosed Porous Medium

As the simplest configuration of simultaneous heat and mass transfer in an enclosed porous medium consider the two-dimensional system defined in Fig. 9.7. The uniform temperature and concentration are maintained at different levels along the two side walls. The main engineering challenge is the calculation of the overall heat and mass transfer rates expressed by Eq. (9.44).

Relative to the single-wall problem (Fig. 9.5) the present phenomenon depends on the geometric aspect ratio L/H as an additional dimensionless group next to N and Le. These groups account for the many distinct heat and mass transfer regimes that can exist. Trevisan and Bejan (1985) identified these regimes on the basis of scale analysis and numerical experiments. Figure 9.8 shows that in the case of heat transfer-driven flows ($|N| \ll 1$) there are five distinct regimes, which are labeled I–V. The proper Nu and Sh scales are listed directly on the [Le, $(L/H)^2$Ra] subdomain occupied by each regime.

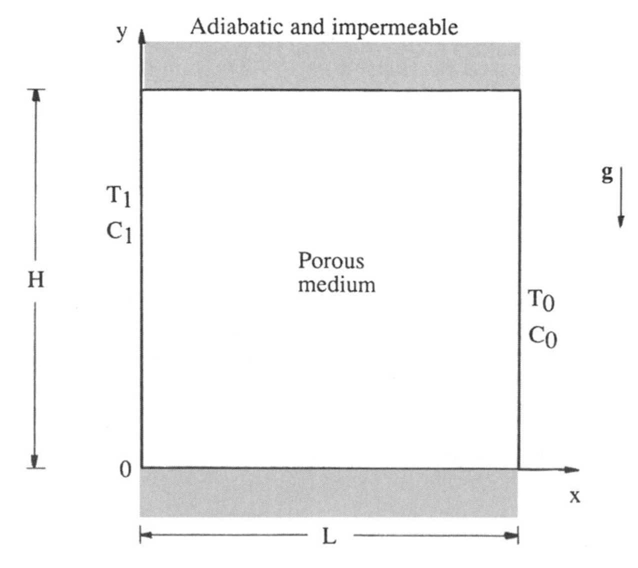

Figure 9.7. Enclosed porous medium subjected to heat and mass transfer in the horizontal direction.

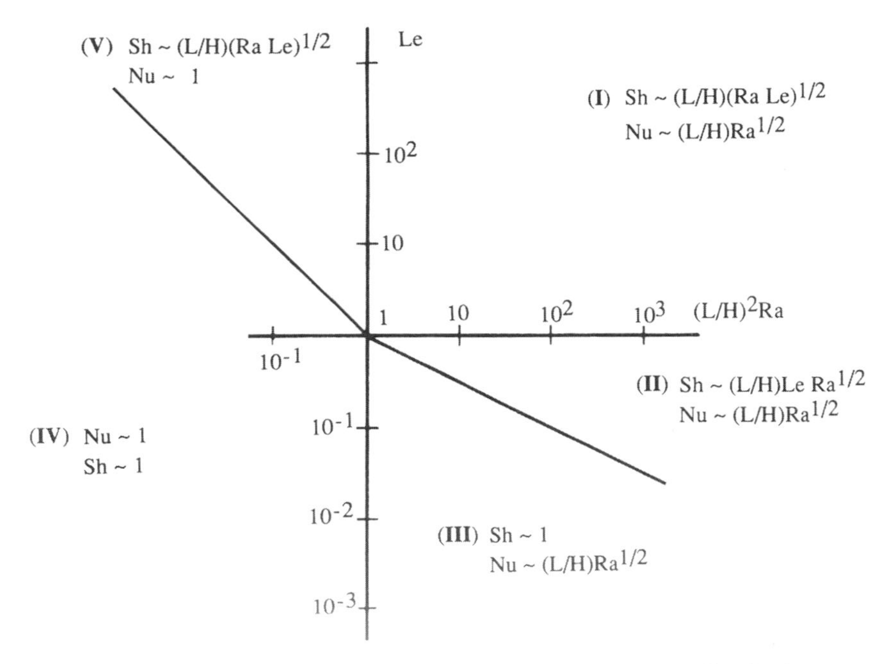

Figure 9.8. The heat and mass transfer regimes when the buoyancy effect in the system of Fig. 9.7 is due mainly to temperature gradients, $|N| \ll 1$ (Trevisan and Bejan, 1985).

Five distinct regimes also are possible in the limit of mass transfer driven flows, $|N| \gg 1$. Figure 9.9 shows the corresponding Nusselt and Sherwood number scales and the position of each regime in the plane [Le, $(L/H)^2$ Ra$|N|$]. Had we used the plane [Le^{-1}, $(L/H)^2$ Ra$|N|$ Le] then the symmetry with Fig. 9.8 would have been apparent. The Nu and Sh scales reported in Figs. 9.8 and 9.9 are correct within a numerical factor of order 1. Considerably more accurate results have been developed numerically and reported in Trevisan and Bejan (1985).

The most striking effect of varying the buoyancy ratio N between the extremes represented by Figs. 9.8 and 9.9 is the suppression of convection in the vicinity of $N = -1$. In this special limit, the temperature and concentration buoyancy effects are comparable in size but have opposite signs. Indeed, the flow disappears completely if Le $= 1$ and $N = -1$. This dramatic effect is illustrated in Fig. 9.10, which shows how the overall mass transfer rate approaches the pure diffusion level (Sh $= 1$) as N passes through the value -1.

When the Lewis number is smaller or greater than 1, the passing of N through the value -1 is not accompanied by the total disappearance of the flow. This aspect is illustrated by the sequence of streamlines, isotherms, and concentration lines displayed in Fig. 9.11. The figure shows that when N is algebraically greater than approximately -0.85, the natural convection pattern resembles the one that would be expected in a porous layer in which the opposing buoyancy effect is not the dominant driving force. The circulation is reversed at N values lower

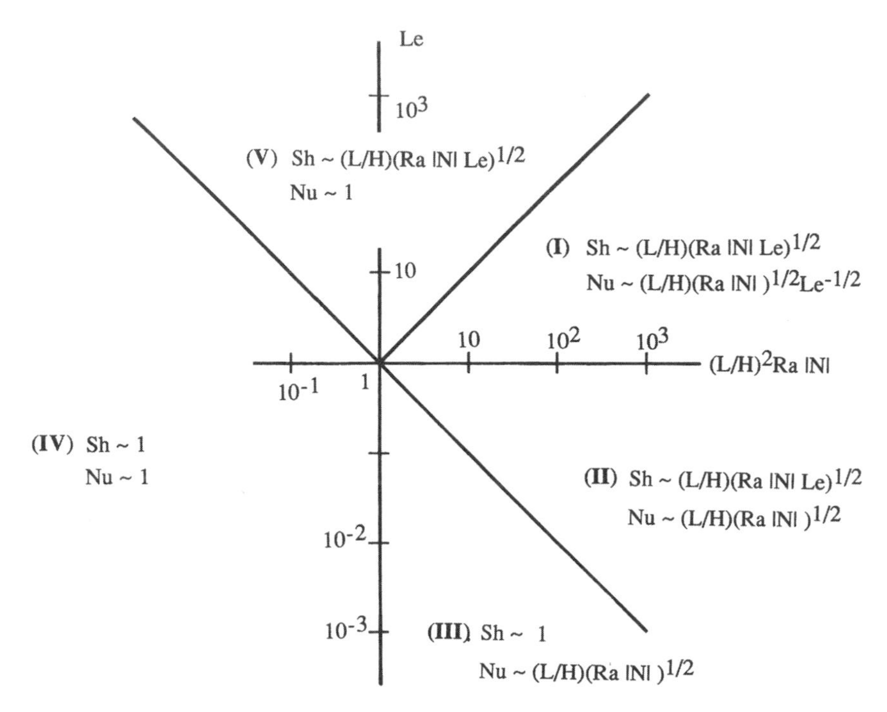

Figure 9.9. The heat and mass transfer regimes when the buoyancy effect in the system of Fig. 9.7 is due mainly to concentration gradients, $|N| \gg 1$ (Trevisan and Bejan, 1985).

than approximately -1.5. The flow reversal takes place rather abruptly around $N = -0.9$, as is shown in Fig. 9.11b. The core, which exhibited temperature and concentration stratification at N values sufficiently above and below -0.9, is now dominated by nearly vertical constant T and C lines. This feature is consistent with the tendency of both Nu and Sh to approach their pure diffusion limits (e.g., Fig. 9.10).

A compact analytical solution that documents the effect of N on both Nu and Sh was developed in a subsequent paper by Trevisan and Bejan (1986). This solution is valid strictly for Le = 1 and is based on the constant-flux model according to which both sidewalls are covered with uniform distributions of heat flux and mass flux. The overall Nusselt number and Sherwood number expressions for the high Rayleigh number regime (distinct boundary layers) are

$$\mathrm{Nu} = \mathrm{Sh} = \frac{1}{2}\left(\frac{H}{L}\right)^{1/5} \mathrm{Ra}_*{}^{2/5}(1+N)^{2/5}, \tag{9.52}$$

where Ra_* is the heat-flux Rayleigh number defined by $\mathrm{Ra}_* = g\beta K H^2 q''/\nu\alpha_m k_m$. These theoretical Nu and Sh results agree well with numerical simulations of the heat and mass transfer phenomenon.

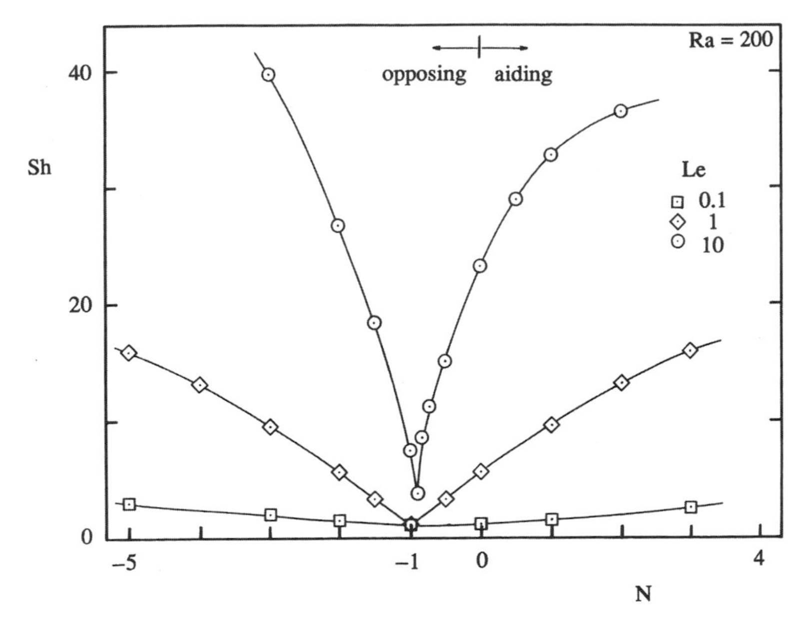

Figure 9.10. The effect of the buoyancy ratio on the overall mass transfer rate through the enclosed porous medium shown in Fig. 9.7 (Ra = 200, $H/L = 1$) (Trevisan and Bejan, 1985).

Another theoretical result has been developed by Trevisan and Bejan (1986) for the large Lewis numbers limit in heat transfer driven flows ($|N| \ll 1$). In this limit the concentration boundary layer can be described by means of a similarity solution, leading to the following expression for the overall Sherwood number:

$$\mathrm{Sh} = 0.665 \left(\frac{L}{H}\right)^{1/10} \mathrm{Le}^{1/2} \mathrm{Ra}_*^{3/10}. \tag{9.53}$$

The mass flux j used in the Sh definition, $\mathrm{Sh} = jH/D_m \Delta C$, is constant, while ΔC is the resulting concentration-temperature difference between the two sidewalls. Equation (9.53) is also in good agreement with numerical experiments.

It has been shown that the constant-flux expressions (9.50) and (9.53) can be recast in terms of dimensionless groups (Ra, Nu, Sh) that are based on temperature and concentration differences. This was done in order to obtain approximate theoretical results for the configuration of Fig. 9.7, in which the sidewalls have constant temperature and concentrations (Trevisan and Bejan, 1986). Similarly, appropriately transformed versions of these expressions can be used to anticipate the Nu and Sh values in enclosures with mixed boundary conditions, that is, constant T and j, or constant q'' and C on the same wall. Numerical simulations of the convective heat and mass transfer across enclosures with mixed boundary conditions are reported by Trevisan and Bejan (1986).

Figure 9.11. Streamlines, isotherms, and isosolutal lines for natural convection in the enclosed porous medium of Fig. 9.7, showing the flow reversal that occurs near $N = -1$ ($\mathrm{Ra} = 200$, $\mathrm{Le} = 10$, $H/L = 1$). (a) $N = -0.85$; (b) $N = -0.9$; and (c) $N = -1.5$ (Trevisan and Bejan, 1985).

An analytical and numerical study of convection in vertical slots due to prescribed heat flux at the vertical boundaries was made by Alavyoon (1993), whose numerical results showed that of any value of $\mathrm{Le} > 1$ there exists a minimum aspect ratio A below which the concentration field in the core region is rather uniform and above which it is linearly stratified in the vertical direction. For $\mathrm{Le} > 1$ the thermal layers at the top and bottom of the enclosure are thinner than their solutal counterparts. In the boundary layer regime and for sufficiently large A the thicknesses of the vertical boundary layers of velocity, concentration, and temperature were found to be equal. The case of opposing fluxes was studied by Alavyoon *et al.* (1994). They found that at sufficiently large values of Ra, Le, and A there is a domain of N in which one obtains oscillating convection, while outside this domain the solution approaches steady-state convection.

Numerical simulations based on an extension to the Brinkman model for the case of cooperating thermal and solutal buoyancy forces in the domain of positive N and for $\text{Le} > 1$ were reported by Goyeau *et al.* (1996a). The Brinkman model was also employed by Mamou *et al.* (1998a).

The studies reviewed in this subsection are based on the homogeneous and isotropic porous-medium model. The effect of medium heterogeneity on the heat and mass transfer across an enclosure with constant-flux boundary conditions is documented by Mehta and Nandakumar (1987). They show numerically that the Nu and Sh values can differ from the values anticipated based on the homogeneous porous-medium model.

For the case $N = -1$, a purely diffusive solution exists for suitable geometry and boundary conditions. Charrier-Mojtabi *et al.* (1997, 1998) have studied this case for a rectangular slot with constant temperature imposed on the side walls. The onset of convection, for which $\gamma = \text{Le}\,\theta$ occurs when $\text{Ra}\,|\text{Le} - 1|$ exceeds a certain critical value, depending on the aspect ratio A. The critical value is 184.06 for a square cavity ($A = 1$) and 105.33 for a vertical layer of infinite extent; the corresponding critical wavenumber has the value 2.51. For $A = 1$, they also performed numerical simulations, the results of which confirmed the linear instability results. They observed that the bifurcation to convection was of the transcritical type and that the bifurcation diagrams indicated the existence of both symmetrical and asymmetrical subcritical and supercritical solutions.

A numerical study for a square cavity, comparing the Darcy, Forchheimer, and Brinkman models, was made by Karimi-Fard *et al.* (1997). They found that Nu and Sh increase with Da and decrease with increase of a Forchheimer parameter. The quadratic drag effects are almost negligible, but the boundary effect is important. A further numerical study, for the case of opposing buoyancy effects, was reported by Angirasa and Peterson (1997a). Effects of porosity variation were emphasized in the numerical study by Nithiarasu *et al.* (1996). Three-dimensional convection in a cubic or rectangular enclosure with opposing horizontal gradients of temperature and concentration was studied numerically by Sezai and Mohamad (1999) and Mohamad and Sezai (2002). A numerical treatment with a random porosity model was reported by Fu and Ke (2000).

The various studies for the case $N = -1$ have demonstrated that there exists a threshold for the onset of monotonic convection, such that oscillatory convection occurs in a narrow range of values of Le (close to 1, applicable for many gases) depending on the normalized porosity. For the case of an infinite layer, the wavelength at the onset of stationary convection is independent of the Lewis number but this is not so for overstability. When the Lewis number is close to unity the system remains conditionally stable provided that the normalized porosity is less than unity. For a vertical enclosure with constant heat and solute fluxes, the particular case $N = -1 + \varepsilon$ case (where ε is a very small positive number) was studied by Amahmid *et al.* (2000). In this situation multiple unicellurar convective flows are predicted.

A non-Newtonian fluid was studied theoretically and numerically by Getachew *et al.* (1998) and by Benhadji and Vasseur (2003). An electrochemical experimental method was demonstrated by Chen *et al.* (1999). An inverse method, leading to the determination of an unknown solute concentration on one wall given known conditions for temperature and concentration on the remaining faces, was reported by Prud'homme and Jiang (2003). A numerical study of the effect of thermal stratification on convection in a square enclosure was made by Ratish Kumar *et al.* (2002).

Analytical and numerical studies of convection in a vertical layer were reported by Amahmid *et al.* (1999b,c, 2000, 2001), Bennacer *et al.* (2001b), Mamou *et al.* (1998a) and Mamou (2002a). The effect of evaporation was added by Asbik *et al.* (2002) in their numerical treatment. Convection in a square cavity, or a horizontal layer with the Soret effect included, under crossed heat and mass fluxes was studied analytically and numerically by Bennacer *et al.* (2001a, 2003b). Convection in a vertically layered system, with a porous layer between two clear layer, was studied by Mharzi *et al.* (2000). Anisotropic cavities were studied analytically and numerically by Tobbal and Bennacer (1998). Bera *et al.* (1998, 2000), and Bera and Khalili (2002a). Three algebraic analytical solutions were presented by Cai *et al.* (2003).

9.2.3. Transient Effects

Another basic configuration in which the net heat and mass transfer occurs in the horizontal direction is the time-dependent process that evolves from a state in which two (side-by-side) regions of a porous medium have different temperatures and species concentrations. In time, the two regions share a counterflow that brings both regions to a state of thermal and chemical equilibrium. The key question is how parameters such as N, Le, and the height-length ratio of the two-region ensemble affect the time scale of the approach to equilibrium. These effects have been documented both numerically and on the basis of scale analysis by Zhang and Bejan (1987).

As an example of how two dissimilar adjacent regions come to equilibrium by convection, Fig. 9.12 shows the evolution of the flow, temperature, and concentration fields of a relatively high Rayleigh number flow driven by thermal buoyancy effects ($N = 0$). As the time increases, the warm fluid (initially on the left-hand side) migrates into the upper half of the system. The thermal barrier between the two thermal regions is smoothed gradually by thermal diffusion. Figures 9.12c and 9.12d show that as the Lewis number decreases the sharpness of the concentration dividing line disappears, as the phenomenon of mass diffusion becomes more pronounced.

Figure 9.12. The horizontal spreading and layering of thermal and chemical deposits in a porous medium ($N = 0$, Ra $=$ 1000, $H/L = 1$, $\varphi/\sigma = 1$). (a) Streamlines; (b) isotherms, or isosolutal lines for Le $= 1$; (c) isosolutal lines for Le $= 0.1$; and (d) isosolutal lines for Le $= 0.01$ (Zhang and Bejan, 1987).

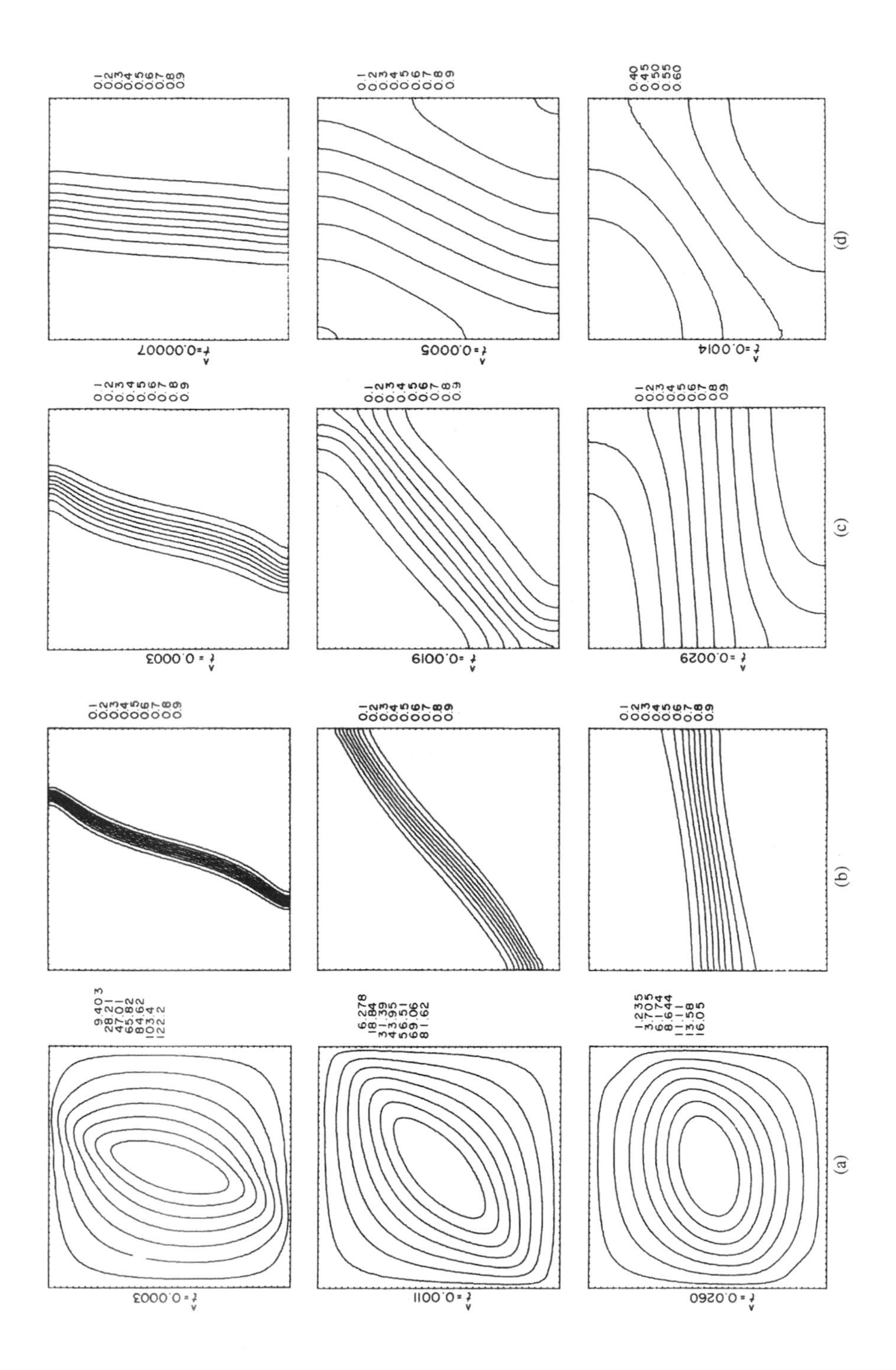

t = 0.0003
9.403
28.21
47.01
65.82
84.62
103.4
122.2
t = 0.0011
6.278
18.84
31.39
43.95
56.51
69.06
81.62
t = 0.0260
1.235
3.705
6.174
8.644
11.11
13.58
16.05
(a)
0.1
0.2
0.3
0.4
0.5
0.6
0.7
0.8
0.9
(b)
t = 0.0003
t = 0.0019
t = 0.0029
(c)
t = 0.00007
t = 0.0005
t = 0.0014
0.40
0.45
0.50
0.55
0.60
(d)

In the case of heat transfer driven flows, the time scale associated with the end of convective mass transfer in the horizontal direction is

$$\hat{t} = \frac{\varphi}{\sigma}\left(\frac{L}{H}\right)^2 \mathrm{Ra}^{-1} \quad \text{if} \quad \mathrm{LeRa} > \frac{\varphi}{\sigma}\left(\frac{L}{H}\right)^2, \tag{9.54}$$

$$\hat{t} = \frac{\varphi}{\sigma}\left(\frac{L}{H}\right)^2 \mathrm{Le} \quad \text{if} \quad \mathrm{LeRa} < \frac{\varphi}{\sigma}\left(\frac{L}{H}\right)^2. \tag{9.55}$$

The dimensionless time $\hat{t}$ is defined as

$$\hat{t} = \frac{\alpha_m t}{\sigma H^2}. \tag{9.56}$$

Values of $\hat{t}$ are listed also on the side of each frame of Fig. 9.12. The time criteria (9.54)–(9.56) have been tested numerically along with the corresponding time scales for approach to thermal equilibrium in either heat transfer driven or mass transfer-driven flows.

The transient problem for the case of a vertical plate, with a simultaneous step change in wall temperature and wall concentration, was treated numerically using a Brinkman-Forchheimer model by Jang *et al.* (1991). They found that the time to reach steady state decreases with increase of Da or magnitude of the buoyancy ratio *N*, increases with increase of the inertia coefficient c_F, and passes through a minimum as Le increases through the value 1. Earlier Pop and Herwig (1990) had shown that when just the concentration was suddenly changed at an isothermal vertical plate, the local Sherwood number decreases with time and approaches its steady-state value. Cheng (2000b) analyzed a problem involving transient heat and mass transport from a vertical plate on which the temperature and concentration are power functions of the streamwise coordinate.

9.2.4. Stability of Flow

The stability of the steady Darcy flow driven by differential heating of the isothermal walls bounding an infinite vertical slab with a stabilizing uniform vertical salinity gradient was studied independently by Gershuni *et al.* (1976, 1980) and Khan and Zebib (1981). Their results show disagreement in some respects. We believe that Gershuni *et al.* are correct. The flow is stable if $|\mathrm{Ra}_D|$ is less than $\mathrm{Ra}_{D1} = 2.486$ and unstable if $|\mathrm{Ra}_D| > \mathrm{Ra}_{D1}$. The critical wavenumber α_c is zero for $\mathrm{Ra}_{D1} < |\mathrm{Ra}_D| < \mathrm{Ra}_{D2}$ where $\mathrm{Ra}_{D2} \approx 52$ for the case $N = 100$, $\sigma = 1$, and nonzero for $|\mathrm{Ra}_D| > \mathrm{Ra}_{D2}$. As $|\mathrm{Ra}_D| \to \infty$; either monotonic or oscillatory instability can occur depending on the values of *N* and σ. If, as in the case of aqueous solutions, *N* and N/σ are fairly large and of the same order of magnitude, then monotonic instability occurs and the critical values are

$$\mathrm{Ra}_c = \frac{2\pi^{1/2}}{|N-1|}|\mathrm{Ra}_D|^{3/4}, \quad \alpha_c = \left(\frac{\pi}{2}\right)^{1/2}|\mathrm{Ra}_D|^{1/4}. \tag{9.57}$$

Mamou *et al.* (1995a) have demonstrated numerically the existence of multiple steady states for convection in a rectangular enclosure with vertical walls. Mamou *et al.* (1995b) studied analytically and numerically convection in an inclined slot. Again multiple solutions were found.

Two-dimensional convection produced by an endothermic chemical reaction and a constant heat flux was examined by Basu and Islam (1996). They identified various routes to chaos. The onset of convection in a rectangular cavity with balanced heat and mass fluxes applied to the vertical walls was analyzed by Marcoux *et al.* (1999a). An analytical and numerical study of a similar situation was reported by Mamou *et al.* (1998d).

9.3. Concentrated Heat and Mass Sources

9.3.1. Point Source

Poulikakos (1985a) considered the transient flow as well as the steady flow near a point source of heat and mass in the limit of small Rayleigh numbers based on the heat source strength q[W], $\tilde{\text{Ra}} = g\beta K q/\nu\alpha_m k_m$. The relative importance of thermal and solutal buoyancy effects is described by the "source buoyancy ratio"

$$N_s = \frac{\beta_C m/D_m}{\beta q/k_m}, \tag{9.58}$$

in which m[kg/s] is the strength of the mass source.

Figure 9.13 shows Poulikakos' (1985a) pattern of streamlines for the time-dependent regime. The curves correspond to constant values of the special group

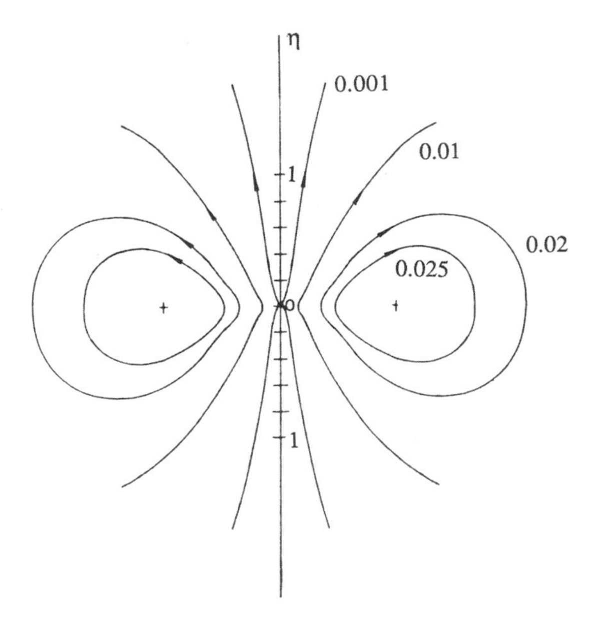

Figure 9.13. The time-dependent flow field around a suddenly placed point source of heat and mass ($A = 1$) (Poulikakos, 1985a, with permission from Pergamon Press).

$\psi_* t_*^{-1/2}(1 - N_s)$, in which

$$\psi_* = \frac{\psi}{\alpha_m} K^{-1/2}, \quad t_* = \frac{\alpha_m t}{\sigma K}, \tag{9.59}$$

and where $\psi[\mathrm{m}^3/\mathrm{s}]$ is the dimensional streamfunction. The radial coordinate η is defined by

$$\eta = \frac{r}{2}\left(\frac{\sigma}{\alpha_m t}\right)^{1/2}, \tag{9.60}$$

showing that the flow region expands as $t^{1/2}$. Figure 9.13 represents the special case $A = 1$, where A is shorthand for

$$A = \left(\frac{\varphi}{\sigma}\mathrm{Le}\right)^{1/2}. \tag{9.61}$$

Poulikakos (1985a) showed that the A parameter has a striking effect on the flow field in cases where the two buoyancy effects oppose one another ($N_s > 0$ in his terminology). Figure 9.14 illustrates this effect for the case $N = 0.5$ and $A = 0.1$; when A is smaller than 1, the ring flow that surrounds the point source (seen also

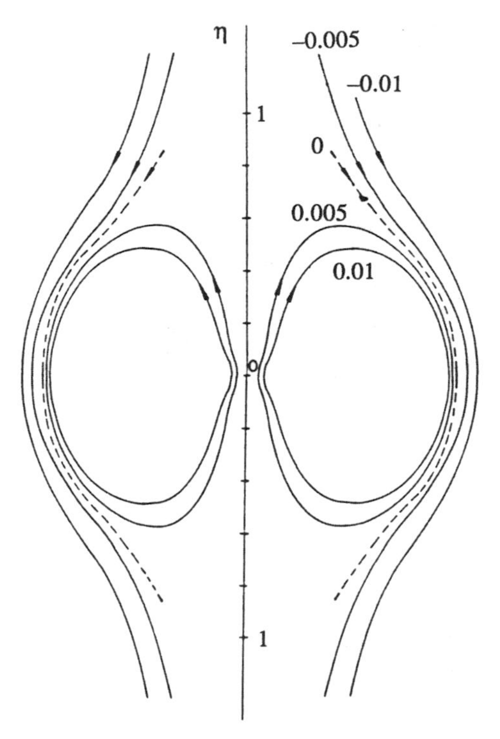

Figure 9.14. The effect of a small Lewis number (or small A) on the transient flow near a point source of heat and mass ($N = 0.5$, $A = 0.1$) (Poulikakos, 1985a, with permission from Pergamon Press).

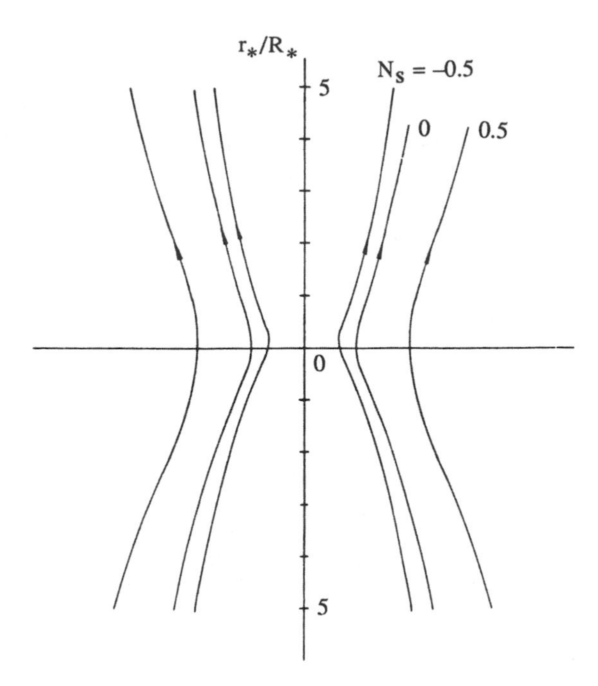

Figure 9.15. The steady-state flow near a point source of heat and mass ($\tilde{\text{Ra}} = 5$, Le = 1), and the effect of the source buoyancy ratio (Poulikakos, 1985a, with permission from Pergamon Press).

in Fig. 9.13) is engulfed by a far-field unidirectional flow. The lines drawn on Fig. 9.14 correspond to constant values of the group $2\pi\psi_* t_*^{-1/2}$.

In the steady state and in the same small-$\tilde{\text{Ra}}$ limit, the flow, temperature, and concentration fields depend only on $\tilde{\text{Ra}}$, N_s, and Le. Figure 9.15 shows the migration of one streamline as the buoyancy ratio N_s increases from -0.5 to 0.5, that is, as the buoyancy effects shift from a position of cooperation to one of competition. When the buoyancy effects oppose one another, $N = 0.5$, the vertical flow field is wider and slower. The curves drawn in Fig. 9.15 correspond to $\psi_* = \text{Ra}R_*/8\pi$, where $R_* = R/K^{1/2}$ and R is a reference radial distance. Asymptotic analytical solutions for the steady-state temperature and concentration fields also are reported by Poulikakos (1985a). Ganapathy (1994a) treated the same problem using the Brinkman model. For the case of large Rayleigh numbers, a boundary layer analysis was carried out by Nakayama and Ashizawa (1996). They showed that for large Le the solute diffuses some distance from the plume centerline and the mass transfer influences both velocity and temperature profiles over a wide range. For large Le the solute diffuses within a narrow region along the centerline. A strongly peaked velocity profile then appears for positive buoyancy ratio N, while a velocity defect emerges along the centerline for negative N.

A finite element model for a leaking third species migration from a heat source buried in a porous medium was demonstrated by Nithiarasu (1999). An inverse problem, the determination from temperature measurement of an unknown volumetric heat source that is a function of the solute concentration, was discussed by Prud'homme and Jasmin (2003) and Jasmin and Prud'homme (2005). Hill (2005) has considered the linear and nonlinear stability of a layer in which there is a concentration-dependent internal volumetric heat source.

9.3.2. Horizontal Line Source

The corresponding heat and mass transfer processes in the vicinity of a horizontal line source were analyzed by Larson and Poulikakos (1986). The source buoyancy ratio in this case is

$$N_s' = \frac{\beta_C m'/D_m}{\beta q'/k_m}, \tag{9.62}$$

where q' [W/m] and m' [kg/m/s] are the heat and mass source strengths. All the features described in the preceding sections also are present in the low Rayleigh number regime of the line source configuration. The Rayleigh number for the line source is based on the heat source strength q',

$$\hat{R}a = \frac{g\beta K^{3/2} q'}{\nu \alpha_m k_m}. \tag{9.63}$$

In addition to developing asymptotic solutions for the transient and steady states, Larson and Poulikakos (1986) illustrated the effect of a vertical insulated wall situated in the vicinity of the horizontal line source. An analysis using the Brinkman model was reported by Ganapathy (1994b).

The high Rayleigh number regime was studied by Lai (1990a). He obtained a similarity solution and made calculations for a range of Le and N values. For the special case Le = 1 he obtained a closed form solution analogous to that given by Eqs. (5.192)–(5.196). The study of Nakayama and Ashizawa (1996) mentioned in the previous section covered the case of a line source also.

9.4. Other Configurations

The double-diffusive case of natural convection over a sphere was analyzed by Lai and Kulacki (1990a), while Yücel (1990) has similarly treated the flow over a vertical cylinder and Lai *et al.* (1990b) the case of a slender body of revolution. Non-Darcy effects on flow over a two-dimensional or axisymmetric body were treated by Kumari *et al.* (1988a,b), and Kumari and Nath (1989c,d) have dealt with the case where the wall temperature and concentration vary with time. A numerical study of convection in an axisymmetric body was reported by Nithiarasu *et al.* (1997b). Flow over a horizontal cylinder, with the concentration gradient being produced by transpiration, was studied by Hassan and Mujumdar (1985). Natural convection in a horizontal shallow layer induced by a finite source of chemical constituent was given a numerical treatment by Trevisan and Bejan (1989). Convection in a vertical annulus was studied analytically and numerically by Marcoux *et al.* (1999b); numerically by Beji *et al.* (1999), who analyzed the effect of curvature on the value of N necessary to pass from clockwise to anticlockwise rolls; and Bennacer (2000). The effect of thermal diffusion for this case was studied numerically and analytically by Bennacer and Lakhal (2005). An analytical and numerical study of the separation of the components in a binary mixture in a vertical annulus with uniform heat fluxes at the walls was conducted by Bahloul *et al.* (2004b). Convection in a partly porous vertical annulus was studied numerically by Benzeghiba *et al.* (2003). A problem involving a vertical enclosure with two isotropic or anisotropic porous layers was studied numerically by Bennacer *et al.* (2003a), while convection in a partly filled rectangular enclosure was studied numerically by Goyeau and Gobin (1999), Singh *et al.* (1999), and Younsi *et al.* (2001).

The onset of convection in an inclined layer has been studied using linear stability analysis numerically by Karimi-Fard *et al.* (1998, 1999), who obtained parameter ranges for which the first primary bifurcation is a Hopf bifurcation (oscillatory convection). The same problem was studied numerically by Mamou *et al.* (1998c) and Mamou (2004) using a finite element method and by Chamkha and Al-Naser (2001) using a finite-difference method.

The composite fluid layer over a porous substrate was studied theoretically by Chen (1990), who extended to a range of Ra_m [the thermal Rayleigh number in the porous medium as defined in Eq. (6.167)], the calculations initiated by Chen and Chen (1988) for the salt-finger situation. For small Ra_m ($= 0.01$) there is a jump in α_c as the depth ratio $\hat{d} = d_f/d_m$ increases (the jump is fivefold as $\hat{d}$ increases between 0.2 and 0.3). For large $\mathrm{Ra}_m (= 1)$ there is no sudden jump. Convection occurs primarily in the fluid layer if $\hat{d}$ is sufficiently large. When this is so, multicellular convection occurs for sufficiently large Ra_m. The cells are superposed and their number increases with increase of Ra_m. For $\hat{d} < 0.1$, the critical Ra_{Dm} (the solutal Rayleigh number for the porous medium layer) and α_{cm} decrease as $\hat{d}$ increases, but when multicellular convection occurs the critical Ra_{Dm} remains almost constant as $\hat{d}$ is increased for fixed Ra_m. Zhao and Chen (2001) returned to the same problem but used a one-equation model rather than a two-equation model. They found that the two models predicted quantitative differences in the critical conditions and flow streamlines at the onset of convection, and they noted that carefully conducted experiments were needed to determine which model gave the more realistic results.

Goyeau *et al.* (1996b) studied numerically for $\mathrm{N} > 0$ the effect of a thin layer of low permeability medium, which suppresses the convective mass transfer. Further numerical studies were reported by Gobin *et al.* (1998, 2005).

Transient double-diffusive convection in a fluid/porous layer composite was studied by Kazmierczak and Poulikakos (1989, 1991) numerically and then experimentally. The system considered was one containing a linear stabilizing salt distribution initially and suddenly heated uniformly from below at constant flux.

In the experiments it was possible to visually observe the flow in the fluid layer but not in the porous layer. In all the experiments $\hat{d} = 1$, and most of the convective flow took place in the fluid layer. In general, a series of mixed layers formed in turn, starting with one just above the porous layer as time increased, as one would expect if the porous matrix was absent. A corresponding numerical study, with the system cooled through its top boundary (adjacent to the solid layer), was conducted by Rastogi and Poulikakos (1993). A numerical study involving two layers of contrasting permeabilities was conducted by Saghir and Islam (1999). A transient problem involving double-diffusive convection from a heated cylinder buried in a saturated porous medium was studied numerically by Chaves *et al.* (2005).

An experimental study with a clear liquid layer below a layer at porous medium was performed by Rastogi and Poulikakos (1997). They took the initial species concentration of the porous layer to be linear and stable and that in the clear fluid uniform and the system initially isothermal and then cooled from above.

Sandner (1986) performed experiments, using salt water and glass beads in a vertical cylindrical porous bed. In his experiments the salt concentration was initially uniform. When the system was heated at the bottom, a stabilizing salinity gradient developed, due to the Soret effect. Some related work is discussed in Section 10.5.

Natural convection in an anisotropic trapezoidal enclosure was studied numerically by Nguyen *et al.* (1997a). A forced convection flow around a porous medium layer placed downstream on a flat plate was studied numerically and experimentally by Lee and Howell (1991). Convection in a parallelogramic enclosure was studied numerically by Costa (2004). A transient problem, involving a smaller rectangular cavity containing initially cold fresh fluid located in the corner of a larger one containing hot salty fluid, was studied numerically by Saghir (1998).

9.5. Inclined and Crossed Gradients

The effects of horizontal gradients on thermosolutal stability, for the particular case where the horizontal thermal and solutal gradients compensate each other as far as density is concerned, was studied theoretically by Parvathy and Patil (1989) and Sarkar and Phillips (1992a,b). The more general case for arbitrary inclined thermal and solutal gradients was treated by Nield *et al.* (1993) and independently but in a less detailed manner by Parthiban and Patil (1994). Even when the gradients are coplanar the situation is complex. The effect of the horizontal gradients may be to either increase or decrease the critical vertical Rayleigh number, and the favored mode may be oscillatory or nonoscillatory and have various inclinations to the plane of the applied gradients according to the signs of the gradients. The horizontal gradients can cause instability even in the absence of any vertical gradients. The noncoplanar case was also treated by Nield *et al.* (1993). A nonlinear stability analysis was presented by Guo and Kaloni (1995a). Their main theorem was proved for the coplanar case. Kaloni and Qiao (2000) extended this analysis to the case of horizontal mass flow. A linear instability analysis for the extension where there is net horizontal mass flow was reported by Manole *et al.* (1994).

The case of horizontal temperature and vertical solutal gradients was investigated numerically by Mohamad and Bennacer (2001, 2002) and both analytically and numerically by Kalla *et al.* (2001b). Bennacer *et al.* (2004, 2005) analyzed convection in a two-layer medium with the lower one thermally anisotropic and submitted to a uniform horizontal heat flux and a vertical mass flux.

Mansour *et al.* (2004) studied numerically the Soret effect on multiple solutions in a square cavity with a vertical temperature gradient and a horizontal concentration gradient. Bourich *et al.* (2004a) showed that the multiplicity of solutions is eliminated if the buoyancy ratio N exceeds some critical value that depends on Le and Ra. A similar problem with a partly heated lower wall was treated by Bourich *et al.* (2004b). A vertical slot heated from below and with horizontal concentration gradients was studied analytically and numerically by Bahloul *et al.* (2004a).

9.6. Mixed Double-Diffusive Convection

Similarity solutions also can be obtained for the double-diffusive case of Darcy mixed convection from a vertical plate maintained at constant temperature and concentration (Lai, 1991a). The relative importance of buoyancy and forcing effects is critically dependent on the values of Le and N. Kumari and Nath (1992) studied convection over a slender vertical cylinder, with the effect of a magnetic field included. Another study of mixed convection was made by Yücel (1993). Darcy-Forchheimer convection over a vertical plate was studied by Jumar *et al.* (2001), and a similar problem with double dispersion was analyzed by Murthy (2000). For convection about a vertical cylinder, the entire mixed convection regime was covered by Yih (1998g). The effect of transpiration on mixed convection past a vertical permeable plate or vertical cylinder was treated numerically by Yih (1997, 1999h). For thermally assisted flow, suction increases the local surface heat and mass transfer rates. Mixed convection in an inclined layer has been analyzed by Rudraiah *et al.* (1987).

Mixed convection over a vertical plate, a wedge, or a cone with variable wall temperature and concentration was analyzed by Yih (1998c,f, 1999b,c, 2000b). Similar studies for MHD convection and a vertical plate were reported by Chamkha and Khaled (1999, 2000a,b) and Chamkha (2000). The effects of variable viscosity and thermal conductivity on mixed convection over a wedge, for the cases of uniform heat flux and uniform mass flux, were analyzed by Hassanien *et al.* (2003a). The influence of lateral mass flux on mixed convection over inclined surfaces was analyzed by Singh *et al.* (2002). Mixed convection over a vertical plate with viscosity variation was analyzed by Chamkha and Khanafer (1999). The case of a vertical plate with transverse spatially periodic suction that produces a three-dimensional flow was analyzed by Sharma (2005).

A numerical study of mixed convection with opposing flow in a rectangular cavity with horizontal temperature and concentration gradients was reported by Younsi *et al.* (2002a,b), who noted that for a certain combination of Ra, Le, and N values the flow has a multicellular structure. Mixed convection driven by a moving lid of a square enclosure was studied numerically by Khanafer and Vafai (2002) for the case of insulated vertical walls and horizontal at different constant temperature and concentration.

10
Convection with Change of Phase

In the examples of forced and natural convection discussed until now, the fluid that flowed through the pores did not experience a change of phase, no matter how intense the heating or cooling effect. In the present chapter we turn our attention to situations in which a change of phase occurs, for example, melting or evaporation upon heating and solidification or condensation upon cooling. These convection problems constitute a relatively new and active area in the field of convection in porous media.

10.1. Melting

10.1.1. Enclosure Heated from the Side

The first analysis of melting dominated by natural convection in a porous matrix saturated with a phase-change material and heated from the side was performed by Kazmierczak *et al.* (1986). Their study was based on a simple model in which (a) the liquid flow was assumed to be slow enough to conform to the Darcy regime, and (b) the melting front that separates the region saturated with solid from the region saturated with liquid was modeled as a surface (i.e., as a region of zero thickness and at the melting point).

These modeling assumptions also have been made in the simplest studies of the geometry illustrated in Fig. 10.1 (Jany and Bejan, 1988a), in which the porous medium is confined by an impermeable boundary and is heated through one of its side walls. On the problem considered by Kazmierczak *et al.* (1986) we will focus in Section 10.1.5, because that problem is in one way more general than the configuration addressed in this section.

Consider the two-dimensional system illustrated schematically in Fig. 10.1. Initially, the walls are all insulated and the cavity is filled with porous medium (*PM*) and phase change material (*PCM*) in the solid state, both at the fusion temperature T_f. For times $t = 0$ the left vertical wall is heated and maintained at constant temperature, T_w, so that $T_w > T_f$. In the domain occupied by liquid *PCM* the conservation of mass, momentum, and energy is governed by the equations

$$\frac{\partial u}{\partial x} + \frac{\partial v}{\partial y} = 0, \tag{10.1}$$

$$u = -\frac{K}{\mu}\frac{\partial P}{\partial x}, \tag{10.2}$$

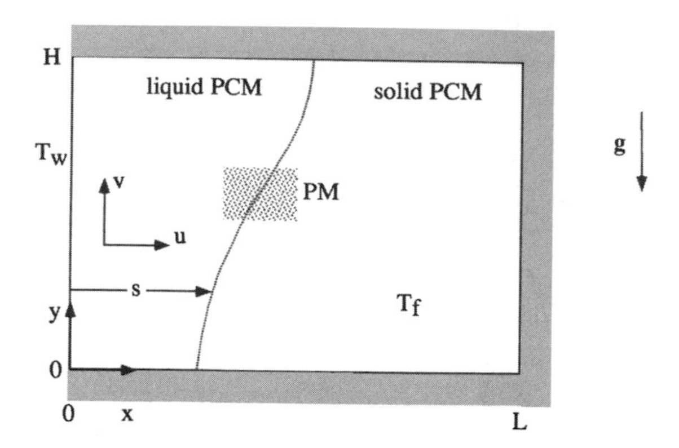

Figure 10.1. Melting in a two-dimensional porous medium heated from the side (after Jany and Bejan, 1988a).

$$v = -\frac{K}{\mu}\left(\frac{\partial P}{\partial y} + \rho g\left[1 - \beta(T - T_f)\right]\right), \tag{10.3}$$

$$\sigma\frac{\partial T}{\partial t} + u\frac{\partial T}{\partial x} + v\frac{\partial T}{\partial y} = \alpha_m\left(\frac{\partial^2 T}{\partial x^2} + \frac{\partial^2 T}{\partial y^2}\right). \tag{10.4}$$

Equations (10.1)–(10.2) are based on the following assumptions: (1) two-dimensional flow, (2) Darcy flow model [see also assumption (a) above], (3) local thermodynamic equilibrium between *PCM* and *PM*, (4) negligible viscous dissipation, (5) isotropic *PM*, and (6) constant thermophysical properties, with the exception of the assumed linear relation between density and temperature in the buoyancy term of Eq. (10.3) (the Oberbeck-Boussinesq approximation). The boundary conditions for Eqs. (10.1)–(10.4) are:

$$y = 0; \quad y = H: \quad v = 0, \quad \frac{\partial T}{\partial y} = 0, \tag{10.5}$$

$$x = 0: \quad u = 0, \quad T = T_w, \tag{10.6}$$

$$x = L: \quad u = 0, \quad \frac{\partial T}{\partial x} = 0, \tag{10.7}$$

$$x = s(<L): \quad u = 0, \quad T = T_f, \tag{10.8}$$

$$\frac{\partial s}{\partial t} = -\frac{\alpha_m c_P}{h_{sf}}\left(\frac{\partial T}{\partial x} - \frac{\partial s}{\partial y}\frac{\partial T}{\partial y}\right), \tag{10.9}$$

where h_{sf} is the latent heat of melting of the phase-change material. Equation (10.9) represents the energy balance at the interface between the liquid and solid saturated regions, while neglecting the difference between the densities of liquid and solid at the melting point.

The melting process was simulated numerically by Jany and Bejan (1988a), based on the streamfunction formulation $u = \partial\psi/\partial y$, $v = -\partial\psi/\partial x$ and in terms of the following dimensionless variables:

$$\Theta = \frac{T - T_f}{T_w - T_f}, \quad X = \frac{x}{H}, \quad Y = \frac{y}{H}, \tag{10.10}$$

$$S = \frac{s}{H}, \quad U = u\frac{H}{\alpha_m}, \quad V = v\frac{H}{\alpha_m}, \tag{10.11}$$

$$\Psi = \frac{\psi}{\alpha_m}, \quad \text{Fo} = \frac{\alpha_m t}{H^2}. \tag{10.12}$$

The transformed, dimensionless equations involve the Fourier number Fo, the aspect ratio L/H, and the Rayleigh and Stefan numbers defined below,

$$\text{Ra} = \frac{g\beta K H(T_w - T_f)}{\nu\alpha_m}, \quad \text{Ste} = \frac{c_P(T_w - T_f)}{h_{sf}} \tag{10.13}$$

They assumed that in the case of small Stefan numbers the interface moves relatively slowly, so that $\partial S/\partial \text{Fo} \ll U, V$. Therefore, it was reasonable to assume that the liquid flow is not disturbed by the interface motion. Said another way, the interface motion results from a fully developed state of natural convection in the liquid. This "quasistationary front" approximation implies a fixed melting domain $[S = S(Y)]$ during each time interval, hence a stepwise motion of the interface. Details of the finite-difference numerical procedure are presented in Jany and Bejan (1988a,b).

Figure 10.2 shows the evolution of the melting front in a square cavity. Because of the quasistationary front assumption, the Stefan and Fourier numbers appear always as a product, Ste Fo. The two-graph sequence of Fig. 10.2 illustrates the strong influence of natural convection on the melting velocity and on the melting front shape. The deviation from the pure heat conduction (vertical interfaces) increases with the dimensionless time (Ste Fo) and with the Rayleigh number.

The transition from a heat transfer regime dominated by conduction to one dominated by convection is illustrated in Fig. 10.3. Isotherms are plotted for a square domain for each of the Rayleigh numbers, 12.5 and 800. The existence of distinct boundary layers is evident in Fig. 10.3 (right), while the nearly equidistant isotherms of Fig. 10.3 (left) suggest a heat transfer mechanism dominated by conduction.

For the same values of Rayleigh number, Fig. 10.4 shows the transition of the flow field from the conduction-dominated regime to the boundary layer (convection) regime. The flow pattern is qualitatively similar to what is found in cavities without porous matrices. However, the velocity and flow rate scales depend greatly on the properties of the fluid-saturated porous medium. These scales are addressed in the next subsection.

An important quantitative measure of the intensity of the flow and heat transfer process is the overall Nusselt number, which is defined as

$$\text{Nu} = \frac{q'}{k_m(T_w - T_f)} = -\int_0^1 \left(\frac{\partial\Theta}{\partial X}\right)_{x=0} dY. \tag{10.14}$$

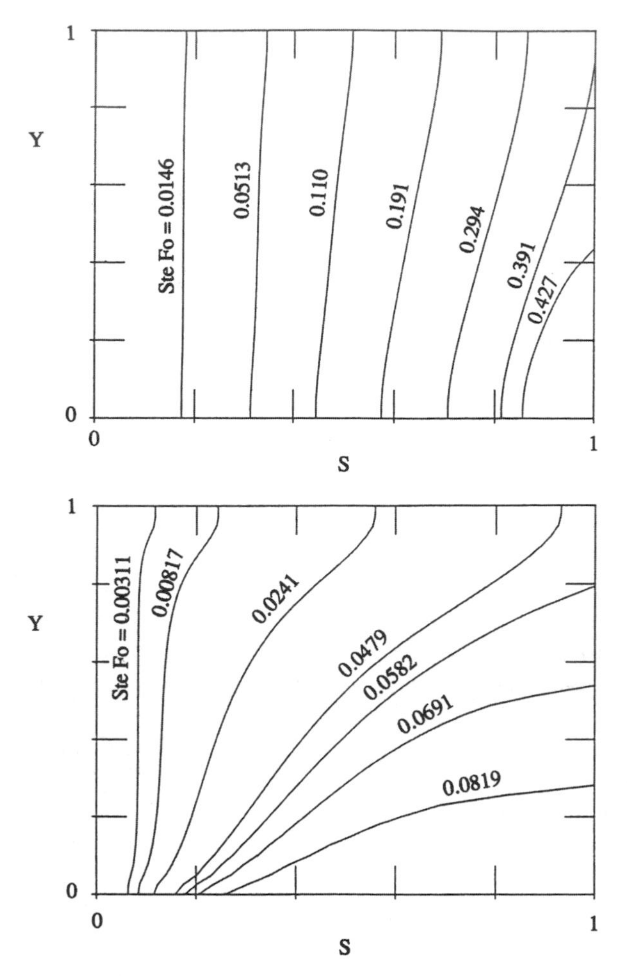

Figure 10.2. The evolution of the melting front ($L/H = 1$). Top: Ra $= 12.5$; bottom: Ra $= 800$ (Jany and Bejan, 1988a).

The numerator in this definition, q', is the heat transfer rate per unit length measured in the direction perpendicular to the (x, y) plane. The results of this calculation are shown in Fig. 10.5 as Nu versus the time number Ste Fo for different Rayleigh numbers and $L/H = 1$. The "knee" point marked on each curve represents the first arrival of the liquid-solid interface at the right vertical wall. This figure shows that the Nusselt number departs significantly from the pure conduction solution (Ra $= 0$) as the Rayleigh number increases above approximately 50. At Ra values of order 200 and higher, the Nu(Ste Fo) curve has a minimum at "short times," i.e., before the melting front reaches the right wall.

Another overall measure of the evolution of the melting process is the melt fraction or the mean horizontal dimensionless position of the melting

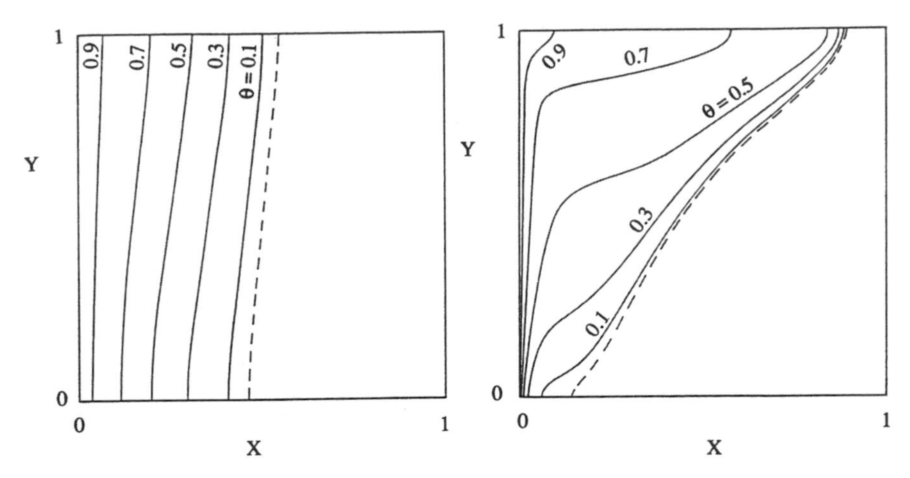

Figure 10.3. Patterns of isotherms in the melting process of Fig. 10.1 ($L/H = 1$). Left: Ra = 12.5, Ste Fo = 0.125; right: Ra = 800, Ste Fo = 0.0452 (Jany and Bejan, 1988a).

front:

$$S_{av} = \int_0^1 S\,dY. \tag{10.15}$$

This quantity is also a measure of the total energy storage and is related to *Nu* by

$$\frac{dS_{av}}{d(\text{Ste Fo})} = \text{Nu}. \tag{10.16}$$

Numerical S_{av} results are presented in Fig. 10.6 for a square cavity at five different Ra values. The melting process is accelerated as Ra increases. On the other hand, the S_{av}(Ste Fo) curves collapse onto a single curve as Ste Fo approaches zero.

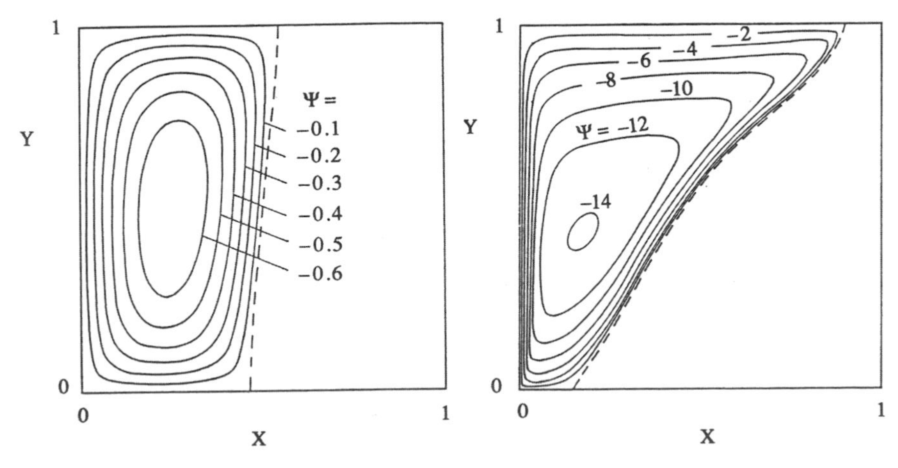

Figure 10.4. Patterns of streamlines in the melting process of Fig. 10.1 ($L/H = 1$). Left: Ra = 12.5, Ste Fo = 0.125; right: Ra = 800, Ste Fo = 0.0452 (Jany and Bejan, 1988a).

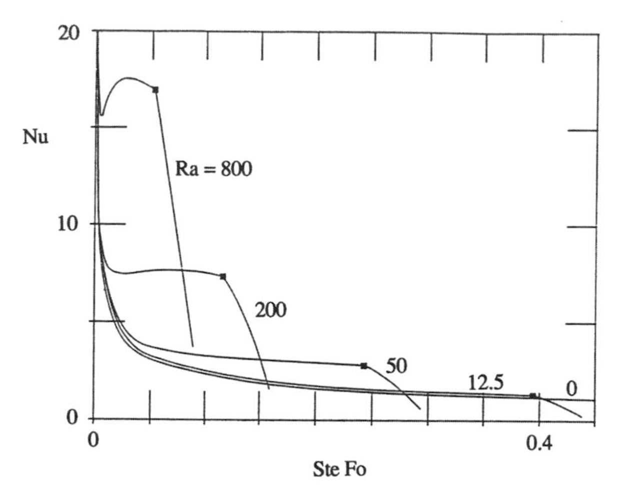

Figure 10.5. The Nusselt number as a function of time and Rayleigh number ($L/H = 1$) (Jany and Bejan, 1988a).

Similar results are revealed by calculations involving rectangular cavities. Figure 10.7 shows the evolution of the melting front in a shallow space ($L/H = 4$) for the Ra values 12.5 and 800. For example, it is evident that the Ra = 12.5 solution represents a case dominated by conduction. Also worth noting is the severe tilting of the liquid-solid interface during the convection-dominated case Ra = 800.

10.1.2. Scale Analysis

The numerical results have features that are similar to those encountered in the classical problem of melting in a cavity without a porous matrix (Jany and Bejan,

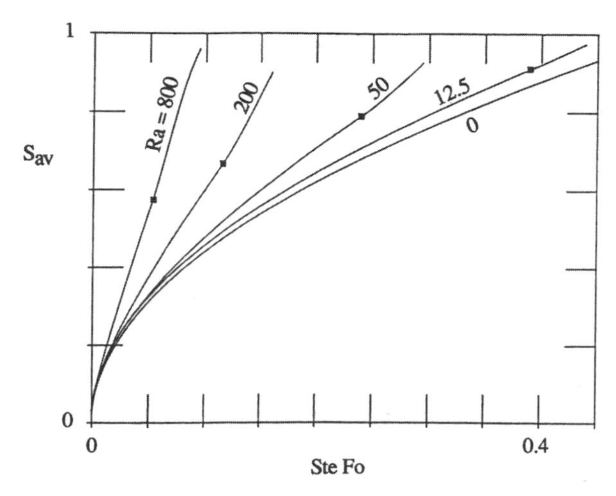

Figure 10.6. The average melting front location as a function of time and Rayleigh number ($L/H = 1$) (Jany and Bejan, 1988a).

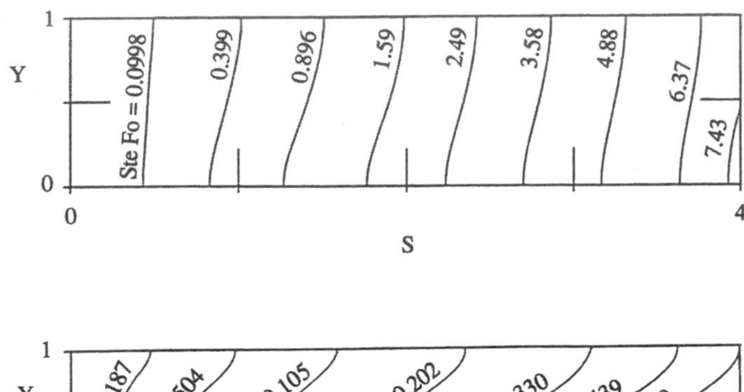

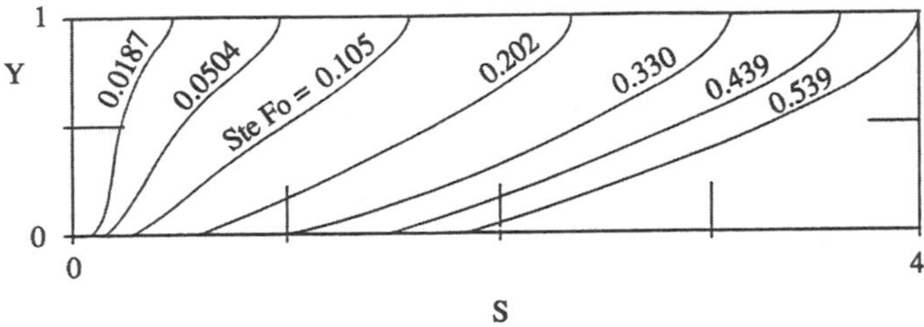

Figure 10.7. The evolution of the melting front in a shallow rectangular porous medium ($L/H = 4$). Top: Ra = 12.5; bottom: Ra = 800 (Jany and Bejan, 1988a).

1988b). In the present problem it is convenient to identify first the four regimes I–IV whose main characteristics are sketched in Fig. 10.8. The "conduction" region (I) is ruled by pure thermal diffusion and covered by the classical Neumann solution

$$\Theta(\text{Fo}) = 1 - \frac{\text{erf}(X/2\text{Fo}^{1/2})}{\text{erf}(C)}, \; S(\text{Fo}) = 2C\text{Fo}^{1/2}, \tag{10.17}$$

where C is the root of the equation

$$\frac{\text{Cerf(C)}}{\exp(-C^2)} = \frac{\text{Ste}}{\pi^{1/2}}. \tag{10.18}$$

The "transition" regime (II) is where the flow carves its own convection-dominated zone in the upper part of the liquid region, while the lower part remains

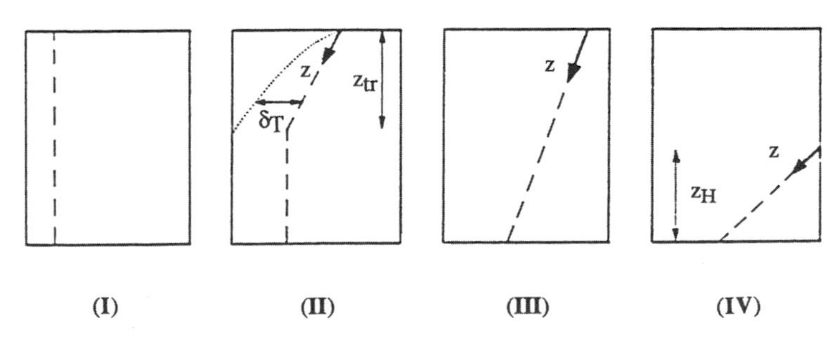

Figure 10.8. The four regimes for the scale analysis of melting in a porous medium heated from the side (Jany and Bejan, 1988a).

ruled by conduction. The "convection" regime (III) begins when the convection-dominated zone of the preceding regime fills the entire height H. Finally, the arrival of the liquid-solid interface at the right vertical wall marks the beginning of the "variable-height" regime (IV).

The scales of regimes I and II become apparent if we focus on the transition regime II, where z_{tr} is the height of the convection-dominated upper zone. The boundary layer thickness scale in this upper zone is (e.g., Bejan, 1984, p. 392)

$$\delta_T \sim z_{tr}\,\mathrm{Ra}_{z_{tr}}^{-1/2} \sim z_{tr}\left(\mathrm{Ra}\frac{z_{tr}}{H}\right)^{-1/2}. \tag{10.19}$$

where $\mathrm{Ra}_{ztr} = g\beta K z_{tr}(T_w - T_f)/\upsilon\alpha_m$. The convection-dominated zone is such that at its lower extremity δ_T is of the same order as the width of the conduction-dominated zone of height $(H - z_{tr})$, in other words,

$$z_{tr}\left(\mathrm{Ra}\frac{z_{tr}}{H}\right)^{-1/2} \sim H(\mathrm{Ste\,Fo})^{1/2}, \tag{10.20}$$

which means that $z_{tr} \sim H$ Ra Ste Fo.

The scale of the overall Nusselt number is obtained by adding the conduction heat transfer integrated over the height $(H - z_{tr})$ to the convection heat transfer integrated over the upper portion of height z_{tr}. The result is

$$\mathrm{Nu} \sim (H - z_{tr})s^{-1} + \int_0^{z_{tr}} \delta_T^{-1} dz \sim (\mathrm{Ste\,Fo})^{-1/2} + \mathrm{Ra}(\mathrm{Ste\,Fo})^{1/2} \tag{10.21}$$

or in terms of the average melting front location [Eqs. (10.15) and (10.16)],

$$S_{av} \sim (\mathrm{Ste\,Fo})^{1/2} + \mathrm{Ra}(\mathrm{Ste\,Fo})^{3/2}. \tag{10.22}$$

The transition regime II expires when z_{tr} becomes of order H, i.e., at a time of order Ste Fo $\sim \mathrm{Ra}^{-1}$.

The most striking feature of this set of scaling results is the Nu minimum revealed by Eq. (10.21). Setting $\partial\mathrm{Nu}/\partial(\mathrm{SteFo}) = 0$, we find that the minimum occurs at a time of order:

$$(\mathrm{Ste\,Fo})_{min} \sim \mathrm{Ra}^{-1}, \tag{10.23}$$

and that the minimum Nusselt number scale is

$$\mathrm{Nu}_{min} \sim \mathrm{Ra}^{1/2}. \tag{10.24}$$

The Nu_{min} scale is supported very well by the heat transfer data of Fig. 10.5, in which the actual values obey the relationship $\mathrm{Nu}_{min} \cong 0.54\mathrm{Ra}^{1/2}$ in the Ra range 200–1200 (Jany and Bejan, 1988a).

In the convection regime III the heat transfer and the melting front progress are controlled by the two thermal resistances of thickness δ_T,

$$Nu \sim \int_0^H \delta_T^{-1} dz \sim \mathrm{Ra}^{1/2}, \tag{10.25}$$

$$S_{av} \sim \mathrm{Ra}^{1/2}\,\mathrm{Ste\,Fo}. \tag{10.26}$$

The convection regime begins at a time of order Ste Fo $\sim$ Ra^{-1} and expires when the melting front reaches the right wall (at the "knee" points in Figs. 10.5 and 10.6). In the entire Ra domain 12.4 – 800, the Nu/Ra$^{1/2}$ ratio during the convection regime is roughly equal to 0.5. It is interesting that the value of Nu/Ra$^{1/2}$ is extremely close to what we expect in the convection regime in a rectangular porous medium, namely 0.577 (Weber, 1975b).

The scales of melting and natural convection during the variable-height regime IV are discussed in Jany and Bejan (1988a).

10.1.3. Effect of Liquid Superheating

In this section we review a theoretical solution to the problem of melting in the presence of natural convection in a porous medium saturated with a phase-change material (Bejan, 1989). The porous medium is held in a rectangular enclosure, which is being heated from the side (Fig. 10.9 or Fig. 10.1). The porous medium is initially saturated with solid phase-change material; its initial temperature is uniform and equal to the melting point of the phase-change material. The heating from the side consists of suddenly raising the side wall temperature and maintaining it at a constant level above the melting point.

We begin with the analysis of the convection-dominated regime. The main features of the temperature distribution in the liquid space are the two distinct

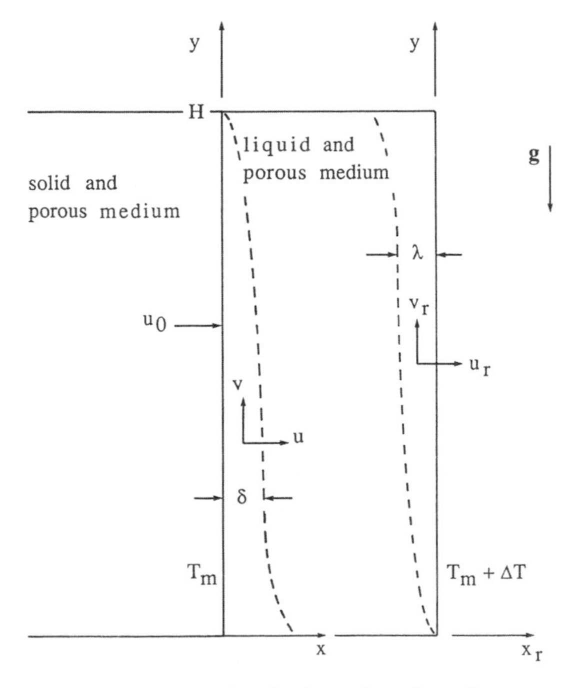

Figure 10.9. The boundary layer regime in the melt region of a porous medium heated from the right (Bejan, 1989).

boundary layers that line the heated wall and the solid-liquid interface. The core region of the liquid space is thermally stratified: its temperature is represented by the unknown function $T_c(y)$. The horizontal boundary layers that line the top and bottom walls and the details of the flow in the four corners are being neglected.

The analysis consists of first obtaining temperature and flow field solutions for the two vertical boundary layer regions and then meshing these solutions with a third (unique) solution for the core region. The key results of the analytical solution are

$$\tilde{\delta} = A(1 - \tau)(1 + \tau^2 \mathrm{Ste})^{-1/2}, \tag{10.27}$$

$$\tilde{\lambda} = A\tau(1 + \tau^2 \mathrm{Ste})^{-1/2}, \tag{10.28}$$

$$\tilde{y} = \frac{A^2}{4\mathrm{Ste}}\left[\frac{\tau(1 + \tau\,\mathrm{Ste})}{1 + \tau^2\,\mathrm{Ste}} - \frac{\tan^{-1}(\tau\,\mathrm{Ste}^{1/2})}{\mathrm{Ste}^{1/2}}\right], \tag{10.29}$$

where A depends only on the Stefan number,

$$A = 2\,\mathrm{Ste}^{1/2}\left[1 - \frac{\tan^{-1}(\mathrm{Ste}^{1/2})}{\mathrm{Ste}^{1/2}}\right]^{-1/2}. \tag{10.30}$$

The dimensionless variables $\tilde{\delta}$, $\tilde{\lambda}$, and τ represent the thickness of the cold boundary layer, the thickness of the warm boundary layer, and the temperature in the core region (cf. Fig. 10.9),

$$(\tilde{\delta}, \tilde{\lambda}) = \frac{(\delta, \lambda)}{H}\mathrm{Ra}^{1/2}, \tag{10.31}$$

$$\tau = \frac{T_c - T_m}{\Delta T}, \qquad \tilde{y} = \frac{y}{H}. \tag{10.32}$$

The left-hand side of Fig. 10.10 shows the solution obtained for the cold boundary layer thickness. The function $\tilde{\delta}(\tilde{y})$ increases monotonically in the flow direction (downward); its bottom value $\tilde{\delta}(0)$ is finite. The cold boundary layer thickness increases substantially as the Stefan number increases.

Figure 10.11 illustrates the manner in which the core temperature distribution responds to changes in the Stefan number. The core temperature distribution is symmetric about the midheight level only when $\mathrm{Ste} = 0$. The core temperature decreases at all levels as Ste increases above zero. Said another way, the average core temperature in the melting and natural convection problem (finite Ste) is always lower than the average core temperature in the pure natural convection problem ($\mathrm{Ste} = 0$).

The thickness of the warm boundary layer has been plotted on the right-hand side of Fig. 10.10. We learn in this way that the warm boundary layer becomes thinner as the Stefan number increases. The Ste effect on the warm layer, however, is less pronounced than on the boundary layer that descends along the solid-liquid interface.

The useful feature of this analytical solution is the ability to predict the rate at which the melting and natural convection process draws heat from the right wall

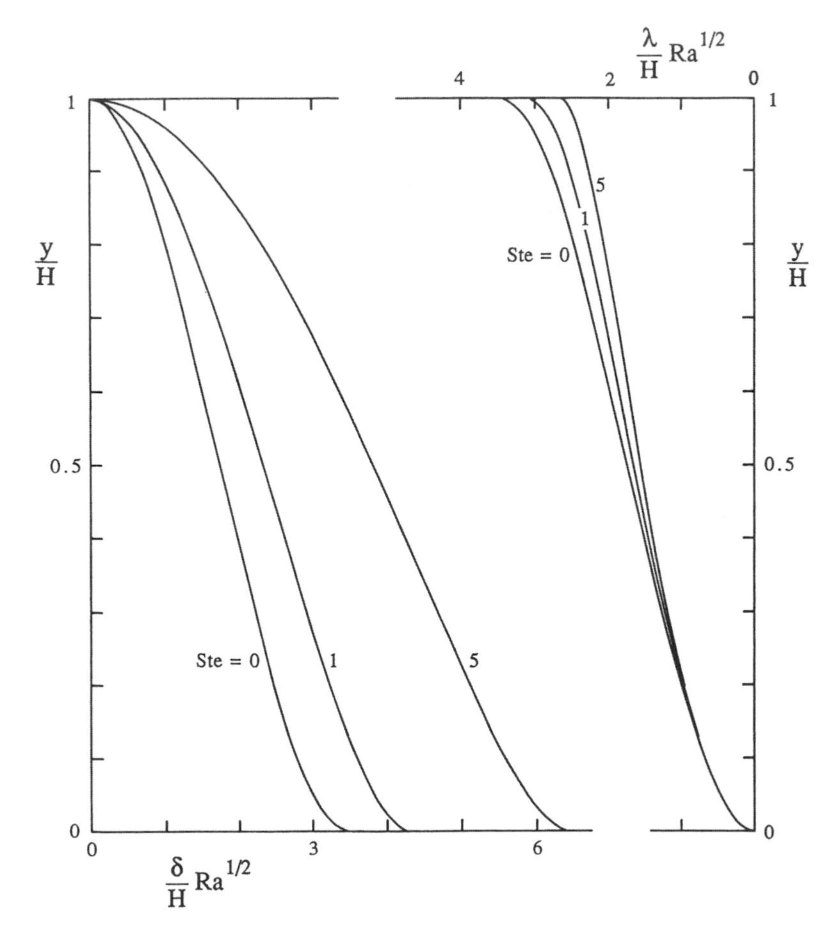

Figure 10.10. The thicknesses of the cold (left) boundary layer and the warm (right) boundary layer (Bejan, 1989).

of the system. This heat transfer rate through the right-hand side of Fig. 10.9 is

$$q_r{}' = k_m \int_0^H \left(\frac{\partial T}{\partial x}\right)_{x_r=0} dy \tag{10.33}$$

or, as an overall Nusselt number,

$$\mathrm{Nu}_r = \frac{q_r'}{k_m \Delta T} = \mathrm{Ra}^{1/2} F_r(\mathrm{Ste}) \tag{10.34}$$

with

$$F_r = \int_0^1 \frac{1-\tau}{\tilde{\lambda}} d\tilde{y} = \frac{\mathrm{Ste}^{3/4}}{\left[\mathrm{Ste}^{1/2} - \tan^{-1}(\mathrm{Ste}^{1/2})\right]^{1/2}} \times \left\{\frac{(\mathrm{Ste}-1)(\mathrm{Ste}+1)^{1/2} - 2\mathrm{Ste}}{\mathrm{Ste}(\mathrm{Ste}+1)} + \mathrm{Ste}^{-3/2} \ln\left[\mathrm{Ste}^{1/2} + (\mathrm{Ste}+1)^{1/2}\right]\right\}. \tag{10.35}$$

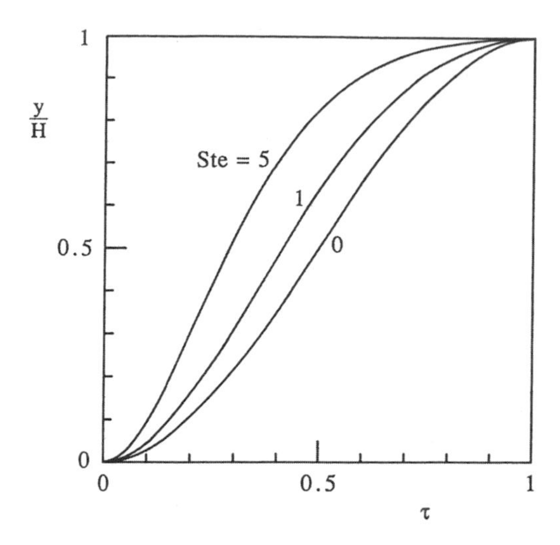

Figure 10.11. The core temperature distribution in the boundary layer regime (Bejan, 1989).

The approximate proportionality $Nu_r \sim Ra^{1/2}$ that is revealed by Eq. (10.34) is expected from the scale analysis shown in the preceding section. The new aspect unveiled by the present solution is the effect of the Stefan number. Representative F_r values constitute the top curve in Fig. 10.12. These values show that the heat transfer rate in the quasisteady regime increases gradually as the Stefan number increases.

One quantity of interest on the cold side of the liquid-saturated region is the overall heat transfer rate into the solid-liquid interface,

$$q' = k_m \int_0^H \left(\frac{\partial T}{\partial x}\right)_{x=0} dy \tag{10.36}$$

or the left-hand side Nusselt number

$$\mathrm{Nu} = \frac{q'}{k_m \Delta T} = \mathrm{Ra}^{1/2} F(\mathrm{Ste}) \tag{10.37}$$

with

$$\begin{aligned} F = \int_0^1 \frac{\tau}{\tilde{\delta}} d\tilde{y} &= \frac{\mathrm{Ste}^{-3/4}}{\left[\mathrm{Ste}^{1/2} - \tan^{-1}(\mathrm{Ste}^{1/2})\right]^{1/2}} \\ &\times \left\{ \ln\left[\mathrm{Ste}^{1/2} + (\mathrm{Ste}+1)^{1/2}\right] - \left(\frac{\mathrm{Ste}}{\mathrm{Ste}+1}\right)^{1/2} \right\}. \end{aligned} \tag{10.38}$$

The behavior of $F(\mathrm{Ste})$ is illustrated in Fig. 10.12. We see that the left-hand side Nusselt number decreases dramatically as the Stefan number increases.

In summary, the effect of increasing the Stefan number is to accentuate the difference between the heat transfer administered to the right wall (Nu_r) and the heat transfer absorbed by the solid-liquid interface (Nu). The difference between

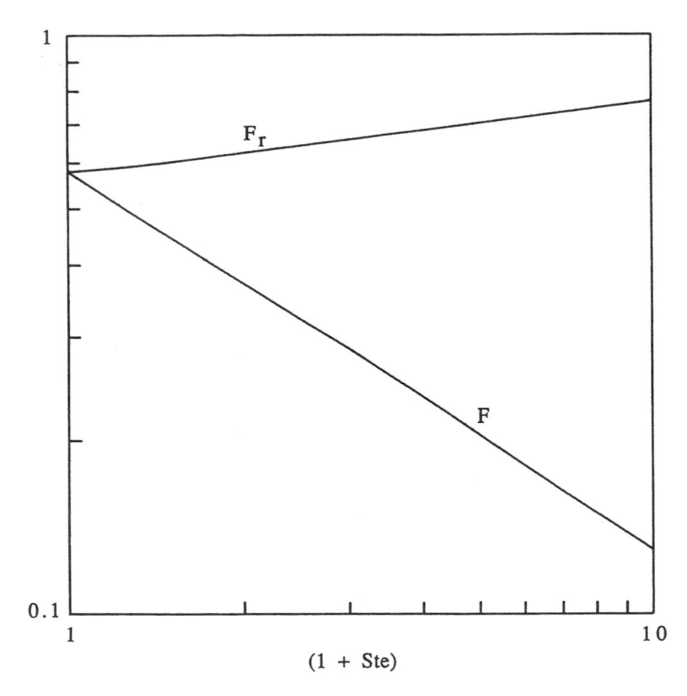

Figure 10.12. The effect of liquid superheating on melting in the convection-dominated regime (Bejan, 1989).

the two heat transfer rates is steadily being spent on raising the temperature of the newly created liquid up to the average temperature of the liquid-saturated zone.

Another quantity that can be anticipated based on this theory is the average melting rate. Writing u_0 for the local rate at which the solid-liquid interface migrates to the left in Fig. 10.9 and $\tilde{u}_0$ for the nondimensional counterpart,

$$\tilde{u}_0 = \frac{u_0}{\alpha_m / H} \mathrm{Ra}^{-1/2} = \mathrm{Ste} \frac{\tau}{\tilde{\delta}} \tag{10.39}$$

leads to

$$\tilde{u}_{0,av} = \mathrm{Ste} \int_0^1 \frac{\tau}{\tilde{\delta}} d\tilde{y} = \mathrm{Ste} F. \tag{10.40}$$

The function $\tilde{u}_{0,av}$ depends only on the Stefan number, as is shown by Fig. 10.12.

In closing, it is worth commenting on the use of (1 + Ste) as abscissa in Fig. 10.12. This choice has the effect of making the F and F_r curves appear nearly straight in the logarithmic plane, improving in this way the accuracy associated with reading numerical values directly off Fig. 10.12. This observation leads to two very simple formulas,

$$F_r \cong 3^{-1/2}(1 + 1.563 \text{ Ste})^{0.107} \tag{10.41}$$

$$F \cong 3^{-1/2}(1 + 0.822 \text{ Ste})^{-0.715}, \tag{10.42}$$

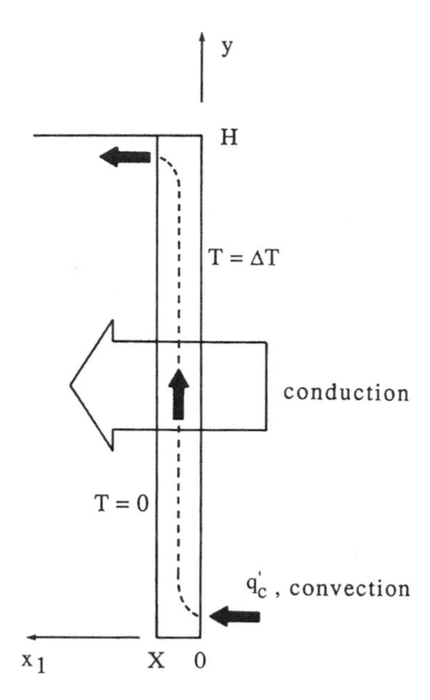

Figure 10.13. Combined conduction and convection during the earliest stages of melting due to heating from the side (Bejan, 1989).

which approach within 0.5 percent the values calculated based on Eqs. (10.35) and (10.38).

In the very beginning of the melting process the liquid-saturated region is infinitely slender and the heat transfer mechanism is that of pure conduction. With reference to the slender liquid zone sketched in Fig. 10.13, the history of the thickness X is described by the well-known Neumann solution [Eqs. (10.17) and (10.18)], which can be written here as

$$X = 2\Lambda(\alpha_m f)^{1/2}, \qquad \frac{\exp(-\Lambda^2)}{\operatorname{erf}(\Lambda)} = \pi^{1/2}\frac{\Lambda}{\mathrm{Ste}}. \tag{10.43}$$

According to the same solution, the excess temperature of the liquid-saturated porous medium depends on t and x_1 (and not on y), where x_1 is chosen such that it increases toward the left in Fig. 10.13 (note that here $T = 0$ on the melting front),

$$T = \Delta T\left[1 - \frac{1}{\operatorname{erf}(\Lambda)}\operatorname{erf}\frac{x_1}{2(\alpha_m t)^{1/2}}\right]. \tag{10.44}$$

The overall heat transfer rate delivered through the heated wall (q'_r, or Nu_r) is also well known. For example, in the limit $\mathrm{Ste} = 0$ the overall Nusselt number decays as

$$\mathrm{Nu}_r = 2^{-1/2}\tau^{-1/2}, \qquad \tau = \frac{\alpha_m t}{H^2}\mathrm{Ste}. \tag{10.45}$$

Bejan (1989) showed that it is possible to develop an analytical transition from the short-times Nusselt number (10.45) to the long-times expression of the quasi-steady regime (10.34). In other words, it is possible to develop a heat transfer theory that holds starting with $\tau = 0$ and covers the entire period during which the heat transfer mechanism is, in order, pure conduction, conduction and convection, and finally convection.

This theoretical development is based on the observation that even in the limit $\tau \to 0$ when the liquid region approaches zero thickness, there is liquid motion in the liquid saturated region. The incipient convective heat transfer contribution is

$$q'_c = \int_0^X \rho c_f v T dx = \rho c_f \frac{g\beta K \Delta T}{\nu} \Delta T X B, \tag{10.46}$$

where the function $B(\mathrm{Ste})$ is the integral

$$B(\mathrm{Ste}) = \int_0^1 \left[\frac{\int_0^\Lambda \mathrm{erf}(m)dm}{\Lambda \mathrm{erf}(\Lambda)} - \frac{\mathrm{erf}(n\Lambda)}{\mathrm{erf}(\Lambda)} \right] \left[1 - \frac{\mathrm{erf}(n\Lambda)}{\mathrm{erf}(\Lambda)} \right] dn. \tag{10.47}$$

This function was evaluated numerically. In the conduction regime the effect of q'_c on the overall heat transfer rate is purely additive, because the top and bottom ends of the liquid-zone temperature field (the only patches affected by the flow are negligible in height when compared with the rest of the system (height H). Therefore, the instantaneous total heat transfer rate through the right wall is

$$q'_r = k_m H \left(-\frac{\partial T}{\partial x_1} \right)_{x_1 = 0} + q'_c \tag{10.48}$$

where the first term on the right-hand side accounts for the dominant conduction contribution. Employing the Nu_r notation defined in Eq. (10.34), expression (10.48) translates into

$$\mathrm{Nu}_r = G_0 \tau^{-1/2} + G_c \,\mathrm{Ra}\, \tau^{1/2}. \tag{10.49}$$

The functions G_0 and G_c depend only on the Stefan number,

$$G_0 = \frac{\mathrm{Ste}^{1/2}}{\pi^{1/2} \mathrm{erf}(\Lambda)}, \; G_c = 2\Lambda B \mathrm{Ste}^{-1/2}, \tag{10.50}$$

and are presented in Fig. 10.14. The Stefan number has a sizeable effect on both G_0 and G_c. For fixed values of τ and Ra, the effect of increasing the Stefan number is to diminish the relative importance of the convection contribution to the overall Nusselt number.

In view of the reasoning on which Eq. (10.49) is based, we must keep in mind that this Nu_r expression cannot be used beyond the moment τ when the second (convection) term begins to outweigh the first (conduction) term. This condition,

$$G_0 \tau^{-1/2} > G_c \mathrm{Ra}\, \tau^{1/2}, \tag{10.51}$$

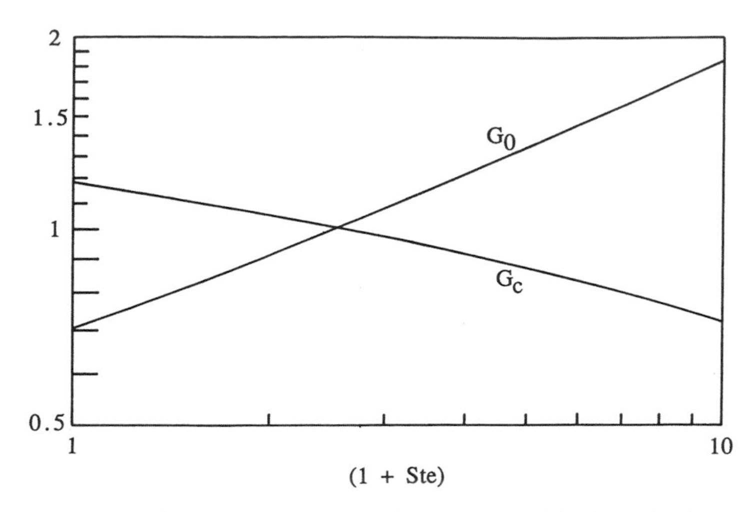

Figure 10.14. The effect of liquid superheating on the combined conduction and convection regime (Bejan, 1989).

yields the following time criterion for the domain of validity of Eq. (10.49),

$$\tau \mathrm{Ra} < \frac{G_0}{G_c}, \tag{10.52}$$

The solid lines of Fig. 10.15 show the Nusselt number history predicted by Eq. (10.49) all the way up to the time limit (10.52). That limit, or the point of expiration of each solid curve, is indicated by a circle. Plotted on the ordinate is the group $\mathrm{Nu}_r \mathrm{Ra}^{-1/2}$: this group was chosen in order to achieve a Ra correlation of the Nusselt number in the convection limit.

The horizontal dash lines of Fig. 10.15 represent the Nusselt number values that prevail at long times in the boundary layer regime, Eq. (10.34). It is remarkable that two different and admittedly approximate theories [Eqs. (10.34) and (10.49)] provide a practically continuous description for the time variation of the overall Nusselt number. Only when Ste increases above 5 does a mismatch of a few percentage points develop between the $\mathrm{Nu}_r \mathrm{Ra}^{-1/2}$ values predicted by the two theories at the transition time (10.52).

10.1.4. Horizontal Liquid Layer

In the convection-dominated regime, the melting front acquires a characteristic shape, the dominant feature of which is a horizontal layer of melt that grows along the top boundary of the phase-change system. The slenderness of the horizontal layer increases with the time and Rayleigh number (Figs. 10.2–10.4, 10.7).

With these images in mind, the liquid-saturated region can be viewed as the union of two simpler regions, an upper zone that is a horizontal intrusion layer and

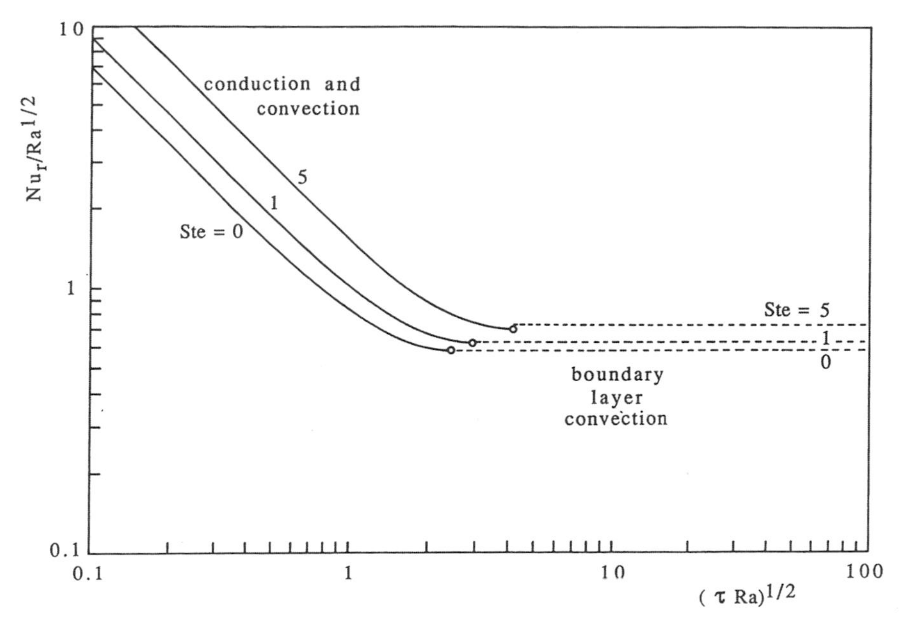

Figure 10.15. The evolution of the Nusselt number during the conduction, mixed, and convection regimes (Bejan, 1989).

a lower zone that houses a vertical counterflow (as in Fig. 10.9). These two zones are labeled A_2 and A_1 in Fig. 10.16.

It is possible to describe the shape and propagation of the horizontal intrusion layer by means of a similarity solution of the boundary layer type (Bejan *et al.*, 1990). In addition to the features of the Darcy flow model described in Section 10.1.1, this similarity solution was based on the assumption that the depth of the intrusion layer (δ) is considerably smaller than the distance of horizontal penetration of the leading edge (L). The melting speed $U = dL/dt$ was assumed small relative to the horizontal velocity in the liquid-saturated region: this particular assumption holds in the limit Ste $\ll 1$. Finally, it was assumed that the melting front shape is preserved in time, i.e., in a frame attached to the leading edge of the intrusion.

The main result of the intrusion layer analysis is the theoretical formula

$$\frac{L}{H} = 0.343\text{Ra}^{1/2}(\text{Ste Fo})^{3/4} \tag{10.53}$$

that describes the evolution of the length of horizontal penetration $L(t)$. This formula agrees very well with the $L(t)$ read off numerical plots such as those of Fig. 10.7 (bottom), in the Ra range 200–800.

Another result of the intrusion layer analysis is that the volume (area A_2 in Fig. 10.16) of the upper region of the liquid-saturated porous medium increases

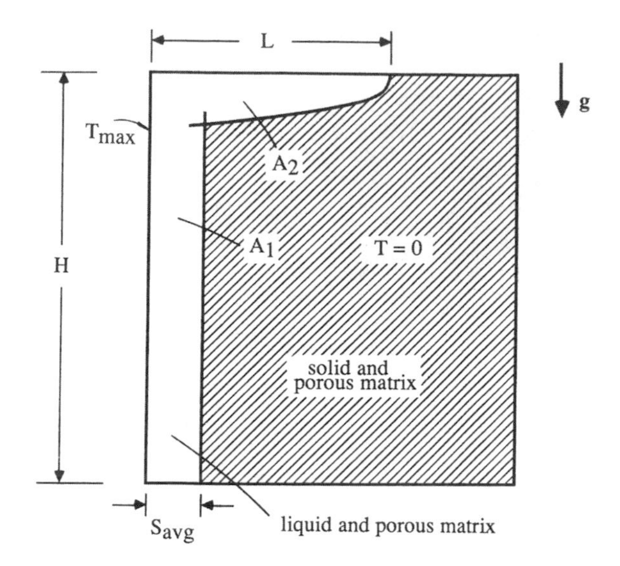

Figure 10.16. Two-zone model ($A_1 + A_2$) for the melt region of a rectangular system heated from the side (Bejan *et al.*, 1990).

with both Ra and Ste Fo as

$$\frac{A_2}{H^2} = 0.419\,\mathrm{Ra}^{1/2}(\mathrm{Ste\,Fo})^{5/4}. \tag{10.54}$$

This estimate can be added to the one for area A_1, which accounts for the regime of boundary layer convection in the vertical slot, cf. Eq. (10.34) and Fig. 10.12 at Ste = 0,

$$\frac{A_1}{H^2} = 0.577\,\mathrm{Ra}^{1/2}\mathrm{Ste\,Fo} \tag{10.55}$$

in order to calculate the total cross-sectional area of the region saturated by liquid:

$$\frac{A_2 + A_1}{H^2} = 0.577\,\mathrm{Ra}^{1/2}\mathrm{Ste\,Fo}\left[1 + 0.725(\mathrm{Ste\,Fo})^{1/4}\right]. \tag{10.56}$$

The relative effect of the horizontal intrusion layer on the size of the melt region is described by the group $(\mathrm{Ste\,Fo})^{1/4}$. When the order of magnitude of the group $(\mathrm{Ste\,Fo})^{1/4}$ is greater than 1, the size of the melt fraction is ruled by the horizontal intrusion layer. When this group is less than 1 (as in the numerical experiments of Section 10.1.1), the melt fraction is dominated by the boundary layer convection that erodes the nearly vertical portion of the two-phase interface (area A_1).

10.1.5. Vertical Melting Front in an Infinite Porous Medium

Kazmierczak *et al.*'s (1986) analysis of melting with natural convection applies to the configuration shown on the left-hand side of Fig. 10.17. The melting front is vertical and at the melting point T_m. The coordinate system $x - y$ is attached

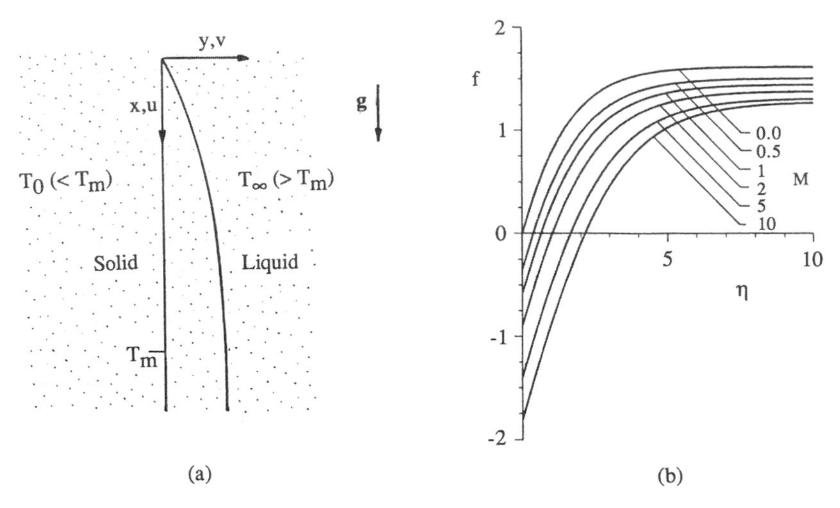

Figure 10.17. The dimensionless streamfunction for boundary layer convection on the liquid side of a vertical melting front in a porous medium (Kazmierczak *et al.*, 1986, with permission from Hemisphere Publishing Corporation).

to the melting front: in it, the porous medium flows to the right, with a melting (or blowing) velocity across the x axis. The melting front is modeled as a vertical plane.

The geometry of Fig. 10.17 is more general than in the systems analyzed until now, because the temperature of the solid region is below the melting point, $T_0 < T_m$. On the right-hand side of the melting front, the liquid is superheated, $T_\infty > T_m$. A vertical boundary layer flow on the liquid side smooths the transition from T_m to T_∞. Because of the presence of solid subcooling, the Stefan number Ste of Eq. (10.13) is now replaced by the "superheating and subcooling" number

$$M = \frac{c_f(T_\infty - T_m)}{h_{sf} + c_s(T_m - T_0)}, \tag{10.57}$$

where c_f and c_s are the specific heats of the liquid and solid.

The flow and temperature field on the liquid side of the melting front was determined in the form of a similarity solution. Figure 10.17 shows the dimensionless streamfunction profile $f(\eta)$, which is defined by

$$\psi = \alpha_m \, \mathrm{Ra}_x^{1/2} f(\eta), \quad \eta = \frac{y}{x} \, \mathrm{Ra}_x^{1/2}, \tag{10.58}$$

and $\mathrm{Ra}_x = g\beta K(T_\infty - T_m)x/\nu\alpha_m$. The streamfunction is defined in the usual way, by writing $u = \partial\psi/\partial y$ and $v = -\partial\psi/\partial x$. The figure shows that the number M can have a sizeable impact on the flow. The limit $M = 0$ corresponds to the case of natural convection near a vertical impermeable plate embedded in a fluid-saturated porous medium (Cheng and Minkowycz, 1977), discussed earlier in Section 5.1.

The superheating and subcooling parameter M also has an effect on the local heat transfer flux through the melting front (q_x'') and on the melting rate $v(x, y) = 0$. The two are related by

$$\frac{x}{\alpha_m} v(x, 0) = M \, \mathrm{Nu}_x, \tag{10.59}$$

where Nu_x is the local Nusselt number $q_x'' x / k_m (T_\infty - T_m)$. It was found that the Nusselt number varies in such a way that the ratio $\mathrm{Nu}_x / \mathrm{Ra}_x^{1/2}$ is a function of M only. Originally that function was calculated numerically and tabulated in Kazmierczak *et al.* (1986). It was shown more recently (Bejan, 1989) that the same numerical results are correlated within 1 percent by an expression similar to Eqs. (10.41) and (10.42):

$$\frac{\mathrm{Nu}_x}{\mathrm{Ra}_x^{1/2}} = 0.444(1 + 0.776M)^{-0.735}. \tag{10.60}$$

Combining Eqs. (10.59) and (10.60), we note that the melting velocity $v(x, 0)$ increases with M and that its rate of increase decreases as M becomes comparable with 1 or greater.

Kazmierczak *et al.* (1986) also treated the companion phenomenon of boundary layer natural convection melting near a perfectly horizontal melting front in an infinite porous medium. They demonstrated that the same parameter M has a significant effect on the local heat flux and melting rate.

10.1.6. A More General Model

An alternative to the Darcy flow model (outlined in Section 10.1.1 and used in all the studies discussed until now) was developed by Beckermann and Viskanta (1988a). One advantage of this general model is that the resulting governing equations apply throughout the porous medium, i.e., in both the liquid-saturated region and the solid region. Because of this feature, the same set of equations can be solved in the entire domain occupied by the porous medium, even in problems with initial solid subcooling (i.e., time-dependent conduction in the solid). Another advantage of this model is that it can account for the inertia and boundary friction effects in the flow of the liquid through the porous matrix.

The model is based on the volume averaging of the microscopic conservation equations. In accordance with Fig. 10.18a, the saturated porous medium is described by three geometric parameters, two of which are independent:

$$\varepsilon = \frac{V_f}{V}, \text{ pore fraction in volume element (previously labeled } \varphi) \tag{10.61}$$

$$\gamma(t) = \frac{V_1(t)}{V_f}, \text{ liquid fraction in pore space} \tag{10.62}$$

$$\delta(t) = \frac{V_1(t)}{V} \; \varepsilon\gamma, \text{ liquid fraction in volume element.} \tag{10.63}$$

Next, the melting "front" actually can have a finite width even when the phase-change substance has a well-defined melting point T_m, because the phase-change

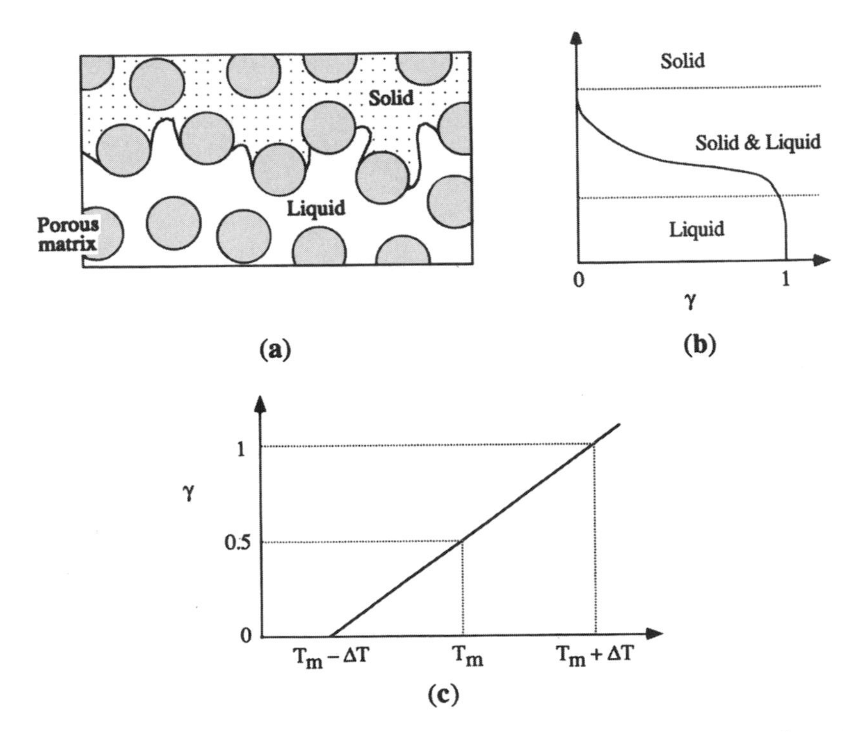

Figure 10.18. A more general model for melting in a saturated porous medium: (a) element used for volume averaging; (b) the coexistence of liquid and solid in the pores in the phase-change region; and (c) the assumed variation of the liquid fraction with the local temperature (after Beckermann and Viskanta, 1988a).

region can be inhabited at the same time by solid and liquid in the pores (Fig. 10.18b). The liquid fraction γ varies from 0 to 1 across this region, while the average temperature of the saturated porous medium in this zone is T_m, Fig. 10.18c.

In addition to these ideas, Beckermann and Viskanta's (1988a) model is based on the assumptions that the flow and temperature fields are two-dimensional, the properties of the solid matrix and the phase-change material (liquid, or solid) are homogeneous and isotropic, local thermal equilibrium prevails, the porous matrix and the solid phase-change material are rigid, the liquid is Boussinesq incompressible and the properties of the liquid and solid phases are constant, the dispersion fluxes due to velocity fluctuations are negligible, and the solid and liquid phases of the phase-change material have nearly the same density ρ. Under these circumstances, the volume-averaged equations for mass and momentum conservation become

$$\nabla \cdot \mathbf{u} = 0, \tag{10.64}$$

$$\frac{\rho}{\delta}\frac{\partial \mathbf{u}}{\partial t} + \frac{\rho}{\delta^2}(\mathbf{u}\cdot\nabla)\mathbf{u} = -\nabla P + \frac{\mu_l}{\delta}\nabla^2\mathbf{u} - \left(\frac{\mu_l}{K} + \frac{\rho c_F}{K^{1/2}}|\mathbf{u}|\right)\mathbf{u} - \rho g\beta(T - T_{ref}) \tag{10.65}$$

where $\mathbf{u}$ is the Darcian velocity $\mathbf{u} = \delta \mathbf{u}_l$, and $\mathbf{u}_l$ is the average liquid velocity through the pore.

The third group on the right-hand side of Eq. (10.65) accounts for the Darcy term and the Forchheimer inertia correction, in which $c_F \cong 0.55$ (Ward, 1964). For a bed of spherical beads of diameter d, the permeability K can be calculated with the Kozeny-Carman relation (1.16), in which $d_p = d$, and $\varphi = \delta$. The permeability is therefore equal to $K(\delta = \varepsilon)$ in the liquid-saturated region, $K(\delta = 0) = 0$ in the solid region, and takes in-between values in the phase-change region (Fig. 10.18b).

The volume-averaged equation for energy conservation is (Beckermann and Viskanta, 1988a)

$$\overline{\rho c}\frac{\partial T}{\partial t} + \rho c_l(\mathrm{u} \cdot \nabla T) = \nabla \cdot (k_{eff} \nabla T) - \varepsilon \rho \Delta h \frac{\partial \gamma}{\partial t}, \tag{10.66}$$

in which c_l is the liquid specific heat, Δh is the latent heat of melting (labeled h_{sf} in the preceding sections), and $\overline{\rho c}$ is the average thermal capacity of the saturated porous medium,

$$\overline{\rho c} = \varepsilon \rho [\gamma c_l + (1 - \gamma) c_s] + (1 - \varepsilon)(\rho c)_m. \tag{10.67}$$

The subscript $(\ )_m$ refers to properties of the solid matrix. The effective thermal conductivity k_{eff} can be estimated using Veinberg's (1967) model,

$$k_{eff} + \varepsilon k_{eff}^{1/3} \frac{k_m - k_{ls}}{k_{ls}^{1/3}} - k_m = 0, \tag{10.68}$$

where k_{ls} is the average conductivity of the phase-change material (liquid and solid phases):

$$k_{ls} = \gamma k_1 + (1 - \gamma) k_s. \tag{10.69}$$

The above model was used by Beckermann and Viskanta (1988a) in the process of numerically simulating the evolution of the melting process in the porous medium geometry shown in Fig. 10.19. The two side walls are maintained at different temperatures, T_h and T_c. Because of the mixed region recognized in Fig. 10.18b, the melting front is a region of finite thickness in Fig. 10.19. These numerical simulations agreed with a companion set of experimental observations in a system consisting of spherical glass beads ($d = 6$ mm) and gallium ($T_m = 29.78°$C). The numerical runs were performed for conditions in which the Rayleigh number Ra varied from 9.22 to 11.52. Because of the low Ra range, the calculated shape of the melting region was nearly plane and vertical, resembling the melting front shapes exhibited here in Fig. 10.2 (top). In the same numerical runs, the Darcy term dominated the Forchheimer and Brinkman terms on the right-hand side of Eq. (10.65).

10.1.7. Further Studies

Kazmierczak *et al.* (1988) analyzed the melting process in a porous medium in which the frozen phase-change material (*PCM*) is not the same substance as the warmer liquid that saturates the melt region. They considered a vertical melting

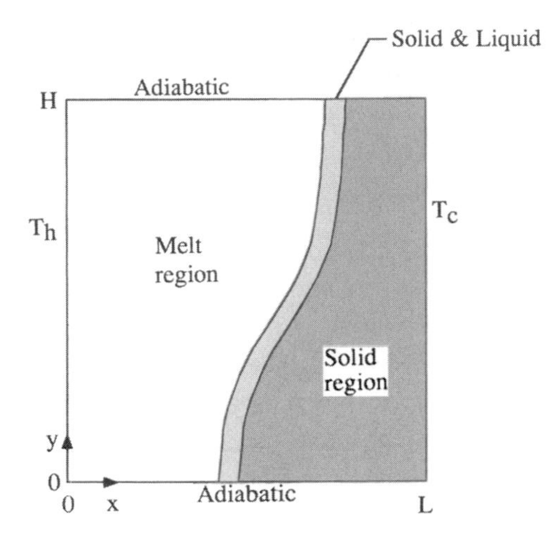

Figure 10.19. The finite thickness of the melting front according to the model of Fig. 10.18 (after Beckermann and Viskanta, 1988a).

front and showed the formation of a liquid counterflow along the melting front. Adjacent to the solid is the liquid formed as the *PCM* melts: in this first layer the liquid rises along the solid-liquid interface. The second layer bridges the gap between the first layer (liquid *PCM*) and the warmer dissimilar liquid that drives the melting process. In this outer boundary layer, the dissimilar liquid flows downward. The corresponding problem in which the heat transfer in the dissimilar fluid is by forced convection was considered by Kazmierczak *et al.* (1987).

Zhang (1993) performed a numerical study on the Darcy model of an ice-water system in a rectangular cavity heated laterally, using the Landau transformation to immobilize the interface and a finite-difference technique. He reported that local maximum and minimum average Nusselt numbers occur at heating temperatures of 5°C and 8°C, respectively. If the heating temperature is less than 8°C the melt region is wider at the bottom than at the top, while the reverse is true for higher heating temperatures. The numerical study of Sasaguchi (1995) was concerned with a cavity with one heated sidewall and three insulated walls, a transient problem. The further numerical study by Zhang *et al.* (1997) dealt with the case of anisotropic permeability with the principal axes oriented at an angle θ to the gravity vector. The effect of a magnetic field on melting from a vertical plate was treated by Tashtoush (2005) using the Forchheimer model.

The research discussed so far in this chapter has dealt with heating from the side. X. Zhang *et al.* (1991) have made a theoretical investigation of the melting of ice in a cavity heated from below. They found that the convection that arises in the unstable layer can penetrate into the stable region but cannot reach the melting front, and this results in a flatter solid-liquid interface than that produced in the absence of a stable layer. They also found that, in transition from onset to final state the convection pattern passes through several intermediate forms, each change being accompanied by a sudden increase (which is followed by a subsequent decline) in the heat transfer rate and in the displacement velocity of the solid-liquid interface. Zhang and Nguyen (1990, 1994) have found that melting from above is more effective than melting from below when the heating temperature is between 0 and 8°C; convection arises earlier, the melting process is faster, and the total melt at steady state is thicker. The time for the onset of convection is a minimum and the heat transfer rate is a maximum when the upper boundary is at 6°C, and at this temperature the heat transfer rate is a maximum. Hguyen (*sic*) and Zhang (1992) studied the penetrative convection that occurs during the melting of a layer of ice heated either above or below. They found that convection starts to play an increasingly important role as the melt thickness attains a certain value corresponding to the critical Rayleigh number for the onset of convection. The new convection cells have an approximately square form. As time passes these cells become more slender and suddenly break up sequentially. The breaking up process is quite short and is associated with a sharp jump in the curve of Nusselt number versus time.

The melting of ice has also been considered by Kazmierczak and Poulilkakos (1988). They dealt with both vertical and horizontal interfaces. Plumb (1994a) developed a simple model for convective melting of particles in a packed bed with throughflow and solved it numerically in one dimension to predict melting rates for a single substance and a system in which the liquid phase at elevated temperature enters a packed bed of the solid phase at the melting temperature. He found that the thickness of the melting zone increases with Péclet number and Prandtl number for systems dominated by convection.

Melting around a horizontal cylinder was studied numerically on the Darcy model by Christopher and Wang (1994). They found that heat transfer from the cylinder is minimized at some value of the burial depth that is a function of Ra and the dimensionless phase change temperature. Chang and Yang (1996) studied numerically, on the Brinkman-Forchheimer model, the melting of ice in a rectangular enclosure. They noted that as time goes on, heat transfer on the hot side decreases and that on the cold side increases.

Ellinger and Beckerman (1991) reported an experimental study of melting of a pure substance (n-Octadecane) in a rectangular enclosure that is partially occupied by horizontal or vertical layers of a relatively high thermal conductivity medium (glass or aluminum beads). They found that though such a porous layer may cause a faster movement of the solid-liquid interface, the effect of low permeability causes a reduction in melting and heat transfer rates compared with the case without the porous layer. Pak and Plumb (1997) studied numerically and experimentally the melting of a mixture that consists of melting and nonmelting components, with heat applied to the bottom of the bed.

Mixed convection with melting from a vertical plate was analyzed by Bakier (1997) and Gorla *et al.* (1999a). They noted that the melting phenomenon decreases the local Nusselt number at the solid-liquid surface. Horizontal forced and mixed convection with local thermal nonequilibrium melting was studied experimentally and theoretically by Hao and Tao (2003a,b). The topic of local thermal nonequilibrium melting was further addressed by Harris *et al.* (2001) and Agwu Nnanna *et al.* (2004).

A related problem, involving a phase-change front at the interface between a diminishing solid volume and an increasing fluid volume, has been treated by Rocha *et al.* (2001) and Bejan *et al.* (2004). This involves a layer of porous medium impregnated by solid methane hydrate material. The clathrate (endowed with a lattice) hydrates are solid crystals of water and methane at sufficiently high pressures and low temperatures. When the layer is depressurized suddenly on its lower plane, the methane hydrate material progressively dissociates into methane gas plus liquid water.

10.2. Freezing and Solidification

10.2.1. Cooling from the Side

10.2.1.1. Steady State

In a study that deals with both freezing and melting, Oosthuizen (1988a) considered the steady state in the two-dimensional configuration of Fig. 10.20. The porous medium is heated from the left and cooled from the right in such a way that the melting point of the phase-change material falls between the temperatures of the two side walls, $T_h > T_m > T_c$.

In the steady state, the freezing front takes up a stationary position and the freezing and melting at the front ceases. This is why in the steady state the latent heat of the phase-change material (h_{sf}) does not play a role in the heat transfer process or in deciding the position and shape of the melting front. The heat transfer from T_h to T_c is one of conjugate convection and conduction: specifically, convection through the zone saturated with liquid and conduction through the zone with pores filled by solid phase-change material.

Oosthuizen (1988a) relied on the finite element method in order to simulate the flow and heat transfer through the entire $H \times L$ domain of Fig. 10.20. The porous medium model was the same as the one outlined in the first part of Section 10.1.1. The parametric domain covered by this study was $0 = \text{Ra} = 500$, $0.5 = H/L = 2$, and $1 = k_F/k_U = 3$. The thermal conductivities k_F and k_U refer to the frozen and the unfrozen zones. They are both of type k_m, i.e., thermal conductivities of the saturated porous medium. The Rayleigh number is defined as Ra $= g\beta K H(T_h - T_c)/\nu\alpha_m$.

Besides Ra, H/L, and K_U, the fourth dimensionless group that governs the steady state is the dimensionless temperature difference ratio,

$$\theta_c = \frac{T_m - T_c}{T_h - T_c}, \tag{10.70}$$

which describes the position of T_m relative to T_h and T_c. Figure 10.20b shows the effect of increasing the Rayleigh number when $k_F = k_U$ and $\theta_c = 0.5$. In this case, in the absence of natural convection (Ra = 0) the melting front constitutes the

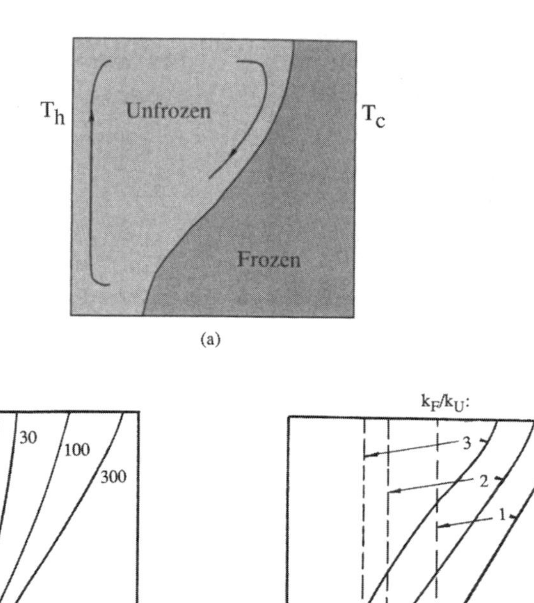

Figure 10.20. (a) Steady-state convection and heat transfer in a porous medium with differentially heated side walls. (b) The effect of Ra on the freezing front ($\theta_c = 0.5$, $k_F = k_U$). (c) The effect of k_F/k_U on the freezing front ($\theta_c = 0.5$) (Oosthuizen, 1988a).

vertical midplane of the $H \times L$ cross section. The melting front becomes tilted, S-shaped, and displaced to the right as Ra increases. The effect of natural convection is important when Ra exceeds approximately 30.

The effect of the conductivity ratio k_F/k_U is illustrated in Fig. 10.20c, again for the case when T_m falls right in the middle of the temperature interval $T_c - T_h$ (i.e., when $\theta_c = 0.5$). The figure shows that when the conductivity of the frozen zone is greater than that of the liquid-saturated zone ($k_F/k_U > 1$), the frozen zone occupies a greater portion of the $H \times L$ cross section. The effect of the k_F/k_U ratio is felt at both low and high Rayleigh numbers.

The melting-point parameter θ_c has an interesting effect, which is illustrated in Fig. 10.21. The ordinate shows the value of the overall Nusselt number, which is the ratio of the actual heat transfer rate to the pure-conduction estimate, $\mathrm{Nu} = q'/[k_U(T_h - T_c)/L]$. On the abscissa, the θ_c parameter decreases from $\theta_c = 1$ (or $T_m = T_c$) to $\theta_c = 0$ (or $T_m = T_c$). The figure shows that when $k_F/k_U > 1$, there exists an intermediate θ_c value for which the overall heat transfer rate is minimum. This effect is particularly evident at high Rayleigh numbers, where convection plays an important role in the unfrozen zone.

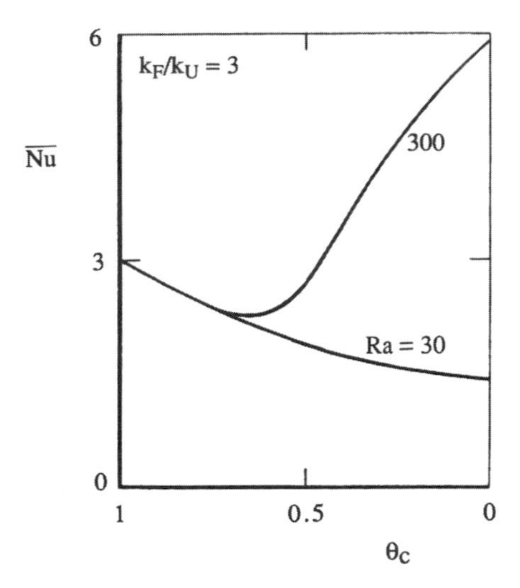

Figure 10.21. The effect of the melting-point parameter θ_c on the overall heat transfer rate through the system of Fig. 10.20a (Oosthuizen, 1988a).

10.2.1.2. Other Studies

Weaver and Viskanta (1986) experimented with a cylindrical capsule (7.3 cm diameter, 15.9 cm length) filled with spherical beads and distilled water. Freezing was initiated by cooling the outer wall of the capsule. Experiments were conducted using either glass beads or aluminum beads, with the capsule oriented vertically or horizontally. Weaver and Viskanta (1986) complemented their measurements with a computational solution in which the heat transfer process was modeled as one-dimensional pure conduction. The computed thickness of the frozen zone agreed well with the experimental data for the combination of glass beads and distilled water, in which the difference between thermal conductivities is small. The agreement was less adequate when the aluminum beads were used. These observations lead to the conclusion that the effective porous-medium thermal conductivity model is adequate when the solid matrix and pore material have similar conductivities and that the local thermal equilibrium model breaks down when the two conductivities differ greatly.

The breakdown of the local thermal equilibrium assumption was studied further by Chellaiah and Viskanta (1987, 1989a). In the first of these experimental studies, Chellaiah and Viskanta examined photographically the freezing of water or water-salt solutions around aluminum spheres aligned inside a tube surrounded by a pool of the same phase-change material. They found that the freezing front advances faster inside the tube. When water was used, they found that the leading aluminum sphere is covered at first by a thin layer of ice of constant thickness. This layer was not present when the phase-change material was a water-salt solution.

In their second study of freezing of water, Chellaiah and Viskanta (1989a) showed that the water is supercooled (i.e., its temperature falls below the freezing point) before freezing is initiated. The degree of supercooling was considerably smaller than the one observed in the freezing of water in the absence of the porous matrix (glass, or aluminum beads).

Chellaiah and Viskanta (1989b, 1990) found good agreement between calculations using the Brinkman-Forchheimer equation and experiments using water and glass beads in a rectangular enclosure suddenly cooled from the side. They investigated the effects of imposed temperature difference and the superheat defined by $S = c_P(T_h - T_f)/h_{sf}$, where T_f is the fusion temperature. For small S the flow is weak and the interface is almost planar. The larger S convection modifies the shape of the interface. Further numerical results for lateral transient freezing were reported by Sasaki *et al.* (1990). A further numerical and experimental study was performed by Sasaki and Aiba (1992).

A boundary layer solution, appropriate for high Rayleigh number, for freezing on the exterior of a vertical cylinder was obtained numerically by Wang *et al.* (1990). Transient freezing about a horizontal cylinder was studied numerically by Bian and Wang (1993). Experiments with an inclined bed of packed spheres were performed by Yang *et al.* (1993a,b).

A generalized formulation of the Darcy-Stefan problem, one valid for irregular geometries with irregular subregions and not requiring the smoothness of the temperature, was proposed by Rodrigues and Urbano (1999). A comprehensive theoretical and experimental study of lateral freezing with an aqueous salt solution as the fluid, and taking into account anisotropy and the formation of dendrite arrays, was made by Song and Viskanta (2001). They found that the porous matrix phase affected the freezing of the aqueous salt solution by offering an additional resistance to the motion of the fluid and migration of separate crystals. The amount of macrosegregation was found to be mainly controlled by the porous matrix permeability in the direction of gravity, while macrosegregation was decreased when the permeabilities of the porous matrix phase and/or dendrite arrays were decreased.

10.2.2. Cooling from Above

Experiments on layers cooled from above were performed by Sugawara *et al.* (1988). They employed water and beads of either glass or steel. Their main interest was in predicting the onset of convection. Experimental and numerical work was reported by Lein and Tankin (1992b). The experimental work involved visualization. The authors reported that the convection process is controlled by the mean Rayleigh number and weakens as the freezing process proceeds. They examined results for various aspect ratios and they found that these agreed reasonably well with the formula of Beck (1972), Fig. 6.20. A nonlinear stability analysis was presented by Karcher and Müller (1995). The analysis shows that due to the kinematic conditions at the solid/liquid interface, hexagons having upflow in the center are stable near the onset of convection, but for sufficiently supercritical

Rayleigh numbers rolls are the only stable mode. The transition from hexagons to rolls is characterized by a hysteresis loop. A numerical study of a superheated fluid saturated porous medium in a rectangular cavity, with the bottom and side walls insulated and the top wall maintained at a constant temperature below the freezing point, was reported by Zhang and Nguyen (1999). A substantial numerical and experimental study was reported by Kimura (2005).

10.2.3. Solidification of Binary Alloys

When a binary mixture solidifies from a solid boundary, the planar solidification front often becomes unstable due to constitutional undercooling and the result is a mushy layer, separating the completely liquid phase from the completely solid phase. The mushy layer has been modeled as a reactive porous medium. A feature of the mushy zone is that it contains columnar solid dendrites, and so the porous medium is anisotropic. One principal axis for the anisotropic permeability is commonly, but not necessarily, approximately aligned with the temperature gradient.

The solidification of aqueous solutions of binary substances (notably ammonium chloride) is analogous in many ways to the solidification of metallic alloys, so experiments are often done with aqueous solutions. A pioneering study of solidification in a vertical container was carried out by Beckermann and Viskanta (1988b). Fundamental experimental work on solidification produced by cooling from the side in a rectangular cavity has been performed by Choi and Viskanta (1993) and Matsumoto *et al.* (1993, 1995), while Cao and Poulikakos (1991a,b) and Choi and Viskanta (1992) observed solidification with cooling from above and Song *et al.* (1993) observed cooling from below. Okada *et al.* (1994) did experiments on solidification around a horizontal cylinder.

The simplest model for the momentum equation, Darcy's law, was introduced in this context by Mehrabian *et al.* (1970). Subsequent modeling has been based on either a mixture theory in which the mushy zone is viewed as an overlapping continuum (e.g., Bennon and Incropera, 1987) or on volume averaging (e.g., Beckermann and Viskanta, 1988b; Ganesan and Poirier, 1990—the latter were more explicit about underlying assumptions). The second approach requires more work, but in relating macroscopic effects to microscopic effects it leads to greater insight about the physical processes involved. The averaging approach also allows the incorporation of the effects of thermal or chemical nonequilibrium or a moving solid matrix (Ni and Beckermann, 1991b). Felicelli *et al.* (1991) investigated the effect of spatially varying porosity, but found that that had no significant effect on the convection pattern. They did find that the effect of remelting in part of the mushy zone was important. Poirier *et al.* (1991) showed that for relatively large solidification rate and/or thermal gradients, the effects of heat of mixing need to be incorporated in the energy equation. Using the mixture continuum model modified to include the effect of shrinkage induced flow, Chiang and Tsai (1992) analyzed solidification in a two-dimensional rectangular cavity with riser. For the same geometry, Schneider and Beckermann (1995) used numerical simulation to compare two types (Scheil and lever-rule) of microsegregation models; the predicted macrosegregation patterns were found to be similar although the predicted eutectic fraction is significantly higher with the Scheil model. They noted that the predicted pattern is sensitive to the permeability function assumed in the model.

Ni and Incropera (1995a,b) produced a new model that retains the computational convenience of the mixture continuum model while allowing for the inclusion of important features of the volume-average two-phase model. They relaxed several assumptions inherent in the original formulation of the two-phase model, making it possible to account for the effects of solutal undercooling, solidification shrinkage, and solid movement.

The effect of anisotropy of permeability has been investigated by Sinha *et al.* (1992, 1993) and Yoo and Viskanta (1992). A three-phase model, in which the release of dissolved gas from the alloy is taken into account, was developed by Kuznetsov and Vafai (1995a).

Prescott and Incropera (1995) introduced the effect of turbulence in the context of stirring produced by an oscillating magnetic field. Their results indicate that turbulence decreases the propensity for channel development and macrosegregation by enhancing mixing and reducing the effective Lewis number from a large value to near unity. For modeling the turbulence, they employed an isotropic low-Reynolds number k-ε model. The turbulence is produced via a shear-production source term. They carried out numerical calculations for comparison with experiments with a lead-tin alloy. The turbulence occurs in the liquid and near the liquidus interface; it is strongly dampened in the mushy zone. Prescott and Incropera remark that turbulence can survive in the mush only in regions with porosity about 0.99 or higher, and there slurry conditions are likely to occur in practice. However, this assumption may be an artifact of an assumption of the model (Lage, 1996), and turbulence may penetrate further into the mushy layer than this model predicts.

Compositional convection can occur in a mushy layer cooled from below when unstable density gradients are formed as a result of rejection of the lighter component of the mixture upon solidification. There is an interaction among convection, heat transfer, and solidification that can lead to the formation of "chimneys," or localized channels devoid of solid through which buoyant liquid rises. An analytical investigation of chimneys was made by Roberts and Loper (1983), who used equations formulated by Hills *et al.* (1983). Observations of chimneys led to stability analyses. Fowler (1985) modeled the mushy layer as a nonreacting porous layer, while the linear stability analysis of Worster (1992) included the effects of the interaction of convection and solidification. Linear stability analysis had been applied previously by Nandapurkar *et al.* (1989). Worster identified two direct modes of convective instability: one driven from a narrow compositional boundary layer about the mush-liquid interface and the other driven from the interior of the mushy layer. The graph of Rayleigh number versus wavenumber has two minima. The boundary-layer mode results in fine-scale convection in the melt above the mushy layer and leaves the interstitial fluid in the mushy layer virtually stagnant. The mushy-layer model causes perturbations to the solid fraction of the mushy layer that are indicative of a tendency to form chimneys. Good quantitative agreement was found with the experimental results of Tait and Jaupart (1992) for the onset of the mushy-layer mode of convection. These authors and Tait *et al.* (1992) discussed geophysical implications of their experimental results.

The linear stability analysis of Emms and Fowler (1994) involved a time-dependent basic state that included the effect of finger-type convection in the liquid. However, their analysis indicated that the onset of convection in the mushy layer is little affected by vigorous convection in the melt.

Worster's (1991, 1992) analysis was extended by Chen *et al.* (1994) to the case of oscillatory modes. They found that when stabilizing thermal buoyancy is present in the liquid, the two steady modes of convection can separate by way of an oscillatory instability. They noted that the oscillatory instability occurred only when the buoyancy ratio (thermal to solutal) in the liquid region was nonzero, so they associated the oscillatory instability with the interaction of the double-diffusive convection in the liquid region with the mushy-layer convective mode. Their results showed that the steady modes became unstable before the oscillatory mode. Chen *et al.* (1994) also performed experiments with ammonium chloride solution which confirmed that during the progress of solidification the melt in the mush is in a thermodynamic equilibrium state except at the melt-mush interface where most of the solidification occurs.

A weakly nonlinear analysis based on the assumption that the mushy layer is decoupled from the overlying liquid layer and the underlying solid layer was performed by Amberg and Homsy (1993). They made progress by considering the case of small growth Péclet number, small departures from the eutectic point, and infinite Lewis number. Their analysis, which revealed the structure of possible nonlinear, steady convecting states in the mushy layer, was extended by Anderson and Worster (1995) to include additional physical effects and interactions in the mushy layer. They employed a near-eutectic approximation and considered the limit of large far-field temperature, so that their model involved small deviations from the classic HRL problem. The effects of asymmetries in the basic state and the nonuniform permeability lead to transcritically bifurcating convection with hexagonal planform. They produced a set of amplitude equations that described the evolution of small-amplitude convecting states associated with direct modes of instability. Analysis of these revealed that either two-dimensional rolls or hexagons can be stable, depending on the relative strengths of different physical mechanisms. They determined how to adjust the control parameters to minimize the degree of subcriticality of the bifurcation, and hence render the system more stable globally. Moreover, their work suggested the possibility of an oscillatory mode of instability despite the lack of any stabilizing thermal buoyancy, in contrast with the results of Chen *et al.* (1994).

The linear instability analysis of Anderson and Worster (1996) was designed to investigate this new oscillatory instability. Their model contained no double-diffusive effects and no region in which a statically stable density gradient exists. They considered the limit of large Stefan number, which incorporates a key balance for the existence of the oscillatory instability. They discovered that the mechanism underlying the oscillatory instability involves a complex interaction between heat transfer, convection, and solidification. Further work on the oscillatory modes of nonlinear convection has been reported by Riahi (2002b). The modes take the form of two- and three-dimensional traveling and standing waves. For most of the parameter range studied, supercritical simple traveling waves are stable. Riahi (1998a) examined the structure of an unsteady convecting mushy layer. He identified four regimes corresponding to high or low Prandtl number melt and strongly or weakly dependent flow. He found that strongly time-dependent flow can lead to nonvertical chimneys and for weakly time-dependent flow of a low Prandtl number melt vertical chimneys are possible only when the chimneys have small radius.

Some of the experimental results reported by Chen (1995) confirm the theoretical predictions, while others reveal phenomena not observed hitherto.

The effects of rotation about a vertical axis were included in the linear stability analysis of Lu and Chen (1997). They noted that very high rotation rates were necessary to significantly increase the critical Rayleigh number, but smaller rates could change the most critical convection mode. They found their results to be sensitive to the value of a buoyancy ratio defined as $\Gamma\alpha_T/(\alpha_s - \Gamma\alpha_T)$, where α_T, α_s are the thermal, solutal expansion coefficients, respectively, and Γ is the slope of the solidus. The effect of rotation also was studied by Riahi (1993b, 1997), Sayre and Riahi (1996, 1997) and Riahi and Sayre (1996). The latter investigated nonlinear natural convection under a high gravity environment, where the rotation axis is inclined to the high gravity vector. They found that for some particular moderate rotation range, the vertical velocity in the chimneys decreases rapidly with increasing rotation rate and appears to have opposite signs across some rotation-dependent vertical level.

The study by Guba (2001) concentrated on the way rotation controls the bifurcating convection with various planforms. Govender and Vadasz (2002a,b) have reported a weak nonlinear analysis of moderate Stefan number stationary convection in rotating layers. Further linear stability studies have been made by Govender and Vadasz (2002c) and Govender (2003b, 2005a–c). The results show that generally the oscillatory mode is the most dangerous mode for intermediate values of the Stefan number at sufficiently large Taylor number values, while the stationary mode is the most dangerous for very small and very large values of the Stefan number. Further finite amplitude studies of convection have been carried out by Govender (2003d,e, 2004c) to consider factors such as large Stefan number or small variations in retardability. Other studies by Riahi (2003a,b) on effects of rotation have dealt with oscillatory modes of convection and with nonlinear steady convection. Some aspects of the topic were reviewed by Riahi (1998b, 2002a).

A numerical study of the effects of rotation was made by Neilson and Incropera (1993). They found that slow, steady rotation had insignificant effect on channel formation, but with intermittent rotation corresponding to successive spin-up and spin-down of the mold in their numerical study channel nucleation was confined to the centerline and outer radius of the casting. They attributed the elimination of channels from the core of the casting to the impulsive change in angular frequency associated with spin-up and its effect on establishing an Ekman layer along the liquidus front, the front being washed by flow within the layer, thereby eliminating the perturbations responsible for channel nucleation.

A flow-focusing instability, driven by expansion or contraction upon solidification, was analyzed by Chiareli and Worster (1995), and comparisons were made with acid-etching instabilities in porous rocks. They concluded that though the potential for instability exists, it is unlikely to occur in practice.

For the case of unidirectional solidification, Krane and Incropera (1996) performed a scaling analysis that showed that Darcy's law was adequate in the mushy zone except in the region near the liquidus isotherm, and that advection dominates the solute transport throughout the mush, though in the denser regions of the solid-liquid region the liquid velocities are so small as to have a negligible effect of macrosegregation.

The review by Worster (1997) contains a summary of a theory of an ideal mushy layer. When use is made of the linear liquidus relationship

$$T = T_E + \Gamma(C - C_E), \tag{10.71}$$

where Γ is a constant and the subscript E refers to the eutectic point, the equation of state (9.1) reduces to

$$\rho_f = \rho_0 + \beta^*(C - C_0), \tag{10.72}$$

where

$$\beta^* = -\beta\Gamma - \beta_C. \tag{10.73}$$

Consequently an appropriate Rayleigh number is

$$\mathrm{Ra}_m = \frac{\rho_0 g \beta^* K_0 \Delta C}{\mu V} = \frac{\rho_0 g \beta * K_0 L \Delta C}{\mu \alpha_m} \tag{10.74}$$

where K_0 is a reference permeability and V is the rate of solidification and the thermal length scale L is defined by $L = \alpha_m / V$. Convection in the ideal mushy layer is governed by Ra_m together with a Stefan number and a compositional ratio. Experimental results such as those by Bergman *et al.* (1997) confirm that Ra_m is indeed a governing parameter.

Worster's (1997) review also includes a discussion of explanations of why chimneys may or not form. The explanation of Worster and Kerr (1994) is that interfacial undercooling causes a strengthening of the boundary-layer mode of convection, which retards growth of the mushy layer, increases its solid fraction, and decreases the compositional contract across it. These three effects combine to reduce Ra_m, and as time progresses Ra_m may reach a maximum less than that required for chimneys to form. Worster (1997) also mentions experiments related to the formation of a mushy zone in sea ice (Wettlaufer *et al.*, 1997), as well as applications to solidifying magmas and the molten outer core of the Earth. The development of chimneys has been further studied numerically by Schulze and Worster (1998, 1999) and by Emms (1998). An alternative model for mush-chimney convection was proposed by Loper and Roberts (2001).

Further work on plume formation in mushy layers has been reported by Chung and Chen (2000) and Chung and Worster (2002). The effect of initial solutal concentration on the evolution of the convection pattern during the solidification of a binary mixture was examined experimentally by Skudarnov *et al.* (2002). An experimental study of the solidification of a ternary alloy was reported by Thompson *et al.* (2003). A model for the diffusion-controlled solidification of ternary alloys was described by Anderson (2003). A morphological instability due to a forced flow in the melt was analyzed by Feltam and Worster (1999) and Chung and Chen (2001). An alternative hybrid model of a mushy zone has been proposed by Mat and Ilegbusi (2002). An experimental study of the suppression of natural convection by an additive to increase the viscosity of the fluid was reported by Nishimura and Wakamatsu (2000).

Further complexities of alloy solidification are discussed in the reviews by Beckermann and Viskanta (1993), Beckermann and Wang (1995), and Prescott and Incropera (1996). Experimental work has been reported by Solomon and Hartley (1998). A numerical investigation of the macrosegregation during the thin strip casting of carbon steel was made by Kuznetsov (1998a). Another convective instability problem involving solidification was analyzed using linear stability theory by Hwang (2001). An expository article on the solidification of fluids was presented by Worster (2000). Adnani and Hsiao (2005) have reviewed transport phenomena in liquid composites modeling processes and their roles in process control and optimization.

Roberts *et al.* (2003) have considered the convective instability of a plane mushy layer that advances as heat is withdrawn at a uniform rate from the bottom of an alloy. They assumed that the solid that forms is composed entirely of the denser constituent, making the residual liquid compositionally buoyant, and thus prone to convective motion. They focused on the large-scale mush mode of instability, quantified the minimum critical Rayleigh number, and determined the structure of the convective modes of motion within the mush and the associated deflections of the mush-melt and mush-solid boundaries.

A related problem involving dissolution-driven convection was investigated by Hallworth *et al.* (2005). They considered experimentally and theoretically the heating from above of an initially homogeneous layer of solid crystals, saturated liquid, and glass ballotini. The heat flux causes crystals at the top of the layer to dissolve, forming liquid that, being more concentrated, drives convection in the lower layer. Mixing of this concentrated liquid into the lower layer leads to precipitation, thereby releasing latent heat that raises the temperature of the lower layer. There results a three-layered system: clear fluid, clear fluid plus close-packed ballotini, and a mixture of solid crystals, ballotini, and saturated liquid. The theoretical model used is based on the concept that the heat supplied from above is used entirely for the dissolution of solid crystals at the upper boundary of the lower layer. The resulting compositional convection redistributes the dissolved salt uniformly through the lower layer where it partly recrystallizes to restore chemical equilibrium. The crystallization leads to a gradual and uniform increase in both the solid fraction and the temperature of the lower layer.

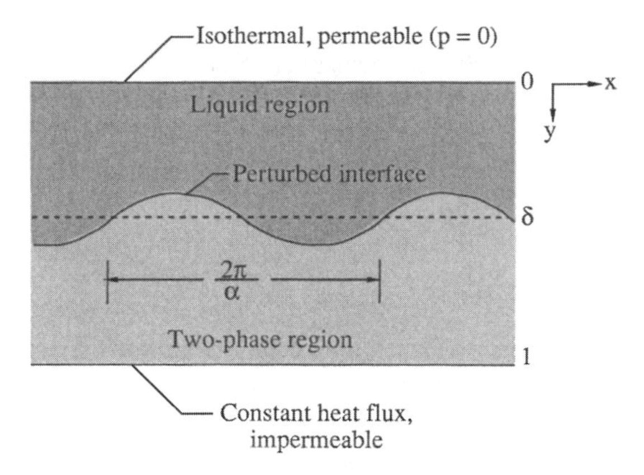

Figure 10.22. Definition sketch for boiling produced by heating from below.

10.3. Boiling and Evaporation

10.3.1. Boiling Produced by Heating from Below

When boiling begins in a fluid-saturated porous medium heated from below, a two-layer system is formed with a liquid region overlying a two-phase region, as sketched in Fig. 10.22. Experiments by Sondergeld and Turcotte (1977) and Bau and Torrance (1982a) have shown that the liquid regime temperature profile may be conductive or convective, but the two-phase region is essentially isothermal at the saturation temperature. The two-phase region may be liquid-dominated or vapor-dominated. Heat is transported across the two-phase region by vertical counterpercolation of liquid and vapor; liquid evaporates on the heating surface and vapor condenses at the interface between the liquid and two-phase regions. Experiments have indicated that thermal convection in the liquid region may occur before the onset of boiling or after the onset of boiling. Visualization experiments (Sondergeld and Turcotte, 1978) reveal that after the onset of convection the liquid region streamlines penetrate the two-phase region. The convection in the liquid region is in the form of polyhedral cells whose dimensions vary with the heat flux.

With the liquid region overlying the two-phase region there are two mechanical mechanisms for instability: buoyancy and gravitational instability, the latter due to the heavier liquid region overlying the lighter two-phase region. The gravitational instability differs from the classical Rayleigh-Taylor instability of superposed fluids because the interface is now permeable and therefore permits both heat and mass transfer across it. Schubert and Straus (1977) noted that convection also can be driven by a phase-change instability mechanism. If steam and water stay in thermal equilibrium, then thermal perturbations lead to pressure variations that tend to move the fluid against the frictional resistance of the medium. Because of conservation of mass, horizontal divergence is accompanied by vertical contraction and phase change takes place so that the vertical forces stay in balance. In a porous medium containing saturated liquid or a liquid-vapor mixture, convection occurs more readily by the phase-change mechanism than it would with ordinary liquid driven by buoyancy. Phase-change-driven convection is concentrated toward the bottom of the porous layer and the cells are narrow in comparison with their depth. The model used by Schubert and Straus (1977) is valid only for a mixture with small amounts of steam.

Schubert and Straus (1980) also considered the stability of a vapor-dominated system with a liquid region overlying a dry vapor region. Their analysis predicts that such systems are stable provided that the permeability is sufficiently small. The stability arises because when liquid penetrates the interface, that interface is distorted so the system remains on the Clapeyron curve, and this results in a pressure gradient that acts to restore equilibrium.

O'Sullivan (1985b) described some numerical experiments modeling a geothermal reservoir in which the level of heat input at the base of a layer is varied. As the heat input is increased the flow changes from conduction to single-phase convection, then to convection with an increasingly larger boiling zone, and finally to an irregular oscillatory two-phase convection.

The onset of two-dimensional roll convection in the configuration of Fig. 10.22 was studied using linear stability analysis by Ramesh and Torrance (1990). They assumed that the relative permeabilities of liquid and vapor were linear functions of the liquid saturation S. Their analysis reveals that the important parameters are the Rayleigh numbers Ra and $\mathrm{Ra}_{2\varphi}$ in the liquid and two-phase regions and the dimensionless heat flux Q_b at the lower boundary. The parameters are defined by

$$\mathrm{Ra} = \frac{g\beta_l KH(T_s - T_0)}{\nu_l \alpha_{ml}}, \quad \mathrm{Ra}_{2\varphi} = \frac{(1-\bar{\rho}_v)KH}{\nu_l \alpha_{ml}}, \quad Q_b = \frac{q_b'' H}{k_m(T_s - T_0)}, \tag{10.75}$$

where q_b'' is the heat flux at the lower boundary, T_s is the saturation temperature, T_0 is the temperature at the top boundary, $\bar{\rho}_v$ is the ratio of vapor to liquid densities, and $\bar{\mu}_l$ is the ratio of liquid to vapor viscosities, while λ (see Fig. 10.23) is defined to be $h_{fg}/[c_{Pl}(T_s - T_0)]$, where h_{fg} is the latent heat.

For sufficiently large Q_b there is dryout of the liquid phase region in the two-phase region. For smaller values of Q_b there are two S values for each value of Q_b (Fig. 10.23). The smaller value ($S < 0.17$ for water) corresponds to a vapor-dominated system and the larger value to a liquid-dominated system. For a liquid-dominated system, the solution map (for water) is shown in Fig. 10.24. The picture is approximate because it is based on a single wavenumber, $\alpha = \pi$. We are primarily interested in values $Q_b = 1$ because $1/Q_b$ is the ratio of the mean depth of the interface to the total depth of the medium.

The onset of boiling is indicated by the curve *ABE*. For Q_b values above this curve, boiling occurs with a liquid layer overlying a two-phase zone. For Q_b values below *ABE* boiling does not occur. The onset of convection in the liquid is denoted by the curve *CBD*; convection occurs only to the right of this curve. Its

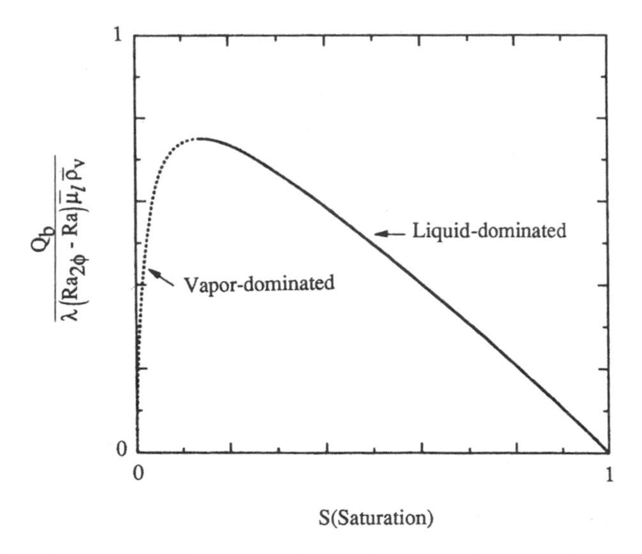

Figure 10.23. The relationship between bottom heat transfer rate and saturation for the basic state in a steam-water system (Ramesh and Torrance, 1990, with permission from Pergamon Press).

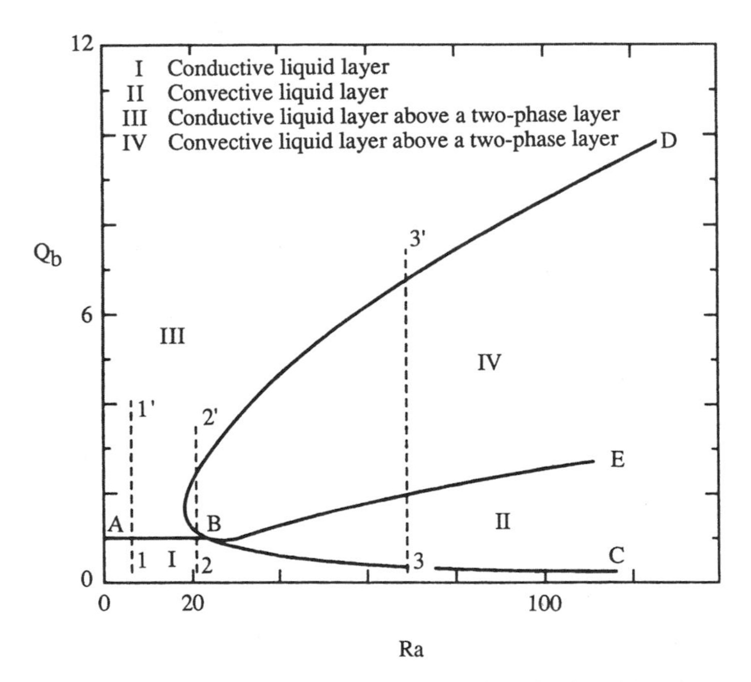

Figure 10.24. Map of conductive and convective solutions in (Ra, Q_b) parameter space for liquid-dominated two-phase systems, for the nondimensional wavenumber $\alpha = \pi$ (Ramesh and Torrance, 1990, with permission from Pergamon Press).

nose defines the critical Rayleigh number as Q_b varies for $\alpha = \pi$. (As Q_b and α both vary, the minimum value of Ra is 14.57, attained at $Q_b = 1.35$, $\alpha = 1.9$.) In laboratory experiments boiling occurs when the temperature at the bottom reaches the saturation temperature T_s. The branch AB corresponds to $Q_b = 1$ and represents the onset of boiling before the onset of convection, while the branch BE represents the onset of boiling after convection already exists within a liquid-filled layer.

We consider experiments conducted on a porous medium with constant properties by varying the bottom heat flux. At low Ra (as indicated by line 1-1′) the liquid region is conductive before and after the onset of boiling. This is consistent with the experiments of Bau and Torrance (1982a) on low permeability porous beds ($K = 11 \times 10^{-12}\text{m}^2$). At higher Ra (as indicated by line 2-2′), the liquid region is conductive before the onset of boiling but becomes convective almost immediately when boiling starts, which is in agreement with the observations of Sondergeld and Turcotte (1977), ($K = 70 \times 10^{-12}\text{m}^2$). For large Ra (as indicated by line 3-3′) the liquid region becomes convective before the onset of boiling and stays convective after the onset of boiling, which is consistent with the experiments of Bau and Torrance (1982c) on high permeability beds ($K = 1600 \times 10^{-12}\text{m}^2$). They observed that at large heat fluxes the liquid region reverts back to a conductive state, which is consistent with Fig. 10.24.

For vapor-dominated systems the density difference between the liquid and two-phase regions is large, and as we noted above we can expect gravitational instability to dominate over buoyancy effects. If the buoyancy effects are negligible (Ra = 0), the stability diagram shown in Fig. 10.25 is obtained. This applies for water

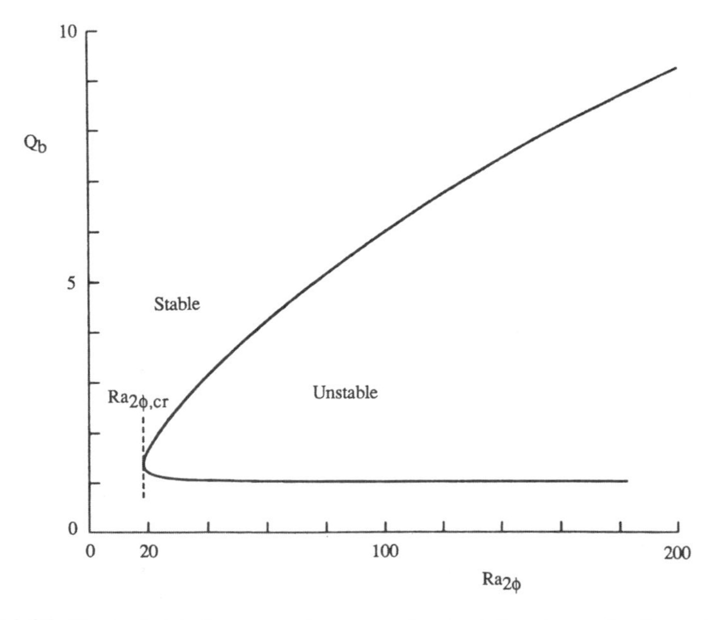

Figure 10.25. Neutral stability curve for vapor-dominated systems, for Ra = 0, $\alpha = \pi$ (Ramesh and Torrance, 1990, with permission from Pergamon Press).

with $T_0 = 30°C$, $T_s = 100°C$. For $\alpha = \pi$ the minimum value of $Ra_{2\varphi}$ is 18.95, occurring for $Q_b = 1.4$.

The minimum value of S on the curve BD in Fig. 10.24 is approximately equal to 0.98. The maximum value of S on the curve in Fig. 10.25 is about 0.02. We conclude that if the rest-state value of S lies in the range 0.02 to 0.98, then the rest state is stable according to linear theory.

However, the numerical study of Ramesh and Torrance (1993) indicates that finite-amplitude instability is possible in this range. This study involved convection and boiling in a two-dimensional rectangular region with length-to-height aspect ratio equal to 2. In order to model experiments in a Hele-Shaw cell, a volumetric cooling term (to take account of heat losses from the front and back walls of the cell) was allowed for in equations for the temperature and saturation. The results indicate three solution regimes: conduction-dominated, steady convection-dominated, and oscillatory convection. In some cases the solutions exhibit a dependence on initial conditions and perturbations. As Figure 10.26 indicates, the finite amplitude solutions agree with the linear stability analysis.

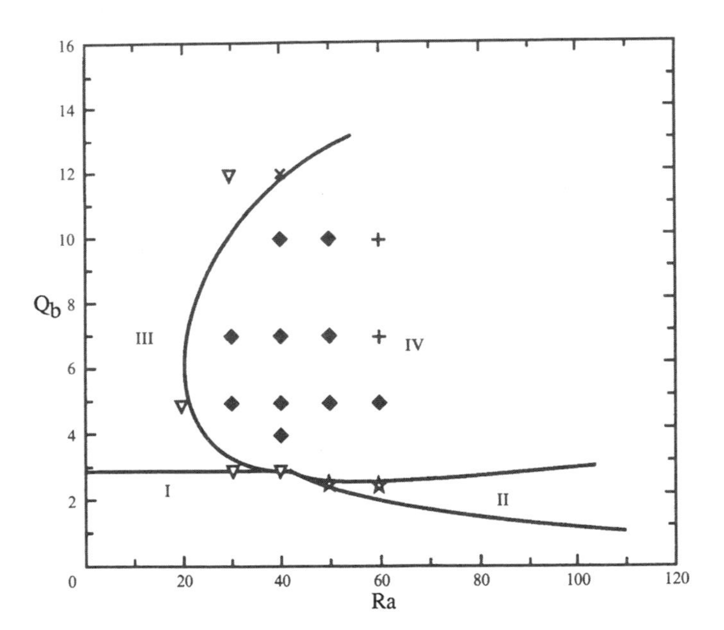

Figure 10.26. Comparison of numerical solutions (symbols) and linear stability theory (solid lines) in Ra - Q_b parameter space. I, II, III, and IV denote four solution regimes: I, a conductive liquid layer—no boiling; II, a convective liquid layer—no boiling; III, a conductive liquid layer overlying a two-phase layer; IV, a convective liquid layer overlying a two-phase layer. The numerically observed solutions are: *, steady convective liquid layer—no boiling, ∇, steady conductive liquid layer over a two-phase layer; ♦, steady convective liquid layer overlying a two-phase layer; +, steady or oscillatory convective liquid layer, overlying a two-phase layer; $\times$, steady conductive or steady convective liquid layer overlying a two-phase layer (Ramesh and Torrance, 1993, with permission from Cambridge University Press).

Ramesh and Torrance (1993) also reported that their numerical results agree with prior laboratory experiments, including those of Echaniz (1984) on the oscillatory convection, which is observed for high-permeability beds (i.e., high Ra). Such solutions are generated numerically for high Ra by introducing asymmmetric perturbations into a one-dimensional initial conduction field (initial symmetric disturbances lead to steady-state solutions). The time period in oscillations decreases with increase of Q_b. Heat transfer rates are drastically increased by the onset of oscillatory convection. Echaniz (1984) concluded that the oscillations are caused by thermals (pairs of small vortices) that originate at the heating surface where the cold fluid descends, grow, and then disappear either at the top boundary or in the two-phase region.

Ramesh and Torrance (1993) also showed that when steady convection had its onset after the onset of boiling the preferred computed convective mode is two cells symmetric about the centerline. The interface moves up as the heat flux is increased and is depressed in the center (indicating downflow of cold fluid there) and raised at the sides (or vice versa). The center of the cell lies in the liquid region, where the buoyancy production term is present. When the onset of convection precedes that of boiling, the stable two-cell convection pattern is retained after boiling if Ra is low, but at larger Ra a transition from a two-cell to a four-cell structure occurs, in qualitative agreement with the experiment of Tewari (1982). [The stable three cells also observed by Tewari (1982), not replicated in the computations, may have been due to experimental nonuniformities.] The steady-state heat flux Q_{top} for the numerical two-cell solutions was found to vary with heat-flux Rayleigh number $\mathrm{Ra}_f(= \mathrm{Ra}Q_b)$ according to $Q_{top} \propto \mathrm{Ra}_f^{0.6}$, in approximate accord with the experimental correlation $\mathrm{Nu} \propto \mathrm{Ra}_f^{0.5}$ reported by Echaniz (1984).

In connection with the testing of a new two-phase mixture model introduced by Wang and Beckermann (1993), Wang *et al.* (1994a,b) have made a numerical study of boiling in a layer of a capillary porous medium heated from below. Their numerical procedure employs a fixed grid and avoids tracking explicitly the moving interface between the liquid and two-phase regions. Also on the new mixture model, Wang and Beckerman (1995) performed a two-phase boundary layer analysis and Easterday *et al.* (1995) studied numerically and experimentally two-phase flow and heat transfer in a horizontal porous formation with horizontal water throughflow and partial heating from below. The latter found that the resulting two-phase structure and flow patterns are strongly dependent on the water inlet velocity and the bottom heat flux. They reported qualitative agreement between numerical and experimental results. Wang *et al.* (1994a) studied numerically transient natural convection and boiling in a square cavity heated from below. They observed boiling-induced natural convection, flow transition from a unicellular to a bicellular pattern with the onset of boiling, and flow hysteresis as the bottom heat flux first increases and then decreases. This subject has been reviewed by Wang (1998a). A numerical study of boiling with mixed convection in a vertical porous layer was made by Najjari and Ben Nasrallah (2002), while Najjari and Ben Nasrallah (2005) similarly studied the effect of aspect ratio on natural convection in a rectangular cavity.

Stemmelen *et al.* (1992) noted that large-amplitude oscillations are observed in a boiling porous medium with high heat fluxes and they presented a simplified linear stability analysis that they carried out to determine the stability criterion.

For discussion of some wider aspects of boiling and two-phase flow in porous media, the reader is referred to the reviews by Dhir (1994, 1997).

10.3.2. Film Boiling

It was observed by Parmentier (1979) that because of the nature of the (P, T) phase diagram, the thin film of water vapor that forms adjacent to a vertical surface is separated from the liquid water by a sharp interface with no mixed region in between. The assumption that the vapor and liquid form adjacent boundary layers (as in Fig. 10.26), with a stable smooth interface, is mathematically convenient and has been adopted in most theoretical studies of film boiling. In reality the interface may be wavy or unsteady, due to the formation and detachment of bubbles.

If one assumes, following Cheng and Verma (1981), that the Oberbeck-Boussinesq approximation and Darcy's law are applicable and variables are defined as in Fig. 10.27, then the governing equations for the region saturated with superheated vapor (subscript v), $y < \delta_v$, are

$$\frac{\partial u_v}{\partial x} + \frac{\partial v_v}{\partial y} = 0, \tag{10.76}$$

$$u_v = -\frac{K}{\mu}(\rho_v - \rho_\infty)g, \tag{10.77}$$

$$u_v \frac{\partial T_v}{\partial x} + v_v \frac{\partial T}{\partial y} = \alpha_m \frac{\partial^2 T_v}{\partial y^2}, \tag{10.78}$$

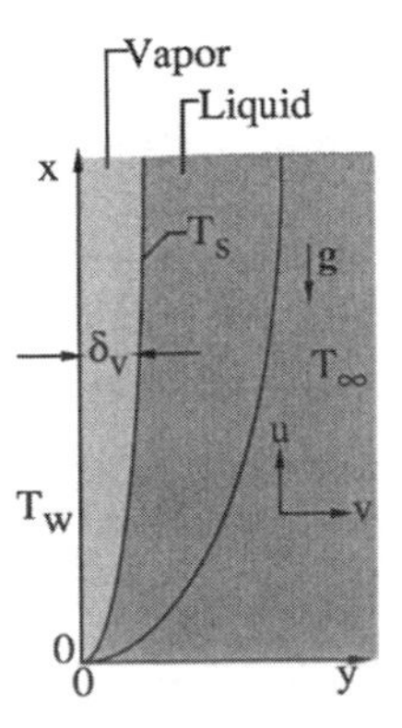

Figure 10.27. Definition sketch for film boiling.

while those for the region filled with subcooled liquid (subscript l), $y > \delta_v$, are

$$\frac{\partial u_l}{\partial x} + \frac{\partial v_l}{\partial y} = 0, \tag{10.79}$$

$$u_l = \frac{\rho g \beta_{l\infty} K (T_l - T_\infty)}{\mu_l}, \tag{10.80}$$

$$u_l \frac{\partial T_l}{\partial x} + v_l \frac{\partial T_l}{\partial y} = \alpha_{ml} \frac{\partial^2 T_l}{\partial y^2}. \tag{10.81}$$

The boundary conditions are

$$y = 0: \quad v_v = 0, \quad T_v = T_w, \tag{10.82}$$

$$y \to \infty: \quad u_l = 0, \quad T_l = T_\infty, \tag{10.83}$$

where the saturation temperature T_s satisfies $T_w > T_s = T_\infty$. At the vapor-liquid interface $y = \delta_v$, we have

$$T_v = T_s = T_l, \tag{10.84}$$

$$\rho_v \left(v_v - u_v \frac{d\delta_v}{dx} \right) = \rho_l \left(v_l - u_l \frac{d\delta_v}{dx} \right) = \dot{m}_\delta, \tag{10.85}$$

$$-k_{mv} \frac{\partial T_v}{\partial y} = \dot{m}_\delta h_{fv} - k_{ml} \frac{\partial T_l}{\partial y}, \tag{10.86}$$

where k_m is the effective thermal conductivity of the porous medium and h_{fv} is the latent heat of vaporization of the liquid at T_s. Equation (10.86) states that the energy crossing the interface is partly used to evaporate liquid at a rate $\dot{m}_\delta$.

We introduce the streamfunctions ψ_v, ψ_l defined by

$$u_v = \frac{\partial \psi_v}{\partial y}, \quad v_v = -\frac{\partial \psi_v}{\partial x}, \tag{10.87}$$

$$u_l = \frac{\partial \psi_l}{\partial y}, \quad v_l = -\frac{\partial \psi_l}{\partial x}, \tag{10.88}$$

and the similarity variables defined by

$$\eta_v = (\mathrm{Ra}_{xv})^{1/2} y/x, \eta_l = (\mathrm{Ra}_{xl})^{1/2} (y - \delta_v)/x, \tag{10.89}$$

$$\psi_v = \alpha_{mv} (\mathrm{Ra}_{xv})^{1/2} f_v(\eta_v), \psi_l = \alpha_{ml} (\mathrm{Ra}_{xl})^{1/2} f_l(\eta_l), \tag{10.90}$$

$$T_v - T_s = (T_w - T_s)\theta_v(\eta_v), T_l - T_s = (T_s - T_\infty)\theta_l(\eta_l), \tag{10.91}$$

where

$$\mathrm{Ra}_{xv} = \frac{(\rho - \rho_v) g K x}{\mu_v \alpha_{mv}}, \quad \mathrm{Ra}_{xl} \frac{\rho g \beta_l K (T_s - T_\infty) x}{\mu_l \alpha_{ml}}. \tag{10.92}$$

We then have

$$f_v' = 1, f_l' = \theta, \tag{10.93}$$

$$2\theta_v'' + f_v \theta_v' = 0, 2\theta_l'' + f_l \theta_l' = 0, \tag{10.94}$$

$$f_v(0) = 0, f_l'(\infty) = 0, \tag{10.95}$$

$$\theta_v(0) = 1, \theta_l(\infty) = 0 \tag{10.96}$$

and at the interface, which is given by $y = \delta_v$, and therefore by

$$\eta_v = \eta_{v\delta} = \mathrm{Ra}_{xv}^{1/2}\delta_v/x, \quad \eta_l = 0, \tag{10.97}$$

we have

$$\theta_v(\eta_{v\delta}) = 0, \quad \theta_l(0) = 0, \tag{10.98}$$

$$f_1(0) = -\frac{\dot{m}_\delta 2x^{1/2}}{\rho[\alpha_{ml}\rho_\infty g\beta_1 K(T_s - T_\infty)/\mu_1]^{1/2}} = \frac{R}{\mathrm{Sc}^{1/2}}\eta_{v\delta}, \tag{10.99}$$

$$\mathrm{Sh}\theta_v'(\eta_{v\delta}) = \frac{\mathrm{Sc}^{3/2}}{R} \quad \theta_l'(0) - \frac{\eta_{v\delta}}{2}. \tag{10.100}$$

Here

$$\mathrm{Sc} = c_{Pl}(T_s - T_\infty)/h_{fv}, \ \mathrm{Sh} = c_{Pv}(T_w - T_s)/h_{fv} \tag{10.101}$$

are "Jakob numbers" measuring, respectively, the degree of subcooling of the fluid and the superheating of the vapor, and R is defined by

$$R = \frac{\rho_v}{\rho_\infty}\left[\frac{\mu_l\alpha_{mv}(\rho_\infty - \rho_v)c_{Pl}}{\mu_v\alpha_{ml}\rho_\infty\beta_l h_{fv}}\right]^{1/2}. \tag{10.102}$$

Equation (10.99), which is related to the rate of evaporation, determines $\eta_{v\delta}$. The remaining equations in f_v, θ_v, f_l, and θ_l constitute a sixth-order eigenvalue problem. Those in f_v, θ_v have the exact solution

$$f_v = \eta_v, \quad \theta_v = 1 - \frac{\mathrm{erf}(\eta_v/2)}{\mathrm{erf}(\eta_{v\delta}/2)} \tag{10.103}$$

while those in f_l, θ_l reduce to the problem discussed in Section 5.1.2 if the values of $\eta_{v\delta}$, R, and Sc are prescribed.

We define the local Nusselt number Nu_x in terms of the wall heat flux q_w'', so

$$\mathrm{Nu}_x = \frac{q_w'' x}{k_{mv}(T_w - T_s)}, \tag{10.104}$$

and then

$$\frac{\mathrm{Nu}_x}{\mathrm{Ra}_{xv}^{1/2}} = -\theta_v'(0) = \frac{1}{\pi^{1/2}\mathrm{erf}(\eta_{v\delta}/2)}. \tag{10.105}$$

The value of $\theta_v'(0)$ can be obtained numerically, and results are shown in Fig. 10.27. In particular we have the asymptotic result

$$\frac{\mathrm{Nu}_x}{\mathrm{Ra}_{xv}^{1/2}} \to 0.564 \quad \text{as} \quad \mathrm{Sh} \to \infty. \tag{10.106}$$

Results for other geometrical configurations are readily attained (Cheng *et al.*, 1982). For example, for a horizontal cylinder of diameter D, we have Eq. (5.120), modified by the replacement of the coefficient 0.628 with the expression $2^{1/2}[-\theta_v'(0)]$. Likewise Eq. (5.122), similarly modified, applies for a sphere of diameter D. For a cone of half-angle α with axis vertical and vertex downward,

$$\frac{\mathrm{Nu}_x}{\mathrm{Ra}_{xv}^{1/2}} = 3^{1/2}[-\theta_v'(0)] \tag{10.107}$$

where now $g \cos\alpha$ replaces g in the definition of Ra_{xv}, while for a wedge the same applies except that the factor $3^{1/2}$ is absent.

Nakayama *et al.* (1987) have extended the boundary layer theory to general two-dimensional and axisymmetric bodies. They show that an accurate approximate formula is

$$\frac{\mathrm{Nu}_x}{(\mathrm{Ra}/I)^{1/2}} = \left\{\pi^{-1} + \left[\left(2\mathrm{Sh} + \left(0.444\frac{\mathrm{Sc}}{R_n}\right)^2\right)^{1/2} - 0.444\frac{\mathrm{Sc}}{R_n}\right]^{-2}\right\}^{1/2} \tag{10.108}$$

where

$$R_n = \frac{\rho_v \alpha_{mv}}{\rho_l \alpha_{ml}}\left[\frac{\alpha_{ml}\upsilon_l(\rho_l - \rho_v)}{\alpha_{mv}\upsilon_v\rho_v\beta_v(T_s - T_\infty)}\right]^{1/2}, \tag{10.109}$$

$$I(x) = \frac{\int_0^x g_x r^{*2} dx}{g_x r^{*2} x}, \tag{10.110}$$

$$r^* = \begin{cases} 1 \text{ for plane flow,} \\ \mathrm{r(x)} \text{ for axisymmetric flow,} \end{cases} \tag{10.111}$$

$$g_x = g\left[1 - \left(\frac{dr}{dx}\right)^2\right]^{1/2}. \tag{10.112}$$

Here $r(x)$ defines the surface, where x is measured along the surface from a stagnation point. Thus, for example, $I = 1$ for a vertical plate and $I = 1/3$ for a vertical cone pointing downward.

Subcooled forced convection film boiling over a vertical plate was analyzed by Nakayama and Koyama (1988b), and similarity solutions for the vertical plate, horizontal circular cylinder, and sphere were found by Nakayama and Koyama (1988a). A theoretical and experimental study of film boiling over a sphere or a horizontal cylinder was performed by Orozco *et al.* (1988). Film boiling of a binary mixture over a vertical plate was studied analytically and experimentally (with good agreement between the results) by Essome and Orozco (1991). A theoretical study of mixed convection film boiling of a binary mixture over a horizontal cylinder was reported by Orozco and Zhu (1993).

10.4. Condensation

Several authors have used a one-dimensional model to analyze condensation in porous media. For example, Vafai and Sarkar (1986, 1987) have reported a transient analysis of moisture migration and condensation in porous and partially

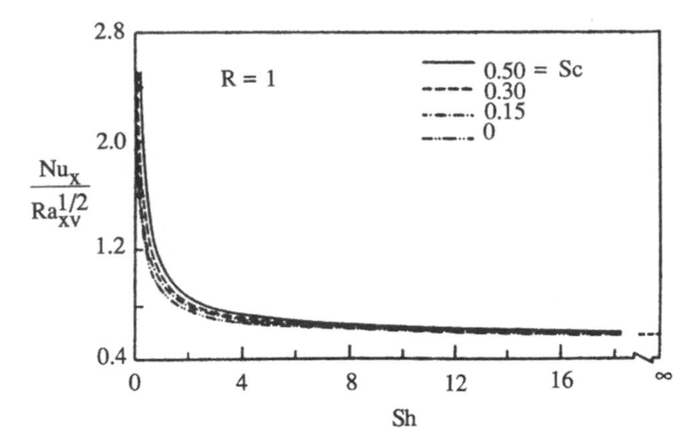

Figure 10.28. Heat transfer results for film boiling (Cheng and Verma, 1981, with permission from Pergamon Press).

porous enclosures, and Sözen and Vafai (1990) have analyzed the transient forced convective condensing flow of a gas through a packed bed, with quadratic drag effects incorporated. A two-dimensional transient model was employed by Vafai and Whitaker (1986) to study the accumulation and migration of moisture in an insulation material; this involved a porous slab.

The only problem that has been studied in depth is that of film condensation. This problem is analogous to that of film boiling, discussed in the previous section. The roles of the liquid and the vapor are reversed and heating is replaced by cooling, but the mathematical analysis is the same provided that the liquid/vapor interface remains sharp, i.e., there is no intervening two-phase region, provided that capillary effects are negligible. In the literature the analysis has been developed in parallel with that discussed in Chapter 5. Hence our discussion will be brief.

The original study by Cheng (1981b) for steady condensation outside a wedge or cone embedded in a porous medium filled with a dry saturated vapor was extended by Cheng and Chui (1984) to the transient situation. Liu *et al.* (1984) extended the analysis to treat general two-dimensional and axisymmetric bodies and to allow for the effect of lateral mass flux.

White and Tien (1987) employed the Brinkman equation to account for boundary friction and also the effect of variable porosity at the wall. Lai and Kulacki (1989b) allowed for the effect of temperature-dependent viscosity; this can significantly increase the heat transfer rate if the wall temperature is close to the saturation temperature. Ebinuma and Nakayama (1990a,b, 1997) have included the effect of quadratic drag for the transient problem (the additional drag increases the time required to reach the steady state) and the transient problem with lateral mass flux. Li and Wang (1998) investigated analytically the influence of an effective thermal conductivity change adjacent to the cooling wall. The effect of a transient suction effect at the porous layer interface was studied by Ma and Wang (1998). The effect of suction on condensation on a finite-sized horizontal flat medium was studied

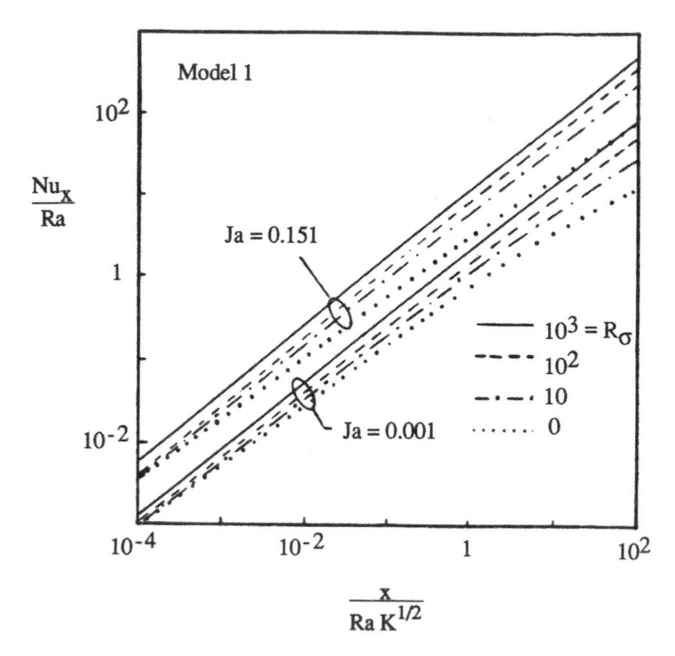

Figure 10.29. Heat transfer results for film condensation (Majumdar and Tien, 1990).

theoretically by Wang *et al.* (2003d). A further study incorporating non-Darcian effects was reported by Masoud *et al.* (2000).

The effects of surface tension on film condensation were analyzed by Majumdar and Tien (1990). Now the thermodynamics of phase equilibria requires the existence of a two-phase zone lying between the liquid and vapor regions. In this zone solutions of the conservation equations indicate a boundary layer profile for the capillary pressure. Majumdar and Tien considered various models for the boundary conditions. They concluded that the best results are attained if one assumes that there is no shear at the interface between liquid and the two-phase zone. Results obtained using this model are shown in Fig. 10.29. The parameter R_σ, the Rayleigh number Ra, and the Jakob number Ja are defined here by

$$R_\sigma = \frac{\sigma^*(K\varphi)^{1/2}}{\mu_l \alpha_m}, \quad \mathrm{Ra} = \frac{g(\rho_l - \rho_v)K^{3/2}}{\mu_l \alpha_m}, \quad \mathrm{Ja} = \frac{c_P(T_s - T_w)}{h_{fg}}, \tag{10.113}$$

where σ^* is the surface tension and the other quantities are as in Section 10.3.2.

Condensation on a vertical surface was investigated experimentally and numerically by Chung *et al.* (1992). Their numerical model assumed a distinct two-phase zone existing between liquid and vapor zones and included the effect of vapor flow in that two-phase zone. Their experiments were performed for steam condensing in packed beds of glass beads of three different sizes. They reported good agreement between numerical and experimental results. They found that the calculated liquid film thicknesses are of the order of the diameter of the glass beads.

Nakayama (1991) used the Forchheimer model in his analytical treatment of film condensation in the presence of both gravity and externally forced flow. He introduced a similarity transformation involving a modified Péclet number based on the resultant velocity of the condensate. Microscale Grashof and Reynolds numbers based on the square root of the permeability govern the delineation of four limiting regions, namely (i) Darcy forced convection, (ii) Forchheimer forced convection, (iii) Darcy natural convection, and (iv) Forchheimer natural convection.

An experimental and numerical investigation on the Brinkman model of condensation of a downward flowing vapor on a horizontal cylinder embedded in a vapor-saturated porous medium was carried out by Orozco (1992). Good agreement was found between predicted and measured values of Nu and condensate thickness.

Renken and Aboye (1993a,b) have reported numerical and experimental studies of film condensation within thin inclined porous coatings. The experiments involved a condensate region overlaying metallic permeable coating adhered to an isothermal copper block. Reduced gravity measurements were obtained by condensing saturated steam containing small concentrations of noncondensables on surfaces with effective body forces between 0.3 and 1 g. They also investigated the effects of surface subcooling. The presence of the coating enhanced the heat transfer substantially. The previous work of Renken *et al.* (1989) involved a numerical study of a porous coating on a vertical surface. The subsequent work by Renken *et al.* (1994) involved further numerical investigation on the Brinkman-Forchheimer model or coatings on inclined surfaces. Experiments on forced convection past porous coatings placed parallel to saturated steam flow were reported by Renkin and Raich (1996).

Wang and Beckerman (1995) performed a two-phase boundary layer analysis based on a two-phase mixture model for buoyancy-driven two-phase flow (condensing or boiling) in capillary porous media. They used the solution to reveal the capillary effect.

For film condensation on a vertical plate, Al-Nimr and Alkam (1997a) obtained closed-form expressions for the condensate film thickness and flow rate and for the convective heat transfer coefficient. They found that the liquid film thickness is proportional to $x^{1/4}$ in a thin porous domain as the permeability tends to infinity, but it is proportional to $x^{1/2}$ in a thick porous domain as the permeability tends to zero. Masoud *et al.* (2004) extended this analysis to a transient problem.

Char and Lin (2001) and Char *et al.* (2001) treated conjugate film condensation in natural and mixed convection between two porous media separated by a vertical plate. A further conjugate problem was studied by Mosaad (1999). Heat and mass transfer with condensation in a fibrous insulation slab was studied experimentally and analytically by Murata (1995). Forced convection film condensation on a vertical porous-layer-coated surface was studied analytically by Toda *et al.* (1998) and by Asbik *et al.* (2003).

10.5. Spaces Filled with Fluid and Fibers Coated with a Phase-Change Material

It has been shown that polyethylene glycols (polyols) can be bonded stably on fibrous materials and that the resulting composites—the "thermally active" fibers—exhibit reproducible energy storage and release properties (e.g., Vigo and Bruno, 1987). The energy storage and release is due to the large latent heat of melting and crystallization of the polyols affixed to the fibers. A fundamental model for heat transfer through a space filled with polyol-coated fibers surrounded by air was described by Lim *et al.* (1993), who also reviewed the applications of this new class of materials. In this model the fibers *and* the phase-change material (polyol, liquid, or solid) constitute the matrix of the porous medium, while air is the fluid that flows through the interstitial spaces.

It is worth noting that this model differs fundamentally from the one used in earlier studies of melting and solidification in porous media (e.g., Section 10.1.1). In the earlier studies the melted phase-change material was the fluid that filled the pores, and therefore there was no flow through regions saturated with solid phase-change material. In the model for spaces filled with thermally active fibers, the fluid (air) flows through the entire matrix regardless of whether the polyol coatings are liquid or solid.

The model of Lim *et al.* (1993) is based on the homogeneous porous medium and local thermal equilibrium assumptions. The composition of the porous medium is described by the porosity, φ (about 80 percent), and the fraction of the matrix occupied by polyol, ε (about 20 percent). This means that a unit volume is distributed in the following proportions: φ = air, $(1-\varphi)$ = matrix (fibers and polyol), $(1-\varphi)\varepsilon$ = polyol, and $(1-\varphi)(1-\varepsilon)$ = fibers. The average heat capacity of the porous medium is

$$(\rho c)_m = \varphi(\rho c_P)_a + (1-\varphi)[\varepsilon(\rho c)_P + (1-\varepsilon)(\rho c)_f], \tag{10.114}$$

in which the subscripts m, a, p, and f refer to the averaged porous medium, air, polyol, and fibers.

Lim *et al.* (1993) applied the model to melting and freezing in three configurations, which were analyzed numerically: one-dimensional conduction, one-dimensional convection, and two-dimensional natural convection due to heating or cooling from the side. In each case the focus was on the relation between the time of complete melting or solidification of the polyol coatings and the various dimensions and external parameters of the enclosure. For example, in a two-dimensional space with time-dependent melting by natural convection (Fig. 10.1) the time-dependent flow and heat transfer is ruled by four independent groups: $\mathrm{Ra} = g\beta K H(T_h - T_i)/\upsilon\alpha_m$, H/L, $S = (1-\varphi)\varepsilon\rho_p\lambda/(\rho c)_m(T_h - T_i)$ and $\theta_m = (T_m - T_i)/(T_h - T_i)$ where T_h, T_i, T_m and λ are the temperature of the heated side wall, the uniform initial temperature of the system, the melting temperature,

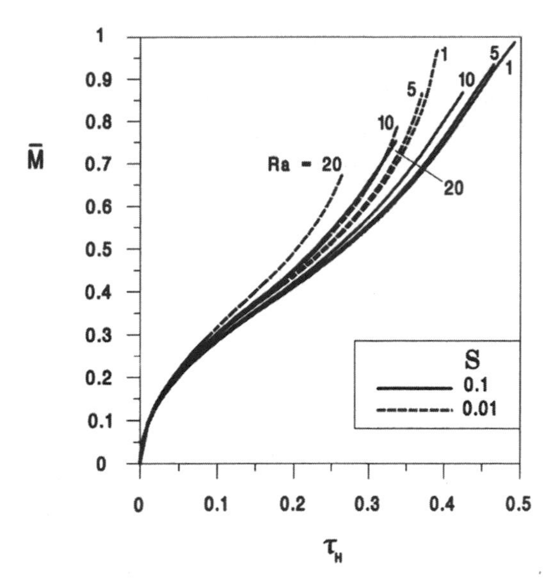

Figure 10.30. The effect of Ra and the latent heat on the evolution of the average melting front position in a space filled with fibers coated by a phase-change material (Lim *et al.*, 1993).

and the latent heat of melting. Note that in the corresponding configuration of Section 10.1.1 the phenomenon was described by only two independent groups, Ra and H/L.

The shape and evolution of the melting front has the same features as in Fig. 10.2. Several effects are presented in condensed form in Fig. 10.30, which shows the average position of the melting front

$$\overline{M}(\tau_H) = \frac{1}{H}\int_0^H \frac{s}{L}dy \tag{10.115}$$

versus the dimensionless time $\tau_H = \alpha_m t/\sigma H^2$, where σ is the heat capacity ratio $\sigma = (\rho c)_m/(\rho C_P)_a$. The dimensions $s(y, t)$, H, and L are defined in Fig. 10.1. Each of the curves plotted in Fig. 10.30 is terminated at the time when the melting front has traveled the distance L along the top of the enclosure. The inflection of each curve is considerably more pronounced than in Fig. 10.6.

The effect of the latent heat parameter S is also shown in Fig. 10.30. A larger latent heat (larger *S*) means a longer time until the coating melts on the fibers located the farthest from the heated wall. The melting times decrease sensibly as the Rayleigh number becomes greater than approximately 5. The effects of changing θ_m and H/L are further documented in Lim *et al.* (1993).

The solidification process in the same two-dimensional configuration is analogous to the melting process discussed until now. In solidification, the $H \times L$ region is initially isothermal (T_i) and all the fibers are coated with liquid polyol, $T_i > T_m$. The temperature of one of the side walls is lowered suddenly to T_c, which is lower than T_m. The movement of the solidification front is similar to that of Fig. 10.2: the shape of the front can be visualized by imagining the mirror image of Fig. 10.2, where the role of mirror is played by one of the horizontal walls. Figure 10.30 continues to be valid subject to the new definitions $\mathrm{Ra} = g\beta KH(T_i - T_c)/\nu\alpha_m$ and $\theta_m = (T_i - T_m)/(T_i - T_c)$.

11
Geophysical Aspects

Most of the studies of convection in porous media published before 1970 were motivated by geophysical applications and many published since have geophysical ramifications; see, for example, the reviews by Cheng (1978, 1985b). On the other hand, geothermal reservoir modeling involves several features that are outside the scope of this book. Relevant reviews include those by Donaldson (1982), Grant (1983), O'Sullivan (1985a), Bodvarsson *et al.* (1986), Bjornsson and Stefansson (1987), McKibbin (1998, 2005), and O'Sullivan *et al.* (2000, 2001). In this chapter we discuss a number of topics that involve additional physical processes or have led to theoretical developments beyond those that we have already covered.

11.1. Snow

It is not uncommon for an unstable air density gradient to be found in a dry snow cover, because the base is often warmer than the upper surface. The geothermal heat flux, the heat release due to seasonal lag, and the release of heat if the soil freezes are factors that tend to keep the bottom boundary of a snow cover near 0°C. In contrast, the upper boundary is usually near the ambient air temperature, which in cold climates can be below 0°C for long periods of time.

When the unstable air density gradient within the snow becomes sufficiently great, convection occurs and the rate of transport of both heat and vapor increases and the snow undergoes metamorphosis. For example, a strong vertical temperature gradient favors the growth of ice particles. These may grow to 1 or 2 cm in diameter. As the particles increase in size their number decreases so rapidly that the density of the snow decreases, relative to that in the absence of a temperature gradient. At the same time there is a change in the shape of ice crystals. The strength of the snow against shear stresses is lowered and on sloping terrain this can lead to slab avalanches.

Thermal convection has been observed in snow both in laboratory experiments and in the field. These experiments indicate that natural convection should be fairly common under subarctic conditions.

The particular feature of convection in snow that distinguishes it from convection in other porous media is the fact that the energy balance is significantly affected by the phase change due to the transport of water vapor from particle to particle in snow. This has been studied by Palm and Tveitereid (1979). Their analysis was refined by Powers *et al.* (1985). The latter assume that the Boussinesq approximation is valid and that the equation of state for vapor at saturation can be taken as

$$\rho_v = \rho_0 \exp[B(T - T_0)]. \tag{11.1}$$

The heat flux is incremented by $L\mathbf{j}_v$, where L is the latent heat and $\mathbf{j}_v$ is the diffusive flux of vapor, given by $\mathbf{j}_v = -D_{eff}\nabla\rho_v$ where D_{eff} is an effective mass diffusivity. At the same time there is an additional energy transport term resulting from the convection of vapor (for details, see Powers *et al.*, 1985). As a consequence one ends up with an energy equation in the form

$$[L\rho_v B + (\rho c_P)_a]\mathbf{v} \cdot \nabla T = \nabla \cdot [(k_m + LD_{eff}\rho_v B)\, \nabla T], \tag{11.2}$$

where the subscript a denotes air and $\mathbf{v}$ is the mass-averaged seepage velocity (which is approximately equal to the air velocity because the density of vapor is much less than that of air). If the various coefficients in Eq. (11.2) can be approximated by constant values, this takes the form

$$\mathbf{v} \cdot \nabla T = \alpha_e \nabla^2 T, \tag{11.3}$$

where

$$\alpha_e = \alpha_m \left(\frac{1+\gamma}{1+a\gamma}\right), \tag{11.4}$$

where in turn

$$\alpha_m = \frac{k_m}{(\rho c_P)_a}, \quad \gamma = \frac{LD_{eff}}{k_m}\left(\frac{d\rho_v}{dT}\right), \quad a = \frac{\alpha}{D_{eff}}. \tag{11.5}$$

We see that the primary effect of the diffusion of water vapor (which arises from the variation of saturation vapor density with temperature) is to change the value of the effective thermal diffusivity.

To Eq. (11.3) we can add the equations of continuity, momentum, and state:

$$\nabla \cdot \mathbf{v} = 0, \tag{11.6}$$

$$-\nabla P - \frac{\mu}{K}\mathbf{v} + \rho_a \mathbf{g} = 0, \tag{11.7}$$

$$\rho_a = \rho_0[1 - \beta(T - T_0)], \tag{11.8}$$

and appropriate boundary conditions to formulate a variant of the Horton-Rogers-Lapwood problem. Powers *et al.* (1985) solved this system for the two-dimensional case using finite differences and calculated the heat transfer for Rayleigh numbers just above critical. They treated various types of boundary conditions and they briefly discussed the case of inclined layers.

We note that the effect of water vapor is destabilizing if $a > 1$ and stabilizing if $a < 1$. In practice the value of a can vary widely, but typical values are in the range of 0.5 to 2. This means that the critical Rayleigh number is in the range 25 to 35 for the case of an isothermal permeable top and an isothermal impermeable bottom boundary.

Sommerfeld and Rocchio (1993) reported experiments on the permeability of snow. They noted that while calculated Rayleigh numbers have exceeded those thought critical for natural convection in snow, field experiments by Sturm and Johnson (1991) indicate that extreme thermal gradients are necessary for even intermittent convection. Sturm and Johnson, however, concluded that convection occurred almost continuously during two of the three winters during which they made their experiments.

Comparing the results of a numerical model with a field experiment where air was forced through a natural snowpack, Albert (1995) concluded that the air-flow through the pack was sufficient to produce advection-dominated heat transfer throughout most of the pack. Aspects of the convective instability of air in snow cover treated as a two-layered system were discussed by Zhekamukhov and Zhekamukhova (2002). A nonequilibrium treatment of heat and mass transfer in alpine snowcovers was reported by Bartelt *et al.* (2004).

11.2. Patterned Ground

There are many places in arctic or mountainous regions where the surface of the ground takes the form of a regular pattern of circles, stripes, or polygons. These are made prominent because of the segregation of stones and fines resulting from diurnal, seasonal, or other recurrent freeze-thaw cycles in water-saturated soils. These patterns also are found underwater, in shallow lakes, or near shores. The diameter of sorted polygons may vary from 0.1 to 10 m. A variety of photographs is included in the article by Krantz *et al.* (1988).

When frozen soil thaws, the potential for convection exists because of the density inversion for water between 0 and 4°C. More dense water at a few degrees above its freezing point can overlie less dense water at 0°C. But convection currents alone are too weak to move either the stones or the soil.

Ray *et al.* (1983) provided the following explanation of the formation of patterned ground. Once gravitationally induced convection occurs, it typically forms hexagonal cells in horizontal ground and roll cells or helical coils in sloped terrain. These regular cellular flow patterns can then be impressed on the underlying ice front, because in areas of downflow the warmer descending water causes extra melting, whereas in areas of upflow the rising cooler water hinders melting of the ice front. Consequently, the ice level is lowered under descending currents and raised over ascending currents, relative to the mean level. Thus a pattern of regularly spaced peaks and troughs is formed on the underlying ice front that mirrors the cellular convection patterns in the thawed layer. This pattern is transferred to the ground surface through the process of mechanisms such as frost push or frost pull. The width of the flow cell at the onset of convection then determines the width W of the observed stone patterns. The height H of the thawed layer at the onset of convection is assumed to correspond to the sorting depth D. Linear instability theory thus predicts the value of W/D. This tallies well with observations (Gleason *et al.*, 1986). The model also provides an explanation for the transition from polygons on horizontal ground to stripes on sloped terrain.

The direction of fluid circulation determines whether the stones concentrate over the ice troughs or peaks. Gleason *et al.* (1986) report results of weakly nonlinear stability theory that shows that under most conditions the determining temperature-dependent property for convection arising from thawing frozen soil is the coefficient of thermal expansion. This decreases from 3.5×10^{-5} per degree Celsius to zero as the temperature increases from 0 to 4°C, and this decrease implies cell circulation with upflow in the cell center and downflow along the polygonal borders. The underlying ice front then should have isolated ice peaks and continuous polygonal troughs. If stones tend to concentrate over troughs during sorting, this would lead to stone-bordered polygons. In fact these are the most frequently observed patterns. If kinematic viscosity were the dominant temperature-dependent property, then the decrease in kinematic viscosity as the temperature increases would imply the opposite direction of circulation and this would lead to stone pits. These are occasionally observed.

Rock conducts heat better than soil does. Thus if in the freeze following the thaw period wherein the convection was initiated the sorting process moves some stones over the convection-induced ice troughs, then during the next thaw period the conductive heat transfer will be largest in precisely those regions. Thus heat conduction will act to accentuate the previous pattern.

George *et al.* (1989) state that three conditions are believed to be essential for the formation of stone polygons: the existence of freeze-thaw cycles within the soil, the saturation of the soil with water for at least part of the year, and the presence of an impermeable ice barrier underlying the active layer. Once these conditions are satisfied, the formation of polygonal ground follows a five-step process. Stone polygons have been grown in the laboratory by reproducing these five steps, namely: (1) Permeability enhancement as the result of the formation of needle ice and frost heaving in the soil. (2) Onset of buoyancy-driven convection in the water saturated soil. (3) Formation of a tessellated surface in the permafrost. (4) Genesis of polygonal ground through frost heaving. (5) Perpetuation of the hexagonal pattern.

Gleason *et al.* (1986) claimed that the two forms of convection cells that can occur in sloped terrain have widely different width-to-depth ratios, 2.7 for two-dimensional rolls (which occur for small downslope flow) and 3.8 for helical coils (which occur for large downslope flow). They have not published the analysis that leads to these values. We would expect the values to be practically the same. The value 2.7 would correspond to an impermeable conducting bottom and a permeable conducting top surface.

George *et al.* (1989) also have extended the theoretical analysis of the onset of convection in several respects. Whereas Ray *et al.* (1983) approximated the density versus temperature relationship by a linear expression, George *et al.* (1989) worked with a more accurate parabolic expression. George *et al.* (1989) also allowed for a permeability that varies linearly with depth and they contributed a nonlinear analysis based on the method of energy. They found that their theoretical predictions of W/D agreed well with field studies when a constant-flux condition is imposed at the upper boundary and an upwardly stratified permeability is chosen. Theoretical extensions to include the effects of solar radiation, phase change, cubic density law, and overlying water have been made by McKay (1992, 1996) and McKay and Straughan (1991, 1993), respectively. In particular, McKay (1992) presented a linear analysis involving Floquet theory, a nonlinear energy analysis, and extensive numerical results.

Experimental work together with the results of a theoretical investigation of heterogeneity effects were reported by Zimmerman *et al.* (1998). The mathematical aspects of the pattern formation were emphasized in the review by Straughan (2004b). The self-organization aspect of the phenomenon was discussed by Kessler and Werner (2003).

11.3. Thawing Subsea Permafrost

During the ice age (18,000 years ago) the sea level was some 100 m lower than it is at present and the lower ambient temperatures led to substantial permafrost forming around arctic shores. With the rise of sea levels the permafrost has responded to the relatively warm and salty sea, which has created a thawing front and a layer of salty sediments beneath the sea bed. Those off the coast of Alaska have been extensively studied. It is believed that convection is taking place in the layer between the sea bed and the permafrost. (This belief is based on the fact that although conduction appears to be the dominant heat transfer mechanism, the molecular diffusion of salt is too slow to explain the observed rate of thawed layer development. Also the salinity Rayleigh number is supercritical, salinity gradients in the thawed layer are small except for a boundary layer near the bottom, and the pure water pressure is different from hydrostatic.) A buoyancy mechanism is provided by the release of relatively fresh and therefore buoyant water liberated by thawing at the base of the layer.

The analysis of Swift and Harrison (1984) is of interest because of the way in which they were able to replace a moving boundary (Stefan) problem with one essentially on a fixed domain, using the facts that the convection is salt dominated and the climatic interface advance is slow (2 to 5 cm/year). The argument is as follows.

On the moving boundary $z = D$, Stefan conditions hold for the temperature and salinity fields. At $z = D$,

$$L_V \frac{dD}{dt} = -k_m \left.\frac{\partial T}{\partial z}\right|_{D-} \quad \text{and} \quad S(D)\frac{dD}{dt} = -\alpha_s \left.\frac{\partial S}{\partial z}\right|_{D-}, \qquad (11.9\text{a,b})$$

where S is the salinity, k_m the thermal conductivity, L_V is the latent heat per unit volume of the salty thawed layer, and α_s is the diffusivity of salt. Because salinity is the driving mechanism, the temperature profile can be assumed linear throughout, and hence the temperature gradient can be replaced by $[T(D) - T_0]/D$, where T_0

is the sea-bed temperature. The requirement for phase equilibrium is that $S(D)$ is proportional to $-T(D)$, and so we can write $S(D)/S_r = T(D)/T_0$, where S_r is the salinity of water that would begin to freeze at temperature T_0. Here $T(D) < T_0 < 0$ and so $S(D) > S_r$. Now dD/dt can be eliminated from Eqs. (11.9a) and (11.9b), and we end up with the nonlinear boundary condition

$$\frac{\partial S}{\partial z} = \frac{k_m T_0}{L_v \alpha_s D} S \left(\frac{S}{S_r} - 1 \right) \text{ at } z = D. \tag{11.10}$$

The other boundary conditions are the usual ones, and the problem is reduced to a standard linear stability problem on a fixed domain.

Swift and Harrison (1984) went on to solve this problem numerically for solute Rayleigh numbers 1750 and 17500 (which we recall are well in excess of the critical value for the onset of convection, which is about 40). Galdi *et al.* (1987) reexamined this problem, using both linear and nonlinear analysis. They used an energy method to determine a critical Rayleigh number below which convection cannot develop. Payne *et al.* (1988) also have applied an energy method to this problem. They assumed that the downward permafrost interface movement is negligible and they allowed the density to vary quadratically with temperature.

Subsequent studies have shown that salt fingering may play a major role in the thawing of the permafrost. The salt gradient is produced partly by salts rejected during sea ice growth producing concentrated brine near the sea bed and partly by salts rejected during sediment freezing near the sea bed causing the formation of a concentrated brine layer within the deeper and yet unfrozen sediments. Gosink and Baker (1990) report theoretical, laboratory, and field investigations. The theoretical ones are based on timescale balances related to the result of Wooding (1959) that convective instability in the form of fingering takes place when the magnitude of the salinity Rayleigh number exceeds a certain critical value (3.390 in the case of a vertical cylinder, the Rayleigh number being based on the radius of the cylinder; compare Section 6.16.1). The results of Gosink and Baker suggest that downward salt fingering will occur at Prudhoe Bay whenever the density gradient in the thawed subsea sediments exceeds $6.2 < 10^{-5}$ g cm^{-4}. The maximum predicted velocity of fingering is about 2 m/day and this is consistent with estimates made from measurements of pressure gradients and numerical modeling in the thawing permafrost. The energy dissipated by viscous force in the thawed layer balances the energy added to the layer by the salt fingers caused by concentrated brines at the seabed.

Hutter and Straughan (1997) have employed a realistic equation of state and have imposed a linear temperature gradient. For this case they have developed linear and fully nonlinear stability analyses. They found that the refinements to the equation of state led to a reduction in critical Rayleigh number. An unconditional nonlinear stability bound (close to that of linear theory) was found by Budu (2001). A further study was carried out by Hutter and Straughan (1999). Their multiscale perturbation analysis verified the observed thaw rates with a parabolic-in-time phase boundary retreat and enabled an investigation of possible currents induced by the ocean circulation overlying the thawed permafrost layer. Their analysis indicates that the phase boundary beneath the sea bed and below the thawing layer has a parabolic shape, something that is observed in practice. The topic of this section has been reviewed by Straughan (2004b), who concludes that the nonlinear stability thresholds will be extremely close to the linear instability ones for any practical choice of the density equation of state.

11.4. Magma Production and Magma Chambers

In general the flow of magma can be treated like that of a viscous fluid subject to the Navier-Stokes equation, but there are two situations where Darcy's equation is applicable. The first is when crystallization leads to a porous structure near the walls of a magma chamber. The second is when a partial melt is formed during magma genesis and the melt products tend to concentrate along interconnected grain boundaries. Lowell (1985) has applied double-diffusive stability analysis to each of these situations.

The partial melt problem involves a layer whose thickness varies with time, and so the associated boundary condition is of Stefan type. Lowell (1985) obtains as an approximate expression for the critical thermal Rayleigh number

$$\mathrm{Ra}_c = 4\pi^2 \left(1 - \frac{Q^2}{2\pi^2}\right), \tag{11.11}$$

where Q is determined as the root of

$$\pi^{1/2} Q \operatorname{erf}(Q) \exp\left(Q^2\right) = \mathrm{Ste}, \tag{11.12}$$

where the Stefan (or Jakob) number Ste = $\Delta T c_P / \varphi L_h$. Here ΔT is the difference between the basal temperature of the layer and the eutectic temperature (the starting temperature for the melting process), c_P is the specific heat of the solid/melt mixture, φ is the melt fraction (porosity), and L_h the latent heat. In the present context Q is a small parameter, so the dynamics of the melt front can be decoupled from the double-diffusive effects. Thus the basic stability results of Nield (1968) are applicable. The critical thickness can vary from about 800 m to a few centimeters, depending on the composition of the magma. Lowell concluded that convective processes will tend to homogenize the melt before it separates from the source zone, but the vigor of mixing is dependent upon the composition of the source.

Lowell's (1985) other problem concerns the structure of the porous boundary layer that forms as a result of side-wall crystallization in a convecting magma chamber. His examination of the steady-state boundary layer equations shows that the structure may be one of two types. If upon crystallization at the wall the residual melt fraction has negative compositional buoyancy or if the negative thermal buoyancy at the cold wall exceeds the positive compositional buoyancy of the residual melt, then the flow across the whole boundary layer will be downward. Then if the residual melt fraction has negative compositional buoyancy, the magma chamber will become stratified as the result of the accumulation of a layer of dense cold liquid on the floor, while if the melt fraction has positive compositional buoyancy, the boundary layer fluid will tend to be remixed into the interior of the magma chamber. If, on the other hand, the positive compositional buoyancy exceeds the negative thermal buoyancy, counterflowing boundary layers will occur and the compositional buoyancy liquid will tend to be fractionated towards the top of the magma chamber.

11.5. Diagenetic Processes

Diagenetic processes involve reactions between pure water and mineral phases during which unstable minerals are dissolved and more stable phases are precipitated, resulting in changes in porosity and permeability. If fluid flow is involved, then the dissolution and precipitation occur in different parts of the medium. Davis *et al.* (1985) have computed the flow pattern and the resulting diagenetic contours (of $\mathbf{v} \cdot \nabla T$) for convection in a folded porous layer (sand) bounded by an impermeable medium (shale) heated from below and held at a constant temperature above. They assumed that the dip angles are small and the convection is weak, so that the temperature field can be uncoupled from the fluid flow. Their results are shown in Fig. 11.1. The direction of circulation, and hence the region of precipitation, depends on whether the conductivity of the porous medium (k_m) is less than or greater than the conductivity of the impermeable medium (k_s). If $k_m/k_s < 1$, the precipitation of quartz takes place on the lower flanks of the porous layer, because the solubility increases with temperature, and hence the material is leached from the porous matrix in regions where the fluid is being heated and precipitated in regions where it is cooled.

The rate of mass transfer is radically increased if the critical Rayleigh number is exceeded and multicellular convection occurs. Palm (1990) has modified the analysis which Palm and Tveitereid (1979) developed for convection in snow (see Section 11.1) to slightly supercritical two-dimensional flow in a sloping layer in order to determine the rate of change of mean porosity $\bar{\varphi}$ (averaged with respect to the upslope coordinate). Palm (1990) showed that

$$\frac{\partial \bar{\varphi}}{\partial t} = 4\pi \frac{\rho_w}{\rho_s} \frac{dC_s}{dT}\bigg|_{\bar{T}} \alpha_m \frac{\Delta T}{H^2} \frac{\mathrm{Ra} - \mathrm{Ra}_c}{\mathrm{Ra}} \sin\left(\frac{2\pi z}{H}\right), \tag{11.13}$$

where ρ_w is the density of water while C_s is the mass fraction of the transported material in water, in his case silica quartz, and ρ_s is its density. The other quantities are as in Section 7.8. We note that the maximal changes in porosity occur at $z = (1/4)H$ and $z = (3/4)H$. This work was applied to the sedimentary basin under the North Sea by Bjørlykke *et al.* (1988).

The book by Phillips (1991, Ch.7) contains further extensions. Phillips presents detailed analysis of convective flow at small Rayleigh number in submerged banks of slowly varying thickness or in compact platforms or reefs. He also treats flow patterns at intermediate Rayleigh number and scale ratio.

A computation of porosity redistribution resulting form thermal convection in slanted porous layers was made by Gouze *et al.* (1994). Implications for hydrothermal circulation at midocean ridges, resulting from permeability changes due to diagenesis in the fractured crust, were studied by Fontaine *et al.* (2001).

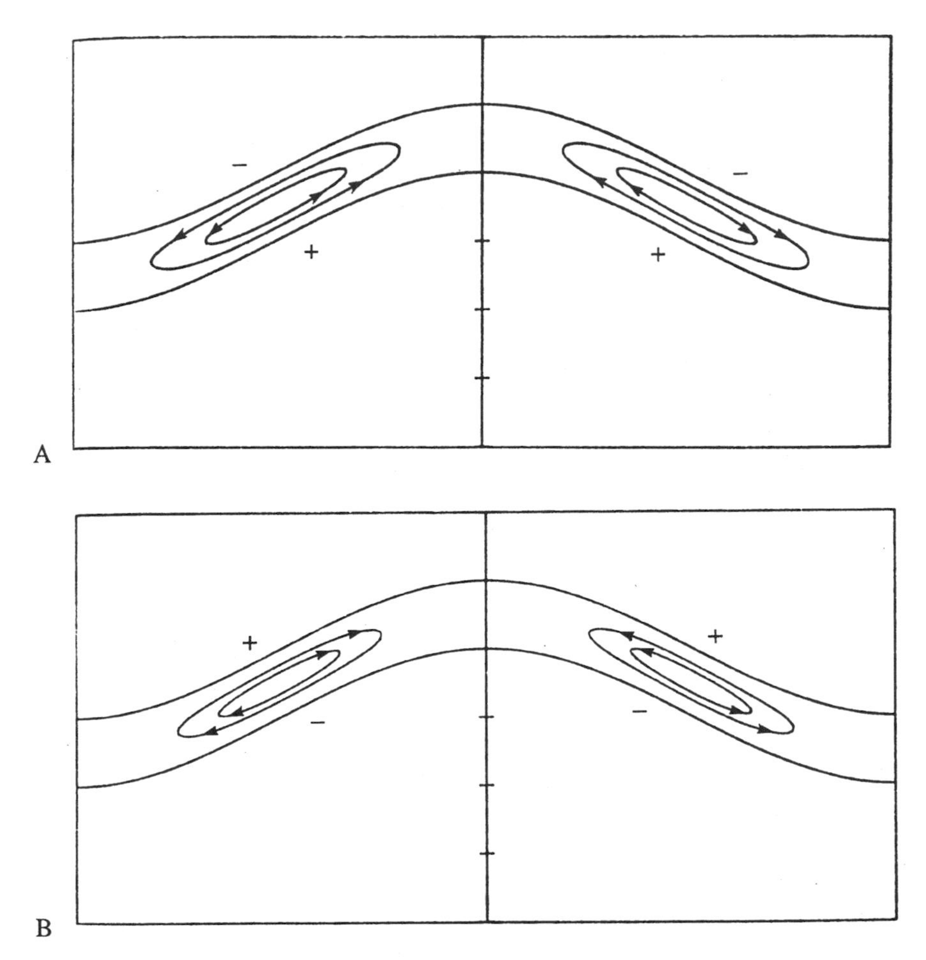

Figure 11.1. Streamlines in a folded porous layer. In A the thermal conductivity ratio k_m/k_s is 0.8, where m refers to the porous layer (sandstone) and s to the surrounding impermeable material (shale). In B the ratio is 1.25. The + signs denote the loci of maximum precipitation of quartz and the − signs the loci of maximum dissolution (Davis *et al.*, 1985, with permission from the *American Journal of Science*).

11.6. Oceanic Crust

11.6.1. Heat Flux Distribution

Measurements of heat flow on the ocean floor near the Galapagos spreading center have revealed a spatial periodicity with a wavelength of about 7 km, peaks of 12 HFU (where 1 HFU $\equiv 1\mu$ cal cm^{-2} s^{-1} is the "heat flux unit") and troughs of 2 HFU, i.e., a peak to trough ratio of 6. Ribando *et al.* (1976) calculated this ratio for various values of a Rayleigh number Ra based on heat flux, for the cases of permeable and impermeable upper boundary, and for exponentially decreasing and constant permeability. The heat flux distributions for permeable and impermeable tops are similar and in the parameter range of interest the peak to trough ratio is not sensitive to whether the permeability is constant or exponentially decreasing, taking the value 6 for Ra = 100. For a cell depth of 3.5 km this corresponds to a permeability of 4.5×10^{-12} cm^2, in accordance with other estimates of the permeability of oceanic basalts.

11.6.2. Topographical Forcing

Convection in oceanic crust has motivated studies of convection initiated by topography giving rise to horizontal temperature gradients and also of the extent to which topography influences the wavelength of convection cells produced by vertical temperature gradients. Lowell (1980) studied the first aspect. He assumed that the topography is two-dimensional, of uniform wavelength L and amplitude d, with $d/L \ll 1$, as shown in Fig. 11.2. This allows the temperature boundary condition to be changed from $T = 0$ at the surface to

$$T = \frac{d\Delta T}{2H}(1 + \cos kx) \text{ at } z = 0, \tag{11.14}$$

where $k = 2\pi/L$. The other boundary conditions are taken as

$$\frac{\partial w}{\partial z} = 0 \text{ at } z = 0, \text{ and } w = T = 0 \text{ at } z = H. \tag{11.15}$$

The linearized momentum and energy equations for steady flow take the form

$$\nabla^2 w = \frac{g\beta K}{\nu}\frac{\partial^2 T}{\partial x^2}, \tag{11.16}$$

$$\nabla^2 T = \frac{\Delta T}{H\alpha_m} w. \tag{11.17}$$

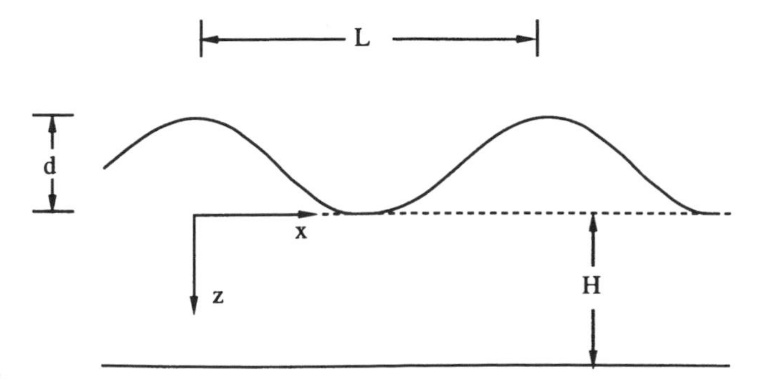

Figure 11.2. Definition sketch for low amplitude, wavelike crustal topography.

This system of equations can be solved iteratively. The first-order solution is

$$T_1 = \frac{d\Delta T}{2H}\left[1 + \cos kx \frac{\sinh k(h-z)}{\sinh kH} - \frac{z}{H}\right], \tag{11.18}$$

$$w_1 = -\frac{g\beta K d\Delta T}{4\nu H \sinh kH} \cos kx[(1 + kH \tanh kh) \sinh k(H-z) - k(H-z)\cosh k(H-z)]. \tag{11.19}$$

The last equation shows that the fluid descends at topographic troughs and ascends beneath topographic peaks as expected. The vertical velocity is proportional to the topographic amplitude d, but the convective heat flux $\rho c_P w_1 T_1$ is proportional to d^2. Lowell (1980) also analyzed the case when the topography is covered with a layer of sediment.

The extent to which boundary topography can control the pattern of convection in a porous layer was examined by Hartline and Lister (1981). Their experiments using a Hele-Shaw cell indicate that for supercritical values of Ra the topography does not control the convection pattern except when the topographic wavelength is comparable to the depth of water penetration, the nondimensional wavenumber $2\pi H/L$ taking values between 2.5 and 4.8. We note that this range brackets π, the critical wavenumber for a slab with planar, isothermal, and impermeable boundaries. Topographies within this range control the circulation pattern perfectly, with downwelling under troughs and upwelling aligned with peaks. Other topographies do not force the pattern, although in some cases the convection wavenumber may be a harmonic of the topographic wavenumber. Unforced convection cells wander and vary in size. Hartline and Lister (1981) conclude that where the submarine circulation correlates with bottom topography it may be because the topographic wavelength is comparable to the depth to which water penetrates the porous crust.

11.7. Geothermal Reservoirs: Injection and Withdrawal

Geothermal reservoir modeling has motivated many numerical studies of problems involving the withdrawal and injection of fluids. It is often convenient to formulate such problems in terms of pressure and temperature. For example, Cheng and Teckchandani (1977) studied the transient response in a liquid-dominated geothermal reservoir resulting from sudden heating and the withdrawal of fluids. They considered a two-dimensional rectangular reservoir confined by caprock at the top, heated by bedrock from below, and recharged continuously through vertical boundaries from the sides, with withdrawal from either a centrally placed line sink or a vertical plane sink. The characteristic feature is the contraction of isotherms in the neighborhood of the sink (see Fig. 11.3). Oscillatory convection starts at Ra = 200, a lower Rayleigh number than in the absence of cold water recharge from the sides.

In other studies the withdrawal and recharge of fluid has been through a permeable top. The numerical results of Horne and O'Sullivan (1974b) showed that fluid withdrawal can increase or decrease the rate of heat transfer from the bottom

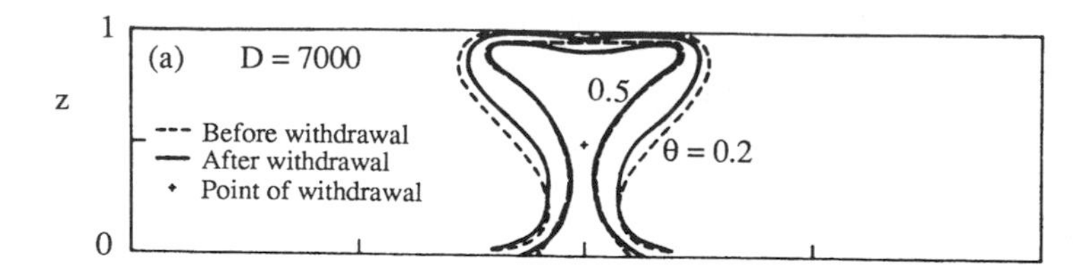

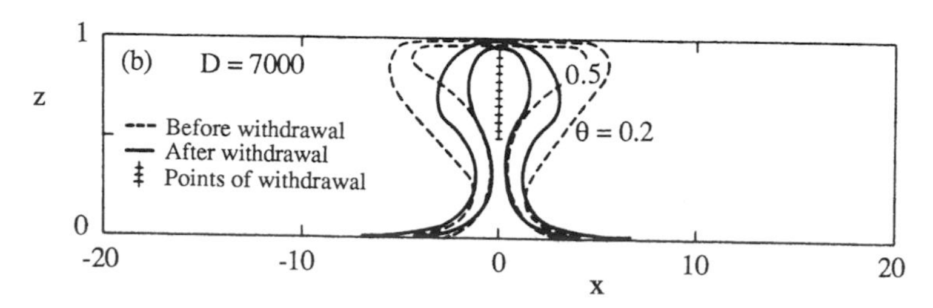

Figure 11.3. Contraction of isotherms in a geothermal reservoir resulting from fluid withdrawal from (a) a point sink and (b) a vertical line sink. Here θ is the nondimensional temperature, and $D = \mathrm{Ra}/\beta\Delta T$ (Cheng and Teckchandani, 1977).

(heated) surface depending on its position relative to the heat source. A two-temperature model was used by Turcotte *et al.* (1977) to simulate hot springs. Fluid is assumed to enter an upper permeable boundary at ambient temperature. That leaving is at a temperature greater than ambient temperature. At large Ra the significant temperature differences between fluid and solid are restricted to a thin layer near the upper boundary. Further work on this topic has been reviewed by Cheng (1978, 1985b).

11.8. Other Aspects of Single-Phase Flow

In the vicinity of the fluid critical point the intensity of natural convective circulation can increase dramatically. Dunn and Hardee (1981) presented laboratory data that show that in a porous medium heat transfer rates can increase by a factor of 70 in the vicinity of the critical point. They also showed that the conditions for this type of superconvection are compatible with expected geological conditions above magma bodies in the Earth's crust. Numerical experiments on convective heat transfer at near-critical conditions were reported by Cox and Pruess (1990). The heat transfer rates obtained in the simulations were considerably smaller than those reported by Dunn and Hardee (1981). Cox and Pruess suggested that possible causes of the discrepancy are the effects of pressure variation, channeling, and vertical asymmetry of the temperature field. Ingebritsen and Hayba (1994) observed that singularities in the equations of state of water at its critical point could be avoided by switching to a pressure-enthalpy formulation. Their numerical simulations showed that there was little near-critical enhancement in heat transfer for systems in which flow is driven by fixed pressure drops. However, in density-driven systems there can be an enhancement of heat transfer by a factor 10^2 or more, with convection occurring in narrow cells, if the permeability is sufficiently high. The restriction to high permeability environments within a fairly narrow pressure-enthalphy window indicates that superconvection may be quite rare in natural near-magma systems.

In order to discuss convective flow patterns in ground water near salt domes, Evans and Nunn (1989) made some calculations of double-diffusive convection. They did not invoke the Boussinesq approximation. They found that along a salt flank the flow can be either up or down, the sense of direction depending mainly on the value of the buoyancy ratio N[defined in Eq. (9.10)] and how sharply the isotherms are pulled up near the salt dome. These factors depend in turn on the regional salinity variation, the time since diaparism, and the thermal conductivity of water-saturated sediments.

A time-dependent numerical model of heat transfer across a thickening conductive boundary layer, between a crystallizing magma chamber and a single-pass hydrothermal system in the ocean crust, was developed by Lowell and Burnell (1991) and applied to sea-floor black smokers. General discussions of submarine hydrothermal systems were presented by Lowell (1991) and Lowell *et al.* (1995), Wilcock (1998), and Jupp and Schultz (2000, 2004). High Rayleigh number convection in an open top porous layer (or Hele-Shaw cell) heated from below was studied by Cherkaoui and Wilcock (1999, 2001).

Convection in a mushy zone at the Earth's inner-outer core boundary was discussed by Bergman and Fearn (1994). They concluded that the magnetic field may be strong enough to act against the tendency for convection to be in the form of narrow chimneys.

The interaction of thermally driven convective circulation in a steeply dipping fault zone and groundwater flow through the surrounding rock that is driven by a regional topographic gradient was examined by López and Smith (1995). Three-dimensional thermoconvection in an anisotropic inclined sedimentary layer was numerically simulated by Ormond and Genthon (1993).

Numerical modeling was used by Mullis (1995) to check the usefulness of the analytical solution given by Eq. (7.100). He found that for a homogeneous aquifer this solution is a good approximation provided that the inclination of the layer is replaced by the inclination of the isotherms. He also numerically modeled convection in wedges and lenses.

A general discussion based on numerical simulation of the patterns of flow induced by geothermal sources in deep ground was presented by Holzbecher and Yusa (1995). A geological thermosyphon, where the convection in a closed loop is coupled to conduction in the surrounding earth, was simulated numerically by Paterson and Schlanger (1992). They found that at a Rayleigh number above 1, convection leads to a temperature reduction near the source.

The problem of confinement of nuclear wastes in places like Yucca mountain in which the temperature and humidity inside emplacement drifts are of interest has led to new numerical simulations by Webb *et al.* (2003) and Itamura *et al.* (2004). An analytical assessment of the impact of covers on the onset of air convection in mine wastes was reported by Lu (2001).

Studies of the successive formation and evolution of layered structures in porous media resulting from heating a compositionally stable stratified fluid from below were made by Schoofs *et al.* (1998, 2000a). Thermochemical convection in and between intracratonic basins was studied by Schoofs *et al.* (2000b). The depletion of a brine layer at the base of ridge-crest hydrothermal systems was simulated by Schoofs and Hansen (2000). Numerical simulations of midocean ridge hydrothermal circulation including the phase separation of sea water were made by Kawada *et al.* (2004). A comprehensive study of NaCl-H_2O convection in the Earth's crust was reported by Geiger *et al.* (2005) who employed a novel finite element-finite volume numerical method. They allowed for phase separation. To characterize the onset of convection with a non-Boussinesq situation they introduced a fluxibility parameter (a scaled energy flux) and a local Rayleigh number.

Using finite-element numerical modeling, Zhao *et al.* (1997, 1998a, 1999c,d, 2000a, 2001a,b) have treated a range of situations. Zhao *et al.* (1998b, 1999a) studied high Rayleigh number steady-state heat transfer in media heated from below. The first paper dealt with the effect of geological inhomogeneity with both heat and mass transfer and the second with the effect of medium thermoelasticity, mineralization, and deformable media. Zhao *et al.* (2003a) transformed a magma solidification problem with a moving boundary into a problem without the moving boundary but with an equivalent heat source.

Steady-state heat transfer through midcrustal vertical cracks with upward throughflow in hydrothermal systems was analyzed by Zhao *et al.* (2002). The onset of convective flow in three-dimensional fluid-saturated faults was analyzed by Zhao *et al.* (2003a,b, 2004a, 2005). Further interesting studies involving layering or plume separation of thermohaline convection have been carried out by Oldenburg and Preuss (1998, 1999).

The influence of free convection on soil salinization in arid regions was studied by Gilman and Bear (1996). A numerical technique useful for such problems was supplied by Payne and Straughan (2000a). Straughan (2004b) notes that a nonlinear energy theory for this problem is lacking, but Payne *et al.* (1999) have used energylike techniques to derives continuous dependence and convergence results for the basic equations arising from the Gilman and Bear (1996) theory. Numerical modeling of reaction-induced cavities in a porous rock was conducted by Ormond and Ortoleva (2000). Solute transport in a peat moss layer produced by buoyancy-driven flow was discussed by Rappoldt *et al.* (2003). Thermal convection in faulted extensional sedimentary basins was simulated by Simms and Garven (2004).

Highly heterogeneous geologic systems recently have received special attention from Simmons *et al.* (2001) and Prasad and Simmons (2003). They pointed out that in many geologic systems, hydraulic properties such as the hydraulic conductivity of the system under consideration can vary by many orders of magnitude and sometimes rapidly over small spatial scales. Geologic systems, characterized by fractured rock environments or lenticular mixes of sand and clay, are common in many hydrogeologic systems. Such heterogeneity occurs over many spatial scales and variable density flow phenomena may be triggered, grow, and decay over a very large mix of different spatial and temporal scales. Dense plume problems in these geologic environments, in general, are expected to be inherently transient in nature and often may involve sharp plume interfaces whose spatiotemporal development is very sensitive to initial conditions. Importantly, the onset of instability in transient, sharp interface problems is controlled by very local conditions in the vicinity of the evolving boundary layer and not by the global layer properties or some average property of that macroscopic layer. Simmons *et al.* (2001) and Prasad and Simmons (2003) pointed out that any averaging process is likely to remove the very structural controls and physics that are important in controlling the onset, growth, and/or decay of instability in a highly heterogeneous system. These authors, together with Schincariol *et al.* (1997), reported that in the case of dense plume migration in highly heterogeneous environments the application of an average global Rayleigh number based upon average hydraulic conductivity of the medium was problematic. In these cases, an average Rayleigh number appears to be unable to predict the onset of instability accurately because the system is characterized by unsteady flows and large amplitude perturbations. For statistically equivalent geologic systems, and hence average global Ra, dense plume behavior was observed by Simmons *et al.* (2001) and Prasad and Simmons (2003) to vary between highly unstable to highly stable.

A number of factors limit the application the of the Rayleigh number in highly heterogeneous geologic environments. These include the invalidity of steady-state flow assumptions and the inability to accurately quantify both time-dependent nondimensionalizing length scales and dispersion in plume problems. In real field settings the critical transition regions in flow and transport behavior in groundwater systems are rarely known, and the idealized boundary conditions assumptions underlying the classic Horton-Rogers-Lapwood problem are not met. Also it is likely that in the case of transient development of fingers elements of Rayleigh-Taylor instability are involved and the effect is accentuated when there is heterogeneity. Thus it is not surprising that the use of a single average Rayleigh number has severe limitations in the situations investigated by Simmons and his colleagues. However, we believe that as a criterion for the onset of instability (as distinct from its subsequent development) in the case of moderately heterogeneous media the average Rayleigh number may do a better job than Simmons and his colleagues have reported. Because they worked in terms of a log permeability field they employed a poor estimate (too low) of the average permeability. Further work is needed to examine the consequences of this.

11.9. Two-Phase Flow

11.9.1. Vapor-Liquid Counterflow

For a geothermal field, the solid (rock) is at rest and the gas is the vapor. With subscript v (for vapor) replacing g, Eqs. (3.77–3.78) reduce to

$$\mathrm{v}_l = -\frac{k_f K}{\mu_l}(\nabla P - \rho_l \mathbf{g}) \tag{11.20}$$

$$\mathrm{v}_v = -\frac{k_v K}{\mu_v}(\nabla P - \rho_v \mathbf{g}) \tag{11.21}$$

Here it is assumed that K is constant. Likewise, Eqs. (3.82)–(3.83) reduce to

$$J_M = \rho_l \mathrm{v}_l + \rho_v \mathrm{v}_v, \tag{11.22}$$

$$J_E = \rho_l h_l v_l + \rho_v h_v v_v - k\nabla T \tag{11.23}$$

Under the two-phase conditions, the pressure P and temperature T are functionally related through the saturation line relation $T = T_{sat}(P)$. It is customary to take the z-axis in the vertically downward direction. In the absence of source terms and with the pressure term there negligible Eqs. (3.85)–(3.86) give for vertical flow,

$$\frac{\partial A_M}{\partial t} + \frac{\partial J_M}{\partial z} = 0, \tag{11.24}$$

$$\frac{\partial A_E}{\partial t} + \frac{\partial J_E}{\partial z} = 0, \tag{11.25}$$

where $A_M(P, S)$, $A_E(P, S)$, $J_M(P, \partial P/\partial z, S)$, and $J_E(P, \partial P/\partial z, S)$ and S is the liquid saturation. The relative permeabilities $k_l(S)$ and $k_v(S)$ are assumed to be monotonic increasing and decreasing, respectively, and to satisfy the conditions

$$k_l(S) = 0 \quad \text{for} \quad 0 < S < S_* \tag{11.26}$$

$$k_v(S) = 0 \quad \text{for} \quad S^* < S < 1, \tag{11.27}$$

where S_* and $1 - S^*$ denote the residual liquid saturation and vapor saturation, respectively. From Eqs. (11.20–11.24) it follows that for the case of negligible conduction,

$$J_M = -F + G_M, \tag{11.28}$$

$$J_E = -hF + G_E, \tag{11.29}$$

where the gravitational terms are

$$G_M = \left(\frac{\rho_l^2 K k_l}{\mu_l} + \frac{\rho_v^2 K k_v}{\mu_v} \right) g, \tag{11.30}$$

$$G_E = \left(\frac{\rho_l^2 K k_l h_l}{\mu_l} + \frac{\rho_v^2 K k_v h_v}{\mu_v} \right) g, \tag{11.31}$$

The mass mobility F is given by

$$F = K \left(\frac{\rho_l k_l}{\mu_l} + \frac{\rho_v k_v}{\mu_v} \right), \tag{11.32}$$

and the flowing enthalpy h is given by

$$h\,(P, S) = \frac{\rho_l k_l h_l/\mu_l + \rho_v k_v h_v/\mu_v}{\rho_l k_l/\mu_l + \rho_v k_v/\mu_v}. \tag{11.33}$$

Substituting Eqs. (11.28)–(11.29) into Eqs. (11.24)–(11.25) and eliminating second derivatives of the pressure, one obtains a first-order wave equation of the form

$$\frac{\partial S}{\partial t} + c\frac{\partial S}{\partial z} = f_l(S, P, \partial P/\partial t, \partial P/\partial z)\,, \tag{11.34}$$

where f_1 is a forcing term and the wave-speed c (whose reciprocal is an eigenvalue of the differential system) is given by

$$c = \frac{1}{E_S}\left[\frac{\partial h}{\partial S}J_M - \frac{\partial G}{\partial S}\right], \tag{11.35}$$

where in turn, for the case of negligible conduction,

$$G = hG_M - G_E, \tag{11.36}$$

$$E_S = \frac{\partial A_E}{\partial S} - h\frac{\partial A_M}{\partial S}. \tag{11.37}$$

Equation (11.34) may be analyzed by the standard method of characteristics. Rankine-Hugoniot equations, expressing conservation of mass and energy, relate the shock velocity to changes in densities and flows:

$$U = \frac{[J_M]}{[A_M]} = \frac{[J_E]}{[A_E]}, \tag{11.38}$$

where [ ] denotes a jump across the shock. It can be verified that for the case of zero conduction the second equality in the last equation is equivalent to the continuity of a volumetric flux vector J_Q given by

$$J_Q = -\frac{K}{\mu}\left(\frac{\partial P}{\partial z} - \rho g\right), \tag{11.39}$$

where μ and ρ are defined by

$$\frac{1}{\mu} = \frac{k_l}{\mu_l} + \frac{k_v}{\mu_v}, \qquad \frac{\rho}{\mu} = \frac{k_l\rho_l}{\mu_l} + \frac{k_v\rho_v}{\mu_v}. \tag{11.40}$$

On the basis of analysis and numerical simulations, Kissling *et al.* (1992b) concluded that although the phases can travel in opposite directions (counterflow), information travels either up or down, depending on the sign of the wave-speed c. Wave-speed, saturation, and other quantities are defined on a two-sheeted surface over the mass-energy flow plane, with the sheets overlapping in the counterflow region. [For counterflow, there are either two or zero solutions of Eq. (11.34), for the case of zero conduction.] Most saturations are of the wetting type, i.e. they leave the environment more saturated after their passage. In fact, when the flow is horizontal all shocks are wetting, but in passage. In fact, when the flow is horizontal all shocks are wetting, but in vertical two-phase flow there also exist drying shocks for sufficiently small mass and energy flows.

A general analytical treatment of three-dimensional flow was given by Weir (1991). He showed that when both phases were mobile the generalization of Eq. (11.34) is of the form

$$\frac{\partial S}{\partial t} + \mathbf{c}\cdot\nabla S = f, \tag{11.41}$$

where

$$\mathbf{c} = \frac{1}{E_S}\left(\frac{\partial h}{\partial S}\mathbf{J}_M - \frac{\partial G}{\partial S}\mathbf{k}\right), \tag{11.42}$$

where $\mathbf{k}$ is the unit vector in the z-direction. Weir (1991) showed that at each point in space, flows are essentially two-dimensional, in the sense that $\mathbf{J}_M$, $\mathbf{J}_E$, $\mathbf{J}_Q$, and $\mathbf{c}$ all lie in a vertical plane. Here $\mathbf{J}_Q$ is the vector generalization of the scalar in Eq. (11.39). Further, gravity establishes a vertical hierarchy; the volumetric, energy, and mass flux vectors (listed in descending order) can never point below a lower member of this triple.

For a one-dimensional horizontal two-phase flow, Eqs. (11.41–11.42) give, analogous to Eqs. (11.34–11.35) with zero gravity,

$$\frac{\partial S}{\partial t} + c\frac{\partial S}{\partial x} = f_1(S, P, \partial P/\partial t, \partial P/\partial x), \tag{11.43}$$

where

$$c = -\frac{F}{E_S}\frac{\partial h}{\partial S}\frac{\partial P}{\partial x}. \tag{11.44}$$

Equation (11.44) is formally similar to the Buckley-Leverett equation (of oil recovery theory) describing isothermal flow of a two-component single-phase fluid in a porous medium when capillarity can be ignored. However, in the present situation the saturation equation (11.44) is strongly coupled to the nonlinear diffusion equation, for P, obtained by eliminating $\partial S/\partial t$ from the conservation equations:

$$\frac{\partial P}{\partial t} - D\frac{\partial^2 P}{\partial x^2} = f_2(S, P, \partial S/\partial x, \partial P/\partial x), \tag{11.45}$$

where

$$D = -\frac{E_S F}{\dfrac{\partial M}{\partial S}\dfrac{\partial E}{\partial P} - \dfrac{\partial E}{\partial S}\dfrac{\partial M}{\partial P}}. \tag{11.46}$$

Kissling *et al.* (1992a) solved Eqs. (11.45) and (11.43) in turn under the assumption that pressure disturbances diffuse to steady state faster than saturation changes convect. They performed numerical simulations for a block of porous material with pressure and saturation given constant values at the ends of the block. When pressure diffusion occurs much faster than saturation convection, the numerical results can be described in terms of either saturation expansion fans or isolated saturation shocks. When pressure diffusion and saturation convection occur on the same timescale, initial simple shock profiles evolve into multiple shocks.

In the work discussed so far conduction has been neglected. Weir (1994a) has shown that this is certainly valid for sufficiently high temperatures and sufficiently high permeabilities. Young (1993b) has shown that even when conduction has been included, the geothermal saturation wave-speed is formally identical to the Buckley-Leverett wave-speed when the latter is written as the saturation derivative of a volumetric flow.

For the case of two-phase brine mixtures, one has to add an equation expressing conservation of salt. Young (1993a) presented a model in which the flows are described by a parabolic equation for the pressure with a derivative coupling to a pair of equations for saturation and salt concentration. He showed that the wave-speed matrix for the hyperbolic part of the coupled system is formally identical to the corresponding matrix in the polymer flood model for oil recovery. Indeed, for a class of strongly diffusive hot brine models, the wave phenomena in geothermal reservoirs can be predicted from the polymer flood model.

The two-phase geothermal theory has been extended by Weir (1994b) to the case where nonreacting chemical transport (of CO_2, for example) is added. He derived a natural factorization of the system of equations into diffusive and wave equations. Each wave equation allows for the corresponding variable to be discontinuous or equivalently for shock propagation to occur. In general, there now are more that the usual two (vapor and liquid dominated) saturations for a given mass, energy, and chemical flux in steady flow.

A further extension of the theory to the case of withdrawal of fluid at a constant rate was made by Young and Weir (1994). They defined a parameter α,

$$\alpha = \frac{\mu_v W}{K g \rho_l (\rho_l - \rho_v)}, \tag{11.47}$$

where W is the rate of withdrawal (mass per unit area per unit time). They concluded that for large α fluid withdrawal is a mining process, a vapor-dominated zone spreads out from the production level, and production enthalpies tend toward steam values. For small α gravity predominates and buoyancy forces can lead to the formation of a steam bubble that escapes from the production boundary and rises toward the surface. Then production enthalpy may remain at the liquid value over long periods. In addition, certain saturation ranges at the sink may be forbidden as a consequence of the constant rate boundary condition and then saturation shocks will form at the production boundary and travel out from the sink. Internally generated shocks also may occur.

A more general study of vapor-liquid counterflow is that of Satik, Parlar, and Yortsos (1991). They considered a situation in which the counterflow is inclined to the vertical and their analysis included capillarity, heat conduction, and Kelvin effects (the lowering of the vapor pressure due to capillarity). They treated a three-zone model in which the counterflow zone is sandwiched between two zones (one containing mainly vapor and one containing mainly liquid) in which there is no flow. They found that the critical heat flux (above which dryout occurs) increases with decreasing permeability and that a threshold permeability exists below which steady states may not exist. In this context, the critical heat flux is dependent on the pressure and the temperature and so is not precisely defined. As special cases of their general theory, they considered what they called the "heat pipe" and "geothermal" problems. In the former the flow is driven by capillary pressure and the Kelvin effects are of significance only over a narrow boundary layer at the vapor-phase boundary. In the latter the flow is driven by gravity.

The effect of capillary heterogeneity induced by variation in permeability was analyzed by Stubos *et al.* (1993b). They found that the heterogeneity acts as a spatially varying body force that may enhance or diminish gravity effects on heat pipes. A detailed numerical investigation of a transient problem involving a self-heated porous bed was conducted by Stubos *et al.* (1997). Another investigation of a heterogeneous medium, one involving oscillatory instability, was made by Xu and Lowell (1998).

For the axially symmetric problem of constant-strength heat source embedded in an infinite homogeneous medium with uniform initial conditions, Doughty and Pruess (1990, 1992) obtained a similarity solution in terms of the variable $r/t^{1/2}$. In their second paper they included an air component and investigated vapor-pressure lowering, pore-level phase change effects, and an effective continuum representation of fractured porous media.

A model taking into account latent heat, vertical flow, and heat conduction terms, and so involving a new parameter representing a combination of those quantities, was presented by Pestov (1997, 1998).

11.9.2. Heat Pipes

A heat pipe is a system in which a very efficient heat transfer process is effected by vapor-liquid counterflow and associated evaporation and condensation effects with transfer of latent heat. Vapor and liquid may flow in opposite directions due to gravity or capillary action, or both. If heat is injected into such a system, the liquid phase will vaporize, causing pressurization of the vapor phase and vapor flow away from the heat source. In cooler regions the vapor condenses and deposits its latent heat. In the case of a heat pipe depending on capillary action, this sets up a saturation profile, with liquid-phase saturations increasing away from the heat source and capillary forces then cause backflow of the liquid toward the heat source.

For a vertical heat pipe, McGuinness *et al.* (1993) showed that the steady-state values of J_E and $\partial P/\partial z$ are given by

$$\frac{J_E}{K}\left(\frac{\mu_l}{\rho_l k_l}+\frac{\mu_v}{\rho_v k_v}\right)=\frac{J_M}{K}\left(\frac{\mu_l h_l}{\rho_l k_l}+\frac{\mu_v h_v}{\rho_v k_v}\right)-g(h_v-h_l)(\rho_l-\rho_v), \tag{11.48}$$

$$\frac{\partial P}{\partial z}\left(\frac{\rho_l k_l}{\mu_l}+\frac{\rho_v k_v}{\mu_v}\right)=-\frac{J_M}{K}+\left(\frac{\rho_l^2 k_l}{\mu_l}+\frac{\rho_v^2 k_v}{\mu_v}\right). \tag{11.49}$$

If the simplification $k_l+k_v=1$ is assumed (this is a good approximation in many situations), the value of the wave-speed that appears in Eq. (11.34) is, for the case $J_M=0$ (which is appropriate for a heat pipe),

$$c=A\left(\frac{\mu_l k_v^2}{\rho_l}-\frac{\mu_v k_l^2}{\rho_v}\right), \tag{11.50}$$

where A, defined by

$$A=\frac{K(\partial k_l/\partial S)g(\partial_l-\partial_v)}{(k_l\mu_v+k_v\mu_l)(k_l\mu_v/\rho_l+k_v\mu_i/\rho_l)}, \tag{11.51}$$

is always positive. Hence c is normally negative for a steady liquid-dominated pipe ($k_v \approx 0$) and normally positive for a steady vapor-dominated pipe ($k_l \approx 0$). This fixes the direction of information flow, and hence tells one at which end of the pipe one should impose flux values in numerical simulations of geothermal systems and at which end one should specify the saturation and pressure. For the vapor-dominated solution the pressure and saturation should be fixed at depth and the heat and mass flux specified at the top. These boundary conditions are appropriate for a laboratory heat pipe but they are questionable for geothermal systems.

An extension of this work was made by McGuinness (1996), who pointed out that the three-zone model used by Satik *et al.* (1991) limits the possible range of heat flow values through the heat pipe and also limits solutions to those with a smooth transition from pure vapor to pure liquid. The single-zone model of McGuinness allowed these restrictions to be removed. He used a singular perturbation approach (valid for $K > 10^{-15}$ m^2, so that the heat flow is convection dominated), allowing for capillary boundary layers in the temperature-saturation phase-plane. He found that in the geothermal context and with heat flow that is dominated by convection phase-plane trajectories of temperature versus saturation track zero-capillarity (gravity driven) solutions (one liquid-dominated and one vapor-dominated) when they exist. Which of the two solutions is selected depends on the boundary conditions. In the case of bottom heating it is the liquid-dominated solution that should be selected. Whereas the work of Satik *et al.* (1991) suggested that only the vapor-dominated solution is typically obtained, the results of McGuinness (1996) explain why Bau and Torrance (1982a) and others obtained only liquid-dominated solutions in their laboratory experiments. McGuinness (1996) also calculated bounds (maxima) for the lengths of heat pipes in cases where previous work had predicted unbounded lengths.

The theory of two-phase convection, and in particular the theory of heat pipes, is currently controversial. The work of Satik *et al.* (1991) and Stubos *et al.* (1993a,b) was developed in the context of laboratory experiments, and care needs to be exercised in extending their theory to geothermal systems. For further discussion, the reader is referred to Young (1996a,b, 1998a,b).

The quadratic drag (Forchhiemer) effect was included in the analysis by Zhu and Vafai (1999). The dynamics of submarine geothermal heat pipes was investigated by Bai *et al.* (2003). A further study of the stability of heat pipes in vapor-dominated systems was reported by Amili and Yortsos (2003).

11.9.3. Other Aspects

A numerical investigation of two-phase fluid flow and heat transfer in a porous medium heated from the side was conducted by Waite and Amin (1999). A general local thermal nonequilibrium model for two-phase flows with phase change in porous media was proposed by Duval *et al.* (2004). Two-phase flow in porous-channel heat sinks was studied by Peterson and Chang (1997, 1998). Buoyancy effects together with phase change have been discussed by Zhao *et al.* (1999e,

2000b). A review of several aspects of liquid and vapor flow in superheated rock was made by Woods (1999).

11.10. Cracks in Shrinking Solids

The earth's crust is a cracked porous medium with multiple scales, which result from erosion and from periodic shrinking due to volumetric cooling and drying. In spite of the apparent diversity of crack sizes and locations, there is pattern. For example, wet soil exposed to the sun and the wind becomes drier, shrinks superficially, and develops a network of cracks. The loop in the network has a characteristic length scale. The loop is round, more like a hexagon or a square, not slender. The loop is smaller (i.e., cracks are denser) when the wind blows harder, that is, when the drying rate is higher.

The characteric scales of cracks in volumetrically shrinking solids were recently deduced from constructal theory (Bejan *et al.*, 1998; Bejan, 2000). They were deduced by invoking the constructal law: the maximization of access for the mass transfer from wet and cracked soil to the ambient. In Bejan *et al.* (1998), model was a heat transfer analog in which a one-dimensional solid slab of thickness L is initially at the high temperature T_H and has the property of shrinking on cooling. The coolant is a single-phase fluid of temperature T_L.

The question is how to maximize the thermal contact between the solid and the fluid or how to minimize the overall cooling time. This objective makes it necessary to allow the fluid to flow through the solid. In Fig. 11.4 the cracks are spaced uniformly, but their spacing R is arbitrary. The channel width D increases in time, as each solid piece R shrinks. The fluid is driven by the pressure difference ΔP, which is maintained across the solid thickness L. The imposed ΔP is an essential aspect of the channel spacing selection mechanism. For example, in the air cooling of a hot solid layer the scale of ΔP is set at $(1/2)\rho_f U_\infty^2$, where ρ_f and U_∞ are the density and the free-stream velocity, respectively, of the external air flow.

To examine the effect of the channel spacing R on the time needed for cooling the solid we consider the asymptotes $R \to 0$ and $R \to \infty$. The approach is known as the intersection of asymptotes method (Lewins, 2003). When the number of channels per unit length is large, the spacing R is small and so is the eventual shrinkage that is experienced by each R element. This means that when $R \to 0$ we can expect $D \to 0$ and laminar flow through each D-thin channel, such that the channel mass flow rate is $\dot{m}' = \rho_f DU \sim \rho_f D^3 \Delta P/(\mu L)$. In the same limit, R is small enough so that the solid conduction is described by the lumped thermal capacitance model. The solid piece R has a single temperature T, which decreases in time from the initial level T_H to the inlet temperature of the fluid T_L. This cooling effect is governed by the energy balance $\rho c R L (dT/dt) = -q'$, where ρ and c are the density and the specific heat, respectively, of the solid. The cooling effect (q') provided by the flow through the channel is represented well by $q' = \dot{m}' c_p (T - T_L)$, where c_p is the specific heat of the coolant. We obtain the

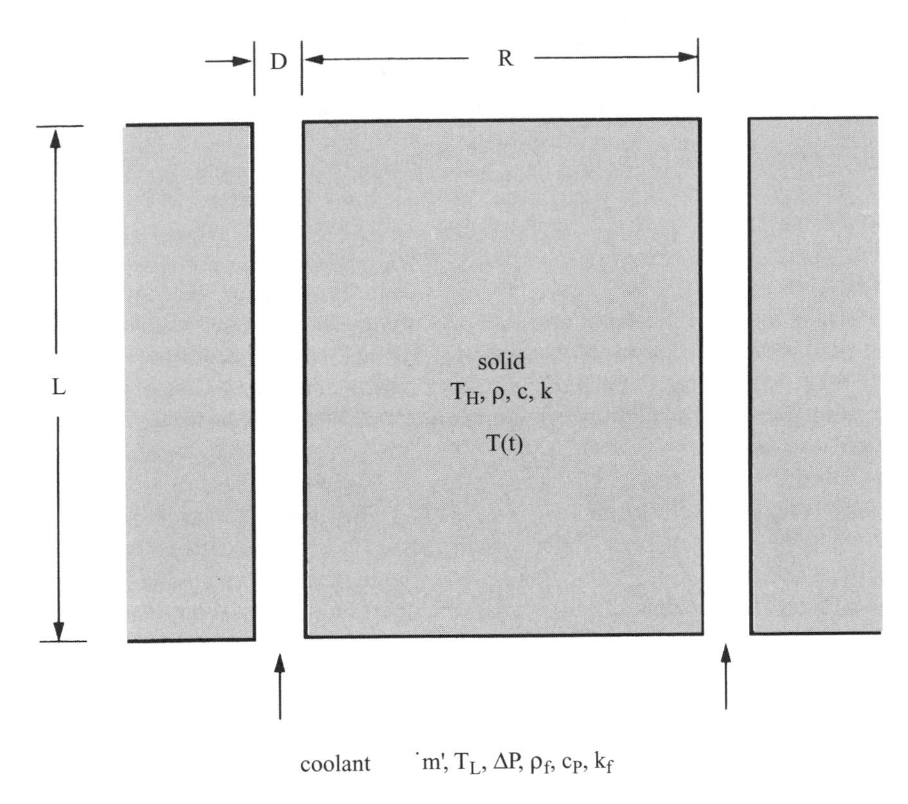

Figure 11.4. Cracks in a shrinking solid cooled by forced convection (Bejan *et al.*, 1998).

order-of-magnitude statement $\rho c RL(\Delta T/t) \sim \dot{m}' c_p \Delta T$, where ΔT is the scale of the instantaneous solid excess temperature $T - T_L$. Finally, by using the scale, we find the cooling time scale:

$$t \sim \frac{\rho c}{\rho_f c_p} \frac{\mu R L^2}{D^3 \Delta P} \qquad (R \to 0) \tag{11.52}$$

In the opposite limit, R is large and the shrinkage (the channel width D) is potentially very large in proportion to R. The fluid present at one time in the channel is mainly isothermal at the inlet temperature T_L. The cooling of each solid side of the crack is ruled by one-dimensional thermal diffusion into a semiinfinite medium. The cooling time in this regime is the same as the time of thermal diffusion

$$t \sim \frac{R^2}{\alpha} \qquad (R \to \infty) \tag{11.53}$$

where $\alpha = k/(\rho c)$ and k is the thermal conductivity of the solid.

To summarize, in the limit $R \to 0$, the cooling time is proportional to R/D^3 or R^{-2} because we expect a proportionality between D and R, namely,

$D/R \sim \beta \Delta T \ll 1$, where $\Delta T \sim T_H - T_L$ and β is the coefficient of thermal contraction of the solid. In the opposite limit, $R \to \infty$, the cooling time is proportional to R^2. Put together, these proportionalities suggest that the cooling time possesses a sharp minimum with respect to R or the channel density. Intersecting the two asymptotes, we find that the optimal crack distance for fastest cooling

$$R_{opt} \sim \left[\frac{k}{k_f} \frac{\alpha_f \nu L^2}{U_\infty^2 (\beta \Delta T)^3} \right]^{1.4}. \quad (11.54)$$

The optimal crack distance decreases as the external pressure (or flow) is intensified. This is in accord with observations that mud cracks become denser when the wind speed increases. The R_{opt} result predicts a higher density of cracks (a smaller R_{opt}) as the solid excess temperature ΔT increases, again in agreement with observations.

An important geometric aspect of the R_{opt} scale is that the optimal distance between consecutive cracks must increase as $L^{1/2}$. This is relevant to predicting the length scale of the lattice of vertical cracks formed in a horizontal two-dimensional surface cooled (or dried) from above, under the influence of external forced convection. As the air flow direction changes locally from time to time and as the material (its graininess) is such that cracks may propagate in more than one direction, we arrive at the problem of cooling a two-dimensional terrain (area A, when seen from above) with cracks of length L and associated area elements of width R_{opt}.

Figure 11.5 shows the two extremes in which L may find itself in relation to R_{opt}. First, when L is considerably shorter than R_{opt} it is impossible to cover the

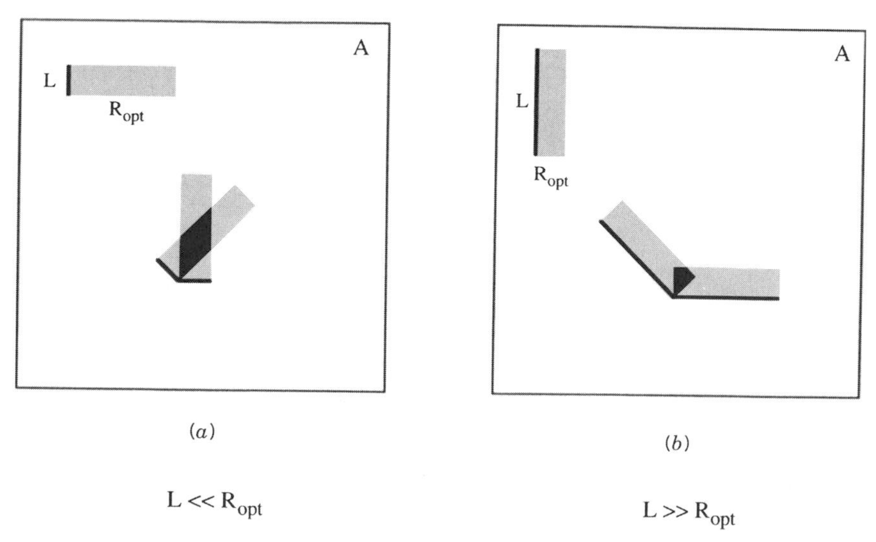

Figure 11.5. Two extremes in covering a two-dimensional solid (A) with cracks (L) and optimally shaped volume elements ($L \times R_{opt}$) (Bejan *et al.*, 1998).

area A exclusively with patches of size $L \times R_{opt}$. The reason is that when two cracks of length L are joined at an angle, the elemental area $\sim L^2$ trapped between them is too small to accommodate the amount of ideally cooled solid material. When L is considerably longer than R_{opt}, any lattice of cracks will fail to cover the area A completely. Now the trapped elemental area ($\sim L^2$) is considerably larger than the amount of ideally cooled solid ($\sim L R_{opt}$): most of the interior of the area element of size L^2 would require a cooling time that is considerably longer than the minimum time determined in the preceding analysis.

In conclusion, maximum access for the global heat current is achieved by covering the A cross section with $L \times R_{opt}$ elements, in which $L \sim R_{opt}$. The optimal pattern is one with relatively round or square loops, not slender loops. Combining $L \sim R_{opt}$ with the R_{opt} expression, we find the optimal length scale of the loop in the network of cracks that will minimize the cooldown time: $R_{opt} \sim (\alpha_f \nu k/k_f)^{1/2}/[U_\infty(\beta \Delta T)^{3/2}]$. Once again, in agreement with observations, the lattice length scale R_{opt} must decrease as the wind speed and the initial excess temperature increase.

Further geophysical applications of constructal theory are explored in Bejan *et al.* (2005).

References

Abel, M. S., Khan, S. K. and Prasad, K. V. 2001 Convective heat and mass transfer in a viscoelastic fluid flow through a porous medium over a stretching sheet. *Int. J. Numer. Meth. Heat. Fluid Flow* **11**, 779–792. [9.2.1]

Aboubi, K., Robillard, L. and Bilgen, E. 1995a Convective heat transfer in an annular porous layer with centrifugal force field. *Numer. Heat Transfer A* **28**, 375–388. [7.3.3]

Aboubi, K., Robillard, L. and Bilgen, E. 1995b Natural convection in horizontal annulus filled with an anisotropic porous medium. *Proc. ASME/JSME Thermal Engineering Joint Conf.* Vol. 3, 415–422. [7.3.3]

Aboubi, K., Robillard, L. and Vasseur, P. 1998 Natural convection in horizontal annulus filled with an anisotropic porous medium. *Int. J. Numer. Meth. Heat Fluid Flow* **8**, 689–702. [6.16.2]

Abu-Hijleh, B. A. K. 1997 Convection heat transfer from a laminar flow over a 2-D backward facing step with asymmetric and orthotropic porous floor segments. *Numer. Heat Transfer A* **31**, 325–335. [4.11]

Abu-Hijleh, B. A. K. 2000 Heat transfer from a 2D backward facing step with isotropic porous floor segments. *Int. J. Heat Mass Transfer* **43**, 2727–2737. [4.11]

Abu-Hijleh, B. A. K. 2001a Natural convection heat transfer from a cylinder covered with an orthotropic porous layer in cross-flow. *Numer. Heat Transfer A* **40**, 767–782. [5.12.3]

Abu-Hijleh, B. A. K. 2001b Laminar forced convection heat transfer from a cylinder covered with an orthotropic porous layer in cross-flow. *Int. J. Numer. Meth. Heat Fluid Flow* **11**, 106–120. [4.11]

Abu-Hijleh, B. A. K. 2002 Entropy generation due to cross-flow heat transfer from a cylinder covered with an orthotropic porous layer. *Heat Mass Transfer* **39**, 27–40. [4.11]

Abu-Hijleh, B. A. K. 2003 Enhanced forced convection heat transfer from a cylinder using permeable fins. *ASME J. Heat Transfer* **125**, 804–811. [4.11]

Abu-Hijleh, B. A. K. and Al-Nimr, M. A. 2001 The effect of the local inertial term on the fluid flow in channels partially filled with porous material. *Int. J. Heat Mass Transfer* **44**, 1565–1572. [4.11]

Abu-Hijleh, B. A. K., Al-Nimr, M. A. and Hader, M. A. 2004 Thermal equilibrium in transient forced convection porous channel flow. *Transport Porous Media* **57**, 49–58. [4.10]

Acharya, M., Dash, G. C. and Singh, L. P. 2000 Magnetic effects on the free convection and mass transfer flow through porous medium with constant suction and constant heat. *Indian J. Pure Appl. Math.* **31**, 1–18. [9.2.1]

Achenbach, E. 1995 Heat and flow characteristics of packed beds. *Expt. Thermal Fluid Science* **10**, 17–27. [1.5.2]

Adnani, P., Catton, I. and Abdou, M. A. 1995 Non-Darcian forced convection in porous media with anisotropic dispersion. *ASME J. Heat Transfer* **117**, 447–451. [4.9]

Adnani, S. G. and Hsiao, K. T. 2005 Transport phenomena in liquid composites molding processes and their roles in process control and optimization. *Handbook of Porous Media* (ed. K. Vafai), 2nd ed., Taylor and Francis, New York, pp. 573–606. [10.2.3]

Afzal, N. and Salam, M. Y. 1990 Natural convection from point source embedded in Darcian porous medium. *Fluid Dyn. Res.* **6**, 175–184. [5.8]

Afifi, R. I. and Berbish, N. S. 1999 Experimental investigation of forced convection heat transfer over a horizontal flat plate in a porous medium. *J. Engng. Appl. Sci.* **46**, 693–710. [4.1]

Afzal, N. 1985 Two-dimensional buoyant plume in porous media: higher-order effects. *Int. J. Heat Mass Transfer* **28**, 2029–2041. [5.10.1.1]

Agwu Nnanna, A. G., Haji-Sheikh, A. and Harris, K. T. 2004 Experimental study of local thermal non-equilibrium phenomena during phase change in porous media. *Int. J. Heat Mass Transfer* **43**, 4365–4375. [10.1.7]

Aidun, C. K. 1987 Stability of convection rolls in porous media. *ASME HTD* **94**, 31–36. [6.8]

Aidun, C. K. and Steen, P. H. 1987 Transition of oscillatory convective heat transfer in a fluid-saturated porous medium. *AIAA J. Thermophys. Heat Transfer* **1**, 268–273. [6.8]

Aithal, S. M., Aldemir, T. and Vafai, K. 1994 Assessment of the impact of neutronic/thermal-hydraulic coupling in the design and performance of nuclear reactors on the design and performance of nuclear reactors for space propulsion. *Nuclear Tech.* **106**, 187–202. [4.16.2]

Alagoa, K. D., Tay, G. and Abbey, T. M. 1999 Radiative and free-convection effects of a MHD flow through porous medium between infinite parallel plates with time-dependent suction. *Astrophys. Space Science* **260**, 455–468. [7.3.7]

Al-Amiri, A. M. 2002 Natural convection in porous enclosures: The application of the two-energy model. *Numer. Heat Transfer A* **41**, 817–834. [7.6.2]

Alavyoon, F. 1993 On natural convection in vertical porous enclosures due to prescribed fluxes of heat and mass at the vertical boundaries. *Int. J. Heat Mass Transfer* **36**, 2479–2498. [9.2.2]

Alavyoon, F., Masuda, Y. and Kimura, S. 1994 On the natural convection in vertical porous enclosures due to opposing fluxes of heat and mass prescribed at the vertical walls. *Int. J. Heat Mass Transfer* **37**, 195–206. [9.2.2]

Alazmi, B. and Vafai, K. 2000 Analysis of variants within the porous media transport models. *ASME J. Heat Transfer* **122**, 303–326. [7.6.2]

Alazmi, B. and Vafai, K. 2001 Analysis of fluid flow and heat transfer interfacial conditions between a porous medium and a fluid layer. *Int. J. Heat Mass Transfer* **44**, 1735–1749. [1.6]

Alazmi, B. and Vafai, K. 2002 Constant wall heat flux boundary conditions in porous media under local thermal non-equilibrium conditions. *Int. J. Heat Mass Transfer* **45**, 3071–3087. [2.2.2, 4.10]

Alazmi, B. and Vafai, K. 2004 Analysis of variable porosity, thermal dispersion and local thermal nonequilibrium on free surface flows through porous media. *ASME J. Heat Mass Transfer* 126, 389–399. [4.10]

Albert, M. R. 1995 Advective-diffusive heat transfer in snow. *ASME Int. Mech. Cong. Expos.*, San Francisco, Paper 95-Wa/HT-44. [11.1]

Alchaar, S., Vasseur, P. and Bilgen, E. 1995a Effects of a magnetic field on the onset of convection in a porous medium. *Heat Mass Transfer* **30**, 259–267. [6.21]

Alchaar, S., Vasseur, P. and Bilgen, E. 1995b Hydromagnetic natural convection in a tilted rectangular porous enclosure. *Numer. Heat Transfer A* **27**, 107–127. [7.8]

Aldoss, T. K. 1996 MHD mixed convection from a vertical cylinder embedded in a porous medium. *Int. Comm. Heat Mass Transfer* **23**, 517–530. [8.1.3]

Aldoss, T. K. and Ali, Y. D. 1997 MHD mixed convection from a horizontal cylinder in a porous medium. *JSME Int. J. Ser. B* **40**, 290–295. [8.1.3]

Aldoss, T. K., Alkam, M. and Shatarah, M. 2004 Natural convection from a horizontal annulus partially filled with a porous medium. *Int. Comm. Heat Mass Transfer* **31**, 441–452. [7.7]

Aldoss, T. K., Alnimr, M. A., Jarrahr, M. A. and Alshaer, B. J. 1995 Magnetohydrodynamic mixed convection from a vertical plate embedded in a porous medium. *Numer. Heat Transfer A* **28**, 635–645. [8.1.1]

Aldoss, T. K., Chen, T. S. and Armaly, B. F. 1993a Nonsimilarity solutions for mixed convection from horizontal surfaces in a porous medium—variable surface heat flux. *Int. J. Heat Mass Transfer* **36**, 463–470. [8.1.2]

Aldoss, T. K., Chen, T. S. and Armaly, B. F. 1993b Nonsimilarity solutions for mixed convection from horizontal surfaces in a porous medium—variable wall temperature. *Int. J. Heat Mass Transfer* **36**, 471–477. [8.1.2]

Aldoss, T. K., Chen, T. S. and Armaly, B. F. 1994a Mixed convection over nonisothermal horizontal surfaces in a porous medium—the entire regime. *Numer. Heat Transfer A* **25**, 685–701. [8.1.2]

Aldoss, T. K., Jarrah, M. A. and Al-Sha'er, B. J. 1996 Mixed convection from a vertical cylinder embedded in a porous medium: non-Darcy model. *Int. J. Heat Mass Transfer* **39**, 1141–1148. [8.1.3]

Aldoss, T. K., Jarrah, M. A. and Duwairi, H. M. 1994b Wall effect on mixed convection flow from horizontal surfaces with a variable heat flux. *Canad. J. Chem. Engng.* **72**, 35–42. [8.1.2]

Alex, S. M. and Patil, P. R. 2000a Thermal instability in an anisotropic rotating porous medium. *Heat Mass Transfer* **36**, 159–163. [6.22]

Alex, S. M. and Patil, P. R. 2000b Thermal instability in an inclined isotropic porous medium. *J. Porous Media* **3**, 207–216. [7.8]

Alex, S. M. and Patil, P. R. 2001 Effect of variable gravity field on Soret driven thermosolutal convection in a porous medium. *Int. Comm. Heat Mass Transfer* **28**, 509–518. [9.1.6.4]

Alex, S. M. and Patil, P. R. 2002a Effect of a variable gravity field on convection in an isotropic porous medium with internal heat source and inclined temperature gradient. *ASME J. Heat Transfer* **124**, 144–150. [7.9]

Alex, S. M. and Patil, P. R. 2002b Effect of a variable gravity field on thermal instability in a porous medium with inclined temperature gradient and vertical throughflow. *J. Porous Media* **5**, 137–147. [7.9]

Alex, S. M., Patil, P. R. and Venakrishnan, K. S. 2001 Variable gravity effects on thermal instability in a porous medium with internal heat source and inclined temperature gradient. *Fluid Dyn. Res.* **29**, 1–6. [7.9]

Al-Hadhrami, A. K., Elliot, L. and Ingham, D. B. 2002 Combined free and forced convection in vertical channels of porous media. *Transport in Porous Media* **49**, 265–289. [8.3.1]

Al-Hadhrami, A. K., Elliot, L. and Ingham, D. B. 2003 A new model for viscous dissipation in porous media across a range of permeability values. *Transport in Porous Media* **53**, 117–122. [2.2.2]

Al-Hadhrami, A. K., Elliot, L., Ingham, D. B. and Wen, X. 2001a Fluid flows through two-dimensional channels of composite materials. *Transport Porous Media* **45**, 283–302. [4.11]

Al-Hadhrami, A. K., Elliot, L., Ingham, D. B. and Wen, X. 2001b Analytical solutions of fluid flows through composite channels. *J. Porous Media* **4**, 149–163. [4.11]

Alkam, M. K. and Al-Nimr, M. A. 1998 Transient non-Darcian forced convection flow in a pipe partially filled with a porous material. *Int. J. Heat Mass Transfer* **41**, 347–356. [4.6.4]

Alkam, M. K. and Al-Nimr, M. A. 1999a Improving the performance of double-pipe heat exchangers by using porous substrates. *Int. J. Heat Mass Transfer* **42**, 3609–3618. [4.11]

Alkam, M. K. and Al-Nimr, M. A. 1999b Solar collectors with tubes partially filled with porous substrates. *ASME J. Solar Energy* **121**, 20–24. [4.11]

Alkam, M. K. and Al-Nimr, M. A. 2001 Transient flow hydrodynamics in circular channels partially filled with a porous material. *Heat Mass Transfer* **37**, 133–137. [4.11]

Alkam, M. K., Al-Nimr, M. A. and Hamdan, M. O. 2001 Enhancing heat transfer in parallel-plate channels by using porous media. *Int. J. Heat Mass Transfer* **44**, 931–938. [4.11]

Alkam, M. K., Al-Nimr, M. A. and Hamdan, M. O. 2002 On forced convection in channels partially filled with porous substrates. *Heat Mass Transfer* **38**, 337–342. [4.11]

Alkam, M. K., Al-Nimr, M. A. and Mousa, Z. 1998 Forced convection of non-Newtonian fluids in porous concentric annuli. *Int. J. Numer. Meth. Heat. Fluid Flow* **8**, 703–716. [4.16.3]

Allain, C., Cloitre, M. and Mongruel, A. 1992 Scaling in flows driven by heat and mass convection in a porous medium. *Europhys. Lett.* **20**, 313–318. [9.2.1]

Allan, F. M. and Hamdan, M. H. 2002 Fluid mechanics of the interface region between two porous layers. *Appl. Math. Comput.* **128**, 37–43. [1.6]

Al-Nimr, M. A. and Abu-Hijleh, B. A. 2002 Validation of thermal equilibrium assumption in transient forced convection flow in porous channel. *Transport in Porous Media* **49**, 127–138. [4.10]

Al-Nimr, M. A. and Alkam, M. K. 1997a Film condensation on a vertical plate embedded in a porous medium. *J. Appl. Energy* **56**, 47–57. [10.4]

Al-Nimr, M. A. and Alkam, M. K. 1997b Unsteady non-Darcian forced convection analysis in an annulus partially filled with a porous material. *ASME J. Heat Transfer* **119**, 799–804. [4.11]

Al-Nimr, M. A. and Alkam, M. K. 1998a A modified tubeless solar collector partially filled with porous substrate. *Renewable Energy* **13**, 165–173. [4.11]

Al-Nimr, M. A. and Alkam, M. K. 1998b Unsteady non-Darcian fluid flow in parallel-plates channels partially filled with porous materials. *Heat Mass Transfer* **33**, 315–318. [4.11]

Al-Nimr, M. A. and Darabseh, T. T. 1995 Analytical solution for transient laminar free convection in open-ended vertical concentric porous annuli. *ASME J. Heat Transfer* **117**, 762–764. [7.5]

Al-Nimr, M. A. and Haddad, O. M. 1999a Fully developed free convection in open-ended vertical channels partially filled with porous material. *J. Porous Media* **2**, 179–204. [7.7]

Al-Nimr, M. A. and Hader, M. A. 1999b MHD free convection flow in open-ended vertical porous channels. *Chem. Engng. Sci.* **54**, 1883–1889. [7.7]

Al-Nimr, M. A. and Khadrawi A. F. 2003 Transient free convection fluid flow in domains partially filled with porous media. *Transport in Porous Media* **51**, 157–172. [7.7]

Al-Nimr, M. A. and Kiwan, S. 2002 Examination of the thermal equilibrium assumption in periodic forced convection in a porous channel. *J. Porous Media* **5**, 35–40. [4.10]

Al-Nimr, M. A. and Massoud, S. 1998 Unsteady free convection flow over a vertical flat plate immersed in a porous medium. *Fluid Dyn. Res.* **23**, 153–160. [5.1.9.9]

Al-Nimr, M. A., Aldoss, T. and Abuzaid, M. M. 2005 Effect of the macroscopic local inertial term on the non-Newtonian free-convection flow in channels filled with porous materials. *J. Porous Media* **8**, 421–430. [7.5]

Al-Nimr, M. A., Aldoss, T. and Naji, M. I. 1994a Transient forced convection in the entrance region of a porous tube. *Canad. J. Chem. Engng.* **72**, 249–255. [4.6.4]

Al-Nimr, M. A., Aldoss, T. and Naji, M. I. 1994b Transient forced convection in the entrance region of porous concentric annuli. *Canad. J. Chem. Engng.* **72**, 1092–1096. [4.6.4]

Al-Odat, M. Q. 2004a Nonsimilar solutions for unsteady free convection from a vertical plate embedded in a non-Darcy porous medium. *Int. Comm. Heat Mass Transfer* **31**, 377–386. [5.1.7.2]

Al-Odat, M. Q. 2004b Transient non-Darcy mixed convection along a vertical surface in porous medium with suction or injection. *Appl. Math. Comput.* **156**, 679–694. [8.1.1]

Altevogt, A. S., Rolston, D. E. and Whitaker, S. 2003 New equations for binary gas transport in porous media; part 1: equation development. *Adv. Water Res.* **26**, 695–715. [1.4.1]

Alvarez, G., Bournet, P. E. and Flick, D. 2003 Two-dimensional simulation of turbulent flow and transfer through stacked spheres. *Int. J. Heat Mass Transfer* **46**, 2459–2469. [1.8]

Alves, L. S. de B. and Cotta, R. M. 2000 Transient natural convection inside porous cavities: hybrid numerical-analytical solution and symbolic-numerical computation. *Num. Heat Transfer A* **38**, 89–110. [7.5]

Alves, L. S. de B., Cotta, R. M. and Pontes, J. 2002 Stability analysis of natural convection in porous cavities through integral transforms. *Int. J. Heat Mass Transfer* **45**, 1185–1195. [6.15.1].

Aly, E. H., Elliott, L. and Ingham, D. B. 2003 Mixed convection boundary-layer flow over a vertical surface embedded in a porous medium. *European J. Mech. B/Fluids* **22**, 529–543. [8.1.1]

Amahmid, A., Hasnaoui, M. and Douamna, S. 2001 Analytic and numerical study of double-diffusive parallel flow induced in a vertical porous layer subjected to constant heat and mass fluxes. *Strojniski Vestnik-J. Mech. Engng.* **47**, 501–505. [9.2.2]

Amahmid, A., Hasnaoui, M. and Vasseur, P. 1999c Étude analytique et numérique de la convection naturelle dans une couche poreuse de Brinkman doublement diffusive. *Int. J. Heat Mass Transfer* **42**, 2991–3005. [9.2.2]

Amahmid, A., Hasnaoui, M., Mamou, M. and Vasseur, P. 1999a Double-diffusive parallel flow induced in a horizontal Brinkman porous layer subjected to constant heat and mass fluxes: analytical and numerical studies. *Heat Mass Transfer* **35**, 409–421. [9.1.6.3]

Amahmid, A., Hasnaoui, M., Mamou, M. and Vasseur, P. 1999b Boundary layer flows in a vertical porous enclosure induced by opposing buoyancy forces. *Int. J. Heat Mass Transfer* **42**, 3599–3608. [9.2.2]

Amahmid, A., Hasnaoui, M., Mamou, M. and Vasseur, P. 2000 On the transition between aiding and opposing double-diffusive flows in a vertical porous cavity, *J. Porous Media* **3**, 123–137. [9.2.2]

Amara, T., Slimi, K. and Ben Nasrallah, S. 2000 Free convection in a vertical cylindrical enclosure. *Int. J. Thermal Sci.* **39**, 616–634. [7.3.3]

Amari, B., Vasseur, P. and Bilgen, E. 1994 Natural convection of non-Newtonian fluids in a horizontal porous layer. *Wärme-Stoffübertrag.* **29**, 185–193. [6.23]

Amberg, G. and Homsy, G. M. 1993 Nonlinear analysis of buoyant convection in binary solidification with application to channel formation. *J. Fluid Mech.* **252**, 79–98. [10.2.3]

Ames, K. A. and Cobb, S. S. 1994 Penetrative convection in a porous medium with internal heat sources. *Int. J. Engng. Sci.* **32**, 95–106. [6.11.2]

Amili, P. and Yortsos, Y. C. 2003 Stability of heat pipes in vapor-dominated systems. *Int. J. Heat Mass Transfer* **47**, 1233–1246. [11.9.2]

Amiri, A. and Vafai, K. 1994 Analysis of dispersion effects and non-thermal equilibrium non-Darcian, variable porosity incompressible flow through porous media. *Int. J. Heat Mass Transfer* **37**, 939–954. [4.6.4]

Amiri, A. and Vafai, K. 1998 Transient analysis of incompressible flow through a packed bed. *Int. J. Heat Mass Transfer* **41**, 4259–4279. [4.6.4]

Amiri, A., Vafai, K. and Kuzay, T. M. 1995 Effects of boundary conditions on non-Darcian heat transfer through porous media and experimental comparisons. *Numer. Heat Transfer A* **27**, 651–664. [4.6.4]

Anderson, D. M. 2003 A model for diffusion-controlled solidification of ternary alloys in mushy layers. *J. Fluid Mech.* **483**, 165–197. [10.2.3]

Anderson, D. M. and Worster, M. G. 1995 Weakly nonlinear analysis of convection in mushy layers during solidification of binary alloys. *J. Fluid Mech.* **302**, 307–331. [10.2.3]

Anderson, D. M. and Worster, M. G. 1996 A new oscillatory instability in a mushy layer during the solidification of binary alloys. *J. Fluid Mech.* **307**, 245–267. [10.2.3]

Anderson, P. and Glassner, D. 1990 Thermal convection and surface temperatures in porous media. *Int. J. Heat Mass Transfer* **33**, 1321–1330. [7.11]

Angirasa, D. 2002a Forced convection heat transfer in metallic fibrous materials. *ASME J. Heat Transfer* **124**, 739–745. [4.9]

Angirasa, D. 2002b Experimental investigation of forced convection heat transfer augmentation with metallic fibrous materials. *Int. J. Heat Mass Transfer* **45**, 919–922. [4.9]

Angirasa, D. and Peterson, G. P. 1997a Combined heat and mass transfer by natural convection with opposing buoyancy effects in a fluid saturated porous medium. *Int. J. Heat Mass Transfer* **40**, 2755–2733. [9.2.2]

Angirasa, D. and Peterson, G. P. 1997b Natural convection heat transfer from an isothermal vertical surface to a fluid saturated thermally stratified porous medium. *Int. J. Heat Mass Transfer* **40**, 4329–4335. [5.1.4]

Angirasa, D. and Peterson, G. P. 1998a Upper and lower Rayleigh number bounds for two-dimensional natural convection over a finite horizontal surface situated in a fluid-saturated porous medium. *Numer. Heat Transfer A* **33**, 477–493. [5.2]

Angirasa, D. and Peterson, G. P. 1998b Natural convection below a downward facing heated horizontal surface in a fluid-saturated porous medium. *Numer. Heat Transfer A* **34**, 301–311. [5.2]

Angirasa, D. and Peterson, G. P. 1999 Forced convection heat transfer augmentation in a channel with a localized heat source using fibrous materials. *ASME J. Electr. Pack.* **121**, 1–7. [4.16.2]

Angirasa, D., Peterson, G. P. and Pop, I. 1997 Combined heat and mass transfer by natural convection in a saturated thermally stratified porous medium. *Numer. Heat Transfer A* **31**, 255–272. [9.2.1]

Ansari, A. and Daniels, P. G. 1993 Thermally driven tall cavity flows in porous media. *Proc. Roy. Soc. Lond. A.* **433**, 163–181. [7.1.2]

Ansari, A. and Daniels, P. G. 1994 Thermally driven tall cavity flows in porous media: the convective regime. *Proc. Roy. Soc. Lond. A.* **444**, 375–388. [7.1.2]

Antohe, B. V. and Lage, J. L. 1994 A dynamic thermal insulator: inducing resonance within a fluid saturated porous medium heated periodically from the side. *Int. J. Heat Mass Transfer* **37**, 771–782. [7.10]

Antohe, B. V. and Lage, J. L. 1996 Amplitude effect on convection induced by tile-periodic heating. *Int. J. Heat Mass Transfer* **39**, 1121–1133. [7.10]

Antohe, B. V. and Lage, J. L. 1997a The Prandtl number effect on the optimum heating frequency of an enclosure filled with fluid or with a saturated porous medium. *Int. J. Heat Mass Transfer* **40**, 1313–1323. [7.10]

Antohe, B. V. and Lage, J. L. 1997b A general two-equation macroscopic turbulence model for incompressible flow in porous media. *Int. J. Heat Mass Transfer* **40**, 3013–3024. [1.8]

Artem'eva, E. L. and Stroganova, E. V. 1987 Stability of a nonuniformly heated fluid in a porous horizontal layer. *Fluid Dyn.* **21**, 845–848. [6.10]

Asako, Y., Nakamura, H., Yamaguchi, Y. and Fagri, M. 1992 Three-dimensional natural convection in a vertical porous layer with a hexagonal honeycomb core. *ASME J. Heat Transfer* **114**, 924–927. [7.3.7]

Asbik, M., Chaynane, R., Boushaba, H., Zeghmati, B. and Khmou, A. 2003 Analytic investigation of forced convection film condensation on a vertical porous-layer coated surface. *Heat Mass Transfer* **40**, 143–155. [10.4]

Asbik, M., Sadki, H., Hajar, M., Zeghmati, B. and Khmou, A. 2002 Numerical study of laminar mixed convection in a vertical saturated porous enclosure: The combined effect of double diffusion and evaporation. *Numer. Heat Transfer A* **41**, 403–420. [9.2.2]

Assato, M., Pedras, M. H. J. and de Lemos, M. J. S. 2005 Numerical solution of turbulent channel flow past a backward-facing step with a porous insert using linear and nonlinear k-ε models. *J. Porous Media* **8**, 13–29. [1.8]

Auriault, J.-L. 1999 Comments on the paper "Local and global transitions to chaos and hysteresis in a porous layer heated from below," by P. Vadasz. *Transport in Porous Media* **37**, 247–249. [6.4]

Auriault, J.-L., Geindreau, C. and Royer, P. 2002a Coriolis effects on filtration law in rotating porous media. *Transport in Porous Media* **48**, 315–330. [6.22]

Auriault, J.-L., Royer, P. and Geindreau, C. 2002b Filtration law for power-law fluids in anisotropic porous media. *Int. J. Engng. Sci.* **40**, 1151–1163. [1.5.4]

Avramenko, A. A. and Kuznetsov, A. V. 2004 Stability of a suspension of gyrotactic microorganisms in superposed fluid and porous layers. *Int. Comm. Heat Mass Transfer* **31**, 1057–1066. [6.25]

Avramenko, A. A. and Kuznetsov, A. V. 2005 Linear stability analysis of a suspension of oxytactic bacteria in superposed fluid and porous layers. *Transport Porous Media* **61**, 157–176. [6.25]

Avroam, D. G. and Payatakes, A. C. 1995 Flow regimes and relative permeabilities during steady-state two-phase flow in porous media. *J. Fluid Mech.* **293**, 207–236. [3.5.4]

Azoumah, Y., Mazet, N. and Neveu, P. 2004 Constructal network for heat and mass transfer in a solid-gas reactive porous medium. *Int. J. Heat Mass Transfer* **47**, 2961–2970. [4.18]

Badr, H. M. and Pop, I. 1988 Combined convection from an isothermal horizontal rod buried in a porous medium. *Int. J. Heat Mass Transfer* **31**, 2527–2541. [8.1.3]

Badr, H. M. and Pop, I 1992 Effect of flow direction on mixed convection from a horizontal rod embedded in a porous medium. *Trans. Can. Soc. Mech. Engng.* **16**, 267–290. [8.1.3]

Bae, J. H., Hyun, J.M. and Kim, J. W. 2004 Mixed convection in a channel with porous multiblocks under imposed thermal modulation. *Numer. Heat Transfer A* **46**, 891–908. [8.4]

Bagai, S. 2003 Similarity solutions of free convection boundary layers over a body of arbitrary shape in a porous medium with internal heat generation. *Int. Comm. Heat Mass Transfer* **30**, 997–1003. [5.9]

Bagai, S. 2004 Effect of variable viscosity on free convection over a non-isothermal axisymmetric body in a porous medium with internal heat generation. *Acta Mech.* **169**, 187–194. [5.9]

Bahloul, A., Boutana, N. and Vasseur, P. 2003 Double-diffusive and Soret-induced convection in a shallow horizontal porous layer. *J. Fluid Mech.* **491**, 325–352. [9.1.4]

Bahloul, A., Kalla, L., Bennacer, R., Beji, H. and Vasseur, P. 2004a Natural convection in a vertical porous slot heated from below and with horizontal concentration gradients. *Int. J. Thermal Sci.* **43**, 653–663. [9.5]

Bahloul, A., Yahiaoui, M. A., Vasseur, P. and Robillard, L. 2004b Thermogravitational separation in a vertical annular porous layer. *Int. Comm. Heat Mass Transfer* **31**, 783–794. [9.4]

Bai, M. and Roegiers, J. C. 1994 Fluid flow and heat flow in deformable fractured media. *Int. J. Engng. Sci.* **32**, 1615–1663. [1.9]

Bai, M., Ma, Q. and Roegiers, J. C. 1994a Dual porosity behaviour of naturally fractured reservoirs. *Int. J. Num. Anal. Mech. Geomech.* **18**, 359–376. [1.9]

Bai, M., Ma, Q. and Roegiers, J. C. 1994b Nonlinear dual-porosity model. *Appl. Math. Model.* **18**, 602–610. [1.9]

Bai, M., Roegiers, J. C. and Inyang, H. F. 1996 Contaminant transport in nonisothermal fractured porous media. *J. Env. Engng.* **122**, 416–423. [1.9]

Bai, W. M., Xu, W. Y. and Lowell, R. P. 2003 The dynamics of submarine geothermal heat pipes. *Geophsy. Res. Lett.* **30**, art. no. 1108. [11.9.2]

Bakier, A. Y. 1997 Aiding and opposing mixed convection flow in melting from a vertical flat plate embedded in a porous medium. *Transport in Porous Media* **29**, 127–139. [10.1.7]

Bakier, A. Y. 2001a Thermal radiation effect on mixed convection from vertical surfaces in saturated porous media. *Int. Comm. Heat Mass Transfer* **28**, 119–126. [8.1.1]

Bakier, A. Y. 2001b Thermal radiation effect on mixed convection from vertical surfaces in saturated porous media. *Indian J. Pure Appl. Math.* **32**, 1157–1163. [8.1.1]

Bakier, A. Y. and Gorla, R. S. R. 1996 Thermal radiation effect on mixed convection from horizontal surfaces in saturated porous media. *Transport Porous Media* **23**, 357–363. [8.1.2]

Bakier, A. Y., Mansour, M. A. , Gorla, R. S. R. and Ebiana, A. B. 1997 Nonsimilar solutions for free convection from a vertical plate in porous media. *Heat Mass Transfer* **33**, 145–148. [5.1.9.4]

Balakotaiah, V. and Portalet, P. 1990a Natural convection effects on thermal ignition in a porous medium. I. Semenov model. *Proc. Roy. Soc. Lond. A.* **429**, 533–554. [3.4]

Balakotaiah, V. and Portalet, P. 1990b Natural convection effects on thermal ignition in a porous medium. II. Lumped thermal model-I. *Proc. Roy. Soc. Lond. A.* **429**, 555–567. [3.4]

Balasubramanian, R. and Thangaraj, R. P. 1998 Thermal convection in a fluid layer sandwiched between two porous layers of different permeabilities. *Acta Mech.* **130**, 81–93. [6.19.1]

Bankvall, C. G. 1974 Natural convection in a vertical permeable space. *Wärme-Stoffübertrag.* **7**, 22–30. [7.1.2]

Bansod, V. J. 2003 The Darcy model for boundary layer flows in a horizontal porous medium induced by combined buoyancy forces. *J. Porous Media* **6**, 273–281. [9.2.1]

Bansod, V. J. 2005 The effects of blowing and suction on double diffusion by mixed convection over inclined permeable surfaces. *Transport Porous Media* **60**, 301–317. [9.6]

Bansod, V. J., Singh, P. and Ratishkumar, B. V. 2002 Heat and mass transfer by natural convection from a vertical surface to the stratified Darcian fluid. *J. Porous Med.* **5**, 57–66. [9.2.1]

Bansod, V. J., Singh, P. and Ratishkumar, B. V. 2005 Laminar natural convection heat and mass transfer from a horizontal in non-Darcy porous media. *J. Porous Med.* **5**, 57–66. [9.2.1]

Banu, N. and Rees, D. A. S. 2000 The effect of inertia on vertical free convection boundary layer flow from a heated surface in porous media with suction. *Int. Comm. Heat Mass Transfer* **27**, 775–783. [5.1.7.2]

Banu, N. and Rees, D. A. S. 2001 Effect of a traveling thermal wave on weakly nonlinear convection in a porous layer. *J. Porous Media* **4**, 225–239. [6.4]

Banu, N. and Rees, D. A. S. 2002 Onset of Darcy-Bénard convection using a thermal non-equilibrium model. *Int. J. Heat Mass Transfer* **45**, 2221–2228. [6.5]

Banu, N., Rees, D. A. S. and Pop, I. 1998 Steady and unsteady free convection in porous cavities with internal heat generation. *Heat Transfer 1998, Proc. 11th IHTC*, **4**, 375–380. [6.17]

Barak, A. Z. 1987 Comments on "High velocity flow in porous media" by Hassanizadeh and Gray. *Transport in Porous Media* **2**, 533–535. [1.5.2]

Barbosa Mota, J. P. and Saatdjian, E. 1994 Natural convection in a porous, horizontal cylindrical annulus. *ASME J. Heat Transfer* **116**, 621–626. [7.3.3]

Barbosa Mota, J. P. and Saatdjian, E. 1995 Natural convection in porous cylindrical annuli. *Int. J. Numer. Methods Heat Fluid Flow* **5**, 3–17. [7.3.3]

Barbosa Mota, J. P. and Saatdjian, E. 1997 Reduction of natural convection heat transfer in horizontal eccentric annuli containing saturated porous media. *Int. J. Numer. Methods Heat Fluid Flow* **7**, 401–416. [6.16.2]

Barbosa Mota, J. P., Le Provost, J. F., Puons, E. and Saatdjian, E. 1994 Natural convection in porous, horizontal eccentric annuli. *Heat Transfer 1994*, Inst. Chem. Engrs, Rugby, Vol. 5, 435–440. [7.3.3]

Bardan, G. and Mojtabi, A. 2000 On the Horton-Rogers-Lapwood convective instability with vertical vibration: onset of convection. *Phys. Fluids* **12**, 2723–2731. [6.24]

Bardan, G., Pinaud, G. and Mojtabi, A. 2001 Onset of convection in porous cell submitted to high-frequency vibrations. *C. R. Acad. Sci. II B* **329**, 283–286. [6.24]

Bardan, G., Razi, Y. P. and Mojtabi, A. 2004 Comments on the mean flow averaged model. *Phys. Fluids* **16**, 4535–4538. [6.24]

Barenblatt, G. I., Entov, V. M. and Ryzhik, V. M. 1990 *Theory of Fluid Flow through Natural Rocks*, Kluwer Academic Publishers, Dordrecht. [1.9]

Bartelt, P., Buser, O. and Sokratov, S. A. 2004 A nonequilibrium treatment of heat and mass transfer in alpine snowcovers. *Cold Regions Sci. Tech.* **39**, 219–242. [11.1]

Bartlett, R. F. and Viskanta, R. 1996 Enhancement of forced convection in an asymmetrically heated duct filled with high thermal conductivity porous media. *J. Enhanced Heat Transfer* **3**, 291–299. [4.9]

Bassom, A. P. and Rees, D. A. S. 1995 The linear vortex instability of flow induced by a horizontal heated surface in a porous medium. *Quart. J. Mech. Appl. Mech.* **48**, 1–19. [5.4]

Bassom, A. P. and Rees, D. A. S. 1996 Free convection from a heated vertical cylinder embedded in a fluid-saturated porous medium. *Acta Mech.* **116**, 139–151. [5.7]

Basu, A. and Islam, M. R. 1996 Instability in a combined heat and mass transfer problem in porous media. *Chaos, Solitons, Fractals* **7**, 109–123. [9.2.4]

Basu, A. J. and Khalili, A. 1999 Computation of flow through a fluid-sediment interface in a benthic chamber. *Phys. Fluids* **11**, 1395–1405. [6.19.3]

Batchelor, G. K. 1967 *An Introduction to Fluid Dynamics*, Cambridge University Press. [1.5.1]

Bau, H. H. 1984a Low Rayleigh number thermal convection in a saturated porous medium bounded by two horizontal, eccentric cylinders. *ASME J. Heat Transfer* **106**, 166–175. [7.3.3]

Bau, H. H. 1984b Convective heat losses from a pipe buried in a semi-infinite porous medium. *Int. J. Heat Mass Transfer* **27**, 2047–2056. [7.11]

Bau, H. H. 1984c Thermal convection in a horizontal, eccentric annulus containing a saturated porous medium—an extended perturbation expansion. *Int. J. Heat Mass Transfer* **27**, 2277–2287. [7.3.3]

Bau, H.H. 1986 Estimation of heat losses from flows in buried pipes. *Handbook of Heat and Mass Transfer* (ed. N.P. Cheremisinoff), Gulf Publishing Co., Houston, TX, Vol. 1, 1009–1024. [7.11]

Bau, H. H. 1993 Controlling chaotic convection. *Theoretical and Applied Mechanics 1992* (eds. S.R. Bodner, *et al.*), Elsevier, Amsterdam, 187–203. [6.11.3]

Bau, H. H. and Torrance, K. E. 1981 Onset of convection in a permeable medium between vertical coaxial cylinders. *Phys. Fluids* **24**, 382–385. [6.16.1]

Bau, H. H. and Torrance, K. E. 1982a Boiling in low permeability porous materials. *Int. J. Heat Mass Transfer* **25**, 45–55. [11.9.2]

Bau, H. H. and Torrance, K. E. 1982b Low Rayleigh number thermal convection in a vertical cylinder filled with porous materials and heated from below. *ASME J. Heat Transfer* **104**, 166–172. [6.16.1]

Bau, H. H. and Torrance, K. E. 1982c Thermal convection and boiling in a porous medium. *Lett. Heat Mass Transfer* **9**, 431–333. [10.3.1]

Baytas, A. C. 2000 Entropy generation for natural convection in an inclined porous cavity. *Int. J. Heat Mass Transfer* **43**, 2089–2099. [7.8]

Baytas, A. C. 2003 Thermal non-equilibrium natural convection in a square enclosure filled with a heat-generating solid phase, non-Darcy porous medium. *Int. J. Energy Res.* **27**, 975–988. [6.5]

Baytas, A. C. 2004 Thermal non-equilibrium free convection in a cavity filled with a non-Darcy porous medium. In *Emerging Technologies and Techniques in Porous Media* (D. B. Ingham, A. Bejan, E. Mamut and I. Pop, eds), Kluwer Academic, Dordrecht, pp. 247–258. [6.5]

Baytas, A. C. and Baytas, A. F. 2005 Entropy generation in porous media. In *Transport Phenomena in Porous Media III* (eds. D. B. Ingham and I. Pop), Elsevier, Oxford, pp. 201–224. [7.8]

Baytas, A. C. and Pop, I. 1999 Free convection in oblique enclosures filled with a porous medium. *Int. J. Heat Mass Transfer* **42**, 1047–1057. [7.8]

Baytas, A. C. and Pop, I. 2001 Natural convection in a trapezoidal enclosure filled with a porous medium. *Int. J. Engng. Sci.* **39**, 125–134. [7.3.7]

Baytas, A. C. and Pop, I. 2002 Free convection in a square porous cavity using a thermal nonequilibrium model. *Int. J. Thermal Sci.* **41**, 861–870. [7.6.2]

Baytas, A. C., Baytas, A. F. and Pop, I. 2004 Free convection in a porous cavity filled with pure or saline water. In *Applications of Porous Media (ICAPM 2004)*, (eds. A. H. Reis and A. F. Miguel), Évora, Portugal, pp. 121–125. [7.3.5]

Baytas, A. C., Grosan, T. and Pop, I. 2002 Free convection in spherical annular sectors filled with a porous medium. *Transport Porous Media* **49**, 191–207. [7.3.4]

Baytas, A. C., Liaqat, A., Grosan, T. and Pop, I. 2001 Conjugate natural convection in a square porous cavity. *Heat Mass Transfer* **37**, 467–473. [7.1.5]

Bear, J. and Bachmat, Y. 1990 *Introduction to Modeling of Transport Phenomena in Porous Media*, Kluwer Academic, Dordrecht. [1.1, 1.5.3]

Beavers, G. S. and Joseph, D. D. 1967 Boundary conditions at a naturally permeable wall. *J. Fluid Mech.* **30**, 197–207. [1.6]

Beavers, G. S., Sparrow, E. M. and Magnuson, R. A. 1970 Experiments on coupled parallel flows in a channel and a bounding medium. *ASME J. Basic Engng.* **92**, 843–848. [1.6]

Beavers, G. S., Sparrow, E. M. and Masha, B. A. 1974 Boundary conditions at a porous surface which bounds a fluid flow. *AIChE J.* **20**, 596–597. [1.6]

Beavers, G. S., Sparrow, E. M. and Rodenz, D. E. 1973 Influence of bed size on the flow characteristics and porosity of randomly packed beds of spheres. *J. Appl. Mech.* **40**, 655–660. [1.5.2]

Beck, J. L. 1972 Convection in a box of porous material saturated with fluid. *Phys. Fluids* **15**, 1377–1383. [1.5.1, 6.15.1]

Becker, S. M., Kuznetsov, A. V. and Avramenko, A. A. 2004 Numerical modeling of a falling bioconvection plume in a porous medium. *Fluid Dyn. Res.* **33**, 323–339. [6.25]

Beckermann, C. and Viskanta, R. 1987 Forced convection boundary layer flow and heat transfer along a flat plate embedded in a porous medium. *Int. J. Heat Mass Transfer* **30**, 1547–1551. [4.8]

Beckermann, C. and Viskanta, R. 1988a Natural convection solid/liquid phase change in porous media. *Int. J. Heat Mass Transfer* **31**, 35–46. [10.1.6]

Beckermann, C. and Viskanta, R. 1988b Double-diffusive convection during dendritic solidification of a binary mixture. *Physicochemical Hydrodyn.* **10**, 195–213. [10.2.2, 10.2.3]

Beckermann, C. and Viskanta, R. 1993 Mathematical modeling of transport phenomena during alloy solidification. *Appl. Mech. Rev.* **46**, 1–27. [10.2.3]

Beckermann, C. and Wang, C. Y. 1995 Multiphase/-scale modeling of alloy solidification. *Ann. Rev. Heat Transfer* **6**, 115–198. [10.2.3]

Beckermann, C., Viskanta, R. and Ramadhyani, S. 1986 A numerical study of non-Darcian natural convection in a vertical enclosure filled with a porous medium. *Numer. Heat Transfer* **10**, 557–570. [7.6.2]

Beckermann, C., Viskanta, R. and Ramadhyani, S. 1988 Natural convection in vertical enclosures containing simultaneously fluid and porous layers. *J. Fluid Mech.* **186**, 257–284. [7.7]

Beg, O. A., Takhar, H. S., Soundalgekar, V. M. and Prasad, V. 1998 Thermoconvective flow in a saturated, isotropic, homogeneous porous medium using Brinkman's model—numerical study. *Int. J. Numer. Methods Heat Fluid Flow* **8**, 559–589. [5.1.7.1]

Beithou, N., Aybar, H. S., Albayrak, K. and Erenay, O. 2001 Free convection flow of Newtonian fluid along a vertical plate embedded in a double layer porous medium. *JSME Int. J. Ser B* **44**, 255–261. [5.1.9.2]

Bejan, A. 1978 Natural convection in an infinite porous medium with a concentrated heat source. *J. Fluid Mech.* **89**, 97–107. [5.11.2]

Bejan, A. 1979 On the boundary layer regime in a vertical enclosure filled with a porous medium. *Lett. Heat Mass Transfer* **6**, 93–102. [7.1.2, 7.2]

Bejan, A. 1980 A synthesis of analytical results for natural convection heat transfer across rectangular enclosures. *Int. J. Heat Mass Transfer* **23**, 723–726. [7.1.3, 7.4, 7.4.2]

Bejan, A. 1981 Lateral intrusion of natural convection into a horizontal porous structure. *ASME J. Heat Transfer* **103**, 237–241. [7.4, 7.4.1]

Bejan, A. 1983a Natural convection heat transfer in a porous layer with internal flow obstructions. *Int. J. Heat Mass Transfer* **26**, 815–822. [7.3.1]

Bejan, A. 1983b The boundary layer regime in a porous layer with uniform heat flux from the side. *Int. J. Heat Mass Transfer* **26**, 1339–1346. [7.2]

Bejan, A. 1984 *Convection Heat Transfer*, Wiley, New York. [1.1, 4.1, 4.2, 4.5, 4.17, 5.1.4, 5.11.1, 6.9.2, 7.1.1, 7.1.2, 7.3.3, 7.4.2, 9.2.1, 10.1.2]

Bejan, A. 1985 The method of scale analysis: natural convection in porous media. *Natural Convection: Fundamentals and Applications* (ed. S. Kakaç, *et al.*), Hemisphere, Washington, DC, 548–572. [7.1.1]

Bejan, A. 1987 Convective heat transfer in porous media. *Handbook of Single-Phase Convective Heat Transfer* (eds. S. Kakaç, R. K. Shah, and W. Aung), Chapter 16, Wiley, New York. [6.20, 7.3.3, 7.3.4, 7.3.5]

Bejan, A. 1989 Theory of melting with natural convection in an enclosed porous medium. *ASME J. Heat Transfer* **111**, 407–415. [10.1.3, 10.1.5]

Bejan, A. 1990a Theory of heat transfer from a surface covered with hair. *ASME J. Heat Transfer* **112**, 662–667. [4.14]

Bejan, A. 1990b Optimum hair strand diameter for minimum free-convection heat transfer from a surface covered with hair. *Int. J. Heat Mass Transfer* **33**, 206–209. [5.13]

Bejan, A. 1992a Comments on "Coupled heat and mass transfer by natural convection from vertical surfaces in porous media." *Int. J. Heat Mass Transfer* **35**, 3498. [9.2.1]

Bejan, A. 1992b Surfaces covered with hair: optimal strand diameter and optimal porosity for minimum heat transfer. *Biomimetics* **1**, 23–38. [4.14]

Bejan, A. 1993 *Heat Transfer*, 2nd ed., Wiley, New York. [4.4, 4.15]

Bejan, A. 1995 The optimal spacing for cylinders in cross flow forced convection. *J. Heat Transfer* **117**, 767–770. [4.15]

Bejan, A. 1996a *Entropy Generation Minimization*, CRC Press, Boca Raton, FL. [4.15]

Bejan, A. 1996b Street network theory of organization in nature. *J. Adv. Transportation* **30**, 85–107. [4.18]

Bejan, A. 1997a Constructal-theory network of conducting paths for cooling a heat generating volume. *Int. J. Heat Mass Transfer* **40**, 815. [4.18]

Bejan, A. 1997b Constructal tree network for fluid flow between a finite-size volume and one source or sink. *Rev. Gén. Thermique* **36**, 592–604. [4.18]

Bejan, A. 1997c *Advanced Engineering Thermodynamics*, 3rd ed., Wiley, New York, [4.19, 6.26]

Bejan, A. 2000 *Shape and Structure, from Engineering to Nature*, Cambridge University Press, Cambridge, UK. [1.5.2, 4.18, 6.2, 6.26, 11.10]

Bejan, A. 2004a *Convection Heat Transfer*, 3rd ed., Wiley, New York. [1.5.2, 2.1, 4.17, 4.18, 4.20]

Bejan, A. 2004b Designed porous media: maximal heat transfer density at decreasing length scales. *Int. J. Heat Mass Transfer* **47**, 3073–3083. [4.15]

Bejan, A. and Anderson, R. 1981 Heat transfer across a vertical impermeable partition imbedded in a porous medium. *Int. J. Heat Mass Transfer* **24**, 1237–1245. [5.1.5, 7.3.1]

Bejan, A. and Anderson, R. 1983 Natural convection at the interface between a vertical porous layer and an open space. *ASME J. Heat Transfer* **105**, 124–129. [5.1.5]

Bejan, A. and Fautrelle, Y. 2003 Constructal multi-scale structure for maximal heat transfer density. *Acta Mech.* **163**, 39–49. [4.19]

Bejan, A. and Khair, K. R. 1985 Heat and mass transfer by natural convection in a porous medium. *Int. J. Heat Mass Transfer* **28**, 909–918. [9.2.1]

Bejan, A. and Lage, J. L. 1991 Heat transfer from a surface covered with hair. *Convective Heat and Mass Transfer in Porous Media* (eds. S. Kakaç, *et al.*), Kluwer Academic, Dordrecht, 823–845. [1.2, 4.14]

Bejan, A. and Lorente, S. 2004 The constructal law and the thermodynamics of flow systems with configuration. *Int. J. Heat Mass Transfer* **47**, 3203–3214. [4.18]

Bejan, A. and Morega, A. M. 1993 Optimal arrays of pin fins and plate fins in laminar forced convection. *J. Heat Transfer* **115**, 75–81. [4.15]

Bejan, A. and Nield, D. A. 1991 Transient forced convection near a suddenly heated plate in a porous medium. *Int. Comm. Heat Mass Transfer* **18**, 83–91. [4.6]

Bejan, A. and Poulikakos, D. 1984 The non-Darcy regime for vertical boundary layer natural convection in a porous medium. *Int. J. Heat Mass Transfer* **27**, 717–722. [5.1.7.2]

Bejan, A. and Sciubba, E. 1992 The optimal spacing of parallel plates cooled by forced convection. *Int. J. Heat Mass Transfer* **35**, 3259–3264. [4.15]

Bejan, A. and Tien, C. L. 1978 Natural convection in a horizontal porous medium subjected to an end-to-end temperature difference. *ASME J. Heat Transfer* **100**, 191–198; errata **105**, 683–684. [7.1.3, 7.3.2, 7.3.3]

Bejan, A. and Tien, C. L. 1979 Natural convection in horizontal space bounded by two concentric cylinders with different end temperatures. *Int. J. Heat Mass Transfer* **22**, 919–927. [7.3.3]

Bejan, A., Dincer, I., Lorente, S., Miguel, A. F. and Reis, A. H. 2004 *Porous and Complex Flow Structures in Modern Technologies*. Springer, New York. [1.1, 1.5.2, 2.1, 3.3, 3.7, 4.18, 4.19, 6.26, 10.1.7]

Bejan, A., Ikegami, Y. and Ledezma, G. A. 1998 Constructal theory of natural crack pattern formation for fastest cooling. *Int. J. Heat Mass Transfer* **41**, 1945–1954. [11.10]

Bejan, A., Lorente, S., Miguel, A. F. and Reis, A. H. 2005 *Along with Constructal Theory*, Faculty of Geosciences and the Environment, University of Lausanne, Switzerland. [11.10]

Bejan, A., Rocha, L. A. O. and Cherry, R. S. 2002 Methane hydrates in porous layers: gas formation and convection. In *Transport Phenomena in Porous Media II* (D. B. Ingham and I. Pop, eds.) Elsevier, Oxford, pp. 365–396. [10.1.7]

Bejan, A., Zhang, Z. and Jany, P. 1990 The horizontal intrusion layer of melt in a saturated porous medium. *Int. J. Heat Fluid Flow* **11**, 284–289. [10.1.4]

Beji, H. and Gobin, P. 1992 The effect of thermal dispersion on natural dispersion heat transfer in porous media. *Numer. Heat Transfer A* **23**, 487–500. [7.6.2]

Beji, H., Bennacer, R. and Duval, R. 1999 Double-diffusive natural convection in a vertical porous annulus. *Numer. Heat Transfer A* **36**, 153–170. [9.4]

Belhachmi, Z., Brighi, B. and Taous, K. 2000 Similarity solutions for a boundary layer problem in porous media. *C. R. Acad. Sci. II B* **328**, 407–410. [5.1.9.1]

Belhachmi, Z., Brighi, B. and Taous, K. 2001 On a family of differential equations for boundary layer approximations in porous media. *Eur. J. Appl. Math.* **12**, 513–528. [5.1.9.1]

Belhachmi, Z., Brighi, B., Sac-Eppe, J. M. and Taous, K. 2003 Numerical simulations of free convection about a vertical flat plate embedded in a porous medium. *Comp. Geosci.* **7**, 137–166. [5.1.9.1]

Bello-Ochende, T. and Bejan, A. 2004 Plates with multiple lengths in forced convection. *Int. J. Thermal Sci.* **43**, 4300–4306. [4.19]

Bello-Ochende, T. and Bejan, A. 2005 Constructal multi-scale cylinders in cross-flow. *Int. J. Heat Mass Transfer* **48**, 1373–1383. [4.19]

Ben Nasrallah, S., Mara, A. and Du Peuty, M. A. 1997 Transient free convection in a vertical cylinder filled with a granular product: Open at the extremities and heated with heat flux density : validity of the local thermal equilibrium hypothesis. *Int. J. Heat Mass Transfer* **40**, 1155–1168. [7.5]

Benano-Melly, L. B., Caltagirone, J.-P., Faissat, B., Montel, F. and Costeseque, P. 2001 Modelling Soret coefficient measurement experiments in porous media considering thermal and solutal convection. *Int. J. Heat Mass Transfer* **44**, 1285–1297. [9.1.4]

Benhadji, K. and Vasseur P. 2001 Double-diffusive convection in a shallow porous cavity filled with a non-Newtonian fluid. *Int. Comm. Heat Mass Transfer* **28**, 763–772. [9.2.2]

Benhadji, K., Robillard, L. and Vasseur P. 2003 Convection in a porous cavity saturated with water near 4 degrees C and subject to Dirichlet-Neumann thermal boundary conditions. *Int. Comm. Heat Mass Transfer* **29**, 897–906. [7.3.5]

Bennacer, R. 2000 The Brinkman model for thermosolutal convection in a vertical annular porous layer. *Int. Comm. Heat Mass Transfer* **27**, 69–80. [9.4]

Bennacer, R. 2004 Natural convection in anisotropic hetereogeneous porous medium. In *Emerging Technologies and Techniques in Porous Media* (D. B. Ingham, A. Bejan, E. Mamut and I. Pop, eds), Kluwer Academic, Dordrecht, pp. 271–284. [9.1.6.2]

Bennacer, R. and Lakhal, A. 2005 Numerical and analytical analysis of the thermosolutal convection in an annular field: effect of thermodiffusion. In *Transport Phenomena in Porous Media III*, (eds. D. B. Ingham and I. Pop), Elsevier, Oxford, pp. 341–365. [9.4]

Bennacer, R., Beji, H. and Mohamad, A. A. 2003a Double diffusive convection in a vertical enclosure inserted with two saturated porous layers containing a fluid layer. *Int. J. Thermal Sci.* **42**, 141–151. [9.4]

Bennacer, R., Beji, H., Oueslati, F. and Belgith, A. 2001a Multiple natural convection solution in porous media under cross temperature and concentration gradients. *Numer. Heat Transfer A* **39**, 553–567. [9.2.2]

Bennacer, R., El Ganaoui, M. and Fauchais, P. 2004 On the thermal anisotropy affecting transfers in multiplayer porous medium. *Comptes Rendus Mécanique* **332**, 539–546. [9.5]

Bennacer, R., Mahidjibo, A., Vasseur, P., Beji, H. and Duval, R. 2003b The Soret effect on convection in a horizontal porous domain under cross-temperature and concentration gradients. *Int. J. Numer. Meth. Heat Fluid Flow* **13**, 199–215. [9.2.2]

Bennacer, R., Mohamad, A. A. and El Ganaoui, M. 2005 Analytical and numerical investigation of double diffusion in thermally anisotropic multilayer porous medium. *Heat Mass Transfer* **41**, 298–305. [9.5]

Bennacer, R., Tobbal, A., Beji, H. and Vasseur, P. 2001b Double diffusive convection in a vertical enclosure filled with anisotropic media. *Int. J. Thermal Sci.* **40**, 30–41. [9.2.2]

Bennethum, L. S. and Giorgi, T. 1997 Generalized Forchheimer equation for two-phase flow based on hybrid mixture theory. *Transport Porous Media* **26**, 261–275. [1.5.2]

Bennon, W. D. and Incropera, F. P. 1987 A continuum model for momentum, heat and species transport in binary-phase change systems. I. Model formulation. *Int. J. Heat Transfer* **40**, 2161–2170. [10.2.3]

Benzeghiba, M., Chikh, S. and Campo, A. 2003 Thermosolutal convection in a partly porous vertical annular cavity. *ASME J. Heat Transfer* **125**, 703–715. [9.4]

Bera, P. and Khalili, A. 2002a Double-diffusive natural convection in an anisotropic porous cavity with opposing buoyancy forces: multi-solutions and oscillations. *Int. J. Heat Mass Transfer* **45**, 3205–3222. [9.2.2]

Bera, P. and Khalili, A. 2002b Stability of mixed convection in an anisotropic vertical porous channel. *Phys. Fluids* **14**, 1617–1630. [8.3.1]

Bera, P., Eswaran, V. and Singh, P. 1998 Numerical study of heat and mass transfer in an anisotropic porous enclosure due to constant heating and cooling. *Numer. Heat Transfer A*, **34**, 887–905. [9.2.2]

Bera, P., Eswaran, V. and Singh, P. 2000 Double-diffusive convection in slender anisotropic porous enclosures. *J. Porous Media* **3**, 11–29. [9.2.2]

Bergman, M. I., Fearn, D. R., Bloxam, J. and Shannon, M. C. 1997 Convection and channel formation in solidifying Pb-Sn alloys. *Metall. Mat. Trans. A* **28**, 859–866. [10.2.3]

Bergman, M. I. and Fearn, D. R. 1994 Chimneys on the Earth's inner-outer core boundary? *Geophys. Res. Lett.* **21**, 477–480. [6.21,11.8.2]

Bessonov, O.A. and Brailovskaya, V. A. 2001 Three-dimensional model of thermal convection in an anisotropic porous medium bounded by two horizontal coaxial cylinders. *Fluid Dyn.* **36**, 130–138. [7.3.3]

Beukema, K. J. and Bruin, S. 1983 Three-dimensional natural convection in a confined porous medium with internal heat generation. *Int. J. Heat Mass Transfer* **26**, 451–458. [6.15.3]

Bhattacharjee, S. and Grosshandler, W. L. 1988 The formation of a wall jet near a high temperature wall under a microgravity environment, *ASMR HTD* **96**, 711–716. [4.15]

Bhattacharya, A. and Mahajan, R. L. 2002 Finned metal foam heat sinks for electronics cooling in forced convection. *ASME J. Electronic Packaging* **124**, 155–163. [4.16.2]

Bhattacharyya, S., Pal, A. and Pop, I. 1998 Unsteady mixed convection on a wedge in a porous medium. *Int. Comm. Heat Mass Transfer* **25**, 743–752. [8.1.4]

Bian, W. and Wang, B. X. 1993 Transient freezing and natural convection around a cylinder in saturated porous media. *Proc. 6th Int. Sympos. Transport Phenomena in Thermal Engineering*, Seoul, Korea, pp. 79–84. [10.2.1.2]

Bian, W., Vasseur, P. and Bilgen, E. 1994a Natural convection of non-Newtonian fluids in an inclined porous layer. *Chem. Engng. Commun.* **129**, 79–97. [7.8]

Bian, W., Vasseur, P. and Bilgen, E. 1994b Boundary-layer analysis for natural convection in a vertical porous layer filled with a non-Newtonian fluid. *Int. J. Heat Fluid Flow* **15**, 384–391. [7.8]

Bian, W., Vasseur, P. and Bilgen, E. 1996a Effect of an external magnetic field on buoyancy driven flow in a shallow porous cavity. *Numer. Heat Transfer A* **29**, 625–638. [6.21]

Bian, W., Vasseur, P., Bilgen, E. and Meng, F. 1996b Effect of an electromagnetic field on natural convection in an inclined porous layer. *Int. J. Heat Fluid Flow*, **17**, 36–44. [7.8]

Biggs, M. J. and Humby, S. J. 1998 Lattice-gas automata methods for engineering. *Chem. Engng. Res. Design* **76**, 162–174. [2.6]

Bilgen, E. and Mbaye, M. 2001 Bénard cells in fluid-saturated porous enclosures with lateral cooling. *Int. J. Heat Fluid Flow* **22**, 561–570. [6.8]

Bjørlykke, K., Mo, A. and Palm, E. 1988 Modelling of thermal convection in sedimentary basins and its relevance to diagenetic reactions. *Mar. Petr. Geol.* **5**, 338–351. [11.5]

Bjornsson, S. and Stefansson, V. 1987 Heat and mass transport in geothermal reservoirs. *Advances in Transport Phenomena in Porous Media* (eds. J. Bear and M. Y. Corapcioglu), Martinus Nijhoff, The Netherlands, 145–153. [11]

Blake, K. R., Bejan, A. and Poulikakos, D. 1984 Natural convection near 4°C in a water saturated porous layer heated from below. *Int. J. Heat Mass Transfer* **27**, 2355–2364. [6.20]

Blythe, P. A. and Simpkins, P. G. 1981 Convection in a porous layer for a temperature dependent viscosity. *Int. J. Heat Mass Transfer* **24**, 497–506. [6.7]

Blythe, P. A., Daniels, P. G. and Simpkins, P. G. 1982 Thermally driven cavity flows in porous media. I. The vertical boundary layer structure near the corners. *Proc. Roy. Soc. London A* **380**, 119–136. [7.1.2]

Blythe, P. A., Daniels, P. G. and Simpkins, P. G. 1985a Convection in a fluid saturated porous medium due to internal heat generation. *Int. Comm. Heat Mass Transfer* **12**, 493–504. [6.17]

Blythe, P. A., Daniels, P. G. and Simpkins, P. G. 1985b Limiting behaviours in porous media cavity flows. *Natural Convection: Fundamentals and Applications* (eds. S. Kakaç, W. Aung, and R. Viskanta), Hemisphere, Washington, DC, pp. 600–611. [7.1.3]

Blythe, P. A., Simpkins, P. G. and Daniels, P. G. 1983 Thermal convection in a cavity filled with a porous medium: a classification of limiting behaviours. *Int. J. Heat Mass Transfer* **26**, 701–708. [7.1.1]

Bodvarsson, G. S., Pruess, K. and Lippmann, M. J. 1986 Modeling of geothermal systems. *J. Petrol. Tech.* **38**, 1007–1021. [11]

Boomsma, K., Poulikakos, D. and Zwick, F. 2003 Metal foams as compact high performance heat exchangers. *Mech. Mater.* **35**, 1161–1176. [4.15]

Bories, S. 1970a Sur les méchanismes fondamentaux de la convection naturelle en milieu poreux. *Rev. Gén. Therm.* **9**, 1377–1401. [6.9.1]

Bories, S. 1970b Comparaison des prévisions d'une théorie non linéaire et des résultats expérimentaux en convection naturelle dans une couche poreuse saturée horizontale. *C. R. Acad. Sci. Paris B* **271**, 269–272. [6.9.1]

Bories, S. 1993 Convection naturelle dans une couche porous inclineé. Sélection du nombre d'onde des configurations d'écoulements. *Compt. Rend. Acad. Sci., Paris, Sér.* II, **316**, 151–156. [7.8]

Bories, S. A. 1987 Natural convection in porous media. *Advances in Transport Phenomena in Porous Media* (eds. J. Bear and M. Y. Corapcioglu), Martinus Nijhoff, The Netherlands, 77–141. [6.3, 6.5]

Bories, S. A. 1991 Fundamentals of drying of capillary-porous bodies. *Convective Heat and Mass Transfer in Porous Media* (eds. S. Kakac, *et al.*), Kluwer Academic, Dordrecht, 391–434. [3.6]

Bories, S. A. and Combarnous, M. A. 1973 Natural convection in a sloping porous layer. *J. Fluid Mech.* **57**, 63–79. [7.8]

Bories, S. and Deltour, A. 1980 Influence des conditions aux limites sur la convection naturelle dans un volume poreux cylindrique. *Int. J. Heat Mass Transfer* **23**, 765–771. [6.16.1]

Bories, S. and Thirriot, C. 1969 Échanges thermiques et tourbillons dans une couche poreuse horizontale. *La Houille Blanche* **24**, 237–245. [6.9.1]

Borkowska-Pawlak, B. and Kordylewski, W. 1982 Stability of two-dimensional natural convection in a porous layer. *Q. J. Mech. Appl. Math.* **35**, 279–290. [6.15.1]

Borkowska-Pawlak, B. and Kordylewski, W. 1985 Cell-pattern sensitivity to box configuration in a saturated porous medium. *J. Fluid Mech.* **150**, 169–181. [6.15.1]

Bortolozzi, R. A. and Deibler, J. A. 2001 Comparison between two-field and one-field models for natural convection in porous media. *Chem. Engng. Sci.* **56**, 157–172. [7.3.3]

Bouddour, A., Auriault, J. L. and Mhamdi-Alaoui, M. 1998 Heat and mass transfer in wet porous media in presence of evaporation. *Int. J. Heat Mass Transfer* **41**, 2263–2277. [3.6]

Bourich, M., Amahmid, A. and Hasnaoui M. 2004a Double diffusive convection in a porous enclosure submitted to cross gradients of temperature and concentration. *Energy Conv. Management* **45**, 1655–1670. [9.5]

Bourich, M., Hasnaoui, M. and Amahmid, A. 2004b Double-diffusive natural convection in a porous enclosure partially heated from below and differentially salted. *Int. J. Heat Fluid Flow* **25**, 1034–1046. [9.5]

Bourich, M., Hasnaoui, M., Amahmid, A. 2004c A scale analysis of thermosolutal convection in a saturated porous enclosure submitted to vertical temperature and horizontal concentration gradients. *Energy Conv. Management* **45**, 2795–2811. [9.1.3]

Bourich, M., Hasnaoui, M., Amahmid, A. and Mamou, M. 2002 Soret driven thermosolutal convection in a shallow porous enclosure. *Int. Comm. Heat Mass Transfer* **29**, 717–728. [9.1.4]

Bourich, M., Hasnaoui, M., Amahmid, A. and Mamou, M. 2004d Soret convection in a shallow porous cavity submitted to uniform fluxes of heat and mass. *Int. Comm. Heat Mass Transfer* **31**, 773–782. [9.1.4]

Bourich, M., Hasnaoui, M., Amahmid, A. and Mamou, M. 2005 Onset of convection and finite amplitude flow due to Soret effect within a sparsely packed porous enclosure heated from below. *Int. J. Heat Fluid Flow* **26**, 513–525. [9.1.4]

Bourich, M., Hasnaoui, M., Mamou, M. and Amahmid, A. 2004e Soret effect inducing subcritical and Hopf bifurcations in a shallow enclosure filled with a clear binary fluid or a saturated porous medium: A comparative study. *Phys. Fluids* **16**, 551–568. [9.1.4]

Bourich, M., Mamou, M., Hasnaoui, M. and Amahmid, A. 2004f On stability analysis of Soret convection within a horizontal porous layer. In *Emerging Technologies and Techniques in Porous Media* (D. B. Ingham, A. Bejan, E. Mamut and I. Pop, eds.), Kluwer Academic, Dordrecht, pp. 221–234. [9.1.4]

Boussinesq, J. 1903 *Théorie Analytique de la Chaleur*, Vol. 2, Gauthier-Villars, Paris. [2.3]

Boutana, N., Bahlooul, N., Vasseur, P. and Joly, F. 2004 Soret and double diffusive convection in a porous cavity. *J. Porous Media* **7**, 41–57. [9.1.4]

Bradean, R., Heggs, P. J., Ingham, D, B, and Pop, I. 1998a Convective heat flow from suddenly heated surfaces embedded in porous media. *Transport Phenomena in Porous Media* (eds. D. B. Ingham and I. Pop), Elsevier, Oxford, pp. 411–438. [5.5.1]

Bradean, R., Ingham, D. B., Heggs, P. and Pop, I. 1997b Unsteady free convection adjacent to an impulsively heated horizontal circular cylinder in porous media. *Numer. Heat Transfer A* **31**, 325–346. [5.5.1]

Bradean, R., Ingham, D. B., Heggs, P. J. and Pop, I. 1995a Buoyancy induced flow adjacent to a periodically heated and cooled horizontal surface in porous media. *Int. J. Heat Mass Transfer* **39**, 615–630. [5.2]

Bradean, R., Ingham, D. B., Heggs, P. J. and Pop, I. 1995b Free convection fluid flow due to a periodically heated and cooled vertical plate embedded in a porous media. *Int. J. Heat Mass Transfer* **39**, 2545–2557. [5.1.9]

Bradean, R., Ingham, D. B., Heggs, P. J. and Pop, I. 1996 Unsteady free convection from a horizontal surface embedded in a porous media. *Proc. 2nd European Thermal-Sciences and 14th UIT Nat. Heat Transfer Conference*, Edizioni ETS, Pisa Vol. 1, 329–335. [5.2]

Bradean, R., Ingham, D. B., Heggs, P. J. and Pop, I. 1997a The unsteady penetration of free convection flows caused by heating and cooling flat surfaces in a porous media. *Int. J. Heat Mass Transfer* **40**, 665–687. [5.1.9, 5.2]

Bradean, R., Ingham, D. B., Heggs, P. J. and Pop, I. 1998b Mixed convection adjacent to a suddenly heated horizontal circular cylinder embedded in a porous medium. *Transport Porous Media* **32**, 329–355. [8.1.3]

Bradshaw, P. 2001 Shape and structure, from engineering to nature. *AAIA J.* **39**, 983. [4.19]

Bradshaw, S., Glasser, D. and Brooks, K. 1991 Self-ignition and convection patterns in an infinite coal layer. *Chem. Engng. Comm.* **105**, 255–278. [6.11.2]

Braester, C. and Vadasz, P. 1993 The effect of a weak heterogeneity of a porous medium on natural convection. *J. Fluid Mech.* **254**, 345–362. [6.9.1, 6.13.4].

Braga, E. J. and de Lemos, M. J. S. 2004 Turbulent natural convection in a porous square cavity computed with a macroscopic k-ε model. *Int. J. Heat Mass Transfer* **47**, 5639–5663. [1.8]

Brand, H. and Steinberg, V. 1983a Convective instabilities in binary mixtures in a porous medium. *Physica A* **119**, 327–338. [9.1.3, 9.1.4]

Brand, H. and Steinberg, V. 1983b Nonlinear effects in the convective instability of a binary mixture in a porous medium near threshold. *Phys. Lett. A* **93**, 333–336. [9.1.3, 9.1.4]

Brand, H. R., Hohenberg, P. C., and Steinberg, V. 1983 Amplitude equation near a polycritical point for the convective instability of a binary fluid mixture in a porous medium. *Phys. Rev. A* **27**, 591–594. [9.1.3]

Bratsun, D. A. and Lyubimov, D. V. 1995 Co-symmetry breakdown in problems of thermal convection in porous medium. *Physica D* **82**, 398–417. [6.16.2]

Bresch, D. and Sy, M. 2003 Convection in rotating porous media: The planetary geostrophic equations, used in geophysical fluid dynamics, revisited. *Cont. Mech. Thermodyn.* **15**, 247–263. [6.22]

Brinkman, H. C. 1947a A calculation of the viscous force exerted by a flowing fluid on a dense swarm of particles. *Appl. Sci. Res. A* **1**, 27–34. [1.5.3]

Brinkman, H. C. 1947b On the permeability of media consisting of closely packed porous particles. *Appl. Sci. Res. A* **1**, 81–86. [1.5.3]

Budu, P. 2001 Stability results for convection in thawing subsea permafrost. *Cont. Mech. Thermodyn.* **13**, 269–285. [11.3]

Buikis, A. and Ulanova, N. 1996 Modelling of non-isothermal gas flow through a heterogeneous medium. *Int. J. Heat Mass Transfer* **39**, 1743–1748. [2.6]

Buonanno, G. and Carotenuto, A. 1997 The effective thermal conductivity of a porous medium with interconnected particles. *Int. J. Heat Mass Transfer* **40**, 393–405. [2.2.1]

Buretta, R. J. and Berman, A. S. 1976 Convective heat transfer in a liquid saturated porous layer. *ASME J. Appl. Mech.* **43**, 249–253. [6.3, 6.9.1, 6.9.2, 6.11.2]

Burns, A. S. and Stewart, W. E. 1992 Convection in heat generating porous media in a concentric annulus with a permeable outer boundary. *Int. Comm. Heat Mass Transfer* **19**, 127–136. [6.17]

Burns, P. J. and Tien, C. L. 1979 Natural convection in porous media bounded by concentric spheres and horizontal cylinders. *Int. J. Heat Mass Transfer* **22**, 929–939. [7.3.3, 7.3.4]

Burns. P. J., Chow, L. C. and Tien, C. L. 1977 Convection in a vertical slot filled with porous insulation. *Int. J. Heat Mass Transfer* **20**, 919–926. [7.1.7]

Busse, F. H. 1985 Transition to turbulence in Rayleigh-Bénard convection. *Hydrodynamic Instabilities and the Transition to Turbulence* (eds. H. L. Swinney and J. Gollub), 2nd ed., Springer, Berlin, 97–137. [6.4]

Busse, F. H. and Joseph, D. D. 1972 Bounds for heat transport in a porous layer. *J. Fluid Mech.* **54**, 521–543. [6.3]

Cai, R. X., Zhang, N. and Liu, W. W. 2003 Algebraically explicit analytical solutions of two-buoyancy natural convection in porous media. *Prog. Nat. Sci.* **13**, 848–851. [9.2.2]

Calmidi, V. V. and Mahajan, R. L. 2000 Forced convection in high porosity metal foams. *ASME J. Heat Transfer* **122**, 557–565. [4.9]

Caltagirone, J. P. 1975 Thermoconvective instabilities in a horizontal porous layer. *J. Fluid Mech.* **72**, 269–287. [6.8]

Caltagirone, J. P. 1976a Stabilité d'une couche poreuse horizontale soumise a des conditions aux limites périodiques. *Int. J. Heat Mass Transfer* **19**, 815–820. [6.11.3]

Caltagirone, J. P. 1976b Thermoconvective instabilities in a porous medium bounded by two concentric horizontal cylinders. *J. Fluid Mech.* **76**, 337–362. [7.3.3]

Caltagirone, J. P. 1980 Stability of a saturated porous layer subject to a sudden rise in surface temperature: comparison between the linear and energy methods. *Q. J. Mech. Appl. Math.* **33**, 47–58. [6.11.3]

Caltagirone, J. P. and Bories, S. 1985 Solutions and stability criteria of natural convective flow in an inclined porous layer. *J. Fluid Mech.* **155**, 267–287. [7.8]

Caltagirone, J. P. and Fabrie, P. 1989 Natural convection in a porous medium at high Rayleigh numbers. Part 1—Darcy's model. *Eur. J. Mech. B* **8**, 207–227. [6.8]

Caltagirone, J. P., Cloupeau, M. and Combarnous, M. 1971 Convection naturelle fluctuante dans une couche poreuse horizontale. *C. R. Acad. Sci. Paris B* **273**, 833–836. [6.9.1]

Caltagirone, J. P., Fabrie, P. and Combarnous, M. 1987 De la convection naturelle oscillante en milieu poreux au chaos temporel? *C. R. Acad. Sci. Paris, Sér. II* **305**, 549–553. [6.8]

Caltagirone, J. P., Meyer, G. and Mojtabi, A. 1981 Structurations thermoconvectives tridimensionnelles dans une couche poreuse horizontale. *J. Mécanique* **20**, 219–232. [6.8, 6.15.1]

Campos, H., Morales, J. C., Lacoa, U. and Campo, A. 1990 Thermal aspects of a vertical annular enclosure divided into a fluid region and a porous region. *Int. Comm. Heat Mass Transfer* **17**, 343–354. [7.3.7]

Cao, W. Z. and Poulikakos, D. 1991a Solidification of a binary mixture saturating a bed of glass spheres. *Convective Heat and Mass Transfer in Porous Media* (eds. S. Kakaç, *et al.*), Kluwer Academic, Dordrecht, 725–772. [10.2.2]

Cao, W. Z. and Poulikakos, D. 1991b Freezing of a binary alloy saturating a packed bed of spheres. *AIAA J. Thermophys. Heat Transfer* **5**, 46–53. [10.2.3]

Capone, F. 2001 On the onset of convection in porous media: temperature depending viscosity. *Bull. Unione Mat. Ital.* **4B**, 143–156. [6.7]

Capone, F. and Rionero, S. 2003 Nonlinear stability of a convective motion in a porous layer driven by a horizontally periodic temperature gradient. *Cont. Mech. Thermodyn.* **15**, 529–538. [6.14]

Carr, M. 2003 Unconditional nonlinear stability for temperature-dependent density flow in a porous medium. *Math. Mod. Meth. Appl. Sci.* **13**, 207–220. [6.4]

Carr, M. 2004 Penetrative convection in a superposed porous-medium—fluid layer via internal heating. *J. Fluid Mech.* **509**, 305–329.

Carr, M. and de Putter, S. 2003 Penetrative convection in a horizontally isotropic porous layer. *Cont. Mech. Thermodyn.* **15**, 33–43. [6.11.4]

Carr, M. and Straughan, B. 2003 Penetrative convection in a fluid overlying a porous layer. *Adv. Water Res.* **26**, 263–276. [6.11.4]

Carrillo, L. P. 2005 Convective heat transfer for viscous fluid flow through a metallic packed bed. *Interciencia* **30**, 81–86 and 109–111. [2.2.2]

Carson, J. K., Lovatt, S. J., Tanner, D. J. and Cleland, A. C. 2005 Thermal conductivity bounds for isotropic porous materials. *Int. J. Heat Mass Transfer* **48**, 2150–2158. [2.2.1]

Castinel, G. and Combarnous, M. 1975 Natural convection in an anisotropic porous layer. *Rev. Gén. Therm.* **168**, 937–947. English translation, *Int. Chem. Eng.* **17**, 605–614 (1977). [6.12]

Catton, I. 1985 Natural convection heat transfer in porous media. *Natural Convection: Fundamentals and Applications* (eds. S. Kakaç, W. Aung, and R. Viskanta), Hemisphere, Washington, DC, 514–547. [6.9.2, 6.19.2]

Catton, I. and Chung, M. 1991 An experimental study of steam injection into a uniform water flow through porous media. *Wärme-Stoffübertrag.* **27**, 29–31. [11.9.5]

Catton, I. and Travkin, V. S. 1996 Turbulent flow and heat transfer in high permeability porous media. *Proceedings of the International Conference on Porous Media and their Applications in Science, Engineering and Industry*, Kona, Hawaii, June 1996, (K.Vafai, editor), Engineering Foundation, New York, pp. 333–368. [1.8]

Catton, I., Georgiadis, J. G. and Adnani, P. 1988 The impact of nonlinear convective processes in transport phenomena in porous media. *ASME HTD* **96**, Vol. 1, 767–777. [2.2.3]
Chakrabarti, A. and Gupta, A. S. 1981 Nonlinear theromohaline convection in a rotating porous medium. *Mech. Res. Comm.* **8**, 9–22. [9.1.6.4]
Chamkha, A. J. 1996 Non-Darcy hydromagnetic free convection from a cone and a wedge in porous media. *Int. Comm. Heat Mass Transfer* **23**, 875–887. [5.8]
Chamkha, A. J. 1997a Solar radiation assisted natural convection in uniform porous medium supported by a vertical flat plate. *ASME J. Heat Transfer* **119**, 89–96. [5.1.9.4]
Chamkha, A. J. 1997b A note on unsteady hydromagnetic free convection from a vertical fluid saturated porous medium channel. *ASME J. Heat Transfer* **119**, 638–641. [7.5]
Chamkha, A. J. 1997c Similarity solutions for buoyancy-induced flow of a power-law fluid over a horizontal surface immersed in a porous medium. *Int. Comm. Heat Mass Transfer* **24**, 805–814. [5.2]
Chamkha, A. J. 1997d Hydromagnetic flow and heat transfer of a heat-generating fluid over a surface in a porous medium. *Int. Comm. Heat mass Transfer* **24**, 815–825. [5.1.9.4]
Chamkha, A. J. 1997e Hydromagnetic natural convection from an isothermal inclined surface adjacent to a thermally stratified porous medium. *Int. J. Engng. Sci.* **35**, 975–986. [5.3]
Chamkha, A. J. 1997f Non-Darcy fully developed mixed convection in a porous medium channel with heat generation/absorption and hydromagnetic effects. *Numer. Heat Transfer A* **32**, 653–675. [8.3.1]
Chamkha, A. J. 1997g MHD-free convection from a vertical plate embedded in a thermally stratified porous medium with Hall effects. *Appl. Math. Modelling* **21**, 603–609. [5.1.4]
Chamkha, A. J. 1998 Mixed convection flow along a vertical permeable plate embedded in a porous medium in the presence of a transverse magnetic field. *Numer. Heat Transfer A* **34**, 93–103. [8.1.1]
Chamkha, A. J. 1999 Magnetohydrodynamic mixed convection from a rotating cone embedded in a porous medium with heat generation. *J. Porous Media* **2**, 87–105. [8.1.4]
Chamkha, A. J. 2000 Non-similar solutions for heat and mass transfer by hydromagnetic mixed convection flow over a plate in porous media with surface suction or injection. *Int. J. Numer. Meth. Heat Fluid Flow* **10**, 142–162. [9.6]
Chamkha, A. J. 2001 Unsteady laminar hydromagnetic flow and heat transfer in porous channels with temperature-dependent properties. *Int. J. Numer. Meth. Heat Fluid Flow* **11**, 430–448. [4.16.2]
Chamkha, A. J. 2002 Double-diffusive convection in a porous enclosure with cooperating temperature and concentration gradients and heat generation or absorption effects. *Numer. Heat Transfer A* **41**, 65–87. [9.1.6.4]
Chamkha, A. J. 2004 Unsteady MHD convective heat mass transfer past a semi-infinite vertical permeable moving plate with heat absorption. *Int. J. Engng. Sci.* **42**, 217–230. [9.2.1]
Chamkha, A. J. and Al-Naser, H. 2001 Double-diffusive convection in an inclined porous enclosure with opposing temperature and concentration gradients. *Int. J. Therm. Sci.* **40**, 227–244. [9.4]
Chamkha, A. J. and Al-Naser, H. 2002 Hydromagnetic double-diffusive convection in a rectangular enclosure with opposing temperature and concentration gradients. *Int. J. Heat Mass Transfer* **45**, 2465–2483. [9.1.6.4]
Chamkha, A. J. and Khaled, A. R. A. 1999 Nonsimilar hydomagnetic simultaneous heat and mass transfer by mixed convection from a vertical plate embedded in a uniform porous medium. *Numer. Heat Transfer A* **36**, 327–344. [9.6]

Chamkha, A. J. and Khaled, A. R. A. 2000a Hydromagnetic simultaneous heat and mass transfer by mixed convection from a vertical plate embedded in a stratified porous medium with thermal dispersion effects. *Heat Mass Transfer* **36**, 63–70. [9.6]

Chamkha, A. J. and Khaled, A. R. A. 2000b Similarity solutions for hydromagnetic mixed convection heat and mass transfer for Hiemenz flow through porous media. *Int. J. Numer. Meth. Heat Fluid Flow* **10**, 94–115. [9.6]

Chamkha, A. J. and Khaled, A. R. A. 2000c Hydromagnetic combined heat and mass transfer by natural convection from a permeable surface embedded in a fluid-saturated porous medium. *Int. J. Numer. Meth. Heat Fluid Flow* **10**, 445–476. [9.1.6.4]

Chamkha, A. J. and Khaled, A. R. A. 2000d Hydromagnetic coupled heat and mass transfer by natural convection from a permeable constant heat flux surface in porous media. *J. Porous Media* **3**, 259–266. [9.2.1]

Chamkha, A. J. and Khanafer, K. 1999 Nonsimilar combined convection flow over a vertical surface embedded in a variable porosity medium. *J. Porous Media* **2**, 231–249. [9.6]

Chamkha, A. J. and Pop, I. 2004 Effect of thermophoresis particle deposition in free convection boundary layer from a vertical flat plate embedded in a porous medium. *Int. Comm. Heat Mass Transfer* **31**, 421–430. [9.2.1]

Chamkha, A. J. and Quadri, M. M. A. 2001 Heat and mass transfer from a permeable cylinder in a porous medium with magnetic field and heat generation/adsorption effects. *Numer. Heat Transfer A*. **40**, 387–401. [9.2.1]

Chamkha, A. J. and Quadri, M. M. A. 2002 Combined heat and mass transfer by hydromagnetic natural convection over a cone embedded in a non-Darcian porous medium with heat generation/absorption effects. *Heat Mass Transfer* **38**, 487–495. [9.2.1]

Chamkha, A. J. and Quadri, M. M. A. 2003 Simultaneous heat and mass transfer by natural convection from a plate embedded in a porous medium with thermal dispersion effects. *Heat Mass Transfer* **39**, 561–569. [9.2.1]

Chamkha, A. J., Issa, C. and Khanafer, K. 2001 Natural convection due to solar radiation from a vertical plate embedded in a porous medium with variable porosity. *J. Porous Media* **4**, 69–77. [5.1.9.4]

Chamkha, A. J., Issa, C. and Khanafer, K. 2002 Natural convection from an inclined plate embedded in a variable porosity porous medium due to solar radiation. *Int. J. Therm. Sci.* **41**, 73–78. [5.3]

Chamkha, A. J., Khaled, A. R. A. and Al-Hawaj, O. 2000 Simultaneous heat and mass transfer by natural convection from a cone and a wedge in porous media. *J. Porous Media* **3**, 155–164. [9.2.1]

Chan, B. K. C., Ivey, C. M. and Barry, J. M. 1970 Natural convection in enclosed porous media with rectangular boundaries. *ASME J. Heat Transfer* **92**, 21–27. [7.6.2, 7.8]

Chan, Y. T. and Banerjee, S. 1981 Analysis of transient three-dimensional natural convection in porous media. *ASME J. Heat Transfer* **103**, 242–248. [6.9.2]

Chandrasekhara, B. C. 1985 Mixed convection in the presence of horizontal impermeable surfaces in saturated porous media with variable permeability. *Wärme-Stoffübertrag.* **19**, 195–201. [8.1.2]

Chandrasekhara, B. C. and Nagaraju, P. 1988 Composite heat transfer in the case of a steady laminar flow of a gray fluid with small optical density past a horizontal plate embedded in a saturated porous medium. *Wärme-Stoffübertrag.* **23**, 343–352. [8.1.2]

Chandrasekhara, B. C. and Nagaraju, P. 1993 Composite heat transfer in a variable porosity medium bounded by an infinite flat plate. *Wärme-Stoffübertrag.* **28**, 449–456. [5.1.9.9]

Chandrasekhara, B. C. and Namboodiri, P. M. S. 1985 Influence of variable permeability on combined free and forced convection about inclined surfaces in porous media. *Int. J. Heat Mass Transfer* **28**, 199–206. [8.1.1]

Chandrasekhara, B. C., Radha, N. and Kumari, M. 1992 The effect of surface mass transfer on buoyancy induced flow in a variable porosity medium adjacent to a vertical heated plate. *Wärme-Stoffübertrag.* **27**, 157–166. [5.1.9.9]

Chang, I. D. and Cheng, P. 1983 Matched asymptotic expansions for free convection about an impermeable horizontal surface in a porous medium. *Int. J. Heat Mass Transfer* **26**, 163–173. [5.1.6]

Chang, M. H. 2004 Stability of convection induced by selective absorption of radiation in a fluid overlying a porous layer. *Phys. Fluids* **16**, 3690–3698. [6.19.1]

Chang, M. H. 2005 Thermal convection in superposed fluid and porous layers subjected to a horizontal plane Couette flow. *Phys. Fluids* **17**, Art. No. 064016. [6.19.3]

Chang, P. Y., Shiah, S. W. and Fu, M. N. 2004 Mixed convection in a horizontal square packed-sphere channel under uniform heating peripherally uniform wall temperature. *Numer. Heat Transfer A* **45**, 791–809. [6.19.3]

Chang, W. J. and Chang, W. L. 1995 Mixed convection in a vertical tube partially filled with porous medium. *Numer. Heat Transfer A* **28**, 739–754. [8.3.3]

Chang, W. J. and Chang, W. L. 1996 Mixed convection in a vertical parallel-plate channel partially filled with porous media of high permeability. *Int. J. Heat Mass Transfer* **39**, 1331–1342. [8.3.1].

Chang, W. J. and Hsiao, C. F. 1993 Natural convection in a vertical cylinder filled with anisotropic porous media. *Int. J. Heat Mass Transfer* **36**, 3361–3367. [7.3.3]

Chang, W. J. and Jang, J. Y. 1989a Non-Darcian effects on vortex instability of a horizontal natural convection flow in a porous medium. *Int. J. Heat Mass Transfer* **32**, 529–539. [5.4]

Chang, W. J. and Jang, J. Y. 1989b Inertia effects on vortex instability of a horizontal natural convection flow in a saturated porous medium. *Int. J. Heat Mass Transfer* **32**, 541–550. [5.4]

Chang, W. J. and Lin, H. C. 1994a Wall heat conduction effect on natural convection in an enclosure filled with a non-Darcian porous medium. *Numer. Heat Transfer A* **25**, 671–684. [7.1.5]

Chang, W. J. and Lin, H. C. 1994b Natural convection in a finite wall rectangular cavity filled with an anisotropic porous medium. *Int. J. Heat Mass Transfer* **37**, 303–312. [7.3.2]

Chang, W. J. and Wang, C. I. 2002 Heat and mass transfer in porous material. In *Transport Phenomena in Porous Media II* (D. B. Ingham and I. Pop, eds.) Elsevier, Oxford, pp. 257–275. [3.6]

Chang, W. J. and Yang, D. F. 1995 Transient natural convection of water near its density extremum in a rectangular cavity filled with porous medium. *Numer. Heat Transfer A* **28**, 619–633. [6.20]

Chang, W. J. and Yang, D. F. 1996 Natural convection for the melting of ice in porous media in a rectangular enclosure. *Int. J. Heat Mass Transfer* **39**, 2333–2348. [10.1.7]

Chao, B. H., Wang, H. and Cheng, P. 1996 Stagnation point flow of a chemical reactive fluid in a catalytic porous bed. *Int. J. Heat Mass Transfer* **39**, 3003–3019. [3.4]

Char, M. I. and Lee, G. C. 1998 Maximum density effects on natural convection in a vertical annulus filled with a non-Darcy porous medium. *Acta Mech.* **128**, 217–231. [8.3.3]

Char, M. I. and Lin, J. D. 2001 Conjugate film condensation and natural convection between two porous media separated by a vertical plate. *Acta Mech.* **148**, 1–15. [10.4]

Char, M. I, Lin, J. D. and Chen, H. T. 2001 Conjugate mixed convection laminar non-Darcy film condensation along a vertical plate in a porous medium. *Int. J. Engng. Sci.* **39**, 897–912. [10.4]

Charrier-Mojtabi, M. C. 1997 Numerical simulation of two- and three-dimensional free convective flows in a horizontal porous annulus using a pressure and temperature formulation. *Int. J. Heat Mass Transfer* **40**, 1521–1533. [7.3.3]

Charrier-Mojtabi, M. C. and Mojtabi, A. 1994 Numerical simulation of three-dimensional free convection in a horizontal porous annulus. *Heat Transfer 1994*, Inst. Chem. Engrs. Rugby. pp. 319–324. [7.3.3]

Charrier-Mojtabi, M. C. and Mojtabi, A. K. 1998 Natural convection in a horizontal porous annulus. *Transport Phenomena in Porous Media* (eds. D. B. Ingham and I. Pop), Elsevier, Oxford, pp. 155–178. [7.3.3]

Charrier-Mojtabi, M. C., Karimi-Fard, M. and Mojtabi, A. 1997 Onset of thermosolutal convective regimes in rectangular porous cavity. *C. R. Acad. Sci. II B* **324**, 9–17. [9.2.2]

Charrier-Mojtabi, M. C., Karimi-Fard, M., Azaiez, M. and Mojtabi, A. 1998 Onset of a double-diffusive convective regime in a rectangular porous cavity. *J. Porous Media* **1**, 104–118. [9.2.2]

Charrier-Mojtabi, M. C., Mojtabi, A., Azaiez, M. and Labrosse, G. 1991 Numerical and experimental study of multicellular free convection flows in an annular porous pipe. *Int. J. Heat Mass Transfer* **34**, 3061–3074. [6.16.3]

Charrier-Mojtabi, M. C., Maliwan, K., Razi, Y. P., Bardan, G. and Mojtabi, A. 2003 Contrôle des ecoulements thermoconvectifs au moyen de vibration. *J. Méc. Ind.* **4**, 545–549. [6.24]

Charrier-Mojtabi, M. C., Razi, Y. P., Maliwan, K. and Mojtabi, A. 2004 Influence of vibration on Soret-driven convection in porous media.. *Numer. Heat Transfer* **46**, 981–993. [9.1.6.4]

Charrier-Mojtabi, M. C., Razi, Y. P., Maliwan, K. and Mojtabi, A. 2005 Effect of vibration on the onset of double-diffusive convection in porous media. In *Transport Phenomena in Porous Media III*, (eds. D. B. Ingham and I. Pop), Elsevier, Oxford, pp. 261–286. [9.1.6.4]

Chaudary, R. C. and Sharma, P. K. 2003 Three dimensional unsteady convection and mass transfer flow through porous medium. *Heat Mass Transfer* **39**, 765–770. [5.1.9.7]

Chaudhary, M. A., Merkin, J. H. and Pop, I. 1995a Similarity solutions in free-convection boundary-layer flows adjacent to vertical permeable surfaces in porous media. 1. Prescribed surface temperature. *Europ. J. Mech. B.* **14**, 217–237. [5.1.2]

Chaudhary, M. A., Merkin, J. H. and Pop. I. 1995b Similarity solutions in free convection boundary-layer flows adjacent to vertical permeable surfaces in porous media: II prescribed surface heat flux. *Heat Mass Transfer* **30**, 341–347. [5.1.2]

Chaudhary, M. A., Merkin, J. H. and Pop, I. 1996 Natural convection from a horizontal permeable surface in a porous medium—numerical and asymptotic solutions. *Transport in Porous Media* **22**, 327–344. [5.2]

Chaves, C. A., Camargo, J. R., Cardoso, S. and de Macedo, A. G. 2005 Transient natural convection heat transfer by double-diffusion from a heated cylinder buried in a saturated porous medium. *Int. J. Thermal Sci.* **44**, 720–725. [9.4]

Chelghoum, D. E., Weidman, P. D. and Kassoy, D. R. 1987 Effect of slab width on the stability of natural convection in confined saturated porous media. *Phys. Fluids* **30**, 1941–1947. [6.15.1]

Chellaiah, S. and Viskanta, R. 1987 Freezing of water and water-salt solutions around aluminum spheres. *Int. Comm. Heat Mass Transfer* **14**, 437–446. [10.2.1.2]

Chellaiah, S. and Viskanta, R. 1989a On the supercooling during freezing of water saturated porous media. *Int. Comm. Heat Mass Transfer* **16**, 163–172. [10.2.1.2]

Chellaiah, S. and Viskanta, R. 1989b Freezing of water-saturated porous media in the presence of natural convection: experiments and analysis. *ASME J. Heat Transfer* **111**, 424–432; errata 648. [10.2.1.2]

Chellaiah, S. and Viskanta, R. 1990 Natural convection melting of a frozen porous medium. *Int. J. Heat Mass Transfer* **33**, 887–899. [10.2.1.2]

Chen, B., Wang, B. and Fang, Z. 1999 Electrochemical experimental method for natural convective heat and mass transfer in porous media. *Heat Transfer Asian Res.* **28**, 266–277. [9.2.2]

Chen, C. F. 1995 Experimental study of convection in a mushy layer during directional solidification. *J. Fluid Mech.* **293**, 81–98. [10.2.3]

Chen, C. H. 1996 Non-Darcy mixed convection from a horizontal surface with variable surface heat flux in a porous medium. *Num. Heat Transfer A* **30**, 859–869. [8.1.2]

Chen, C. H. 1997a Non-Darcy mixed convection over a vertical flat plate in porous media with variable wall heat flux. *Int. Comm. Heat Mass Transfer* **24**, 427–437. [8.1.1]

Chen, C. H. 1997b Analysis of non-Darcian mixed convection along a vertical plate embedded in a porous medium. *Int. J. Heat Mass Transfer* **40**, 2993–2997. [8.1.2]

Chen, C. H. 1998a Mixed convection heat transfer from a horizontal plate with variable surface heat flux in a porous medium. *Heat Mass Transfer* **34**, 1–7. [8.1.2]

Chen, C. H. 1998b Nonsimilar solutions for non-Darcy mixed convection from a non-isothermal horizontal surface in a porous medium. *Int. J. Engng. Sci.* **36**, 251–263. [8.1.2]

Chen, C. H. and Chen, C. K. 1990a Non-Darcian mixed convection along a vertical plate embedded in a porous medium. *Appl. Math. Modelling* **14**, 482–488. [8.1.1]

Chen, C. H. and Chiou, J. S. 1994 Conjugate free convection heat transfer analysis of a vertical plate fin embedded in non-Darcian porous media. *Int. J. Engng. Sci.* **32**, 1703–1716. [5.12.1]

Chen, C. H. and Horng, J. H. 1999 Natural convection from a vertical cylinder in a thermally stratified porous medium. *Heat Mass Transfer* **34**, 423–428. [5.7]

Chen, C. H., Chen, T. S. and Chen, C. K. 1996 Non-Darcy mixed convection along non-isothermal vertical surfaces in porous media. *Int. J. Heat Mass Transfer* **39**, 1157–1164. [8.1.1]

Chen, C. K. and Chen, C. H. 1990b Nonuniform porosity and non-Darcian effects on conjugate mixed convection heat transfer from a plate fin in porous media. *Int. J. Heat Fluid Flow* **11**, 65–71. [8.1.4]

Chen, C. K. and Chen, C. H. 1991 Non-Darcian effects on conjugate mixed convection about a vertical circular pin in a porous medium. *Comput. Struct.* **38**, 529–535. [8.1.3]

Chen, C. K. and Lin, C. R. 1995 Natural convection from an isothermal vertical surface embedded in a thermally stratified high-porosity medium. *Int. J. Engng. Sci.* **33**, 131–138. [5.1.4]

Chen, C. K., Chen, C. H., Minkowitz, W. J. and Gill, U. S. 1992 Non-Darcian effects on mixed convection about a vertical cylinder embedded in a saturated porous medium. *Int. J. Heat Mass Transfer* **35**, 3041–3046. [8.1.3]

Chen, C. K., Hsiao, S. W. and Cheng, P. 1990 Transient natural convection in an eccentric porous annulus between horizontal cylinders. *Numer. Heat Transfer A.* **17**, 431–448. [7.3.3]

Chen, C. K., Hung, C. I. and Horng, H. C. 1987 Transient natural convection on a vertical flat plate embedded in a high-porosity medium. *ASME J. Energy Res. Tech.* **109**, 112–118. [5.1.3]

Chen, F. 1990 On the stability of salt-finger convection in superposed fluid and porous layers. *ASME J. Heat Transfer* **112**, 1088–1092. [9.4]

Chen, F. 1991 Throughflow effects on convective instability in superposed fluid and porous layers. *J. Fluid Mech.* **231**, 113–133. [6.19.1.2]

Chen, F. 1992 Salt-finger instability in an anisotropic and inhomogeneous porous substrate underlying a fluid layer. *J. Appl. Phys.* **71**, 5222–5236. [9.1.6.2]

Chen, F. and Chen, C. F. 1988 Onset of finger convection in a horizontal porous layer underlying a fluid layer. *ASME J. Heat Transfer* **110**, 403–409. [6.19.1.1, 6.19.1.2, 9.4]

Chen, F. and Chen, C. F. 1989 Experimental investigation of convective stability in a superposed fluid and porous layer when heated from below. *J. Fluid Mech.* **207**, 311–321. [6.19.1.1, 6.19.1.2]

Chen, F. and Chen, C. F. 1992 Convection in superposed fluid and porous layers. *J. Fluid Mech.* **234**, 97–119. [1.6, 6.19.1.2]

Chen, F. and Chen, C. F. 1993 Double-diffusive fingering convection in a porous medium. *Int. J. Heat Mass Transfer* **36**, 793–807. [9.1.6.3]

Chen, F. and Chen, C. F. 1996 Analysis of heat transfer regulation and modification using intermittently emplaced porous cavities—Analysis of flow and heat transfer over an external boundary covered with a porous substrate. *ASME J. Heat Transfer* **118**, 266–267. [5.1.9.5]

Chen, F. and Hsu, L. H. 1991 Onset of thermal convection in an anisotropic and inhomogeneous porous layer underlying a fluid layer. *J. Appl. Phys.* **69**, 6289–6301. [6.19.1.2]

Chen, F. and Lu, J. W. 1991 Influence of viscosity variation on salt-finger instability in a fluid layer, a porous layer, and their superposition. *J. Appl. Phys.* **70**, 4121–4131. [9. 4]

Chen, F. and Lu, J. W. 1992a Variable viscosity effects on convective instability in superposed fluid and porous layers. *Phys. Fluids* **4**, 1936–1944. [6.19.1.2]

Chen, F. and Lu, J. W. 1992b Onset of salt-finger convection in anisotropic and inhomogeneous porous media. *Int. J. Heat Mass Transfer* **35**, 3451–3464. [6.19.1.2, 9.1.6.2]

Chen, F. and Wang, C. Y. 1993a Convective instability in a porous enclosure with a horizontal conducting baffle. *ASME J. Heat Transfer* **115**, 810–813. [7.3.7]

Chen, F. and Wang, C. Y. 1993b Convective instability in saturated porous enclosures with a vertical insulating baffle. *Int. J. Heat Mass Transfer* **36**, 1897–1904. [7.3.7]

Chen, F., Chen, C. F. and Pearlstein, A. J. 1991 Convective instability in superposed fluid and anisotropic porous layers. *Phys. Fluids* A **3**, 556–565. [6.19.1.2]

Chen, F., Lu, J. W. and Yang, T. L. 1994 Convective instability in ammonium chloride solution directionally solidified from below. *J. Fluid Mech.* **276**, 163–187. [10.2.3]

Chen, G. and Hadim, H. A. 1995 Numerical study of forced convection of a power-law fluid in a porous channel. *ASME HTD* -**309**, 65–72. [4.16.3]

Chen, G. and Hadim, H. A. 1998a Forced convection of a power-law fluid in a porous channel—numerical solutions. *Heat Mass Transfer* **34**, 221–228. [4.16.3]

Chen, G. and Hadim, H. A. 1998b Numerical study of non-Darcy forced convection in a packed bed saturated with a power-law fluid. *J. Porous Media* **1**, 147–157. [4.16.3]

Chen, G. and Hadim, H. A. 1999a Forced convection of a power-law fluid in a porous channel—integral solutions. *J. Porous Media* **2**, 59–69. [4.16.3]

Chen, G. and Hadim, H. A. 1999b Numerical study of three dimensional non-Darcy forced convection in a square porous duct. *Int. J. Numer. Meth. Heat Fluid Flow* **9**, 151–169. [4.9]

Chen, H. T. and Chen, C. K. 1987 Natural convection of non-Newtonian fluids about a horizontal surface in a porous medium. *ASME J. Energy Res. Tech.* **109**, 119–123. [5.2]

Chen, H. T. and Chen, C. K. 1988a Free convection flows of non-Newtonian fluids along a vertical plate embedded in a porous medium. *ASME J. Heat Transfer* **110**, 257–259. [5.1.9.2]

Chen, H. T. and Chen, C. K. 1988b Natural convection of a non-Newtonian fluid about a horizontal cylinder and a sphere in a porous medium. *Int. Comm. Heat Mass Transfer* **15**, 605–614. [5.5.1, 5.6.1].

Chen, H., Besant, R. W. and Tao, Y. X. 1998 Numerical modeling of heat transfer and water vapor transfer and frosting within a fiberglass filled cavity during air infiltration. *Heat Transfer 1998, Proc. 11th IHTC*, **4**, 381–386. [3.6]

Chen, K. S. and Ho, J. R. 1986 Effects of flow inertia on vertical natural convection in saturated porous media. *Int. J. Heat Mass Transfer* **29**, 753–759. [5.1.7.2]

Chen, L., Li, Y. and Thorpe, G. 1998 High Rayleigh-number natural convection in an enclosure containing a porous layer. *Heat Transfer 1998, Proc. 11th IHTC*, **4**, 423–428. [1.8, 7.7]

Chen, Q. S., Prasad, V. and Chatterjee, A. 1999 Modeling of fluid flow and heat transfer in a hydrothermal crystal growth system: use of fluid-superposed porous layer theory. *ASME J. Heat Transfer* **121**, 1049–1058. [6.19.2]

Chen, S. C. and Vafai, K. 1996 Analysis of free surface momentum and energy transport in porous media *Numer. Heat Transfer* **29**, 281–296. [2.2.2]

Chen, Y. C. 2004 Non-Darcy flow stability of mixed convection in a vertical channel filled with a porous medium. *Int. J. Heat Mass Transfer* **47**, 1257–1266. [8.3.1]

Chen, Y. C. and Chung, J. N. 1998 Stability of shear flow in a vertical heated channel filled with a porous medium. *Heat Transfer 1998, Proc. 11th IHTC*, **4**, 435–440. [8.3.1]

Chen, Y. C., Chung, J. N., Wu, C. S. and Lue, Y. F. 2000a Non-Darcy mixed convection in a vertical channel filled with a porous medium. *Int. J. Heat Mass Transfer* **43**, 2421–2429. [8.3.1]

Chen, Y. H. and Lin, H. T. 1997 Natural convection in an inclined enclosure with a fluid layer and a heat-generating porous bed. *Heat Mass Transfer* **33**, 247–255. [7.8]

Chen, Z. Q., Cheng, P. and Hsu, C. T. 2000 A theoretical and experimental study on stagnant thermal conductivity of bi-dispersed porous media. *Int. Comm. Heat Mass Transfer* **27**, 601–610. [4.16.4]

Chen, Z., Lyons, S. L. and Qin, G. 2001 Derivation of the Forchheimer equation via homogenization. *Transport Porous Media* **44**, 325–335. [1.5.2]

Cheng, C. Y. 1999 Effect of a magnetic field on heat and mass transfer by natural convection from vertical surfaces in porous media—an integral approach. *Int. Comm. Heat Mass Transfer* **26**, 935–943. [9.2.1]

Cheng, C. Y. 2000a An integral approach for heat and mass transfer by natural convection from truncated cones in porous media with variable wall temperature and concentration. *Int. Comm. Heat Mass Transfer* **27**, 537–548. [9.2.1]

Cheng, C. Y. 2000b Transient heat and mass transfer by natural convection from vertical surfaces in porous media. *J. Phys. D* **33**, 1425–1430. [9.2.3]

Cheng, C. Y. 2000c Natural convection heat and mass transfer near a wavy cone with constant wall temperature and concentration in a porous medium. *Mech. Res. Comm.* **27**, 613–620. [9.2.1]

Cheng, C. Y. 2000d Natural convection heat and mass transfer near a vertical wavy surface with constant wall temperature and concentration in a porous medium. *Int. Comm. Heat Mass Transfer* **27**, 1143–1154. [9.2.1]

Cheng, C. Y. 2005 An integral approach for hydromagnetic natural convection heat and mass transfer form vertical surfaces with power-law variation in wall temperature and concentration in porous media. *Int. Comm. Heat Mass Transfer* **32**, 204–213. [9.2.1]

Cheng, G. J., Yu, A. B. and Zulli, P. 1999 Evaluation of effective thermal conductivity from the structure of a packed bed. *Chem. Eng. Sci.* **54**, 4199–4209. [2.2.1]

Cheng, J., Liao, S. J. and Pop, I. 2005 Analytic series solution for unsteady mixed convection boundary layer flow near the stagnation point on a vertical surface in a porous medium. *Transport Porous Media* **61**, 365–379. [8.1.1]

Cheng, L. P. and Kuznetsov, A. V. 2005 Heat transfer in a laminar flow in a helical pipe filled with a saturated porous medium. *Int. J. Thermal Sci.* **44**, 787–798. [4.5]

Cheng, P. 1977a Constant surface heat flux solutions for porous layer flows. *Lett. Heat Mass Transfer* **4**, 119–128. [5.1.2]

Cheng, P. 1977b The influence of lateral mass flux on free convection boundary layers in a saturated porous medium. *Int. J. Heat Mass Transfer* **20**, 201–206. [5.1.2]

Cheng, P. 1977c Combined free and forced boundary layer flows about inclined surfaces in a porous medium. *Int. J. Heat Mass Transfer* **20**, 807–814. [4.1, 8.1.4]

Cheng, P. 1977d Similarity solutions for mixed convection from horizontal impermeable surfaces in saturated porous media. *Int. J. Heat Mass Transfer* **20**, 893–898. [8.1.2]

Cheng, P. 1978 Heat transfer in geothermal systems. *Adv. Heat Transfer* **14**, 1–105. [3.5, 6.9.1, 6.26, 11, 11.7]

Cheng, P. 1981a Thermal dispersion effects in non-Darcian convective flows in a saturated porous medium. *Lett. Heat Mass Transfer* **8**, 267–270. [5.1.7.3, 5.1.8]

Cheng, P. 1981b Film condensation along an inclined surface in a porous medium. *Int. J. Heat Mass Transfer* **24**, 983–990. [10.4]

Cheng, P. 1982 Mixed convection about a horizontal cylinder and a sphere in a fluid saturated porous medium. *Int. J. Heat Mass Transfer* **25**, 1245–1247. [4.3, 8.1.3]

Cheng, P. 1985a Natural convection in a porous medium: external flows. *Natural Convection: Fundamentals and Applications* (eds. S. Kakaç, W. Aung, and R. Viskanta), Hemisphere, Washington, DC, 475–513. [5, 5.1.7.3, 5.1.8, 5.3, 5.6.1, 5.10.1.2]

Cheng, P. 1985b Geothermal heat transfer. *Handbook of Heat Transfer Applications* (eds. W. M. Rohsenow *et al.*), 2nd edition, McGraw-Hill, New York, Chapter 11. [11, 11.7]

Cheng, P. 1987 Wall effects on fluid flow and heat transfer in porous media. *Proc. 1987 ASME JSME Thermal Engineering Joint Conf.* **2**, 297–303. [4.8]

Cheng, P. and Ali, C. L. 1981 An experimental investigation of free convection about an inclined surface in a porous medium. *ASME 20th National Heat Transfer Conference*, Paper No. 81–HT-85. [5.3]

Cheng, P. and Chang, I. D. 1976 On buoyancy induced flows in a saturated porous medium adjacent to impermeable horizontal surfaces. *Int. J. Heat Mass Transfer* **19**, 1267–1272. [5.2, 6.26]

Cheng, P. and Chang, I. D. 1979 Convection in a porous medium as a singular perturbation problem. *Lett. Heat Mass Transfer* **6**, 253–258. [5.1.6]

Cheng, P. and Chau, W. C. 1977 Similarity solutions for convection of groundwater adjacent to horizontal surface with axisymmetric temperature distribution. *Water Resour. Res.* **13**, 768–772. [5.2]

Cheng, P. and Chui, D. K. 1984 Transient film condensation on a vertical surface in a porous medium. *Int. J. Heat Mass Transfer* **27**, 795–798. [10.4]

Cheng, P. and Hsu, C. T. 1984 Higher-order approximations for Darcian free convective flow about a semi-infinite vertical flat plate. *ASME J. Heat Transfer* **106**, 143–151. [5.1.6]

Cheng, P. and Hsu, C. T. 1986a Fully developed, forced convective flow through an annular packed-sphere bed with wall effects. *Int. J. Heat Mass Transfer* **29**, 1843–1853. [4.9]

Cheng, P. and Hsu, C. T. 1986b Applications of Van Driest's mixing length theory to transverse thermal dispersion in forced convective flow through a packed bed. *Int. Comm. Heat Mass Transfer* **13**, 613–626. [4.9]

Cheng, P. and Hsu, C. T. 1998 Heat conduction. *Transport Phenomena in Porous Media I* (eds. D. B. Ingam and I. Pop), Elsevier, Oxford, pp. 57–76. [2.2.1]

Cheng, P. and Hsu, C. T. 1999 The effective stagnant thermal conductivity of porous media with periodic structure. *J. Porous Media*. **2**, 19–38. [2.2.1, 4.16.4]

Cheng, P. and Minkowycz, W. J. 1977 Free convection about a vertical flat plate embedded in a porous medium with application to heat transfer from a dike. *J. Geophys. Res.* **82**, 2040–2044. [5.1.1, 10.1.5]

Cheng, P. and Pop, I. 1984 Transient free convection about a vertical flat plate imbedded in a porous medium. *Int. J. Engng. Sci.* **22**, 253–264. [5.1.3]

Cheng, P. and Teckchandani, L. 1977 Numerical solutions for transient heating and fluid withdrawal in a liquid-dominated geothermal reservoir. *The Earth's Crust* (ed. J. G. Heacock), Amer. Geophys. Union, Washington, DC, 705–721. [11.7]

Cheng, P. and Verma, A. K. 1981 The effect of subcooled liquid on film boiling about a vertical heated surface in a porous medium. *Int. J. Heat Mass Transfer* **24**, 1151–1160. [10.3.2]

Cheng, P. and Vortmeyer, D. 1988 Transverse thermal dispersion and wall channeling in a packed bed with forced convective flow. *Chem. Engng. Sci.* **43**, 2523–2532. [4.9]

Cheng, P. and Wang, C. Y. 1996 Multiphase mixture model for multiphase, multicomponent transport in capillary porous media—II. Numerical simulation of the transport of organic compounds in the subsurface. *Int. J. Heat Mass Transfer* **39**, 3619–3632. [3.6.2]

Cheng, P. and Zheng, T. M. 1986 Mixed convection in the thermal plume above a horizontal line source of heat in a porous medium of infinite extent. *Heat Transfer 1986*, Hemisphere, Washington, DC, **5**, 2671–2675. [8.1.4]

Cheng, P. and Zhu, H. 1987 Effects of radial dispersion on fully-developed forced convection in cylindrical packed tubes. *Int. J. Heat Mass Transfer* **30**, 2373–2383. [4.9]

Cheng, P., Ali, C. L. and Verma, A. K. 1981 An experimental study of non-Darcian effects in free convection in a saturated porous medium. *Lett. Heat Mass Transfer* **8**, 261–265. [5.1.8]

Cheng, P., Chowdhury, A. and Hsu, C. T. 1991 Forced convection in packed tubes and channels with variable porosity and thermal dispersion effects. *Convective Heat and Mass Transfer in Porous Media*, (eds. S. Kakaç, *et al.*), Kluwer Academic, Dordrecht, 625–653. [1.7, 4.9]

Cheng, P., Chui, D. K. and Kwok, L. P. 1982 Film boiling about two-dimensional and axisymmetric isothermal bodies of arbitrary shape in a porous medium. *Int. J. Heat Mass Transfer* **25**, 1247–1249. [10.3.2]

Cheng, P., Hsu, C. T. and Chowdhury, A. 1988 Forced convection in the entrance region of a packed channel with asymmetric heating. *ASME J. Heat Transfer* **110**, 946–954. [4.9]

Cheng, P., Le, T. T. and Pop, I. 1985 Natural convection of a Darcian fluid about a cone. *Int. Comm. Heat Mass Transfer* **12**, 705–717. [5.8]

Cheng, W. T. and Lin, H. T. 2002 Unsteady forced convection heat transfer on a flat plate embedded in the fluid-saturated porous medium with inertia effect and thermal dispersion. *Int. J. Heat Mass Transfer*, **45**, 1563–1569. [4.6.3]

Cherkaoui, A.S.M. and Wilcock, W.S.D. 1999 Characteristics of high Rayleigh number two-dimensional convection in an open-top porous layer heated from below. *J. Fluid Mech.* **394**, 241–260. [11.8]

Cherkaoui, A.S.M. and Wilcock, W.S.D. 2001 Laboratory studies of high Rayleigh number circulation in an open-top Hele-Shaw cell: an analogue to mid-ocean ridge hydrothermal systems. *J. Geophys. Res.* **106**, 10983–11000. [11.8]

Chevalier, S., Bernard, D. and Grenet, J. P. 1996 Ètude expérimentale de la convection naturelle dans une couche poreuse inclinée limitée pardes frontiéres conductrices de la chaleur. *C.R. Acad. Sci. Paris*, Sér. II, **322**, 305–311. [7.8]

Chevalier, S., Bernard, D. and Joly, N. 1999 Natural convection in a porous layer bounded by impervious domains: from numerical approaches to experimental realization. *Int. J. Heat Mass Transfer* **42**, 581–597. [7.8]

Chhuon, B. and Caltagirone, J. P. 1979 Stability of a horizontal porous layer with timewise periodic boundary conditions. *ASME J. Heat Transfer* **101**, 244–248. [6.11.3]

Chiang, K. C. and Tsai, H. L. 1992 Interaction between shrinkage-induced fluid flow and natural convection during alloy solidification. *Int. J. Heat Mass Transfer*, **35**, 1771–1778. [10.2.3]

Chiareli, A. O. P. and Worster, M. G. 1995 Flow focusing instability in a solidifying mushy layer. *J. Fluid Mech.* **297**, 293–305. [10.2.3]

Chiem, K. S. and Zhao, Y. 2004 Numerical study of steady/unsteady flow and heat transfer in porous media using a characteristics-based matrix-free implicit FV method on unstructured grids. *Int. J. Heat Fluid Flow* **25**, 1015–1033. [4.11]

Chikh, S., Boumedien, A. Bouhadef, K. and Lauriat, G. 1998 Analysis of fluid flow and heat transfer in a channel with intermittent heated porous disks. *Heat Mass Transfer* **33**, 405–413. [4.11]

Chikh, S., Boumedien, A., Bouhadef, K. and Lauriat, G. 1995a Analytical solution of non-Darcian forced convection in an annular duct partially filled with a porous medium. *Int. J. Heat Mass Transfer* **38**, 1543–1551. [4.12.1]

Chikh, S., Boumedien, A., Bouhadef, K. and Lauriat, G. 1995b Non-Darcian forced convection analysis in an annulus partially filled with a porous material *Numer. Heat Transfer A* **28**, 707–722. [4.11]

Choi, C. Y. and Kulacki, F. A. 1990 Non-Darcian effects on mixed convection in a vertical porous annulus. *Heat Transfer 1990*, Hemisphere, New York, **5**, 271–276. [8.3.4]

Choi, C. Y. and Kulacki, F. A. 1992a Mixed convection in vertical porous channels and annuli. *Heat and Mass Transfer in Porous Media*, (ed. M. Quintard and M. Todorovic), Elsevier, Amsterdam, 61–98. [8.3]

Choi, C. Y. and Kulaki, F. A. 1992b Mixed convection through vertical porous annuli locally heated from the inner cylinder. *ASME J. Heat Transfer* **114**, 143–151. [8.3.3]

Choi, C. Y. and Kulaki, F. A. 1993 Non-Darcian effects on mixed convection in a vertical packed-sphere annulus. *ASME J. Heat Transfer* **115**, 506–510. [8.3.3]

Choi, C. Y., Lai, F. C. and Kulacki, F. A. 1989 Mixed convection in vertical porous annuli. *AIChE Sympos. Ser.* **269**, 356–361. [8.3.4]

Choi, E., Chamkha, A. and Nandakumar, K. 1998 A bifurcation study of natural convection in porous media with internal heat sources: the non-Darcy effects. *Int. J. Heat Mass Transfer* **41**, 383–392. [6.11.2]

Choi, J. and Viskanta, R. 1992 Freezing of aqueous sodium chloride solution saturated packed bed from above. *ASME HTD* **206**, Vol. 2, 159–166. [10.2.3]

Choi, J. and Viskanta, R. 1993 Freezing of aqueous sodium chloride solution saturated packed bed from a vertical wall of a rectangular cavity. *Int. J. Heat Mass Transfer* **36**, 2805–2813. [10.2.3]

Chou, F. C. and Chung, P. Y. 1995 Effect of stagnant conductivity on non-Darcian mixed convection in horizontal square packed channels. *Numer. Heat Transfer A* **27**, 195–209. [8.2.1]

Chou, F. C., Cheng, C. J. and Lien, W. Y. 1992a Analysis and experiment of non-Darcian convection in horizontal square packed-sphere channels—2. Mixed convection. *Int. J. Heat Mass Transfer* **35**, 1197–1207. [8.2.1]

Chou, F. C., Chung, P. Y. and Cheng, C.J . 1992b Effects of stagnant and dispersion conductivities on non-Darcian forced convection in square packed-sphere channels. *Canad. J. Chem. Engng.*, **69**, 1401–1407. [4.9]

Chou, F. C., Lien, W. Y. and Lin, S. H. 1992c Analysis and experiment of non-Darcian convection in horizontal square packed-sphere channels—1. Forced convection. *Int. J. Heat Mass Transfer* **35**, 195–205. [4.9]

Chou, F. C., Su, J. H. and Lien, S. S. 1994 A reevaluation of non-Darcian forced and mixed convection in cylindrical packed tubes. *ASME J. Heat Transfer*, **116**, 513–516. [4.9, 8.2.1]

Christopher, D. M. and Wang, B. X. 1993 Non-Darcy natural convection around a horizontal cylinder buried near the surface of a fluid-saturated porous medium. *Int. J. Heat Mass Transfer* **36**, 3663–3669. [7.11]

Christopher, D. M. and Wang, B. X. 1994 Natural convection melting around a horizontal cylinder buried in frozen water-saturated porous media. *Heat Transfer 1994*, Inst. Chem. Engrs, Rugby, Vol. 4, 19–24. [10.1.7]

Chung, C. A. and Chen., F. 2000 Onset of plume convection in mushy layers. *J. Fluid Mech.* **408**, 53–82. [10.2.3]

Chung, C. A. and Chen., F. 2001 Morphological instability in a directionally solidifying binary solution with an imposed shear flow. *J. Fluid Mech.* **436**, 85–106. [10.2.3]

Chung, C. A. and Worster, M. G. 2002 Steady-state chimneys in a mushy layer. *J. Fluid Mech.* **455**, 387–411. [10.2.3]

Chung, J. N., Plumb, O. A. and Lee, W. C. 1992 Condensation in a porous region bounded by a cold vertical surface. *ASME J. Heat Transfer*, **114**, 1011–1018. [10.4]

Chung, K., Lee, K. S. and Kim, W. S. 2003 Modified macroscopic turbulence modeling for the tube with channel geometry in porous media. *Numer. Heat Transfer A* **43**, 659–668. [1.8]

Chung, T. J., Park, J. H., Choi, C. K. and Yoon, D. Y. 2002 The onset of vortex instability in a laminar forced convection flow through a horizontal porous channel. *Int. J. Heat Mass Transfer* **45**, 3061–3064. [8.2.1]

Cieszko, M. and Kubik, J. 1999 Derivation of matching conditions at the contact surface between fluid-saturated porous solid and bulk fluid. *Transp. Por. Media* **34**, 319–336. [1.6]

Clarksean, R., Kwendakwema, N. and Boehm, R. 1988 A study of mixed convection in a porous medium between vertical concentric cylinders. *ASME HTD* **96**, Vol. 2, 339–344. [8.3.3]

Close, D. J. 1986 A general correlation for natural convection in liquid-saturated beds of spheres. *ASME J. Heat Transfer* **108**, 983–985. [6.9.2]

Close, D. J., Symons, J. G. and White, R. F. 1985 Convective heat transfer in shallow gas-filled porous media: experimental investigation. *Int. J. Heat Mass Transfer* **28**, 2371–2378. [6.9.1]

Collins, R. E. 1961 *Flow of Fluids through Porous Materials*, Reinhold, New York. [1.1]

Combarnous, M. 1970 Convection naturelle et convection mixte dans une couche poreuse horizontale. *Rev. Gén. Therm.* **9**, 1355–1375. [6.9.1]

Combarnous, M. 1972 Description du transfert de chaleur par convection naturelle dans une couche poreuse horizontale à l'aide d'un coefficient de transfert solide-fluide. *C. R. Acad. Sci. Paris A* **275**, 1375–1378. [6.5, 6.9.2]

Combarnous, M. and Bia, P. 1971 Combined free and forced convection in porous media. *Soc. Petrol. Eng. J.* **11**, 399–405. [6.9.1, 8.2.1]

Combarnous, M. and Bories, S. 1974 Módelization de la convection naturelle au sein d'une couche poreuse horizontale à l'aide d'un coefficient de transfert solide-fluide. *Int. J. Heat Mass Transfer* **17**, 505–515. [6.9.2]

Combarnous, M. A. and Bories, S. A. 1975 Hydrothermal convection in saturated porous media. *Adv. Hydroscience* **10**, 231–307. [6.9.1, 6.9.2]

Combarnous, M. and Le Fur, B. 1969 Transfert de chaleur par convection naturelle dans une couche poreuse horizontale. *C. R. Acad. Sci. Paris B* **269**, 1009–1012. [6.3, 6.9.1]

Cooper, C. A., Glass, R. J. and Tyler, S. W. 1997 Experimental investigation of the stability boundary for double-diffusive finger convection in a Hele-Shaw cell. *Water Resources Res.* **33**, 517–526. [2.5, 9.1.1]

Cooper, C. A., Glass, R. J. and Tyler, S. W. 2001 Effect of buoyancy ratio on the development of double-diffusive finger convection in a Hele-Shaw cell. *Water Resources Res.* **37**, 2323–2332. [9.1.1]

Corey, A. T., Rathjens, C. H., Henderson, J. H. and Wyllie, M. R. J. 1956 Three-phase relative permeability. *Trans. Am. Inst. Min. Metall. Pet. Eng.* **207**, 349–351. [3.5.4]

Costa, V. A. F. 2003 Unified streamline, heatline and massline methods for visualization of two-dimensional heat and mass transfer in anisotropic media. *Int. J. Heat Mass Transfer* **46**, 1309–1320. [4.17]

Costa, V. A. F. 2004 Double-diffusive natural convection in parallelogrammic enclosures filled with fluid-saturated porous media. *Int. J. Heat Mass Transfer* **47**, 2913–2926. [9.4]

Costa, V. A. F. 2005a Thermodynamics of natural convection in enclosures with viscous dissipation. *Int. J. Heat Mass Transfer* **47**, 2913–2926. [2.2.2]

Costa, V. A. F. 2005b *Appl. Mech. Rev.*, to appear. [4.17]

Cotta, R. M., Luz Neto, H., Alves, L. S. de B., and Quaresima, J. N. N. 2005 Integral transforms for natural convection in cavities filled with porous media. In *Transport Phenomena in Porous Media III*, (eds. D. B. Ingham and I. Pop), Elsevier, Oxford, pp. 97–119. [7.5]

Costa, V. A. F. , Oliveira, L. A. and Sousa, A. C. M. 2004b Simulation of coupled flows in adjacent porous and open domains using a control volume finite-element method. *Energy Conversion and Management* **45**, 2795–2811. [1.6]

Costa, V. A. F. , Oliveira, L. A., Baliga, B. R. and Sousa, A. C. M. 2004a Simulation of coupled flows in adjacent porous and open domains using a control volume finite-element method. *Numer. Heat Transfer A* **45**, 675–697. [1.6]

Costa, V. A. F. 2006 Bejan's heatlines and masslines for convection visualization and analysis. *Appl. Mech. Rev.* **59**, in press. [4.17]

Coulaud, O., Morel, P. and Caltagirone, J. P. 1988 Numerical modelling of nonlinear effects in laminar flow through a porous medium. *J. Fluid Mech.* **190**, 393–407. [1.5.2]

Coussot, P. 2000 Scaling approach to the convective drying of a porous medium. *European Phys. J. B* **15**, 557–566. [3.6]

Cox, B. L. and Pruess, K. 1990 Numerical experiments on convective heat transfer in water-saturated porous media at near-critical conditions. *Transport in Porous Media* **5**, 299–323. [11.8]

Cserepes, L. and Lenkey, L. 2004 Forms of hydrothermal and hydraulic flow in a homogeneous unconfined aquifer. *Geophys. J. Inter.* **158**, 785–797. [7.8]

Cui, C., Huang, X. Y. and Liu, C. Y. 2001 Forced convection in a porous channel with discrete heat sources. *ASME J. Heat Transfer* **123**, 404–407. [4.11]

Daniels, P. G. and Punpocha, M. 2004 Cavity flow in a porous medium driven by differential heating. *Int. J. Heat Mass Transfer* **47**, 3013–3030. [7.3.7]

Daniels, P. G. and Punpocha, M. 2005 On the boundary-layer structure of cavity flow in a porous medium driven by differential heating. *J. Fluid Mech.* **532**, 321–344. [7.3.7]

Daniels, P. G. and Simpkins, P. G. 1984 The flow induced by a heated vertical wall in a porous medium. *Q. J. Mech. Appl. Math.* **37**, 339–354. [5.12.2]

Daniels, P. G., Blythe, P. A. and Simpkins, P. G. 1982 Thermally driven cavity flows in porous media. II. The horizontal boundary layer structure. *Proc. Roy. Soc. London Ser. A* **382**, 135–154. [7.1.1]

Daniels, P. G., Blythe, P. A. and Simpkins, P. G. 1986 Thermally driven shallow cavity flows in porous media: the intermediate regime. *Proc. Roy. Soc. London Ser. A* **406**, 263–285. [7.1.3]

Daniels, P. G., Simpkins, P. G. and Blythe, P. A. 1989 Thermally driven shallow cavity flows in porous media: the merged layer regime. *Proc. Roy. Soc. London Ser. A* **426**, 107–124. [7.1.3]

Darcy, H. P. G. 1856 *Les Fontaines Publiques de la Ville de Dijon*. Victor Dalmont, Paris. [1.4.1]

Darnault, C. J. G., Steenhuis, T. S., Parlange, J. Y., Baveye, P. and Montemagno, C. 2004 Flow and transport phenomena in porous media: visualization, measurements and modeling techniques. *Emerging Technologies and Techniques in Porous Media* (D. B. Ingham, A. Bejan, E. Mamut and I. Pop, eds.), Kluwer Academic, Dordrect, 97–105.

Das, S. and Morsi, Y. 2003 Natural convection in domed porous enclosures: non-Darcian flow. *J. Porous Media* **6**, 159–175. [7.3.7]

Das, S. and Morsi, Y. S. 2005 A non-Darcian numerical modeling in domed enclosures filled with heat-generating porous media. *Numer. Heat Transfer A* **48**, 149–164. [7.3.7]

Das, S. and Sahoo, R. K. 1999 Effect of Darcy, fluid Rayleigh and heat generation parameters on natural convection in a square enclosure: a Brinkman-extended Darcy model. *Int. Comm. Heat Mass Transfer* **26**, 569–578. [7.3.8]

Das, S., Sahoo, R. K. and Morsi, Y. S. 2003 Natural convection in heat generating porous enclosures: a non-Darcian model. *Canad. J. Chem. Engng.* **81**, 289–296. [6.17]

da Silva Miranda, B. M. and Anand, N. K. 2004 Convective heat transfer in a channel with porous baffles. *Numer. Heat Transfer A* **46**, 425–452, [4.11]

da Silva, A. K. and Bejan, A. 2005 Constructal multi-scale structure for maximal heat transfer density in natural convection. *Int. J. Heat Fluid Flow* **26**, 34–44. [4.19]

Daurelle, J. V. , Topin, F. and Occelli, R. 1998 Modeling of coupled heat and mass transfers with phase change in a porous medium; application to superheated steam drying. *Numer. Heat Transfer A* **33**, 39–63. [3.6]

David, E., Lauriat, G. and Cheng, P. 1988 Natural convection in rectangular cavities filled with variable porosity media. *ASME HTD* **96**, Vol. 1, 605–612. [7.6.2]

David, E., Lauriat, G. and Cheng, P. 1991 A numerical solution of variable porosity effects on natural convection in a packed-sphere cavity. *ASME J. Heat Transfer* **113**, 391–399. [7.6.2]

David, E., Lauriat, G. and Prasad, V. 1989 Non-Darcy natural convection in packed-sphere beds between concentric vertical cylinders. *AIChE Sympos. Ser.* **269**, 90–95. [7.6.2]

Davis, A.M.J. and James, D.F. 1996 Slow flow through a model porous medium. *Int. J. Multiphase Flow* **22**, 969–989. [1.4.2]

Davis, S. H., Rosenblat, S., Wood, J. R. and Hewitt, T. A. 1985 Convective fluid flow and diagenetic patterns in domed sheets. *Amer. J. Sci.* **285**, 207–223. [11.5]

Dawood, A. S. and Burns, P. J. 1992 Steady three-dimensional convective heat transfer in a porous box via multigrid. *Numer. Heat Transfer A* **22**, 167–198. [7.2]

Dayan, A., Flesh, J. and Saltiel, C. 2004 Drying of a porous spherical rock for compressed air energy storage. *Int. J. Heat Mass Transfer* **47**, 4459–4468. [3.6]

Debeda, V., Caltagirone, J. P. and Watremez, P. 1995 Local multigrid refinement method for natural convection in fissured porous media. *Numer. Heat Transfer* **28**, 455–467. [6.13.4]
Degan, G. and Vasseur, P. 1995 The non-Darcy regime for boundary layer natural convection from a point source of heat in a porous medium. *Int. Comm. Heat Mass Transfer* **22**, 381–390. [5.11.1]
Degan, G. and Vasseur, P. 1996 Natural convection in a vertical slot filled with an anisotropic porous medium with oblique principal axes. *Numer. Heat Transfer A* **30**, 397–412. [7.3.2]
Degan, G. and Vasseur, P. 1997 Boundary-layer regime in a vertical porous layer with anisotropic permeability and boundary effects. *Int. J. Heat Fluid Flow* **18**, 334–343. [7.3.2]
Degan, G. and Vasseur, P. 2002 Aiding mixed convection through a vertical anisotropic porous channel with oblique axes. *Int. J. Engng. Sci.* **40**, 193–209. [8.3.1]
Degan, G. and Vasseur, P. 2003 Influence of anisotropy on convection in porous media with nonuniform thermal gradient. *Int. J. Heat Mass Transfer* **46**, 781–789. [6.12]
Degan, G., Beji, H., and Vasseur, P. 1998b Natural convection in a rectangular cavity filled with an anisotropic porous medium. *Heat Transfer 1998, Proc. 11th IHTC*, **4**, 441–446. [7.3.2]
Degan, G., Beji, H., Vasseur, P. and Robillard, L. 1998a Effect of anisotropy on the development of convective boundary layer flow in porous media. *Int. Comm. Heat Mass Transfer* **25**, 1159–1168. [7.3.2]
Degan, G., Vasseur, P. and Bilgen, E. 1995 Convective heat transfer in a vertical anisotropic porous layer. *Int. J. Heat Mass Transfer* **38**, 1975–1987. [7.3.2]
Degan, G., Zohoun, S. and Vasseur, P. 2002 Forced convection in horizontal porous channels with hydrodynamic anisotropy. *Int. J. Heat Mass Transfer* **45**, 3181–3188. [4.16.2]
Degan, G., Zohoun, S. and Vasseur, P. 2003 Buoyant plume above a line source of heat in an anisotropic porous medium. *Heat Mass Transfer* **39**, 209–213. [5.10.1.l]
Deibler, J. A. and Bortolozzi, R. A. 1998 A two-field model for natural convection in a porous annulus at high Rayleigh numbers. *Chem. Engng. Sci.* **53**, 1505–1516. [7.3.3]
Delache, A., Ouarzazi, N. and Néel, M. C. 2002 Pattern formation of mixed convection in a porous medium confined laterally and heated from below: effect of inertia. *Comptes Rendus Mécanique* **330**, 885–891. [6.10]
De la Torre Juarez, M. and Busse, F. H. 1995 Stability of two-dimensional convection in a fluid-saturated porous medium. *J. Fluid Mech.* **292**, 305–323. [6.8]
de Lemos, M. J. S. 2004 Turbulent heat and mass transfer in porous media. In *Technologies and Techniques in Porous Media* (D. B. Ingham, A. Bejan, E. Mamut and I. Pop, eds), Kluwer Academic, Dordrecht, pp. 157–168. [1.8].
de Lemos, M. J. S. 2005a Turbulent kinetic energy distribution across the interface between a porous medium and a clear region. *Int. Comm. Heat Mass Transfer* **32**, 107–115. [1.8]
de Lemos, M. J. S. 2005b Mathematical modeling and applications of turbulent heat and mass transfer in porous media. *Handbook of Porous Media* (ed. K. Vafai), 2nd ed., Taylor and Francis, New York, pp. 409–454. [1.8]
de Lemos, M. J. S. 2005c The double-decomposition concept for turbulent transport in porous media. In *Transport Phenomena in Porous Media III*, (eds. D. B. Ingham and I. Pop), Elsevier, Oxford, pp. 1–33. [1.8]
de Lemos, M. J. S. and Braga, E. J. 2003 Modeling of turbulent natural convection in porous media. *Int. Comm. Heat Mass Transfer* **30**, 615–624. [1.8]

de Lemos, M. J. S. and Mesquita, M. S. 2003 Modeling of turbulent natural convection in porous media. *Int. Comm. Heat Mass Transfer* **30**, 105–113. [1.8]

de Lemos, M. J. S. and Pedras, M. H. J. 2000 On the definitions of turbulent kinetic energy for flow in porous media. *Int. Comm. Heat Mass Transfer* **27**, 211–220. [1.8]

de Lemos, M. J. S. and Pedras, M. H. J. 2001 Recent mathematical models for turbulent flow in saturated rigid porous media. *ASME J. Fluids Engng.* **123**, 935–940. [1.8]

de Lemos, M. J. S. and Rocamora, F. D. 2002 Turbulent transport modeling for heat flow in rigid porous media. *Heat Transfer 2002, Proc. 12th Int. Heat Transfer Conf.*, Elsevier, Vol. 2, pp. 791–796. [1.8]

de Lemos, M. J. S. and Tofaneli, L. A. 2004 Modelling of double-diffusive turbulent natural convection in porous media. *Int. J. Heat Mass Transfer* **47**, 4233–4241. [1.8]

Delmas, A. and Arquis, E. 1995 Early initiation of natural convection in an open porous layer due to the presence of solid conductive inclusions. *ASME J. Heat Transfer* **117**, 733–739. [6.13.2]

Demirel, Y. and Kahraman, R. 1999 Entropy generation in a rectangular packed duct with wall heat flux. *Int. J. Heat Mass Transfer* **42**, 2337–2344. [4.9]

Demirel, Y. and Kahraman, R. 2000 Thermodynamic analysis of convective heat transfer in an annular packed bed. *Int. J. Heat Fluid Flow* **21**, 442–448. [4.16.2]

Demirel, Y., Abu-Al-Saud, B. A., Al-Ali, H. H. and Makkawi, Y. 1999 Packing size and shape effects on forced convection in large rectangular packed ducts with asymmetrical heating. *Int. J. Heat Mass Transfer* **42**, 3267–3277. [4.16.2]

Demirel, Y., Sharma, R. N. and Al-Ali, H. H. 2000 On the effective heat transfer parameters in a packed bed. *Int. J. Heat Mass Transfer* **43**, 327–332. [4.16.2]

Desaive, T., Hennenberg, M. and Lebon, G. 2002 Thermal instability of a rotating saturated porous medium heated from below and submitted to rotation. *Eur. Phys. J.* B **29**, 641–647. [6.22]

Desaive, T., Lebon, G. and Hennenberg, M. 2001 Coupled capillary and gravity-driven instability in a liquid film overlying a porous layer. *Phys. Rev. E.* **64**, art. no. 066304 Part 2. [6.19.3]

Desrayaud, G. and Lauriat, G. 1991 Thermal convection around a line heat source buried in a porous medium. *ASME HTD* **172**, 17–24. [7.11]

Dessaux, A. 1998 Analytical and numerical solutions to a problem of convection in a porous media with lateral mass flux. *Int. Comm. Heat Mass Transfer* **25**, 641–650. [5.1.9.9]

Devi, S.N.S., Nagaraju, P. and Hanumanthappa, A. R. 2002 Effect of radiation on Rayleigh-Bénard convection in an anisotropic porous medium. *Indian J. Engng. Mater. Sci.* **9**, 163–171. [6.12]

Dhanasekharan, M. R., Das, S. K. and Venkateshan, S. P. 2002 Natural convection in a cylindrical enclosure filled with heat generating anisotropic porous medium. *ASME J. Heat Transfer* **124**, 203–207. [6.16.1]

Dharma Rao, V., Naidu, S. V. and Sarma, P. K. 1996 Non-Darcy effects in natural convection heat transfer in a vertical porous annulus. *ASME J. Heat Transfer* **118**, 502–505. [7.3.3]

Dhir, V. K. 1994 Boiling and two-phase flow in porous media. *Ann. Rev. Heat Transfer* **5**, 303–350. [10.3.1]

Dhir, V. K. 1997 Heat transfer from heat-generating pools and particulate beds. *Adv. Heat Transfer* **29**, 1–57. [10.3.1]

Dickey, J. T. and Peterson, G. P. 1997 High heat flux absorption utilizing porous material with two-phase heat transfer. *ASME J. Energy Res. Tech.* **119**, 171–179. [3.6]

Diersch, H. J. G. 2000 Note to the opposing flow regime at mixed convection around a heated cylinder in a porous medium. *Transport in Porous Media* **38**, 345–352. [8.1.3]

Diersch, H. J. G. and Kolditz, O. 2002 Variable-density flow and transport in porous media: approaches and challenges. *Adv. Water Res.* **25**, 899–944. [9]

Dixon, A. G. and Cresswell, D. L. 1979 Theoretical predictions of effective heat transfer mechanisms in regular shaped packed beds. *AIChE Journal* **25**, 663–676. [2.2.2]

Doering, C. R. and Constantin, P. 1998 Bounds for heat transport in a porous layer. *J. Fluid Mech.* **376**, 263–296. [6.3]

Dona, C. L. G. and Stewart, W. E. 1989 Variable property effects on convection in a heat generating porous medium. *ASME J. Heat Transfer* **111**, 1100–1102. [6.17]

Donaldson, I. G. 1962 Temperature gradients in the upper layers of the Earth's crust due to convective water flows. *J. Geophys. Res.* **67**, 3449–3459. [6.13.1]

Donaldson, I. G. 1982 Heat and mass circulation in geothermal systems. *Ann. Rev. Earth Planet Sci.* **10**, 377–395. [11]

Doughty, C. and Pruess, K. 1990 A similarity solution for the two-phase fluid and heat flow near high-level nuclear waste packages emplaced in porous media. *Int. J. Heat Mass Transfer* **33**, 1205–1222. [11.9.1]

Doughty, C. and Pruess, K. 1992 A similarity solution for two-phase water, air and heat flow near a linear heat source in a porous medium. *J. Geophys. Res*, **97**, 1821–1838. [11.9.1]

du Plessis, J. P. 1994 Analytical quantification of coefficients in the Ergun equation for fluid friction in a packed bed. *Transport in Porous Media* **16**, 189–207. [1.5.2]

Du, J. H. and Wang, B.X. 1999 Mixed convection in porous media between vertical concentric cylinders. *Heat Transfer Asian Res.* **28**, 95–101. [8.3.3]

Du, Z. G. and Bilgen, E. 1990 Natural convection in vertical cavities with partially filled heat-generating porous media. *Numer. Heat Transfer A* **18**, 371–386. [7.7]

Du, Z. G. and Bilgen, E. 1992 Natural convection in vertical cavities with internal heat generating porous medium. *Wärme-Stoffübertrag.* **27**, 149–155. [7.3.8]

Dufour, F. and Néel, M. C. 1998 Numerical study of instability in a horizontal porous channel with bottom heating and forced horizontal flow. *Phys. Fluids* **10**, 2198–2207. [6.10]

Dufour, F. and Néel, M. C. 2000 Time-periodic convection patterns in a horizontal porous layer with through-flow. *Quart. Appl. Math.* **58**, 265–281. [6.10]

Dullien, F. A. L. 1992 *Porous Media: Fluid Transport and Pore Structure*, Academic, New York, 2nd Edit. [1.4.2]

Dunn, J. C. and Hardee, H. C. 1981 Superconvecting geothermal zones. *J. Volcanol. Geotherm. Res.* **11**, 189–201. [11.8]

Dupuit, A. J. E. J. 1863 *Études Théoriques et Pratiques sur le Mouvement des aux dans les Canaux Découverts et a Travers les Terrains Perméables.* Victor Dalmont, Paris. [1.5.2]

Durlofsky, L. and Brady, J. F. 1987 Analysis of the Brinkman equation as a model for flow in porous media. *Phys. Fluids* **30**, 3329–3341. [1.5.3]

Dutta, P. and Seetharamu, K. N. 1993 Free convection in a saturated porous medium adjacent to a vertical impermeable wall subjected to a non-uniform heat flux. *Wärme-Stoffübertrag.* **28**, 27–32. [5.1.9]

Duval, F., Fichet, F. and Quintard, M. 2004 A local thermal non-equilibrium model for two-phase flows with phase change in porous media. *Int. J. Heat Mass Transfer* **47**, 613–639. [11.9.3]

Duwairi, H. M., Aldoss, T. K. and Jarrah, M. A. 1997 Nonsimilarity solutions for non-Darcy mixed convection from horizontal surfaces in a porous medium. *Heat Mass Transfer* **33**, 149–156. [8.1.2]

Dwiek, Y. J., Rabadi, N. J. and Ghazzawi, N. A. 1994 Effect of lateral mass flux on free convection from an inclined plate embedded in a saturated porous medium. *Arab. J. Sci. Engng.* **14**, 449–460. [5.3]

Dybbs, A. and Edwards, R. V. 1984 A new look at porous media fluid mechanics–Darcy to turbulent. *Fundamentals of Transport Phenomena in Porous Media* (eds. J. Bear and V. Carapcioglu), Martinus Nijhoff, The Netherlands, 199–256. [7.6.1]

Easterday, O. T., Wang, C. Y. and Cheng, P. 1995 A numerical and experimental study of two-phase flow and heat transfer in a porous formation with localized heating from below. *ASME HTD*-**321**, 723–732. [10.3.2]

Ebinuma, C. D. and Nakayama, A. 1990a Non-Darcy transient and steady film condensation in a porous medium. *Int. Comm. Heat Mass Transfer* **17**, 49–58. [10.4]

Ebinuma, C. D. and Nakayama, A. 1990b An exact solution for transient film condensation in a porous medium along a vertical surface with lateral mass flux. *Int. Comm. Heat Mass Transfer* **17**, 105–111. [4.6.2, 10.4]

Ebinuma, C. D. and Nakayama, A. 1997 Approximate solution for non-Darcy transient film condensation in a porous medium. *J. Brazilian Soc. Mech. Sci.* **19**, 496–503. [10.4]

Echaniz, H. L. 1984 Oscillatory convection with boiling in a water-saturated porous medium. MS thesis, Cornell University. [10.3.1]

Egorov, S. D. and Poleshaev, V. I. 1993 Thermal convection in anisotropic porous insulation. *Heat Transfer Res.* **25**, 968–990. [7.3.2]

Ekholm, T. C. 1983 Studies of convection using a Hele-Shaw cell. Project Report, School of Engineering, University of Auckland. [6.13.2]

El-Amin, M. F. 2003a Combined effect of magnetic field and viscous dissipation on a power-law fluid over plate with variable surface heat flux embedded in a porous medium. *J. Magn. Magn, Mater.* **261**, 228–237. [5.1.9.2]

El-Amin, M. F. 2003b Combined effect of viscous dissipation and Joule heating on MHD forced convection over a non-isothermal horizontal cylinder embedded in a fluid saturated porous medium.. *J. Magn. Magn. Mater.* **263**, 337–343. [4.3]

El-Amin, M. F. 2004a Double dispersion effects on natural convection heat and mass transfer in non-Darcy porous medium. *Appl. Math. Comput.* **156**, 1–17. [9.2.1]

El-Amin, M. F. 2004b Non-Darcy free convection from a vertical plate with time-periodic surface oscillations. *J. Porous Media* **7**, 331–338. [5.1.9.7]

El-Amin, M. F., El-Hakiem, M. A. and Mansour, M. A. 2003 Effects of viscous dissipation on a power-law fluid over plate embedded in a porous medium. *Heat Mass Transfer* **39**, 807–813. [5.1.9.2]

El-Amin, M. F., El-Hakiem, M. A. and Mansour, M. A. 2004 Combined effect of magnetic field and lateral mass transfer on non-Darcy axisymmetric free convection in a power-law fluid saturated porous medium. *J. Porous Media* **7**, 65–71. [5.2]

Elbashbeshy, E. M. A. 2001 Laminar mixed convection over horizontal flat plate embedded in a non-Darcian porous medium with suction and injection. *Appl. Math. Comput.* **121**, 123–128. [8.1.2]

Elbashbeshy, E. M. A. 2003 The mixed convection along a vertical plate embedded in a non-Darcian porous medium with suction and injection. *Appl. Math. Comput.* **136**, 139–149. [8.1.1]

Elbashbeshy, E. M. A. and Bazid, M. A. 2000a Heat transfer over a continuously moving plate embedded in a non-Darcian porous medium. *Int. J. Heat Mass Transfer* **43**, 3087–3092. [8.1.4]

Elbashbeshy, E. M. A. and Bazid, M. A. 2000b The mixed convection along a vertical plate with variable heat flux embedded in porous medium. *Appl. Math. Comput.* **125**, 317–324. [8.1.1]

Elder, J. W. 1967a Steady free convection in a porous medium heated from below. *J. Fluid Mech.* **27**, 29–48. [2.5, 6.9.1, 6.18]

Elder, J. W. 1967b Transient convection in a porous medium. *J. Fluid Mech.* **27**, 609–623. [6.9.1, 6.18]

El-Hakiem, M. A. 2000 MHD oscillatory flow on free convection-radiation through a porous medium with constant suction velocity. *J. Magn. Magn. Mater.* **220**, 271–276. [5.1.9.4]

El-Hakiem, M. A. 2001a Thermal dispersion effects on combined convection in non-Newtonian fluids along a non-isothermal vertical plate in a porous medium. *Transport in Porous Media* **45**, 29–40. [8.1.1]

El-Hakiem, M. A. 2001b Combined convection in non-Newtonian fluids along a nonisothermal vertical plate in a porous medium with lateral mass flux. *Heat Mass Transfer* **37**, 379–385. [8.1.1]

El-Hakiem, M. A. and El-Amin, M. F. 2001a Thermal radiation effect on non-Darcy natural convection with lateral mass transfer. *Heat Mass Transfer* **37**, 161–165. [5.1.9.4]

El-Hakiem, M. A. and El-Amin, M. F. 2001b Mass transfer effects on the non-Newtonian fluids past a vertical plate embedded in a porous medium with non-uniform surface heat flux. *Heat Mass Transfer* **37**, 293–297. [5.1.9.2]

El-Khatib, G. and Prasad, V. 1987 Effects of stratification on thermal convection in horizontal porous layers with localized heating from below. *ASME J. Heat Transfer* **109**, 683–687. [6.18]

Ellinger, E. A. and Beckerman, C. 1991 On the effect of porous layers on melting heat transfer in an enclosure. *Exp. Therm. Fluid Sci.* **4**, 619–629. [10.1.7]

El-Shaarawi, M. A. I., Al-Nimr, M. A. and Al Yah, M. M. K. 1999 Transient conjugate heat transfer in a porous medium in concentric annuli. *Int. J. Numer. Meth. Heat Fluid Flow* **9**, 444–460. [4.6.4]

Emms, P. W. 1998 Freckle formation in a solidifying binary alloy. *J. Engng. Math.* **33**, 175–200. [10.2.3]

Emms, P. W. and Fowler, A. C. 1994 Compositional convection in the solidification of binary alloys. *J. Fluid Mech.* **262**, 111–139. [10.2.3]

Ene, H. I. 1991 Effects of anisotropy on the free convection from a vertical plate embedded in a porous medium. *Transport in Porous Media* **6**, 183–194. [5.1.9.5]

Ene, H. I. 2004 Modeling the flow through porous media. In *Emerging Technologies and Techniques in Porous Media* (D. B. Ingham, A. Bejan, E. Mamut and I. Pop, eds), Kluwer Academic, Dordrecht, pp. 25–41. [1.4.3]

Ene, H. I. and Poliševski, D. 1987 *Thermal Flow in Porous Media*, Reidel, Dordrecht. [1.4.3, 5.6.2, 5.11.2]

Ene, H. J. and Sanchez-Palencia. E. 1982 On thermal equation for flow in porous media. *Int. J. Engng. Sci.* **20**, 623–630. [2.2.2]

Ennis-King, J. and Paterson, L. 2005 Role of convective mixing in the long-term storage of carbon dioxide in deep saline formations. *SPR J.* **10**, 349–356. [6.11.3]

Ennis-King, J., Preston, I. and Paterson, J. 2005 Onset of convection in anisotropic porous media subject to a rapid change of boundary conditions. *Phys. Fluids* **17**, Art. No. 084107. [6.11.3]

Epherre, J. F. 1975 Criterion for the appearance of natural convection in an anisotropic porous layer. *Rev. Gén. Therm.* **168**, 949–950. English translation, *Int. Chem. Engng.* **17**, 615–616, 1977. [6.12]

Ergun, S. 1952 Fluid flow through packed columns. *Chem. Engrg. Prog.* **48**, 89–94. [1.5.2, 6.9.2]

Er-Raki, M., Hasnaoui, M., Amahmid, A. and Bourich, M. 2005 Soret driven thermosolutal convection in a shallow porous layer with a stress-free upper surface. *Engng. Comput.* **22**, 186–205. [9.1.4]

Essome, G. R. and Orozco, J. 1991 An analysis of film boiling on a binary mixture in a porous medium. *Int. J. Heat Mass Transfer* **34**, 757–766. [10.3.2]

Estebe, J. and Schott, J. 1970 Concentration saline et cristallisation dans un milieu poreux par effet thermogravitationnel. *C. R. Acad. Sci. Paris* **271**, 805–807. [9.1.6.4]

Ettefagh, J., Vafai, K. and Kim, S. J. 1991 Non-Darcian effects in open ended cavities filled with a porous medium. *ASME J. Heat Transfer* **113**, 747–756. [5.12.3]

Evans, D. G. and Nunn, J. A. 1989 Free thermohaline convection in sediments surrounding a salt column. *J. Geophys. Res.* **94**, 12413–12422. [11.8]

Evans, G. H. and Plumb, O. A. 1978 Natural convection from a vertical isothermal surface imbedded in a saturated porous medium. *AIAA-ASME Thermophysics and Heat Transfer Conf.*, Paper 78-HT -55, Palo Alto, California. [5.1.7.1, 5.1.8]

Facas, G. N. 1994 Reducing the heat transfer from a hot pipe buried in a semi-infinite, saturated, porous medium. *ASME J. Heat Transfer*, **116**, 473–476. [7.11]

Facas, G. N. 1995a Natural convection from a buried pipe with external baffles. *Numer. Heat Transfer A*, **27**, 595–609. [7.11]

Facas, G. N. 1995b Natural convection from a buried elliptic heat source. *Int. J. Fluid Flow* **16**, 519–526. [7.11]

Facas, G. N. and Farouk, B. 1983 Transient and steady-state natural convection in a porous medium between two concentric cylinders. *ASME J. Heat Transfer* **105**, 660–663. [7.3.3]

Fand, R. M. and Phan, R. T. 1987 Combined forced and natural convection heat transfer from a horizontal cylinder embedded in a porous medium. *Int. J. Heat Mass Transfer* **30**, 1351–1358. [8.1.3]

Fand, R. M. and Yamamoto, L. H. 1990 Heat transfer by natural convection from a horizontal cylinder embedded in a porous medium: the wall effect. *Heat Transfer 1990*, Hemisphere, New York, **5**, 183–188. [5.5.2]

Fand, R. M., Steinberger, T. E. and Cheng, P. 1986 Natural convection heat transfer from a horizontal cylinder embedded in a porous medium. *Int. J. Heat Mass Transfer* **29**, 119–133. [5.5.2]

Fand, R. M., Varahasamy, M. and Greer, L. S. 1993 Empirical correlation equations for heat transfer by forced convection from cylinders embedded in porous media that account for the wall effect and dispersion. *Int. J. Heat Mass Transfer* **36**, 4407–4418. [4.8]

Fand, R. M., Varahasamy, M. and Yamamoto, L. M. 1994 Heat transfer by natural convection from horizontal cylinders embedded in porous media whose matrices are composed of spheres: viscous dissipation. *Heat Transfer 1994*, Inst. Chem. Engrs., Rugby, vol. 5, pp. 237–242. [5.5.1]

Farouk, B. and Shayer, H. 1988 Natural convection around a heated cylinder in a saturated porous medium. *ASME J. Heat Transfer* **110**, 642–648. [7.11]

Faruque D, Saghir, M. Z., Chacha, M. and Ghorayeb, K. 2004 Compositional variation considering diffusion and convection for a binary mixture in a porous medium. *J. Porous Media* **7**, 73–91. [9.1.4]

Farr, W. W., Gabito, J. F., Luss, D. and Balakotaiah, V. 1991 Reaction-driven convection in a porous medium: Part 1. Linear stability analysis. *AIChE J.* **37**, 963–975. [3.4]

Felicelli, S. D., Heinrich, J. C. and Poirier, D. R. 1991 Simulation of freckles during vertical solidification of binary alloys. *Metall. Trans. B* **22**, 847–859. [10.2.3]

Feltham, D. L and Worster, M. G. 1999 Flow-induced morphological instability of a mushy layer. *J. Fluid Mech.* **391**, 337–357. [10.2.3]

Feng, Z. G. and Michaelides, E. E. 1999 Unsteady mass transport from a sphere immersed in a porous medium at finite Peclet numbers. *Int. J. Heat Mass Transfer* **42**, 535–546. [8.1.3]

Fernandez, R. T. and Schrock, V. E. 1982 Natural convection from cylinders embedded in a liquid-saturated porous medium. *Heat Transfer 1982*, Elsevier, Amsterdam, **2**, 335–340. [7.11].

Figueiredo, J. R. and Llagostera, T. 1999 Comparative study of the unified finite approach exponential-type scheme (UNIFAES) and its application to natural convection in a porous cavity. *Numer. Heat Transfer B* **35**, 347–367. [6.8]

Figus, C., Le Bray, Y., Bories, S. and Prat, M. 1998 Heat transfer in porous media considering phase change, capillarity and gravity. Application to capillary evaporator. *Heat Transfer 1998, Proc. 11th IHTC*, **4**, 393–398. [3.6]

Firdaouss, M., Gurmond, J. L. and Le Quéré, P. 1997 Nonlinear corrections to Darcy's law at low Reynolds Number. *J. Fluid Mech.* **343**, 331–350. [1.5.2]

Flavin, J. N. and Rionero, S. 1999 Nonlinear stability for a thermofluid in a vertical porous slab. *Cont. Mech. Thermodyn.* **11**, 173–179. [7.1.4]

Flick, D., Leslous, A. and Alvarez, G. 2003 Semi-empirical modeling of turbulent fluid flow and heat transfer in porous media. *Int. J. Refrig.* **26**, 349–359. [1.8]

Fomin, S., Shimizu, A. and Hashida, T. 2002 Mathematical modeling of convection heat transfer in a geothermal reservoir of fractal geometry. *Heat Transfer 2002, Proc. 12th Int. Heat Transfer Conf.*, Elsevier, Vol. 2, pp. 809–814. [2.6]

Fontaine, F. J., Rabinowicz, M. and Boulegue J. 2001 Permeability changes due to mineral diagenesis in fractured crust: implications for hydrothermal circulation at mid-ocean ridges. *Earth Planet. Sci. Lett.* **184**, 407–425. [11.5]

Forchheimer, P. 1901 Wasserbewegung durch Boden. *Zeitschrift des Vereines Deutscher Ingenieure* **45**, 1736–1741 and 1781–1788. [1.5.2]

Fourar, M., Lenormand, R., Karimi-Fard, M. and Horne, R. 2005 Inertia effects in high-rate flow through heterogeneous porous media. *Transport Porous Media* **60**, 353–370. [1.5.2]

Fourie, J. G. and Du Plessis, J. P. 2003a A two-equation model for heat conduction in porous media. (I. Theory) *Transport Porous Media* **53**, 145–161. [2.2.2]

Fourie, J. G. and Du Plessis, J. P. 2003b A two-equation model for heat conduction in porous media. (II. Application) *Transport Porous Media* **53**, 163–174. [2.2.2]

Fowler, A. C. 1985 The formation of freckles in binary alloys. *IMA J. Appl. Maths.* **35**, 159–174. [10.2.3]

Fowler, A. J. and Bejan, A. 1994 Forced convection in banks of inclined cylinders at low Reynolds numbers. *Int. J. Heat Fluid Flow* **15**, 90–99. [4.14]

Fowler, A. J. and Bejan, A. 1995 Forced convection from a surface covered with flexible fibers. *Int. J. Heat Mass Transfer* **38**, 767–777. [1.9, 4.14]

Francis, N. D. and Wepfer, J. W. 1996 Jet impingement drying of a moist porous solid. *Int. J. Heat Mass Transfer* **35**, 469–480. [3.6]

Frei, K. M., Cameron, D. and Stuart, P. R. 2004 Novel drying process using forced aeration through a porous biomass matrix. *Drying Tech.* **22**, 1191–1215. [3.6]

Friedrich, R. 1983 Einfluss der Prandtl-Zahl auf die Zellularkonvektion in einem rotierenden, mit Fluid gesättigten porösen Medium. *ZAMM* **63**, T246–T249. [6.22]

Frizon, F., Lorente, S., Ollivier, J. P. and Thouvenot, P. 2003 Transport model for the nuclear decontamination of cementitious materials. *Comput. Materials Sci.* **27**, 507–516. [3.7]

Fu, H. L., Leong, K. C., Huang, X. Y. and Liu, C. Y. 2001 An experimental study of heat transfer of a porous channel subjected to oscillating flow. *ASME J. Heat Transfer* **123**, 162–170 (erratum p. 1194). [4.6.4]

Fu, W. S. and Chen, S. F. 2002 A numerical study of heat transfer of a porous block with the random porosity model in a channel flow. *Heat Mass Transfer* **38**, 695–704. [4.16.1]

Fu, W. S. and Huang, H. C. 1999 Effects of random porosity model on heat transfer performance of porous media. *Int. J. Heat Mass Transfer* **42**, 13–25. [1.7]

Fu, W. S. and Ke, W. W. 2000 Effects of random porosity model on double-diffusive natural convection in a porous medium enclosure. *Int. Comm. Heat Mass Transfer* **27**, 119–132. [9.2.2]

Fu, W. S., Huang, H. C. and Liou, W. Y. 1996 Thermal enhancement in laminar channel flow with a porous block. *Int. J. Heat Mass Transfer* **39**, 2165–2175. [4.11]

Fu, W. S., Wang, K. N. and Ke, W. W 2001 Heat transfer of porous medium with random porosity model in a laminar channel flow. *J. Chinese Inst. Engrs.* **24**, 431–438. [4.16.2]

Fu, X., Viskanta, R. and Gore, J. P. 1998 Prediction of effective thermal conductivity for cellular ceramics. *Int. Comm. Heat Mass Transfer* **25**, 151–161. [2.2.1]

Fujii, Y., Ohita, K. and Hijikata, A. 1994 Unsteady heat transfer around a periodically-heated cylinder embedded in saturated porous media. *Heat Transfer 1994*, Inst. Chem. Engrs, Rugby, Vol. 5, 249–254. [4.8]

Gabito, J. F. and Balakotaiah, V. 1991 Reaction-driven convection in a porous medium: Part 2. Numerical bifurcation analysis. *AIChE J.* **37**, 976–985. [3.4]

Galdi, G. P., Payne, L. E., Proctor, M. R. E. and Straughan, B. 1987 Convection in thawing subsea permafrost. *Proc. Roy. Soc. London Ser. A* **414**, 83–102. [11.3]

Ganapathy, R. 1992 Thermal convection in an infinite porous medium due to a source in a sphere. *Fluid Dyn. Res.* **9**, 223–234. [5.11.2]

Ganapathy, R. 1994a Free convective heat and mass transfer flow induced by an instantaneous point source in an infinite porous medium. *Fluid Dyn. Res.* **14**, 313–329. [5.11.2]

Ganapathy, R. 1994b Free convection flow induced by a line source in a sparsely packed porous medium. *Adv. Water Resources* **17**, 251–258, Corrigendum **19**, 255–257. [9.3.2]

Ganapathy, R. 1997 Thermal convection in an infinite porous medium induced by a heated sphere. *ASME J. Heat Transfer* **119**, 647–650. [5.6.2]

Ganapathy, R. and Purushothaman, R. 1990 Thermal convection from an instantaneous point heat source in a porous medium. *Int. J. Engng. Sci.* **28**, 907–918. [5.11.2]

Ganapathy, R. and Purushothaman, R. 1992 Free convection in a saturated porous medium due to a traveling thermal wave. *Z. Angew. Math. Mech.* **72**, 142–145. [6.11.3]

Ganesan, S. and Poirier, D. R. 1990 Conservation of mass and momentum for the flow of interdendritic liquid during solidification. *Mettal. Trans. B* **21**, 173–181. [10.2.3]

Gartling, D. K. and Hickox, C. E. 1985 Numerical study of the applicability of the Boussinesq approximation for a fluid-saturated porous medium. *Int. J. Numer. Methods Fluids* **5**, 995–1013. [6.7]

Gartling, D. K., Hickox, C. E. and Givler, R. C. 1996 Simulation of coupled viscous and porous flow problems. *Comp. Fluid Dyn.* **7**, 23–48. [1.6]

Gasser, R. D. and Kazimi, M. S. 1976 Onset of convection in a porous medium with internal heat generation. *ASME J. Heat Transfer* **98**, 49–54. [6.11.2]

Gatica, J. E., Viljoen, H. J., and Hlavacek, V. 1989 Interaction between chemical reaction and natural convection in porous media. *Chem. Eng. Sci.* **44**, 1853–1870. [3.4]

Geiger, S., Driesner, T., Heinrich, C. A. and Matthäi, S. K. 2005 On the dynamics of NaCl-H_2O fluid convection in the Earths's crust. *J. Geophys. Res.* **110**, Art. No. B07101. [11.8]

George, J. H., Gunn, R. D., and Straughan, B. 1989 Patterned ground formation and penetrative convection in porous media. *Geophys. Astrophys. Fluid Dyn.* **46**, 135–158. [11.2]

Georgiadis, J. G. 1991 Effect of randomness on heat and mass transfer in porous media. *Convective Heat and Mass Transfer in Porous Media* (eds. S. Kakaç, *et al.*), Kluwer Academic, Dordrecht, 499–524. [1.1]

Georgiadis, J. G. and Catton, I. 1985 Free convective motion in an infinite vertical porous slot: the non-Darcian regime. *Int. J. Heat Mass Transfer* **28**, 2389–2392. [7.1.4]

Georgiadis, J. G. and Catton, I. 1986 Prandtl number effect on Bénard convection in porous media. *ASME J. Heat Transfer* **108**, 284–290. [6.9.2]

Georgiadis, J. G. and Catton, I. 1987 Stochastic modeling of unidirectional fluid transport in uniform and random packed beds. *Phys. Fluids* **30**, 1017–1022. [1.1, 1.7]

Georgiadis, J. G. and Catton, I. 1988 Dispersion in cellular thermal convection in porous media. *Int. J. Heat Mass Transfer* **31**, 1081–1091. [1.1, 6.6]

Gerritsen, M. G., Chen, T. and Chen, Q. 2005 Stanford University, private communication.

Gershuni, G. Z. and Lyubimov, D. V. 1998 *Thermal Vibration Convection*. Wiley, New York. [6.24]

Gershuni, G. Z., Zhukhovitskii, E, M. and Lyubimov, D. V. 1976 Thermo-concentration instability of a mixture in a porous medium. *Dokl. Akad, Nauk. SSSR* 229, 575-578 (English translation *Sov. Phys. Dokl.* **21**, 375–377.) [9.2.4]

Gershuni, G. Z., Zhukhovitskii, E. M. and Lyubimov, D. V. 1980 Stability of stationary convective flow of a mixture in a vertical porous layer. *Fluid Dynamics* **15**, 122–127. [9.2.4]

Getachew, D., Minkowycz, W. J. and Lage, J. L. 2000 A modified form of the κ-ε model for turbulent flows of an incompressible fluid in porous media. *Int. J. Heat Mass Transfer* **43**, 2909–2915. [1.8]

Getachew, D., Minkowycz, W. J. and Poulikakos, D. 1996 Natural convection in a porous cavity saturated with a non-Newtonian fluid. *J. Thermophys. Heat Transfer* **10**, 640–651. [7.1.6]

Getachew, D., Poulikakos, D. and Minkowycz, W. J. 1998 Double diffusion in porous cavity saturated with non-Newtonian fluid. *J. Thermophys. Heat Transfer* **12**, 437–446. [9.2.2]

Ghafir, R. and Lauriat, G. 2001 Forced convection heat transfer with evaporation in a heat generating porous medium. *J. Porous Media* **4**, 309–322. [10.4]

Ghorayeb, K. and Firoozabadi, A. 2000a Numerical study of natural convection and diffusion in fractured porous media. *SPE Journal* **5**, 12–20. [1.9]

Ghorayeb, K. and Firoozabadi, A. 2000b Modeling multicomponent diffusion and convection in porous media. *SPE Journal* **5**, 158–171. [1.9]

Ghorayeb, K. and Firoozabadi, A. 2001 Features of convection and diffusion in porous media for binary systems. *J. Canad. Petrol. Tech.* **40**, 21–28. [1.9]

Gibson, P. W. and Charmchi, M. 1997 Modeling convection/diffusion processes in porous textiles with inclusion of humidity-dependent air permeability. *Int. Comm. Heat Mass Transfer* **24**, 709–724. [3.6]

Gill, A. E. 1969 A proof that convection in a porous vertical slab is stable. *J. Fluid Mech.* **35**, 545–547. [7.1.4]

Gill, U. S. and Minkowycz, W. J. 1988 Boundary and inertia effects on conjugate mixed convection heat transfer from a vertical plate fin in a high-porosity porous medium. *Int. J. Heat Mass Transfer* **31**, 419–427. [5.12.1]

Gill, U. S., Minkowycz, W. J., Chen, C. K. and Chen, C. H. 1992 Boundary and inertia effects on conjugate mixed convection-conduction heat transfer from a vertical circular fin embedded in a porous medium. *Numer. Heat Transfer A* **21**, 423–441. [8.1.4]

Gilman, A. and Bear, J. 1996 The influence of free convection on soil salinization in arid regions. *Transport Porous Media* **23**, 275–301. [9.1.6.4, 11.8]

Giorgi, T. 1997 Derivation of the Forchheimer law via matched asymptotic expansions. *Transport Porous Media* **29**, 191–206. [1.5.2]

Givler, R. C. and Altobelli, S. A. 1994 A determination of the effective viscosity for the Brinkman-Forchheimer flow model. *J. Fluid Mech.***258**, 355–370. [1.5.3]

Gjerde, K. M. and Tyvand, P. A. 1984 Thermal convection in a porous medium with continuous periodic stratification. *Int. J. Heat Mass Transfer* **27**, 2289–2295. [6.13.2]

Gleason, K. J., Krantz, W. B., Caine, N., George, J. H. and Gunn, R. D. 1986 Geometrical aspects of sorted patterned ground in recurrently frozen soil. *Science* **232**, 216–220. [11.2]

Gobin, D., Goyeau, B. and Songbe, J. P. 1998 Double diffusive natural convection in a composite porous layer. *ASME J. Heat Transfer* **120**, 234–242. [9.4]

Gobin, D., Goyeau, B. and Neculae, A. 2005 Convective heat and solute transfer in partially porous cavities. *Int. J. Heat Mass Transfer* **48**, 1898–1908. [9.4]

Goel, A. K. and Agrawal, S. C. 1998 A numerical study of the hydromagnetic thermal convection in a visco-elastic dusty fluid in a porous medium. *Indian J. Pure Appl. Math.* **29**, 929–940. [6.21]

Goering, D. J. 2003 Passively cooled railway embankments for use in permafrost areas. *J. Cold Reg. Engng.* **17**, 119–133. [7.3.7]

Goering, D. J. and Kumar, P. 1996 Winter-time convection in open-graded embankments. *Cold Reg. Sci. Tech.* **24**, 57–74. [7.3.7]

Goharzadeh, A., Khalili, A. and Jorgensen, B. B. 2005 Transition layer thickness at a fluid-porous interface. *Physics Fluids* **17**, Art. No. 057102. [1.6]

Goldstein, R. E., Pesci, A. I. and Shelley, M. J. 1998 Instabilities and singularities in Hele-Shaw flow. *Phys. Fluids* **10**, 2701–2723. [2.5]

Golfier, F., Quintard, M. and Whitaker, S. 2002a Heat and mass transfer in tubes: an analysis using the method of volume averaging. *J. Porous Media* **5**, 169–185. [2.2.2]

Gorin, A. V., Nakoryakov, V. E., Khoruzhenko, A. G. and Tsoi, O. N. 1988 Heat transfer during mixed convection on a vertical surface in a porous medium with deviation from Darcy's law. *J. Appl. Mech. Tech. Phys.* **29**, 133–139. [8.1.1]

Gorin, A. V., Sikovsky, D. P., Mikhailova, T. N. and Mukhin, V. A. 1998 Forced convection heat and mass transfer from a circular cylinder in a Hele-Shaw cell. *Heat Transfer 1998, Proc. 11th IHTC*, **3**, 109–114. [2.5]

Gorla, R. S. R. and Kumari, M. 1996 Mixed convection in non-Newtonian fluids along a vertical plate in a porous medium. *Acta Mech.* **118**, 55–64. [8.1.1]

Gorla, R. S. R. and Kumari, M. 1998 Nonsimilar solutions for mixed convection in non-Newtonian fluids along a vertical plate in a porous medium. *Transport Porous Media* **33**, 295–307. [8.1.1]

Gorla, R. S. R. and Kumari, M. 1999a Mixed convection in non-Newtonian fluids along a vertical plate with a variable surface heat flux in a porous medium. *Heat Mass Transfer* **35**, 221–227. [8.1.1]

Gorla, R. S. R. and Kumari, M. 1999b Nonsimilar solutions for mixed convection in non-Newtonian fluids along a wedge with variable surface temperature in a porous medium. *Int. J. Numer. Meth. Heat Fluid Flow* **9**, 601–611. [8.1.1]

Gorla, R. S. R. and Kumari, M. 1999c Nonsimilar solutions for free convection in non-Newtonian fluids along a vertical plate in a porous medium. *Int. J. Numer. Meth. Heat Fluid Flow* **9**, 847–859. [8.1.1]

Gorla, R. S. R. and Kumari, M. 2000 Nonsimilar solutions for mixed convection in non-Newtonian fluids along a wedge with variable surface heat flux in a porous medium. *J. Porous Media* **3**, 181–184. [8.1.4]

Gorla, R. S. R. and Kumari, M. 2003 Free convection in non-Newtonian fluids along a horizontal plate in a porous medium. *Heat Mass Transfer* **39**, 101–106. [5.2]

Gorla, R. S. R. and Tornabene, R. 1988 Free convection from a vertical plate with nonuniform surface heat flux and embedded in a porous medium. *Transport in Porous Media* **3**, 95–105. [5.1.9.9]

Gorla, R. S. R. and Zinalabedini, A. H. 1987 Free convection from a vertical plate with nonuniform surface temperature and embedded in a porous medium. *ASME J. Energy Res. Tech.* **109**, 27–30. [5.1.9.9]

Gorla, R. S. R., Bakier, A. Y. and Byrd, L. 1996 Effects of thermal dispersion and stratification on combined convection on a vertical surface embedded in a porous medium. *Transport Porous Media* **25**, 275–282. [8.1.1]

Gorla, R. S. R., Mansour, M. A. and Sarhar, M. G. 1999b Natural convection from a vertical plate in a porous medium using Brinkman's model. *Transport Porous Media* **36**, 357–371. [5.1.7.1]

Gorla, R. S. R., Mansour, M. A., Hasssanien, I. A. and Bakier, A. Y. 1999a Mixed convection effect on melting from a vertical plate in a porous medium. *Transport Porous Media* **36**, 245–254. [10.1.7]

Gorla, R. S. R., Shanmugam, K. and Kumari, M. 1998 Mixed convection in non-Newtonian fluids along nonisothermal horizontal surfaces in porous media. *Heat Mass Transfer* **33**, 281–286. [8.1.2]

Gorla, R. S. R., Slaouti, A. and Takhar, H. S. 1997 Mixed convection in non-Newtonian fluids along a vertical plate in porous media with surface mass transfer. *Int. J. Numer. Methods Heat Fluid Flow* **7**, 598–608. [8.1.1]

Gosink, J. P. and Baker, G. C. 1990 Salt fingering in subsea permafrost: some stability and energy considerations. *J. Geophys. Res.* **95**, 9575–9583. [11.3]

Gounot, J. and Caltagirone, J. P. 1989 Stabilité et convection naturelle au sein d'une couche poreuse non homogène. *Int. J. Heat Mass Transfer* **32**, 1131–1140. [6.13.4]

Gouze, P., Coudrain-Ribstein, A. and Dominique, B. 1994 Computation of porosity redistribution resulting from thermal convection in slanted porous layers. *J. Geophys. Res.* **99**, 697–706. [11.5]

Govender, S. 2003a Oscillating convection induced by gravity and centrifugal forces in a rotating porous layer distant from the axis of rotation. *Int. J. Engng. Sci.* **41**, 539–545. [6.22]

Govender, S. 2003b On the linear stability of large Stephan number convection in rotating mushy layers for a new Darcy equation formulation. *Transport Porous Media* **51**, 173–189. [10.2.3]

Govender, S. 2003c Coriolis effect on the linear stability of convection in a porous layer placed far away from the axis of rotation. *Transport Porous Media* **51**, 315–326. [6.22]

Govender, S. 2003d Finite amplitude analysis of convection in rotating mushy layers during solidification of binary alloys. *J. Porous Media* **6**, 137–147. [10.2.3]

Govender, S. 2003e Moderate time scale finite amplitude analysis of large Stephan number convection in rotating mushy layers. *Transport Porous Media* **53**, 357–366. [10.2.3]

Govender, S. 2004a Three-dimensional convection in an inclined porous layer heated from below and subjected to gravity and Coriolis effects. *Transport Porous Media* **55**, 103–112. [7.8]

Govender, S. 2004b Stability of convection in a gravity modulated porous layer heated from below. *Transport Porous Media* **57**, 113–123. [6.24]

Govender, S. 2004c Finite amplitude analysis of convection in rotating mushy layers for small variations in retardability. *J. Porous Media* **7**, 227–238. [10.2.3]

Govender, S. 2005a Moderate Stefan number convection in rotating mushy layers: A new Darcy number formulation. *Transport Porous Media* **59**, 127–137. [10.2.3]

Govender, S. 2005b Stefan number effect on the transition from stationary to oscillatory convection in a solidifying mushy layer subjected to rotation; Response to reviewer's comment. *Transport Porous Media* **58**, 361–369. [10.2.3]

Govender, S. 2005c Destabilizing a fluid saturated gravity modulated porous layer heated from above. *Transport Porous Media* **59**, 215–225. [6.24]

Govender, S. 2005d Linear stability and convection in a gravity modulated porous layer heated from below: Transition from synchronous to subharmonic oscillations. *Transport Porous Media* **59**, 227–238. [6.24]

Govender, S. 2005e Stability analysis of a porous layer heated from below and subjected to low frequency vibration: Frozen time analysis. *Transport Porous Media*, in press. [6.24]

Govender, S. 2005f Weak non-linear analysis of convection in a gravity modulated porous layer. *Transport Porous Media* **60**, 33–42. [6.24]

Govender, S. and Vadasz, P. 1995 Centrifugal and gravity driven convection in rotating porous media—an analogy with the inclined porous layer. *ASME HTD* **309**, 93–98. [6.22, 7.1.4]

Goyeau, B. and Gobin, D. 1999 Heat transfer by thermosolutal natural convection in a vertical composite fluid-porous cavity. *Int. Comm. Heat Mass Transfer* **26**, 1115–1126. [9.4]

Govender, S. and Vadasz, P. 2002a Weak nonlinear analysis and moderate Stephan number oscillatory convection in rotating mushy layers. *Transport Porous Media* **48**, 353–372. [10.2.3]

Govender, S. and Vadasz, P. 2002b Weak nonlinear analysis and moderate Stephan number stationary convection in rotating mushy layers. *Transport Porous Media* **49**, 247–263. [10.2.3]

Govender, S. and Vadasz, P. 2002c Moderate time scale linear stability of moderate Stefan number convection in rotating mushy layers. *J. Porous Media* **5**, 113–121. [10.2.3]

Goyeau, B., Lhuillier, D. , Gobin, D. and Velarde, M. G. 2003 Momentum transport at a fluid-porous interface. *Int. J. Heat Mass Transfer* **46**, 4071–4081. [1.6]

Goyeau, B., Lhuillier, D. and Gobin, D. 2002 Momentum transfer at a fluid/porous interface. *Heat Transfer 2002, Proc. 12th Int. Heat Transfer Conf.*, Elsevier, Vol. 3, pp. 147–152. [1.6]

Goyeau, B., Lhuillier, D., Gobin, D. and Velarde, M. G. 2003 Momentum transport at a fluid-porous interface. *Int. J. Heat Mass Transfer* **46**, 4071–4081. [1.6]

Goyeau, B., Mergui, S., Songe, J. P. and Gobin, D. 1996b Convection thermosolutale en cavité partiellement occupée par une couch poreuse faibllement perméable. *C.R. Acad. Sci. Paris, Sér. II*, **323**, 447–454. [9.4]

Goyeau, B., Songe, J. P. and Gobin, D. 1996a Numerical study of double-diffusive natural convection in a porous cavity using the Darcy-Brinkman formulation. *Int. J. Heat Mass Transfer* **39**, 1363–1378. [9.2.2]

Graham, M. D. and Steen, P. H. 1991 Structure and mechanism of oscillatory convection in a cube of fluid-saturated porous material heated from below. *J. Fluid Mech.*, **232**, 591–609. [6.15.1]

Graham, M. D. and Steen, P. H. 1992 Strongly interacting traveling waves and quasiperiodic dynamics in porous medium convection. *Physica D* **54**, 331–350. [6.15.1]

Graham, M. D. and Steen, P. H. 1994 Plume formation and resonant bifurcations in porous-media convection. *J. Fluid Mech.* **272**, 67–89. [6.15.1]

Graham, M. D., Steen, P. H. and Titi,. S. 1993 Computational efficiency and approximate inertial manifolds for a Bénard convection system. *J. Nonlinear Sci.* **3**, 153–167. [6.15.1]

Grangeot, G., Quintard, M. and Whitaker, S. 1994 Heat transfer in packed beds: interpretation of experiments in terms of one- and two-equation models. *Heat Transfer 1994*, Inst. Chem. Engrs, Rugby, vol. 5, pp. 291–296. [2.2.2]

Grant, M. A. 1983 Geothermal reservoir modeling. *Geothermics* **12**, 251–263. [11]

Gratton, L. J., Travkin, V. S. and Catton, I. 1996 Influence of morphology upon two-temperature statements for convective transport in porous media. *J. Enhanced Heat Transfer* **3**, 129–145. [1.8]

Gray, W. C. and O'Neill, K. 1976 On the general equations for flow in porous media and their reduction to Darcy's law. *Water Resources Res.* **12**, 148–154. [3.5.2]

Green, T. 1984 Scales for double-diffusive fingering in porous media. *Water Resources Res.* **20**, 1225–1229. [9.1.3]

Green, T. 1990 The momentary instability of a saturated porous layer with a time-dependent temperature, and the most unstable disturbance. *Water Resources Res.* **26**, 2015–2021. [6.11.3]

Green, T. and Freehill, R. L. 1969 Marginal stability in inhomogeneous porous media. *J. Appl. Phys.* **40**, 1759–1762. [6.13.1]

Greenkorn, R. A. 1983 *Flow Phenomena in Porous Media*, Marcel Dekker, New York. [2.2.3]

Griffiths, R. W. 1981 Layered double-diffusive convection in porous media. *J. Fluid Mech.* **102**, 221–248. [9.1.3]

Grundmann, M. and Mojtabi, A. 1995 Solution asymptotique du problème de la convection naturelle dans la cavité carrée poreuse chaufée par le bas. *C.R. Acad. Sci. Paris, Sér.* II, **321**, 401–406. [6.3]

Grundmann, M., Mojtabi, A. and vant Hof, B. 1996 Asymptotic solution of natural convection problem in a square cavity heated from below. *Int. J. Numer. Meth. Heat Fluid Flow* **6**, 29–36. [6.3]

Guba, P. 2001 On the finite-amplitude steady convection in rotating mushy layers, *J. Fluid Mech.* **347**, 337–365. [10.2.3]

Guo, J. and Kaloni, P. N. 1995a Nonlinear stability of convection induced by inclined thermal and solutal gradients. *Z. Angew. Math. Phys.* **46**, 645–654. [9.5]

Guo, J. and Kaloni, P. N. 1995b Double-diffusive convection in a porous-medium, nonlinear stability, and the Brinkman effect. *Stud. Appl. Math.* **94**, 341–358. [9.1.6.3]

Guo, Z. L. and Zhao, T. S. 2005a A lattice Boltzmann model for convective heat transfer in porous media. *Numer. Heat Transfer B* **47**, 157–177. [2.6]

Guo, Z. L. and Zhao, T. S. 2005b Lattice Boltzmann simulation of natural convection with temperature-dependent viscosity in a porous cavity. *Prog. Comput. Fluid Dyn.* **5**, 110–117. [2.6]

Guo, Z., Kim, S. Y. and Sung, H. J. 1997a Pulsating flow and heat transfer in a pipe

Guo, Z., Sung, H. J. and Hyan, J. M. 1997b Pulsating flow and heat transfer in an annulus partially filled with porous media. *Numer. Heat Transfer A* **31**, 517–527. [4.11]

Gupta, V. P. and Joseph, D. D. 1973 Bounds for heat transport in a porous layer. *J. Fluid Mech.* **57**, 491–514. [6.3, 6.9.1, 6.22]

Gustafson, M. R. and Howle, L. E. 1999 Effects of anisotropy and boundary plates on the critical values of a porous medium heated from below. *Int. J. Heat Mass Transfer* **42**, 3419–3430. [6.12]

Haajizadeh, M. and Tien, C. L. 1983 Natural convection in a rectangular porous cavity with one permeable endwall. *ASME J. Heat Transfer* **105**, 803–808. [7.3.2]

Haajizadeh, M. and Tien, C. L. 1984 Combined natural and forced convection in a horizontal porous channel. *Int. J. Heat Mass Transfer* **27**, 799–813. [8.2.4]

Haajizadeh, M., Ozguc, A. F. and Tien, C. L. 1984 Natural convection in a vertical porous enclosure with internal heat generation. *Int. J. Heat Mass Transfer* **27**, 1893–1902. [6.17]

Haber, S. and Mauri, R. 1983 Boundary conditions for Darcy's flow through porous media. *Int. J. Multiphase Flow* **9**, 561–574. [1.6]

Haddad, O.M., Al-Nimr, M. A. and Abu-Ayyad, M. A. 2002 Numerical simulation of forced convection flow past a parabolic cylinder embedded in porous media. *Int. J. Numer. Meth. Heat Fluid Flow* **12**, 6–28. [4.3]

Haddad O.M., Al-Nimr, M. A. and Al-Khateeb, A. N. 2004 Validation of the local thermal equilibrium assumption in natural convection from a vertical plate embedded in porous medium. *Int. J. Heat Mass Transfer* **47**, 2037–2042. [5.1.9.3]

Haddad O.M., Al-Nimr, M. A. and Al-Khateeb, A. N. 2005 Validity of the local thermal equilibrium assumption in natural convection from a vertical plate embedded in porous medium. *J. Porous Media* **8**, 85–95. [5.1.9.3]

Hadim, A. 1994a Forced convection in a porous channel with localized heat sources. *ASME J. Heat Transfer* **116**, 465–472. [4.11]

Hadim, A. 1994b Numerical study of non-Darcy mixed convection in a vertical porous channel. *AIAA J. Thermophys. Heat Transfer* **8**, 371–373. [8.3.1]

Hadim, A. and Burmeister, L. C. 1988 Onset of convection in a porous medium with internal heat generation and downward flow. *AIAA J. Thermophys. Heat Transfer* **2**, 343–351. [6.11.2]

Hadim, A. and Burmeister, L. C. 1992 Conceptual design of a downward-convecting solar pond filled with a water-saturated, porous medium. *ASME J. Solar Energy Engng.* **114**, 240–245. [6.11.2]

Hadim, A. and Chen, G. 1994 Non-Darcy mixed convection in a vertical porous channel with discrete heat sources at the walls. *Int. Comm. Heat Mass Transfer* **21**, 377–387. [8.3.2]

Hadim, A. and Govindarajan, S. 1988 Development of laminar mixed convection in a vertical porous channel. *ASME HTD* **105**, 145–153. [8.3.1]

Hadim, H. A. and Chen, G. 1994 Non-Darcy mixed convection in a vertical porous channel with asymmetric wall heating. *AIAA J. Thermophys. Heat Transfer* **8**, 805–808. [8.3.2]

Hadim, H. A. and Chen, G. 1995 Numerical study of non-Darcy natural convection of a power-law fluid in a porous cavity. *ASME HTD*-**317**, 301–307. [7.1.6]

Hadim, H. and North, A. 2005 Forced convection in a sintered porous channel with inlet and outlet slots. *Int. J. Thermal Sci.* **44**, 33–42. [4.16.2]

Hadim, H.A. and Bethancourt, A. 1995 Numerical study of forced convection in a partially porous channel with discrete heat sources. *ASME J. Electron. Packaging* **117**, 46–51. [4.11]

Hady, F. M. and Ibrahim, F. S. 1997 Forced convection heat transfer on a flat plate embedded in porous media for power-law fluids. *Transport Porous Media* **28**, 125–134. [4.16.3]

Haji-Sheikh, A. 2004 Estimation of average and local heat transfer in parallel plates and circular ducts filled with porous materials. *ASME J. Heat Transfer* **126**, 400–409. [4.5]

Haji-Sheik, A. and Vafai, K. 2004 Analysis of flow and heat transfer in porous media imbedded inside various shaped ducts. *Int. J. Heat Mass Transfer* **47**, 1889–1905. [4.13]

Haji-Sheikh, A., Minkowycz W. J. and Sparrow E. M. 2004b Green's function solution of temperature field for flow in porous passages. *Int. J. Heat Mass Transfer* **47**, 4685–4695. [4.13]

Haji-Sheikh, A., Minkowycz W. J. and Sparrow E. M. 2005 Heat transfer to flow through porous passages using extended weighted residual method–a Green's function solution. *Int. J. Heat Mass Transfer* **48**, 1330–1349. [4.13]

Haji-Sheikh, A., Sparrow, E. M. and Minkowycz, W. J. 2004a A numerical study of heat transfer to fluid flow through circular porous passages. *Numer. Heat Transfer A* **46**, 929–955. [4.5]

Hall, M. J. and Hiatt, J. P. 1996 Measurements of pore scale flows within and exiting ceramic foams. *Expt. Fluids* **20**, 433–440. [1.9]

Hallworth, M. A., Huppert, H. E. and Woods, A. W. 2005 Dissolution-driven convection in a reactive porous medium. *J. Fluid Mech.* **535**, 255–285. [10.2.3]

Hamdan, M. O., Al-Nimr, M. A. and Alkam, M. K. 2000 Enforcing forced convection by inserting porous substrate in the core of a parallel-plate channel. *Int. J. Numer. Meth. Heat Fluid Flow* **10**, 502–517. [4.11]

Hanamura, K. and Kaviany, M. 1995 Propagation of condensation front in steam injection into dry porous media. *Int. J. Heat Mass Transfer* **38**, 1377–1386. [3.6]

Handley, D. and Heggs, P. J. 1968 Momentum and heat transfer mechanisms in regular shaped packings. *Trans. Inst. Chem. Engrs.* **46**, T251–T264. [2.2.2]

Hao, Y. L. and Tao, Y. X. 2003a Non-equilibrium melting of granular packed bed in horizontal forced convection. Part I: Experiment. *Int. J. Heat Mass Transfer* **46**, 5017–5030. [10.1.7]

Hao, Y. L. and Tao, Y. X. 2003b Non-equilibrium melting of granular packed bed in horizontal forced convection. Part II: Numerical simulation. *Int. J. Heat Mass Transfer* **46**, 5031–5044. [10.1.7]

Haq, S. and Mulligan, J. C. 1990a Transient free convection about a vertical flat plate embedded in a saturated porous medium. *Numer. Heat Transfer A* **18**, 227–242. [5.1.3]

Haq, S. and Mulligan, J. C. 1990b Transient free convection from a vertical plate to a non-Newtonian fluid in a porous medium. *J. Non-Newtonian Fluid Mech.* **36**, 395–440. [5.1.9.2]

Hardee, H. C. and Nilson, R. H. 1977 Natural convection in porous media with heat generation. *Nucl. Sci. Engng.* **63**, 119–132. [6.11.2]

Harris, K.T., Haji-Sheikh, A. and Nhanna, A. G. A. 2001 Phase change phenomena in porous media–a non-local thermal equilibrium model. *Int. J. Heat Mass Transfer* **44**, 1619–1625. [10.1.7]

Harris, S. D. and Ingham, D. B. 2004 Techniques for solving the boundary-layer equations. In *Emerging Technologies and Techniques in Porous Media* (D. B. Ingham, A. Bejan, E. Mamut and I. Pop, eds), Kluwer Academic, Dordrecht, pp. 43–64. [5.1.3]

Harris, S. D., Ingham, D. B. and Pop, I. 1996 Transient free convection from a vertical plate subjected to a change in surface heat flux in porous media. *Fluid Dyn. Res.* **18**, 313–324. [5.1.3]

Harris, S. D., Ingham, D. B. and Pop, I. 1997a Free convection from a vertical plate in a porous media subjected to a sudden change in surface temperature. *Int. Comm. Heat Mass Transfer* **24**, 543–552. [5.1.3]

Harris, S. D., Ingham, D. B. and Pop, I. 1997b Free convection from a vertical plate in a porous medium subjected to a sudden change in surface heat flux. *Transport in Porous Media* **26**, 205–224. [5.1.3]

Harris, S. D., Ingham, D. B. and Pop, I. 1998 Transient mixed convection from a vertical surface in a porous medium. In *Mathematics of Heat Transfer* (G. E. Tupholme and A. S. Wood, eds.) Clarendon Press, Oxford, pp. 157–164. [8.1.1]

Harris, S. D., Ingham, D. B. and Pop, I. 1999 Unsteady mixed convection boundary layer flow on a vertical surface in a porous medium. *Int. J. Heat Mass Transfer* **42**, 357–372. [8.1.1]

Harris, S. D., Ingham, D. B. and Pop, I. 2000 Transient free convection from a horizontal surface in a porous medium subjected to a sudden change in surface heat flux. *Transport Porous Media* **39**, 97–117. [5.2]

Harris, S. D., Ingham, D. B. and Pop, I. 2002 Thermal capacity effect on transient free convection adjacent to a fixed surface in a porous medium. *Transport Porous Media* **46**, 1–18. [8.1.1]

Hartline, B. K. and Lister, C. R. B. 1977 Thermal convection in a Hele-Shaw cell. *J. Fluid Mech.***79**, 379–389. [2.5]

Hartline, B. K. and Lister, C. R. B. 1981 Topographic forcing of supercritical convection in a porous medium such as the oceanic crust. *Earth Planet. Sci. Lett.* **55**, 75–86. [11.6.2]

Hasnaoui, M., Vasseur, P., Bilgen, E. and Robillard, L. 1995 Analytical and numerical study of natural convection heat transfer in a vertical porous annulus. *Chem. Engng. Comm.* **131**, 141–159. [7.3.3]

Hassan, M. and Mujumdar, A. S. 1985 Transpiration-induced buoyancy effect around a horizontal cylinder embedded in a porous medium. *Int. J. Energy Res.* **9**, 151–163. [9.4]

Hassanien, I. A. 2003 Variable permeability effects on mixed convection along a vertical wedge embedded in a porous medium with variable heat flux. *Appl. Math. Comp.* **138**, 41–59. [8.1.4]

Hassanien, I. A. and Allah, M. H. O. 2002 Oscillatory hydromagnetic flow through a porous medium with variable permeability in the presence of free convection and mass transfer flow. *Int. Comm. Heat Mass Transfer* **29**, 567–575. [9.2.1]

Hassanien, I. A. and Omer, G.M. 2002 Nonsimilarity solutions for mixed convection flow along nonisothermal vertical plate embedded in porous media with variable permeability. *J. Porous Media* **5**, 159–167. [8.1.1]

Hassanien, I. A. and Omer, G. M. 2005 Mixed-convection flow adjacent to a horizontal surface in a porous medium with variable permeability and surface heat flux. *J. Porous Media* **8**, 225–235. [8.1.2]

Hassanien, I. A., Bakier, A. Y. and Gorla, R. S. R. 1998 Effects of thermal dispersion and stratification on non-Darcy mixed convection from a vertical plate in a porous medium. *Heat Mass Transfer* **34**, 209–212. [8.1.1]

Hassanien, I. A., Essawy, A. H. and Moursy, N. M. 2003a Variable viscosity and thermal conductivity effects on combined heat and mass transfer in mixed convection over a UHF/UMF wedge in porous media: the entire regime. *Appl. Math. Comput.* **145**, 667–682. [9.6]

Hassanien, I. A., Essawy, A. H. and Moursy, N. M. 2003b Variable viscosity and thermal conductivity effects on heat transfer by natural convection from a cone and a wedge in porous media. *Arch. Mech.***55**, 345–356. [5.8]

Hassanien, I. A., Essawy, A. H. and Moursy, N. M. 2004 Natural convection flow of micropolar fluid from a permeable uniform heat flux surface in porous medium. *Appl. Math. Comput.* **152**, 3232–335. [5.1.9.2]

Hassanien, I. A., Ibrahim, F. S. and Omer, G. M. 2005 The effect of variable permeability and viscous dissipation on a non-Darcy natural-convection regime with thermal dispersion. *J. Porous Media* **8**, 237–256. [5.1.9.4]

Hassanien, I. A., Salama, A. A. and Elaiw, A. M. 2003c Variable permeability effect on vortex instability of mixed convection flow in a semi-infinite porous medium bounded by a horizontal surface. *Appl. Math. Comput.* **146**, 829–847. [8.1.2]

Hassanien, I. A., Salama, A. A. and Elaiw, A. M. 2004a The onset of longitudinal vortices in mixed convection flow over an inclined surface in a porous medium with variable permeability. *Appl. Math. Comput.* **154**, 313–333. [8.1.2]

Hassanien, I. A., Salama, A. A. and Elaiw, A. M. 2004b Variable permeability effect on vortex instability of a horizontal natural convection flow in a saturated porous medium with a variable wall temperature. *Zeit. Angew. Math. Mech.* **84**, 39–47. [5.4]

Hassanien, I. A., Salama, A. A. and Moursy, N. M. 2004c Non-Darcian effects on vortex instability of mixed convection over horizontal plates porous medium with surface mass flux. *Int. Comm. Heat Mass Transfer* **31**, 231–240. [8.1.2]

Hassanien, I. A., Salama, A. A. and Moursy, N. M. 2004d Inertia effect on vortex instability of horizontal natural convection flow in a saturated porous medium with surface mass flux. *Int. Comm. Heat Mass Transfer* **31**, 741–750. [5.4]

Hassanizadeh, S. M. and Gray, W. G. 1988 Reply to comments by Barak on "High velocity flow in porous media" by Hassanizadeh and Gray. *Transport in Porous Media* **3**, 319–321. [1.5.2]

Hassanizadeh, S. M. and Gray, W. G. 1993 Toward an improved description of the physics of two-phase flow. *Adv. Water Resources* **16**, 53–67. [3.5.4]

Haugen, K. B. and Tyvand, P. A. 2003 Onset of thermal convection in a vertical porous cylinder with conducting wall. *Phys. Fluids* **15**, 2661–2667. [7.3.3]

Havstad, M. A. and Burns, P. J. 1982 Convective heat transfer in vertical cylindrical annuli filled with a porous medium. *Int. J. Heat Mass Transfer* **25**, 1755–1766. [7.3.3]

Hayes, R. E. 1990 Forced convection heat transfer at the boundary layer of a packed bed. *Transport in Porous Media* **5**, 231–245. [4.8]

Hayes, R. E., Afacan, A., Boulanger, B. and Shenoy, A. V. 1996 Modelling the flow of power law fluids in a packed bed using a volume-averaged equation of motion. *Transport in Porous Media* **23**, 175–196. [1.5.4]

He, X. S. and Georgiadis, J. G. 1990 Natural convection in porous media: effect of weak dispersion on bifurcation. *J. Fluid Mech.* **216**, 285–298. [6.6, 6.11.2]

He, X. S. and Georgiadis, J. G. 1992 Direct numerical solution of diffusion problems with intrinsic randomness. *Int. J. Heat Mass Flow* **35**, 3141–3151. [2.6]

Helmy, K. A. 1998 MHD unsteady free convection flow past a vertical porous plate. *Z. Angew. Math. Mech.* **78**, 255–270. [5.1.9.9]

Hennenberg, M., Saghir, M. Z., Rednikov, A. and Legros, J. C. 1997 Porous media and the Bénard-Marangoni problem. *Transport Porous Media* **27**, 327–355. [6.19.3]

Herron, I. H. 2001 Onset of convection in a porous medium with heat sources and variable gravity. *Int. J. Engng. Sci.* **39**, 201–208. [6.24]

Herwig, H. and Koch, M. 1990 An asymptotic approach to natural convection momentum and heat transfer in saturated highly porous media. *ASME J. Heat Transfer* **112**, 1085–1088. [5.1.9.9, 9.7]

Hguyen (*sic*), T. H. and Zhang, X. 1992 Onset and evolution of penetrative convection during the melting process in a porous medium. *Heat and Mass Transfer in Porous Media* (ed. M. Quintard and M. Todorovic), Elsevier, Amsterdam, 381–392. [10.1.7]

Hickox, C. E. 1981 Thermal convection at low Rayleigh number from concentrated sources in porous media. *ASME J. Heat Transfer* **103**, 232–236. [5.11.2]

Hickox, C. E. and Chu, T. Y. 1990 A numerical study of convection in a layered porous medium heated from below. *ASME HTD*-149, 13–21. [6.13.2]

Hickox, C. E. and Gartling, D. K. 1981 A numerical study of natural convection in a horizontal porous layer subjected to an end-to-end temperature difference. *ASME J. Heat Transfer* **103**, 797–802. [7.1.3]

Hickox, C. E. and Gartling, D. K. 1985 A numerical study of natural convection in a vertical annular porous layer. *Int. J. Heat Mass Transfer* **28**, 720–723. [7.3.3]

Hickox, C. E. and Watts, H. A. 1980 Steady thermal convection from a concentrated source in a porous medium. *ASME J. Heat Transfer* **102**, 248–253. [5.11.3]

Higuera, F.J. 1997 Conjugate natural convection heat transfer between two porous media separated by a horizontal wall. *Int. J. Heat Mass Transfer* **40**, 3157–3161. [5.1.5]

Higuera, F.J. and Pop, I. 1997 Conjugate natural convection heat transfer between two porous media separated by a vertical wall. *Int. J. Heat Mass Transfer*, **40**, 123–129. [5.1.5]

Higuera, F. J. and Weidman, P. D. 1995 Natural convection beneath a downward facing heated plate in a porous medium. *Eur. J. Mech. B/Fluids* **14**, 29–40. [5.2]

Higuera, F. J. and Weidman, P. D. 1998 Natural convection far downstream of a heat source on a solid wall. *J. Fluid Mech.***361**, 225–39. [5.11.1]

Hill, A. A. 2003 Convection due to the selective absorption of radiation in a porous medium. *Cont. Mech. Thermodyn.* **15**, 451–462, erratum 629. [6.11.2]

Hill, A. A. 2004a Convection induced by the selective absorption of radiation for the Brinkman model. *Cont. Mech. Thermodyn.* **16**, 43–52. [6.11.2]

Hill, A. A. 2004b Conditional and unconditional nonlinear stability for convection induced by absorption of radiation in a non-Darcy porous medium. *Cont. Mech. Thermodyn.* **16**, 305–318. [6.11.2]

Hill, A. A. 2005 Double-diffusive convection in a porous medium with concentration based internal heat source. *Proc. Roy. Soc. Lond. A* **461**, 561–574. [9.3.1]

Hills, R. N., Loper, D. E. and Roberts, P. H. 1983 A thermodynamically consistent model for flow through dendrites. *Quart. J. Mech. Appl. Math.* **36**, 505–539. [10.2.3]

Himasekhar, K. and Bau, H. H. 1986 Large Rayleigh number convection in a horizontal, eccentric annulus containing saturated porous media. *Int. J. Heat Mass Transfer* **29**, 703–712. [7.3.3]

Himasekhar, K. and Bau, H. H. 1987 Thermal convection associated with hot/cold pipes buried in a semi-infinite, saturated porous medium. *Int. J. Heat Mass Transfer* **30**, 263–273. [7.11]

Himasekhar, K. and Bau, H. H. 1988a Thermal convection around a heat source embedded in a box containing a saturated porous medium. *ASME J. Heat Transfer* **110**, 649–654. [5.12.3]

Himasekhar, K. and Bau, H. H. 1988b Two-dimensional bifurcation phenomena in thermal convection in horizontal concentric annuli containing saturated porous media. *J. Fluid Mech.* **187**, 267–300. [7.3.3]

Hoffman, M. R. and van der Meer, F. M. 2002 A simple space-time averaged porous media model for flow in densely vegetated channels. In *Computational Methods in Water Resources* (eds. S. M. Hassanizadeh, R. J. Schotting, W, G, Gray and G. F. Pinder.) Vol. 2., Elsevier, Amsterdam. [1.8]

Hollard, S., Layadi, M., Boisson, H. and Bories, S. 1995 Ecoulements stationnaires et instationnaires de convection naturelle en milieu poreux. *Rev. Gen. Therm.* **34**, S499–S514. [7.8]

Holst, P. H. and Aziz, K. 1972 Transient three-dimensional natural convection in confined porous media. *Int. J. Heat Mass Transfer* **15**, 73–90. [6.8]

Holzbecher, E. 1997 Numerical studies on thermal convection in cold groundwater. *Int. J. Heat Mass Transfer* **40**, 605–612. [6.20]

Holzbecher, E. 2001 On the relevance of oscillatory convection regimes in porous media–review and numerical experiments. *Computers and Fluids* **30**, 189–209. [6.4]

Holzbecher, E. 2004a The mixed convection number for porous media flow. In *Emerging Technologies and Techniques in Porous Media* (D. B. Ingham, A. Bejan, E. Mamut and I. Pop, eds), Kluwer Academic, Dordrecht, pp. 169–181. [8.1.1]

Holzbecher, E. 2004b Free convection in open-top enclosures filled with a porous medium heated from below. *Numer. Heat Transfer A* **46**, 241–254. [6.4]

Holzbecher, E. 2004c Free convection induced by oscillating conditions at the top. In *Applications of Porous Media* (ICAPM 2004), (eds. A. H. Reis and A. F. Miguel), Evora, pp. 147–152. [6.11.3]

Holzbecher, E. 2005a Free and forced convection in porous media open at the top. *Heat Mass Transfer* **41**, 606–614. [6.4]

Holzbecher, E. 2005b Groundwater flow pattern in the vicinity of a salt lake. *Hydrobiologica* **532**, 233–242. [9.1.6.4]

Holzbecher, E. and Yusa, Y. 1995 Numerical experiments on free and forced convection in porous media. *Int. J. Heat Mass Transfer* **38**, 2109–2115. [11.8.2]

Homsy, G. M. and Sherwood, A. E. 1976 Convective instabilities in porous media with throughflow. *AIChE J.* **22**, 168–174. [6.10]

Hong, J. T. and Tien, C. L. 1987 Analysis of thermal dispersion effect on vertical-plate natural convection in porous media. *Int. J. Heat Mass Transfer* **30**, 143–150. [5.1.7.3]

Hong, J. T., Tien, C. L. and Kaviany, M. 1985 Non-Darcian effects on vertical-plate natural convection in porous media with high porosities. *Int. J. Heat Mass Transfer* **28**, 2149–2157. [5.1.7.2]

Hong, J. T., Yamada, Y. and Tien, C. L. 1987 Effects of non-Darcian and nonuniform porosity on vertical-plate natural convection in porous media. *ASME J. Heat Transfer* **109**, 356–362. [5.1.7.1, 5.1.7.2]

Hooman, K. and Ranjbar-Kani, A. A. 2003 Forced convection in a fluid-saturated porous-medium tube with iso-flux wall. *Int. Comm. Heat Mass Transfer* **30**, 1015–1026. [4.9]

Hooman, K. and Ranjbar-Kani, A. A. 2004 A perturbation based analysis to investigate forced convection in a porous saturated tube. *J. Comp. Appl. Math.* **162**, 411–419. [4.9]

Hooman, K., Ranjbar-Kani, A. A. and Ejlali, A. 2003 Axial conduction effects on thermally developing forced convection in a porous medium: circular tube with uniform wall temperature. *Heat Transfer Research* **34**, 34–40. [4.13]

Hooper, W. B., Chen, T. S. and Armaly, B. F. 1994a Mixed convection along an isothermal vertical cylinder in porous media. *AIAA J. Thermophys. Heat Transfer* **8**, 92–99. [8.1.3]

Hooper, W. B., Chen, T. S. and Armaly, B. F. 1994b Mixed convection from a vertical plate in porous media with surface injection or suction. *Numer. Heat Transfer A* **25**, 317–329. [8.1.1]

Horne, R. N. 1979 Three-dimensional natural convection in a confined porous medium heated from below. *J. Fluid Mech.* **92**, 751–766. [6.15.1]

Horne, R. N. and Caltagirone, J. P. 1980 On the evaluation of thermal disturbances during natural convection in a porous medium. *J. Fluid Mech.* **100**, 385–395. [6.8, 6.15.1]

Horne, R. N. and O'Sullivan, M. J. 1974a Oscillatory convection in a porous medium heated from below. *J. Fluid Mech.* **66**, 339–352. [6.8, 6.9.1, 6.18]

Horne, R. N. and O'Sullivan, M. J. 1974b Oscillatory convection in a porous medium: the effect of through flow. *Proc. 5th Australasian Conf. Hydraulics Fluid Mech.*, Univ. Canterbury, Christchurch, New Zealand, vol.2, pp.234–237. [11.7]

Horne, R. N. and O'Sullivan, M. J. 1978a Origin of oscillatory convection in a porous medium heated from below. *Phys. Fluids* **21**, 1260–1264. [6.8, 6.15.1]

Horne, R. N. and O'Sullivan, M. J. 1978b Convection in a porous medium heated from below: the effect of temperature dependent viscosity and thermal expansion coefficient. *ASME J. Heat Transfer* **100**, 448–452. [6.18]

Horton, C. W. and Rogers, F. T. 1945 Convection currents in a porous medium. *J. Appl. Phys.* **16**, 367–370. [6.1, 6.26]

Hossain, M. A. and Nakayama, A. 1993 Non-Darcy free convective flow along a vertical cylinder embedded in a porous medium with surface mass flux. *Int. J. Heat Fluid Flow* **14**, 385–390. [4.3]

Hossain, M. A. and Nakayama, A. 1994 Nonsimilar free convection boundary layer in non-Newtonian fluid saturated porous media. *AIAA J. Thermophys. Heat Transfer* **8**, 107–112. [5.1.9.2]

Hossain, M. A. and Pop, I. 1997 Radiation effect on Darcy free convection flow along an inclined surface placed in porous media. *Heat Mass Transfer* **32**, 223–227. [5.3]

Hossain, M. A. and Pop, I. 2001 Radiation effects on free convection over a vertical flat plate embedded in a porous medium with high porosity. *Int. J. Therm. Sci.* **40**, 289–295. [5.1.9.4]

Hossain, M. A. and Rees, D. A. S. 1997 Non-Darcy free convection along a horizontal heated surface. *Transport Porous Media* **29**, 309–321. [5.2]

Hossain, M. A. and Rees, D. A. S. 2003 Natural convection flow of a viscous incompressible fluid in a rectangular porous cavity heated from below with cold sidewalls. *Heat Mass Transfer* **39**, 657–663. [6.4]

Hossain, M. A. and Wilson, M. 2002 Natural convection flow in a fluid-saturated porous medium enclosed by non-isothermal walls with heat generation. *Int. J. Therm. Sci.* **41**, 447–454. [7.3.8]

Hossain, M. A., Banu, N. and Nakayama, A. 1994 Non-Darcy forced convection boundary layer flow over a wedge embedded in a saturated porous medium. *Numer. Heat Transfer A* **26**, 399–414. [4.16.2]

Hossain, M. A., Banu, N., Rees, D. A. S. and Nakayama, A. 1996 Unsteady forced convection boundary layer flow through a saturated porous medium. *Proceedings of the International Conference on Porous Media and their Applications in Science, Engineering and Industry*, (K. Vafai and P.N. Shivakumar, eds), Engineering Foundation, New York, 85–101. [4.8]

Hossain, M. A., Pop, I. and Rees, D. A. S. 2000 The effect of time-periodic surface oscillations on free convection from a vertical surface in a porous medium. *Transport Porous Media* **39**, 119–130. [5.1.9.7]

Hossain, M. A., Pop, I. and Vafai, K. 1999a Combined free-convection heat and mass transfer above a near-horizontal surface in a porous medium. *Hybrid Methods Engng.* **1**, 87–102. [9.2.1]

Hossain, M. A., Vafai, K. and Khanafer, K. M. N. 1999b Non-Darcy natural convection heat and mass transfer along a vertical permeable cylinder embedded in a porous medium. *Rev. Gén. Therm.* **38**, 854–862. [9.2.1]

Hossain, M.A., Nakayama, A. and Pop, I. 1995 Conjugate free convection of non-Newtonian fluids about a vertical cylindrical fin in porous media. *Heat Mass Transfer* **30**, 149–153. [5.12.1]

Howell, J. R. 2000 Radiative transfer in porous media. *Handbook of Porous Media* (K. Vafai, ed.) Marcel Dekker, New York, pp. 663–698. [2.6]

Howells, I. D. 1998 Drag on fixed beds of fibres in slow flow. *J. Fluid Mech.* **355**, 163–192. [1.5.3]

Howle, L. E. 2002 Convection in ordered and disordered porous layers. In *Transport Phenomena in Porous Media II* (D. B. Ingham and I. Pop, eds.) Elsevier, Oxford, pp. 155–176. [6.9.1]

Howle, L. and Georgiadis, J. G. 1994 Natural convection in porous media with anisotropic dispersive thermal conductivity. *Int. J. Heat Mass Transfer* **37**, 1081–1094. [6.12]

Howle, L E., Behringer, R. P. and Georgiadis, J. G. 1997 Convection and flow in porous media. Part 2. Visualization by shadowgraph. *J. Fluid Mech.* **332**, 247–262 [6.9.1]

Hsiao, K. T. and Advani, S. G. 1999 Modified effective thermal conductivity due to heat dispersion in fibrous porous media. *Int. J. Heat Mass Transfer* **42**, 1237–1254. [2.1.1]

Hsiao, S. W. 1995 A numerical study of transient natural convection about a corrugated plate embedded in an enclosed porous medium. *Int. J. Num. Meth. Heat Fluid Flow* **5**, 629–646. [7.11]

Hsiao, S. W. 1998 Natural convection in an inclined porous cavity with variable porosity and thermal dispersion effects. *Int. J. Numer. Methods Heat Fluid Flow* **8**, 97–117. [7.8]

Hsiao, S. W. and Chen, C. K. 1994 Natural convection heat transfer from a corrugated plate in an enclosed porous medium. *Numer. Heat Transfer A* **25**, 331–345. [7.11]

Hsiao, S. W., Chen, C. K. and Cheng, P. 1994 A numerical solution for natural convection in an inclined porous cavity with a discrete heat source on one wall. *Int. J. Heat Mass Transfer* **37**, 2193–2201. [7.8]

Hsiao, S. W., Cheng, P. and Chen, C.K . 1992 Non-uniform porosity and thermal dispersion effects on natural convection about a heated horizontal cylinder in an enclosed porous medium. *Int. J. Heat Mass Transfer* **35**, 3407–3418. [7.11]

Hsieh, J. C., Chen, T. S. and Armaly, B. F. 1993a Nonsimilarity solutions for mixed convection from vertical surfaces in porous media: variable surface temperature or heat flux. *Int. J. Heat Mass Transfer* **36**, 1485–1493. [8.1.1]

Hsieh, J. C., Chen, T. S. and Armaly, B. F. 1993b Mixed convection along a non isothermal vertical flat plate embedded in a porous medium: the entire regime. *Int. J. Heat Mass Transfer* **36**, 1819–1825. [8.1.1]

Hsieh, W. H. and Lu, S. F. 1998 Heat-transfer analysis of thermally developing region of annular porous media. *Heat Transfer 1998, Proc. 11th IHTC* **4**, 447–452. [4.13]

Hsieh, W. H. and Lu, S. F. 2000 Heat-transfer analysis and thermal dispersion in thermally-developing region of a sintered porous metal channel. *Int. J. Heat Mass Transfer* **43**, 3001–3011. [2.2.3]

Hsieh, W. H., Wu, J. Y., Shih, W. H. and Chiu, W. C. 2004 Experimental investigation of heat-transfer characteristics of aluminum-foam heat sinks. *Int. J. Heat Mass Transfer* **47**, 5149–5157. [4.10]

Hsu, C. T. 1999 A closure model for transient heat conduction in porous media. *ASME J. Heat Transfer* **121**, 733–739. [2.2.1]

Hsu, C. T. 2000 Heat conduction in porous media. *Handbook of Porous Media* (K Vafai, ed.) Marcel Dekker, New York, pp. 171–200. [2.2.1]

Hsu, C. T. 2005 Dynamic modeling of convective heat transfer in porous media. *Handbook of Porous Media* (K Vafai, ed.), 2nd ed., Taylor and Francis, New York, pp. 39–80. [2.6S]

Hsu, C. T. and Cheng, P. 1979 Vortex instability in buoyancy-induced flow over inclined heated surfaces in porous media. *ASME J. Heat Transfer* **101**, 660–665. [5.4]

Hsu, C. T. and Cheng, P. 1980a Vortex instability of mixed convection flow in a semi-infinite porous medium bounded by a horizontal surface. *Int. J. Heat Mass Transfer* **23**, 789–798. [8.1.2]

Hsu, C. T. and Cheng, P. 1980b The onset of longitudinal vortices in mixed convective flow over an inclined surface in a porous medium. *ASME J. Heat Transfer* **102**, 544–549. [8.1.2]

Hsu, C. T. and Cheng, P. 1985a Effects of upstream geometry on natural convection of a Darcian fluid about a semi-infinite inclined heated surface. *ASME J. Heat Transfer* **107**, 283–292. [5.12.2]

Hsu, C. T. and Cheng, P. 1985b The Brinkman model for natural convection about a semi-infinite vertical flat plate in a porous medium. *Int. J. Heat Mass Transfer* **28**, 683–697. [5.1.7.1]

Hsu, C. T. and Cheng, P. 1988 Closure schemes of the macroscopic energy equation for convective heat transfer in porous media. *Int. Comm. Heat Mass Transfer* **15**, 689–703. [4.9]

Hsu, C. T. and Cheng, P. 1990 Thermal dispersion in a porous medium. *Int. J. Heat Mass Transfer* **33**, 1587–1597. [1.5.3, 2.2.3, 4.9]

Hsu, C. T. and Fu, H. 1998 Reacting stagnation flows in catalytic porous beds. *Int. J. Heat Mass Transfer* **41**, 2335–2346. [3.4]

Hsu, C. T., Cheng, P. and Homsy, G. M. 1978 Instability of free convective flow over a horizontal impermeable surface in a porous medium. *Int. J. Heat Mass Transfer* **21**, 1221–1228. [5.4]

Hsu, C. T., Cheng, P. and Wong, K. W. 1994 Modified Zehner-Schundler models for stagnant thermal conductivity of porous media. *Int. J. Heat Mass Transfer* **37**, 2751–2759. [2.2.1]

Hsu, C. T., Cheng, P. and Wong, K. W. 1995 A lumped-parameter model for stagnant thermal conductivity of spatially periodic media. *ASME J. Heat Transfer* **117**, 264–269. [2.2.1]

Hsu, C. T., Fu, H. L. and Cheng, P. 1999 On pressure-velocity correlation of steady and oscillating flows in generators made of wire screens. *ASME Fluids Engng.* **121**, 52–56. [1.5.2]

Hu, X. J., Du, J. H. , Lei, S. Y. and Wang, B. X. 2001 A model for the thermal conductivity of unconsolidated porous media based on capillary pressure-saturation relation. *Int. J. Heat Mass Transfer* **44**, 247–251 (corrigendum 1267–1268). [2.2.1]

Huang, M. J. and Chen, C. K. 1985 Effects of surface mass transfer on free convection flow over vertical cylinder imbedded in a saturated porous medium. *ASME J. Energy Resources Tech.* **107**, 394–396. [5.7]

Huang, M. J., Yih, K. A., Chou, Y. L. and Chen, C. K. 1986 Mixed convection over a horizontal cylinder or a sphere embedded in a saturated porous medium. *ASME J. Heat Transfer* **108**, 469–471. [8.1.3]

Huang, P. C. and Vafai, K. 1993 Flow and heat transfer control over an external surface using a porous block array arrangement. *Int. J. Heat Mass Transfer* **36**, 4019–4032. [4.11]

Huang, P. C. and Vafai, K. 1994a Passive alteration and control of convective heat transfer utilizing alternate porous cavity-block wafers. *Int. J. Heat Fluid Flow* **15**, 48–61. [4.11]

Huang, P. C. and Vafai, K. 1994b Internal heat transfer augmentation in a channel using alternate set of porous cavity-block obstacles. *Numer. Heat Transfer A* **25**, 519–539. [4.11]

Huang, P. C. and Vafai, K. 1994c Analysis of flow and heat transfer over an external boundary covered with a porous substrate. *ASME J. Heat Transfer* **116**, 768–771. [4.11]

Huang, P. C. and Vafai, K. 1994d Analysis of forced convection enhancement in a channel using porous blocks. *AIAA J. Thermophys. Heat Transfer* **8**, 563–573. [4.11]

Huang, P. C., Yang, C. F. and Chang, S. Y. 2004a Mixed convection cooling of heat sources mounted with porous blocks. *J. Thermophys. Heat Transfer* **18**, 464–475. [8.3.1]

Huang, P. C., Yang, C. F., Hwang, J. J. and Cjiu, M. T. 2004b Enhancement of forced-convection cooling of multiple heated blocks in a channel using porous covers. *Int. J. Heat Mass Transfer* **48**, 674–664. [4.11]

Huenefeld, J. S. and Plumb, O. A. 1981 A study of non-Darcy natural convection from a vertical heated surface in a saturated porous medium. *ASME 20th National Heat Transfer Conf.* Paper 81-HT-45. [5.1.8]

Hung, C. I. 1991 Note on conjugate natural convection-conduction heat transfer for a vertical plate fin embedded in high-porosity medium. *Int. J. Non-linear Mech.* **26**, 135–140. [5.12.1]

Hung, C. I. and Chen, C. B. 1997 Non-Darcy free convection in a thermally stratified porous medium along a vertical plate with variable heat flow. *Heat Mass Transfer* **33**, 101–107. [5.1.4]

Hung, C. I., Chen, C. H. and Chen, C. B. 1999 Non-Darcy free convection along a non-isothermal vertical surface in a thermally stratified porous medium. *Int. J. Engng. Sci.* **37**, 477–495. [5.1.4]

Hung, C. I., Chen, C. K. and Cheng, P. 1989 Transient conjugate natural convection heat transfer along a vertical plate fin in a high-porosity medium. *Numer. Heat Transfer* **15**, 133–148. [5.12.1]

Hunt, M. L. and Tien, C. L. 1988a Effects of thermal dispersion on forced convection in fibrous media. *Int. J. Heat Mass Transfer* **31**, 301–309. [4.9]

Hunt, M. L. and Tien, C. L. 1988b Non-Darcian convection in cylindrical packed beds. *ASME J. Heat Transfer* **110**, 378–384. [4.9]

Hunt, M. L. and Tien, C. L. 1990 Non-Darcian flow, heat and mass transfer in catalytic packed-bed reactors. *Chem. Engng. Sci.* **45**, 55–63. [6.19.2]

Hutter, K. and Straughan, B. 1997 Penetrative convection in thawing subsea permafrost. *Cont. Mech. Thermodyn.* **9**, 259–272. [11.3]

Hutter, K. and Straughan, B. 1999 Models for convection in thawing porous media in support for the subsea permafrost equations. *J. Geophys. Res.* **104**, 29249–29260. [11.3]

Hwang, G. and Chao, C. H. 1992 Effects of wall conduction and Darcy number on laminar mixed convection in a horizontal square porous channel. *ASME J. Heat Transfer* **114**, 614–621. [8.2.5]

Hwang, G. J. and Chao, C. H. 1994 Heat transfer measurements and analysis for sintered porous channels. *ASME J. Heat Transfer* **116**, 456–464. [4.9]

Hwang, G. J., Cai, Y. and Cheng, P. 1992 An experimental study of forced convection in a packed channel with asymmetric heating. *Int. J. Heat Mass Transfer* **35**, 3029–3039. [4.5]

Hwang, G. J., Wu, C. C. and Chao, C. H. 1995 Investigation on non-Darcian forced convection in an asymmetrically heated sintured porous channel. *ASME J. Heat Transfer* **117**, 725–732. [4.16.2]

Hwang, I. G. 2001 Convective instability in porous media during solidification. *AIChE J.* **47**, 1698–1700. [10.2.3]

Hwang, J. J., Hwang, G. J., Yeh, R. H. and Chao, C. H. 2002 Measurement of interstitial convective heat transfer and frictional drag for flow across metal foams. *ASME J. Heat Transfer* **124**, 120–124. [4.9]

Ibrahim, F. S. and Hassanien, I. A. 2000 Influence of variable permeability on combined convection along a nonisothermal wedge in a saturated porous medium. *Transport Porous Media* **39**, 57–71. [8.1.4]

Ibrahim, F. S., Abdel-Gaid, S. M. and Gorla, R. S. R. 2000 Non-Darcy mixed convection flow along a vertical plate embedded in a non-Newtonian fluid saturated porous medium with surface mass transfer. *Int. J. Numer. Meth. Heat Fluid Flow* **10**, 397–408. [8.1.1]

Ibrahim, F. S., Mansour, M. A. and Abdel-Gaid, S. M. 2005 Radiative and thermal dispersion effects on non-Darcy natural convection with lateral mass flux for non-Newtonian fluid from a vertical flat plate in a porous medium. *Transport Porous Media* **61**, 45–57. [5.1.9.2]

Imadojemu, H. and Johnson, R. 1991 Convective heat transfer from a heated vertical plate surrounded by a saturated porous medium. *Proc. ASME & JSME Thermal Engineering Joint Conference–1991*. vol.4, pp. 203–212. [5.1.8]

Imadojemu, H. E. and Porter, L. H. 1995 Effective thermal conductivity of a saturated porous medium. *AIAA J. Thermophys. Heat Transfer* **9**, 573–575. [2.2.1]

Imhoff, P. T. and Green, T. 1988 Experimental investigation of double-diffusive groundwater fingers. *J. Fluid Mech.* **188**, 363–382. [9.1.3]

Impey, M. D. and Riley, D. S. 1991 On exchanges between convective modes in a slightly tilted porous cavity. *Math. Proc. Camb. Phil. Soc.* **110**, 395–416. [7.8]

Impey, M. D., Riley, D. S. and Winters, K. H. 1990 The effect of sidewall imperfections on pattern formation in Lapwood convection. *Nonlinearity* **3**, 197–230. [6.15.1]

Inaba, H., Ozaki, K. and Nozu, S. 1993 Convective heat transfer of a horizontal-spherical particle layer heated from below and cooled from above. *Heat Transfer Japan Res.* **22**, 573–595. [4.5]

Inaba, H., Ozaki, K. and Nozu, S. 1994 Mixed convection heat transfer in an open shallow cavity heated from below and packed with a one-step arrangement of spherical particles. *Heat Transfer Japan Res.* **23**, 66–85. [4.5, 8.2.1]

Inaba, H., Sugawara, M. and Blumenberg, J. 1988 Natural convection heat transfer in an inclined porous layer. *Int. J. Heat Mass Transfer* **31**, 1365–1372. [7.8]

Ingebritsen, S. E. and Hayba, D. O. 1994 Fluid-flow and heat-transport near the critical-point of H_2O. *Geophys. Res. Lett.* **21**, 2199–2202. [11.8]

Ingham, D. B. 1986 The non-Darcy free convection boundary layer on axi-symmetric and two-dimensional bodies of arbitrary shape. *Int. J. Heat Mass Transfer* **29**, 1759–1760. [5.9]

Ingham, D. B. 1988 An exact solution for non-Darcy free convection from a horizontal line source. *Wärme-Stoffübertrag.* **22**, 125–127. [5.10.1.1, 5.10.1.2]

Ingham, D. B. and Brown, S. N. 1986 Flow past a suddenly heated vertical plate in a porous medium. *Proc. Roy. Soc. London Ser. A* **403**, 51–80. [5.1.3]

Ingham, D. B. and Merkin, J. H. 1982 Flow past a suddenly cooled vertical flat surface in a saturated porous-medium. *Int. J. Heat Mass Transfer* **25**, 1916–1919. [5.1.3]

Ingham, D. B. and Pop, I. 1986a Free-forced convection from a heated longitudinal horizontal cylinder embedded in a porous medium. *Wärme-Stoffübertrag.* **20**, 283–289. [8.1.3]

Ingham, D. B. and Pop, I. 1986b A horizontal flow past a partially heated infinite vertical cylinder embedded in a porous medium. *Int. J. Engng. Sci.* **24**, 1351–1363. [8.1.3]

Ingham, D. B. and Pop, I. 1987a Darcian free convective flow about an impermeable horizontal surface bounded by a vertical wall. *Int. J. Engng. Sci.* **25**, 373–383. [5.12.2]

Ingham, D. B. and Pop, I. 1987b Free convection from a semi-infinite vertical surface bounded by a horizontal wall in a porous medium. *Int. J. Heat Mass Transfer* **30**, 1615–1622. [5.12.2]

Ingham, D. B. and Pop, I. 1987c Natural convection about a heated horizontal cylinder in a porous medium. *J. Fluid Mech.* **184**, 157–181. [5.5.2]

Ingham, D. B. and Pop, I. 1988 Higher-order effects in natural convection flow over uniform flux inclined flat plates in porous media. *Wärme-Stoffübertrag.* **22**, 239-242. [5.3]

Ingham, D. B. and Pop, I. 1991 Mixed convection about a cylinder embedded to a wedge in porous media. *AIAA J. Thermophys. Heat Transfer* **5**, 117–120. [8.1.4]

Ingham, D. B., Merkin, J. H. and Pop, I. 1982 Flow past a suddenly cooled vertical flat surface in a saturated porous medium. *Int. J. Heat Mass Transfer* **25**, 1916–1919. [5.1.3]

Ingham, D. B., Merkin, J. H. and Pop, I. 1983 The collision of free-convection boundary layers on a horizontal cylinder embedded in a porous medium. *Quart. J. Mech. Appl. Math.* **36**, 313–335. [5.5.1]

Ingham D. B., Merkin, J. H. and Pop, I. 1985a Flow past a suddenly cooled horizontal flat surface in a saturated porous medium. *Acta Mech.* **57**, 183–202. [5.3]

Ingham, D. B., Merkin, J. H. and Pop, I. 1985b Natural convection from a semi-infinite flat plate inclined at a small angle to the horizontal in a saturated porous medium. *Acta Mech.* **57**, 183–202. [5.3]

Ingham, D. B., Pop, I. and Cheng, P. 1990 Combined free and forced convection in a porous medium between two vertical walls with viscous dissipation. *Transport in Porous Media* **5**, 381–398. [8.3.1]

Inoue, M. and Nakayama, A. 1998 Numerical modeling of non-Newtonian fluid flow in a porous medium using a three-dimensional periodic array. *ASME J. Fluids Engng.* **120**, 131–135. [1.5.4]

Irmay, S. 1958 On the theoretical derivation of Darcy and Forchheimer formulas. *Eos, Trans. AGU* **39**, 702–707. [1.5.2]

Islam, M. R. 1992 Evolution of oscillatory and chaotic flows in mixed convection in porous media in the non-Darcy regime. *Chaos, Solitons, Fractals*, **2**, 51–71. [8.2.1]

Islam, M. R. 1993 Route to chaos in chemically enhanced thermal convection in porous media. *Chem. Engng. Comm.* **124**, 77–95. [6.17]

Islam, M. R. and Nandakumar, K. 1990 Transient convection in saturated porous layers. *Int. J. Heat Mass Transfer* **33**, 151–161. [6.17]

Islam, R. M. and Nandakumar, K. 1986 Multiple solutions for buoyancy-induced flow in saturated porous media for large Peclet numbers. *ASME J. Heat Transfer* **108**, 866–871. [8.2.1]

Islam, R. M. and Nandakumar, K. 1988 Mixed convection heat transfer in porous media in the non-Darcy regime. *Canad. J. Chem. Engng.* **66**, 68–74. [8.2.1]

Israel-Cookey, C., Ogulu, A. and Omubo-Pepple, Ichimiya, K. 2003 Influence of viscous dissipation and radiation on unsteady MHD free-convection flow past an infinite heated vertical plate in a porous medium with time-dependent suction. *Int. J. Heat Mass Transfer* **46**, 2305–2311. [5.1.9.4]

Itamura, M. T., Francis, N. D., Webb, S. W. and James, D. L. 2004 In-drift natural convection analysis of the low-temperature operating mode design. *Nuclear Tech.* **148**, 115–124. [11.8]

Iyer, S. V. and Vafai, K. 1999 Passive heat transfer augmentation in a cylindrical annulus utilizing a porous perturbation. *Numer. Heat Transfer A* **36**, 115–128. [7.3.3]

Izadpanah, M. R., Müller-Steinhagen, H. and Jamialahmadi, M. 1998 Experimental and theoretical studies of convective heat transfer in a cylindrical porous medium. *Int. J. Heat Fluid Flow* **19**, 629–635. [4.9]

Jadhav, R. S. and Pillai, K. M. 2003 Numerical study of heat transfer during unsaturated flow in dual-scale porous media. *Numer. Heat Transfer A* **43**, 385–407. [3.6]

Jaffrenou, J. Y., Bories, S. A. and Combarnous, M. A. 1974 Natural convective flows and mean heat transfer in a sloping porous layer. *Heat Transfer 1974*, Japan Soc. Mech. Engs. and Soc. Chem. Engs. Japan, Tokyo. Vol. 5, pp. 83–87. [7.8]

Jaiswal, B. S. and Soundalgekar, V. M. 2001 Oscillating plate temperature effects on a flow past an infinite vertical porous plate with constant suction and embedded in a porous medium. *Heat Mass Transfer* **37**, 125–131. [5.1.9.7]

James, D. F. and Davis, A. M. 2001 Flow at the interface of a model fibrous porous medium. *J. Fluid Mech.* **426**, 47–72. [1.6]

Jamet, P., Frague, D., Costesèque, P., de Marsily, G. and Cernes, A. 1992 The thermo-gravitational effect in porous media: A modeling approach. *Transport Porous Media* **9**, 223–240. [9.1.4]

Jamialahmadi, M., Muller-Steinhagen, H. and Izadpanah, M. R. 2005 Pressure drop, gas hold-up and heat transfer during single and two-phase flow through porous media. *Int. J. Heat Fluid Flow* **26**, 156–172. [3.5]

Jang, J. Y. and Chang, W. J. 1987 Vortex instability of inclined buoyant layer in porous media saturated with cold water. *Int. Comm. Heat Mass Transfer* **14**, 405–416. [5.4]

Jang, J. Y. and Chang, W. J. 1988a Vortex instability of buoyancy-induced inclined boundary layer flow in a saturated porous medium. *Int. J. Heat Mass Transfer* **31**, 759–767. [5.4]

Jang, J. Y. and Chang, W. J. 1988b The flow and vortex instability of horizontal natural convection in a porous medium resulting from combined heat and mass buoyancy effects. *Int. J. Heat Mass Transfer* **31**, 769–777. [9.2.1]

Jang, J. Y. and Chang, W. J. 1988c Buoyancy-induced inclined boundary layer flow in a porous medium resulting from combined heat and mass buoyancy effects. *Int. Comm. Heat Mass Transfer* **15**, 17–30. [9.2.1]

Jang, J. Y. and Chang, W. J. 1988d Buoyancy-induced inclined boundary layer flow in a saturated porous medium. *Comput. Methods Appl. Mech.* **68**, 333–344. [5.3]

Jang, J. Y. and Chang, W. J. 1989 Maximum density effects on vortex instability of horizontal and inclined buoyancy-induced flows in porous media. *ASME J. Heat Transfer* **111**, 572–574. [5.3]

Jang, J. Y. and Chen, C. N. 1989 Natural convection in an inclined porous enclosure with an off-centre diathermal partition. *Wärme-Stoffübertrag.* **24**, 117–123. [7.8]

Jang, J. Y. and Chen, J. L. 1992 Forced convection in a parallel plate channel partially filled with a high porosity medium. *Int. Comm. Heat Mass Transfer* **14**, 263–273. [4.11]

Jang, J. Y. and Chen, J. L. 1993a Thermal dispersion and inertia effects on vortex instability of a horizontal mixed convection flow in a saturated porous medium. *Int. J. Heat Mass Transfer* **36**, 383–389. [5.4]

Jang, J. Y. and Chen, J. L. 1993b Variable porosity effect on vortex instability of a horizontal mixed convection flow in a saturated porous medium. *Int. J. Heat Mass Transfer* **36**, 1573–1583. [5.4]

Jang, J. Y. and Chen, J. L. 1994 Variable porosity and thermal dispersion effects on vortex instability of a horizontal natural convection flow in a saturated porous medium. *Wärme-Stoffübertrag.* **29**, 153–160. [5.4]

Jang, J. Y. and Leu, J. S. 1992 Buoyancy-induced boundary layer flow of liquid in a porous medium with temperature-dependent viscosity. *Int. Comm. Heat Mass Transfer* **19**, 435–444. [5.1.9.9]

Jang, J. Y. and Leu, J. S. 1993 Variable viscosity effects on the vortex instability of free convection boundary layer flow over a horizontal surface in a porous medium. *Int. J. Heat Mass Transfer* **36**, 1287–1294. [5.4]

Jang, J. Y. and Lie, K. N. 1992 Vortex instability of mixed convection flow over horizontal and inclined surfaces in a porous medium. *Int. J. Heat Mass Transfer* **35**, 2077–2085. [5.4]

Jang, J. Y. and Ni, J. R. 1989 Transient free convection with mass transfer from an isothermal vertical flat plate embedded in a porous medium. *Int. J. Heat Fluid Flow* **10**, 59–65. [9.2.1]

Jang, J. Y. and Ni, J. R. 1992 Mixed convection adjacent to inclined flat surfaces embedded in a porous medium. *Wärme-Stoffübertrag.* **27**, 103–108. [8.1.1]

Jang, J. Y. and Shiang, C. T. 1997 The mixed convection plume along a vertical adiabatic surface embedded in a non-Darcian porous medium. *Int. J. Heat Mass Transfer* **40**, 1693–1699. [8.1.4]

Jang, J. Y., Lie, K. N. and Chen, J. L. 1995 Influence of surface mass flux on vortex instability of a horizontal mixed convection flow in a saturated porous medium. *Int. J. Heat Mass Transfer* **38**, 3305–3311. [8.1.2]

Jang, J. Y., Tzeng, D. J. and Shaw, H. J. 1991 Transient free convection with mass transfer on a vertical plate embedded in a high porosity medium. *Numer. Heat Transfer A* **20**, 1–18. [9.2.3]

Jannot, M., Naudin, P. and Vianny, S. 1973 Convection mixte en milieu poreux. *Int. J. Heat Mass Transfer* **16**, 395–410. [8.1.4]

Jany, P. and Bejan, A. 1988a Scales of melting in the presence of natural convection in a rectangular cavity filled with porous medium. *ASME J. Heat Transfer* **110**, 526–529. [10.1.1, 10.1.2]

Jany, P. and Bejan, A. 1988b Scaling theory of melting with natural convection in an enclosure. *Int. J. Heat Mass Transfer* **31**, 1221–1235 [10.1.1]

Jasmin, S. and Prud'homme, M. 2005 Inverse determination of a heat source from a solute concentration generation model in porous medium. *Int. Comm. Heat Mass Transfer* **32**, 43–53. [9.3.1]

Jecl, R. and Skerget, L. 2000 Natural convection in a vertical porous cavity. *Zeit. Angew. Math. Mech.* **80**, S691–S693. [7.1.2]

Jecl, R. and Skerget, L. 2003 Boundary element method for natural convection in non-Newtonian fluid saturated square porous cavity. *Engng. Anal. Bound. Elem.* **27**, 963–975. [7.1.6]

Jecl, R. and Škerget, L. 2004 Comparison between Forchheimer and the Brinkman model for natural convection in porous media by the boundary element method. In *Applications of Porous Media (ICAPM 2004)*, (eds. A. H. Reis and A. F. Miguel), Évora, Portugal, pp. 113–120. [7.6.2]

Jecl, R., Škerget, L. and Petresin, E. 2001 Boundary domain integral method for transport phenomena in porous media. *Int. J. Numer. Meth. Fluids* **35**, 39–54. [7.6.2]

Jen, T. C. and Yan, T. Z. 2005 Developing flow and heat transfer in a channel partially filled with a porous medium. *Int. J. Heat Mass Transfer* **48**, 3995–4009. [4.11]

Jha, B. K. 1997 Transient convection through vertical porous medium. *Heat Mass Transfer* **33**, 261–263. [7.5]

Jha, B. K. 2005 Free convection flow through an annular porous medium. *Heat Mass Transfer* **41**, 675–679. [7.3.3]

Jiang, C. G., Sahgir, M. Z. and Kawaji, M. 2004a Thermo-solutal convection in heterogeneous porous media. In *Applications of Porous Media (ICAPM 2004)*, (eds. A. H. Reis and A. F. Miguel), Évora, Portugal, pp. 287–292. [9.1.4]

Jiang, C. G., Saghir, M. Z., Kawaji, M. and Ghorayeb, K. 2004b Two-dimensional numerical simulation of thermo-gravitational convection in a vertical porous column filled with a binary mixture. *Int. J. Therm. Sci.* **43**, 1057–1065. [9.1.4]

Jiang, C. G., Saghir, M. Z., Kawaji, M. and Ghorayeb, K. 2004c Contribution of the thermal and molecular diffusion to convection in a vertical porous cavity. In *Emerging Technologies and Techniques in Porous Media* (D. B. Ingham, A. Bejan, E. Mamut and I. Pop, eds), Kluwer Academic, Dordrecht, pp. 307–320. [9.1.4]

Jiang, F., Liu, S. Wang, G. and Chen, H. Z. 2004d The air flow and heat transfer in gravel embankment in permafrost regions. *Science in China D* **47**, 142–151. [7.3.7]

Jiang, P. X. and Ren, Z. 2001 Numerical investigation of forced convection heat transfer in porous media using a thermal non-equilibrium model. *Int. J. Heat Fluid Flow* **22**, 102–110. [4.10]

Jiang, P. X., Fan, M. H., Si, G. S. and Ren. Z. P. 2001 Thermal-hydraulic performance of small scale microchannel and porous-media heat exchangers. *Int. J. Heat Mass Transfer* **44**, 1039–1051. [4.15]

Jiang, P. X., Li, M., Ma, Y. C. and Ren, Z. P. 2004e Boundary conditions and wall effect for forced convection heat transfer in sintered porous channels. *Int. J. Heat Mass Transfer* **47**, 2073–2083. [4.9]

Jiang, P. X., Li, M., Ma, Y. C. and Ren, Z. P. 2004f Experimental research on convection heat transfer in sintered porous channels. *Int. J. Heat Mass Transfer* **47**, 2085–2096. [4.9]

Jiang, P. X., Li, M., Lu, T. J., Ren, Z. P. and Sun, X. J. 2002 Convection heat transfer in sintered porous plate channels. *Heat Transfer 2002, Proc. 12th Int. Heat Transfer Conf.*, Elsevier, Vol. 2, pp. 803–808. [4.10]

Jiang, P. X., Ren, Z. and Wang, B. X. 1999a Numerical simulation of forced convection heat transfer in porous plate channels using thermal equilibrium and non-thermal equilibrium models. *Numer. Heat Transfer A* **35**, 99–113. [4.10]

Jiang, P. X., Si, G. S., Li. M. and Ren, Z. P. 2004g Experimental and numerical investigation of forced convection heat transfer of air in non-sintered porous media. *Exp. Thermal Fluid Sci.* **28**, 545–555. [4.9]

Jiang, P. X., Wang, B. X. and Ren, Z. P. 1994 A numerical investigation of mixed convection in a vertical porous annulus. *Heat Transfer 1994*, Inst. Chem. Engrs, Rugby, vol. 5, pp. 303–308. [8.3.3]

Jiang, P. X., Wang, B. X., Juo, D. A. and Ren, Z. P. 1996 Fluid flow and convective heat transfer in a vertical annulus. *Numer. Heat Transfer A* **30**, 305–320. [8.3.3]

Jiang, P. X., Wang, Z. and Ren, Z. 1998 Fluid flow and concentration heat transfer in a plate channel filled with solid particles. *Heat Transfer 1998, Proc. 11th IHTC* **4**, 405–410. [4.10]

Jiang, P. X., Wang, Z., Ren, Z. and Wang, B. X. 1999b Experimental research of fluid flow and convection heat transfer in plate channels filled with glass or metallic particles. *Exp. Thermal Fluid Sci.* **20**, 45–54. [4.9]

Jiang, P. X., Xu, Y. J. and Li, M. 2004h Experimental investigation of convection heat transfer in min-fin structures and sintered porous media. *J. Enhanced Heat Transfer* **11**, 391–405. [4.9]

Jiang, P. X., Xu, Y. J. and Shi, R. F. 2004i Experimental investigation of convection heat transfer of CO at supercritical pressures in a porous tube. In *Applications of Porous Media (ICAPM 2004)*, (eds. A. H. Reis and A. F. Miguel), Évora, Portugal, pp. 173–181. [4.5]

Jiang, P. X., Xu, Y. J., Lu, J., Shi, R. F., He, S. and Jackson, J. D. 2004j Experimental investigation of convection heat transfer of CO_2 at supercritical pressure in vertical mini-tubes and porous media. *Appl. Thermal Engng.* **24**, 1255–1270. [4.5]

Jiang, Y. Y. and Shoji, M. 2002 Thermal convection in a porous toroidal thermosyphon. *Int. J. Heat Mass Transfer* **45**, 3459–3470. [7.3.7]

Jiménez-Islas, H., López-Isunza, F. and Ochoa-Tapia, J. A. 1999 Natural convection in a cylindrical porous cavity with internal heat source: a numerical study with Brinkman-extended Darcy model. *Int. J. Heat Mass Transfer* **42**, 4185–4195. [7.3.8]

Jiménez-Islas, H., Navarrete-Bolanos, J. L. and Botello-Alvarez, E. 2004 Numerical study of the natural convection of heat and 2-D mass of grain stored in cylindrical silos. *Agrosciencia* **38**, 325–342. [6.11.2]

Johannsen, K. 2003 On the validation of the Bousinesq approximation for the Elder problem. *Comput. Geosci.* **7**, 169–182. [2.3]

Johnson, C. H. and Cheng, P. 1978 Possible similarity solutions for free convection boundary layers adjacent to flat plates in porous media. *Int. J. Heat Mass Transfer* **21**, 709–718. [5.1.9.9]

Joly, F., Vasseur, P. and Labrosse, G. 2001 Soret instability in a vertical Brinkman porous enclosure. *Numer. Heat Transfer A* **39**, 339–359. [9.1.4]

Joly, N. and Bernard, D. 1995 Critère d'apparition de la convection naturelle dans une couch poreuse horizontale limitée par des frontièrs conductrices de la chaleur: effet des anisotropies. *C.R. Acad. Sci. Paris, Ser. II*, **320**, 573–579. [6.12]

Joly, N., Bernard, D. and Menegazzi, P. 1996 ST2D3D: An FE program to compute stability criteria for natural convection in complex porous structures. *Numer. Heat Transfer B* **29**, 91–112. [1.9]

Jones, I. P. 1973 Low Reynolds number flow past a porous spherical shell. *Proc. Camb. Phil. Soc.* **73**, 231–238. [1.6]

Jones, M. C. and Persichetti, J. M. 1986 Convective instability in packed beds with through-flow. *AIChE J.* **32**, 1555–1557. [6.10]

Jonsson, T. and Catton, I. 1987 Prandtl number dependence of natural convection in porous media. *ASME J. Heat Transfer* **109**, 371–377. [6.9.2]

Joseph, D. D. 1976 *Stability of Fluid Motions II*, Springer, Berlin. [2.3, 6.3, 6.4]

Joseph, D. D., Nield, D. A. and Papanicolaou, G. 1982 Nonlinear equation governing flow in a saturated porous medium. *Water Resources Res.* **18**, 1049–1052 and **19**, 591. [1.5.2]

Joshi, V. and Gebhart, B. 1984 Vertical natural convection flows in porous media: calculations of improved accuracy. *Int. J. Heat Mass Transfer* **27**, 69–75. [5.1.6]

Joshi, Y. and Gebhart, B. 1985 Mixed convection in porous media adjacent to a vertical uniform heat flux surface. *Int. J. Heat Mass Transfer* **28**, 1783–1786. [8.1.1]

Jou, J. J., Kung, K. Y. and Hsu, C. H. 1996 Thermal stability of horizontally superposed porous fluid layers in a rotating system. *Int. J. Heat Mass Transfer* **39**, 1847–1857. [6.19.1.2]

Joulin, A. and Ouarzazi, M. N. 2000 Mixed convection of a binary mixture in a porous medium. *C. R. Acad. Sci. Paris II* **B 328**, 311–316. [9.1.6.4]

Jounet, A. and Bardan, G. 2001 Onset of thermohaline convection in the presence of vertical vibration. *Phys. Fluids* **13**, 3234–3246. [9.1.6.4]

Jue, T. C. 2000 Analysis of heat and fluid flow in partially divided fluid saturated porous cavities. *Heat Mass Transfer* **36**, 285–294. [7.3.7]

Jue, T. C. 2001a Analysis of oscillatory flow with thermal convection in a rectangular cavity filled with a porous medium. *Int. Comm. Heat Mass Transfer* **27**, 985–994. [6.15.3]

Jue, T. C. 2001b Analysis of Bénard convection in rectangular cavities filled with a porous medium. *Acta Mech.* **146**, 21–29. [6.8, 6.15.3]

Jue, T. C. 2003 Analysis of thermal convection in a fluid-saturated porous cavity with internal heat generation. *Heat Mass Transfer* **40**, 83–89. [7.3.8]

Jumah, R. Y. and Majumdar, A. S. 2000 Free convection heat mass transfer of non-Newtonian power law fluids with yield stress from a vertical flat plate in saturated porous media. *Int. Comm. Heat Mass Transfer* **27**, 485–494. [9.2.1]

Jumah, R. Y. and Majumdar, A. S. 2001 Natural convection heat mass transfer from a vertical flat plate with variable wall temperature and concentration to power-law fluids with yield stress in a porous medium . *Chem. Engng. Comm.* **185**, 165–182. [9.2.1]

Jumar, R. J., Banat, F. A. and Abu-Al-Rub, F. 2001 Darcy-Forchheimer mixed convection heat and mass transfer in fluid saturated porous media. *Int. J. Numer. Meth. Heat Fluid Flow* **11**, 600–618. [9.2.1]

Jupp, T. E. and Schultz, A. 2000 A thermodynamic explanation for black smoker temperatures. *Nature* **403**, 880–883. [11.8]

Jupp, T. E. and Schultz, A. 2004 Physical balances in subseafloor hydrothermal convection cells. *J. Geophys. Res.–Solid Earth* **109**, Art # B05101. [11.8]

Kacur, J. and Van Keer, R. 2003 Numerical approximation of a flow and transport system in unsaturated porous media. *Chem Engng. Sci.* **58**, 4805–4813. [3.6]

Kalla, L., Mamou, M., Vasseur, P. and Robillard, L. 1999 Multiple steady states for natural convection in shallow porous cavity subject to uniform heat fluxes. *Int. Comm. Heat Mass Transfer* **26**, 761–770. [7.2]

Kalla, L., Mamou, M., Vasseur, P. and Robillard, L. 2001a Multiple solutions for double-diffusive convection in a shallow porous cavity with vertical fluxes of heat and mass. *Int. J. Heat Mass Transfer* **44**, 4493–4504. [9.1.6.4]

Kalla, L., Vasseur, P., Benacer, R., Beji, H and Duval, R. 2001b Double-diffusive convection within a horizontal porous layer salted from the bottom and heated horizontally. *Int. Comm. Heat Mass Transfer* **28**, 1–10. [9.5]

Kaloni, P. N. and Qiao, Z. 1997 Non-linear stability of convection in a porous medium with inclined temperature gradient. *Int. J. Heat Mass Transfer* **40**, 1611–1615. [7.9]

Kaloni, P. N. and Qiao, Z. 2000 Nonlinear convection induced by inclined thermal and solutal gradient with mass flow. *Cont. Mech. Thermodyn.* **12**, 185–194. [9.5]

Kaloni, P. N. and Qiao, Z. 2001 Nonlinear convection in a porous medium with inclined temperature gradient and variable gravity effects. *Int. J. Heat Mass Transfer* **44**, 1585–1591. [7.9]

Kamel, M. H. 2001 Unsteady MHD convection through porous medium with combined heat and mass transfer with heat source/sink. *Energy Conv. Management* **42**, 393–405. [9.2.1]

Kamiuto, K. and Saitoh, S. 1994 Fully developed forced-convection heat transfer in cylindrical packed beds with constant wall temperatures. *JSME Int. J. Series B* **37**, 554–559. [4.5]

Kaneko, T., Mohtadi, M. F., and Aziz, K. 1974 An experimental study of natural convection in inclined porous media. *Int. J. Heat Mass Transfer* **17**, 485–496. [6.9.1, 7.8]

Karasozen, B. and Tsybulin, V. G. 2004 Cosymmetric families of steady states in Darcy convection and their collision. *Phys. Lett. A.* **323**, 67–76. [6.4]

Karcher, C. and Müller, U. 1995 Convection in a porous medium with solidification. *Fluid Dyn. Res.* **15**, 25–42. [10.2.2]

Karimi-Fard, M., Charrier-Mojtabi, M. C. and Mojtabi, A. 1997 Non-Darcy effects on double-diffusive convection within a porous medium. *Numer. Heat Transfer A* **31**, 837–852. [9.2.2]

Karimi-Fard, M., Charrier-Mojtabi, M. C. and Mojtabi, A. 1998 Analytical and numerical simulation of double-diffusive convection in a tilted cavity filled with porous medium. *Heat Transfer 1998, Proc. 11th IHTC*, **4**, 453–458. [9.4]

Karimi-Fard, M., Charrier-Mojtabi, M. C. and Mojtabi, A. 1999 Onset of stationary and oscillatory convection in a tilted porous cavity saturated with a binary fluid: Linear stability analysis. *Phys. Fluids* **11**, 1346–1358. [9.4]

Kassoy, D. R. and Cotte, B. 1985 The effects of sidewall heat loss on convection in a saturated porous vertical slab. *J. Fluid Mech.* **152**, 361–378. [6.15.2]

Kassoy, D. R. and Zebib, A. 1978 Convection fluid dynamics in a model of a fault zone in the earth's crust. *J. Fluid Mech.* **88**, 769–792. [6.15.2]

Katto, Y. and Masuoka, T. 1967 Criterion for the onset of convective flow in a fluid in a porous medium. *Int. J. Heat Mass Transfer* **10**, 297–309. [6.9.1]

Kauffman, S. A. 1993 *The Origins of Order: Self-Organization and Selection and Evolution*, Oxford University Press, London. [4.18]

Kaviany, M. 1984a Thermal convective instabilities in a porous medium. *ASME J. Heat Transfer* **106**, 137–142. [6.11.2, 6.11.3]

Kaviany, M. 1984b Onset of thermal convection in a saturated porous medium: experiment and analysis. *Int. J. Heat Mass Transfer* **27**, 2101–2110. [6.11.3]

Kaviany, M. 1985 Laminar flow through a porous channel bounded by isothermal parallel plates. *Int. J. Heat Mass Transfer* **28**, 851–858. [4.9]

Kaviany, M. 1986 Non-Darcian effects on natural convection in porous media confined between horizontal cylinders. *Int. J. Heat Mass Transfer* **29**, 1513–1519. [7.6.2]

Kaviany, M. 1987 Boundary-layer treatment of forced convection heat transfer from a semi-infinite flat plate embedded in porous media. *ASME J. Heat Transfer* **109**, 345–349. [4.8]

Kaviany, M. 1995 *Principles of Heat Transfer in Porous Media*, Second Edition, Springer, New York. [1.5.2, 6.10]

Kaviany, M. and Mittal, M. 1987 Natural convection heat transfer from a vertical plate to high permeability porous media: an experiment and an approximate solution. *Int. J. Heat Mass Transfer* **30**, 967–978. [5.1.7.2, 5.1.8]

Kawada, Y., Yoshida, S. and Watanabe, S. 2004 Numerical simulations of mid-ocean ridge hydrothermal circulation including the phase separation of sea water. *Earth Planets Space* **56**, 193–215. [11.8]

Kazmierczak, M. and Muley, A. 1994 Steady and transient natural convection experiments in a horizontal porous layer: The effects of a thin top fluid layer and oscillating bottom wall temperature. *Int. J. Heat Fluid Flow* **15**, 30–41. [6.9.1, 6.19.1.2]

Kazmierczak, M. and Poulikakos, D. 1988 Melting of an ice surface in a porous medium. *AIAA J. Thermophys. Heat Transfer* **2**, 352–358. [10.1.7]

Kazmierczak, M. and Poulikakos, D. 1989 Numerical simulation of transient double diffusion in a composite porous/fluid layer heated from below. *AIChE Sympos. Ser.* **269**, 108–114. [9.4]

Kazmierczak, M. and Poulikakos, D. 1991 Transient double diffusion in a fluid layer extending over a permeable substrate. *ASME J. Heat Transfer* **113**, 148–157. [9.4]

Kazmierczak, M., Poulikakos, D. and Pop, I. 1986 Melting from a flat plate embedded in a porous medium in the presence of steady natural convection. *Numer. Heat Transfer* **10**, 571–582. [10.1.1, 10.1.5]

Kazmierczak, M., Poulikakos, D. and Sadowski, D. 1987 Melting of a vertical plate in porous medium controlled by forced convection of a dissimilar fluid. *Int. Comm. Heat Mass Transfer* **14**, 507–517. [10.1.7]

Kazmierczak, M., Sadowski, D. and Poulikakos, D. 1988 Melting of a solid in a porous medium induced by free convection of a warm dissimilar fluid. *ASME J. Heat Transfer* **110**, 520–523. [10.1.7]

Kessler, M. A. and Werner, B. T. 2003 Self-organization of sorted patterned ground. *Science* **299**, 380–383. [11.2]

Khadrawi, A. F. and Al-Nimr, M. A. 2003a The effect of the local inertial term on the free-convection fluid flow in vertical channels partially filled with porous media. *J. Porous Media* **6**, 59–70. [7.7]

Khadrawi, A. F. and Al-Nimr, M. A. 2003b Examination of the thermal equilibrium assumption in transient natural convection flow in porous channel. *Transport Porous Media* **53**, 317–329. [7.5]

Khadrawi, A. F. and Al-Nimr, M. A. 2005 The effect of the local inertial term on the transient free convection from a vertical plate inserted in a semi-infinite domain partly filled with porous material. *Transport Porous Media*, in press. [5.1.3]

Khaled, A. R. A. and Chamkha, A. J. 2001 Variable porosity and thermal dispersion effects on coupled heat and mass transfer by natural convection from a surface embedded in a non-metallic porous medium. *Int. J. Numer. Meth. Heat Fluid Flow* **11**, 413–429. [9.2.1]

Khaled, A. R. A. and Vafai, K. 2003 The role of porous media in modeling flow and heat transfer in biological tissues. *Int. J. Heat Mass Transfer* **46**, 4989–5003. [1.9, 2.6]

Khalili, A. and Shivakumara, I. S. 1998 Onset of convection in a porous layer with net through-flow and internal heat generation. *Phys. Fluids* **10**, 315–317. [6.10]

Khalili, A. and Shivakumara, I. S. 2003 Non-Darcian effects on the onset of convection in a porous layer with throughflow. *Transport Porous Media* **53**, 245–263. [6.10]

Khalili, A., Shivakumara, I. S. and Huettel, M. 2002 Effects of throughflow and internal heat generation on convective instabilities in an anisotropic porous layer. *J. Porous Media* **5**, 187–198. [6.10]

Khalili, A., Shivakumara, I. S. and Suma, S. P. 2003 Convective instability in superposed fluid and porous layers with vertical throughflow. *Transport Porous Media* **51**, 1–18. [6.19.3]

Khallouf, H., Gershuni, G. Z. and Mojtabi, A. 1996 Some properties of convective oscillations in porous medium. *Numer. Heat Transfer A* **30**, 605–618. [6.15.3]

Khan, A. A. and Zebib, A. 1981 Double diffusive instability in a vertical layer of porous medium. *ASME J. Heat Transfer* **103**, 179–181. [9.2.4]

Khanafer, K. and Vafai, K. 2001 Isothermal surface production and regulation for high heat flux applications using porous inserts. *Int. J. Heat Mass Transfer* **44**, 2933–2947. [4.11]

Khanafer, K. and Vafai, K. 2002 Double-diffusive mixed convection in a lid-driven enclosure filled with a fluid-saturated porous medium. *Numer. Heat Transfer A* **42**, 465–486. [9.6]

Khanafer, K. and Vafai, K. 2005 Transport through porous media–a synthesis of the state of the art for the past couple of decades. *Ann. Rev. Heat Transfer*, in press. [4.11]

Khanafer, K. M. and Chamkha, A. J. 1998 Hydromagnetic natural convection from an inclined porous square enclosure with heat generation. *Numer. Heat Transfer A* **33**, 891–910. [7.8]

Khanafer, K. M. and Chamkha, A. J. 1999 Mixed convection flow in a lid-driven enclosure filled with a fluid-saturated porous medium. *Int. J. Heat Mass Transfer* **42**, 2465–2481. [8.2.1]

Khanafer, K. M. and Chamkha, A. J. 2003 Mixed convection with a porous bed heat generating horizontal annulus. *Int. J. Heat Mass Transfer* **46**, 1725–1735. [8.2.3]

Khanafer, K. M., Al-Najem, N. M. and El-Refaee, M. M. 2000 Natural convection in tilted porous enclosures in the presence of a transverse magnetic field. *J. Porous Media* **3**, 79–91. [7.8]

Khanafer, K., Vafai, K. and Kangarlu, A. 2003 Computational modeling of cerebral diffusion–application to stroke imaging. *Magnet. Reson. Imag.* **21**, 651–661. [2.6]

Khare, H. C. and Sahai, A. K. 1993 Thermosolutal convection in a heterogeneous fluid layer in porous medium in the presence of a magnetic field. *Int. J. Engng. Sci.* **31**, 1507–1517. [9.1.6.4]

Khashan, S. A. and Al-Nimr, M. A. 2005 Validation of the local thermal equilibrium assumption in forced convection in non-Newtonian fluids through porous channels. *Transport Porous Media* **61**, 291–305. [4.10]

Khashan, S. A., Al-Amiri, A. M. and Al-Nimr, M. A. 2005 Assessment of the local thermal non-equilibrium condition in developing forced convection flows through fluid-saturated porous tubes. *Appl. Therm. Engng.* **25**, 1429–1445. [4.10]

Kim, G. B. and Hyun, J. M. 2004 Buoyant convection of a power-law fluid in an enclosure filled with heat-generating porous media. *Numer. Heat Transfer A* **45**, 569–582. [6.11.2]

Kim, G.B., Hyun. J. M. and Kwak, H. S. 2001 Buoyant convection in a square cavity partially filled with a heat-generating porous medium. *Numer. Heat Transfer A* **40**, 601–618. [6.19.3]

Kim, K. H., Kim, S. J. and Hyun, J. M. 2004 Development of boundary layers in transient convection about a vertical plate in a porous medium. *J. Porous Media* **7**, 249–259. [5.1.3]

Kim, M. C. and Kim, S. 2005 Onset of convective stability in a fluid-saturated porous layer subjected to time-dependent heating. *Int. Comm. Heat Mass Transfer* **32**, 416–424. [6.11.3]

Kim, M. C, Kim, S. and Choi, C. K. 2002a Convective instability in fluid-saturated porous layer under uniform volumetric heat sources. *Int. Comm. Heat Mass Transfer* **29**, 919–928. [6.11.2]

Kim, M. C, Kim, S., Chung, B. J. and Choi, C. K. 2003a Convective instability in a horizontal porous layer saturated with oil and a layer of gas underlying it. *Int. Comm. Heat Mass Transfer* **30**, 225–234. [6.19.3]

Kim, M. C., Kim, K.Y. and Kim, S. 2004 The onset of transient convection in fluid-saturated porous layer. *Int. Comm. Heat Mass Transfer* **31**, 53–62. [6.11.3]

Kim, M. C., Lee, S. B., Chung, B. J. and Kim, S. 2002b Heat transfer correlation in fluid-saturated porous layer under uniform volumetric heat sources. *Int. Comm. Heat Mass Transfer* **29**, 1089–1097. [6.11.2]

Kim, M. C., Lee, S. B., Kim. and Chung, B. J. 2003b Thermal instability of viscoelastic fluids in porous media. *Int. J. Heat Mass Transfer* **46**, 5065–5072. [6.23]

Kim, S. and Kim, M. C. 2002 A scale analysis of turbulent heat transfer driven by buoyancy in a porous layer with homogeneous heat sources. *Int. Comm. Heat Mass Transfer* **29**, 127–134. [6.11.2]

Kim, S. J, Kang, B. H. and Kim, J. H. 2001 Forced convection from aluminum foam materials in an asymmetrically heated channel. *Int. J. Heat Mass Transfer* **44**, 1451–1454. [4.9]

Kim, S. J. and Choi, C. Y. 1996 Convective heat transfer in porous and overlying fluid layers heated from below. *Int. J. Heat Mass Transfer* **39**, 319–329. [6.19.1.2, 6.19.2]

Kim, S. J. and Hyun, J. M. 2005 A porous medium aproach for the thermal analysis of heat transfer devices. In *Transport Phenomena in Porous Media III*, (eds. D. B. Ingham and I. Pop), Elsevier, Oxford, pp. 120–146. [4.16.1]

Kim, S. J. and Jang, S. P. 2002 Effects of the Darcy number, the Prandtl number and the Reynolds number on local thermal non-equilibrium. *Int. J. Heat Mass Transfer* **45**, 3885–3896. [4.10]

Kim, S. J. and Kim, D. 1999 Forced convection in microstructure for electronic equipment cooling. *ASME J. Heat Transfer* **121**, 639–645. [4.9]

Kim, S. J. and Kim, D. 2000 Discussion: "Heat transfer measurement and analysis for sintered porous channels" (Hwang, G. J. and Chao, C. H., 1994, ASME J. Heat Transfer, 116, pp. 456–469). *ASME J. Heat Transfer* **122**, 632–633. [4.9]

Kim, S. J. and Kim, D. 2001 Thermal interaction at the interface between a porous medium and an impermeable wall. *ASME J. Heat Transfer* **123**, 527–533. [2.4]

Kim, S. J. and Vafai, K. 1989 Analysis of natural convection about a vertical plate embedded in a porous medium. *Int. J. Heat Mass Transfer* **32**, 665–677. [5.1.7.1]

Kim, S. J., Kim, D. and Lee, D. Y. 2000 On the local thermal equilibrium of microchannel heat sinks. *Int. J. Heat Mass Transfer* **43**, 1735–1748. [4.10]

Kim, S. J., Yoo, J. W. and Jang, S. P. 2002 Thermal optimization of a circular-sectored finned tube using a porous medium approach. *ASME J. Heat Transfer* **124**, 1026–1033. [4.16.2]

Kim, S. Y. and Kuznetsov, A. V. 2003 Optimization of pin-fin heat sinks using anisotropic local thermal non-equilibrium porous model in a jet impinging channel. *Numer. Heat Transfer A* **44**, 771–787. [4.16.2]

Kim, S. Y., Kang, B. H. and Hyun, J. M. 1994 Heat transfer from pulsating flow in a channel filled with porous media. *Int. J. Heat Mass Transfer* **37**, 2025–2033. [4.12.2]

Kim, S. Y., Koo, J. M. and Kuznetsov, A. V. 2001 Effect of anisotropy in permeability and effective thermal conductivity on thermal performance of an aluminum foam heat sink. *Numer. Heat Transfer A* **40**, 21–36. [4.16.2]

Kim, S. Y., Paek, J. W. and Kang, B. H. 2000 Flow and heat transfer correlations for porous fin in a plate-fin heat exchanger. *ASME J. Heat Transfer* **122**, 572–578. [4.16.2]

Kim, W. T., Hong, K. H., Jhon, M. S., Van Osdol, J. G. and Smith, D. H. 2003 Forced convection in a circular pipe with a partially filled porous medium. *KSME Int. J.* **17**, 1583–1595. [4.11]

Kim, Y. J. 2001a Unsteady MHD convection flow of polar fluids past a vertical moving porous plate in a porous medium. *Int. J. Heat Mass Transfer* **44**, 2791–2799. [5.1.9.2]

Kim, Y. J. 2001b Unsteady convection flow of micropolar fluids past a vertical porous plate embedded in a porous medium. *Acta Mech.* **148**, 105–116. [5.1.9.2]

Kim, Y. J. 2004 Heat and mass transfer in MHD micropolar flow over a vertical moving porous plate in a porous medium. *Transport Porous Media* **56**, 17–37. [9.2.1]

Kimmich, R., Klemm, A. and Weber, M. 2001 Flow, diffusion, and thermal convection in percolation clusters: NMR experiments and numerical FEM/FVM simulations. *Magnet. Reson. Imag.* **19**, 353–361. [6.9.1]

Kimura, S. 1988a Forced convection heat transfer about an elliptic cylinder in a saturated porous medium. *Int. J. Heat Mass Transfer* **31**, 197–199. [4.3]

Kimura, S. 1988b Forced convection heat transfer about a cylinder placed in porous media with longitudinal flows. *Int. J. Heat Fluid Flow* **9**, 83–86. [4.3]

Kimura, S. 1988c Transient heat transfer from a circular cylinder with constant heat flux in a saturated porous layer; application to underground water velocimetry. *Int. Symp. Geothermal Energy*, Kumamoto and Beppu, Japan, November 10–14. [4.6]

Kimura, S. 1989a Transient forced convection heat transfer from a circular cylinder in a saturated porous medium. *Int. J. Heat Mass Transfer* **32**, 192–195. [4.6]

Kimura, S. 1989b Transient forced and natural convection heat transfer about a vertical cylinder in a porous medium. *Int. J. Heat Mass Transfer* **32**, 617–620. [4.6.4, 5.7]

Kimura, S. 1992 Time-dependent phenomena in porous media convection. *Heat and Mass Transfer in Porous Media* (eds. M. Quintard and M. Todorovic), Elsevier, Amsterdam, 277–292. [7.1.4]

Kimura, S. 1998 Onset of oscillatory convection in a porous medium. *Transport Phenomena in Porous Media* (eds. D. B. Ingham and I. Pop), Elsevier, Oxford, pp. 77–102. [6.15.11]

Kimura, S. 2003 Heat transfer through a vertical partition separating porous-porous or porous-fluid reservoirs at different temperatures. *Int. J. Energy Res.* **27**, 891–905. [7.1.5]

Kimura, S. 2005 Dynamic solidification in a water-saturated porous medium cooled from above. In *Transport Phenomena in Porous Media III*, (eds. D. B. Ingham and I. Pop), Elsevier, Oxford, pp. 399–417. [10.2.2]

Kimura, S. and Bejan, A. 1983 The "heatline" visualization of convective heat transfer. *J. Heat Transfer* **105**, 916–919. [4.17]

Kimura, S. and Bejan, A. 1985 Natural convection in a stably heated corner filled with porous medium. *ASME J. Heat Transfer* **107**, 293–298. [5.12.2]

Kimura, S. and Nigorinuma, H. 1991 Heat transfer from a cylinder in a porous medium subjected to axial flow. *Heat Transfer Jap. Res.* **20**, 368–375. [4.3]

Kimura, S. and Okajima, A. 2000 Natural convection heat transfer in an anisotropic porous cavity heated form the side: Part 1. Theory. *Heat Transfer—Asian Research* **29**, 373–384. [7.3.2]

Kimura, S. and Pop, I. 1991 Non-Darcian effects on conjugate natural convection between horizontal concentric cylinders filled with a porous medium. *Fluid Dyn. Res.* **7**, 241–253. [7.3.3]

Kimura, S. and Pop, I. 1992a Conjugate natural convection between horizontal concentric cylinders filled with a porous medium. *Wärme-Stoffübertrag.* **27**, 85–91. [7.3.3]

Kimura, S. and Pop, I. 1992b Conjugate free convection from a circular cylinder in a porous medium. *Int. J. Heat Mass Transfer* **35**, 3105–3113. [5.5.1]

Kimura, S. and Pop, I. 1994 Conjugate free convection from a sphere in a porous medium. *Int. J. Heat Mass Transfer* **37**, 2187–2192. [5.6.1]

Kimura, S., Bejan, A., and Pop, I. 1985 Natural convection near a cold plate facing upward in a porous medium. *ASME J. Heat Transfer* **107**, 819–825. [5.2]

Kimura, S., Kiwata, T., Okajima, A. and Pop, I. 1997 Conjugate natural convection in porous media. *Adv. Water Resource*, **20**, 111–126. [5.12.3]

Kimura, S., Masuda, Y. and Hayashi, K. 1993 Natural convection in an anisotropic porous medium heated from the side. *Heat Transfer Japanese Research* **22**, 139–153. [7.3.2]

Kimura, S., Okajima, A. and Kiwata, T. 2000 Natural convection heat transfer in an anisotropic porous cavity heated from the side (2nd report, experiment by Hele-Shaw cell). *Trans. Jap. Soc. Mech. Engrs. B* **66**, 2950 et seq. [7.3.2]

Kimura, S., Schubert, G. and Straus, J. M. 1986 Route to chaos in porous-medium thermal convection. *J. Fluid Mech.* **166**, 305–324. [6.8]

Kimura, S., Schubert, G. and Straus, J. M. 1987 Instabilities of steady, periodic and quasi-periodic modes of convection in porous media. *ASME J. Heat Transfer* **109**, 350–355. [6.8]

Kimura, S., Schubert, G. and Straus, J. M. 1989 Time-dependent convection in a fluid-saturated porous cube heated from below. *J. Fluid Mech.* **207**, 153–189. [6.15.1]

Kimura, S., Vynnycky, M. and Alavyoon, F. 1995 Unicellular natural circulation in a shallow horizontal porous layer heated from below by a constant flux. *J. Fluid Mech.* **294**, 231–257. [6.8]

Kimura, S., Yoneya, M., Ikeshoji, T. and Shiraishi, M. 1994 Heat transfer to ultralarge-scale heat pipes placed in a geothermal reservoir (3rd report)—Effects of natural convection. *Geotherm. Sci. Technol.* **4**, 77–96. [8.1.4]

Kissling, W., McGuinness, M. J., McNabb, A., Weir, G., White, S. and Young, R. 1992a Analysis of one-dimensional horizontal two-phase flow in geothermal reservoirs. *Transport in Porous Media* **7**, 223–253. [11.9.1]

Kissling, W., McGuinness, M.J., Weir, G., White, S. and Young, R. 1992b Vertical two-phase flow in porous media. *Transport in Porous Media* **8**, 99–131. [11.9.1]

Kladias, N. and Prasad, V. 1989a Convective instabilities in horizontal porous layers heated from below: effects of grain size and its properties. *ASME HTD* **107**, 369–379. [6.9.2]

Kladias, N. and Prasad, V. 1989b Natural convection in horizontal porous layers: effects of Darcy and Prandtl numbers. *ASME J. Heat Transfer* **111**, 926–935. [6.8, 6.9.2]

Kladias, N. and Prasad, V. 1990 Flow transitions in buoyancy-induced non-Darcy convection in a porous medium heated from below. *ASME J. Heat Transfer* **112**, 675–684. [6.8, 6.9.2]

Kladias, N. and Prasad, V. 1991 Experimental verification of Darcy-Brinkman-Forchheimer model for natural convection in porous media. *AIAA J. Thermophys. Heat Transfer* **5**, 560–576. [6.9.2]

Klarsfeld, S. 1970 Champs de température associés aux movements de convection naturelle dans un milieu poreux limité. *Rev. Gén. Therm.* **9**, 1403–1424. [7.1.2]

Knobloch, E. 1986 Oscillatory convection in binary mixtures. *Phys. Rev. A* **34**, 1538–1549. [9.1.3]

Knupp, P. M. and Lage, J. L. 1995 Generalization of the Forchheimer-extended Darcy flow model to the tensor permeability case via a variational principle. *J. Fluid Mech.* **299**, 97–104. [1.5.2]

Kodah, Z. H. and Al-Gasem, A. M. 1998 Non-Darcy mixed convection from a vertical plate in saturated porous media—variable surface heat flux. *Heat Mass Transfer* **33**, 377–382. [8.1.1]

Kodah, Z. H. and Duwairi, H. M. 1996 Inertia effects on mixed convection for vertical plates with variable wall temperature in saturated porous media. *Heat Mass Transfer* **31**, 333–338.[8.1.1]

Koh, J. C. Y. and Colony, R. 1974 Analysis of cooling effectiveness by porous material in coolant passage. *ASME J. Heat Transfer* **96**, 324–330. [4.11]

Koh, J. C. Y. and Stevens, R. L. 1975 Enhancement of cooling effectiveness by porous material in coolant passage, *ASME J. Heat Transfer* **97**, 309–311. [4.11]

Kolesnikov, A. K. 1979 Concentration-dependent convection in a horizontal porous bed containing a chemically active liquid. *J. Engng. Phys.* **36**, 97–101. [3.4]

Koponen, E., Kandhai, D., Hellén, E., Alava, M., Hoekstra, A., Kataja, M, Niskasen, K., Sloot, P. and Timonen, J. 1998 Permeability of three-dimensional random fiber web. *Phys. Rev. lett.* **80**, 716–719. [1.5.2]

Kordylewski, W. and Borkowska-Pawlak, B. 1983 Stability of nonlinear thermal convection in a porous medium. *Arch. Mech.* **35**, 95–106. [6.15.1]

Kordylewski, W., Borkowska-Pawlak, B. and Slany, J. 1987 Stability of three-dimensional natural convection in a porous layer. *Arch. Mech.* **38**, 383–394. [6.15.1]

Kou H. S. and Huang, D. K. 1996a Some transformations for natural convection on a vertical flat plate embedded in porous media with prescribed wall temperature. *Int. Comm. Heat Transfer* **23**, 273–286. [5.1.9.9]

Kou, H. S. and Huang, D. K. 1996b Possible transformations for natural convection on a vertical flat plate embedded in porous media with prescribed wall heat flux. *Int. Comm. Heat Transfer* **23**, 1031–1042. [5.1.9.9]

Kou, H. S. and Huang, D. K. 1997 Fully developed laminar mixed convection through a vertical annular duct filled with porous media. *Int. Comm. Heat Mass Transfer* **24**, 99–110. [8.3.3]

Kou, H. S. and Lu, K. T. 1993a Combined boundary and inertia effects for fully developed mixed convection in a vertical channel embedded in porous media. *Int. Comm. Heat Mass Transfer* **20**, 333–345. [8.3.1]

Kou, H. S. and Lu, K. T. 1993b The analytical solution of mixed convection in a vertical channel embedded in porous media with asymptotic wall heat fluxes. *Int. Comm. Heat Mass Transfer* **20**, 737–750. [8.3.1]

Kozak, R., Saghir, M. Z. and Viviani, A. 2004 Marangoni convection in a liquid layer overlying a porous layer with evaporation at the free surface. *Acta Astronautica* **55**, 189–197. [6.19.3]

Krane, M. J. M. and Incropera, F. P. 1996 A scaling analysis of the unidirectional solidification of a binary alloy. *Int. J. Heat Mass Transfer* **39**, 3567–3579. [10.2.3]

Krantz, W. B., Gleason, K. J. and Caine, N. 1988 Patterned ground. *Sci. Amer.* **159**, 44–50. [11.2]

Krishna, C. V. S. 2001 Effects of non-inertial acceleration on the onset of convection in a second-order fluid-saturated porous medium. *Int. J. Engng. Sci* **39**, 599–609. [6.22]

Krishnan, S. Murthy, J. Y. and Garaimella, S. V. 2004 A two-temperature model for the analysis of passive thermal control systems. *AMSE J. Heat Transfer* **126**, 628–637. [7.5]

Kubitscheck, J. P. and Weidman, P. D. 2003 Stability of fluid-saturated porous medium heated from below by forced convection. *Int. J. Heat Mass Transfer* **46**, 3697–3705. [6.10]s enclosures. *Math. Comp. Sim.* **65**, 221–229. [7.3.6]

Kulacki, F. A. and Freeman, R. G. 1979 A note on thermal convection in a saturated, heat generating porous layer. *ASME J. Heat Transfer* **101**, 169–171. [6.11.2]

Kulacki, F. A. and Rajen, G. 1991 Buoyancy-induced flow and heat transfer in saturated fissured media. *Convective Heat and Mass Transfer in Porous Media*,(eds. S. Kakaç, B. Kilkis, F. A. Kulacki and F. Arinç), Kluwer Academic Publishers, Dordrecht, 465–498. [1.9]

Kulacki, F. A. and Ramchandani, R. 1975 Hydrodynamic instability in porous layer saturated with heat generating fluid. *Wärme-Stoffübertrag.* **8**, 179–185. [6.11.2]

Kumar, P. 1999 Thermal convection in Walters B′ viscoelastic fluid permeated with suspended particles in porous medium. *Indian J. Pure Appl. Math.* **30**, 1117–1132. [6.23]

Kumar, B. R. and Kumar, B. P. 2004 Parallel computation of natural convection in trapezoidal porous enclosures. *Math. Comput. Simul.* **65**, 221–229. [7.3.6]

Kumari, M. 2001a Effect of variable viscosity on non-Darcy free or mixed convection flow on the horizontal surface in a saturated porous medium. *Int. Comm. Heat Mass Transfer* **28**, 723–732. [5.2, 8.1.2]

Kumari, M. 2001b Variable viscosity effects on free and mixed convection boundary layer flow from a horizontal surface in a saturated porous medium–variable heat flux. *Mech. Res. Commun.* **28**, 339–348. [5.2]

Kumari, M. and Gorla, R. S. R. 1996 Combined convection in power-law fluids along a nonisothermal vertical plate in a porous medium. *Transport in Porous Media* **24**, 157–166. [8.1.1]

Kumari, M. and Gorla, R. S. R. 1997 Combined convection along a non-isothermal wedge in a porous medium. *Heat Mass Transfer* **32**, 393–398. [8.1.4]

Kumari, M. and Jayanthi, S. 2004 Non-Darcy non-Newtonian free convection flow over a horizontal cylinder in a saturated porous medium. *Int. Comm. Heat Mass Transfer* **31**, 1219–1226. [5.5.1]

Kumari, M. and Jayanthi, S. 2005 Uniform lateral mass flux on natural-convection flow over a vertical cone embedded in a porous medium saturated with a non-Newtonian fluid. *J. Porous Media* **8**, 73–84. [5.8]

Kumari, M. and Nath, G. 1989a Non-Darcy mixed convection boundary layer flow on a vertical cylinder in a saturated porous medium. *Int. J. Heat Mass Transfer* **32**, 183–187. [8.1.3]

Kumari, M. and Nath, G. 1989b Unsteady mixed convection with double diffusion over a horizontal cylinder and a sphere within a porous medium. *Wärme-Stoffübertrag.* **24**, 103–109. [8.1.3]

Kumari, M. and Nath, G. 1989c Double diffusive unsteady free convection on two-dimensional and axisymmetric bodies in a porous medium. *Int. J. Energy Res.* **13**, 379–391. [9.4]

Kumari, M. and Nath, G. 1989d Double diffusive unsteady mixed convection flow over a vertical plate embedded in a porous medium. *Int. J. Energy Res.* **13**, 419–430. [9.4]

Kumari, M. and Nath, G. 1990 Non-Darcy mixed convection flow over a nonisothermal cylinder and sphere embedded in a saturated porous medium. *ASME J. Heat Transfer* **112**, 518–523. [8.1.3]

Kumari, M. and Nath, G. 1992 Simultaneous heat and mass transfer under unsteady mixed convection along a vertical slender cylinder embedded in a porous medium. *Wärme-Stoffübertrag.* **28**, 97–105. [9.6]

Kumari, M. and Nath, G. 2004a Non-Darcy mixed convection in power-law fluids along a non-isothermal horizontal surface in a porous medium *Int. J. Engng. Sci.* **42**, 353–369. [8.1.2]

Kumari, M. and Nath, G. 2004b Radiation effect on mixed convection from a horizontal surface in a porous medium. *Mech. Res. Comm.* **31**, 483–491. [8.1.2]

Kumari, M., Gorla, R. S. R., and Byrd, L. 1997 Mixed convection in non-Newtonian fluids along a horizontal plate in a porous medium. *ASME J. Energy Resources Tech.* **119**, 35–37. [8.1.2]

Kumari, M., Nath, G. and Pop, I. 1990c Non-Darcian effects on forced convection heat transfer over a flat plate in a highly porous medium. *Acta Mech.* **84**, 201–207. [4.8]

Kumari, M., Nath, G. and Pop, I. 1993 Non-Darcy mixed convection flow with thermal dispersion on a vertical cylinder in a saturated porous medium. *Acta Mech.* **100**, 69–77. [8.1.3]

Kumari, M., Pop, I. and Nath, G. 1985 Non-Darcy natural convection from a heated vertical plate in saturated porous media with mass transfer. *Int. Comm. Heat Mass Transfer* **12**, 337–346. [5.1.7.2]

Kumari, M., Pop, I. and Nath, G. 1986 Non-Darcy natural convection on a vertical cylinder in a saturated porous medium. *Wärme-Stoffubertrag..* **20**, 33–37. [5.7]

Kumari, M., Pop, I. and Nath, G. 1987 Mixed convection boundary layer over a sphere in a saturated porous medium. *Z. Ang. Math. Mech.* **67**, 569–571. [8.1.3]

Kumari, M., Pop, I. and Nath, G. 1988 Darcian mixed convection plumes along vertical adiabatic surfaces in a saturated porous medium. *Wärme-Stoffubertrag..* **22**, 173–178 [8.1.1].

Kumari, M., Pop, I. and Nath, G. 1990a Nonsimilar boundary layers for non-Darcy mixed convection flow about a horizontal surface in a saturated porous medium. *Int. J. Engng. Sci.* **28**, 253–263. [8.1.2]

Kumari, M., Pop, I. and Nath, G. 1990b Natural convection in porous media above a near horizontal uniform heat flux surface. *Wärme-Stoffübertrag.*, **25**, 155–159. [5.3]

Kumari, M., Takhar, H. S. and Nath, G. 1988a Non-Darcy double-diffusive mixed convection from heated vertical and horizontal plates in saturated porous media. *Wärme-Stoffübertrag.* **23**, 267–273. [9.4]

Kumari, M., Takhar, H. S. and Nath, G. 1988b Double-diffusive non-Darcy free convection from two-dimensional and axisymmetric bodies of arbitrary shape in a saturated porous medium. *Indian J. Tech.* **26**, 324–328. [9.4]

Kumari, M., Takhar, H. S. and Nath, G. 2001 Mixed convection flow over a vertical wedge embedded in a highly porous medium. *Heat Mass Transfer* **37**, 139–146. [8.1.4]

Kurdyumov, V. N. and Liñán, A. 2001 Free and forced convection around line sources of heat and heated cylinders in porous media. *J. Fluid Mech.* **427**, 389–409. [5.5.1]

Kuwahara, F. and Nakayama, A. 1998 Numerical modelling of non-Darcy convective flow in a porous medium. *Heat Transfer 1998, Proc. 11th IHTC*, **4**, 411–416. [1.5.2, 1.8]

Kuwahara, F. and Nakayama, A. 1999 Numerical determination of thermal dispersion coefficients using periodic porous structure. *ASME J. Heat Transfer* **121**, 160–163. [2.2.3]

Kuwahara, F. and Nakayama, A. 2005 Three-dimensional flow and heat transfer within highly anisotropic porous media. *Handbook of Porous Media* (K Vafai, ed.), 2nd ed., Taylor and Francis, New York, pp. 235–266. [2.2.3]

Kuwahara, F., Kameyama, Y., Yamashita, S. and Nakayama, A. 1998 Numerical modeling of turbulent flow in porous media using a spatially periodic array. *J. Porous Media* **1**, 47–55. [1.8]

Kuwahara, F., Nakayama, A. and Koyama, H. 1994 Numerical modelling of heat and fluid flow in a porous medium. *Heat Transfer 1994*, Inst. Chem. Engrs, Rugby, vol. 5, pp. 309–314. [2.6]

Kuwahara, F., Nakayama, A. and Koyama, H. 1996 A numerical study of thermal dispersion in porous media. *ASME J. Heat Transfer* **118**, 756–761. [2.2.3]

Kuwahara, F., Shirota, M. and Nakayama, A. 2001 A numerical study of interfacial convective heat transfer coefficient in two-energy model for convection in porous media. *Int. J. Heat Mass Transfer* **44**, 1153–1159. [2.2.2]

Kuznetsov, A. V. 1994 An investigation of a wave of temperature difference between solid and fluid phases in a porous packed bed. *Int. J. Heat Mass Transfer* **37**, 3030–3033. [4.6.4]

Kuznetsov, A. V. 1995a Comparisons of the waves of temperature difference between the solid and fluid phases in a porous slab and in a semi-infinite porous body. *Int. Comm. Heat Mass Transfer* **22**, 499–506. [4.6.4]

Kuznetsov, A. V. 1995b An analytical solution for heating a two-dimensional porous packed bed by a non-thermal equilibrium fluid flow. *Appl. Sci. Res.* **55**, 83–93. [4.6.4]

Kuznetsov, A. V. 1996a Analytical investigation of the fluid flow in the interface region between a porous medium and a clear fluid in channels partially filled with a porous medium. *Appl. Sci. Res.* **56**, 53–67. [1.6]

Kuznetsov, A. V. 1996b Analysis of a non-thermal equilibrium fluid flow in a concentric tube annulus filled with a porous medium. *Int. Comm. Heat Mass Transfer* **23**, 929–938. [4.6.4]

Kuznetsov, A. V. 1996c Stochastic modeling of heating of a 1D porous slab by a flow of incompressible fluid. *Acta Mech.* **114**, 39–50. [4.6.4]

Kuznetsov, A. V. 1996d A perturbation solution for a nonthermal equilibrium fluid flow through a three-dimensional sensible heat storage packed bed. *ASME J. Heat Transfer* **118**, 508–510. [4.6.4]

Kuznetsov, A. V. 1996e Investigation of a non-thermal equilibrium flow of an incompressible fluid in a cylindrical tube filled with porous media. *Z. Angew. Math. Mech.* **76**, 411–418. [4.6.4]

Kuznetsov, A. V. 1996f Analysis of heating a three-dimensional porous bed utilizing the two energy equation model. *Heat Mass Transfer* **31**, 173–177. [4.6.4]

Kuznetsov, A. V. 1997a Determination of the optimal initial temperature distribution in a porous bed. *Acta Mech.* **120**, 61–69. [4.6.4]

Kuznetsov, A. V. 1997b Influence of the stress jump condition at the porous-medium/clear-fluid interface on a flow at a porous wall. *Int. Comm. Heat Mass Transfer* **24**, 401–410. [1.6]

Kuznetsov, A. V. 1997c Optimal control of the heat storage in a porous slab. *Int. J. Heat Mass Transfer* **40**, 1720–1723. [4.15]

Kuznetsov, A. V. 1997d Thermal nonequilibrium, non-Darcian forced convection in a channel filled with a fluid saturated porous medium—a perturbation solution. *Appl. Sci. Res.* **57**, 119–131. [4.10]

Kuznetsov, A. V. 1998a Numerical investigation of the macrosegregation during thin strip casting of carbon steel. *Numer. Heat Transfer A* **33**, 515–532. [10.2.3]

Kuznetsov, A. V. 1998b Analytical study of fluid flow and heat transfer during forced convection in a composite channel partly filled with a Brinkman-Forchheimer porous medium. *Flow, Turbulence and Combustion* **60**, 173–192. [4.11]

Kuznetsov, A. V. 1998c Analytical investigation of heat transfer in Couette flow through a porous medium utilizing the Brinkman-Forchheimer-extended Darcy model. *Acta Mech.* **129**, 13–24. [4.9]

Kuznetsov, A. V. 1998d Analytical investigation of Couette flow in a composite channel partially filled with a porous medium and partially with a clear fluid. *Int. J. Heat Mass Transfer* **41**, 2556–2560. [4.16.1]

Kuznetsov, A. 1998e Non-thermal equilibrium forced convection in porous media. *Transport Phenomena in Porous Media* (eds. D. B. Ingham and I. Pop), Elsevier, Oxford, pp.103–130. [2.2.2, 4.6.4]

Kuznetsov, A. V. 1999a Fluid mechanics and heat transfer in the interface region between a porous medium and a fluid layer: A boundary layer solution. *J. Porous Media* **2**, 309–321. [4.11]

Kuznetsov, A. V. 1999b Analytical investigation of forced convection from a flat plate enhanced by a porous substrate. *Acta Mech.* **137**, 211–223. [4.11]

Kuznetsov, A. V. 1999c Forced convection heat transfer in a parallel-plate channel with a porous core. *Appl. Mech. Engng.* **4**, 271–290. [4.11]

Kuznetsov, A. V. 2000a Analytical studies of forced convection in partly porous configurations. *Handbook of Porous Media* (K. Vafai, ed.), Marcel Dekker, New York, pp. 269–312. [4.11]

Kuznetsov, A. V. 2000b Fluid flow and heat transfer analysis of Couette flow in a composite duct. *Acta Mech.* **140**, 163–170. [4.11]

Kuznetsov, A. V. 2000c Investigation of the effect of transverse thermal dispersion on forced convection in porous media. *Acta Mech.* **145**, 35–43. [4.9]

Kuznetsov, A. V. 2001 Influence of thermal dispersion on forced convection in a composite parallel-plate channel. *Z. Angew. Math. Phys.* **52**, 135–150. [4.11]

Kuznetsov, A. V. 2004a Effect of turbulence on forced convection in a composite tube partly filled with a porous medium. *J. Porous Media* **7**, 59–64. [4.11]

Kuznetsov, A. V. 2004b Numerical modeling of turbulent flow in a composite porous/fluid duct utilizing a two-layer k-epsilon model to account for interface roughness. *Int. J. Therm. Sci.* **43**, 1047–1056. [1.8]

Kuznetsov, A. V. 2005 Modeling bioconvection in porous media. *Handbook of Porous Media* (ed. K. Vafai), 2nd ed., Taylor and Francis, New York, pp. 645–686. [6.25]

Kuznetsov, A. V. and Avramenko, A. A. 2002 A 2D analysis of stability of bioconvection in a fluid saturated porous medium—Estimation of the critical permeability value. *Int. Comm. Heat Mass Transfer* **29**, 175–184. [6.25]

Kuznetsov, A. V. and Avramenko, A. A. 2003a The effect of deposition and declogging on the critical permeability in bioconvection in a porous medium. *Acta Mech.* **160**, 113–125. [6.25]

Kuznetsov, A. V. and Avramenko, A. A. 2003b Stability analysis of bioconvection of gyrotactic motile microorganisms in a fluid saturated porous medium. *Transport Porous Media* **53**, 95–104. [6.25]

Kuznetsov, A. V. and Avramenko, A. A. 2005 Effect of fouling on stability of bioconvection of gyrotactic micro-organisms in a porous medium. *J. Porous Media* **8**, 45–53. [6.25]

Kuznetsov, A. V. and Becker, S. M. 2004 Effect of the interface roughness on turbulent convective heat transfer in a composite porous/fluid duct. *Int. Comm. Heat Mass Transfer* **31**, 11–20. [4.11]

Kuznetsov, A. V. and Jiang, N. 2001 Numerical investigation of bioconvection of gravitactic microorganisms in an isotropic porous medium. *Int. Comm. Heat Mass Transfer* **28**, 877–866. [6.25]

Kuznetsov, A. V. and Jiang, N. 2003 Bioconvection of negatively geotactic microorganisms in a porous medium: The effect of cell deposition and declogging. *Int. J. Numer. Meth. Heat Fluid Flow* **13**, 341–364. [6.25]

Kuznetsov, A. V. and Nield, D. A. 2001 Effects of heterogeneity in forced convection in a porous medium: triple layer or conjugate problem. *Numer Heat Transfer A* **40**, 363–385. [4.12]

Kuznetsov, A.V. and Nield, D. A. 2005a Thermally developing forced convection in a channel occupied by a porous medium saturated by a non-Newtonian fluid. *International Journal of Heat and Mass Transfer*, to appear. [4.13]

Kuznetsov, A.V. and Nield D.A. 2005b Thermally developing forced convection in a bi-disperse porous medium. *J. Porous Media*, to appear. [4.16.4]

Kuznetsov, A. V. and Vafai, K. 1995a Development and investigation of three-phase model of the mushy zone for analysis of porosity formation in solidifying castings. *Int. J. Heat Mass Transfer* **38**, 2557–2567. [10.2.3]

Kuznetsov, A. V. and Vafai, K. 1995b Analytical comparison and criteria for heat and mass transfer models in metal hydride packed beds. *Int. J. Heat Mass Transfer* **38**, 2873–2884. [3.4]

Kuznetsov, A. V. and Xiong, M. 1999 Limitation of the single-domain approach for computation of convection in composite channels: comparisons with exact solutions. *Hybrid Methods in Engineering* **1**, 249–264. [4.11]

Kuznetsov, A. V. and Xiong, M. 2000 Numerical simulation of the effect of thermal dispersion on forced convection in a circular duct partly filled with a Brinkman-Forchheimer porous medium. *Int. J. Numer. Meth. Heat Fluid Flow* **10**, 488–501. [4.11]

Kuznetsov, A. V. and Xiong, M. 2003 Development of an engineering approach to computations of turbulent flows in composite porous/fluid domains. *Int. J. Therm. Sci.* **42**, 9123–919. [1.8]

Kuznetsov, A. V., Avramenko, A. A. and Geng. P. 2003a A similarity solution for a falling plume in bioconvection of oxytactic bacteria in a porous medium. *Int. Comm. Heat Mass Transfer* **30**, 37–46. [6.25]

Kuznetsov, A. V., Avramenko, A. A. and Geng. P. 2004 Analytical investigation a falling plume caused by bioconvection of oxytactic bacteria in a fluid saturated porous medium. *Int. J. Engng. Sci.* **42**, 557–569. [6.25]

Kuznetsov, A. V., Cheng, L. and Xiong, M. 2002 Effects of thermal dispersion and turbulence in forced convection in a composite parallel-plate channel: Investigation of constant wall heat flux and constant wall temperature cases. *Numer. Heat Transfer A* **42**, 365–383. [4.11]

Kuznetsov, A. V., Cheng, L. and Xiong, M. 2003b Investigation of turbulence effects on forced convection in a composite porous/fluid duct: Constant wall heat flux and constant wall temperature cases. *Heat Mass Transfer* **39**, 613–623. [4.11]

Kuznetsov, A.V., Xiong, M. and Nield, D. A. 2003c Thermally developing forced convection in a porous medium: circular duct with walls at constant temperature, with longitudinal conduction and viscous dissipation effects. *Transport Porous Media* **53**, 331–345. [4.13]

Kvernvold, O. 1979 On the stability of nonlinear convection in a Hele-Shaw cell. *Int. J. Heat Mass Transfer* **22**, 395–400. [2.5]

Kvernvold, O. and Tyvand, P. A. 1979 Nonlinear thermal convection in anisotropic porous media. *J. Fluid Mech.* **90**, 609–624. [6.12]

Kvernvold, O. and Tyvand, P. A. 1980 Dispersion effects on thermal convection in porous media. *J. Fluid Mech.* **99**, 673–686. [6.6]

Kvernvold, O. and Tyvand, P. A. 1981 Dispersion effects on thermal convection in a Hele-Shaw cell. *Int. J. Heat Mass Transfer* **24**, 887–990. [2.5]

Kwendakwema, N.J. and Boehm, R.F. 1991 Parametric study of mixed convection in a porous medium between vertical concentric cylinders. *ASME J. Heat Transfer* **113**, 128–134. [8.3.3]

Kwok, L. P. and Chen, C. F. 1987 Stability of thermal convection in a vertical porous layer. *ASME J. Heat Transfer* **109**, 889–893. [7.1.4]

Laakkonen, K. 2003 Method to model dryer fabrics in paper machine scale using small-scale simulations and porous medium model. *Int. J. Heat Fluid Flow* **24**, 114–121. [1.8]

Lage, J. L. 1992 Effect of the convective inertia term on Bénard convection in a porous medium. *Numer. Heat Transfer A* **22**, 469–485. [1.5.2]

Lage, J. L. 1993a Natural convection within a porous medium cavity: predicting tools for flow regime and heat transfer. *Int. Comm. Heat Mass Transfer* **20**, 501–513. [1.5.2, 6.6]

Lage, J. L. 1993b On the theoretical prediction of transient heat transfer within a rectangular fluid-saturated porous medium enclosure. *ASME J. Heat Transfer*,**115**, 1069–1071. [7.5]

Lage, J. L. 1996 Comments on "The effect of turbulence on solidification of a binary metal alloy with electromagnetic stirring." *ASME J. Heat Transfer* **118**, 996–997. [10.2.3]

Lage, J. L. 1997 Contaminant clean-up in a single rock fracture with porous obstructions. *ASME J. Fluids Engng.* **119**, 180–187. [1.4.1, 1.9]

Lage, J. L. 1998 The fundamental theory of flow through permeable media: from Darcy to turbulence. *Transport Phenomena in Porous Media* (eds. D.B. Ingham and I. Pop), Elsevier, Oxford, pp.1–30. [1.5.2, 1.8]

Lage, J. L. and Antohe, B. V. 2000 Darcy's experiments and the deviation to nonlinear flow regime. *ASME J. Fluids Engng.* **122**, 619–625. [1.5.2]

Lage, J. L. and Bejan, A. 1990 Numerical study of forced convection near a surface covered with hair. *Int. J. Heat Fluid Flow* **11**, 242–248. [5.13]

Lage, J. L. and Bejan, A. 1991 Natural convection from a vertical surface covered with hair. *Int. J. Heat Fluid Flow* **12**, 46–53. [5.13]

Lage, J. L. and Bejan, A. 1993 The resonance of natural convection in an enclosure heated periodically from the side. *Int. J. Heat Mass Transfer* **36**, 2027–2038. [7.10]

Lage, J. L. and Narasimhan, A. 2000 Porous media enhanced forced convection fundamentals. *Handbook of Porous Media* (K. Vafia, ed.), Marcel Dekker, New York, pp. 357–394. [4]

Lage, J. L. and Nield, D. A. 1997 Comments on "numerical studies of forced convection heat transfer from a cylinder from a cylinder embedded in a packed bed. *Int. J. Heat Mass Transfer* **40**, 1725–1726. [4.8]

Lage, J. L. and Nield, D. A. 1998 Convection induced by inclined gradients in a shallow porous medium layer. *Journal of Porous Media* **1**, 57–69. [7.9]

Lage, J. L., Antohe, B. V. and Nield, D. A. 1997 Two types of nonlinear pressure-drop versus flow rate relation observed for saturated porous media. *ASME J. Fluids Engng.* **119**, 701–706. [1.5.2]

Lage, J. L., Bejan, A. and Georgiadis, J. G. 1992 The Prandtl number effect near the onset of Bénard convection in a porous medium. *Int. J. Heat Fluid Flow* **13**, 408–411. [6.9.2, 6.12]

Lage, J. L., de Lemos, M. J. S. and Nield, D. A. 2002 Modeling turbulence in porous media. In *Transport Phenomena in Porous Media II* (D. B. Ingham and I. Pop, eds.) Elsevier, Oxford, pp. 198–230. [1.8]

Lage, J. L., Krueger, P. S. and Narasimhan, A. 2005 Protocol for measuring permeability and form coefficient of porous media. *Phys. Fluids* **17**, Art. No. 088101. [1.5.2]

Lage, J. L., Merrikh, A. A. and Kulish, V.V. 2004a A porous medium model to investigate the red cell distribution effect on alveolar respiration. In *Emerging Technologies and Techniques in Porous Media* (D. B. Ingham, A. Bejan, E. Mamut and I. Pop, eds), Kluwer Academic, Dordrecht, pp. 381–407. [1.9]

Lage, J. L., Narasimhan, A., Porneala, P.C. and Price, D. C. 2004b Experimental study of forced convection through microporous enhanced heat sinks. In *Technologies and Techniques in Porous Media* (D. B. Ingham, A. Bejan, E. Mamut and I. Pop, eds), Kluwer Academic, Dordrecht, pp. 433–452. [4.5]

Lage, J. L., Weinert, A. K., Price, D. C. and Weber, R. M. 1996 Numerical study of a low permeability microporous heat sink for cooling phased-array radar systems. *Int. J. Heat Mass Transfer*, **39**, 3633–3647. [4.9]

Lai, C. H., Bodvarsson, G. S. and Truesdell, A. H. 1994 Modeling studies of heat transfer and phase distribution in two-phase geothermal reservoirs. *Geothermics* **23**,3–20. [11]

Lai, F. C. 1990a Coupled heat and mass transfer by natural convection from a horizontal line source in saturated porous medium. *Int. Comm. Heat Mass Transfer* **17**, 489–499. [9.3.2]

Lai, F.C. 1990b Natural convection from a concentrated heat source in a saturated porous medium. *Int. Comm. Heat Mass Transfer* **17**, 791–800. [5.11.1]

Lai, F. C. 1991a Coupled heat and mass transfer by mixed convection from a vertical plate in a saturated porous medium. *Int. Comm. Heat Mass Transfer* **18**, 93–106. [9.6]

Lai, F. C. 1991b Non-Darcy natural convection from a line source of heat in a saturated porous medium. *Int. Comm. Heat Mass Transfer* **18**, 445–457. [5.10.1.2]

Lai, F. C. 1991c Non-Darcy mixed convection from a line source of heat in a saturated porous medium. *Int. Comm. Heat Mass Transfer* **18**, 875–887. [8.1.4]

Lai, F. C. 1993a Improving effectiveness of pipe insulation by using radial baffles to suppress natural convection. *Int. J. Heat Mass Transfer* **36**, 899–908. [7.3.7]

Lai, F. C. 1993b Natural convection in a horizontal porous annulus with mixed type of radial baffles. *Int. Comm. Heat Mass Transfer* **20**, 347–359. [7.3.7]

Lai, F. C. 1994 Natural convection in horizontal porous annuli with circumferential baffles. *AIAA J. Thermophys. Heat Transfer* **8**, 376–378. [7.3.7]

Lai, F. C. 2000 Mixed convection in saturated porous media. *Handbook of Porous Media* (K. Vafia, ed.), Marcel Dekker, New York, pp. 605–661. [8]

Lai, F. C. and Kulacki, F. A. 1987 Non-Darcy convection from horizontal impermeable surfaces in saturated porous media. *Int. J. Heat Mass Transfer* **30**, 2189–2192. [4.7, 8.1.2]

Lai, F. C. and Kulacki, F. A. 1988a Effects of flow inertia on mixed convection along a vertical surface in a saturated porous medium. *ASME HTD* **96**, Vol. 1, 643–652. [8.1.1]

Lai, F. C. and Kulacki, F. A. 1988b Transient mixed convection in horizontal porous layer locally heated from below. *ASME HTD* **96**, Vol. 2, 353–364. [8.2.2]

Lai, F. C. and Kulacki, F. A. 1988c Natural convection across a vertical layered porous cavity. *Int. J. Heat Mass Transfer* **31**, 1247–1260. [7.3.2]

Lai, F. C. and Kulacki, F. A. 1989a Thermal dispersion effects on non-Darcy convection over horizontal surfaces in saturated porous media. *Int. J. Heat Mass Transfer* **32**, 971–976. [4.7, 8.1.2]

Lai, F. C. and Kulacki, F. A. 1989b Effects of variable fluid viscosity on film condensation along an inclined surface in saturated porous medium. *ASME HTD* **127**, 7–12. [10.4]

Lai, F. C. and Kulacki, F. A. 1990a Coupled heat and mass transfer from a sphere buried in an infinite porous medium. *Int. J. Heat Mass Transfer* **33**, 209–215. [9.4]

Lai, F. C. and Kulacki, F. A. 1990b The influence of surface mass flux on mixed convection over horizontal plates in saturated porous media. *Int. J. Heat Mass Transfer* **33**, 576–579. [8.1.2]

Lai, F. C. and Kulacki, F. A. 1990c The effect of variable viscosity on convective heat transfer along a vertical surface in a saturated porous medium. *Int. J. Heat Mass Transfer* **33**, 1028–1031. [8.1.1]

Lai, F. C. and Kulacki, F. A. 1990d The influence of lateral mass flux on mixed convection over inclined surfaces in saturated porous media. *ASME J. Heat Transfer* **112**, 515–518. [8.1.1]

Lai, F.C. and Kulacki, F.A. 1991a Non-Darcy mixed convection along a vertical wall in a saturated porous medium. *ASME J. Heat Transfer* **113**, 252–255. [8.1.1]

Lai, F.C. and Kulacki, F.A. 1991b Experimental study of free and mixed convection in horizontal porous layers locally heated from below. *Int. J. Heat Mass Transfer* **34**, 525–541. [8.2.2]

Lai, F.C. and Kulacki, F.A. 1991c Oscillatory mixed convection in horizontal porous layers locally heated from below. *Int. J. Heat Mass Transfer* **34**, 887–890. [8.2.2]

Lai, F.C. and Kulacki, F.A. 1991d Coupled heat and mass transfer by natural convection from vertical surfaces in porous media. *Int. J. Heat Mass Transfer* **34**, 1189–1194. [9.2.1]

Lai, F. C. and Kulacki, F. A. 1991e Experimental study in horizontal layers with multiple heat sources. *AIAA J. Thermophys. Heat Transfer* **5**, 627–630. [6.18]

Lai, F. C., Choi, C. Y. and Kulacki, F. A. 1990a Free and mixed convection in horizontal porous layers with multiple heat sources. *AIAA J. Thermophys. Heat Transfer* **4**, 221–227. [6.18, 8.2.2]

Lai, F. C., Choi, C. Y. and Kulacki, F. A. 1990b Coupled heat and mass transfer by natural convection from slender bodies of revolution in porous media. *Int. Comm. Heat Mass Transfer* **17**, 609–620. [9.4]

Lai, F. C., Kulacki, F. A. and Prasad, V. 1987a Mixed convection in horizontal porous layers: effects of thermal boundary conditions. *ASME HTD* **84**, 91–96. [8.2.2]

Lai, F. C., Kulacki, F. A. and Prasad, V. 1991a Mixed convection in saturated porous media. *Convective Heat and Mass Transfer in Porous Media* (eds. S. Kakaç, *et al.*), Kluwer Academic, Dordrecht, 225–287. [8.1.1]

Lai, F. C., Pop, I and Kulacki, F. A. 1991b Natural convection from isothermal plates embedded in thermally stratified porous media. *AIAA J. Thermophys. Heat Transfer* **4**, 533–535. [5.1.4]

Lai, F. C., Pop, I. and Kulacki, F. A. 1990c Free and mixed convection from slender bodies of revolution in porous media. *Int. J. Heat Mass Transfer* **33**, 1767–1769. [5.9, 8.1.4]

Lai, F. C., Prasad, V. and Kulacki, F. A. 1987b Effects of the size of heat source on mixed convection in horizontal porous layers heated from below. *Proc. ASME JSME Thermal Engineering Joint Conference* vol. 2, pp. 413–419. [8.2.2]

Lai, F. C., Prasad, V. and Kulacki, F. A. 1988 Aiding and opposing mixed convection in a vertical porous layer with a finite wall heat source. *Int. J. Heat Mass Transfer* **31**, 1049–1061. [8.3.2]

Lai, Y. M., Li, J. J., Niu, F. J. and Yu, W. B. 2003 Nonlinear thermal analysis for Qing-Tibet railway embankments in cold regions. *Cold Regions Engng.* **17**, 171–184. [4.16.2, 7.3.7]

Lai, Y. M., Zhang, S. J. and Mi, L. 2004 Effect of climatic warming on the temperature fields of embankments in cold regions and a countermeasure. *Numer. Heat Transfer* **45**, 191–210. [4.16.2, 7.3.7]

Lai, Y. M., Zhang, S. J., Zhang, L. X. and Xiao, J. Z. 2004 Adjusting temperature distribution under the south and north slopes of embankment in permafrost regions by the ripped-rock revetment. *Cold Regions Sci. Tech.* **39**, 67–79. [4.16.2, 7.3.7]

Lan, X. K. and Khodadadi, J. M. 1993 Fluid flow and heat transfer through a porous medium channel with permeable walls. *Int. J. Heat Mass Transfer* **36**, 2242–2245. [4.9]

Landman, K. A., Pel., L. and Kaassschieler, E. F. 2001 Analytical modeling of drying of porous materials. *Math. Engng. Ind.* **8**, 89–122. [3.6]

Lapwood, E. R. 1948 Convection of a fluid in a porous medium. *Proc. Camb. Phil. Soc.* **44**, 508–521. [6.1, 6.26]

Larbi, S., Bacon, G. and Bories, S. A. 1995 Diffusion d'air humide avec condensation de vapeur d'eau en milieu poreux. *Int. J. Heat Mass Transfer* **38**, 2411–2416. [3.6]

Larson, S. E. and Poulikakos, D. 1986 Double diffusion from a horizontal line source in an infinite porous medium. *Int. J. Heat Mass Transfer* **29**, 492–495. [9.3.2]

Laurent, S. and Ollivier, J. P. 2005 Scale analysis of electrodiffusion through porous media. *J. Porous Media* **8**, to appear. [3.7]

Lauriat, G. and Ghafir, R. 2000 Forced convective heat transfer in porous media. *Handbook of Porous Media* (K. Vafia, ed.), Marcel Dekker, New York, pp. 201–267. [4]

Lauriat, G. and Prasad, V. 1987 Natural convection in a vertical porous cavity: a numerical study for Brinkman-extended Darcy formulation. *ASME J. Heat Transfer* **109**, 688–696. [7.6.2]

Lauriat, G. and Prasad, V. 1989 Non-Darcian effects on natural convection in a vertical porous layer. *Int. J. Heat Mass Transfer* **32**, 2135–2148. [7.6.2]

Lauriat, G. and Prasad, V. 1991 Natural convection in a vertical porous annulus. *Convective Heat and Mass Transfer in Porous Media* (eds. S. Kakaç, B. Kilkis̄, F. A. Kulacki, and F. Arinc̄), Kluwer Academic, Dordrecht, pp. 143–172. [7.6.2]

Lauriat, G. and Vafai, K. 1991 Forced convection flow and heat transfer through a porous medium exposed to a flat plate or a channel. *Convective Heat and Mass Transfer in Porous Media* (eds. S. Kakaç, *et al.*), Kluwer Academic, Dordrecht, pp. 289–327. [4.8, 4.9]

Lawson, M. L. and Yang, W. J. 1975 Thermal instability of binary gas mixtures in a porous medium. *ASME J. Heat Transfer* **97**, 378–381. [9.1.4]

Lawson, M. L., Yang, W. J. and Bunditkul, S. 1976 Theory of thermal stability of binary gas mixtures in porous media. *ASME J. Heat Transfer* **98**, 35–41. [9.1.4]

Layeghi, M. and Nouri-Borujerdi, A. 2004 Fluid flow and heat transfer around circular cylinders in the presence or no-presence of porous media. *J. Porous Media* **7**, 239–247. [4.11]

Layton, W., Schieweck, F. and Yotov, I. 2003 Coupling fluid flow with porous media flow. *SIAM J. Numer. Anal.* **40**, 2195–2218. [1.6]

Le Breton, P., Caltagirone, J.P. and Arquis, E. 1991 Natural convection in a square cavity with thin porous layers on its vertical walls. *ASME J. Heat Transfer* **113**, 892–898. [7.7]

Lebon, G. and Cloot, A. 1986 A thermodynamical modelling of fluid flows through porous media: application to natural convection. *Int. J. Heat Mass Transfer* **29**, 381–390. [6.6]

Ledezma, G. A., Bejan, A. and Errera, M. R. 1997, Constructal tree networks for heat transfer. *J. Appl. Phys.* **82**, 89–100. [4.18].

Ledezma, G., Morega, A. M. and Bejan, A. 1996 Optimal spacing between fins with impinging flow. *J. Heat Transfer* **118**, 570–577. [4.15]

Lee, D. H., Yoon, D. Y. and Choi, C. K. 2000 The onset of vortex instability in laminar convection flow over an inclined plate embedded in a porous medium. *Int. J. Heat Mass Transfer* **43**, 2895–2908. [5.4]

Lee, D. Y. and Vafai, K. 1999 Analytical characterization and conceptual assessment of solid and fluid temperature differentials in porous media. *Int. J. Heat Mass Transfer* **42**, 423–435 (erratum 4077). [4.10]

Lee, H. M. 1983 An experimental study of natural convection about an isothermal downward-facing inclined surface in a porous medium. M.S. Thesis, University of Hawaii. [5.3]

Lee, J. S. and Ogawa, K. 1994 Pressure drop through packed bed. *J. Chem. Engrg Japan* **27**, 691–693. [1.5.2]

Lee, K. B. and Howell, J. R. 1991 Theoretical and experimental heat and mass transfer in highly porous media. *Int. J. Heat Mass Transfer* **34**, 2123–2132. [1.8]

Lee, S. L. and Yang, J. H. 1997 Modelling of Darcy-Forchheimer drag for fluid flow across a bank of circular cylinders. *Int. J. Heat Mass Transfer* **40**, 3149–3155. [1.5.2]

Lee, S. L. and Yang, J. H. 1998 Modelling of effective thermal conductivity for a nonhomogeneous anisotropic porous medium. *Int. J. Heat Mass Transfer* **41**, 931–937. [2.2.1]

Lein, H. and Tankin, R. S. 1992a Natural convection in porous media—I. Nonfreezing. *Int. J. Heat Mass Transfer* **35**, 175–186. [6.9.1]

Lein, H. and Tankin, R. S. 1992b Natural convection in porous media—II. Freezing. *Int. J. Heat Mass Transfer* **35**, 187–194. [10.2.2]

Leong, J. C. and Lai, F. C. 2001 Effective permeability of a layered porous cavity. *ASME J. Heat Transfer* **123**, 512–519. [6.13.2]

Leong, J. C. and Lai, F. C. 2004 Natural convection in rectangular layers porous cavities. *J. Thermophys. Heat Transfer* **18**, 457–463. [6.13.2]

Leong, K. C. and Jin, L. W. 2004 Heat transfer of oscillating and steady flows in a channel filled with porous media. *Int. Comm. Heat Mass Transfer* **31**, 63–72. [4.16.5]

Leong, K. C. and Jin, L. W. 2005 An experimental study of oscillatory flow through a channel filled with an aluminum foam. *Int. J. Heat Mass Transfer* **48**, 243–253. [4.16.5]

Leppinen, D. M. and Rees, D. A. S. 2004 Sidewall heating in shallow cavities near the density maximum. In *Emerging Technologies and Techniques in Porous Media* (D. B. Ingham, A. Bejan, E. Mamut and I. Pop, eds), Kluwer Academic, Dordrecht, pp. 183–194. [7.1.6]

Leppinen, D. M., Pop, I., Rees, D. A. S. and Storesletten, L. 2004 Free convection in a shallow annular cavity filled with a porous medium. *J. Porous Media* **7**, 289–302. [7.3.3]

Lesnic, D. and Pop, I. 1998a Free convection in a porous medium adjacent to horizontal surfaces. *Zeit. Angew. Math. Mech.* **78**, 197–205. [5.2]

Lesnic, D. and Pop, I. 1998b Mixed convection over a horizontal surface embedded in a porous medium. In *Mathematics of Heat Transfer* (G. E. Tupholme and A. S. Wood, eds.) Clarendon Press, Oxford, pp. 219–224. [8.1.2]

Lesnic, D., Ingham, D. B. and Pop, I. 1995 Conjugate free convection from a horizontal surface in a porous medium. *Z. Angew. Math. Mech.* **75**, 715–722. [5.2]

Lesnic, D., Ingham, D. B. and Pop, I. 1999 Free convection boundary-layer flow along a vertical surface in a porous medium with Newtonian heating. *Int. J. Heat Mass Transfer* **42**, 2621–2627. [5.1.9.8]

Lesnic, D., Ingham, D. B. and Pop, I. 2000 Free convection from a horizontal surface in a porous medium with Newtonian heating. *J. Porous Media* **3**, 227–235. [5.2]

Lesnic, D., Ingham, D. B., Pop, I. and Storr, C. 2004 Free convection boundary-layer flow above a nearly horizontal surface in a porous medium with Newtonian heating. *Heat Mass Transfer* **40**, 665–672. [5.3]

Leu, J. S. and Jang, J. Y. 1994 The wall and free plumes above a horizontal line source in non-Darcian porous media. *Int. J. Heat Mass Transfer* **37**, 1925–1933. [5.10.1]

Leu, J. S. and Jang, J. Y. 1993 Variable viscosity effects on the vortex instability of the convective boundary layer flow over a horizontal uniform heat flux surface in a saturated porous medium. *Proc. 6th Int. Sympos. Transport Phenomena and Thermal Engineering*, Seoul, Korea. Vol.1, pp. 203–208. [5.4]

Leu, J. S. and Jang, J. Y. 1995 The natural convection from a point heat source embedded in a non-Darcian porous medium. *Int. J. Heat Mass Transfer* **38**, 1097–1104. [5.11.1]

Levy, A., Levi-Hevroni, D., Sorek, S. and Ben-Dor, G. 1999 Derivation of Forchheimer terms and their verification by applications to waves propagated in porous media. *Int. J. Multiphase Flows* **25**, 683–704. [1.5.2]

Levy, T. 1981 Loi de Darcy ou loi de Brinkman? *C. R. Acad. Sci. Paris, Sér. II* **292**, 872–874. [1.5.3]

Levy, T. 1990 Écoulement dans un milieu poreux avec fissures unidirectionelles. *C.R. Acad. Sci. . Paris, Sér. II*, 685–690. [1.9]

Lewins, J. 2003 Bejan's constructal theory of equal potential distribution. *Int. J. Heat Mass Transfer* **46**, 1541–1543. [4.18, 4.20, 6.26, 11.10]

Lewis, S., Bassom, A. P. and Rees, D. A. S. 1995 The stability of vertical thermal boundary-layer flow in a porous medium. *Eur. J. Mech. B/Fluids* **14**, 395–407. [5.1.9.9]

Lewis, S., Rees, D. A. S. and Bassom, A. P. 1997 High wavenumber convection in tall porous containers heated from below. *Quart. J. Mech. Appl. Math.* **50**, 545–563. [6.15.2]

Li, C. T. and Lai, F. C. 1998 Re-examination of double diffusive natural convection from horizontal surfaces in porous media. *J. Thermophys. Heat Transfer* **12**, 449–452. [9.2.1]

Li, J. M. and Wang, B. X. 1998 Investigation on wall effect of condensation in porous media. *Heat Transfer 1998, Proc. 11th IHTC*, **4**, 459–463. [10.4]

Li, L. and Kimura, S. 2005 Mixed convection around a heated vertical cylinder embedded in porous medium. *Prog. Nat. Sci.* **15**, 661–664. [8.1.3]

Li, Y. and Park, C. W. 1998 Permeability of packed beds filled with polydiverse spherical particles. *Ind. Eng. Chem Res*. **37**, 2005–2011. [1.4.2]

Liao, S. J. and Pop, I. 2004 Explicit analytic solution for similarity boundary layer equations. *Int. J. Heat Mass Transfer* **47**, 75–85. [5.1.9.1]

Libera, J. and Poulikakos, D. 1990 Parallel-flow and counter-flow conjugate convection from a vertical insulated pipe. *AIAA J. Thermophys. Heat Transfer* **4**, 400–404. [5.7]

Lie, K. N. and Jang, J. Y. 1993 Boundary and inertia effects on vortex instability of a horizontal mixed convection flow in a porous medium. *Numer. Heat Transfer A* **23**, 361–378. [5.4]

Lienhard, J. H. 1973 On the commonality of equations for natural convection from immersed bodies. *Int. J. Heat Mass Transfer* **16**, 2121–2123. [4.3]

Lim, J. S., Fowler, A. J. and Bejan, A., 1993, Spaces filled with fluid and fibers coated with phase-change material. *J. Heat Transfer* **115**, 1044–1050. [10.5]

Lin, D. K. 1992 Unsteady natural convection heat and mass transfer in a saturated porous enclosure. *Wärme-Stoffübertrag.* **28**, 49–56.[9.1.6.4]

Lin, D. S. and Gebhart, B. 1986 Buoyancy induced flow adjacent to a horizontal surface submerged in porous medium saturated with cold water. *Int. J. Heat Mass Transfer* **29**, 611–623. [5.2]

Lin, G., Zhao, C. B., Hobbs, B. E., Ord, A. and Muhlhaus, H. B. 2003 Theoretical and numerical analyses of convective instability in porous media with temperature-dependent viscosity. *Comm. Numer. Meth. Engng.* **19**, 787–799. [6.7]

Lin, S., Chao, J. T., Chen, T. F. and Chen, D. K. 1998 Analytical and experimental study of drying process in a porous medium with a non-penetrating surface. *J. Porous Media* **1**, 159–166. [3.6]

Ling, J. X. and Dybbs, A. 1992 The effect of variable viscosity on forced convection over a flat plate submersed in a porous medium. *ASME J. Heat Transfer* **114**, 1063–1065. [4.2]

Lister, C. R. B. 1990 An explanation for the multivalued heat transport found experimentally for convection in a porous medium. *J. Fluid Mech.* **214**, 287–320. [6.4, 6.9.3]

Liu, C. Y. and Guerra, A. C. 1985 Free convection in a porous medium near the corner of arbitrary angle formed by two vertical plates. *Int. Comm. Heat Mass Transfer* **12**, 431–440. [5.12.2]

Liu, C. Y. and Ismail, K. A. R. 1980 Asymptotic solution of free convection near a corner of two vertical porous plates embedded in a porous medium. *Lett. Heat Mass Transfer* **7**, 457–463. [5.12.2]

Liu, C. Y., Ismail, K. A. R. and Ibinuma, C. D. 1984 Film condensation with lateral mass flux about a body of arbitrary shape in a porous medium. *Int. Comm. Heat Mass Transfer* **11**, 377–384. [10.4]

Liu, C. Y., Lam, C. Y. and Guerra, A. C. 1987a Free convection near a corner formed by two vertical plates embedded in a porous medium. *Int. Comm. Heat Mass Transfer* **14**, 125–136. [5.12.2]

Liu, J. Y. and Minkowycz, W. J. 1986 The effect of Prandtl number on conjugate heat transfer in porous media. *Int. Comm. Heat Mass Transfer* **13**, 439–448. [5.12.1]

Liu, J. Y., Shih, S. D. and Minkowycz, W. J. 1987b Conjugate natural convection about a vertical cylindrical fin with lateral mass flux in a saturated porous medium. *Int. J. Heat Mass Transfer* **30**, 623–630. [5.12.1]

Liu, P. C. 2003 Temperature distribution in a porous medium subjected to solar radiative incidence and downward flow: convective boundaries. *ASME J. Solar Energy Engng.* **125**, 190–194. [6.11.2]

Liu, S. and Masliyah, J. H. 1998 On non-Newtonian fluid flow in ducts and porous media. *Chem. Engng. Sci.* **53**, 1175–1201. [1.5.4]

Liu, S. and Masliyah, J. H. 2005 Dispersion in porous media. *Handbook of Porous Media* (ed. K. Vafai), 2nd ed., Taylor and Francis, Boca, Raton, FL, pp. 81–140. [1.5.1, 2.2.3]

Liu, S., Afacan, A. and Masliyah, J. 1994 Steady incompressible laminar flow in porous media. *Chem. Engng. Sci.* **49**, 3565–3586. [1.4.2]

Liu, W., Shen, S. and Riffat, S. B. 2002 Heat transfer and phase change of liquid in an inclined enclosure packed with unsaturated porous media. *Int. J. Heat Mass Transfer* **45**, 5209–5219. [3.6]

Llagostera, J. and Figueiredo, J. R. 1998 Natural convection in porous cavity: application of UNIFAES discretization scheme. *Heat Transfer 1998, Proc. 11th IHTC*, **4**, 465–470. [6.8]

Llagostera, J. and Figueiredo, J. R. 2000 Application of the UNIFAES discretization scheme to mixed convection in a porous layer with a cavity, using the Darcy model. *J. Porous Media* **3**, 139–154. [8.2.1]

Lombardo, S. and Mulone, G. 2002 Necessary and sufficient conditions of global nonlinear stability of rotating double-diffusive convection in a porous medium. *Cont. Mech. Thermodyn.* **14**, 527–540. [9.1.6.4]

Lombardo, S. and Mulone, G. 2003 Nonlinear stability and convection for laminar flows in a porous medium with Brinkman law. *Math. Meth. Appl. Sci.* **26**, 453–462. [6.10]

Lombardo, S., Mulone, G. and Straughan, B. 2001 Nonlinear stability in the Bénard problem for a double-diffusive mixture in a porous medium. *Math. Meth. Appl. Sci.* **24**, 1229–1246. [9.1.3]

Loper, D. E. and Roberts, P. H. 2001 Mush-chimney convection. *Stud. Appl. Math.* **106**, 187–227. [10.2.3]

Lopez, D. L. and Smith, L. 1995 Fluid flow at fault zones: analysis of the interplay of convective circulation and topographically driven groundwater flow. *Water Resour. Res.* **31**, 1489–1503. [11.8]

Lorente, S. and Bejan, A. 2002 Combined "flow and strength" geometric optimization: internal structure in a vertical insulating wall with air cavities and prescribed strength. *Int. J. Heat Mass Transfer* **45**, 3313–3320. [7.7]

Lorente, S. and Ollivier, J. P. 2005 Scale analysis of electrodiffusion through porous media. *J. Porous Media* **8**(4) in press [3.7].

Lorente, S., Petit, M. and Javelas, R. 1996 Simplified analytical model for thermal transfer in a vertical hollow brick. *Energy Build.* **24**, 95–103. [7.7]

Lorente, S., Petit, M. and Javelas, R. 1998 The effects of temperature conditions on the thermal resistance of walls made of different shapes vertical hollow bricks. *Energy Build.* **28**, 237–240. [7.7]

Lowell, R. P. 1980 Topographically driven subcritical hydrothermal convection in the oceanic crust. *Earth Planet. Sci. Lett.* **49**, 21–28. [11.4, 11.6.2]

Lowell, R. P. 1985 Double-diffusive convection in partially molten silicate systems: its role during magma production and in magma chambers. *J. Volcanol. Geotherm. Res.* **26**, 1–24. [11.4]

Lowell, R. P. 1991 Modeling continental and submarine hydrothermal systems. *Rev. Geophys.* **29**, 457–476. [11.8.1]

Lowell, R. P. and Burnell, D. K. 1991 Mathematical modeling of conductive heat-transfer from a freezing, convecting magma chamber to a single pass hydrothermal system—Implications for sea-floor black smokers. *Earth Planet. Phys.* **104**, 59–69. [11.8]

Lowell, R. P. and Hernandez, H. 1982 Finite amplitude convection in a porous container with fault-like geometry: effect of initial and boundary conditions. *Int. J. Heat Mass Transfer* **25**, 631–641. [6.15.2]

Lowell, R. P. and Shyu, C. T. 1978 On the onset of convection in a water-saturated porous box: effect of conducting walls. *Lett. Heat Mass Transfer* **5**, 371–378. [6.15.2]

Lowell, R. P., Rona, P. A. and von Herzen, R. P. 1995 Seafloor hydrothermal systems. *J. Geophys. Res.* **100**, 327–352. [11.8]

Lu, J. W. and Chen, F. 1997 Rotation effects on the convection of binary alloys unidirectionally solidified from below. *Int. J. Heat Mass Transfer* **40**, 237–246. [10.2.3]

Lu, N. 2001 An analytical assessment on the impact of covers on the onset of air convection in mine wastes. *Int. J. Numer. Anal. Meth. Geomech.* **25**, 347–364. [11.8]

Lu, N. and Zhang, Y. 1997 Onset of thermally induced gas convection in mine wastes. *Int. J. Heat Mass Transfer* **40**, 2621–2636. [6.11.2]

Lu, N., Zhang, Y. and Ross, B. 1999 Onset of gas convection in a moist porous layer with the top boundary open to the atmosphere. *Int. Comm. Heat Mass Transfer* **26**, 33–44. [6.2]

Lu, T. J., Stone, H. A. and Ashby, M. F. 1998 Heat transfer in open-cell metal foams. *Acta Mater.* **46**, 3619–3635. [4.15]

Ludvigsen, A., Palm, E. and McKibbin, R. 1992 Convective momentum and mass transport in sloping porous layers. *J. Geophys. Res.* **97**, 12315–12326. [7.8]

Luna, E., Medina, A., Pérez-Rosales, C. and Treviño, C. 2004 Convection and dispersion in a naturally fractured reservoir. *J. Porous Media* **7**, 303–316. [7.8]

Luna, N. and Mendez, F. 2005 Forced convection on a heated horizontal flat plate with finite thermal conductivity in a non-Darcian porous medium. *Int. J. Thermal Sci.* **44**, 656–664. [4.8]

Lundgren, T. S. 1972 Slow flow through stationary random beds and suspensions of spheres. *J. Fluid Mech.* **51**, 273–299. [1.5.3]

Ly, H. V. and Titi, E. S. 1999 Global Gevrey regularity for the Bénard convection in a porous medium with zero Darcy-Prandtl number. *J. Nonlinear Sci.* **9**, 333–362. [6.15.1]

Lyubimov, D. V. 1975 Convective motions in a porous medium heated from below. *J. Appl. Mech. Tech. Phys.* **16**, 257–261. [6.16.2]

Lyubimov, D. V. 1993 Dynamical properties of thermal convection in porous medium. *Instabilities in Multiphase Flows* (G. Gouesbet and A. Berlement, eds.) Plenum, New York, 289–295. [7.3.3]

Ma, H. and Ruth, D. W. 1993 The microscopic analysis of high Forchheimer number flow in porous media. *Transport Porous Media* **13**, 139–160. [1.5.2]

Ma, X. H. and Wang, B. X. 1998 Suction effect of a vertical coated plain porous layer on film condensation heat transfer enhancement. *Heat Transfer 1998, Proc. 11th IHTC* **5**, 387–391. [10.4]

Macdonald, I. F., El-Sayed, M. S., Mow, K. and Dullien, F. A. L. 1979 Flow through porous media: The Ergun equation revisited. *Ind. Chem. Fundam.* **18**, 199–208. [1.5.2]

MacDonald, M. J., Chu, C. F., Guillot, P. P. and Ng, K. M. 1991 A generalized Blake-Kozeny equation for multisized spherical particles. *AIChE J.* **37**, 1583–1588. [1.4.2]

Macedo, H. H., Costa, U. M. S. and Almeido, M. P. 2001 Turbulent effects on fluid flow through disordered porous media. *Physica A* **299**, 371–377. [1.8]

Mackie, C. 2000 Thermal convection in a sparsely packed porous layer saturated with suspended particles. *Int. Comm. Heat Mass Transfer* **27**, 315–324. [6.23]

Mackie, C., Desai, P. and Myers, C. 1999 Rayleigh-Bénard stability of a solidifying porous medium. *Int. J. Heat Mass Transfer* **42**, 3337–3350. [10.2.2]

Magyari, E. and Keller, B. 2000 Exact analytic solutions for free convection boundary layers on a heated vertical plate with lateral mass flux embedded in a saturated porous medium. *Heat Mass Transfer* **36**, 109–116. [5.1.9.1]

Magyari, E. and Keller, B. 2002 Note on 'A two-equation analysis of convection heat transfer in porous media' by H. Y. Zhang and X. Y. Huang. *Transport Porous Media* **46**, 109–112. [4.10]

Magyari, E. and Keller, B. 2003a The opposing effect of viscous dissipation allows for a parallel free convection boundary-layer flow along a cold vertical flat plate. *Transport Porous Media* **51**, 227–230. [5.1.9.4]

Magyari, E. and Keller, B. 2003b Effect of viscous dissipation on a quasi-parallel free convection boundary-layer flow over a vertical plate in a porous medium. *Transport Porous Media* **51**, 231–236. [5.1.9.4]

Magyari, E. and Keller, B. 2003c Buoyancy sustained by viscous dissipation. *Transport Porous Media* **53**, 105–115. [5.9.4]

Magyari, E. and Keller, B. 2004a The free convection boundary-layer flow induced in a fluid-saturated porous medium by a non-isothermal vertical cylinder approaches the shape of Schlicting's round jet as the porous radius tends to zero. *Transport Porous Media* **54**, 265–271. [5.7]

Magyari, E. and Keller, B. 2004b Backward free convection boundary layers in porous media. *Transport Porous Media* **55**, 285–300. [5.1.9.1]

Magyari, E. and Rees, D. A. S. 2005 Effect of viscous dissipation on the Darcy free convection boundary-layer flow over a vertical plate with exponential temperature distribution in a porous medium. *Fluid Dyn. Res.* (submitted). [5.1.9.4]

Magyari, E., Keller, B. and Pop, I. 2001a, Exact analytic solutions of forced convection flow in a porous medium. *Int. Comm. Heat Mass Transfer* **28**, 233–241. [4.1]

Magyari, E., Pop, I. and Keller, B. 2001b Exact dual solutions occurring in the Darcy mixed convection flow. *Int. J. Heat Mass Transfer* **44**, 4563–4566. [8.1.1]

Magyari, E., Pop, I. and Keller, B. 2002 The "missing" self-similar free convection boundary-layer flow over a vertical permeable surface in a porous medium. *Transport Porous Media* **46**, 91–102. [8.1.1]

Magyari, E., Pop, I. and Keller, B. 2003a New similarity solutions for boundary-layer flow on a horizontal surface in a porous medium. *Transport Porous Media* **51**, 123–140. [8.1.2]

Magyari, E., Pop, I. and Keller, B. 2003b Effect of viscous dissipation on the Darcy forced-convection flow past a plane surface. *J. Porous Media* **6**, 111–122. [4.1]

Magyari, E., Pop, I. and Keller, B. 2003c New analytic solutions of a well-known boundary value problem in fluid mechanics. *Fluid Dyn. Res.* **33**, 313–317. [5.1.9.1]

Magyari, E., Pop, I. and Keller, B. 2004 Analytical solutions for unsteady free convection in porous media. *J. Engng. Math.* **48**, 93–104. [5.1.9.1]

Magyari, E., Pop, I. and Keller, B. 2005a Exact solutions for a longitudinal steady mixed convection flow over a permeable vertical thin cylinder in a porous medium. *Int. J. Heat Mass Transfer* **48**, 3435–3442. [8.1.3]

Magyari, E., Rees, D. A. S. and Keller, B. 2005b Effect of viscous dissipation on the flow in fluid saturated porous media. *Handbook of Porous Media* (ed. K. Vafai), 2nd ed., Taylor and Francis, New York, pp. 373–406. [2.2.2, 5.1.9.4, 6.6]

Mahidjiba, A., Mamou, M. and Vasseur, P. 2000a Onset of double-diffusive convection in a rectangular porous cavity subject to mixed boundary conditions. *Int. J. Heat Mass Transfer* **43**, 1505–1522. [9.1.3]

Mahidjiba, A., Robillard, L. and Vasseur, P. 2002 The horizontal anisotropic porous layer saturated with water near 4 degrees C–A peculiar stability problem. *Heat Transfer 2002, Proc. 12th Int. Heat Transfer Conf.*, Elsevier, Vol. 2, pp. 797–802. [6.11.4]

Mahidjiba, A., Robillard, L. and Vasseur, P. 2003 Linear stability of cold water layer saturating an anisotropic porous medium—effect of confinement. *Int. J. Heat Mass Transfer* **46**, 323–332. [6.11.4]

Mahidjiba, A., Robillard, L. and Vasseur, P. 2000b Onset of convection in a horizontal porous layer saturated with water near 4 degrees C. *Int. Comm. Heat Mass Transfer* **27**, 765–774. [6.11.4]

Mahidjiba, A., Robillard, L., Vasseur, P. and Mamou, M. 2000c Onset of convection in an anisotropic porous layer of finite lateral extent. *Int. Comm. Heat Mass Transfer* **27**, 333–342. [6.12]

Mahmood, T. and Merkin, J. H. 1998 The convective boundary-layer flow on a reacting surface in a porous medium. *Transport Porous Media* **32**, 285–298. [3.4, 5.12.3]

Mahmud, S. and Fraser, R. A. 2004a Flow and heat transfer inside porous stack: steady state problem. *Int. Comm. Heat Mass Transfer* **31**, 951–962. [4.9]

Mahmud, S. and Fraser, R. A. 2004b Magnetohydrodynamic free convection and entropy generation in a square porous cavity. *Int. J. Heat Mass Transfer* **47**, 3245–3256. [7.1.6]

Mahmud, S. and Fraser, R. A. 2005a Conjugate heat transfer inside a porous channel. *Heat Mass Transfer* **41**, 568–575. [4.12]

Mahmud, S. and Fraser, R. A. 2005b Flow, thermal, and entropy characteristics inside a porous channel with viscous dissipation. *Int. J. Thermal Sci.* **44**, 21–32. [4.9]

Maier, R. S., Kroll, D. M., Davis, H. T. and Bernard, R. S. 1998 Simulation of flow through bead packs using the lattice Boltzmann method. *Phys. Fluids* **10**, 60–74. [1.9]

Majumdar, A. and Tien, C. L. 1990 Effects of surface tension on film condensation in a porous medium. *ASME J. Heat Transfer* **112**, 751–757. [10.4]

Makuoka, T., Tsurota., Y. , Tanigawa, H. and Izaki, H. 1994b Natural convection in porous insulation layers containing a row of heat pipes. *Heat Transfer 1994 (Proc. 10th Int. Heat Transfer Conf.* Brighton, UK), Inst. Chem. Engrs, Rugby, vol. 6, pp. 385–390. [6.13.2]

Malashetty, M. S. and Basavaraja, D. 2002 Rayleigh-Bénard convection subject to time dependent wall temperature/gravity in a fluid-saturated anisotropic porous medium *Heat Mass Transfer* **38**, 551–563. [6.11.3]

Malashetty, M. S. and Basavaraja, D. 2004 Effect of time-periodic boundary temperatures on the onset of double diffusive convection in a horizontal anisotropic porous layer. *Int. J. Heat Mass Transfer* **47**, 2317–2327. [9.1.6.4]

Malashetty, M. S. and Gaikwad, S. N. 2003 Onset of convective instabilities in a binary liquid mixtures with fast chemical reactions in a porous media. *Heat Mass Transfer* **39**, 415–420. [9.1.6.4]

Malashetty, M. S. and Padmavathi, V. 1997 Effect of gravity modulation on onset of convection in a fluid and porous layer. *Int. J. Engng. Sci.* **35**, 829–840. [6.19.3]

Malashetty, M. S. and Padmavathi, V. 1998 Effect of gravity modulation on the onset of convection in a porous layer. *J. Porous Media* **1**, 219–226. [6.24]

Malashetty, M. S. and Wadi, V. S. 1999 Rayleigh-Bénard convection subject to time-dependent wall temperature in a fluid saturated porous layer. *Fluid Dyn. Res.* **24**, 293–308. [6.11.3]

Malashetty, M. S., Cheng, P. and Chao, B. H. 1994 Convective instability in a horizontal porous layer saturated with a chemically reacting fluid. *Int. J. Heat Mass Transfer* **37**, 2901–2908. [6.11.2]

Malashetty, M. S., Shivakumara, I. S. and Kulkarni, S. 2005a The onset of convection in an anisotropic porous layer using a thermal non-equilibrium model. *Transport Porous Media* **60**, 199–215. [6.5]

Malashetty, M. S., Shivakumara, I. S. and Kulkarni, S. 2005b The onset of Lapwood-Brinkman convection using a non-equilibrium model. *Int. J. Heat Mass Transfer* **48**, 1155–1163. [6.5]

Malashetty, M. S., Umavathi, J. C. and Pratap Kumar, J. 2001 Convective flow and heat transfer in an inclined composite porous medium. *J. Porous Media* **4**, 15–22. [7.8]

Malashetty, M. S., Umavathi, J. C. and Pratap Kumar, J. 2004 Two fluid flow and heat transfer in an inclined channel containing porous and fluid layer. *Heat Mass Transfer* **40**, 871–876. [7.8]

Malkovski, V. I. and Pek, A. A. 1999 Onset conditions of free thermal convection of a single-phase fluid in a horizontal porous layer with depth-dependent permeability. *Fizika Zemli* (12): 27–31. [6.13.1]

Malkovsky, V. I. and Pek, A. A. 2004 Onset of thermal convection of a single-phase fluid in an open vertical fault. *Izv. Phys. Solid Earth* **40**, 672–679. [6.2]

Mamou, M. 2002a Stability analysis of thermosolutal convection in a vertical packed porous enclosure. *Phys. Fluids* **14**, 4302–4314. [9.1.6.4]

Mamou, M. 2002b Stability analysis of double-diffusive convection in porous enclosures. In *Transport Phenomena in Porous Media II* (D. B. Ingham and I. Pop, eds.) Elsevier, Oxford, pp. 113–154. [9]

Mamou, M. 2003 Stability analysis of the perturbed rest state and of the finite amplitude steady double-diffusive convection in a shallow porous enclosure. *Int. J. Heat Mass Transfer* **46**, 2263–2277. [9.1.3]

Mamou, M. 2004 Onset of oscillatory and stationary double-diffusive convection within a tilted porous enclosure. In *Emerging Technologies and Techniques in Porous Media* (D. B. Ingham, A. Bejan, E. Mamut and I. Pop, eds), Kluwer Academic, Dordrecht, pp. 209–219. [9.4].

Mamou, M. and Vasseur, P. 1999 Thermosolutal bifurcation phenomena in porous enclosures subject to vertical temperature and concentration gradients. *J. Fluid Mech.* **395**, 61–87. [9.1.3]

Mamou, M., Hasnaoui, M., Amahmid, A. and Vasseur, P. 1998a Stability analysis of double diffusive convection in a vertical Brinkman porous enclosure. *Int. Comm. Heat Mass Transfer* **25**, 491–500. [9.2.2]

Mamou, M., Mahidjiba, A., Vasseur, P. and Robillard, L. 1998b Onset of convection in an anisotropic porous medium heated from below by a constant heat flux. *Int. Comm. Heat Mass Transfer* **25**, 799–808. [6.12]

Mamou, M., Robillard, L. and Vasseur, P. 1999 Thermoconvective instability in a horizontal porous cavity saturated with cold water. *Int. J. Heat Mass Transfer* **42**, 4487–4500. [6.11.4]

Mamou, M., Robillard, L., Bilgen, E. and Vasseur, P. 1996 Effects of a moving thermal wave on Bénard convection in a horizontal saturated porous layer. *Int. J. Heat Mass Transfer* **39**, 347–354. [6.11.3]

Mamou, M., Vasseur, P. and Bilgen, E. 1995a Multiple solutions for double-diffusive convection in a vertical porous enclosure. *Int. J. Heat Mass Transfer* **38**, 1787–1798. [9.2.4]

Mamou, M., Vasseur, P. and Bilgen, E. 1998c A Galerkin finite-element study of the onset of double-diffusive convection in an inclined porous enclosure. *Int. J. Heat Mass Transfer* **41**, 1513–1529. [9.4]

Mamou, M., Vasseur, P. and Bilgen, E. 1998d Double-diffusive convection instability in a vertical porous enclosure. *J. Fluid Mech.* **368**, 263–289. [9.2.4]

Mamou, M., Vasseur, P., Bilgen, E. and Gobin, D. 1994 Double-diffusive convection in a shallow porous layer. *Heat Transfer 1994*, Inst.Chem. Engrs, Rugby, vol. 5, pp. 339–344. [9.1.3]

Mamou, M., Vasseur, P., Bilgen, E. and Gobin, D. 1995b Double-diffusive convection in an inclined slot filled with porous medium. *Eur. J. Mech. B/Fluids* **14**, 629–652. [9.2.4]

Manole, D. M. and Lage, J. L. 1993 The inertial effect on the natural convection flow within a fluid saturated porous medium. *Int. J. Heat Fluid Flow* **14**, 376–384. [1.5.2]

Manole, D. M. and Lage, J. L. 1995 Numerical simulation of supercritical Hadley circulation, within a porous layer, induced by inclined temperature gradients. *Int. J. Heat Mass Transfer* **38**, 25–83–2593. [7.9]

Manole, D. M., Lage, J. L. and Antohe, B. V. 1995 Supercritical Hadley circulation within a layer of fluid saturated porous medium: bifurcation to traveling wave. *ASME HTD* **309**, 23–29. [7.9]

Manole, D. M., Lage, J. L. and Nield, D. D. 1994 Convection induced by inclined thermal and solutal gradients, with horizontal mass flow, in a shallow horizontal layer of a porous medium. *Int. J. Heat Mass Transfer* **37**, 2047–2057. [9.5]

Mansour, A., Amahmid, A., Hasnaoui, M. and Bourich, M. 2004 Soret effect on double-diffusive multiple solutions in a square porous cavity subject to cross gradients of temperature and concentration. *Int. Comm. Heat Mass Transfer* **31**, 431–440. [9.5]

Mansour, M. A. 1997 Forced convection radiation interaction heat transfer in boundary layer over a flat plate submersed in a porous medium. *Appl. Mech. Engng.***2**, 405–413. [4.16.2]

Mansour, M. A. and El-Shaer, N. A. 2001 Radiative effects on magnetohydrodynamic natural convection flows saturated in porous media. *J. Magn. Magn. Mater.* **237**, 327–341. [5.1.9.4]

Mansour, M. A. and Gorla, R. S. R. 1998 Mixed convection-radiation in power-law fluids along a nonisothermal wedge embedded in a porous medium. *Transport Porous Media* **30**, 113–124. [8.1.4]

Mansour, M. A. and Gorla, R. S. R. 2000a Thermal dispersion effects on non-Darcy natural convection with internal heat generation. *Chem. Engng. Commun.* **177**, 177–181. [5.1.9.4]

Mansour, M. A. and Gorla, R. S. R. 2000b Combined convection in non-Newtonian fluids along a nonisothermal vertical plate in a porous medium. *Int. J. Numer. Meth. Heat Fluid Flow* **10**, 163–178. [8.1.1]

Mansour, M. A. and Gorla, R. S. R. 2000c Radiative and thermal dispersion effects on non-Darcy natural convection. *J. Porous Media* **3**, 267–272. [5.1.9.4]

Mansour, M. A., El-Hakiem, M. A. and El-Gaid, S. A. 1997 Mixed convection in non-Newtonian fluids along an isothermal vertical cylinder in a porous medium. *Transport Porous Media* **28**, 307–317. [8.1.3]

Mansour, M. and El-Shaer, N. 2004 Mixed convection-radiation in power-law fluids along a non-isothermal wedge in a porous medium with variable permeability. *Transport Porous Media* **57**, 333–346. [8.1.4]

Marafie, A. and Vafai, K. 2001 Analysis of non-Darcian effects on temperature differentials in porous media. *Int. J. Heat Mass Transfer* **44**, 4401–4411. [4.10]

Marcondes, F., de Medeiros, J. M., and Gurgel, J. M. 2001 Numerical analysis of natural convection in cavities with variable porosity. *Numer. Heat Transfer A* **40**, 403–420. [7.6.2]

Marcoux, M. and Charrier-Mojtabi, M. C. 1998 Etude paramétrique de la thermogravitation en milieu poreux. *C. R. Acad. Sci., Paris* 326, 539–546. [9.1.4]

Marcoux, M., Charrier-Mojtabi, M. C. and Azaiez, M. 1999b Double-diffusive convection in an annular vertical porous layer. *Int. J. Heat Mass Transfer* **42**, 2313–2325. [9.4]

Marcoux, M., Karimi-Fard, M. and Charrier-Mojtabi, M. C. 1999a Onset of double-diffusive convection in a rectangular porous cavity submitted to heat and mass fluxes at the vertical walls. *Int. J. Thermal Sci.* **38**, 258–266. [9.2.4]

Marpu, D. R. 1993 Non-Darcy flow and axial conduction effects of forced convection in porous material filled pipes. *Wärme- Stoffübertrag.* **29**, 51–58. [4.9]

Marpu, D. R. 1995 Forchheimer and Brinkman extended Darcy flow model on natural convection in a vertical cylindrical porous annulus. *Acta Mech.* **109**, 41–48. [7.3.3]

Marpu, D. R. and Satyamurty, V. V. 1989 Influence of variable fluid density on free convection in rectangular porous media. *ASME J. Energy Res. Tech.* **111**, 214–220. [6.15.3]

Martínez-Suástegui, L., Trevino, C. and Méndez, F. 2003 Natural convection in a vertical strip immersed in a porous medium. *Europ. J. Mech. B/Fluids* **22**, 5454–553. [7.1.5]

Martins-Costa, M. L. 1996 A local model for a packed-bed heat exchanger with a multiphase matrix. *Int. Comm. Heat Mass Transfer* **23**, 1133–1142. [2.6]

Martins-Costa, M. L. and Saldanha da Gama, R.M. 1994 Local model for the heat transfer process in two distinct flow regions. *Int. J. Heat Fluid Flow* **15**, 477–485. [2.6]

Martins-Costa, M. L., Sampaio, R. and Saldanha da Gama, R.M. 1992 Modelling and simulation of energy transfer in a saturated flow through a porous medium. *Appl. Math. Model.* **16**, 589–597. [2.6]

Martins-Costa, M. L., Sampaio, R. and Saldanha da Gama, R.M. 1994 Modeling and simulation of natural convection flow in a saturated porous cavity. *Meccanica* **29**, 1–13. [2.6]

Martys, N., Bentz, D. P. and Garboczi, E. J. 1994 Computer simulation study of the effective viscosity in Brinkman equation. *Phys. Fluids* **6**, 1434–1439. [1.5.3]

Martys, N. S. 2001 Improved approximation of the Brinkman equation using a lattice Boltzmann method. *Phys. Fluids* **13**, 1807–1810. [1.5.3]

Masoud, S. A., Al-Nimr, M. A. and Alkam, M. K. 2000 Transient film condensation on a vertical plate imbedded in porous medium. *Transport Porous Media* **40**, 345–354. [10.4]

Massarotti, N., Nithiarasu, P. and Zienkiewicz, O. C. 2001 Natural convection in porous medium-fluid interface problems—A finite element analysis by using the CBS procedure. *Int. J. Numer. Meth. Heat Fluid Flow* **11**, 473–490. [6.19.1]

Masuda, Y., Yoneda, M., Ikeshoji, T., Kimura, S., Alavyoon, F., Tsukada, T. and Hozawa, M. 2002 Oscillatory double-diffusive convection in a porous enclosure due to opposing heat and mass fluxes on the vertical walls. *Int. J. Heat Mass Transfer* **45**, 1365–1369. [9.1.6.4]

Masuda, Y., Yoneda, M., Sumi, S., Kimura, S. and Alavyoon, F. 1999 Double-diffusive natural convection in a porous medium under constant heat and mass fluxes. *Heat Transfer Asian Res.* **28**, 255–265. [9.1.6.4]

Masuoka, T. 1986 Natural convection in stratified porous media heated from the side. (In Japanese) *Trans. ASME B* **52**, 866–869. [7.3.2]

Masuoka, T. and Takatsu, Y. 1996 Turbulence model for flow through porous media. *Int. J. Heat Mass Transfer* **39**, 2803–2809. [1.8]

Masuoka, T. and Takatsu, Y. 2002 Turbulence characteristics in porous media. In *Transport Phenomena in Porous Media II* (D. B. Ingham and I. Pop, eds.) Elsevier, Oxford, pp. 231–256. [1.8]

Masuoka, T., Kakimoto, Y., Nomura, A. and Ooba, M 2004 Fluid flow through a permeable porous obstacle. In *Applications of Porous Media (ICAPM 2004)*, (eds. A. H. Reis and A. F. Miguel), Évora, Portugal, pp. 107–112. [4.11]

Masuoka, T., Nishimura, T., Kawamotu, S. and Tsuruta, T. 1991 Effects of mid-height cooling on natural convection in a porous layer heated from below. *Proc. ASME, JSME Thermal Engineering Joint Conference—1991*, vol. 4, pp. 325–330. [6.13.2]

Masuoka, T., Shibata, K., Nakumura, H., Tanaka, T. and Tsuruta, T. 1988 Natural convection in porous insulation layers with peripheral gaps. *Proc. 2nd Symposium on Heat Transfer*, Tsinghua Univ. Beijing, China. [6.13.2]

Masuoka, T., Takatsu, Y., Kawamoto, S., Koshino, H. and Tsuruta, T. 1995b Buoyant plume through a permeable porous layer located above a line heat source in an infinite fluid space. *JSME Int. J., Series B* **38**, 79–85. [5.10.1.2]

Masuoka, T., Takatsu, Y., Tsuruta, T. and Nakamura, H. 1994 Buoyancy-driven channeling flow in vertical porous layer. *JSME Int. J., Series B* **37**, 915–923. [7.7]

Masuoka, T., Tanigawa, H., Tsuruta, T. and Izaki, H. 1995a Studies on the improvement of the performances of insulation layers with isothermal screens. *Proc. 3rd. ASME/JSME Thermal Engineering Joint Conference*, vol.3, pp. 333–338. [6.13.2]

Masuoka, T., Tohda, Y., Tsurota, Y. and Yasuda, Y. 1986 Buoyant plume above a concentrated heat source in stratified porous media (in Japanese). *Trans JSME Ser. B*, 2656–2662. [5.10.2]

Masuoka, T., Yokote, Y. and Katsuhara, T. 1981 Heat transfer by natural convection in a vertical porous layer. *Bull. JSME* **24**, 995–1001. [7.1.2]

Mat, M. D. and Ilegbusi, O. J. 2002 Application of a hybrid model of mushy zone to macrosegregation in alloy solidification. *Int. J. Heat Mass Transfer* **45**, 279–289. [10.2.3]

Matsumoto, K., Okada, M., Murakami, M. and Yabushita, Y. 1993 Solidification of porous medium saturated with aqueous solution in a rectangular cell. *Int. J. Heat Mass Transfer* **36**, 2869–2880. [10.2.3]

Matsumoto, K., Okada, M., Murakami, M. and Yabushita, Y. 1995 Solidification of porous medium saturated with aqueous solution in a rectangular cell—II. *Int. J. Heat Mass Transfer* **38**, 2935–2943. [10.2.3]

Mauran, S., Riguad, L. and Coudevylle, O. 2001 Application of the Carman-Kozeny correlation to a high porosity and anisotropic consolidated medium: The compressed expanded natural graphite. *Transport Porous Media* **43**, 355–376. [1.4.2]

Mbaye, M. and Bilgen, E. 1992 Natural convection and conduction in porous wall, solar collector systems with porous absorber. *ASME J. Solar Engng.*, **114**, 41–46. [7.3.1]

Mbaye, M. and Bilgen, E. 1993 Conduction and convection heat transfer in composite solar collector systems with porous absorber. *Wärme-Stoffübertrag.* **28**, 267–274. [7.3.1]

Mbaye, M. and Bilgen, E. 2001 Subcritical oscillatory instability in porous beds. *Int. J. Thermal Sci.* **40**, 595–602. [9.1.3]

Mbaye, M., Bilgen, E. and Vasseur, P. 1993 Natural-convection heat transfer in an inclined porous layer boarded by a finite-thickness wall. *Int. J. Heat Fluid Flow* **14**, 284–291. [7.8]

McCarthy, J. F. 1994 Flow through arrays of cylinders: lattice gas cellular automata simulations. *Phys. Fluids* **6**, 435–437. [2.6]

McGuinness, M. J. 1996 Steady solution selection and existence in geothermal heat pipes—I. The convective case. *Int. J. Heat Mass Transfer* **39**, 259–274. [11.9.2]

McGuinness, M. J., Blakely, M., Pruess, K. and O'Sullivan, M. J. 1993 Geothermal heat pipe stability: solution selection by upstreaming and boundary conditions. *Transport in Porous Media* **11**, 71–100. [11.9.2]

McKay, G. 1992 Patterned ground formation and solar radiation ground heating. *Proc. Roy. Soc. Lond. A* **438**, 249–263. [11.2]

McKay, G. 1996 Patterned ground formation and convection in porous media with phase change. *Continuum Mech. Thermodyn.* **8**, 189–199. [11.2]

McKay, G. 1998a Onset of buoyancy-driven convection in superposed reacting fluid and porous layers. *J. Engng. Math.* **33**, 31–46. [6.19.1]

McKay, G. 1998b Onset of double-diffusive convection in a saturated porous layer with time-periodic surface heating. *Cont. Mech. Thermodyn.* **10**, 241–251. [9.1.6.4]

McKay, G. 2000 Double-diffusive convective motions for a saturated porous layer subject to a modulated surface heating. *Cont. Mech. Thermodyn.* **12**, 69–78. [9.1.6.4]

McKay, G. and Straughan, B. 1991 The influence of a cubic density law on patterned ground formation. *Math. Models Meth. Appl. Sci.* **1**, 27–39. [11.2]

McKay, G. and Straughan, B. 1993 Patterned ground formation under water. *Continuum Mech. Thermodyn.* **5**, 145–162. [11.2]

McKibbin, R. 1983 Convection in an aquifer above a layer of heated impermeable rock. *N. Z. J. Sci.* **26**, 49–64. [6.13.1]

McKibbin, R. 1985 Thermal convection in layered and anisotropic porous media: a review. *Convective Flows in Porous Media* (eds. R. A. Wooding and I. White), Dept. Sci. Indust. Res., Wellington, New Zealand, pp. 113–127. [6.12]

McKibbin, R. 1986a Thermal convection in a porous layer: effects of anisotropy and surface boundary conditions. *Transport in Porous Media* **1**, 271–292. [6.12]

McKibbin, R. 1986b Heat transfer in a vertically-layered porous medium heated from below. *Transport in Porous Media* **1**, 361–370. [6.13.4]

McKibbin, R. 1998 Mathematical models for heat and mass transport in geothermal systems. *Transport Phenomena in Porous Media* (eds. D. B. Ingham and I. Pop), Elsevier, Oxford, pp. 131–154. [11]

McKibbin, R. 2005 Modeling heat and mass transport processes in geothermal systems. *Handbook of Porous Media* (ed. K. Vafai), 2nd ed., Taylor and Francis, New York, pp. 545–571. [11]

McKibbin, R. and O'Sullivan, M. J. 1980 Onset of convection in a layered porous medium heated from below. *J. Fluid Mech.* **96**, 375–393. [6.13.2]

McKibbin, R. and O'Sullivan, M. J. 1981 Heat transfer in a layered porous medium heated from below. *J. Fluid Mech.* **111**, 141–173. [6.13.2]

McKibbin, R. and Tyvand, P. A. 1982 Anisotropic modelling of thermal convection in multilayered porous media. *J. Fluid Mech.* **118**, 315–319. [6.13.3]

McKibbin, R. and Tyvand, P. A. 1983 Thermal convection in a porous medium composed of alternating thick and thin layers. *Int. J. Heat Mass Transfer* **26**, 761–780. [6.13.3]

McKibbin, R. and Tyvand, P. A. 1984 Thermal convection in a porous medium with horizontal cracks. *Int. J. Heat Mass Transfer* **27**, 1007–1023. [6.13.3, 6.13.4]

McKibbin, R., Tyvand, P. A. and Palm, E. 1984 On the recirculation of fluid in a porous layer heated from below. *N. Z. J. Sci.* **27**, 1–13. [6.12]

McNabb, A. 1965 On convection in a porous medium. *Proc. 2nd Australasian Conf. Hydraulics and Fluid Mech.*, University of Auckland, pp. C161–171. [5.12.3]

Medina, A., Luna, E., Perez-Rosales, C. and Higuera, F. J. 2002 Thermal convection in tilted porous fractures. *J. Phys. Condensed Matter* **14**, 2467–2474. [7.8]

Mehrabian, R., Keane, M. and Flemings, M. C. 1970 Interdendritic fluid flow and macrosegregation: influence of gravity. *Metall. Trans. B* **1**, 1209–1220. [10.2.3]

Mehta, K. N. and Nandakumar, K. 1987 Natural convection with combined heat and mass transfer buoyancy effects in non-homogeneous porous medium. *Int. J. Heat Mass Transfer* **30**, 2651–2656. [9.2.2]

Mehta, K. N. and Rao, K. N. 1994 Buoyancy induced flow of non-Newtonian fluids over a non-isothermal horizontal plate embedded in a porous medium. *Int. J. Engng. Sci.* **32**, 521–525. [5.2]

Mehta, K. N. and Sood, S. 1992a Transient free convective flow about a non-isothermal vertical flat plate immersed in a saturated inhomogeneous porous medium. *Int. Comm. Heat Mass Transfer* **19**, 687–699. [5.1.3]

Mehta, K. N. and Sood, S. 1992b Transient free convective flow with temperature dependent viscosity in a fluid saturated porous medium. *Int. J. Engng. Sci.* **30**, 1083–1087. [5.1.9.9]

Mehta, K. N. and Sood, S. 1994 Free convection about axisymmetric bodies immersed in inhomogeneous porous medium. *Int. J. Engng. Sci.* **32**, 945–953. [5.9]

Mei, C. C., Auriault, J. L. and Ng, C. O. 1996 Some applications of the homogenization theory. *Adv. Appl. Mech.* **32**, 278–348. [1.4.4]

Méndez, F., Luna, E., Treviño, C. and Pop, I. 2004 Asymptotic and numerical transient analysis of the free convection cooling of a vertical plate embedded in a porous medium. *Heat Mass Transfer* **40**, 593–602. [5.1.5]

Méndez, F., Treviño, C., Pop, I. and Liñán, A. 2002 Conjugate free convection along a thin vertical plate with internal heat generation in a porous medium. *Heat Mass Transfer* **38**, 631–638. [7.1.5]

Mercier, J. F., Weisman, C., Firdaouss, M. and le Quêré, P. 2002 Heat transfer associated to natural convection flow in a partly porous cavity. *ASME J. Heat Transfer* **124**, 130–143. [7.7]

Merkin, J. H. 1978 Free convection boundary layers in a saturated porous medium with lateral mass flux. *Int. J. Heat Mass Transfer* **21**, 1499–1504. [5.1.2, 5.5.1]

Merkin, J. H. 1979 Free convection boundary layers on axisymmetric and two-dimensional bodies of arbitrary shape in a saturated porous medium. *Int. J. Heat Mass Transfer* **22**, 1461–1462. [5.5.1, 5.6.1, 5.9]

Merkin, J. H. 1980 Mixed convection boundary layer flow on a vertical surface in a saturated porous medium. *J. Engng. Math.* **14**, 301–313. [8.1.1]

Merkin, J. H. 1985 On dual solutions occurring in mixed convection in a porous medium. *J. Engng. Math.* **20**, 171–179. [8.1]

Merkin, J. H. 1986 On dual solutions occurring in mixed convection in a porous medium. *J. Acta Mech.* **62**, 19–28. [5.7]

Merkin, J. H. and Mahmood, T. 1998 Convective flows on reactive surfaces in porous media. *Transport Porous Media* **33**, 279–293. [3.4, 5.12.3]

Merkin, J. H. and Needham, D. J. 1987 The natural convection flow above a heated wall in a saturated porous medium. *Q. J. Mech. Appl. Math.* **40**, 559–574. [5.1.9.9]

Merkin, J. H. and Pop, I. 1987 Mixed convection boundary layer on a vertical cylinder embedded in a porous medium. *Acta Mech.* **66**, 251–262. [8.1.3]

Merkin, J. H. and Pop, I. 1989 Free convection above a horizontal circular disk in a saturated porous medium. *Wärme-Stoffübertrag.* **24**, 53–60. [5.12.3]

Merkin, J. H. and Pop, I. 1997 Mixed convection on a horizontal surface embedded in a porous medium: the structure of a singularity. *Transport Porous Media* **29**, 355–364. [8.1.2]

Merkin, J. H. and Pop, I. 2000 Free convection near a stagnation point in a porous medium resulting from an oscillatory wall temperature. *Int. J. Heat Mass Transfer* **43**, 611–621. [5.5.1]

Merkin, J. H. and Zhang, G. 1990a On the similarity solutions for free convection in a saturated porous medium adjacent to impermeable horizontal boundaries. *Wärme-Stoffübertrag.* **25**, 179–184. [5.2]

Merkin, J. H. and Zhang, G. 1990b Free convection in a horizontal porous layer with a partly heated wall. *J. Engrg Math.* **24**, 125–149. [6.18]

Merkin, J. H. and Zhang, G. 1992 Boundary-layer flow past a suddenly heated vertical surface in a saturated porous medium. *Wärme-Stoffübertrag.* **27**, 299–304. [5.1.3]

Merrikh, A. A. and Lage, J. L. 2005 From continuum to porous continuum: The visual resolution impact on modeling natural convection in heterogeneous media. In *Transport Phenomena in Porous Media* III, (eds. D. B. Ingham and I. Pop), Elsevier, Oxford, pp. 60–96. [2.2.1]

Merrikh, A. A. and Mohamad, A. A. 2000 Transient natural convection in differentially heated porous enclosures. *J. Porous Media* **3**, 165–178 (erratum **4**, 195). [7.5]

Merrikh, A. A. and Mohamad, A. A. 2002 Non-Darcy effects in buoyancy driven flows in an enclosure filled with vertically layered porous media. *Int. J. Heat Mass Transfer* **45**, 4305–4313. [7.3.2]

Merrikh, A. A., Lage, J. L. and Mohamad, A. A. 2002 Comparison between pore-level and porous medium models for natural convection in a nonhomogeneous enclosure. *AMS Contemp. Math.* **295**, 387–396. [2.2.1]

Merrikh, A. A., Lage, J. L. and Mohamad, A. A. 2005a Natural convection in an enclosure with disconnected and conducting solid blocks. *Int. J. Heat Mass Transfer* **46**, 1361–1372. [2.2.1]

Merrikh, A. A., Lage, J. L. and Mohamad, A. A. 2005b Natural convection in non-homogeneous heat generating media: Comparison of continuum and porous-continuum models. *J. Porous Media* **8**, 149–163. [2.2.2]

Mharzi, M., Daguenet, M. and Daoudi, S. 2000 Thermosolutal natural convection in a vertically layered fluid-porous medium heated from the side. *Energy Conv. Management* **41**, 1065–1090. [9.2.2]

Mhimid, A., Ben Nasrallah, S. and Fohr, J. P. 2000 Heat and mass transfer during drying of granular products—simulation with convective and conductive boundary conditions. *Int. J. Heat Mass Transfer* **43**, 2779–2791. [3.6]

Mhimid, A., Fohr, J. P. and Ben Nasrallah, S. 1999 Heat and mass transfer during drying of granular products by combined convection and conduction. *Drying Tech.* **17**, 1043–1063. [3.6]

Min, J. Y. and Kim, S. J. 2005 A novel methodology for thermal analysis of a composite system consisting of a porous medium and an adjacent fluid layer. *ASME J. Heat Transfer* **127**, 648–656. [1.6, 2.4]

Minkowycz, W. J. and Cheng, P. 1976 Free convection about a vertical cylinder embedded in a porous medium. *Int. J. Heat Mass Transfer* **19**, 805–813. [5.7]

Minkowycz, W. J. and Cheng, P. 1982 Local non-similar solutions for free convective flow with uniform lateral mass flux in a porous medium. *Lett. Heat Mass Transfer* **9**, 159–168. [5.1.2]

Minkowycz, W. J., Cheng, P. and Chang, C. H. 1985a Mixed convection about non-isothermal cylinders and spheres in a porous medium. *Numer. Heat Transfer* **8**, 349–359. [8.1.3]

Minkowycz, W. J., Cheng, P. and Hirschberg, R. N. 1984 Non-similar boundary layer analysis of mixed convection about a horizontal heated surface in a fluid-saturated porous medium. *Int. Comm. Heat Mass Transfer* **11**, 127–141. [8.1.2]

Minkowycz, W. J., Cheng, P. and Moalem, F. 1985b Effect of surface mass transfer on buoyancy-induced Darcian flow adjacent to a horizontal heated surface. *Int. Comm. Heat Mass Transfer* **12**, 55–65. [5.2]

Minkowycz, W. J., Haji-Sheik, A. and Vafai. K. 1999 On departure from local thermal equilibrium in porous media due to rapidly changing heat source: the Sparrow number. *Int. J. Heat Mass Transfer* **42**, 3373–3385. [4.10]

Minto, B. J., Ingham, D. B. and Pop, I. 1998 Free convection driven by an exothermic reaction on vertical surface embedded in porous media. *Int. J. Heat Mass Transfer* **41**, 11–23. [5.1.9.9]

Miranda, B. M. D. and Anand, N. K. 2004 Convective heat transfer in a channel with porous baffles. *Numer. Heat Transfer A* **46**, 425–452. [4.11]

Mishra, A. K., Paul, T. and Singh, A. K. 2002 Mixed convection flow in a porous medium bounded by two vertical walls. *Forch. Ingen.* **67**, 198–205. [8.3.1]

Misirlioglu, A., Baytas, A. C. and Pop, I. 2005 Free convection in a wavy cavity filled with a porous medium. *Int. J. Heat Mass Transfer* **48**, 1840–1850. [7.3.7]

Misra, D. and Sarkar, A. 1995 A comparative study of porous media models in a differentially heated square cavity using a finite element method. *Int. J. Numer. Meth. Heat Fluid Flow* **5**, 735–752. [7.6.2]

Miyauchi, H., Kataoka, H. and Kikuchi, T. 1976 Gas film coefficients of mass transfer in low Péclet number region for sphere packed beds. *Chem. Engng. Sci.* **31**, 9–13. [2.2.2]

Mohamad, A. A. 2000 Nonequilibrium natural convection in a differentially heated cavity filled with a saturated porous matrix. *ASME J. Heat Transfer* **122**, 380–384. [7.6.2]

Mohamad, A. A. 2001 Natural convection from a vertical plate in a saturated porous medium: nonequilibrium theory. *J. Porous Media* **4**, 181–186. [5.1.9.3]

Mohamad, A. A. 2003 Heat transfer enhancements in heat exchangers filled with porous media. Part 1: Constant wall temperature. *Int. J. Thermal Sci.* **42**, 385–395. [4.15]

Mohamad, A. A. and Bennacer, R. 2001 Natural convection in a confined saturated porous medium with horizontal temperature and vertical solutal gradients. *Int. J. Thermal Sci.* **40**, 82–93. [9.5]

Mohamad, A. A. and Bennacer, R. 2002 Double diffusion, natural convection in an enclosure filled with saturated porous medium subjected to cross-gradients: stably stratified fluid. *Int. J. Heat Mass Transfer* **45**, 3725–3740. [9.5]

Mohamad, A. A. and Karim, G. A. 2001 Flow and heat transfer with segregated beds of solid particles. *J. Porous Media* **4**, 215–224. [4.6.4]

Mohamad, A. A. and Rees, D. A. S. 2004 Conjugate free convection in a porous medium attached to a wall held at a constant temperature. In *Applications of Porous Media (ICAPM 2004)*, (eds. A. H. Reis and A. F. Miguel), Évora, Portugal, pp. 93–97. [7.1.5]

Mohamad, A, A, and Sezai, I. 2002 Effect of lateral aspect ratio on three-dimensional double diffusive convection in porous enclosures with opposing temperature and concentration gradients. *Heat Transfer Research* **33**, 318–325. [9.2.2]

Mohamad, A. A., Bennacer, R. and Azaiez, J. 2004 Double-diffusion natural convection in a rectangular enclosure filled with binary fluid saturated porous media: the effect of lateral aspect ratio. *Phys. Fluids* **16**, 184–199. [9.1.3]

Mohammadien, A. A. and El-Amin, M. F. 2000 Thermal dispersion-radiation effects on non-Darcy natural convection in a fluid saturated porous medium. *Transport Porous Media* **40**, 153–163. [5.1.9.4]

Mohammadien, A. A. and El-Amin, M. F. 2001 Thermal radiation effects on power-law fluids over a horizontal plate embedded in a porous medium. *Int. Comm. Heat Mass Transfer* **27**, 1025–1035. [5.2]

Mohammadien, A. A. and El-Shaer, N. A. 2004 Influence of variable permeability on combined free and forced convection past a semi-infinite vertical plate in a saturated porous medium. *Heat Mass Transfer* **40**, 341–346. [8.1.1]

Mohammadein, A. A., Mansour, M. A., Abd el Gaied, S. M. and Gorla, R. S. R. 1998 Radiative effect on natural convection flows in porous media. *Transport Porous Media* **32**, 263–283. [5.1.9.4]

Mojtabi, A. 2002 Influence of vibrations on the onset of thermo-convection in porous medium. *Proc. 1st Int. Conf. Applications of Porous Media*, 704–721. [6.24]

Mojtabi, A. and Charrier-Mojtabi, M. C. 1992 Analytic solution of steady natural convection in an annular porous medium evaluated with a symbolic algebra code. *ASME J. Heat Transfer* **114**, 1065–1068. [7.3.3]

Mojtabi, A. and Charrier-Mojtabi, M. C. 2000 Double-diffusive convection in porous media. *Handbook of Porous Media* (K. Vafai, ed.), Marcel Dekker, New York, pp. 559–603. [3.3, 9]

Mojtabi, A. and Charrier-Mojtabi, M. C. 2005 Double-diffusive convection in porous media. *Handbook of Porous Media* (K. Vafai, ed.), 2nd ed., Taylor and Francis, New York, pp. 269–320. [3.3, 9]

Mojtabi, A., Charrier-Mojtabi, M. C., Maliwan, K. and Pedramrazi, Y. 2004 Active control of the onset of convection in a porous medium. In *Emerging Technologies and Techniques in Porous Media* (D. B. Ingham, A. Bejan, E. Mamut and I. Pop, eds), Kluwer Academic, Dordrecht, pp. 195–207. [6.24]

Montillet, A. 2004 Flow through a finite packed bed of spheres: A note on the limit of applicability of the Forchheimer-type equation. *ASME J. Fluids Engng.* **126**, 139–143. [1.5.2]

Morega, A. M. and Bejan, A. 1994 Heatline visualization of convection in porous media. *Int. J. Heat Fluid Flow* **15**, 42–47. [4.17]

Morega, A. M., Bejan, A. and Lee, S. W. 1995 Free stream cooling of a stack of parallel plates. *Int. J. Heat Mass Transfer* **38**, 519–531. [4.15]

Morland, L. W., Zebib, A. and Kassoy, D. R. 1977 Variable property effects on the onset of convection in an elastic porous matrix. *Phys. Fluids* **20**, 1255–1259. [6.7]

Mosaad. M. 1999 Natural convection in a porous medium coupled across an impermeable vertical wall with film condensation. *Heat Mass Transfer* **35**, 177–183. [10.4]

Mota, J. P. B., Esteves, A. A. C., Portugal, C. A. M., Esperança, J. M. S. S. and Saatdjain, E. 2000 Natural convection heat transfer in horizontal eccentric elliptic annuli containing saturated porous media. *Int. J. Heat Mass Transfer* **43**, 4367–4379. [7.3.3]

Moutsopoulos, K. N. and Koch, D. L. 1999 Hydrodynamic and boundary-layer dispersion in bidisperse porous media. *J. Fluid Mech.* **385**, 359–379. [4.16.4]

Moya, R. E. S., Prata, A. T. and Cunha Neto, J. A. B. 1999 Experimental analysis of unsteady heat and moisture transfer around a heated cylinder buried in a porous medium. *Int. J. Heat Mass Transfer* **42**, 2187–2198. [3.6]

Moya, S. L., Ramos, E. and Sen, M. 1987 Numerical study of natural convection in a tilted rectangular porous material. *Int. J. Heat Mass Transfer* **30**, 741–756. [7.8]

Mullis, A. M. 1995 Natural convection in porous, permeable media æ sheets, wedges and lenses. *Marine Petrol. Geol.* **12**, 17–25. [11.8.2]

Muralidhar, K. 1989 Mixed convection flow in a saturated porous annulus. *Int. J. Heat Mass Transfer* **32**, 881–888. [8.2.3, 8.3.3]

Muralidhar, K. 1992 Study of heat transfer from buried nuclear waste canisters. *Int. J. Heat Mass Transfer* **35**, 3493–3495. [7.11]

Muralidhar, K. 1993 Near-field solution for heat and mass transfer from buried nuclear waste canisters. *Int. J. Heat Mass Transfer* **36**, 2665–2674. [7.11]

Muralidhar, K. and Misra, D. 1997 Determination of dispersion coefficients in a porous medium using the frequency response method. *Expt. Heat Transfer* **10**, 109–118. [2.2.3]

Muralidhar, K. and Suzuki, K. 2001 Analysis of flow and heat transfer in a regenerator mesh using a non-Darcy thermally non-equilibrium model. *Int. J. Heat Mass Transfer* **44**, 2493–2504. [4.10]

Muralidhar, K., Baunchalk, R.A. and Kulacki, F.A. 1986 Natural convection in a horizontal porous annulus with a step distribution in permeability. *ASME J. Heat Transfer* **108**, 889–893. [7.3.3]

Murata, K. 1995 Heat and mass transfer with condensation in a fibrous insulation slab bounded on one side by a cold surface. *Int. J. Heat Mass Transfer* **38**, 3253–3262. [10.4]

Murdoch, A. and Soliman, A. 1999 On the slip-boundary condition for liquid flow over planar boundaries. *Proc. Roy. Soc. Lond. A* **455**, 1315–1340. [1.6]

Murphy, H. D. 1979 Convective instabilities in vertical fractures and faults. *J. Geophys. Res.* **84**, 6234–6245. [6.15.2]

Murray, B. T. and Chen, C. F. 1989 Double-diffusive convection in a porous medium. *J. Fluid Mech.* **201**, 147–166. [9.1.3]

Murthy, P. V. S. N. 1998 Thermal dispersion and viscous dissipation effects on a non-Darcy mixed convection in a saturated porous medium. *Heat Mass Transfer* **33**, 295–300. [8.1.1]

Murthy, P. V. S. N. 2000 Effect of double dispersion on mixed convection heat and mass transfer in non-Darcy porous medium. *ASME J. Heat Transfer* **122**, 476–484. [9.2.1]

Murthy, P. V. S. N. 2001 Effect of viscous dissipation on mixed convection in a non-Darcy porous medium. *J. Porous Media* **4**, 23–32. [8.1.1]

Murthy, P. V. S. N. and Singh, P. 1997a Effects of viscous dissipation on a non-Darcy natural convection regime. *Int. J. Heat Mass Transfer* **40**, 1251–1260. [5.1.9.4]

Murthy, P. V. S. N. and Singh, P. 1997b Thermal dispersion effects on non-Darcy natural convection with lateral mass flux. *Heat Mass Transfer* **33**, 1–5. [8.1.1]

Murthy, P. V. S. N. and Singh, P. 1997c Thermal dispersion effects on non-Darcy natural convection over horizontal plate with surface mass flux. *Arch. Appl. Mech.* **67**, 487–495. [8.1.2]

Murthy, P. V. S. N. and Singh, P. 1999 Heat and mass transfer by natural convection in a non-Darcy porous medium. *Acta Mech.* **138**, 243–254. [9.2.1]

Murthy, P. V. S. N. and Singh, P. 2000 Thermal dispersion effects on non-Darcy convection over a cone. *Comp. Math. Appl.* **40**, 1433–1444. [8.1.4]

Murthy, P. V. S. N., Mukherjeee, S., Srinivasacharya, D. and Krishna, P. V. S. S. S. R. 2004a Combined radiation and mixed convection from a vertical wall with suction/injection in a non-Darcy porous medium. *Acta Mech.* **168**, 145–156. [8.1.1]

Murthy, P. V. S. N., Ratish Kumar, B. V. and Singh, P. 1997 Natural convection form a horizontal wavy surface in a porous enclosure. *Numer. Heat Transfer* **31**, 207–221. [6.15.3]

Murthy, P. V. S. N., Srinivasacharya, D. and Krishna, P. V. S. S. S. R. 2004b Effect of double stratification on free convection in a Darcian porous medium. *ASME J. Heat Transfer* **126**, 297–300. [9.2.1]

Murty, V. D., Camden, M. P., Clay, C. L. and Paul, D. B. 1989 Natural convection in porous media between concentric and eccentric cylinders. *AIChE Sympos. Ser.* **269**, 96–101. [7.6.2]

Murty, V. D., Camden, M. P., Clay, C. L. and Paul, D. B. 1990 A study of non-Darcian effects on forced convection heat transfer over a cylinder embedded in a porous medium. *Heat Transfer 1990*, Hemisphere, Washington, DC, vol. 5, pp. 201–206. [4.3]

Murty, V. D., Clay, C. L., Camden, M. P. and Paul, D. B. 1994 Natural convection around a cylinder buried in a porous medium—non-Darcian effects. *Appl. Math. Modell.* **18**, 134–141. [7.11]

Na, T. Y. and Pop, I. 1996 A note to the solution of Cheng-Minkowycz equation arising in free convection in porous media. *Int. Comm. Heat Mass Transfer* **23**, 697–703. [5.1.9.9]

Na, T. Y. and Pop, I. 1999 A note to the solution of Cheng-Chang equation arising in free convection in porous media. *Int. Comm. Heat Transfer* **26**, 145–151. [5.2]

Naidu, S. V., Dharma Rao, V., Sarma, P. K. and Subrahmanyam, T. 2004a Performance of a circular fin in a cylindrical enclosure. *Int. Comm. Heat Transfer* **31**, 1209–1218. [7.3.7]

Naidu, S. V., Rao, V. D., Sarma, P. K. and Subrahmanyam, Y. 2004b Performance of a circular fin in a cylindrical porous enclosure. *Int. Comm. Heat Mass Transfer* **31**, 1209–1218. [7.3.3]

Najjari, M. and Ben Nasrallah, S. 2002 Numerical study of boiling with mixed convection in a vertical porous layer. *Int. J. Therm. Sci.***41**, 936–948. [10.3]

Najjari, M. and Ben Nasrallah, S. 2005 Numerical study of the effects of geometric dimensions on a liquid-vapor phase change and free convection in a rectangular porous cavity. *J. Porous Media* **8**, 1–12. [10.3]

Nakagano, K., Mochida, T. and Ochifuji, K. 2002 Influence of natural convection on forced horizontal flow in saturated porous media for aquifer thermal energy storage. *Appl. Therm. Engng*. **22**, 1299–1311. [6.10]

Nakayama, A. 1991 A general treatment for non-Darcy film condensation within a porous medium in the presence of gravity and forced flow. *Wärme-Stoffübertrag.* **27**, 119–124. [10.4]

Nakayama, A. 1993a A similarity solution for free convection from a point heat source embedded in a non-Newtonian fluid-saturated porous medium. *ASME J. Heat Transfer* **115**, 510–513. [5.11.1]

Nakayama, A. 1993b Free convection from a horizontal line heat source in a power-law fluid-saturated porous medium. *Int. J. Heat Fluid Flow* **14**, 279–283. [5.10.1.2, 8.1.5]

Nakayama, A. 1994 A unified theory for non-Darcy free, forced and mixed convection problems associated with a horizontal line heat source in a porous medium. *ASME J. Heat Transfer* **116**, 508–513. [5.10.1, 8.1.5]

Nakayama, A. 1995 *PC-Aided Numerical Heat Transfer and Convective Flow*. CRC Press, Tokyo. [8.1.5]

Nakayama, A. 1998 Unified treatment of Darcy-Forchheimer boundary-layer flows. *Transport Phenomena in Porous Media* (eds. D. B. Ingham and I. Pop), Elsevier, Oxford, pp. 179–204. [8.1.5]

Nakayama, A. and Ashizawa, T. 1996 A boundary layer analysis of combined heat and mass transfer by natural convection from a concentrated source in a saturated porous medium. *Appl. Sci. Res.* **56**, 1–11. [9.3.1, 9.3.2]

Nakayama, A. and Ebinuma, C. D. 1990 Transient non-Darcy forced convective heat transfer from a flat plate embedded in a fluid-saturated porous medium. *Int. J. Heat Fluid Flow* **11**, 249–263. [4.6, 4.7]

Nakayama, A. and Hossain, M. A. 1994 Free convection in a saturated porous medium beyond the similarity solution. *Appl. Sci. Res.* **52**, 133–145. [5.1.9.9]

Nakayama, A. and Hossain, M. A. 1995 An integrated treatment for combined heat and mass transfer by natural convection in a porous medium. *Int. J. Heat Mass Transfer* **38**, 761–765. [9.2.1]

Nakayama, A. and Koyama, H. 1987a Free convective heat transfer over a nonisothermal body of arbitrary shape embedded in a fluid-saturated porous medium. *ASME J. Heat Transfer* **109**, 125–130. [5.9]

Nakayama, A. and Koyama, H. 1987b A general similarity transformation for combined free and forced-convection flows within a fluid-saturated porous medium. *ASME J. Heat Transfer* **109**, 1041–1045. [8.1.4]

Nakayama, A. and Koyama, H. 1987c Effect of thermal stratification on free convection within a porous medium. *AAIA J. Thermophys. Heat Transfer* **1**, 282–285. [5.1.4]

Nakayama, A. and Koyama, H. 1988a A similarity transformation for subcooled mixed convection film boiling in a porous medium. *Appl. Sci. Res.*, **45**, 129–143. [10.3.2]

Nakayama, A. and Koyama, H. 1988b Subcooled forced convection film boiling over a vertical flat plate embedded in a fluid-saturated porous medium. *Wärme-Stoffübertrag.* **22**, 269–273. [10.3.2]

Nakayama, A. and Koyama, H. 1989 Similarity solutions for buoyancy induced flows over a non-isothermal curved surface in a thermally stratified porous medium. *Appl. Sci. Res.* **46**, 309–323. [5.9]

Nakayama, A. and Koyama, H. 1991 Buoyancy-induced flow of non-Newtonian fluids over a non-isothermal body of arbitrary shape in a fluid-saturated porous medium. *Appl. Sci. Res.* **48**, 55–70. [5.9]

Nakayama, A. and Kuwahara, F. 1999 A macroscopic turbulence model for flow in a porous medium. *ASME J. Fluids Engng*. **121**, 427–433. [1.8]

Nakayama, A. and Kuwahara, F. 2000 Numerical modeling of convective heat transfer in porous media using microscopic structures. *Handbook of Porous Media* (K. Vafai, ed.), Marcel Dekker, New York, pp. 441–488. [1.8]

Nakayama, A. and Kuwahara, F. 2004 Closure to discussion [by B. Yu]. *ASME J. Heat Transfer* **126** in a porous medium. *ASME J. Fluids Engng.* **121**, 427–433. [1.8]

Nakayama, A. and Kuwahara, F. 2005 Three-dimensional numerical models for periodically-developed heat and fluid flows within porous media. In *Transport Phenomena in Porous Media*, 1062. [2.2.3]

Nakayama, A. and Pop, I. 1989 Free convection over a non-isothermal body in a porous medium with viscous dissipation. *Int. Comm. Heat Mass Transfer* **16**, 173–180. [5.9]

Nakayama, A. and Pop, I 1991 A unified similarity transformation for free, forced and mixed convection in Darcy and non-Darcy porous media. *Int. J. Heat Mass Transfer* **34**, 357–367. [8.1.5]

Nakayama, A. and Pop, I. 1993 Momentum and heat transfer over a continuously moving surface in a non-Darcian fluid. *Wärme-Stoffübertrag*, **28**, 177–184. [4.16.2]

Nakayama, A. and Shenoy, A. V. 1992 A unified similarity transformation for Darcy and non-Darcy forced-, free- and mixed-convection heat transfer in non-Newtonian inelastic fluid-saturated porous media. *Chem. Engng. Sci.* **50**, 33–45. [8.1.1]

Nakayama, A. and Shenoy, A. V. 1993a Combined forced and free convection heat transfer in power-law fluid-saturated porous media. *Appl. Sci. Res.* **50**, 83–95. [8.1.1]

Nakayama, A. and Shenoy, A. V. 1993b Non-Darcy forced convective heat transfer in a channel embedded in a non-Newtonian inelastic fluid-saturated porous medium. *Canad. Chem. Engng*. **71**, 168–173. [4.16.3]

Nakayama, A., Kokudai, T. and Koyama, H. 1988a Integral method for non-Darcy free convection over a vertical flat plate and cone embedded in a fluid-saturated porous medium. *Wärme-Stoffübertrag.* **23**, 337–341. [5.8]

Nakayama, A., Kokudai, T. and Koyama, H. 1990a Non-Darcian boundary layer flow and forced convective heat transfer over a flat plate in a fluid-saturated porous medium. *ASME J. Heat Transfer* **112**, 157–162. [4.8, 8.1.5]

Nakayama, A., Kokudai, T. and Koyama, H. 1990b Forchheimer free convection over a nonisothermal body of arbitrary shape in a saturated porous medium. *ASME J. Heat Transfer* **112**, 511–515. [5.9]

Nakayama, A., Koyama, H. and Kuwahara, F. 1987 Two-phase boundary layer treatment for subcooled free-convection film boiling around a body of arbitrary shape in a porous medium. *ASME J. Heat Transfer* **109**, 997–1002. [10.3.2]

Nakayama, A., Koyama, H. and Kuwahara, F. 1988b An analysis on forced convection in a channel filled with a porous medium: Exact and approximate solutions. *Wärme-Stoffubertrag*. **23**, 291–296. [4.9]

Nakayama, A., Koyama, H. and Kuwahara, F. 1989 Similarity solution for non-Darcy free convection from a non-isothermal curved surface in a fluid-saturated porous medium. *ASME J. Heat Transfer* **111**, 807–811. [5.9]

Nakayama, A., Koyama, H. and Kuwahara, F. 1991 A general transformation for transient non-Darcy free and forced convection within a fluid-saturated porous medium. *Proc. ASME/JSME Thermal Engineering Joint Conference—1991*, vol. 4, pp. 287–293. [5.9]

Nakayama, A., Kuwahara, F. and Hayashi, T. 2004 Numerical modeling for three-dimensional heat and fluid flow through a bank of cylinders with yaw. *J. Fluid Mech.* **498**, 139–159. [1.8]

Nakayama, A., Kuwahara, F. and Koyama, H. 1993 Transient non-Darcy free convection between parallel vertical plates in a fluid-saturated porous medium. *Appl. Sci. Res.*, **50**, 29–43. [7.5]

Nakayama, A., Kuwahara, F., Kawamura, Y. and Koyama, H. 1995 Three-dimensional numerical simulation of flow through microscopic porous structure. *Proc. ASME/JSME Thermal Engineering Conf.*, vol. 3, pp. 313–318. [1.5.2]

Nakayama, A., Kuwahara, F., Sugiyama, M. and Xu, G. 2001 A two-energy model for conduction and convection in porous media. *Int. J. Heat Mass Transfer* **44**, 4375–4379. [4.10]

Nakayama, A., Kuwahara, F., Unemoto, T. and Hatashi, T. 2002 Heat and fluid flow within an anisotropic porous medium. *ASME J. Heat Transfer* **124**, 746–753. [4.16.2]

Nandakumar, K. and Weinitschke, H.J. 1992 A bifurcation of chemically driven convection in a porous medium. *Chem. Engng. Sci.* **47**, 4107–4120. [3.4]

Nandakumar, K., Weinitschke, H. J. and Sankar, S. R. 1987 The calculation of singularities in steady mixed convection flow in porous media. *ASME HTD* **84**, 67–73. [8.2.1]

Nandapurkar, P., Poirier, D. R., Heinrich, J. C. and Felicelli, S. 1989 Thermosolutal convection during dendritic solidification of alloys: Part 1. Linear stability analysis. *Metall. Trans. B* **20**, 711–721. [10.2.3]

Narasimhan, A. and Lage, J. L. 2001a Modified Hazen-Dupuit-Darcy model for forced convection of a fluid with temperature-dependent viscosity. *ASME J. Heat Transfer* **123**, 31–38. [4.16.1]

Narasimhan, A. and Lage, J. L. 2001b Forced convection of a fluid with temperature-dependent viscosity through a porous medium channel. *Numer. Heat Transfer A* **40**, 801–820. [4.16.1]

Narasimhan, A. and Lage, J. L. 2002 Inlet temperature influence on the departure from Darcy flow of a fluid with variable viscosity. *Int. J. Heat Mass Transfer* **45**, 2419–2422. [4.16.1]

Narasimhan, A. and Lage, J. L. 2003 Temperature-dependent viscosity effects on the thermohydraulics of heated porous-medium channel flows. *J. Porous Media* **6**, 149–158. [4.16.1]

Narasimhan, A. and Lage, J. L. 2004a Predicting inlet temperature effects on the pressure-drop of heated porous medium channel flows using the M-HDD model. *ASME J. Heat Transfer* **126**, 301–303. [4.16.1]

Narasimhan, A. and Lage, J. L. 2004b Pump power gain for heated porous medium channel flows. *ASME J. Fluids Engng.* **126**, 494–497. [4.16.1]

Narasimhan, A. and Lage, J. L. 2005 Variable viscosity forced convection in porous medium channels. *Handbook of Porous Media* (ed. K. Vafai), 2nd ed., Taylor and Francis, New York, pp. 195–234. [4.16.1]

Narasimhan, A., Lage, J. L. and Nield, D. A. 2001b New theory for forced convection through porous media by fluids with temperature-dependent viscosity. *ASME J. Heat Transfer* **123**, 1045–1051. [4.16.1]

Narasimhan, A., Lage, J. L., Nield, D. A. and Porneala, D. C. 2001a Experimental verification of two new theories predicting temperature-dependent viscosity effects on the forced convection in a porous channel. *ASME J. Fluids Engng.* **123**, 948–951. [4.16.1]

Nasr, K., Ramadhyani, S. and Viskanta, R. 1994 An experimental investigation on forced convection heat transfer from a cylinder embedded in a packed bed. *ASME J. Heat Transfer* **116**, 73–78. [4.3]

Nasr, K., Ramadhyani, S. and Viskanta, R. 1995 Numerical studies of forced convection heat transfer from a cylinder embedded in a packed bed. *Int. J. Heat Mass Transfer* **38**, 2353–2366. [4.8]

Nassehi, V. 1998 Modelling of combined Navier-Stokes and Darcy flows in crossflow membrane filtration. *Chem. Engng. Sci.* **53**, 1253–1265. [1.6]

Natale, M. F. and Santillan Marcus, E. A. 2003 The effect of heat convection on drying of porous semi-infinite space with a heat flux condition on the fixed face $x = 0$. *Appl. Math. Comp.* **137**, 109–129. [3.6]

Naylor, D. and Oosthuizen, P. H. 1995 Free convection in a horizontal enclosure partly filled with a porous medium. *AIAA J. Thermophys. Heat Transfer* **9**, 797–800. [7.7]

Nazar, R. and Pop, I 2004 Unsteady mixed convection boundary layer flow near the stagnation point on a horizontal surface in a porous medium. In *Applications of Porous Media (ICAPM 2004)*, (eds. A. H. Reis and A. F. Miguel), Évora, Portugal, pp. 215–221. [8.1.2]

Nazar, R., Amin, N. and Pop, I. 2003a Unsteady mixed convection boundary layer flow near a stagnation point on a vertical surface in a porous medium. *Int. J. Heat Mass Transfer* **47**, 2681–2688. [8.1.1]

Nazar, R., Amin, N., Filip, D. and Pop, I. 2003b The Brinkman model for mixed convection boundary layer flow past a horizontal circular cylinder in a porous medium. *Int. J. Heat Mass Transfer* **46**, 3167–3178. [8.1.3]

Neale, G. and Nader, W. 1974 Practical significance of Brinkman's extension of Darcy's law: coupled parallel flows within a channel and a bounding porous medium. *Canad. J. Chem. Engng.* **52**, 475–478. [1.6]

Néel, M. C. 1990a Convection in a horizontal porous layer of infinite extent. *Eur. J. Mech. B* **9**, 155–176. [6.15.1]

Néel, M. C. 1990b Convection naturelle dans une couche poreuse horizontale d'extension infinie: chauffage inhomogène. *C. R. Acad. Sci. Paris, Sér. II* **309**, 1863–1868. [6.15.1]

Néel, M. C. 1992 Inhomogeneous boundary conditions and the choice of convective patterns in a porous layer. *Int. J. Engng. Sci.* **30**, 507–521. [6.14]

Néel, M. C. 1998 Driven convection in porous media: deviations from Darcy's law. *C. R. Acad, Sci. Paris II B* **326**, 615–620. [6.10]

Néel, M. C. and Lyubimov, P. 1995 Periodic solutions for differential-equations of order-3, with application to heat-flux induced convection. *Math. Meth. Appl. Sci.* **18**, 1133–1164. [6.4]

Néel, M. C. and Nemrouch, F. 2001 Instabilities in an open top horizontal porous layer subjected to pulsating thermal boundary conditions. *Cont. Mech. Thermodyn.* **13**, 41–58. [6.11.3]

Neichloss, H. and Dagan, G. 1975 Convection currents in a porous layer heated from below: the influence of hydrodynamic dispersion. *Phys. Fluids* **18**, 757–761. [6.6]

Neilson, D. G. and Incropera, F. P. 1993 Effect of rotation on fluid motion and channel formation during unidirectional solidification of a binary alloy. *Int. J. Heat Mass Transfer* **36**, 489–505. [10.2.3]

Nejad, M., Saghar, M. Z. and Islam, M. R. 2001 Role of thermal diffusion on heat and mass transfer in porous media. *Int. J. Comput. Fluid Dyn.* **15**, 157–168. [9.1.4]

Nelson, R. A., Jr, and Bejan, A. 1998 Constructal optimization of internal flow geometry in convection. *ASME J. Heat Transfer* **120**, 357–364. [6.2, 6.26]

Nepf, H. M. 1999 Drag, turbulence, and diffusion in flow through emergent vegetation. *Water Resources Res.* **35**, 479–489. [1.8]

Neto, H. L., Quaresima, J. N. N. and Cotta, R. M. 2004 Transient natural convection in three-dimensional cavities: Reference results via integral transforms. In *Applications of Porous Media (ICAPM 2004)*, (eds. A. H. Reis and A. F. Miguel), Évora, Portugal, pp. 165–172. [7.5]

Neto, H. L., Quaresma, J. N. N. and Cotta, R. M. 2002 Natural convection in three-dimensional porous cavities. *Int. J. Heat Mass Transfer* **45**, 3013–3032. [7.5]

Nganhou, J. 2004 Heat and mass transfer through a thick bed of cocoa beans during drying. *Heat Mass Transfer* **40**, 727–735. [3.6]

Ngo, C. C. and Lai, F. C. 2000 Effective permeability for natural convection in a layered porous annulus, *J. Thermophys. Heat Transfer* **14**, 363–367. [6.13.2]

Ngo, C. C. and Lai, F. C. 2005 Effects of backfill on heat transfer from a buried pipe. *ASME J. Heat Transfer* **127**, 780–784. [7.11]

Nguyen, D. and Balakotaiah, V. 1995 Reaction-driven instabilities in down-flow packed beds. *Proc. Roy. Soc. Lond. A* **450**, 1–21. [3.4]

Nguyan, H. D., Paik, S. and Douglass, R. W. 1994 Study of double-diffusive convection in layered anisotropic porous media. *Numer. Heat Transfer B* **26**, 489–505. [9.1.6.2]

Nguyen, H. D. and Paik, S. 1994 Unsteady mixed convection from a sphere in water-saturated porous media with variable surface temperature—heat flux. *Int. J. Heat Mass Transfer* **37**, 1783–1793. [8.1.3]

Nguyen, H. D., Paik, S. and Douglass, R. W 1997a Double-diffusive convection in a porous trapezoidal enclosure with oblique principal axes. *AAIA J. Thermophysics Heat Transfer* **11**, 309–312. [9.4]

Nguyen, H. D., Paik, S. and Pop, I. 1997b Transient thermal convection in a spherical enclosure containing a fluid core and a porous shell. *Int. J. Heat Mass Transfer*, **40**, 379–392. [7.5]

Nguyen, H. D., Paik, S., Douglass, R. W. and Pop, I. 1996 Unsteady non-Darcy reaction-driven flow from an anisotropic cylinder in porous media. *Chem. Engng. Sci.* **51**, 4963–4977. [5.5.1]

Nguyen, T. H., Lepalec, G., Nguyen-Quang, T. and Bahloul, A. 2004 Bioconvection: spontaneous pattern formation of micro-organisms. In *Applications of Porous Media (ICAPM 2004)*, (eds. A. H. Reis and A. F. Miguel), Évora, Portugal, pp. 521–526. [6.25]

Nguyen-Quang, T., Bahloul, A. and Nguyen, T. H. 2005 Stability of gravitactic microorganisms in a fluid-saturated porous medium. *Int. Comm. Heat Mass Transfer* **32**, 54–63. [6.25]

Ni, J. and Beckermann, C. 1991a Natural convection in a vertical enclosure filled with anisotropic porous media. *ASME J. Heat Transfer* **113**, 1033–1037. [7.3.2]

Ni, J. and Beckermann, C. 1991b A volume-averaged two-phase model for transport phenomena during solidification. *Metall. Trans. B* **22**, 349–361. [10.2.3]

Ni, J. and Incropera, F. P. 1995a Extension of the continuum model for transport phenomena occurring during metal alloy solidification—I. The conservation equations. *Int. J. Heat Mass Transfer* **38**, 1271–1284. [10.2.3]

Ni, J. and Incropera, F. P. 1995b Extension of the continuum model for transport phenomena occurring during metal alloy solidification—II. Microscopic considerations. *Int. J. Heat Mass Transfer* **38**, 1285–1296. [10.2.3]

Nield, D. A. 1968 Onset of thermohaline convection in a porous medium. *Water Resources Res.* **11**, 553–560. [6.2, 9.1.1, 11.4]

Nield, D. A. 1975 The onset of transient convective instability. *J. Fluid Mech.* **71**, 441–454. [6.11.1]

Nield, D. A. 1977 Onset of convection in a fluid layer overlying a layer of a porous medium. *J. Fluid Mech.* **81**, 513–522. [6.19.1.2]

Nield, D. A. 1982 Onset of convection in a porous layer saturated by an ideal gas. *Int. J. Heat Mass Transfer* **25**, 1605–1606. [6.7]

Nield, D. A. 1983 The boundary correction for the Rayleigh-Darcy problem: limitations of the Brinkman equation. *J. Fluid Mech.* **128**, 37-46. Corrigendum **150**, 503. [1.6, 6.19.1.2]

Nield, D. A. 1987a Convective instability in porous media with throughflow. *AIChE J.* **33**, 1222–1224. [6.10]

Nield, D. A. 1987b Convective heat transfer in porous media with columnar structure. *Transport in Porous Media* **2**, 177–185. [6.3, 6.13.4]

Nield, D. A. 1990 Convection in a porous medium with inclined temperature gradient and horizontal mass flow. *Heat Transfer 1990*, Hemisphere, New York, vol. 5, pp. 153–158. [7.9]

Nield, D. A. 1991a Convection in a porous medium with inclined temperature gradient. *Int. J. Heat Mass Transfer* **34**, 87–92. [7.9]

Nield, D. A. 1991b Estimation of the stagnant thermal conductivity of saturated porous media. *Int. J. Heat Mass Transfer* **34**, 1575–1576. [2.2.1]

Nield, D. A. 1993 Correlation formulas for mixed convection heat transfer in a saturated porous medium. *Int. J. Heat Fluid Flow* **14**, 206. [8.3.4]

Nield, D. A. 1994a The effect of channeling on heat transfer across a horizontal layer of a porous medium. *Int. J. Heat Fluid Flow* **15**, 247–248. [6.9.1, 6.19.1.2]

Nield, D. A. 1994b Modelling high speed flow of a compressible fluid in a saturated porous medium. *Transport in Porous Media* **14**, 85–88. [1.5.1]

Nield, D. A. 1994c Estimation of an effective Rayleigh number for convection in a vertically inhomogeneous porous medium or clear fluid. *Int. J. Heat Fluid Flow* **15**, 337–340. [5.4, 6.7, 6.13.2]

Nield, D. A. 1994d Convection in a porous medium with inclined temperature gradient: additional results. *Int. J. Heat Mass Transfer* **37**, 3021–3025. [7.9]

Nield, D. A. 1995 Onset of convection in a porous medium with nonuniform time-dependent volumetric heating. *Int. J. Heat Fluid Flow* **16**, 217–222. [6.11.3]

Nield, D. A. 1996 The effect of temperature-dependent viscosity on the onset of convection in a saturated porous medium. *ASME J. Heat Transfer* **118**, 803–805. [6.7]

Nield, D. A. 1997a Discussion of a discussion by F. Chen and C.F. Chen. *ASME J. Heat Transfer* **119**, 193–194. [1.6]

Nield, D. A. 1997b Notes on convection in a porous medium: (i) an effective Rayleigh number for an anisotropic layer, (ii) the Malkus hypothesis and wavenumber selection. *Transport in Porous Media* **27**, 135–142. [6.4, 6.9.1, 6.12]

Nield, D. A. 1997c Comments on "Turbulence model for flow through porous media". *Int. J. Heat Mass Transfer* **40**, 2499. [1.8]

Nield, D. A. 1998a Effects of local thermal nonequilibrium in steady convective processes in a saturated porous medium: forced convection in a channel. *J. Porous Media* **1**, 181–186. [2.2.2, 4.10]

Nield, D. A. 1998b Convection in a porous medium with inclined temperature gradient and vertical throughflow. *Int. J. Heat Mass Transfer* **41**, 241–243. [7.9]

Nield, D. A. 1998c Modeling the effect of surface tension on the onset of natural convection in a saturated porous medium. *Transport Porous Media* **31**, 365–368. [6.19.3]

Nield, D. A. 1999 Modeling the effects of a magnetic field or rotation on flow in a porous medium: momentum equation and anisotropic permeability analogy. *Int. J. Heat Mass Transfer* **42**, 3715–3718. [6.22]

Nield, D. A. 2000 Resolution of a paradox involving viscous dissipation and nonlinear drag in a porous medium. *Transport Porous Media* **41**, 349–357. [2.2.2]

Nield, D. A. 2001a Comments on 'Fully developed free convection in open-ended vertical channels partially filled with porous material' by M. A. Al-Nimr and O. M. Hadad. *J. Porous Media* **4**, 97–99. [7.7]

Nield, D. A. 2001b Alternative models of turbulence in a porous medium, and related matters. *ASME J. Fluids Engng.* **123**, 928–931. [1.8]

Nield, D. A. 2001c Some pitfalls in the modeling of convective flows in porous media. *Transport Porous Media* **43**, 597–601. [6.7]

Nield, D. A. 2002 A note on the modeling of local thermal non-equilibrium in a structured porous medium. *Int. J. Heat Mass Transfer* **45**, 4367–4368. [4.10]

Nield, D. A. 2003 The stability of flow in a channel or duct occupied by a porous medium. *Int. J. Heat Mass Transfer* **46**, 4351–4354. [1.8]

Nield, D. A. 2004a Comments on 'The onset of transient convection in bottom heated porous media' by K. K. Tan, T. Sam and H. Jamaludin: Rayleigh and Biot numbers. *Int. J. Heat Mass Transfer* **47**, 641–643. [6.11.3]

Nield, D. A. 2004b Comments on 'A new model for viscous dissipation in porous media across a range of permeability values'. *Transport in Porous Media* **55**, 253–254. [2.2.2]

Nield, D. A. 2004c Forced convection in a plane plate channel with asymmetric heating. *Int. J. Heat Mass Transfer* **47**, 5609–5612. [4.9]

Nield, D. A. 2006 A note on a Brinkman-Brinkman forced convection problem. *Transport Porous Media*, to appear. [4.9]

Nield, D. A. and Joseph, D. D. 1985 Effects of quadratic drag on convection in a saturated porous medium. *Phys. Fluids* **28**, 995–997. [6.6]

Nield, D. A. and Kuznetsov, A. V. 1999 Local thermal nonequilibrium effects in forced convection in a porous medium channel: a conjugate problem. *Int. J. Heat Mass Transfer* **42**, 3245–3252. [4.10]

Nield, D. A. and Kuznetsov, A. V. 2000 Effects of heterogeneity in forced convection in a porous medium: parallel plate channel or circular duct. *Int. J. Heat Mass Transfer* **43**, 4119–4134. [4.12]

Nield, D. A. and Kuznetsov, A. V. 2001a Effects of heterogeneity in forced convection in a porous medium: parallel plate channel, asymmetric property variation, and asymmetric heating. *J. Porous Media* **4**, 137–148. [4.12]

Nield, D. A. and Kuznetsov, A. V. 2001b The interaction of thermal nonequilibrium and heterogeneous conductivity effects in forced convection in layered porous channels. *Int. J. Heat Mass Transfer* **44**, 4369–4373. [4.12]

Nield, D. A. and Kuznetsov, A. V. 2003a Effects of gross heterogeneity and anisotropy in forced convection in a porous medium: layered medium analysis. *J. Porous Media* **6**, 51–57. [4.12]

Nield, D. A. and Kuznetsov, A. V. 2003b Effects of temperature-dependent viscosity in forced convection in a porous medium: layered medium analysis. *J. Porous Media* **6**, 213–222. [4.16.6]

Nield, D. A. and Kuznetsov, A. V. 2003c Effects of heterogeneity in forced convection in a porous medium: parallel-plate channel, Brinkman model. *J. Porous Media* **6**, 257–266. [4.12]

Nield, D. A. and Kuznetsov, A. V. 2003d Boundary-layer analysis of forced convection with a plate and a porous substrate. *Acta Mech.* **166**, 141–148. [4.11]

Nield, D. A. and Kuznetsov, A. V. 2004a Interaction of transverse heterogeneity and thermal development of forced convection in a porous medium. *Transport Porous Media* **57**, 103–111. [4.13]

Nield, D. A. and Kuznetsov, A. V. 2004b Forced convection in a helical pipe filled with a saturated porous medium. *Int. J. Heat Mass Transfer* **47**, 5175–5180. [4.5]

Nield, D. A. and Kuznetsov, A. V. 2004c Forced convection in a bi-disperse porous medium channel: a conjugate problem. *Int. J. Heat Mass Transfer* **47**, 5375–5380. [4.16.4]

Nield, D. A. and Kuznetsov, A. V. 2005a Thermally developing forced convection in a channel occupied by a porous medium saturated by a non-Newtonian fluid. *Int. J. Heat Mass Transfer* **48**, 1214–1218. [4.13]

Nield, D. A. and Kuznetsov, A. V. 2005b A two-velocity two-temperature model for a bi-dispersed porous medium: forced convection in a channel. *Transport in Porous Media*, **59**, 325–339. [4.16.4]

Nield, D. A. and Kuznetsov, A. V. 2005c Heat transfer in bidisperse porous media. In *Transport Phenomena in Porous Media III*, (eds. D. B. Ingham and I. Pop), Elsevier, Oxford, pp. 34–59. [4.16.4]

Nield, D. A. and Kuznetsov, A. V. 2006 Forced convection with slip-flow in a channel or *ASME J. Heat Transfer* **119**, 195–197. [1.5.3]

Nield, D. A. and Lage, J. L. 1997 Discussion of a Discussion by K. Vafai and S.J. Kim. *ASME J. Heat Transfer* **119**, 195–197. [1.5.3]

Nield, D. A. and Lage, J. L. 1998 The role of longitudinal diffusion in a fully developed forced convective slug flow in a channel. *Int. J. Heat Mass Transfer* **41**, 4375–4377. [4.5]

Nield, D. A. and White, S. P. 1982 Natural convection in an infinite porous medium produced by a line heat source. *Mathematics and Models in Engineering Science* (ed. A. McNabb *et al.*) Dept. Sci. Indust. Res., Wellington, New Zealand, pp. 121–128. [5.10.2]

Nield, D. A., Junqueira, S. L. M. and Lage, J. L. 1996 Forced convection in a fluid saturated porous medium with isothermal and isoflux boundaries. *J. Fluid Mech.* **322**, 201–214. [4.5]

Nield, D. A., Kuznetsov, A. V. and Avramenko, A. A. 2004c The onset of bioconvection in a horizontal porous medium layer. *Transport in Porous Media* **54**, 335–344. [6.25]

Nield, D. A., Kuznetsov, A. V. and Xiong, M. 2002 Effect of local thermal non-equilibrium on thermally developing forced convection in a porous medium. *Int. J. Heat Mass Transfer* **45**, 4949–4955. [4.13]

Nield, D. A., Kuznetsov, A. V. and Xiong, M. 2003a Thermally developing forced convection in a porous medium: parallel plate channel with walls at uniform temperature, with axial conduction and viscous dissipation effects. *Int. J. Heat Mass Transfer* **46**, 643–651. [4.13]

Nield, D. A., Kuznetsov, A. V. and Xiong, M. 2003b Thermally developing forced convection in a porous medium: parallel-plate channel or circular tube with walls at constant heat flux. *J. Porous Media* **6**, 203–212. [4.13]

Nield, D. A., Kuznetsov, A. V. and Xiong, M. 2004a Thermally developing forced convection in a porous medium: parallel-plate channel or circular tube with isothermal walls. *J. Porous Media* **7**, 19–27. [4.13]

Nield, D. A., Kuznetsov, A. V. and Xiong, M. 2004b Effects of viscous dissipation and flow work on forced convection in a channel filled by a saturated porous medium. *Transport Porous Media* **565**, 351–367. [4.9]

Nield, D. A., Manole, D. M. and Lage, J. L. 1993 Convection induced by inclined thermal and solutal gradients in a shallow layer of a porous medium. *J. Fluid Mech.* **257**, 559–574. [9.5]

Nield, D. A., Porneala, D. C. and Lage, J. L. 1999 A theoretical study, with experimental verification, of the temperature-dependent viscosity effect on the forced convection through a porous medium channel. *ASME J. Heat Transfer* **121**, 500–503. [4.16.1]

Nilsen, T. and Storesletten, L. 1990 An analytical study on natural convection in isotropic and anisotropic porous channels. *ASME J. Heat Transfer* **112**, 396–401. [6.15.3]

Nilson, R. H. 1981 Natural convective boundary layer on two-dimensional and axisymmetric surfaces in high-Pr fluids or in fluid saturated porous media. *ASME J. Heat Transfer* **103**, 803–807. [5.6.1]

Nishimura, T. and Wakamatsu, M. 2000 Natural convection suppression and crystal growth during unidirectional solidification of a binary system. *Heat Transfer Asian Res.* **29**, 120–131. [10.2.3]

Nishimura, T., Kunitsugu, K. and Itoh, T. 1996 Natural convection suppression by azimuthal partitions in a horizontal porous annulus. *Numer. Heat Transfer A* **29**, 65–81. [7.3.3]

Nishimura, T., Takumi, T., Shiraishi, M., Kawamura, Y. and Ozoe, H. 1986 Numerical analysis of natural convection in a rectangular enclosure horizontally divided into fluid and porous regions. *Int. J. Heat Mass Transfer* **29**, 889–898. [7.7]

Nisse, L. and Néel, M. C. 2005 Spectral stability of convective rolls in porous media. *ZAMM* **85**, 366–383. [6.4]

Nithiarasu, P. 1999 Finite element modeling of a leaky third component migration from a heat source buried into a fluid saturated porous medium. *Math. Comput. Model.* **29**, 27–39. [9.3.1]

Nithiarasu, P., Seetharamu, K.N. and Sundararajan, T. 1996 Double-diffusive natural convection in an enclosure filled with a fluid-saturated porous medium: a generalized non-Darcy approach. *Numer. Heat Transfer A* **30**, 413–426. [6.6]

Nithiarasu, P., Seetharamu, K. N. and Sundararajan, T 1999a Numerical investigations of buoyancy driven flow in a fluid saturated non-Darcian porous medium. *Int. J. Heat Mass Transfer* **42**, 1205–1215. [7.1.6]

Nithiarasu, P., Seetharamu, K. N. and Sundararjan, T. 1997a Natural convective heat transfer in a fluid saturated variable porosity medium. *Int. J. Heat Mass Transfer* **40**, 3955–3957. [7.6.2]

Nithiarasu, P., Seetharamu, K. N. and Sundararajan, T. 1997b Non-Darcy double-diffusive natural convection in axisymmetric fluid saturated porous cavities. *Heat Mass Transfer* **32**, 427–433. [9.4]

Nithiarasu, P., Seetharamu, K. N. and Sundararajan, T. 1998 Effect of porosity on natural convective heat transfer in a fluid saturated porous medium. *Int. J. Heat Fluid Flow* **19**, 56–58. [7.6.2]

Nithiarasu, P., Seetharamu, K. N. and Sundararajan, T. 2002 Finite element modeling of flow, heat and mass transfer in fluid saturated porous media. *Arch. Comp. Meth. Engng.* **9**, 3–42. [7.1.6]

Nithiarasu, P., Sujatha, K. S., Sundararajan, T. and Seetharamu, K. N. 1999b Buoyancy driven flow in a non-Darcian, fluid saturated porous enclosure subjected to uniform heat flux—A numerical study. *Comm. Numer. Meth. Engng.* **15**, 765–776. [7.1.6]

Nsofor, E. C. and Adebiyi, G. A. 2003 Forced-convection gas-to-wall heat transfer in a packed bed for high-temperature storage. *Exper. Heat Transfer* **16**, 81–95. [4.9]

Oberbeck, A. 1879 Ueber die Wärmeleitung der Flüssigkeiten bei Berücksichtigung der Strömungen infolge von Temperaturdifferenzen. *Ann. Phys. Chem.* **7**, 271–292. [2.3]

Ochoa-Tapia, J. A. and Whitaker, S. 1995a Momentum transfer at the boundary between a porous medium and a homogeneous fluid—I. Theoretical development. *Int. J. Heat Mass Transfer* **38**, 2635–2646. [1.6]

Ochoa-Tapia, J. A. and Whitaker, S. 1995b Momentum transfer at the boundary between a porous medium and a homogeneous fluid—II. Comparison with experiment. *Int. J. Heat Mass Transfer* **38**, 2647–2655. [1.6]

Ochoa-Tapia, J. A. and Whitaker, S. 1997 Heat transfer at the boundary between a porous medium and a heterogeneous fluid. *Int. J. Heat Mass Transfer* **40**, 2691–2707. [2.4]

Ochoa-Tapia, J. A. and Whitaker, S. 1998 Momentum jump condition at the boundary between a porous medium and a homogeneous fluid: Inertia effects. *J. Porous Media* **1**, 201–207. [1.6]

Okada, M. , Matsumoto, K. and Yabushita, Y. 1994 Solidification around horizontal cylinder in porous medium saturated with aqueous solution. *Heat Transfer 1994*, Inst. Chem. Engrs, Rugby, vol. 4, pp. 109–114. [10.2.3]

Oldenburg, C. M. and Pruess, K. 1998 Layered thermohaline convection in hypersaline geothermal systems. *Transport Porous Media* **33**, 29–63. [11.8]

Oldenburg, C. M. and Pruess, K. 1999 Plume separation by transient thermohaline convection in porous media. *Geophys. Res. Lett.* **26**, 2997–3000. [11.8]

Olek, S. 1998 Heat transfer regimes for free convection in rotating porous media. *Heat Transfer 1998, Proc. 11th IHTC*, **4**, 423–428. [7.12]

Oliveira, L. S. and Haghighi, K. 1998 Conjugate heat and mass transfer in convective drying of porous media. *Numer. Heat Transfer A* **34**, 105–117. [3.6]

Oliver, M. and Titi, E. S. 2000 Gevrey regularity for the attractor of a partially dissipative model of Bénard convection in a porous medium. *J. Differ. Equations* **163**, 292–311. [6.15.1]

Oosthuizen, P. H. 1987 Mixed convection heat transfer from a cylinder in a porous medium near an impermeable surface. *ASME HTD* **84**, 75–82. [8.1.3]

Oosthuizen, P. H. 1988a The effects of free convection on steady-state freezing in a porous medium-filled cavity. *ASME HTD* **96**, Vol. 1, 321–327. [10.2.1]

Oosthuizen, P. H. 1988b Mixed convective heat transfer from a heated horizontal plate in a porous medium near an impermeable surface. *ASME J. Heat Transfer* **110**, 390–394. [8.1.4]

Oosthuizen, P. H. 1995 Heat transfer from a cylinder with a specified surface heat flux buried in a frozen porous medium in an enclosure with a cooled top surface. *Proc. ASME/JSME Thermal Engineering Joint Conf.* vol. 3, pp. 379–386. [7.11]

Oosthuizen, P. H. 2000 Natural convective heat transfer in porous-media-filled enclosures. *Handbook of Porous Media* (K. Vafai, ed.), Marcel Dekker, New York, pp. 489–520. [7]

Oosthuizen, P. H. and Naylor, D. 1996a Natural convective heat transfer from a cylinder in an enclosure partly filled with a porous medium. *Int. J. Numer. Meth. Heat Fluid Flow* **6**, 51–63. [7.11]

Oosthuizen, P. H. and Naylor, D. 1996b Heat transfer from a heated cylinder with a specified surface heat flux buried in a frozen porous medium in an enclosure with non-uniform wall temperatures. *ASME HTD* **331**, 43–52. [7.11]

Oosthuizen, P. H. and Paul, J. T. 1992 Heat transfer from a heated cylinder buried in a frozen porous medium in an enclosure. *Heat and Mass Transfer in Porous Media* (ed. M. Quintard and M. Toderovic), Elsevier, Amsterdam, pp. 315–326. [7.11]

Or, A. C. 1989 The effects of temperature-dependent viscosity and the instabilities in convection rolls of a layer of fluid-saturated porous medium. *J. Fluid Mech.* **206**, 497–515. [6.8]

Ormond, A. and Genthon, P. 1993 3-D thermoconvection in an anisotropic inclined sedimentary layer. *Geophys. J. Int.* **112**, 257–263. [11.8]

Ormond, A. and Ortoleva, P. 2000 Numerical modeling of reaction-induced cavities in a porous rock. *J. Geophys. Res.–Solid Earth* **105**, 16737–16747. [11.8]

Orozco, J. 1992 Condensation of a downward flowing vapor on a horizontal cylinder embedded in a porous medium. *ASME J. Energy Resour. Tech.* **113**, 300–304. [10.4]

Orozco, J. and Zhu, K. H. 1993 Mixed convection film boiling of a binary mixture on a horizontal cylinder embedded in a porous medium. *Chem., Engng. Comm.* **135**, 91–104. [10.3.2]

Orozco, J., Stellman, R. and Gutjahr, M. 1988 Film boiling heat transfer from a sphere and a horizontal cylinder embedded in a liquid-saturated porous medium. *ASME J. Heat Transfer* **110**, 961–967. [10.3.2]

O'Sullivan, M. J. 1985a Geothermal reservoir simulation. *Int. J. Energy Res.* **9**, 319–332. An edited version was reprinted in *Applied Geothermics* (eds. M. Economides and P. Ungemash), Wiley, New York, 1987. [11]

O'Sullivan, M. J. 1985b Convection with boiling in a porous layer. *Convective Flows in Porous Media* (eds. R. A. Wooding and I. White), Dept. Sci. Indust. Res., Wellington, N.Z., pp. 141–155. [10.3.1]

O'Sullivan, M. J. and McKibbin, R. 1986 Heat transfer in an unevenly heated porous layer. *Transport in Porous Media* **1**, 293–312. [6.14]

O'Sullivan, M. J., Pruess, K. and Lippman, M. J. 2000 Geothermal reservoir simulation: the state-of-practice and emerging trends. *Proc. World Geothermal Congr. 2000*, Kyushu-Tohoku, Japan, pp. 4065–4070. [11]

O'Sullivan, M. J., Pruess, K. and Lippman, M. J. 2001 State of the art of geothermal reservoir simulation.. *Geothermics* **30**, 395–429. [11]

Otero, J., Dontcheva, L. A., Johnston, H., Worthing, R. A., Kurganov, A., Petrova, G. and Doering, C. R. 2004 High-Rayleigh number convection in a fluid saturated porous layer. *J. Fluid Mech.* **500**, 263–282. [6.8]

Ouarzazi, M. N. and Bois, P. A. 1994 Convective instability of a fluid mixture in a porous medium with time-dependent temperature gradient. *Eur. J. Mech. B/Fluids* **13**, 275–295. [9.1.6.4]

Ouarzazi, M. N., Bois, P. A. and Taki, M. 1994 Nonlinear interaction of convective instabilities and temporal chaos of a fluid mixture in a porous medium. *Eur. J. Mech. B/Fluids* **13**, 423–438. [9.1.6.4]

Ould-Amer, Y., Chikh, S., Bouhadef, K. and Lauriat, G. 1998 Forced convection cooling enhancement by use of porous materials. *Int. J. Heat Fluid Flow* **19**, 251–258. [4.11]

Ozaki, K. and Inaba, H. 1997 Convective heat transfer of horizontal spherical particle layer heated from below and cooled above—Part II. Effect of thickness of the layer. *Heat Transfer Japan Res.* **26**, 176–192. [6.9.2]

Ozdemir, M. and Ozguc, F. 1997 Forced convection heat transfer in porous medium of wire screen meshes. *Heat Mass Transfer* **33**, 129–136. [4.16.2]

Paek, J. W., Kang, B. H. and Hyun, J. M. 1999a Transient cool-down of a porous medium in pulsating flow. *Int. J. Heat Mass Transfer* **42**, 3523–3527. [4.16.5]

Paek, J. W., Kang, B. H., Kim, S. Y. and Hyun, J. M. 2000 Effective thermal conductivity and permeability of aluminum foam materials. *Int. J. Thermophys.* **21**, 435–464. [2.2.1]

Paek, J. W., Kim, S. Y., Kang, B. H. and Hyun, J. M. 1999b Forced convective heat transfer from anisotropic aluminum foam in a channel flow. *Proc. 33rd Nat. Heat Transfer Conf., NHTC99-158*, 1–8. [4.12]

Paik, S., Nguyen, H. D. and Pop, I. 1998 Transient conjugate mixed convection form a sphere in a porous medium saturated with cold pure or saline water. *Heat Mass Transfer* **34**, 237–245. [8.1.3]

Pak, J. and Plumb, O. A. 1997 Melting in a two-component packed bed. *ASME J. Heat Transfer* **119**, 553–559. [10.1.7]

Palm, E. 1990 Rayleigh convection, mass transport, and change in porosity in layers of sandstone. *J. Geophys. Res.* **95**, 8675–8679. [11.5]

Palm, E. and Tveitereid, M. 1979 On heat and mass flux through dry snow. *J. Geophys. Res.* **84**, 745–749. [11.1, 11.5]

Palm, E. and Tyvand, P. 1984 Thermal convection in a rotating porous layer. *J. Appl. Math. Phys. (ZAMP)* **35**, 122–123. [6.22]

Palm, E., Weber, J. E. and Kvernvold, O. 1972 On steady convection in a porous medium. *J. Fluid Mech.* **54**, 153–161. [6.3, 6.6]

Pan, C. P. and Lai, F. C. 1995 Natural convection in horizontal-layered porous annuli. *AIAA J. Thermophys. Heat Transfer* **9**, 792–795. [7.3.3]

Pan, C. P. and Lai, F. C. 1996 Re-examination of natural convection in a horizontal layered porous annulus. *ASME J. Heat Transfer* **118**, 990–992. [7.3.3]

Papathanasiou, T. D., Markicevic, B. and Dendy, E. D. 2001 A computational evaluation of the Ergun and Forchheimer equations for fibrous porous media. *Phys. Fluids* **13**, 2795–2804. [1.5.2]

Parang, M. and Keyhani, M. 1987 Boundary effects in laminar mixed convection flow through an annular porous medium. *ASME J. Heat Transfer* **109**, 1039–1041. [8.3.3]

Park, H. M., Lee, H. S. and Cho, D. H. 1996 Thermal convective instability in translucent porous media with radiative heat transfer. *Chem. Engng. Comm.* **145**, 155–171. [6.11.2]

Parmentier, E. M. 1979 Two-phase natural convection adjacent to a vertical heated surface in a permeable medium. *Int. J. Heat Mass Transfer* **22**, 849–855. [10.3.2]

Parthiban, C. and Patil, P. R. 1993 Effect of inclined temperature gradient on thermal instability in an anisotropic porous medium. *Wärme-Stoffübertrag.* **29**, 63–69. [7.9]

Parthiban, C. and Patil, P. R. 1994 Effect of inclined gradients on thermohaline convection in porous medium. *Wärme-Stoffübertrag.* **29**, 291–297. [9.5]

Parthiban, C. and Patil, P. R. 1995 Effect of non-uniform boundary temperatures on thermal instability in a porous medium with internal heat source. *Int. Comm. Heat Mass Transfer* **22**, 683–692. [6.11.2, 7.9]

Parthiban, C. and Patil, P. R. 1996 Convection in a porous medium with velocity slip and temperature jump boundary conditions. *Heat Mass Transfer* **32**, 7–31. [6.7]

Parthiban, C. and Patil, P. R. 1997 Thermal instability in an anisotropic porous medium with internal heat source and inclined temperature. *Int. Comm. Heat Mass Transfer* **24**, 1049–1058. [7.9]

Parvathy, C. P. and Patil, P. R. 1989 Effect of thermal diffusion on thermohaline interleaving in a porous medium due to horizontal gradients. *Indian J. Pure Appl. Math.* **20**, 716–727. [9.5]

Pascal, H. 1990 Rheological effects of non-Newtonian fluids on natural convection in a porous medium. [5.1.9.2]

Pascal, J. P. and Pascal, H. 1997 Free convection in a non-Newtonian fluid saturated porous medium with lateral mass flux. *Int. J. Non-Linear Mech.* **32**, 471–482. [5.7]

Paterson, I. and Schlanger, H. P. 1992 Convection in a porous thermosyphon imbedded in a conducting medium. *Int. J. Heat Mass Transfer* **35**, 877–886. [11.8]

Patil, P. R. 1982 Soret driven instability of a reacting fluid in a porous medium. *Israel J. Tech.* **19**, 193–196. [9.1.6.4]

Patil, P. R. and Rudraiah, N. 1973 Stability of hydromagnetic thermoconvective flow through porous medium. *Trans. ASME, J. Appl. Mech. E* **40**, 879–884. [6.21]

Patil, P. R. and Rudraiah, N. 1980 Linear convective stability and thermal diffusion of a horizontal quiescent layer of a two-component fluid in a porous medium. *Int. J. Engng. Sci.* **18**, 1055–1059. [9.1.4]

Patil, P. R. and Subramanian, L. 1992 Soret instability in anisotropic porous medium with temperature dependent viscosity. *Fluid Dyn. Res.* **10**, 159–168. [9.1.6.2]

Patil, P. R. and Vaidyanathan, G. 1981 Effect of variable viscosity on the setting up of convection currents in a porous medium. *Int. J. Engng. Sci.* **19**, 421–426. [6.7]

Patil, P. R. and Vaidyanathan, G. 1982 Effect of variable viscosity on thermohaline convection in a porous medium. *J. Hydrol.* **57**, 147–161. [9.1.6.2]

Patil, P. R. and Vaidyanathan, G. 1983 On setting up of convective currents in a rotating porous medium under the influence of variable viscosity. *Int. J. Engng. Sci.* **21**, 123–130. [6.22]

Patil, P. R., Parvathy, C. P. and Venkatakrishnan, K. S. 1990 Effect of rotation on the stability of a doubly diffusive layer in a porous medium. *Int. J. Heat Mass Transfer* **33**, 1073–1080. [9.1.6.4]

Patil, P. R., Parvathy, C.P. and Venkatakrishnan, K. S. 1989 Thermohaline instability in a rotating anisotropic porous medium. *Appl. Sci. Res.* **46**, 73–88. [9.1.6.4]

Paul, T. and Singh, A. K. 1998 Natural convection between co-axial vertical cylinders partially filled with a porous material. *Forschung Ingen.* **64**, 157–162. [7.7]

Paul, T., Jha, B. K. and Singh, A. K. 1998 Free-convection between vertical walls partially filled with porous medium. *Heat Mass Transfer* **33**, 515–519. [7.7]

Paul, T., Singh, A. K. and Misra, A. K. 2001 Transient natural convection between two vertical walls filled with a porous material having variable porosity. *Math. Engng. Ind.* **8**, 177–185. [7.5]

Paul, T., Singh, A. K., and Thorpe, G. R. 1999 Transient natural convection in a vertical channel partially filled with a porous medium. *Math. Engng. Ind.* **7**, 441–455. [7.7]

Pavel, B. I. and Mohamad, A. A. 2004a An experimental and numerical study on heat transfer enhancement for gas heat exchangers fitted with porous media. *Int. J. Heat Mass Transfer* **47**, 4939–4952. [4.11]

Pavel, B. I. and Mohamad, A. A. 2004b Experimental and numerical investigation of heat transfer enhancement using porous material. In *Applications of Porous Media (ICAPM 2004)*, (eds. A. H. Reis and A. F. Miguel), Évora, Portugal, pp. 331–338. [4.11]

Pavel, B. I. and Mohamad, A. A. 2004c Experimental investigation of the potential of metallic porous inserts in enhancing forced convection heat transfer. *ASME J. Heat Transfer* **126**, 540–545. [4.11]

Payne, L. E. and Song, J. C. 1997 Spatial decay estimates for the Brinkman and Darcy flows in a semi-infinite cylinder. *Contin. Mech. Thermodyn.* **9**, 175–190. [1.5.3]

Payne, L. E. and Song, J. C. 2000 Spatial decay for a model of double diffusive convection in Darcy and Brinkman flow. *Z. Angew. Math. Phys.* **51**, 867–889. [1.5.3]

Payne, L. E. and Song, J. C. 2002 Spatial decay bounds for the Forchheimer equations. *Int. J. Engng. Sci.* **40**, 943–956. [1.5.3]

Payne, L. E. and Straughan, B. 1998a Analysis of the boundary condition at the interface between a viscous fluid and a porous medium and related modelling questions. *J. Math. Pures Appl.* **77**, 317–354. [1.6]

Payne, L. E. and Straughan, B. 1998b Structure stability for the Darcy equations of flow in porous media. *Proc. Roy. Soc. Lond. A* **454**, 1691–1698. [1.5.3]

Payne, L. E. and Straughan, B. 1999 Convergence and continuous dependence for the Brinkman-Forchheimer equations. *Stud. Appl. Math.* **102**, 419–439. [1.5.3]

Payne, L. E. and Straughan, B. 2000a A naturally efficient numerical technique for porous convection stability with non-trivial boundary conditions. *Int. J. Numer. Anal. Meth. Geomech.* **24**, 815–836. [11.8]

Payne, L. E. and Straughan, B. 2000b Unconditional nonlinear stability in temperature-dependent viscosity flow in a porous medium. *Stud. Appl. Math.* **105**, 59–81. [6.7]

Payne, L. E., Rodrigues, J. F. and Straughan, B. 2001 Effect of anisotropic permeability on Darcy's law. *Math. Meth. Appl. Sci.* **24**, 427–438. [1.5.3]

Payne, L. E., Song, J. C. and Straughan, B. 1988 Double-diffusive porous penetrative convection–thawing subsea permafrost. *Int. J. Engng. Sci.* **26**, 797–809. [11.3]

Pedras, M. H. J. and de Lemos, M. J. S. 2000 On the definition of turbulent kinetic energy for flow in porous media. *Int. Comm. Heat Mass Transfer* **27**, 211–220. [1.8]

Pedras, M. H. J. and de Lemos, M. J. S. 2001a Macroscopic turbulence modeling for incompressible flow through undeformable porous media. *Int. J. Heat Mass Transfer* **44**, 1081–1093. [1.8]

Pedras, M. H. J. and de Lemos, M. J. S. 2001b Simulation of turbulent flow in porous media using a spatially periodic array and low Re two-equation closure. *Numer. Heat Transfer A* **39**, 35–59. [1.8]

Pedras, M. H. J. and de Lemos, M. J. S. 2001c On the mathematical description and simulation of turbulent flow in a porous medium formed by an array of elliptical rods. *ASME J. Fluids Engng.* **123**, 941–947. [1.8]

Pedras, M. H. J. and de Lemos, M. J. S. 2003 Computation of turbulent flow in porous media using a low Reynolds number k-ε model and an infinite array of transversely displaced elliptic rods. *Numer. Heat Transfer A* **43**, 585–602. [1.8]

Peirotti, M. B., Giavedoni, M. D. and Deiber, J. A. 1987 Natural convective heat transfer in a rectangular porous cavity with variable fluid properties–validity of the Boussinesq approximation. *Int. J. Heat Mass Transfer* **30**, 2571–2581. [7.3.2]

Peng, C., Zeng, T. and Cheng, G. 1992 Free convection about vertical needles embedded in a saturated porous medium. *AIAA J. Thermophys. Heat Transfer* **6**, 558–561. [5.12.1]

Peng, S. W., Besant, R. W. and Strathdee, G. 2000 Heat and mass transfer in granular potash fertilizer with a surface dissolution reaction. *Cand. J. Chem. Engng.* **78**, 1076–1086. [3.6]

Pestov, I. 1997 Structured geothermal systems: application of dimensional methods. *Mathl. Comput. Modelling* **25**, 43–63. [11.9.1]

Pestov, I. 1998 Stability of vapour-liquid counterflow in porous media. *J. Fluid Mech.* **364**, 273–295. [11.9.1]

Peterson, G. P. and Chang, C. S. 1997 Heat transfer analysis and evaluation for two-phase flow in porous-channel heat sinks. *Numer. Heat Transfer A* **31**, 113–130. [11.9.3]

Peterson, G. P. and Chang, C. S. 1998 Two-phase heat dissipation utilizing porous-channels of high-conductivity material. *ASME J. Heat Transfer* **120**, 243–252. [11.9.3]

Petit, F., Fichot, F. and Quintard, M. 1999 Two-phase flow in porous media: local non-equilibrium model. *Rev Gén. Therm.* **38**, 250–257. [2.2.2]

Petrescu, S. 1994 Comments on the optimal spacing of parallel plates cooled by forced convection. *Int. J. Heat Mass Transfer* **37**, 1283. [4.15, 4.19, 4.20]

Phanikumar, M. S. and Mahajan, R. L. 2002 Non-Darcy natural convection in high porosity metal foams. *Int. J. Heat Mass Transfer* **45**, 3781–3793. [7.3.7]

Philip, J. R. 1982a Free convection at small Rayleigh number in porous cavities of rectangular, elliptical, triangular and other cross sections. *Int. J. Heat Mass Transfer* **25**, 1503–1510. [7.3.7]

Philip, J. R. 1982b Axisymmetric free convection at small Rayleigh number in porous cavities. *Int. J. Heat Mass Transfer* **25**, 1689–1699. [7.3.7]

Philip, J. R. 1988 Free convection in porous cavities near the temperature of maximum density. *Phys. Chem. Hydrodyn.* **10**, 283–294. [7.3.7]

Phillips, O. M. 1991 *Flow and Reactions in Permeable Rocks*, Cambridge Univ. Press, New York. [11.5]

Pien, S. J. and Sen, M. 1989 Hysteresis effects in three-dimensional natural convection in a porous medium. *ASME HTD* **107**, 343–348. [7.8]

Pillatsis, G., Taslim, M. E. and Narusawa, U. 1987 Thermal instability of a fluid-saturated porous medium bounded by thin fluid layers. *ASME J. Heat Transfer* **109**, 677–682. [6.19.1.2]

Platten, J. K. and Costeceque, P. 2004 The Soret coefficient in porous media. *J. Porous Media* **7**, 317–329. [3.3]

Platten, J. K. and Legros, J. C. 1984 *Convection in Liquids*, Springer, New York. [3.3]

Ploude, F. and Prat, M. 2003 Pore network simulations of drying of capillary porous media. Influence of thermal gradient. *Int. J. Heat Mass Transfer* **46**, 1293–1307. [3.6]

Plumb, O. A. 1983 The effect of thermal dispersion on heat transfer in packed bed boundary layers. *Proc. ASME JSME Thermal Engineering Joint Conference* **2**, 17–22. [5.1.7.3]

Plumb, O. A. 1991a Heat transfer during unsaturated flow in porous media. *Convective Heat and Mass Transfer in Porous Media* (eds S. Kakaç, B. Kilkis, F. A. Kulacki and F. Arinç), Kluwer Academic, Dordrecht, 435–464. [3.6]

Plumb, O. A. 1991b Drying complex porous materials–modelling and experiments. *Convective Heat and Mass Transfer in Porous Media* (eds. S. Kakaç, B. Kilkis, F. A. Kulacki and F. Arinç), Kluwer Academic, Dordrecht, 963–984. [3.6]

Plumb, O. A. 1994a Convective melting of packed beds. *Int. J. Heat Mass Transfer* **37**, 829–836. [10.1.7]

Plumb, O. A. 1994b Analysis of near wall heat transfer in porous media. *Heat Transfer 1994*, Inst. Chem. Engrs, Rugby, vol. 5, pp. 363–368. [4.8]

Plumb, O. A. 2000 Transport phenomena in porous media: modeling the drying process. *Handbook of Porous Media* (K. Vafai, ed.), Marcel Dekker, New York, pp. 755–785. [3.6]

Plumb, O. A. and Huenefeld, J. C. 1981 Non-Darcy natural convection from heated surfaces in saturated porous media. *Int. J. Heat Mass Transfer* **24**, 765–768. [5.1.7.2, 5.1.8]

Poirier, D. R., Nandapurkar, P.J . and Ganesan, S. 1991 The energy and solute conservation equations for dendritic solidification. *Metall. Trans. B* **22**, 889–900. [10.2.3]

Poirier, H. 2003 Une théorie explique l'intelligence de la nature. *Science & Vie* **1034**, 44–63. [4.18]

Polyaev, V. M., Mozhaev, A. P., Galitseysky, B. A. and Lozhkin, A. L. 1996 A study of internal heat transfer in nonuniform porous structures. *Expt. Therm. Fluid Sci.* **12**, 426–432. [2.2.2]

Pop, I. 2004 Some boundary-layer problems in convective flows. In *Emerging Technologies and Techniques in Porous Media* (D. B. Ingham, A. Bejan, E. Mamut and I. Pop, eds), Kluwer Academic, Dordrecht, pp. 65–91. [5]

Pop, I. and Cheng, P. 1986 An integral solution for free convection of a Darcian fluid about a cone with curvature effects. *Int. Comm. Heat Mass Transfer* **13**, 433–438. [5.8]

Pop, I. and Gorla, R. S. R. 1991 Horizontal boundary-layer natural convection in a porous medium saturated with a gas. *Transport in Porous Media* **6**, 159–171. [5.2]

Pop, I. and Herwig, H. 1990 Transient mass transfer from an isothermal vertical flat plate embedded in a porous medium. *Int. Comm. Heat Mass Transfer* **17**, 813–821. [9.5]

Pop, I. and Herwig, H. 1992 Free convection from a vertical surface in a porous medium saturated with fluids of variable properties: an asymptotic approach. *Heat and Mass*

Transfer in Porous Media (ed. M. Quintard and M. Todorovic), Elsevier, Amsterdam, 337–348. [5.1.9.9]

Pop, I. and Ingham, D. B. 1990 Natural convection about a heated sphere in a porous medium. *Heat Transfer 1990*, Hemisphere, New York, vol, 2, pp. 567–572. [5.6.2, 5.6.3]

Pop, I. and Ingham, D. B. 2000 Convective boundary layers in porous media: external flows. *Handbook of Porous Media* (K. Vafai, ed.), Marcel Dekker, New York, pp. 313–356. [5]

Pop, I. and Ingham, D. B. 2001 *Convection Heat Transfer: Mathematical and Computational Modelling of Viscous Fluids and Porous Media*, Elsevier, Oxford. [5]

Pop, I. and Merkin, J. H. 1995 Conjugate free convection on a vertical surface in a saturated porous medium. *Fluid Dyn. Res.* **16**, 71–86. [5.1.5]

Pop, I. and Na, N. Y. 1994 Natural convection of a Darcian fluid about a wavy cone. *Int. Comm. Heat Mass Transfer*, **21**, 891–899. [5.8]

Pop, I. and Na, T. Y. 1995 Natural convection over a frustum of a wavy cone in a porous medium. *Mech. Res. Comm.* **22**, 181–190. [5.8]

Pop, I. and Na, T. Y. 1997 Free convection from an arbitrarily inclined plate in a porous medium. *Heat Mass Transfer* **32**, 55–59. [5.3]

Pop, I. and Na, T. Y. 1998 Darcian mixed convection along slender vertical cylinders with variable surface heat flux embedded in a porous medium. *Int. Comm. Heat Mass Transfer* **25**, 251–260. [8.1.3]

Pop, I. and Na, T. Y. 2000 Conjugate free convection over a vertical slender hollow cylinder embedded in a porous medium. *Heat Mass Transfer* **36**, 375–379. [5.7]

Pop, I. and Nakayama, A. 1994 Conjugate free convection from long vertical plate fins in a non-Newtonian fluid-saturated porous medium. *Int. Comm. Heat Mass Transfer* **21**, 297–305. [5.12.1]

Pop, I. and Nakayama, A. 1999 Conjugate free and mixed convection heat transfer from vertical fins embedded in porous media. In *Recent advances in Analysis of Heat Transfer for Fin Type Surfaces* (B. Sunden and P. J. Heggs, eds.), Computational Mechanics Publications, Southampton, pp. 67–96. [5.12.1]

Pop, I. and Yan, B. 1998 Forced convection flow past a circular cylinder and a sphere in a Darcian fluid at large Peclet numbers. *Int. Comm. Heat Mass Transfer* **25**, 261–267. [4.3]

Pop, I., Angirasa, D. and Peterson, G. P., 1997 Natural convection in porous media near L-shaped corners. *Int. J. Heat Mass Transfer* **40**, 485–490. [5.12.2]

Pop, I., Cheng, P. and Le, T. 1989 Leading edge effects on free convection of a Darcian fluid about a semi-infinite vertical plate with uniform heat flux. *Int. J. Heat Mass Transfer* **32**, 493–501. [5.1.6]

Pop, I., Harris, S. D., Ingham, D. B. and Rhea, S. 2003 Three-dimensional stagnation point free convection on a reactive surface in a porous medium. *Int. J. Energy Res.* **27**, 919–939. [5.12.3]

Pop, I., Ingham D. B. and Bradean, R. 1996 Transient free convection about a horizontal circular cylinder in a porous medium with constant surface flux heating, *Acta Mech.* **119**, 79–91. [5.5.1]

Pop, I., Ingham, D. B. and Cheng, P. 1992a Transient natural convection in a horizontal concentric annulus filled with a porous medium. *ASME J. Heat Transfer* **114**, 990–997. [7.3.3]

Pop, I., Ingham, D. B. and Cheng, P. 1993a Transient free convection about a horizontal circular cylinder in a porous medium. *Fluid Dyn. Res.*, **12**, 295–305. [5.5.1]

Pop, I., Ingham, D. B. and Cheng, P. 1993b Transient free convection between two concentric spheres filled with a porous medium. *AIAA J. Thermophys. Heat Transfer* **7**, 724–727. [7.5]

Pop, I., Ingham, D. B. and Merkin, J. H. 1998 Transient convection heat transfer in a porous medium: External flows. *Transport Phenomena in Porous Media* (eds. D.B. Ingham and I. Pop), Elsevier, Oxford, pp. 205–232. [5.1.3]

Pop, I., Ingham, D. B. and Miskin, I. 1995a Mixed convection in a porous medium produced by a line heat source. *Transport in Porous Media* **18**, 1–13. [8.1.4]

Pop, I., Ingham, D. B., Heggs, P. J. and Gardner, D. 1986 Conjugate heat transfer from a downward projecting fin immersed in a porous medium. *Heat Transfer 1986*, Hemisphere, Washington, DC, **5**, 2635–2640. [5.12.1]

Pop, I., Kumari, M. and Nath, G. 1992b Free convection about cylinders of elliptic cross section embedded in a porous medium. *Int. J. Engng. Sci.* **30**, 35–45. [5.5.1]

Pop, I., Lesnic, D. and Ingham, D. B. 1995b Conjugate mixed convection on a vertical surface in a porous medium. *Int. J. Heat Mass Transfer* **38**, 1517–1525. [8.1.1]

Pop, I., Lesnic, D. and Ingham, D. B. 2000 Asymptotic solutions for the free convection boundary layer flow along a vertical surface in a porous medium with Newtonian heating. *Hybrid Methods in Engineering* **2**, 31–40. [5.1.9.8]

Pop, I., Merkin, J. H. and Ingham, D. B. 2002 Chemically driven convection in porous media. In *Transport Phenomena in Porous Media II* (D. B. Ingham and I. Pop, eds.) Elsevier, Oxford, pp. 341–364. [5.12.3]

Pop, I., Rees, D. A. S. and Egbers, C. 2004 Mixed convection flow in a narrow vertical duct filled with a porous medium. *Int. J. Therm. Sci.* **43**, 489–498. [8.3.2]

Pop, I., Rees, D. A. S. and Storesletten, L. 1998 Free convection in a shallow annular cavity filled with a porous medium. *Journal of Porous Media* **1**, 227–241. [7.3.3]

Pop, I., Sunada, J. K., Cheng, P. and Minkowycz, W. J. 1985 Conjugate free convection from long vertical plate fins embedded in a porous medium at high Rayleigh numbers. *Int. J. Heat Mass Transfer* **28**, 1629–1636. [5.12.1]

Postelnicu, A. 2004 Influence of a magnetic field on heat and mass transfer by natural convection from vertical surfaces in porous media considering Soret and Dufour effects. *Int. J. Heat Mass Transfer* **47**, 1467–1472. [9.2.1]

Postelnicu, A. and Pop, I. 1999 Similarity solutions of free convection boundary layers over vertical and horizontal surfaces in porous media with internal heat generation. *Int. Comm. Heat Mass Transfer* **26**, 1183–1191. [5.1.9.4]

Postelnicu, A. and Rees, D. A. S. 2001 The onset of convection in an anisotropic porous layer inclined at a small angle from the horizontal. *Int. Comm. Heat Mass Transfer* **28**, 641–650. [7.8]

Postelnicu, A. and Rees, D. A. S. 2003 The onset of Darcy-Brinkman convection in a porous layer using a thermal nonequilibrium model–Part I: Stress free boundaries. *Int. J. Energy Res.* **27**, 961–973. [6.5]

Postelnicu, A., Grosan, T. and Pop, I. 2000 Free convection boundary-layer over a vertical permeable flat plate in a porous medium with internal heat generation. *Int. Comm. Heat Mass Transfer* **27**, 729–738. [5.1.9.4]

Postelnicu, A., Grosan, T. and Pop, I. 2001 The effect of variable viscosity on forced convection flow past a horizontal flat plate in a porous medium with internal heat generation. *Mech. Res. Comm.* **28**, 331–337. [5.2]

Poulikakos, D. 1984 Maximum density effects on natural convection in a porous layer differentially heated in the horizontal direction. *Int. J. Heat Mass Transfer* **27**, 2067–2075. [7.3.5]

Poulikakos, D. 1985a On buoyancy induced heat and mass transfer from a concentrated source in an infinite porous medium. *Int. J. Heat Mass Transfer* **28**, 621–629. [9.3.1]

Poulikakos, D. 1985b Onset of convection in a horizontal porous layer saturated with cold water. *Int. J. Heat Mass Transfer* **28**, 1899–1905. [6.20]

Poulikakos, D. 1985c The effect of a third diffusing component on the onset of convection in a horizontal porous layer. *Phys. Fluids* **28**, 3172–3174. [9.1.6.4]

Poulikakos, D. 1986 Double diffusive convection in a horizontally sparsely packed porous layer. *Int. Comm. Heat Mass Transfer* **13**, 587–598. [9.1.6.3]

Poulikakos, D. 1987a Buoyancy driven convection in a horizontal fluid layer extending over a porous substrate. *Phys. Fluids* **29**, 3949–3957. [6.19.2]

Poulikakos, D. 1987b Thermal instability in a horizontal fluid layer superposed on a heat-generating porous bed. *Numer. Heat Transfer* **12**, 83–100. [6.19.2]

Poulikakos, D. and Bejan, A. 1983a Natural convection in vertically and horizontally layered porous media heated from the side. *Int. J. Heat Mass Transfer* **26**, 1805–1814. [7.3.2]

Poulikakos, D. and Bejan, A. 1983b Numerical study of transient high Rayleigh number convection in an attic-shaped porous layer. *ASME J. Heat Transfer* **105**, 476–484. [7.3.6]

Poulikakos, D. and Bejan, A. 1983c Unsteady natural convection in a porous layer. *Phys. Fluids* **26**, 1183–1191. [7.5]

Poulikakos, D. and Bejan, A. 1984a Natural convection in a porous layer heated and cooled along one vertical side. *Int. J. Heat Mass Transfer* **27**, 1879–1891. [7.4.3]

Poulikakos, D. and Bejan, A. 1984b Penetrative convection in porous medium bounded by a horizontal wall with hot and cold spots. *Int. J. Heat Mass Transfer* **27**, 1749–1757. [7.4.3]

Poulikakos, D. and Bejan, A. 1985 The departure from Darcy flow in natural convection in a vertical porous layer. *Phys. Fluids* **28**, 3477–3484. [7.6.1]

Poulikakos, D. and Kazmierczak, M. 1987 Forced convection in a duct partially filled with a porous material. *ASME J. Heat Transfer* **109**, 653–662. [4.9]

Poulikakos, D. and Renken, K. 1987 Forced convection in a channel filled with porous medium, including the effects of flow inertia, variable porosity, and Brinkman friction. *ASME J. Heat Transfer* **109**, 880–888. [4.9]

Poulikakos, D., Bejan, A., Selimos, B. and Blake, K. R. 1986 High Rayleigh number convection in a fluid overlying a porous bed. *Int. J. Heat Fluid Flow* **7**, 109–116. [6.19.2]

Powers, D., O'Neill, K. and Colbeck, S. C. 1985 Theory of natural convection in snow. *J. Geophys. Res.* **90**, 10641–10649. [11.1]

Prakash, K. and Kumar, N. 1999a Effects of suspended particles, rotation and variable gravity field on the thermal instability of Rivlin-Ericksen visco-elastic fluid in porous medium. *Indian J. Pure Appl. Math.* **30**, 1157–1166. [6.2.3]

Prakash, K. and Kumar, N. 1999b Thermal instability in Rivlin-Ericksen elastico-viscous fluid in the presence of finite Larmor radius and variable gravity in porous medium. *J. Phys. Soc. Japan* **68**, 1168–1172. [6.23]

Prasad, A. and Simmons, C. T. 2003 Unstable density-driven flow in heterogeneous porous media: A stochastic study of the Elder [1967b] "short heater" problem. *Water Resour. Res.* **39**, Art. No. 1007. [11.8]

Prasad, K. V., Abel, M. S., Khan, S. K. and Datti, P. S. 2002 Non-Darcy forced convective heat transfer in a viscoelastic fluid flow over a non-isothermal sheet. *J. Porous Media* **5**, 41–47. [4.16.3]

Prasad, V. 1986 Numerical study of natural convection in a vertical porous annulus with constant heat flux on the inner wall. *Int. J. Heat Mass Transfer* **29**, 841–853. [7.2]

Prasad, V. 1987 Thermal convection in a rectangular cavity filled with a heat-generating, Darcy porous medium. *ASME J. Heat Transfer* **109**, 697–703. [6.17]

Prasad, V. 1991 Convective flow interaction and heat transfer between fluid and porous layers. *Convective Heat and Mass Transfer in Porous Media* (eds. S. Kakaç, B. Kilkis, F. A. Kulacki and F. Arinç), Kluwer Academic, Dordrecht, pp. 563–615. [6.19.2]

Prasad, V. 1993 Flow instabilities and heat transfer in fluid overlying horizontal porous layers. *Exp. Therm. Fluid Sci.*, **6**, 135–146. [6.19.2]

Prasad, V. and Chui, A. 1989 Natural convection in a cylindrical porous enclosure with internal heat generation. *ASME J. Heat Transfer* **111**, 916–925. [6.17]

Prasad, V. and Kladias, N. 1991 Non-Darcy natural convection in saturated porous media. *Convective Heat and Mass Transfer in Porous Media* (eds. S. Kakaç, B. Kilkis, F. A. Kulacki and F. Arinç), Kluwer Academic, Dordrecht, pp. 173–224. [7.6.2]

Prasad, V. and Kulacki, F. A. 1984a Natural convection in a rectangular porous cavity with constant heat flux on one vertical wall. *ASME J. Heat Transfer* **106**, 152–157. [7.2]

Prasad, V. and Kulacki, F. A. 1984b Convective heat transfer in a rectangular porous cavity. Effect of aspect ratio on flow structure and heat transfer. *ASME J. Heat Transfer* **106**, 158–165. [7.1.3]

Prasad, V. and Kulacki, F. A. 1984c Natural convection in a vertical porous annulus. *Int. J. Heat Mass Transfer* **27**, 207–219. [7.3.3]

Prasad, V. and Kulacki, F. A. 1985 Natural convection in porous media bounded by short concentric vertical cylinders. *ASME J. Heat Transfer* **107**, 147–154. [7.3.3]

Prasad, V. and Kulacki, F. A. 1986 Effects of size of heat source on natural convection in horizontal porous layers heated from below. *Heat Transfer 1986*, Hemisphere, Washington, DC, vol. 5, pp. 2677–2682. [6.18]

Prasad, V. and Kulacki, F. A. 1987 Natural convection in horizontal fluid layers with localized heating from below. *ASME J. Heat Transfer* **109**, 795–796. [6.18]

Prasad, V. and Tian, Q. 1990 An experimental study of thermal convection in fluid-superposed porous layers heated from below. *Heat Transfer 1990*, Hemisphere, Washington, DC, vol. 5, pp. 207–212. [6.19.2]

Prasad, V. and Tuntomo, A. 1987 Inertial effects on natural convection in a vertical porous cavity. *Numer. Heat Transfer* **11**, 295–320. [7.6.1]

Prasad, V., Brown, K. and Tian, G. 1989a Flow visualization and heat transfer experiments in fluid-superposed porous layers heated from below. *ASME HTD* **117**, 75–83. [6.19.2]

Prasad, V., Brown, K. and Tian, Q. 1991 Flow visualization and heat transfer experiments in fluid-superposed packed beds heated from below. *Exp. Therm. Fluid Sci.* **4**, 12–24. [6.19.2]

Prasad, V., Kladias, N., Bandyopadhaya, A. and Tian, Q. 1989b Evaluation of correlations for stagnant thermal conductivity of liquid-saturated porous beds of spheres. *Int. J. Heat Mass Transfer* **32**, 1793–1796. [2.2.1]

Prasad, V., Kulacki, F. A. and Keyhani, M. 1985 Natural convection in porous media. *J. Fluid Mech.* **150**, 89–119. [6.9.2, 7.3.3, 7.6.2]

Prasad, V., Kulacki, F. A. and Kulkarni, A. V. 1986 Free convection in a vertical, porous annulus with constant heat flux on the inner wall—experimental results. *Int. J. Heat Mass Transfer* **29**, 713–723. [7.2]

Prasad, V., Lai, F. C. and Kulacki, F. A. 1988 Mixed convection in horizontal porous layers heated from below. *ASME J. Heat Transfer* **110**, 395–402. [8.2.2]

Prasad, V., Lauriat, G. and Kladias, N. 1992 Non-Darcy natural convection in a vertical porous cavity. *Heat and Mass Transfer in Porous Media* (eds. M. Quintard and M. Todorovic), Elsevier, Amsterdam, pp. 293–314. [7.6.2]

Prats, M. 1966 The effects of horizontal fluid flow on thermally induced convection currents in porous mediums. *J. Geophys. Res.* 71, 4835–4837. [6.10]

Prax, C., Sadat, H and Slagnac, P. 1996 Diffuse approximation method for solving natural convection in porous media. *Transport in Porous Media* **22**, 215–223. [2.6]

Prescott, P. J. and Incropera, F. P. 1995 The effect of turbulence on solidification of binary metal alloy with electromagnetic stirring. *ASME J. Heat Transfer* **117**, 716–724. [1.8, 10.2.3]

Prescott, P. J. and Incropera, F. P. 1996 Convection heat and mass transfer in alloy solidification. *Advances in Heat Transfer* **28**, 231–329. [10.2.3]

Pringle, S. E. and Glass, R. J. 2002 Double-diffusive finger convection: influence of concentration at fixed buoyancy ratio. *J. Fluid Mech.* **462**, 161–183. [9.1.3]

Pringle, S. E., Glass, R. J. and Cooper, C. A. 2002 Double-diffusive finger convection in a Hele-Shaw cell: An experiment exploring the evolution of concentration fields, length scales and mass transfer. *Transport Porous Media* **47**, 195–214. [9.1.6.4]

Prud'homme, M. and Bougherara, H. 2001 Linear stability of free convection in a vertical cavity heated by uniform heat fluxes. *Int. Comm. Heat Mass Transfer* **28**, 783–750. [7.2]

Prud'homme, M. and Jasmin, S. 2001 Component analysis of a steady inverse convection solution in a porous medium. *Int. Comm. Heat Mass Transfer* **28**, 911–921. [7.2]

Prud'homme, M. and Jasmin, S. 2003 Determination of a heat source in porous medium with convective mass diffusion by an inverse method. *Int. J. Heat Mass Transfer* **46**, 2065–2075. [9.3.1]

Prud'homme, M. and Jiang, H. 2003 Inverse determination of concentration in porous medium with thermosolutal convection. *Int. Comm. Heat Mass Transfer* **30**, 303–312. [9.2.3]

Prud'homme, M. and Nguyen, T. H. 2001 Solution of the inverse steady state convection problem in a porous medium by adjoint equations. *Int. Comm. Heat Mass Transfer* **28**, 11–21. [7.2]

Prud'homme, M., Nguyen, T. H. and Bougherara, H. 2003 Stability of convection flow in a horizontal fluid layer heated by uniform heat fluxes. *Int. Comm. Heat Mass Transfer* **30**, 163–172. [7.2]

Pu, W. L., Cheng, P. and Zhao, T. S. 1999 Mixed-convection heat transfer in vertical packed channels. *J. Thermophys. Heat Transfer* **13**, 517–521. [8.3.1]

Purushothaman, R., Ganapathy, R. and Hiremath, P. S. 1990 Free convection in an infinite porous medium due to a pulsating point heat source. *Z. Angew. Math. Mech.* **70**, 41–47. [5.11.2]

Qiao, Z. and Kaloni, P. N. 1997 Convection induced by inclined temperature gradient with mass flow in a porous medium. *ASME J. Heat Transfer* **119**, 366–370. [7.9]

Qiao, Z. and Kaloni, P. N. 1998 Non-linear convection in a porous medium with inclined temperature. *Int. J. Heat Mass Transfer* **41**, 2549–2552. [7.9]

Qin, Y. and Chadam, J. 1995 A nonlinear stability problem for ferromagnetic fluids in a porous medium. *Appl. Math. Lett.* **8**(2), 25–29. [6.21]

Qin, Y. and Chadam, J. 1996 Nonlinear convective stability in a porous medium with temperature-dependent viscosity and inertial drag. *Stud. Appl. Math.* **96**, 273–288. [6.7]

Qin, Y. and Kaloni, P. N. 1993 A nonlinear stability problem of convection in a porous vertical slab. *Phys. Fluids A* **5**, 2067–2069. [7.1.4]

Qin, Y. and Kaloni, P. N. 1994 Convective instabilities in anisotropic porous media. *Stud. Appl. Math.* **91**, 189–204. [6.12]

Qin, Y. and Kaloni, P. N. 1995 Nonlinear stability problem of a rotating porous layer. *Quart. Appl. Math.* **53**, 129–142. [6.22]

Qin, Y., Guo, J. L. and Kaloni, P.N. 1995 Double diffusive penetrative convection in porous media. *Int. J. Engn Sci.* **33**, 303–312. [9.1.6.4]

Quintard, M. and Prouvost, L. 1982 Instabilités de zones de diffusion thermique instationnaires en milieu poreux. *Int. J. Heat Mass Transfer* **25**, 37–44. [6.10]

Quintard, M. and Whitaker, S. 2000 Theoretical modeling of transport in porous media. *Handbook of Porous Media* (K. Vafai, ed.), Marcel Dekker, New York, pp. 1–52. [2.2.2]

Quintard, M. and Whitaker, S. 2005 Coupled, nonlinear mass transfer and heterogeneous reaction in porous media. *Handbook of Porous Media* (K. Vafai, ed.), 2nd ed., Taylor and Francis, New York, pp. 3–38. [3.4]

Quintard, M., Kaviany, M. and Whitaker, S. 1997 Two-medium treatment of heat results in porous media: Numerical results for effective properties. *Adv. Water Resources* **20**, 77–94. [2.4]

Rabadi, N. J. and Hamdan, E. M. 2000 Free convection from inclined permeable walls embedded in variable permeability porous media with lateral mass flux. *J. Petrol. Sci. Engng.* **26**, 241–251. [5.3]

Rabinowicz, M., Sempéré, J. C. and Genthon, P. 1999 Thermal convection in a vertical permeable slot: implications for hydrothermal circulation along mid-ocean ridges. *J. Geophys. Res.* **104**, 29275–29292. [6.15.2]

Rachedi, R. and Chikh, S. 2001 Enhancement of electronic cooling by insertion of foam materials. *Heat Mass Transfer* **37**, 371–378. [4.11]

Raffensperger, J. P. and Vlassopoulos, D. 1999 The potential for free and mixed convection in sedimentary basins. *Hydrol. J.* **7**, 505–520. [6.10]

Rahli, O., Tadrist, L., Miscevic, M. and Santini, R. 1997 Fluid flow through randomly packed monodisperse fibers: the Kozeny-Carman parameter analysis. *ASME J. Fluids Engng.* **119**, 188–192. [1.4.2]

Rahman, S. U. 1999 An experimental study on buoyancy-driven convective mass transfer from spheres embedded in saturated porous media. *Heat Mass Transfer* **35**, 487–491. [5.6.1]

Rahman, S. U. and Badr, H. M. 2002 Natural convection from a vertical wavy surface embedded in saturated porous media. *Ind. Engng. Chem. Res.* **41**, 4422–4429. [5.1.8]

Rahman, S. U., Al-Slaeh, M. A. and Sharma, R. N. 2000 An experimental study on natural convection from heated surfaces embedded in porous media. *Indust. Engng. Chem. Res.* **39**, 214–218. [5.1.8]

Rajamani, R., Srinivas, C., Nithiarasu, P. and Seetharamu, K. N. 1995 Convective heat transfer in axisymmetric porous bodies. *Int. J. Numer. Meth. Heat Fluid Flow* **5**, 829–837. [7.3.3]

Rajen, G. and Kulacki, F. A. 1987 Natural convection in a porous layer locally heated from below—a regional laboratory model for a nuclear waste repository. *ASME HTD* **67**, 19–26. [6.18]

Ramanathan, A. and Surendra, P. 2003 Effect of magnetic field on viscosity of ferroconvection in an anisotropic sparsely distributed porous medium. *Indian J. Pure Appl. Phys.* **41**, 522–526. [6.21]

Ramanathan, A. and Suresh, G. 2004 Effect of magnetic field dependent viscosity and anisotropy of porous medium on ferroconvection. *Int. J. Engng. Sci.* **42**, 411–425. [6.21]

Ramaniah, G. and Malarvizhi, G. 1990 Non-Darcy regime mixed convection on vertical plates in saturated porous media with lateral mass flux. *Acta Mech.* **81**, 191–200. [8.1.1]

Ramaniah, G. and Malarvizhi G. 1991 Note on exact solutions of certain non-linear boundary value problems governing convection in porous media. *Int. J. Non-linear Mech.* **26**, 345–347. [5.1.9.9, 5.2]

Ramaniah, G. and Malarvizhi G. 1994 Note on the boundary value problem arising in free convection in porous media. *Int. J. Engng. Sci.* **32**, 2011–2013. [5.1.9.9]

Ramaniah, G., Malarvizhi, G. and Merkin, J. H. 1991 A unified treatment of mixed convection on a permeable horizontal plate. *Wärme-Stoffübertrag.* **26**, 187–192. [8.1.2]

Ramazanov, M. M. 2000 Convection in a obliquely heated thin porous elliptic ring. *Fluid Dyn.* **35**, 910–917. [7.3.7]

Ramazanov, M. M. 2001 Convection in a obliquely heated thin porous elliptic ring. *Fluid Dyn.* **36**, 279–284. [9.1.6.4]

Ramesh, P. S. and Torrance, K. E. 1990 Stability of boiling in porous media. *Int. J. Heat Mass Transfer* **33**, 1895–1908. [10.3.1]

Ramesh, P. S. and Torrance, K. E. 1993 Boiling in a porous layer heated from below: effects of natural convection and a moving liquid/two-phase surface. *J. Fluid Mech.* **257**, 289–309. [10.3.1]

Ramirez, N. E. and Saez, A. E. 1990 The effect of variable viscosity on boundary-layer heat transfer in a porous medium. *Int. Comm. Heat Mass Transfer* **17**, 477–485. [4.2]

Ranganathan, R. and Viskanta, R. 1984 Mixed convection boundary-layer flow along a vertical surface in a porous medium. *Numer. Heat Transfer* **7**, 305–317. [8.1.1]

Ranjbar-Kani, A. A. and Hooman, K. 2004 Viscous dissipation effects on thermally developing forced convection in a porous medium: circular duct with isothermal wall. *Int. Comm. Heat Mass Transfer* **31**, 897–907. [4.13]

Rao, B. K. 2001 Heat transfer to power-law fluid flows through porous media. *J. Porous Media* **4**, 339–347. [4.16.3]

Rao, B. K. 2002 Internal heat transfer to power-law fluid flows through porous media. *Exper. Heat Transfer* **15**, 73–88. [4.16.3]

Rao, K.N. and Pop, I. 1994 Transient free convection in a fluid saturated porous medium with temperature dependent viscosity. *Int. Comm. Heat Mass Transfer* **21**, 573–591. [5.1.9.9]

Rao, Y. F. and Glakpe, E. K. 1992 Natural convection in a vertical slot filled with porous medium. *Int. J. Heat Fluid Flow* **31**, 97–99. [7.1.2]

Rao, Y. F. and Wang, B. X. 1991 Natural convection in vertical porous enclosures with internal heat generation. *Int. J. Heat Mass Transfer* **34**, 247–252. [7.3.3]

Rao, Y. F., Fukuda, K. and Hasegawa, S. 1987 Steady and transient analyses of natural convection in a horizontal porous annulus with the Galerkin method. *ASME J. Heat Transfer* **109**, 919–927. [7.3.3]

Rao, Y. F., Fukuda, K. and Hasegawa, S. 1988 Numerical study of three dimensional natural convection in a horizontal porous annulus with Galerkin method. *Int. J. Heat Mass Transfer* **31**, 698–707. [7.3.3]

Rappoldt, C. Pieters, G. J. J. M., Adema, E. B., Baaijens, G. J., Grootjans, A. P. and van Duijn, C. J. 2003 Buoyancy-driven flow in a peat moss layer as a mechanism for solute transport. *Proc. Nat. Acad. Sci.* **100**, 14937–14942. [6.11.3, 11.8]

Raptis, A. 1998 Radiation and free convection flow through a porous medium. *Int. Comm. Heat Mass Transfer* **25**, 289–295. [5.1.9.4]

Raptis, A. and Perdikis, C. 2004 Unsteady flow through a highly porous medium in the presence of radiation. *Transport Porous Media* **57**, 171–179 [5.1.9.4]

Raptis, A. and Tzivanidis, G. 1984 Unsteady flow through a porous medium with the presence of mass transfer. *Int. Comm. Heat Mass Transfer* **11**, 97–102. [9.2.1]

Raptis, A., Tzivanidis, G. and Kafousias, N. 1981 Free convection and mass transfer flow through a porous medium bounded by an infinite vertical limiting surface with constant suction. *Lett. Heat Mass Transfer* **8**, 417–424. [9.2.1]

Rashidi, F., Bahrami, A. and Soroush, H. 2000 Prediction of time required for onset of convection in a porous medium saturated with oil and a layer of gas underlying the oil. *J. Petrol. Sci. Engng.* **26**, 311–317. [6.11.3]

Rastogi, S. K. and Poulikakos, D. 1993 Double diffusion in a liquid layer under a permeable solid region. *Numer. Heat Transfer A* **24**, 427–449. [9.4]

Rastogi, S. K. and Poulikakos, D. 1995 Double diffusion from a vertical surface in a porous region. *Int. J. Heat Mass Transfer* **38**, 935–946. [9.2.1]

Rastogi, S. K. and Poulikakos, D. 1997 Experiments on double-diffusion in a composite system comprised of a packed layer of spheres and an underlying layer. *Heat Mass Transfer* **32**, 181–191. [9.4]

Ratish Kumar, B. V. 2000 A study of free convection induced by a vertical wavy surface with heat flux in a porous enclosure. *Numer. Heat Transfer A* **37**, 493–510. [6.14]

Ratish Kumar, B. V. and Shalini 2003 Natural convection in a thermally stratified wavy vertical porous enclosure. *Numer. Heat Transfer A* **43**, 753–776. [6.14]

Ratish Kumar, B. V. and Shalini 2004a Free convection in a thermally stratified non-Darcian porous enclosure. *J. Porous Media* **7**, 261–277. [7.3.7]

Ratish Kumar, B. V. and Shalini 2004b Double-diffusive natural convection in a stratified porous medium. *J. Porous Media* **7**, 279–288. [9.2.1]

Ratish Kumar, B. V. and Shalini 2004c Non-Darcy free convection induced by a vertical wavy surface in a thermally stratified porous medium. *Int. J. Heat Mass Transfer* **47**, 2353–2363. [6.14]

Ratish Kumar, B. V. and Shalini 2005 Combined influence of mass and thermal stratification on double-diffusion on non-Dracian natural convection from a wavy vertical wall to porous media. *ASME J. Heat Transfer* **127**, 637-674. [9.2.1]

Ratish Kumar, B. V. and Singh, P. 1998 Effect of thermal stratification on free convection in a fluid-saturated porous enclosure. *Numer. Heat Transfer A* **34**, 343–356. [5.1.4]

Ratish Kumar, B. V., Singh, P. and Bansod, V. J. 2002 Effect of thermal stratification on free convection in a fluid-saturated porous enclosure. *Numer. Heat Transfer A* **41**, 421–447. [9.2.2]

Ratish Kumar, B. V., Murthy, P. V. S. N. and Singh, P. 1998 Free convection heat-transfer from an isothermal wavy surface in a porous enclosure. *Int. J. Numer. Meth. Fluids* **28**, 633–661. [6.14]

Ratish Kumar, B. V., Singh, P. and Murthy, P. V. S. N. 1997 Effect of surface undulations on natural convection in a porous square cavity. *ASME J. Heat Transfer* **119**, 848–851. [6.14]

Ray, R. J., Krantz, W. B., Caine, T. N. and Gunn, R. D. 1983 A model for sorted ground regularity. *J. Glaciology* **29**, 317–337. [11.2]

Razi, Y. P., Maliwan, K. and Mojtabi, A. 2002 Two different approaches for studying the Horton-Rogers-Lapwood problem under the effect of vertical vibration. *Proc. 1st Int. Conf. Applications of Porous Media*. pp. 479–488. [6.24]

Razi, Y. P., Maliwan, K., Charrier-Mojtabi, M. C. and Mojtabi, A. 2005 Influence of vibrations on buoyancy induced convection in porous media. *Handbook of Porous Media* (ed. K. Vafai), 2nd ed., Taylor and Francis, New York, pp. 321–370. [6.24]

Reda, D. C. 1983 Natural convection experiments in a liquid-saturated porous medium bounded by vertical coaxial cylinders. *ASME J. Heat Transfer* **105**, 795–802. [6.16.1]

Reda, D. C. 1986 Natural convection experiments in a stratified liquid-saturated porous medium. *ASME J. Heat Transfer* **108**, 660–666. [7.3.3]

Reda, D. C. 1988 Mixed convection in a liquid-saturated porous medium. *ASME J. Heat Transfer* **110**, 147–154. [8.1.3, 8.3.4]

Rees, D. A. S. 1988 The stability of Prandtl-Darcy convection in a vertical porous layer. *Int. J. Heat Mass Transfer* **31**, 1529–1534. [7.1.4]

Rees, D. A. S. 1990 The effect of long-wavelength thermal modulations on the onset of convection in an infinite porous layer heated from below. *Q. J. Mech. Appl. Math.* **43**, 189–214. [6.14]

Rees, D. A. S. 1993 Numerical investigation of the nonlinear wave stability of vertical thermal boundary layer flow in a porous medium. *Z. Angew. Math. Phys.* **44**, 306–313. [5.1.9.9]

Rees, D. A. S. 1996a The effect of inertia on free convection from a horizontal surface embedded in a porous medium. *Int. J. Heat Mass Transfer* **39**, 3425–3430. [5.2]

Rees, D. A. S. 1996b The effect of inertia on the stability of convection in a porous layer heated from below. *J. Theor. Appl. Fluid Dyn.* **1**, 154–171. [6.4]

Rees, D. A. S. 1997a The effect of inertia on the onset of mixed convection in a porous layer heated from below. *Int. Comm. Heat Mass Transfer* **24**, 277–283. [6.9.3]

Rees, D. A. S. 1997b Three-dimensional free convective boundary layers in porous media induced by a heated surface with spanwise temperature gradient. *ASME J. Heat Transfer* **119**, 792–798. [5.2]

Rees, D. A. S. 1998 Thermal boundary layer instabilities in porous media: a critical review. *Transport Phenomena in Porous Media* (eds. D.B. Ingham and I. Pop), Elsevier, Oxford, pp. 233–259. [5.4]

Rees, D. A. S. 1999 Free convective boundary-layer flow from a heated surface in a layered porous medium *J. Porous Media* **2**, 39–58. [5.1.9.5]

Rees, D. A. S. 2000 The stability of Darcy-Bénard convection. *Handbook of Porous Media* (K. Vafai, ed.), Marcel Dekker, New York, pp. 521–588. [6.4]

Rees, D. A. S. 2001 Vortex instability from a near-vertical heated surface in a porous medium. I. Linear instability. *Proc. Roy. Soc. Lond. A* **457**, 1721–1734. [5.4]

Rees, D. A. S. 2002a Vortex instability from a near-vertical heated surface in a porous medium. II. Nonlinear evolution. *Proc. Roy. Soc. Lond. A* **458**, 1773–1782. [5.4]

Rees, D. A. S. 2002b The onset of Darcy-Brinkman convection in a porous layer: an asymptotic analysis. *Int. J. Heat Mass Transfer* **45**, 2213–2220. [6.6]

Rees, D. A. S. 2002c Recent advances in the instability of free convective boundary layers in porous media. In *Transport Phenomena in Porous Media II* (D. B. Ingham and I. Pop, eds.), Elsevier, Oxford, pp. 54–81. [5.4]

Rees, D. A. S. 2003 Vertical free convective boundary-layer flow in a porous medium using a thermal nonequilibrium model: elliptical effects. *Z. Angew. Math. Phys.* **54**, 437–448. [5.1.9.3]

Rees, D. A. S. 2004a Convection in a sidewall-heated porous cavity in the presence of viscous dissipation. In *Applications of Porous Media (ICAPM 2004)*, (eds. A. H. Reis and A. F. Miguel), Évora, Portugal, pp. 231–236. [7.6.2]

Rees, D. A. S. 2004b Nonlinear vortex instabilities in free convective boundary layers in porous media. In *Emerging Technologies and Techniques in Porous Media* (D. B. Ingham, A. Bejan, E. Mamut and I. Pop, eds), Kluwer Academic, Dordrecht, pp. 235–245. [5.4]

Rees, D. A. S. and Bassom, A. P. 1991 Some exact solutions for free convective flows over heated semi-infinite surfaces in porous media. *Int. J. Heat Mass Transfer* **34**, 1564–1567. [5.8]

Rees, D. A. S. and Bassom, A. P. 1993 The nonlinear non-parallel wave instability of boundary-layer flow induced by a horizontal heated surface in porous media. *J. Fluid Mech.* **253**, 267–295. [5.2]

Rees, D. A. S. and Bassom, A. P. 1994 The linear wave instability of boundary layer flow induced by a horizontal heated surface in porous media. *Int. Comm. Heat Mass Transfer* **21**, 143–150. [5.2]

Rees, D. A. S. and Bassom, A. P. 1998 The onset of convection in an inclined porous layer heated from below. *Heat Transfer 1998, Proc. 11th IHTC*, **4**, 497–502. [7.8]

Rees, D. A. S. and Bassom, A. P. 2000 Onset of Darcy-Bénard convection in an inclined layer heated from below. *Acta Mech.* **144**, 103–118. [7.8]

Rees, D. A. S. and Hossain, M. A. 1999 Combined effect of inertia and a spanwise pressure gradient on free convection from a vertical surface in porous media. *Numer. Heat Transfer A* **36**, 725–736. [5.1.7.2]

Rees, D. A. S. and Hossain, M. A. 2001 The effect of inertia on free convective plumes in porous media. *Int. Comm. Heat Mass Transfer* **28**, 1137–1142. [5.10.1.2]

Rees, D. A. S. and Lage, J. L. 1996 The effect of thermal stratification on natural convection in a vertical porous medium layer. *Int. J. Heat Mass Transfer* **40**, 111–121. [6.15.2, 7.1.4]

Rees, D. A. S. and Pop, I. 1994a A note on free convection along a vertical wavy surface in a porous medium. *ASME J. Heat Transfer* **116**, 505–508. [5.1.9.6]

Rees, D. A. S. and Pop, I. 1994b Free convection induced by a horizontal wavy surface in a porous medium. *Fluid Dyn. Res.* **14**, 151–166. [5.2]

Rees, D. A. S. and Pop, I. 1995a Free convection induced by a vertical wavy surface with uniform heat flux in a porous medium. *ASME J. Heat Transfer* **117**, 547–550. [5.1.9.6]

Rees, D. A. S. and Pop, I. 1995b Non-Darcy natural convection from a vertical wavy surface in a porous medium. *Transport in Porous Media* **20**, 223–234. [5.1.9.6]

Rees, D. A. S. and Pop, I. 1997 The effect of longitudinal surface waves on free convection from vertical surfaces in porous media. *Int. Comm. Heat Mass Transfer* **24**, 419–425. [5.1.9.6]

Rees, D. A. S. and Pop, I. 1999 Free convective stagnation-point flow in a porous medium using a thermal nonequilibrium model. *Int. Comm. Heat Mass Transfer* **26**, 945–954. [5.1.9.3, 5.12.3]

Rees, D. A. S. and Pop, I. 2000a Vertical free convection in a porous medium with variable permeability effects. *Int. J. Heat Mass Transfer* **43**, 2565–2571. [5.1.9.5]

Rees, D. A. S. and Pop, I. 2000b The effect of g-jitter on vertical free convection boundary-layer flow in porous media. *Int. Comm. Heat Mass Transfer* **27**, 415–424. [5.1.9.7]

Rees, D. A. S. and Pop, I. 2000c Vertical free convective boundary-layer flow in a porous medium using a thermal nonequilibrium model. *J. Porous Media* **3**, 31–44. [5.1.9.3]

Rees, D. A. S. and Pop, I. 2001 The effect of g-jitter on free convection near a stagnation point in a porous medium. *Int. J. Heat Mass Transfer* **44**, 877–883. [5.12.3]

Rees, D. A. S. and Pop, I. 2002 Comments on 'Natural convection from a vertical plate in a saturated porous medium : nonequilibrium theory' by A. A. Mohamad. *J. Porous Media* **5**, 225–227. [5.1.9.3]

Rees, D. A. S. and Pop, I. 2003 The effect of large-amplitude g-jitter vertical free convection boundary-layer flow in porous media. *Int. J. Heat Mass Transfer* **46**, 1097–1102. [5.1.9.7]

Rees, D. A. S. and Pop, I. 2005 Local thermal non-equilibrium in porous media convection. In *Transport Phenomena in Porous Media* III, (eds. D. B. Ingham and I. Pop) , Elsevier, Oxford, pp. 147–173. [5.1.9.3]

Rees, D. A. S. and Postelnicu, A. 2001 The onset of convection in an inclined anisotropic porous layer. *Int. J. Heat Mass Transfer* **44**, 4127–4138. [7.8]

Rees, D. A. S. and Riley, D. S. 1985 Free convection above a near horizontal semi-infinite heated surface embedded in a saturated porous medium. *Int. J. Heat Mass Transfer* **28**, 183–190. [5.3]

Rees, D. A. S. and Riley, D. S. 1986 Free convection in an undulating saturated porous layer. *J. Fluid Mech.* **166**, 503–530. [6.14]

Rees, D. A. S. and Riley, D. S. 1989a The effects of boundary imperfections on convection in a saturated porous layer: near-resonant wavelength excitation. *J. Fluid Mech.* **199**, 133–154. [6.14]

Rees, D. A. S. and Riley, D. S. 1989b The effects of boundary imperfections on convection in a saturated porous layer: non-resonant wavelength excitation. *Proc. Roy. Soc. London Ser. A* **421**, 303–339. [6.14]

Rees, D. A. S. and Riley, D. S. 1990 The three-dimensionality of finite-amplitude convection in a layered porous medium heated from below. *J. Fluid Mech.* **211**, 437–461. [6.13.2]

Rees, D. A. S. and Storesletten, L. 1995 The effect of anisotropic permeability on free convective boundary layer flow in porous media. *Transport in Porous Media* **19**, 79–92. [5.1.9.5]

Rees, D. A. S. and Storesletten, L. 2002 The linear stability of a thermal boundary layer with suction in an anisotropic porous medium. *Fluid Dyn. Res.* **30**, 155–168. [5.3]

Rees, D. A. S. and Tyvand, P. A. 2004a The Helmholtz equation for convection in two-dimensional porous cavities with conducting boundaries. *J. Engrg. Math.* **49**, 181–193. [6.15.1]

Rees, D. A. S. and Tyvand, P. A. 2004b Degeneracy and the time-dependent onset of convection in porous cavities with conducting boundaries. *Emerging Technologies and Techniques in Porous Media* (D. B. Ingham et al., eds.), Kluwer Academic, Dordrect, 459–467. [6.15.1]

Rees, D. A. S. and Tyvand, P. A. 2004c Oscillatory convection in a two-dimensional porous box with asymmetric lateral boundary conditions. *Phys. Fluids* **16**, 3706–3714. [6.15.1]

Rees, D. A. S. and Vafai, K. 1999 Darcy-Brinkman free convection from a heated horizontal surface. *Numer. Heat Transfer A* **35**, 191–204. [5.2]

Rees, D. A. S., Bassom, A. P. and Pop, I. 2003a Forced convection past a heated cylinder in a porous medium using a thermal non-equilibrium model: boundary layer analysis. *European J. Mech. B* **22**, 473–476. [5.1.9.3, 5.7]

Rees, D. A. S., Magyari, E. and Keller, B. 2003b The development of the asymptotic viscous dissipation profile in a vertical free convective boundary layer flow in a porous medium. *Transport Porous Media* **53**, 347–355. [5.1.9.4]

Rees, D. A. S., Magyari, E., Keller, B. 2005a Vortex instability of the asymptotic dissipation profile in porous media. *Transport Porous Media* **61**, 1–14. [5.1.9.4, 6.6]

Rees, D. A. S., Postelnicu, A. and Storesletten, L. 2005b The onset of Darcy-Forchheimer convection in inclined porous layers heated from below. *Int. Comm. Heat Mass Transfer*, to appear. [7.8]

Rees, D. A. S., Storesletten, L. and Bassom, A. P. 2002 Convective plume paths in anisotropic porous media. *Transport Porous Media* **49**, 9–25. [5.10.1.1]

Rehberg, I. and Ahlers, G. 1985 Experimental observation of a codimension-two bifurcation in a binary fluid mixture. *Phys. Rev. Lett.* **55**, 500–503. [9.1.3]

Renken, K. J . and Raich, M. R. 1996 Forced convection steam condensation experiments within thin porous coatings. *Int. J. Heat Mass Transfer* **39**, 2937–2945. [10.4]

Renken, K. J. and Aboye, M. 1993a Analysis of film condensation promotion within thin inclined porous coatings. *Int. J. Heat Fluid Flow* **14**, 48–53. [10.4]

Renken, K. J. and Aboye, M. 1993b Experiments on film condensation promotion within thin porous coatings. *Int. J. Heat Mass Transfer* **36**, 1347–1355. [10.4]

Renken, K. J. and Poulikakos, D. 1988 Experiment and analysis of forced convective heat transport in a packed bed of spheres. *Int. J. Heat Mass Transfer* **31**, 1399–1408. [4.9]

Renken, K. J. and Poulikakos, D. 1989 Experiments on forced convection from a horizontal heated plate in a packed bed of glass spheres. *ASME J. Heat Transfer* **111**, 59–65. [4.8]

Renken, K. J. and Poulikakos, D. 1990 Mixed convection experiments about a horizontal isothermal surface embedded in a water-saturated packed bed of spheres. *Int. J. Heat Mass Transfer* **33**, 1370-1373. [8.1.2]

Renken, K. J., and Meechan, K. 1995 Impact of thermal dispersion during forced convection condensation in a thin porous/fluid composite system. *Chem. Engn. Comm.* **131**, 189–205. [10.4]

Renken, K. J., Carneiro, M. J. and Meechan, K. 1994 Analysis of laminar forced convection condensation within porous coatings. *AIAA J. Thermophys. Heat Transfer* **8**, 303–308. [10.4]

Renken, K. J., Soltykiewicz, D. J. and Poulikakos, D. 1989 A study of laminar film condensation on a vertical surface with a porous coating. *Int. Comm. Heat Mass Transfer* **16**, 181–192. [10.4]

Rhee, S. J., Dhir, V. K. and Catton, I. 1978 Natural convection heat transfer in beds of inductively heated particles. *ASME J. Heat Transfer* **100**, 78–85. [6.11.2]

Riahi, D. N. 1983 Nonlinear convection in a porous layer with finite conducting boundaries. *J. Fluid Mech.* **129**, 153–171. [6.4]

Riahi, D. N. 1989 Nonlinear convection in a porous layer with permeable boundaries. *Int. J. Non-Linear Mech.* **24**, 459–463. [6.10]

Riahi, D. N. 1993a Preferred pattern of convection in a porous layer with a spatially non-uniform boundary temperature. *J. Fluid Mech.* **246**, 529–543. [6.14]

Riahi, D. N. 1993b Effect of rotation on the stability of the melt during the solidification of a binary alloy. *Acta Mech.* **99**, 95–101. [10.2.3]

Riahi, D. N. 1994 The effect of Coriolis force on nonlinear convection in a porous medium. *Int. J. Math. Math. Sci.* **17**, 515–536. [6.22]

Riahi, D. N. 1996 Modal package convection in a porous layer with boundary imperfections. *J. Fluid Mech.* **318**, 107–128. [6.14]

Riahi, D. N. 1997 Effects of centrifugal and Coriolis forces on chimney convection during alloy solidification. *J. Crystal Growth* **179**, 287–296. [10.2.3]

Riahi, D. N. 1998a On the structure of an unsteady convecting mushy layer. *Acta Mech.* **127**, 83–96. [10.2.3]

Riahi, D. N. 1998b High gravity convection in a mushy layer during alloy solidification. In *Nonlinear Instability, Chaos and Turbulence*, Vol. 1 (L. Debnath and D. N. Riahi, eds.) WIT Press, Boston, pp. 301–336. [10.2.3]

Riahi, D. N. 1999 Effect of surface corrugation on convection in a 3D finite box of fluid saturated porous material. *Theor. Comput. Fluid Dyn.* **13**, 189–208. [6.14]

Riahi, D. N. 2002a Effects of rotation on convection in a porous layer during alloy solidification. In *Transport Phenomena in Porous Media II* (D. B. Ingham and I. Pop, eds.) Elsevier, Oxford, pp. 316–340. [10.2.3]

Riahi, D. N. 2002b On nonlinear convection in mushy layers. Part 1. Oscillatory modes of convection. *J. Fluid Mech.* **467**, 331–359. [10.2.3]

Riahi, D. N. 2003a Nonlinear steady convection in rotating mushy layers. *J. Fluid Mech.* **485**, 279–306. [10.2.3]

Riahi, D. N. 2003b On stationary and oscillatory modes of flow instability during alloy solidification. *J. Porous Media* **6**, 177–188. [10.2.3]

Riahi, D. N. and Sayre, T. L. 1996 Effect of rotation on the structure of a convecting mushy layer. *Acta Mech.* **118**, 109–119. [10.2.3]

Ribando, R. J. and Torrance, K. E. 1976 Natural convection in a porous medium: effects of confinement, variable permeability, and thermal boundary conditions. *ASME J. Heat Transfer* **98**, 42–48. [6.13.1]

Ribando, R. J., Torrance, K. E. and Turcotte, D. L. 1976 Numerical models for hydrothermal circulation in the oceanic crust. *J. Geophys. Res.* **81**, 3007–3012. [11.6.1]

Richardson, L. and Straughan, B. 1993 Convection with temperature dependent viscosity in a porous medium; nonlinear stability and the Brinkman effect. *Atti Accad. Naz. Lincei* **4**, 223–230. [6.7]

Richardson, S. 1971 A model for the boundary condition of a porous material. Part 2. *J. Fluid Mech.* **49**, 327–336. [1.6]

Riley, D. S. 1988 Steady two-dimensional thermal convection in a vertical porous slot with spatially periodic boundary imperfections. *Int. J. Heat Mass Transfer* **31**, 2365–2380. [7.1.4]

Riley, D. S. and Rees, D. A. S. 1985 Non-Darcy natural convection from arbitrarily inclined heated surfaces in saturated porous media. *Q. J. Mech. Appl. Math.* **38**, 277–295. [5.12.2]

Riley, D. S. and Winters, K. H. 1989 Modal exchange mechanisms in Lapwood convection. *J. Fluid Mech.* **204**, 325–358. [6.15.1]

Riley, D. S. and Winters, K. H. 1990 A numerical bifurcation study of natural convection in a tilted two-dimensional porous cavity. *J. Fluid Mech.* **215**, 309–329. [7.8]

Riley, D. S. and Winters, K. H. 1991 Time-periodic convection in porous media: the evolution of Hopf bifurcations with aspect ratio. *J. Fluid Mech.* **223**, 457–474. [6.15.1]

Rionera, S. and Straughan, B. 1990 Convection in a porous medium with internal heat source and variable gravity effects. *Int. J. Engng. Sci.* **28**, 497–503. [6.11.2]

Roberts, P. H. and Loper, D. E. 1983 Towards a theory of the structure and evolution of a dendrite layer. *Stellar and Planetary Magnetism* (ed. A. M. Soward), Gordon and Braech, New York, 329–349. [10.2.3]

Roberts, P. H., Loper, D. E. and Roberts, M. F. 2003 Convective instability of a mushy layer: I. Uniform permeability. *Geophys. Astrophys. Fluid Dyn.* **97**, 97–134. [10.2.3]

Robillard, L. and Torrance, K. E. 1990 Convective heat transfer inhibition in an annular porous layer rotating at weak angular velocity. *Int. J. Heat Mass Transfer* **33**, 953–963. [7.3.3]

Robillard, L. and Vasseur, P. 1992 Quasi-steady state natural convection in a tilted porous layer. *Canad. J. Chem. Engng.* **70**, 1094–1100. [7.8]

Robillard, L., Wang, C. H. and Vasseur, P. 1988 Multiple steady states in a confined porous medium with localized heating from below. *Numer. Heat Transfer* **13**, 91–110. [6.18]

Robillard, L., Zhang, X. and Zhao, M. 1993 On the stability of a fluid-saturated porous medium contained in a horizontal circular cylinder. *ASME HTD* **264**, 49–55. [6.16.2]

Rocha, L. A. O., Neagu, M., Bejan, A. and Cherry, R. S. 2001 Convection with phase change during gas formation from methane hydrates via depressurization of porous layers. *J. Porous Media* **4**, 283–295. [10.1.7]

Rocamora, F. D. and de Lemos, M. J. S. 2000 Analysis of convective heat transfer for turbulent flow in saturated porous media. *Int. Comm. Heat Mass Transfer* **27**, 825–834. [1.8]

Rodrigues, J. F. and Urbano, J. M. 1999 Darcy-Stefan problem arising in freezing and thawing of saturated porous media. *Cont. Mech. Thermodyn.* **11**, 181–191. [10.2.1.2]

Rogers, J. A. 1992 Funicular and evaporation front regimes in convective drying of granular beds. *Int. J. Heat Mass Transfer* **35**, 469–480. [3.6]

Rohsenow, W. M. and Choi, H. Y. 1961 *Heat, Mass and Momentum Transfer*. Prentice-Hall, Englewood Cliffs, NJ. [4.5]

Rohsenow, W. M. and Hartnett, J. P. 1973 *Handbook of Heat Transfer*. McGraw-Hill, New York. [4.5]

Romero, L. A. 1994 Low or high Peclet number flow past a sphere in a saturated porous medium. *SIAM J. Appl. Math.* **54**, 42–71. [4.3]

Romero, L. A. 1995a Low or high Peclet number flow past a prolate spheroid in a saturated porous medium. *SIAM J. Appl. Math.* **55**, 952–974. [4.3]

Romero, L. A. 1995b Forced convection past a slender body in a saturated porous medium. *SIAM J. Appl. Math.* **55**, 975–985. [4.16.2]

Rosa, R. N., Reis, A. H. and Miguel, A. F. 2004 *Bejan's Structural Theory of Shape and Structure*, Évora Geophysical Center, University of Évora, Portugal. [4.18]

Rosenberg, N. D. and Spera, F. J. 1990 Role of anisotropic and/or layered permeability in hydrothermal convection. *Geophys. Res. Lett.* **17**, 235–238. [6.13.2]

Rosenberg, N. D. and Spera, F. J. 1992 Thermohaline convection in a porous medium heated from below. *Int. J. Heat Mass Transfer* **35**, 1261–1273. [9.1.5]

Rosenberg, N. D., Spera, F. J. and Haymon, R. M. 1993 The relationship between flow and permeability field in seafloor hydrothermal systems. *Earth Planet. Sci. Lett.* **116**, 135–153. [6.12]

Royer, J. J. and Flores, L. 1994 Two-dimensional natural convection in an anisotropic and heterogeneous porous medium with internal heat generation. *Int. J. Heat Mass Transfer* **37**, 1387–1399. [6.11.2]

Royer, P., Auriault, J. L. and Boutin, C. 1995 Contribution de l'homogénéisation à l'étude de la filtration d'un fluide en mileux poreux fracturé. *Rev. Inst. Franc. Petr.* **50**, 337–352. [1.9]

Rubin, H. 1974 Heat dispersion effect on thermal convection in a porous medium layer. *J. Hydrol.* **21**, 173–184. [2.2.3]

Rubin, H. 1975 Effect of solute dispersion on thermal convection in a porous medium layer. 2. *Water Resources Res.* **11**, 154–158. [9.1.6.1]

Rubin, H. 1976 Onset of thermohaline convection in a cavernous aquifer. *Water Resources Res.* **12**, 141–147. [9.1.6.1]

Rubin, H. 1981 Onset of thermohaline convection in heterogeneous porous media. *Israel J. Tech.* **19**, 110-117. [6.13.1, 9.1.6.2]

Rubin, H. 1982a Thermohaline convection in a nonhomogeneous aquifer. *J. Hydrol.* **57**, 307–320. [9.1.6.2]

Rubin, H. 1982b Application of the aquifer's average characteristics for determining the onset of thermohaline convection in a heterogeneous aquifer. *J. Hydrol.* **57**, 321–336. [9.1.6.2]

Rubin, H. and Roth, C. 1978 Instability of horizontal thermohaline flow in a porous medium layer. *Israel J. Tech.* **16**, 216–223. [9.1.6.1]

Rubin, H. and Roth, C. 1983 Thermohaline convection in flowing groundwater. *Adv. Water Resources* **6**, 146–156. [9.1.6.1]

Rubinstein, J. 1986 Effective equations for flow in random porous media with a large number of scales. *J. Fluid Mech.* **170**, 379–383. [1.5.3]

Rudraiah, N. 1984 Linear and non-linear magnetoconvection in a porous medium. *Proc. Indian Acad. Sci. (Math. Sci.)* **93**, 117–135. [6.21]

Rudraiah, N. 1988 Turbulent convection in porous media with non-Darcy effects. *ASME HTD* **96**, vol. 1, 747–754. [1.8]

Rudraiah, N. and Malashetty, M. S. 1990 Effect of modulation on the onset of convection in a sparsely packed porous layer. *ASME J. Heat Transfer* **112**, 685–689. [6.11.3]

Rudraiah, N. and Prasad, V. 1998 Effect of Brinkman boundary layer on the onset of Marangoni convection in a fluid-saturated porous layer. *Acta Mech.* **127**, 235–246. [6.19.3]

Rudraiah, N. and Siddheshwar, P. G. 1998 A weak nonlinear stability analysis of double-diffusive convection with cross-diffusion in a fluid-saturated porous medium. *Heat Mass Transfer* **33**, 287–293. [9.1.4]

Rudraiah, N. and Srimani, P. K. 1980 Finite-amplitude cellular convection in a fluid-saturated porous layer. *Proc. Roy. Soc. London Ser. A* **373**, 199–222. [6.3]

Rudraiah, N. and Vortmeyer, D. 1978 Stability of finite-amplitude and overstable convection of a conducting fluid through fixed porous bed. *Wärme Stoffübertrag.* **11**, 241–254. [6.21]

Rudraiah, N. and Vortmeyer, D. 1982 The influence of permeability and of a third diffusing component upon the onset of convection in a porous medium. *Int. J. Heat Mass Transfer* **25**, 457–464. [9.1.6.4]

Rudraiah, N., Kaloni, P. N. and Radhadevi, P. V. 1989 Oscillatory convection in a viscoelastic fluid through a porous layer heated from below. *Rheol. Acta*, **28**, 48–53. [6.23]

Rudraiah, N., Sheela, R. and Srimani, P. K. 1987 Mixed thermohaline convection in an inclined porous layer. *ASME HTD* **84**, 97–101. [9.6]

Rudraiah, N., Shivakumara, I. S. and Friedrich, R. 1986 The effect of rotation on linear and nonlinear double-diffusive convection in a sparsely packed, porous medium. *Int. J. Heat Mass Transfer* **29**, 1301–1317. [9.1.6.4]

Rudraiah, N., Siddheshwar, P. G. and Masuoka, T. 2003 Nonlinear convection in porous media: A review. *J. Porous Media* **6**, 1–32. [6.4]

Rudraiah, N., Srimani, P. K. and Friedrich, R. 1982a Finite amplitude convection in a two-component fluid saturated porous layer. *Int. J. Heat Mass Transfer* **25**, 715–722. [9.1.3]

Rudraiah, N., Veerappa, B. and Balachandra Rao, S. 1980 Effects of nonuniform thermal gradient and adiabatic boundaries on convection in porous media. *ASME J. Heat Transfer* **102**, 254–260. [6.11.2]

Rudraiah, N., Veerappa, B. and Balachandra Rao, S. 1982b Convection in a fluid-saturated porous layer with non-uniform temperature gradient. *Int. J. Heat Mass Transfer* **25**, 1147–1156. [6.11.2]

Ruth, D. and Ma, H. 1992 On the derivation of the Forchheimer equation by means of the averaging theorem. *Transport in Porous Media* **7**, 255–264. [1.5.2]

Ryland, D. K. and Nandakumar, K. 1992 Bifurcation study of convective heat transfer in porous media. Part II. Effect of tilt on stationary and nonstationary solutions. *Phys. Fluids A* **4**, 1945–1958. [6.17]

Saatdjian, E., Lam, R. and Mota, J. P. B. 1999 Natural convection heat transfer in the annular region between porous confocal ellipses. *Int. J. Numer. Meth. Fluids* **31**, 513–522. [7.3.3]

Saeid, N. H. 2004 Analysis of mixed convection in a vertical porous layer using non-equilibrium model. *Int. J. Heat Mass Transfer* **47**, 5619–5627. [8]

Saeid, N. H. 2005 Natural convection in porous cavity with sinusoidal bottom wall temperature variation. *Int. Comm. Heat Mass Transfer*, **32**, 454–463. [6.18]

Saeid, N. H. and Mohamad, A. A. 2005a Periodic free convection from a vertical plate in a saturated porous medium, nonequilibrium model. *Int. J. Heat Mass Transfer* **48**, 3855–3863. [5.1.9.7]

Saeid, N. H. and Mohamad, A. A. 2005b Natural convection in a porous cavity with spatial sidewall temperature variation. *Int. J. Numer. Meth. Heat Fluid Flow* **15**, 555–566. [7.2]

Saeid, N. H. and Pop, I. 2004a Transient free convection in a square cavity filled with a porous medium. *Int. J. Heat Mass Transfer* **47**, 1917–1924. [7.5]

Saeid, N. H. and Pop, I. 2004b Viscous dissipation effects on free convection in a porous cavity. *Int. Comm. Heat Mass Transfer* **31**, 723–732. [7.5]

Saeid, N. H. and Pop, I. 2004c Maximum density effects on natural convection from a discrete heater in a cavity filled with a porous medium. *Acta Mech.* **171**, 203–212. [7.3.5]

Saeid, N. H. and Pop, I. 2005a Non-Darcy natural convection in a square cavity filled with a porous medium. *Fluid Dyn. Res.* **36**, 35–43. [7.6.1]

Saeid, N. H. and Pop, I. 2005b Natural convection from a discrete heater in a square cavity filled with a porous medium. *J. Porous Media* **8**, 55–63. [7.3.7]

Saeid, N. H. and Pop, I. 2005c Mixed convection from two themal sources in vertical porous layer. *Int. J. Heat Mass Transfer* **48**, 4150-4160. [8.1.1]

Sáez, A. E., Perfetti, J. C. and Rusinek, I. 1991 Prediction of effective diffusivities in porous media using spatially periodic models. *Transport Porous Media* **11**, 187–199. [1.5.2]

Saffman, P. 1971 On the boundary condition at the surface of a porous medium. *Stud. Appl. Math.* **50**, 93–101. [1.6]

Saghir, M. Z. 1998 Heat and mass transfer in a multiporous cavity. *Int. Comm. Heat Mass Transfer* **25**, 1019–1030. [9.4]

Saghir, M. Z. and Islam, M. R. 1999 Double diffusive convection in dual-permeability dual-porosity porous media. *Int. J. Heat Mass Transfer* **42**, 437–454. [9.4]

Saghir, M. Z., Comi, P. and Mehrvar, M. 2002 Effects of interaction between Rayleigh and Marangoni convection in superposed fluid and porous layers. *Int. J. Thermal Sci.* **41**, 207–215. [6.19.3]

Saghir, M. Z., Jiang, C. G., Chacha, M., Khawaja, M. and Pan, S. 2005a Thermodiffusion in porous media. In *Transport Phenomena in Porous Media* III, (eds. D. B. Ingham and I. Pop) , Elsevier, Oxford, pp. 227–260. [9.1.4]

Saghir, M. Z., Mahendran, P. and Hennenberg, M. 2005b Marangoni and gravity driven convection in a liquid layer overlying a porous layer: lateral and bottom heating conditions. *Energy Sources* **27**, 151–171. [6.19.2, 6.19.3]

Saghir, M. Z., Nejad, M., Vaziri, H. H. and Islam, M. R. 2001 Modeling of heat and mass transfer in a fractured porous medium. *Int. J. Comput. Fluid Dyn.* **15**, 279–292. [1.9]

Sahimi, M. 1993 Flow phenomena in rocks: from continuum models to fractals, percolation, cellular automata, and simulated annealing. *Rev. Mod. Phys.* **65**, 1393–1534. [1.9]

Sahimi, M. 1995 *Flow and Transport in Porous Media and Fractured Rock.* VCH Verlagsgesellschaft, Weinheim. [1.9]

Sahraoui, M. and Kaviany, M. 1992 Slip and no-slip velocity boundary conditions at the interface of porous, plain media. *Int. J. Heat Mass Transfer* **35**, 927–943. [1.6]

Sahraoui, M. and Kaviany, M. 1993 Slip and no-slip temperature boundary conditions at the interface of porous, plain media: conduction. *Int. J. Heat Mass Transfer* **36**, 1019–1033. [2.4]

Sahraoui, M. and Kaviany, M. 1994 Slip and no-slip temperature boundary conditions at the interface of porous, plain media: convection. *Int. J. Heat Mass Transfer* **37**, 1029–1044. [2.4]

Saito, M. B. and de Lemos, M. J. S. 2005 Interfacial heat transfer coefficient in non-equilibrium convective transport in porous media. *Int. Comm. Heat Mass Transfer* **32**, 666–676. [2.2.2]

Salagnac, P., Glouannec, P. and Lecharpentier, D. 2004 Numerical modeling of heat and mass transfer in porous medium during combined hot air, infrared and microwaves drying. *Int. J. Heat Mass Transfer* **44**, 4479–4489. [3.6]

Salinger, A. G., Aris, R. and Derby, J. J. 1994a Finite element formulations for large-scale, coupled flows in adjacent porous and open fluid domains. *Int. J. Numer. Meth. Fluids* **18**, 1185–1209. [1.6]

Salinger, A. G., Aris, R. and Derby, J. J. 1994b Modeling the spontaneous ignition of coal stockpiles. *AIChE J.* **40**, 991–1004. [3.4]

Salt, H. 1988 Heat transfer across a convecting porous layer with flux boundaries. *Transport in Porous Media* **3**, 325–341. [6.3]

Sanchez, F., Higuera, F. J. and Medina, A. 2005 Natural convection in tilted cylindrical cavities embedded in rocks. *Phys. Rev. E* **71**, Art. No. 066308. [7.8]

Sanchez, F., Perez-Rosales, C. and Medina, A. 2005 Natural convection in symmetrically interconnected tilted layers. *J. Phys. Soc. Japan* **74**, 1170–1180. [7.7]

Sandner, H. 1986 Double diffusion effects in a cylindrical porous bed filled with salt water. *Heat Transfer 1986*, Hemisphere, Washington, DC, vol. 5, pp. 2623–2627. [9.4]

Sano, T. 1996 Unsteady forced and natural convection around a sphere immersed in a porous medium. *.J. Engng. Math.* **30**, 515–525. [5.6.2]

Sano, T. and Makizono, K. 1998 Unsteady mixed convection around a sphere in a porous medium at low Peclet numbers. *Fluid Dyn. Res.* **23**, 45–61. [8.1.3]

Sano, T. and Okihara, R. 1994 Natural convection around a sphere immersed in a porous medium at small Rayleigh numbers. *Fluid Dyn. Res.* **13**, 39–44. [5.6.2]

Saravanan, S. and Kandaswamy, P. 2003a Non-Darcian thermal stability of a heat generating fluid in a porous annulus. *Int. J. Heat Mass Transfer* **46**, 4863–4875. [6.17]

Saravanan, S. and Kandaswamy, P. 2003b Convection currents in a porous layer with a gravity gradient. *Heat Mass Transfer* **39**, 693–699. [7.9]

Saravanan, S. and Yamaguchi, H. 2005 Onset of centrifugal convection in a magnetic-fluid-saturated porous medium. *Phys. Fluids* **17**, Art. No. 084105. [7.12]

Sarkar, A. and Phillips, O. M. 1992a Effects of horizontal gradients on thermohaline instabilities in infinite porous media. *J. Fluid Mech.* **242**, 79–98. [9.5]

Sarkar, A. and Phillips, O. M. 1992b Effects of horizontal gradients on thermohaline instabilities in a thick porous layer. *Phys. Fluids A* **4**, 1165–1175. [9.5]

Sarler, B. 2000 DRBEM solution of porous media natural convection problems with internal heat generation. *Z. Angew. Math. Mech.* **80**, S703–S704, Supp. 3. [2.9, 7.3.8]

Sarler, B., Gobin, D., Goyeau, B., Perko, J. and Power, H. 2000a Natural convection in porous media—dual reciprocity boundary element method solution of the Darcy model. *Int. J. Numer. Meth. Fluids* **33**, 279–312. [7.3.8]

Sarler, B., Perko, J. and Chen, C. S. 2004a Radial basis function collocation method of solution of natural convection in porous media. *Int. J. Numer. Meth. Heat Fluid Flow* **14**, 187–212. [2.9, 7.3.8]

Sarler, B., Perko, J., Gobin, D., Goyeau, B. and Power, H. 2000b Dual reciprocity boundary element method of natural convection in Darcy-Brinkman porous media. *Engng.. Anal. Boundary Elements* **28**, 23–41. [2.9, 7.3.8]

Sarler, B., Perko, J., Gobin, D., Goyeau, B. and Power, H. 2004b Dual reciprocity boundary element method of natural convection in Darcy-Brinkman porous media. *Engng. Anal. Boundary Elem.* **28**, 23–41. [2.9, 7.3.8]

Sasaguchi, K. 1995 Effect of density inversion of water on the melting process of frozen porous media. *Proc. ASME/JSME Thermal Engineering Joint Conf.*,. vol. 3, pp.371–378. [10.1.7]

Sasaki, A., and Aiba, S. 1992 Freezing heat transfer in water-saturated porous media in a vertical rectangular vessel. *Wärme-Stoffübertrag.* **27**, 289–298. [10.2.1.2]

Sasaki, A., Aiba, S. and Fukusako, S. 1990 Numerical study on freezing heat transfer in water-saturated porous media. *Numer. Heat Transfer A* **18**, 17–32. [10.2.1.2]

Sathe, S. B. and Tong, T. W. 1988 Measurements of natural convection in partially porous rectangular enclosures of aspect ratio 5. *Int. Comm. Heat Mass Transfer* **15**, 203–212. [7.7]

Sathe, S. B. and Tong, T. W. 1989 Comparison of four insulation schemes for reduction of natural convective heat transfer in rectangular enclosures. *Int. Comm. Heat Mass Transfer* **16**, 795–802. [7.7]

Sathe, S. B., Lin, W. Q. and Tong, T. W. 1988 Natural convection in enclosures containing an insulation with a permeable-fluid-porous interface. *Int. J. Heat Fluid Flow* **9**, 389–395. [7.7]

Sathe, S. B., Tong, T. W. and Faruque, M. A. 1987 Experimental study of natural convection in a partially porous enclosure. *AIAA J. Thermophys. Heat Transfer* **1**, 260–267. [7.7]

Satik, C., Parlar, M. and Yortsos, Y. C. 1991 A study of steady-state, steam-water counterflow in porous media. *Int. J. Heat Mass Transfer*, **34**, 1755–1771. [11.9.1,11.9.2]

Sattar, M. D. A. 1993 Free and forced convection boundary flow through a porous medium with large suction. *Int. J. Energy Res.* **17**, 1–7. [4.162]

Satya Sai, B. V. K., Seetharamu, K. N. and Aswathanarayana, P. A. 1997a Finite element analysis of heat transfer by natural convection in porous media in vertical enclosures: Investigations in Darcy and non-Darcy regimes. *Int. J. Numer. Methods Heat Fluid Flow* **7**, 367–400. [7.3.3]

Satya Sai, B. V. K., Seetharamu, K. N. and Aswathanarayana, P. A. 1997b *Int. J. Numer. Meth. Heat Fluid Flow* **7**, 367–400. [7.6.2]

Sayre, T. L. and Riahi, D. N. 1996 Effect of rotation on flow instabilities during solidification of a binary alloy. *Int. J. Engng. Sci.* **34**, 1631–1645. [10.2.3]

Sayre, T. L. and Riahi, D. N. 1997 Oscillatory instabilities of the liquid and mushy layers during solidification of alloys under rotational constraint. *Acta Mech.* **121**, 143–152. [10.2.3]

Scheidegger, A. E. 1974 *The Physics of Flow through Porous Media*, University of Toronto Press, Toronto. [1.2]

Schincariol, R. A., Schwartz, F. W. and Mendoza, C. A. 1997 Instabilities in variable density flows: stability analyses for homogeneous and heterogeneous media. *Water Resour. Res.* **33**, 31–41. [11.8]

Schneider, K. J. 1963 Investigation on the influence of free thermal convection on heat transfer through granular material. *Proc. 11th Int. Cong. of Refrigeration*, Pergamon Press, Oxford, Paper 11–4, 247–253. [6.9.1, 6.9.2, 7.1.2]

Schneider, M. C. and Beckermann, C. 1995 A numerical study of the combined effects of microsegregation, mushy zone permeability and flow, caused by volume contraction and thermosolutal convection, on macrosegregation and eutectic formation in binary alloy solidification. *Int. J. Heat Mass Transfer* **38**, 3455–3473. [10.2.3]

Schoofs, S. and Hansen, U. 2000 Depletion of a brine layer at the base of ridge-crest hydrothermal systems. *Earth Planet. Sci. Lett.* **180**, 341–353. [11.8]

Schoofs, S. and Spera, F. J. 2003 Transition to chaos and flow dynamics of thermochemical porous medium convection. *Transport Porous Media* **50**, 179–195. [9.1.3]

Schoofs, S., Spera, F. J. and Hansen, U. 1999 Chaotic thermohaline convection in low-porosity hydrothermal systems. *Earth Planet. Sci. Lett.* **174**, 213–229. [9.1.3]

Schoofs, S., Trompert, R. A. and Hansen, U. 1998 The formation and evolution of layered structures in porous media. *J. Geophys. Res.* **103**, 20843–20858. [11.8]

Schoofs, S., Trompert, R. A. and Hansen, U. 2000a The formation and evolution of layered structures in porous media: effects of porosity and mechanical dispersion. *Phys. Earth Planet. Inter.* **118**, 205–225. [11.8]

Schoofs, S., Trompert, R. A. and Hansen, U. 2000b Thermochemical convection in and between intracratonic basins: Onset and effects. *J. Geophys. Res.* **105**, 25567–25585. [11.8]

Schöpf, W. 1992 Convection onset for a binary mixture in a porous medium and in a narrow cell: a comparison. *J. Fluid Mech.* **245**, 263–278. [2.5]

Schubert, G. and Straus, J. M. 1977 Two-phase convection in a porous medium. *J. Geophys. Res.* **82**, 3411–3421. [10.3.1]

Schubert, G. and Straus, J. M. 1979 Three-dimensional and multicellular steady and unsteady convection in fluid-saturated porous media at high Rayleigh numbers. *J. Fluid Mech.* **94**, 25–38. [6.8, 6.15.1]

Schubert, G. and Straus, J. M. 1980 Gravitational stability of water over steam in vapor-dominated geothermal systems. *J. Geophys. Res.* **85**, 6505–6512. [10.3.1]

Schubert, G. and Straus, J. M. 1982 Transitions in time-dependent thermal convection in fluid-saturated porous media. *J. Fluid Mech.* **121**, 301–303. [6.15.1]

Schulenberg, T. and Müller, U. 1984 Natural convection in saturated porous layers with internal heat sources. *Int. J. Heat Mass Transfer* **27**, 677–685. [6.19.2]

Schulze, T. P. and Worster, M. G. 1998 A numerical investigation of steady convection in mushy layers during the directional solidification of binary alloys. *J. Fluid Mech.* **356**, 199–220. [10.2.3]

Schulze, T. P. and Worster, M. G. 1999 Weak convection, liquid inclusions and the formation of chimneys in mushy layers. *J. Fluid Mech.* **388**, 197–215. [10.2.3]

Scurtu, N. D., Postelnicu, A. and Pop, I. 2001 Free convection between two horizontal cylinders filled with a porous medium—a perturbation solution. *Acta Mech.* **151**, 115–125. [7.3.3]

Seddeek, M. A. 2002 Effects of magnetic field and variable viscosity on forced non-Darcy flow about a flat plate with variable wall temperature in porous media in the presence of suction and blowing. *J. Appl. Mech. Tech. Phys.* **43**, 13–17. [4.16.1]

Seddeek, M. A. 2005 Effects of non-Darcian on forced convection heat transfer over a flat plate in a porous medium with temperature dependent viscosity. *Int. Comm. Heat Mass Transfer* **32**, 258–265. [4.16.1]

Seetharamu, K. N. and Dutta, P. 1990 Free convection in a saturated porous medium adjacent to a non-isothermal vertical impermeable wall. *Wärme-Stoffübertrag.* **25**, 9–15. [5.1.9.9]

Seguin, D., Montillet, A., Comiti, J. and Huet, F. 1998 Experimental characterization of flow regimes in various porous media—II: Transition to turbulent regime. *Chem. Engng. Sci.* **53**, 3897–3909. [1.8]

Sekar, R. and Vaidyanathan, G. 1993 Convective instability of a magnetized ferrofluid in a rotating porous medium. *Int. J. Engng. Sci.* **31**, 1139–1150. [6.2.1]

Sekar, R., Ramanathan, A. and Vaidyanathan, G. 1998 Effect of rotation on ferrothermohaline convection saturating a porous medium. *Indian J. Eng. Mater. Sci.* **5**, 445–452. [9.1.6.4]

Selimos, B. and Poulikakos, D. 1985 On double diffusion in a Brinkman heat generating porous layer. *Int. Comm. Heat Mass Transfer* **12**, 149–158. [9.1.6.4]

Sen, A. K. 1987 Natural convection in a shallow porous cavity–the Brinkman model. *Int. J. Heat Mass Transfer* **30**, 855–868. [7.6.2]

Sen, M., Vasseur, P. and Robillard, L. 1987 Multiple steady states for unicellular natural convection in an inclined porous layer. *Int. J. Heat Mass Transfer* **30**, 2097–2113. [7.8]

Sen, M., Vasseur, P. and Robillard, L. 1988 Parallel flow convection in a tilted two-dimensional porous layer heated from all sides. *Phys. Fluids* **31**, 3480–3487. [7.8]

Serkitjis, M. 1995 Natural convection heat transfer in a horizontal thermal insulation layer underlying an air layer. PhD thesis, Chalmers University of Technology, Göteburg. [6.19.2]

Sezai, I. 2005 Flow patterns in a fluid-saturated porous cube. *J. Fluid Mech.* **523**, 393–410. [6.15.1]

Sezai, I. and Mohamad, A. A. 1999 Three-dimensional double-diffusive convection in a porous cubic enclosure due to opposing gradients of temperature and concentration. *J. Fluid Mech.* **400**, 333–353. [9.2.2]

Shah, C. B. and Yortsos, Y. C. 1995 Aspects of flow of power-law fluids in porous media. *AIChE J.* **41**, 1099–1112. [1.5.4]

Shankar, V. and Hagentoft, C. E. 2000 Numerical investigations of natural convection in horizontal porous media heated from below: comparison with experiments. *J. Thermal Envel. Build. Sci.* **23**, 318–338. [7.3.7]

Sharma, P. K. 2005 Simultaneous thermal and mass diffusion on a three-dimensional mixed convection flow through a porous medium. *J. Porous Media* **8**, 409–417. [9.6]

Sharma, R. C. and Gupta, U. 1995 Thermal convection in micropolar fluids in porous medium. *Int. J. Engng. Sci.* **33**, 1887–1892. [6.23]

Sharma, R. C. and Kango, S. K. 1999 Thermal convection in Rivlin-Ericksen elastico-viscous fluid in porous medium in hydromagnetics. *Czech. J. Phys.* **49**, 197–203. [6.23]

Sharma, R. C. and Kumar, P. 1996 Hall effect on thermosolutal instability in a Maxwellian viscoelastic fluid in porous medium. *Arch. Mech.* **48**, 199–209. [9.1.6.4]

Sharma, R. C. and Kumar, P. 1998 Effect of rotation on thermal convection in micropolar fluids in porous medium. *Indian J. Pure Appl. Math.* **29**, 95–104. [6.22]

Sharma, R. C. and Sharma, M. 2004 On couple-stress fluid permeated with suspended particles heated and soluted from below in porous medium. *Indian J. Phys. B* **78**, 189–194. [9.1.6.4]

Sharma, R. C. and Thakur, K. D. 2000 On couple-stress fluid heated from below in porous medium in hydromagnetics. *Czech. J. Phys.* **50**, 753–758. [9.1.6.4]

Sharma, R. C., Sunil and Chand, S. 1998 Thermosolutal instability of Rivlin-Ericksen rotating fluid in porous medium. *Indian J. Pure Appl. Math.* **29**, 433–440. [9.1.6.4]

Sharma, R. C., Sunil and Chand, S. 1999a Thermosolutal instability of Walters rotating fluid (model B′) in porous medium. *Arch. Mech.* **51**, 181–191. [9.1.6.4]

Sharma, R. C., Sunil and Chandel, R. S. 1999b Thermal convection in Walters viscoelastic fluid B′ permeated with suspended particles through porous medium. *Stud. Geotech. Mech.* **21**, 3–14. [9.1.6.4]

Sharma, R. C., Sunil and Pal, M. 2001 Thermosolutal convection in Rivlin-Ericksen rotating fluid in porous medium in hydromagnetics. *Indian J. Pure Appl. Math.* **32**, 143–156. [9.1.6.4]

Sharma, V. and Kishor, K. 2001 Hall effect on thermosolutal instability of Rivlin-Erikson fluid with varying gravity field in porous medium. *Indian J. Pure Appl. Math.* **32**, 1643–1657. [9.1.6.4]

Sharma, V. and Rana, G. C. 2001 Thermal instability of a Walters (model B′) elastico-viscous fluid in the presence of a variable gravity field and rotation in porous medium. *J. Non-Equil. Thermodyn.* **26**, 31–40. [9.1.6.4]

Sharma, V. and Rana, G. C. 2002 Thermosolutal instability of Walters (model B′) visco-elastic rotating fluid permeated with suspended particles and variable gravity field in porous medium. *Indian J. Pure Appl. Math.* **33**, 97–109. [9.1.6.4]

Sharma, V. and Sharma, S. 2000 Thermosolutal convection of micropolar fluids in hydromagnetics in porous medium. *Indian J. Pure Appl. Math.* **31**, 1353–1367. [9.1.6.4]

Shattuck, M. D., Behringer, R. P., Johnson, G. A. and Georgiadis, J. G. 1997 Convection and flow in porous media. Part 1. Visualization by magnetic resonance imaging. *J. Fluid Mech.* **332**, 215–245. [6.9.1]

Shavit, U., Bar-Yosef, G., Rosenzweig, R. and Assouline, S. 2002 Modified Brinkman equation for a free flow problem at the interface of porous surfaces: The Cantor-Taylor brush configuration case. *Water Resour. Res.* **38**, 1320–1334. [1.6]

Shavit, U., Rosenzweig, R. and Assouline, S. 2004 Free flow at the interface of porous surfaces: A generalization of the Taylor brush configuration. *Transport Porous Media.* **54**, 345–360. [1.6]

Shen, S., Liu, W. and Tao, W. Q. 2003 Analysis of field synergy on natural convective heat transfer in porous media. *Int. Comm. Heat Mass Transfer* **30**, 1081–1090. [3.6]

Shenoy, A. V. 1992 Darcy natural, forced and mixed convection heat transfer from an isothermal vertical flat plate in a porous medium saturated with an elastic fluid of constant viscosity. *Int. J. Engng. Sci.* **30**, 455–467. [4.12.3, 5.1.9.2, 8.1.1]

Shenoy, A. V. 1993a Darcy-Forchheimer natural, forced and mixed convection heat transfer in non-Newtonian power-law fluid-saturated porous media. *Transport in Porous Media* **11**, 219–241. [4.16.3, 5.1.9.2, 8.1.1]

Shenoy, A. V. 1993b Forced convection heat transfer to an elastic fluid of constant viscosity flowing through a channel filled with a Brinkman-Darcy medium. *Wärme-Stoffübertrag.* **28**, 295–297. [4.16.3]

Shenoy, A. V. 1994 Non-Newtonian fluid heat transfer in porous media. *Adv. Heat Transfer* **24**, 101–190. [1.5.4]

Sheridan, J., Williams, A. and Close, D. J. 1992 Experimental study of natural convection with coupled heat and mass transfer in porous media. *Int. J. Heat Mass Transfer* **35**, 2131–2143. [9.1.5]

Shih, M. H. and Huang, M. J. 2002 A study of liquid evaporation on forced convection in porous media with non-Darcy effects. *Acta Mech.* **154**, 215–231. [4.16.2]

Shih, M. H., Huang, M. J. and Chen, C. K. 2005 A study of the liquid evaporation with Darcian resistance effect on mixed convection in porous media. *Int. Comm. Heat Mass Transfer* **32**, 685–694. [10.3.2]

Shim, K. I., Yoo, J. W. and Kim, S. J. 2002 Thermal analysis of an internally finned tube using a porous medium approach. *Heat Transfer 2002, Proc. 12th Int. Heat Transfer Conf.*, Elsevier, Vol. 2, pp. 785–790. [4.16.2]

Shin, U. C., Khedari, J. Mbow, C. and Daguenet, M. 1994 Convection naturelle thermique a l'intérieur d'une calotte cylindrique poreuse d'axe horizontal: Etude théorique. *Rev. Gen. Therm.* **33**, 30–37. [7.3.7]

Shiralkar, G. S., Haadjizadeh, M. and Tien, C. L. 1983 Numerical study of high Rayleigh number convection in a vertical porous enclosure. *Numer. Heat Transfer A* **6**, 223–234. [7.1.2]

Shivakumara, I. S. 1999 Boundary and inertia effects on convection in porous media with throughflow. *Acta Mech.* **137**, 151–165. [6.10]

Shivakumara, I. S. and Khalili, A. 2001 On the stability of double diffusive convection in a porous layer with throughflow. *Acta Mech.* **152**, 1–8. [9.1.6.4]

Shivakumara, I. S. and Sumithra, R. 1999 Non-Darcian effects of double diffusive convection in a sparsely packed porous medium. *Acta Mech.* **132**, 113–127. [9.1.6.3]

Shivakumara, I. S., Prasana, B. M. R., Rudraiah, N. and Venkatachalappa, M. 2002 Numerical study of natural convection in a vertical cylindrical annulus using a non-Darcy equation. *J. Porous Media*, **5**, 87–102. [7.3.3]

Shu, J. J. and Pop, I. 1997 Inclined wall plumes in porous media. *Fluid Dyn. Res.* **21**, 303–317. [5.3]

Shu, J. J. and Pop, I. 1998 Transient conjugate free convection from a vertical flat plate in a porous medium subjected to a sudden change in surface heat flux. *Int. J. Engng. Sci.* **36**, 207–214. [5.1.5]

Shu, J. J. and Pop, I. 1999 Thermal interaction between free convection and forced convection along a conducting plate embedded in a porous medium. *Hybrid Meth. Engng.* **1**, 55–66. [8.1.1]

Siddheshwar, P. G. and Krisna, C. V. S. 2003 Linear and non-linear analyses of convection in a micropolar fluid occupying a porous medium. *Int J. Nonlinear Mech.* **38**, 1561–1579. [6.23]

Silva, R. A. and de Lemos, M. J. S. 2003a Numerical analysis of the stress jump interface condition for laminar flow over a porous layer. *Numer. Heat Transfer A* **43**, 603–617. [1.6]

Silva, R. A. and de Lemos, M. J. S. 2003b Turbulent flow in a channel occupied by a porous layer considering the stress jump at the interface. *Int. J. Heat Mass Transfer* **46**, 5113–5136. [1.8]

Simmons, C. T. and Narayan, K. A. 1997 Mixed convection processes below a saline disposal basin. *J. Hydrol.* **194**, 263–285. [11.8]

Simmons, C. T., Fenstemaker, T. R. and Sharp, J. M. 2001 Variable-density groundwater flow and solute transport in heterogeneous porous media: approaches, resolutions and future challenges. *J. Contam. Hydrol.* **52**, 245–275. [11.8]

Simmons, C. T., Narayan, K. A. and Wooding, R. A. 1999 On a test case for density-dependent groundwater flow and solute transport models: The salt lake problem. *Water Resources Res.* **35**, 3607–3620. [9.1.6.4]

Simms, M. A. and Garven, G. 2004 Thermal convection in faulted extensional sedimentary basins: Theoretical results from finite-element modeling. *Geofluids* **4**, 109–130. [11.8]

Simpkins, P. G. and Blythe, P. A. 1980 Convection in a porous layer. *Int. J. Heat Mass Transfer* **23**, 881–887. [7.1.2]

Singh, A. K. and Thorpe, G. R. 1995 Natural convection in a confined fluid overlying a porous layer—a study of different models. *Indian J. Pure Appl. Math.* **26**, 81–95. [7.7]

Singh, A. K., Leonardi, E. and Thorpe, G. R. 1993 Three-dimensional natural convection in a confined fluid overlying a porous layer. *ASME J. Heat Transfer*, **115**, 631–638.[7.7]

Singh, A. K., Paul, T. and Thorpe, G. R. 1999 Natural convection due to heat and mass transfer in a composite system. *Heat Mass Transfer* **35**, 39–48. [9.4]

Singh, A. K., Paul, T. and Thorpe, G. R. 2000 Natural convection in a non-rectangular porous enclosure. *Forsch. Ingenieurwesen* **65**, 301–308. [7.3.7]

Singh, P. and Tewari, K. 1993 Non-Darcy free convection from vertical surfaces in thermally stratified porous media. *Int. J. Engng. Sci.* **31**, 1233–1242. [5.1.4]

Singh, P., Misra, J. K. and Narayan, K. A. 1988 Three-dimensional convective flow and heat transfer in a porous medium. *Indian J. Pure Appl. Math.* **19**, 1130–1135. [5.1.9.9]

Singh, P., Queeny and Sharma, R. N. 2002 Influence of lateral mass flux on mixed convection heat and mass transfer over inclined surfaces in porous media. *Heat Mass Transfer* **38**, 233–242. [9.6]

Sinha, S. K., Sundararajan, T. and Garg, V. K. 1992 A variable property analysis of alloy solidification using the anisotropic porous medium approach. *Int. J. Heat Mass Transfer* **35**, 2865–2877. [10.2.3]

Sinha, S. K., Sundararajan, T. and Garg, V. K. 1993 A study of the effects of macrosegregation and buoyancy-driven flow in binary mixture solidification. *Int. J. Heat Mass Transfer* **36**, 2349–2358. [10.2.3]

Skjetne, E. and Auriault, J. L. 1999a New insights on steady, nonlinear flow in porous media. *Europ. J. Mech. B—Fluids* **18**, 131–145. [1.5.2]

Skjetne, E. and Auriault, J. L. 1999b Homogenization of wall-slip gas flow through porous media. *Transport Porous Media* **36**, 293–306. [1.6]

Skudarnov, P. V., Lin, C. X., Wang, M. H., Pradeep, N. and Ebadian, M. A. 2002 Evolution of convection pattern during the solidification process of a binary mixture: effect of initial solutal concentration. *Int. J. Heat Mass Transfer* **45**, 5191–5200. [10.2.3]

Slimi, K., Ben Nasrallah, S. and Fohr, J. P. 1998 Transient natural convection in a vertical cylinder opened at the extremities and filled with a fluid saturated porous medium: validity of Darcy flow model and thermal boundary layer approximations. *Int. J. Heat Mass Transfer* **41**, 1113–1125. [7.3.3]

Slimi, K., Zili-Ghedira, L., Ben Nasrallah, S. and Mohamad, A. A. 2004 A transient study of coupled natural convection and radiation in a porous vertical channel using the finite-volume method. *Numer. Heat Transfer A* **45**, 451–478. [7.3.3]

Sobera, M. P., Klein, C. R., van den Akker, H. A. and Brasser, P. 2003 Convective heat and mass transfer to a cylinder sheathed by a porous layer. *AIChE J.* **49**, 3018–3028. [4.11]

Sokolov, V. E. 1982 *Mammal Skin*, University of California Press, Berkeley, CA. [4.14]

Solomon, T. H. and Hartley, R. R. 1998 Measurements of the temperature field of mushy and liquid regions during solidification of aqueous ammonium chloride. *J. Fluid Mech.* **358**, 87–106. [10.2.3]

Somerton, C. W. 1983 The Prandtl number effect in porous layer convection. *Appl. Sci. Res.* **40**, 333–344. [6.9.2]

Somerton, C. W. and Catton, I. 1982 On the thermal instability of superimposed porous and fluid layers. *ASME J. Heat Transfer* **104**, 160–165. [1.6, 6.19.1.2, 6.19.2]

Somerton, C. W., McDonough, J. M. and Catton, I. 1984 Natural convection in a volumetrically heated porous layer. *ASME J. Heat Transfer* **106**, 241–244. [6.11.2]

Sommerfeld, R. A. and Rocchio, J. E. 1993 A study of the effects of macrosegregation and buoyancy-driven flow in binary mixture solidification. *Water Resour. Res.* **29**, 2485–2490. [11.1]

Sondergeld, C. H. and Turcotte, D. L. 1977 An experimental study of two-phase convection in a porous medium with applications to geological problems. *J. Geophys. Res.* **82**, 2045–2053. [10.3.1]

Sondergeld, C. H. and Turcotte, D. L. 1978 Flow visualization studies of two-phase thermal convection in a porous layer. *Pure Appl. Geophys.* **117**, 321–330. [10.3.1]

Song, J. 2002 Spatial decay estimates in time-dependent double-diffusive Darcy plane flow. *J. Math. Anal. Appl.* **267**, 76–88. [1.5.3]

Song, M. and Viskanta, R. 1994 Natural convection flow and heat transfer within a rectangular enclosure containing a vertical porous layer. *Int. J. Heat Mass Transfer* **37**, 2425–2438. [7.7]

Song, M. and Viskanta, R. 2001 Lateral freezing of an anisotropic porous medium saturated with an aqueous salt solution. *Int. J. Heat Mass Transfer* **44**, 733–751. [10.2.1.2]

Song, M., Choi, J. and Viskanta, R. 1993 Upward solidification of a binary solution saturated porous medium. *Int. J. Heat Mass Transfer* **36**, 3687–3695. [10.2.3]

Soundalgekar, V. M., Lahuriher, R. M. and Pohanerkar, S. G. 1991 Heat transfer in unsteady flow through a porous medium between two infinite parallel plates in relative motion. *Forschung. Ing.* **57**, 28–31. [4.12.2]

Souto, H. P. A. and Moyne, C. 1997a Dispersion in two-dimensional periodic porous media. 1. Hydrodynamics. *Phys. Fluids* **9**, 2243–2252. [2.2.3]

Souto, H. P. A. and Moyne, C. 1997b Dispersion in two-dimensional periodic porous media. 2. Dispersion tensor. *Phys. Fluids* **9**, 2253–2263. [2.2.3]

Sovran, O., Bardan, G., Mojtabi, A. and Charrier-Mojtabi, M. C. 2000 Finite frequency external modulation in doubly diffusive convection. *Numer. Heat Transfer A* **37**, 877–896. [9.1.6.4]

Sovran, O., Charrier-Mojtabi, M. C. and Mojtabi, A. 2001 Onset of Soret-driven convection in an infinite porous layer. *C. R. Acad. Sci. IIB* **329**, 287–293. [9.1.4]

Sovran, O., Charrier-Mojtabi, M. C., Azaiez, M. and Mojtabi, A. 2002 Onset of Soret-driven convection in porous medium under vibration. *Heat Transfer 2002, Proc. 12th Int. Heat Transfer Conf.*, Elsevier, Vol. 2, pp. 839–844. [9.1.6.4]

Sözen, M. and Kuzay, T. M. 1996 Enhanced heat transfer in round tubes with porous inserts. *Int. J. Heat Fluid Flow*, **17**, 124–129. [4.12.1]

Sözen, M. and Vafai, K. 1990 Analysis of the non-thermal equilibrium condensing flow of a gas through a packed bed. *Int. J. Heat Mass Transfer* **33**, 1247–1261. [4.6.5, 10.4]

Sözen, M. and Vafai, K. 1991 Analysis of oscillating compressible flow through a packed bed. *Int. J. Heat Fluid Flow* **12**, 130–136. [4.9]

Sözen, M. and Vafai, K. 1993 Longitudinal heat dispersion in porous beds with real-gas flow. *AIAA J. Thermophys. Heat Transfer* **7**, 153–157. [4.6.5]

Spaid, M. A. A. and Phelan, F. R. 1997 Lattice Boltzmann methods for modeling microscale flow in fibrous porous media. *Phys. Fluids* **9**, 2468–2474. [2.6]

Spiga, M. and Morini, G. L. 1999 Transient response of nonthermal equilibrium packed beds. *Int. J. Engng. Sci.* **37**, 179–186. [4.10]

Sri Krishna, C. V. 2002 Effects of non-inertial acceleration on the onset of convection in a second-order fluid-saturated porous medium. *Int. J. Engng. Sci.* **39**, 599–609. [6.23]

Srinivasan, V., Vafai, K. and Christenson, R. N. 1994 Analysis pf heat transfer and fluid flow through a spirally fluted tube using a porous substrate approach. *ASME J. Heat Transfer* **116**, 543–551. [4.11]

Stalio, E., Breugem, W, P, and Boersama, B. J. 2004 Numerical study of turbulent heat transfer above a porous wall. In *Applications of Porous Media (ICAPM 2004)*, (eds. A. H. Reis and A. F. Miguel), Évora, Portugal, pp. 191–198. [4.11]

Stamps, D. W., Arpaci, V. S. and Clark, J. A. 1990 Unsteady three-dimensional natural convection in a fluid-saturated porous medium. *J. Fluid Mech.* **213**, 377–396. [6.15.1]

Stanescu, G., Fowler, A. J. and Bejan, A. 1996 The optimal spacing of cylinders in free-stream cross-flow forced convection. *Int. J. Heat Mass Transfer* **39**, 311–317. [4.15]

Stauffer, P.H., Auer, L.H. and Rosenberg, N.D. 1997 Compressible gas in porous media: a finite amplitude analysis of natural convection. *Int. J. Heat Mass Transfer* **40**, 1585–1589. [6.7]

Steen, P. H. 1983 Pattern selection for finite-amplitude convection states in boxes of porous media. *J. Fluid Mech.* **136**, 219–241. [6.15.1]

Steen, P. H. 1986 Container geometry and the transition to unsteady Bénard convection in porous media. *Phys. Fluids* **29**, 925–933. [6.15.1]

Steen, P. H. and Aidun, C. K. 1988 Time-periodic convection in porous media: transition mechanism. *J. Fluid Mech.* **196**, 263–290. [6.8]

Steinbeck, M. J. 1999 Convective drying of porous material containing a partially miscible mixture. *Chem. Engng. Process.* **38**, 487–502. [3.6]

Steinberg, V. and Brand, H. 1983 Convective instabilities of binary mixtures with fast chemical reaction in a porous medium. *J. Chem. Phys.* **78**, 2655–2660. [9.1.6.4]

Stemmelen, D., Moyne, C. and Degiovanni, A. 1992 Thermoconvective instability of a boiling liquid in porous medium. *C.R. Acad. Sci. Paris, Sér. II*, **314**, 769–775. [10.3.1]

Stewart, W. E. and Burns, A. S. 1992 Convection in a concentric annulus with heat generating porous media and a permeable inner boundary. *Int. Comm. Heat Mass Transfer* **19**, 859–868. [6.17]

Stewart, W. E. and Dona, C. L. G. 1988 Free convection in a heat-generating porous medium in a finite vertical cylinder. *ASME J. Heat Transfer* **110**, 517–520. [6.17]

Stewart, W. E., Cai, L. and Stickler, L. A. 1994 Convection in heat-generating porous media with permeable boundaries—natural ventilation of grain storage bins. *ASME J. Heat Transfer* **116**, 1044–1046. [6.17]

Stewart, W. E., Greer, L. A. and Stickler, L. A. 1990 Heat transfer in a partially insulated enclosure of heat generating porous media. *Int. Commm. Heat Mass Transfer* **17**, 597–607. [6.15.3]

Storesletten, L. 1993 Natural convection in a porous layer with anisotropic thermal diffusivity. *Transport in Porous Media* **12**, 19–29. [6.12]

Storesletten, L. 1998 Effects of anisotropy on convective flow through porous media. *Transport Phenomena in Porous Media* (eds. D. B. Ingham and I. Pop), Elsevier, Oxford, pp. 261–184. [6.12]

Storesletten, L. 2004 Effects of anisotropy on convection in horizontal and inclined porous layers. In *Emerging Technologies and Techniques in Porous Media* (D. B. Ingham, A. Bejan, E. Mamut and I. Pop, eds), Kluwer Academic, Dordrecht, pp. 285–306. [6.12]

Storesletten, L. and Pop, I. 1996 Free convection in a vertical porous layer with walls of non-uniform temperature. *Fluid Dyn. Res.* **17**, 107–119. [6.15.2]

Storesletten, L. and Rees, D.A.S. 1997 An analytical study of free convective boundary layer flow in porous media: the effect of anisotropic diffusivity. *Transport in Porous Media* **27**, 289–304. [5.1.9.9]

Storesletten, L. and Rees, D.A.S. 1998 The influence of higher-order effects on the linear instability of thermal boundary layer flow in porous media. *Int. J. Heat Mass Transfer* **41**, 1833–1843. [5.1.9.9]

Storesletten, L. and Rees, D. A. S. 2004 Onset of convection in an inclined porous layer with internal heat generation. In *Applications of Porous Media (ICAPM 2004)*, (eds. A. H. Reis and A. F. Miguel), Évora, Portugal, pp. 139–145. [7.8]

Storesletten, L. and Tveitereid, M. 1987 Thermal convection in a porous medium confined by a horizontal cylinder. *ASME HTD* **92**, 223–230. [6.16.2]

Storesletten, L. and Tveitereid, M. 1991 Natural convection in a horizontal porous cylinder. *Int. J. Heat Mass Transfer* **34**, 1959–1968. [6.16.2]

Storesletten. L. and Tveitereid, M. 1999 Onset of convection in an inclined porous layer with anisotropic permeability. *Appl. Mech. Engng.* **4**, 575–587. [7.8]

Strange, R. and Rees, D. A. S. 1996 The effect of fluid inertia on the stability of free convection in a saturated porous medium heated from below. *Proceedings of the International Conference on Porous Media and their Applications in Science, Engineering and Industry*, (eds. K. Vafai and P.N. Shivakumar), Engineering Foundation, New York, pp. 71–84. [6.6]

Straughan, B. 1988 A nonlinear analysis of convection in a porous vertical slab. *Geophys. Astrophys. Fluid Dyn.* **42**, 269–276. [7.1.4]

Straughan, B. 2001a Porous convection, the Chebyshev tau method, and spurious eigenvalues. In *The K. Hutter 60th Birthday Volume* (B. Straughan, R. Greve, H. Ehrentraut and Y. Wang), Springer, Heidelberg. [6.15.1]

Straughan, B. 2001b Surface-tension driven convection in a fluid overlying a porous layer. *J. Comput. Phys.* **170**, 320–337. [6.19.3]

Straughan, B. 2001c A sharp nonlinear stability threshold in rotation porous convection. *Proc. Roy. Soc. Lond. A* **457**, 87–93. [6.22]

Straughan, B. 2002 Effect of property variation and modeling on convection in a fluid overlying a porous layer. *Int. J. Numer. Anal. Meth. Geomech.* **26**, 75–97. [6.19.1]

Straughan, B. 2004a Resonant porous penetrative convection. *Proc. Roy. Soc. Lond. A* **460**, 2913–2927. [6.11.4]

Straughan, B. 2004b *The Energy Method, Stability, and Nonlinear Convection*, 2nd ed., Springer, New York. [6.4, 11.2]

Straughan, B. and Hutter, K. 1999 A priori bounds and structural stability for double-diffusive convection incorporating the Soret effect. *Proc. Roy. Soc. Lond. A* **445**, 767–777. [9.1.4]

Straughan, B. and Walker, D. W. 1996a Anisotropic porous penetrative convection. *Proc. Roy. Soc. Lond. A* **452**, 97–115. [6.12]

Straughan, B. and Walker, D. W. 1996b Two very accurate and efficient methods for computing eigenvalues and eigenfunctions in porous convection problems. *J. Comput. Phys.* **127**, 128–141. [7.9]

Straus, J. M. 1974 Large amplitude convection in porous media. *J. Fluid Mech.* **64**, 51–63. [6.4, 6.6, 6.8, 6.9.1]

Straus, J. M. and Schubert, G. 1977 Thermal convection of water in a porous medium: effects of temperature- and pressure-dependent thermodynamic and transport processes. *J. Geophys. Res.* **82**, 325–333. [6.7]

Straus, J. M. and Schubert, G. 1979 Three-dimensional convection in a cubic box of fluid-saturated porous material. *J. Fluid Mech.* **91**, 155–165. [6.15.1]

Straus, J. M. and Schubert, G. 1981 Modes of finite-amplitude three-dimensional convection in rectangular boxes of fluid-saturated porous material. *J. Fluid Mech.* **103**, 23–32. [6.15.1]

Stroh, F. and Balakotaiah, V. 1991 Modeling of reaction-induced flow maldistributions in packed beds. *AIChE J.* **37**, 1035–1052. [3.4]

Stroh, F. and Balakotaiah, V. 1992 Stability of uniform flow in packed-bed reactors. *Chem. Engng. Sci.*, **47**, 593–604. [3.4]

Stroh, F. and Balakotaiah, V. 1993 Flow maldistributions in packed beds: effect of reactant consumption. *Chem. Engng. Sci.* **48**, 1629–1640. [3.4]

Stubos, A. K. and Buchlin, J. M. 1993 Analysis and numerical simulation of the thermodynamic behaviour of a heat dissipating debris bed during power transients. *Int. J. Heat Mass Transfer* **36**, 1391–1401. [6.11.2]

Stubos, A. K., Kanellopoulos, V. and Tassopoulos, M. 1994 Aspects of transient modelling of a volumetrically heated porous bed: a combination of microscopic and macroscopic approach. *Heat Transfer 1994*, Inst. Chem. Engrs, Rugby, vol. 5, pp. 387–392. [11.9.1]

Stubos, A. K., Pérez Caseiras, C., Buchlin, J. M. and Kannelopoulos, N. K. 1997 Numerical investigation of vapor-liquid flow and heat transfer in capillary porous media. *Numer. Heat Transfer A* **31**, 143–166. [11.9.1]

Stubos, A. K., Satik, C. and Yortos, Y. C. 1993a Critical heat flux hysteresis in vapor-liquid counterflow in porous media. *Int. J. Heat Mass Transfer* **36**, 227–231. [11.9.2]

Stubos, A. K., Satik, C. and Yortsos, Y. C. 1993b Effects of capillary heterogeneity on vapo-liquid counterflow in porous media. *Int. J. Heat Mass Transfer* **36**, 967–976. [11.9.1]

Sturm, M. and Johnson, J. B. 1991 Natural convection in the subarctic snow cover. *J. Geophys. Res.* **96**, 11657–11671. [11.1]

Subagyo, Standish, N. and Brooks, G. A. 1998 A new model of velocity distribution of a single-phase fluid flowing in packed beds. *Chem. Engng. Sci.* **53**, 1375–1385. [2.6]

Subramanian, L. and Patil, P. R. 1991 Thermohaline convection with coupled molecular diffusion in an anisotropic porous medium. *Indian J. Pure Appl. Math.* **22**, 169–193. [9.1.6.4]

Subramanian, S. 1994 Convective instabilities induced by exothermic reactions occurring in a porous medium. *Phys. Fluids* **6**, 2907–2922. [9.1.6.4]

Subramanian, S. and Balakotaiah, V. 1995 Mode interactions in reaction-driven convection in a porous medium. *Chem. Engrg Sci.* **50**, 1851–1868. [3.4]

Subramanian, S. and Balakotaiah, V. 1997 Analysis and classification of reaction-driven stationary convective patterns in a porous medium. *Phys. Fluids* **9**, 1674–1695. [3.5]

Sugawara, M., Inaba, H. and Seki, N. 1988 Effect of maximum density of water on freezing of a water-saturated horizontal porous layer. *ASME J. Heat Transfer* **110**, 155–159. [10.2.2]

Sun, B. X., Xu, X. Z., Lai, Y. M., and Fang, M. X. 2005 Evaluation of fractured rock layer heights in ballast railway embankment based on cooling effect of natural convection in cold regions. *Cold Regions Sci. Tech.* **42**, 120–144. [7.3.7]

Sun, Z. S., Tien, C. and Yen, Y. C. 1970 Onset of convection in a porous medium containing liquid with a density maximum. *Heat Transfer 1970*, Elsevier, Amsterdam, vol. 4, paper NC 2.11. [6.20]

Sundaravadivelu, K. and Tso, C. P. 2003 Influence of viscosity variations on the forced convection flow through two types of heterogeneous porous media with isoflux boundary condition. *Int. J. Heat Mass Transfer* **46**, 2329–2339. [4.12]

Sundfor, H. O. and Tyvand, P. A. 1996 Transient free convection in a horizontal porous cylinder with a sudden change in wall temperature. *Waves and Nonlinear Processes In Hydrodynamics* (eds. J. Grue, B. Gjevik and J. E. Weber), Kluwer, Dordrecht, pp. 291–302. [7.5]

Sung, H. J., Kim, S. Y. and Hyun, J. M. 1995 Forced convection from an isolated heat source in a channel with porous medium. *Int. J. Heat Fluid Flow* **16**, 527–535. [4.16.2]

Sunil 1994 Thermosolutal hydromagnetic instability of a compressible and partially ionized plasma in a porous medium. *Arch. Mech.* **46**, 819–828. [9.1.6.4]

Sunil 1999 Finite Larmor radius effect on thermosolutal instability of a Hall plasma in a porous medium. *Phys. Plasmas* **6**, 50–56. [9.1.6.4]

Sunil 2001 Thermosolutal instability of a compressible finite Larmor radius, Hall plasma in a porous medium. *J. Porous Media* **4**, 55–67. [9.1.6.4]

Sunil and Singh, P. 2000 Thermal instability of a porous medium with relaxation and inertia in the presence of Hall effects. *Arch. Appl. Mech.* **70**, 649–658. [6.21]

Sunil, Bharti, P. K. , Sharma, D. and Sharma, R. C. 2004a The effect of magnetic field dependent viscosity on the thermal convection in a ferromagnetic fluid in a porous medium. *Z. Naturfor. A* **59**, 397–406. [6.21]

Sunil, Bharti, P. K. and Sharma, R. C. 2003a On Bénard convection in a porous medium in the presence of throughflow and rotation in magnetics. *Arch. Mech.* **55**, 257–274. [6.21]

Sunil, Divya, and Sharma, R. C. 2004b The effect of rotation on ferromagnetic fluid heated and salted from below saturating a porous medium. *J. Geophys. Engng.* **1**, 116–127. [9.1.6.4]

Sunil, Divya, and Sharma, R. C. 2004c The effect of magnetic-field-dependent viscosity on ferroconvection in a porous medium in the presence of dust particles. *J. Geophys. Engng.* **1**, 277–286. [6.21]

Sunil, Divya, and Sharma, R. C. 2005a The effect of magnetic-field-dependent viscosity on thermosolutal convection in a ferromagnetic fluid saturating a porous medium. *Transport Porous Media* **60**, 251–274. [9.1.6.4]

Sunil, Divya, and Sharma, R. C. 2005d Thermosolutal convection in a ferromagnetic fluid saturating a porous medium. *J. Porous Media* **8**, 393–408. [9.1.6.4]

Sunil, Divya, Sharma, R. C. and Sharma, V. 2003b Compressible couple-stress fluid permeated with suspended particles heated and soluted from below in porous medium. *Indian J. Pure Appl. Phys.* **41**, 602–611. [9.1.6.4]

Sunil, Sharma, R. C. and Chandel, R. S. 2004d Effect of suspended particles on couple-stress fluid heated and soluted from below in porous medium. *J. Porous Media* **7**, 9–18. [9.1.6.4]

Sunil, Sharma, A., Kumar, P. and Gupta, U. 2005c The effect of magnetic-field-dependent viscosity and rotation on ferromagnetic convection saturating a porous medium in the presence of dust particles. *J. Geophys. Engng.* **2**, 238–251. [9.1.6.4]

Sunil, Sharma, D. and Sharma, R. C. 2005b Effect of dust particles on thermal convection in ferromagnetic fluid saturating a porous medium. *J. Magnet. Magnet. Mater.* **288**, 183–195. [6.2.1]

Sunil, Sharma, R. C. and Pal, M. 2001 Hall effect on thermosolutal instability of a Rivlin-Ericksen fluid in a porous medium. *Non-Equil. Thermodyn.* **26**, 373–386. [9.1.6.4]

Sunil, Sharma, R. C. and Pal, M. 2002 On a couple-stress fluid heated from below in a porous medium in the presence of a magnetic field and rotation. *J. Porous Media* **5**, 149–158. [6.22]

Suresh, C. S. Y., Krishna, Y. V., Sundararajan, T. and Das, S. K. 2005 Numerical simulation of three-dimensional natural convection inside a heat generating anisotropic porous medium. *Heat Mass Transfer* **41**, 799–809. [7.3.8]

Sutton, F. M. 1970 Onset of convection in a porous channel with net through flow. *Phys. Fluids* **13**, 1931–1934. [6.10]

Swift, D. W. and Harrison, W. D. 1984 Convective transport of brine and thaw of subsea permafrost: result of numerical simulations. *J. Geophys. Res.* **89**, 2080–2086. [11.3]

Tait, S. and Jauport, C. 1992 Compositional convection in a reactive crystalline mush and melt differentiation. *J. Geophys. Res.* **97**, 6735–6756. [10.2.3]

Tait, S., Jahrling, K. and Jaupart, C. 1992 The planform of compositional convection and chimney formation in a mushy layer. *Nature* **359**, 406–408. [10.2.3]

Takata, Y., Fukuda, K., Hasegawa, S., Iwashige, K., Shimomura, H. and Sanokawa, K. 1982 Three-dimensional natural convection in a porous medium enclosed with concentric inclined cylinders. *Heat Transfer 1982*, Hemisphere, Washington, DC, vol. 2, pp. 351–356. [7.8]

Takatsu, Y. and Masuoka, T. 1998 Turbulent phenomena in flow through porous media. *J. Porous Media* 1, 243–251. [1.8]

Takatsu, Y., Masuoka, T., Hashimoto, Y. and Yokota, K. 1997 Natural convection along a vertical porous surface. *Heat Transfer Japan. Res.* **26**, 385–397. [5.1.9.9]

Takhar, H. S. and Beg, O. A. 1997 Non-Darcy effects on convective boundary-layer flow past a semi-infinite vertical plate in saturated porous media. *Heat Mass Transfer* **32**, 33–34. [8.1.1]

Takhar, H. S. and Ram, P. C. 1994 Magnetohydrodynamic free convection flow of water at 4°C through a porous medium. *Int. J. Heat Mass Transfer* **21**, 371–376. [8.2.1]

Takhar, H. S., Chamkha, A. J. and Nath G. 2002 Natural convection on a vertical cylinder embedded in a thermally stratified high-porosity medium. *Int. J. Thermal Sci.* **41**, 83–93. [5.7]

Takhar, H. S., Chamkha, A. J. and Nath G. 2003 Effects of non-uniform temperature or mass transfer in finite sections of an inclined plate on the MHD natural convection flow in a temperature stratified high-porosity porous medium. *Int. J. Thermal Sci.* **42**, 829–836. [5.3]

Takhar, H. S., Kumari, M. and Bèg, O. A. 1998 Computational analysis of coupled radiation-convection dissipative non-gray gas flow in a non-Darcy gas flow in a non-Darcy porous medium using the Keller box implicit difference scheme. *Int. J. Energy Res.* **22**, 141–159. [5.1.9.4]

Takhar, H. S., Roy, S. and Nath, G. 2003 Unsteady free convection flow over an infinite vertical porous plate due to the combined effects of thermal and mass diffusion, magnetic field and Hall currents. *Heat Mass Transfer* **39**, 825–834. [9.2.1]

Takhar, H. S., Soundalgekar, V. M. and Gupta, A. S. 1990 Mixed convection of an incompressible viscous fluid in a porous medium past a hot vertical plate. *Int. J. Non-Linear Mech.* **25**, 723–728. [8.1.1]

Talukdar, P., Mishra, S. C., Trimis, D. and Durts, F. 2004 Combined radiation and convection heat transfer in a porous channel bounded by isothermal parallel plates. *J. Heat Mass Transfer* **47**, 1001–1013. [2.6]

Tam, C. K. W. 1969 The drag on a cloud of spherical particles in low Reynolds number flow. *J. Fluid Mech.* **38**, 537–546. [1.5.3]

Tan, K. K., Sam, T. and Jamaludin, H. 2003 The onset of transient convection in bottom heated porous media. *Int. J. Heat Mass Transfer* **46**, 2857–2873. [6.11.3]

Tang, J. and Bau, H. H. 1993 Feedback control stabilization of the no-motion state of a fluid confined in a horizontal porous layer heated from below. *J. Fluid Mech.* **257**, 485–505. [6.11.3]

Tao, Z., Wu, H., Chen, G. and Deng, H. 2005 Numerical simulation of conjugate heat and mass transfer process within cylindrical porous media with cylindrical dielectric cores in microwave freeze-drying. *Int. J. Heat Mass Transfer* **48**, 561–572. [3.6]

Tashtoush, B. 2000 Analytic solution for the effect of viscous dissipation on mixed convection in saturated porous media. *Transport in Porous Media* **41**, 197–209. [8.1.1]

Tashtoush, B. 2005 Magnetic and buoyancy effects on melting from a vertical plate embedded in saturated porous media. *Energy Conv. Manag.* **46**, 2566–2577. [10.1.7]

Tashtoush, B. and Kodah, Z. 1998 Non slip boundary effects in non-Darcian mixed convection from a vertical wall in saturated porous media. *Heat Mass Transfer* **34**, 35–39. [8.1.1]

Taslim, M. E. and Narusawa, U. 1986 Binary fluid convection and double-diffusive convection. *ASME J. Heat Transfer* **108**, 221–224. [9.1.4]

Taslim, M. E. and Narusawa, U. 1989 Thermal instability of horizontally superposed porous and fluid layers. *ASME J. Heat Transfer* **111**, 357–362. [6.19.1.2]

Taunton, J. W., Lightfoot, E. N. and Green, T. 1972 Thermohaline instability and salt fingers in a porous medium. *Phys. Fluids* **15**, 748–753. [9.1.3]

Tavman, I. H. 1996 Effective thermal conductivity of granular porous materials. *Int. Comm. Heat Mass Transfer* **23**, 169–176. [2.2.1]

Taylor, G. I. 1971 A model for the boundary condition of a porous material, Part 1. *J. Fluid Mech.* **49**, 319–326. [1.6]

Telles, R. S. and Trevisan, O.V. 1993 Dispersion in heat and mass transfer natural convection along vertical boundaries in porous media. *Int. J. Heat Mass Transfer* **36**, 1357–1365. [9.2.1]

Teng, H. and Zhao, H. 2000 An extension of Darcy's law to non-Stokes flow in porous media. *Chem. Eng. Sci.* **55**, 2727–2735. [1.5.1]

Tewari, K. and Singh, P. 1992 Natural convection in a thermally stratified fluid-saturated porous medium. *Int. J. Engng. Sci.* **30**, 1003–1008. [5.1.4]

Tewari, P. K. 1982 A study of boiling and convection in fluid-saturated porous media. MS thesis, Cornell University. [10.3.1]

Tewari, P. K. and Torrance, K. E. 1981 Onset of convection in a box of fluid-saturated porous material with a permeable top. *Phys. Fluids* **24**, 981–983. [6.15.1]

Thangaraj, R. P. 2000 Effect of a nonuniform basic temperature gradient on the convective instability of a fluid-saturated porous layer with general velocity and thermal conditions. *Acta Mech.***141**, 85–97. [6.11.1]

Thevenin, J. 1995 Transient forced convection heat transfer from a circular cylinder embedded in a porous medium. *Int. Comm. Heat Mass Transfer* **22**, 507–516. [4.6.5]

Thevenin, J. and Sadaoui, D. 1995 About enhancement of heat transfer over a circular cylinder embedded in a porous medium. *Int. J. Heat Mass Transfer* **22**, 295–304. [4.3]

Thiele, M. 1997 Heat dispersion in stationary mixed convection flow about horizontal surfaces in porous media. *Heat Mass Transfer* **33**, 7–16. [8.1.2]

Thompson, A. F., Huppert, H. E., Worster, G. M. and Aitta, A. 2003 Solidification and compositional convection of a ternary alloy. *J. Fluid Mech.* **497**, 167–199. [10.2.3]

Tien, C. L. and Hunt, M. L. 1987 Boundary layer flow and heat transfer in porous beds. *Chem. Engng. Proc.* **21**, 53–63. [4.9]

Tien, C. L. and Vafai, K. 1990a Convective and radiative heat transfer in porous media. *Adv. Appl. Mech.* **27**, 225–281. [3.6]

Tien, H. C. and Vafai, K. 1990b Pressure stratification effects on multiphase transport across a vertical slot porous insulation. *ASME J. Heat Transfer* **112**, 1023–1031. [7.1.7]

Tobbal, T. and Bennacer, R. 1998 Heat and mass transfer in anisotropic porous layer. *Trends in Heat, Mass and Momentum Transfer* **3**, 129–137. [9.2.2]

Toda, S., Hsu, W. S., Hashizume, H. and Kawaguchi, T. 1998 Unsteady heat transfer of steam flow with condensation in porous media. *Heat Transfer 1998, Proc. 11th IHTC*, **6**, 463–468. [10.4]

Tong, T. W. and Orangi, S. 1986 A numerical analysis for high modified Rayleigh number natural convection in enclosures containing a porous medium. *Heat Transfer 1986*, Hemisphere, Washington, DC, vol. 5, pp. 2647–2652. [7.6.2]

Tong, T. W. and Subramanian, E. 1985 A boundary-layer analysis for natural convection in vertical porous enclosures—use of the Brinkman-extended Darcy model. *Int. J. Heat Mass Transfer* **28**, 563–571. [7.6.2]

Tong, T. W. and Subramanian, E. 1986 Natural convection in rectangular enclosures partially filled with a porous medium. *Int. J. Heat Fluid Flow* **7**, 3–10. [7.7]

Tong, T. W., Sharatchandra, M. C. and Gdoura, Z. 1993 Using porous inserts to enhance heat transfer in laminar fully-developed flows. *Int. Comm. Heat Mass Transfer* **20**, 761–770. [4.11]

Tournier, C., Genthon, P. and Rabinowicz, M. 2000 The onset of convection in vertical fault planes: consequences for the thermal regime in crystalline basements and for heat recovery experiments. *Geophys. J. Int.* **110**, 500-508. [6.15.2]

Tracey, J. 1996 Multi-component convection-diffusion in a porous medium. *Cont. Mech. Thermodyn.* **8**, 361–381. [9.1.3]

Tracey, J. 1998 Penetrative convection and multi-component diffusion in a porous medium. *Adv. Water Res.* **22**, 399–412. [9.1.6.4]

Travkin, V. S. and Catton, I. 1994 Turbulent transport of momentum, heat and mass in a two-level highly porous medium. *Heat Transfer 1994*, Inst. Chem. Engrs, Rugby, vol. 6, pp. 399–404. [1.8]

Travkin, V. and Catton, I. 1995 A two-temperature model for turbulent flow and heat transfer in a porous layer. *ASME J. Fluids Engng.* **117**, 181–188. [1.8]

Travkin, V. S. and Catton, I. 1998 Porous media transport descriptions—non-local, linear and nonlinear against effective thermal/fluid properties. *Adv. Colloid Interface Sci.* **77**, 389–443. [1.8]

Travkin. V. S. and Catton, I. 1999 Nonlinear effects in multiple regime transport of momentum in longitudinal capillary porous medium morphology. *J. Porous Media* **2**, 277–294. [1.8]

Trevisan, O. V. and Bejan, A. 1985 Natural convection with combined heat and mass transfer buoyancy effects in a porous medium. *Int. J. Heat Mass Transfer* **28**, 1597–1611. [9.2.2]

Trevisan, O. V. and Bejan, A. 1986 Mass and heat transfer by natural convection in a vertical slot filled with porous medium. *Int. J. Heat Mass Transfer* **29**, 403–415. [9.2.2]

Trevisan, O. V. and Bejan, A. 1987a Combined heat and mass transfer by natural convection in a vertical enclosure. *J. Heat Transfer* **109**, 104–109. [9.2.2]

Trevisan, O. V. and Bejan, A. 1987b Mass and heat transfer by high Rayleigh number convection in a porous medium heated from below. *Int. J. Heat Mass Transfer* **30**, 2341–2356. [9.1.5]

Trevisan, O. V. and Bejan, A. 1989 Mass and heat transfer by natural convection above a concentrated source buried at the base of a shallow porous layer. *ASME HTD* **127**, 47–54. [9.4]

Trevisan, O. V. and Bejan, A. 1990 Combined heat and mass transfer by natural convection in a porous medium. *Adv. Heat Transfer* **20**, 315–352. [9]

Trew, M. and McKibbin, R. 1994 Convection in anisotropic inclined porous layers. *Transport in Porous Media* **17**, 271–283. [7.8]

Tung, V. X. and Dhir, V. K. 1993 Convective heat transfer from a sphere embedded in unheated porous media. *ASME J. Heat Transfer* **115**, 503–506. [4.3, 8.1.3]

Turcotte, D. L., Ribando, R. J. and Torrance, K. E. 1977 Numerical calculation of two-temperature thermal convection in a permeable layer with application to the Steamboat Springs thermal system, Nevada. *The Earth's Crust* (ed. J. G. Heacock), Amer. Geophys. Union, Washington, DC, pp. 722–736. [11.7]

Tveitereid, M. 1977 Thermal convection in a horizontal porous layer with internal heat sources. *Int. J. Heat Mass Transfer* **20**, 1045–1050. [6.11.2]

Tyvand, P. A. 1977 Heat dispersion effect on thermal convection in anisotropic porous media. *J. Hydrol.* **34**, 335–342. [2.2.3, 6.12]

Tyvand, P. A. 1980 Thermohaline instability in anisotropic porous media. *Water Resources Res.* **16**, 325–330. [9.1.6.2]

Tyvand, P. A. 1981 Influence of heat dispersion on steady convection in anisotropic porous media. *J. Hydrol.* **52**, 13–23. [2.2.3, 6.12]

Tyvand, P. A. 1995 First-order transient free convection about a horizontal cylinder embedded in a porous medium. *Fluid Dyn. Res.* **15**, 277–290. [5.5.1]

Tyvand, P. A. 2002 Onset of Rayleigh-Bénard convection in porous bodies. In *Transport Phenomena in Porous Media II* (D. B. Ingham and I. Pop, eds.) Elsevier, Oxford, pp. 82–112. [6.4]

Tyvand, P. A. and Storesletten, L. 1991 Onset of convection in an anisotropic porous medium with oblique principal axes. *J. Fluid Mech.* **226**, 371–382. [6.12]

Tzeng, S. C. and Ma, W. P. 2004 Experimental investigation of heat transfer in sintered porous heat sink. *Int. Comm. Heat Mass Transfer* **31**, 827–836. [4.16.2]

Tzeng, S. C., Soong, C. Y. and Wong, S. C. 2004 Heat transfer in rotating channel with open cell porous aluminum foam. *Int. Comm. Heat Mass Transfer* **31**, 261–272. [4.16.2]

Umavarthi, J. C. and Malashetty, M. S. 1999 Oberbeck convection flow of couple stress fluid through a vertical porous stratum. *Int. J. Nonlinear Mech.* **34**, 1037–1045. [7.1.6]

Umavarthi, J. C., Kumar, J. P., Chamkha, A. J. and Pop, I. 2005 Mixed convection in a vertical porous channel. *Transport Porous Media* **61**, 315–335. [8.3.1]

Vadasz, J. J. , Roy-Aikins, J. E. A. and Vadasz, P. 2005 Sudden or smooth transitions in porous media natural convection. *Int. J. Heat Mass Transfer* **48**, 1096–1106. [6.4]

Vadasz, P. 1990 Bifurcation phenomena in natural convection in porous media. *Heat Transfer 1990*, Hemisphere, Washington, DC, vol. 5, pp. 147–152. [6.13.4]

Vadasz, P. 1992 Natural convection in porous media induced by the centrifugal body force: the solution for small aspect ratio. *ASME J. Energy Res. Tech.*, **114**, 250–254. [6.22]

Vadasz, P. 1993 Three-dimensional free convection in a long rotating porous box: analytical solution. *ASME J. Heat Transfer* **115**, 639–644. [7.12]

Vadasz, P. 1994a Stability of free convection in a narrow porous layer subject to rotation. *Int. Comm. Heat Mass Transfer* **21**, 881–890. [7.12]

Vadasz, P. 1994b Centrifugally generated free convection in a rotating porous box. *Int. J. Heat Mass Transfer* **37**, 2399–2404. [6.22]

Vadasz, P. 1995 Coriolis effect on free convection in a long rotating box subject to uniform heat generation. *Int. J. Heat Mass Transfer* **38**, 2011–2018. [7.12]

Vadasz, P. 1996a Stability of free convection in a rotating porous layer distant from the axis of rotation. *Transport in Porous Media* **23**, 153–173. [7.12]

Vadasz, P. 1996b Convection and stability in a rotating porous layer with alternating direction of the cylindrical body force. *Int. J. Heat Mass Transfer* **39**, 1639–1647. [7.12]

Vadasz, P. 1997a Flow in rotating porous media. In *Fluid Transport in Porous Media* (P. Du Plessis, ed.), Computational Mechanics Publications, Southhampton, Chapter 4. [6.22]

Vadasz, P. 1998a Coriolis effect on gravity-driven convection in a rotating porous layer heated from below. *J. Fluid Mech.* **376**, 351–375. [6.22]

Vadasz, P. 1998b Free convection in rotating porous media. *Transport Phenomena in Porous Media* (eds. D. B. Ingham and I. Pop), Elsevier, Oxford, pp. 285–312. [6.22]

Vadasz, P. 1999a Local and global transitions to chaos and hysteresis in a porous layer heated from below. *Transport Porous Media* **37**, 213–245. [6.4]

Vadasz, P. 1999b A note and discussion on J.-L. Auriault's letter "Comments on the paper 'Local and global transitions to chaos and hysteresis in a porous layer heated from below' by P. Vadasz." *Transport Porous Media* **37**, 251–254. [6.4]

Vadasz, P. 2000a Flow and thermal convection in rotating porous media. *Handbook of Porous Media* (K. Vafai, ed.), Marcel Dekker, New York, pp. 395–439. [6.22]

Vadasz, P. 2000b Weak turbulence and transitions to chaos in porous media. *Handbook of Porous Media* (K. Vafai, ed.), Marcel Dekker, New York, pp. 699–754. [6.4]

Vadasz, P. 2001a Heat transfer regimes and hysteresis in porous media convection. *ASME J. Heat Transfer* **123**, 145–156. [6.4]

Vadasz, P. 2001b Equivalent initial conditions for compatibility between analytical and computational solutions of convection in porous media. *Int. J. Nonlinear Mech.* **36**, 197–208. [6.4]

Vadasz, P. 2001c The effect of thermal expansion on porous media convection. Part 1: Thermal expansion solution. *Transport Porous Media* **44**, 421–443. [6.7]

Vadasz, P. 2001d The effect of thermal expansion on porous media convection. Part 2: Thermal convection solution. *Transport Porous Media* **44**, 445–463. [6.7]

Vadasz, P. 2002 Fundamentals of thermal convection in rotating porous media. *Heat Transfer 2002, Proc. 12th Int. Heat Transfer Conf.*, Elsevier, vol.1, pp. 117–128. [6.22]

Vadasz, P. 2003 Small and moderate Prandtl number convection in a porous layer heated from below. *Int. J. Energy Res.* **27**, 941–960. [6.4]

Vadasz, P. 2005 Explicit conditions for local thermal equilibrium in porous media heat conduction. *Transport Porous Media* **59**, 341–355. [2.2.2]

Vadasz, P. and Braester, C. 1992 The effect of imperfectly insulated sidewalls on natural convection in porous media. *Acta Mech.* **91**, 215–233. [6.15.3]

Vadasz, P. and Govender, G. 1997 Two-dimensional convection induced by gravity and centrifugal forces in a rotating porous layer far away from the axis of rotation. *Int. J. Rotating Machinery*, to appear. [7.12]

Vadasz, P. and Govender, S. 2001 Stability and stationary convection induced by gravity and centrifugal forces in a rotation porous layer distant from the axis of rotation. *Int. J. Engng. Sci.* **39**, 715–732. [6.22]

Vadasz, P. and Heerah, A. 1998 Experimental confirmation and analytical results of centrifugally driven free convection in porous media. *J. Porous Media* **1**, 261–272. [7.1.4]

Vadasz, P. and Olek, S. 1998 Transitions and chaos for free convection in a rotating porous layer. *Int. J. Heat Mass Transfer* **41**, 1417–1435. [6.8]

Vadasz, P. and Olek, S. 1999a Weak turbulence and chaos for low Prandtl number gravity driven convection in porous media. *Transport Porous Media.* **37**, 69–91. [6.4]

Vadasz, P. and Olek, S. 1999b Computational recovery of the homoclinic orbit in porous media convection. *Int. J. Nonlinear Mech.* **34**, 1071–1075. [6.4]

Vadasz, P. and Olek, S. 2000a Route to chaos for moderate Prandtl number convection in a porous layer heated from below. *Transport Porous Media.* **41**, 211–239. [6.4]

Vadasz, P. and Olek, S. 2000b Convergence and accuracy of Adomian's decomposition method for the solution of Lorenz equations. *Int. J. Heat Mass Transfer* **43**, 1715–1734. [6.4]

Vadasz, P., Braester, C. and Bear, J. 1993 The effect of perfectly conducting side walls on natural convection in porous media. *Int. J. Heat Mass Transfer* **36**, 1159–1170. [6.15.3]

Vafai, K. 1984 Convective flow and heat transfer in variable-porosity media. *J. Fluid Mech.* **147**, 233–259. [4.8]

Vafai, K. 1986 Analysis of the channeling effect in variable-porosity media. *ASME J. Energy Res. Tech.* **108**, 131–139. [4.8]

Vafai, K. and Amiri, A. 1998 Non-Darcian effects in confined forced convection flows. In *Transport Phenomena in Porous Media* (D. B. Ingham and I. Pop, eds.) Elsevier, Oxford, pp. 313–329. [4.9]

Vafai, K. and Huang, P. C. 1994 Analysis of heat transfer regulation and modification employing intermittently emplaced porous cavities. *ASME J. Heat Transfer* **116**, 604–613. [4.11]

Vafai, K. and Kim, S. J. 1989 Forced convection in a channel filled with a porous medium: an exact solution. *ASME J. Heat Transfer* **111**, 1103–1106. [4.9]

Vafai, K. and Kim, S. J. 1990 Analysis of surface enhancement by a porous substrate. *ASME J. Heat Transfer* **112**, 700–706. [4.8, 4.14]

Vafai, K. and Kim, S. J. 1997 Closure. *ASME J. Heat Transfer* **119**, 197–198. [1.5.3]

Vafai, K. and Sarkar, S. 1986 Condensation effects in a fibrous insulation slab. *ASME J. Heat Transfer* **108**, 667–675. [10.4]

Vafai, K. and Sarkar, S. 1987 Heat and mass transfer in partial enclosures. *AIAA J. Thermophys. Heat Transfer* **1**, 253–259. [10.4]

Vafai, K. and Sözen, M. 1990a Analysis of energy and momentum transport for fluid flow through a porous bed. *ASME J. Heat Transfer* **112**, 690-699. [4.6.5]

Vafai, K. and Sözen, M. 1990b An investigation of a latent heat storage porous bed and condensing flow through it. *ASME J. Heat Transfer* **112**, 1014–1022. [4.6.5]

Vafai, K. and Thiyagaraja, R. 1987 Analysis of flow and heat transfer at the interface region of a porous medium. *Int. J. Heat Mass Transfer* **30**, 1391–1405. [4.8]

Vafai, K. and Tien, C.L. 1981 Boundary and inertia effects on flow and heat transfer in porous media. *Int. J. Heat Transfer* **24**, 195–203. [1.5.3, 4.9]

Vafai, K. and Tien, C. L. 1982 Boundary and inertial effects on convective mass transfer in porous media. *Int. J. Heat Mass Transfer* **25**, 1183–1190. [1.5.3]

Vafai, K. and Tien, C. L. 1989 A numerical investigation of phase change effects in porous materials. *Int. J. Heat Mass Transfer* **32**, 1261–1277. [4.10]

Vafai, K. and Whitaker, S. 1986 Simultaneous heat and mass transfer accompanied by phase change in porous insulation. *ASME J. Heat Transfer* **108**, 132–140. [10.4]

Vafai, K., Alkire, R. L. and Tien, C. L. 1985 An experimental investigation of heat transfer in variable porosity media. *ASME J. Heat Transfer* **107**, 642–647. [4.8]

Vafai, K., Desai, C.P. and Chen, S.C. 1993 An investigation of heat transfer process in a chemically reacting packed bed. *Numer. Heat Transfer A* **24**, 127–142. [3.4]

Vafai, K., Minkowycz, W. J., Bejan, A. and Khanafer, K. 2006 Synthesis of models for turbulent transport through porous media. *Handbook of Numerical Heat Transfer* (eds. W. J. Minkowycz and E. M. Sparrow), Wiley, New York, ch. 12. [1.8]

Vaidyanathan, G., Ramanathan, A. and Maruthamanikandan, S. 2002a Effect of magnetic field dependent viscosity on ferroconvection in a sparsely distributed porous medium. *Indian J. Pure Appl. Phys.* **40**, 166–171. [6.21]

Vaidyanathan, G., Sekar, R. and Balasubramanian, R. 1991 Ferroconvective instability of fluids saturating a porous medium. *Int. J. Engng. Sci.* **29**, 1259–1267. [6.23]

Vaidyanathan, G., Sekar, R. and Ramanathan, A. 2002b Ferroconvection in an anisotropic densely packed porous medium. *Indian J. Chem. Tech.* **9**, 446–449. [6.21]

Vaidyanathan, G., Sekar, R., Vasanthakumari, R. and Ramanathan, A. 2002c Effect of magnetic field dependent viscosity on ferroconvection in a rotating sparsely distributed porous medium. *J. Magn. Magn. Mater.* **250**, 65–76. [6.21]

Valencia-Lopez, J. J. and Ochoa-Tapia, J. A. 2001 A study of buoyancy-driven flow in a confined fluid overlying a porous layer. *Int. J. Heat Mass Transfer* **44**, 4725–4736. [6.19.2]

Valencia-Lopez, J. J., Espinosa-Paredes, G. and Ochoa-Tapia, J. A. 2003 Mass transfer jump condition at the boundary between a porous medium and a homogeneous fluid. *J. Porous Media* **6**, 33–49. [2.4] van Duijn, C. J., Pieters, G. J. M., Wooding, R. A. and van der Ploeg, A. 2002 Stability criteria for the vertical boundary layer formed by throughflow near the surface of a porous medium. *Environmental Mechanics—Water, Mass and Energy Transfer in the Biosphere* (eds. P. A. C. Roots, D. Smiles and A. W. Warrick), American Geophysical Union, pp. 155–169. [6.10]

Van Dyne, D. G. and Stewart, W. E. 1994 Natural convection heat and mass transfer in a semi-cylindrical enclosure filled with a heat generating porous medium. *Int. Comm. Heat Mass Transfer* **21**, 271–281. [6.17]

Vanover, D. E. and Kulacki, F. A. 1987 Experimental study of mixed convection in a horizontal porous annulus. *ASME HTD* **84**, 61–66. [8.2.3]

Varahasamy, M. and Fand, R. M. 1996 Heat transfer by forced convection in pipes packed with porous media whose matrices are composed of spheres. *Int. J. Heat Mass Transfer* **39**, 3931–3947. [4.9]

Vargas, J. V. C., Laursen, T. A. and Bejan, A. 1995 Nonsimilar solutions for mixed convection on a wedge embedded in a porous medium. *Int. J. Heat Fluid Flow* **16**, 211–216. [8.1.4]

Vasantha, R., Pop, I. and Nath, G. 1986 Non-Darcy natural convection over a slender vertical frustum of a cone in a saturated porous medium. *Int. J. Heat Mass Transfer* **29**, 153–156. [5.8]

Vasile, C., Lorente, S. and Perrin, B. 1998 Study of convective phenomena inside cavities coupled with heat and mass transfers through porous media—Application to vertical hollow bricks—a first approach. *Energy Build.* **28**, 229–235. [7.7]

Vasseur, P. and Degan, G. 1998 Free convection along a vertical heated plate in a porous medium with anisotropic permeability. *Int. J. Numer. Methods Heat Fluid Flow* **8**, 43–63. [5.1.9.5]

Vasseur, P. and Robillard, L. 1987 The Brinkman model for boundary layer regime in a rectangular cavity with uniform heat flux from the side. *Int. J. Heat Mass Transfer* **30**, 717–728. [7.6.2]

Vasseur, P. and Robillard, L. 1993 The Brinkman model for natural convection in a porous layer: effects of non-uniform thermal gradient. *Int. J. Heat Mass Transfer* **36**, 4199–4206. [6.11.1]

Vasseur, P. and Robillard, L. 1998 Natural convection in enclosures filled with anisotropic porous media. *Transport Phenomena in Porous Media* (eds. D. B. Ingham and I. Pop), Elsevier, Oxford, pp. 331–356. [7.3.2]

Vasseur, P. and Wang, C. H. 1992 Natural convection heat transfer in a porous layer with multiple partitions. *Chem. Engng. Comm.* **114**, 145–167. [6.13.2]

Vasseur, P., Nguyen, T. H., Robillard, L. and Thi, V. K. T. 1984 Natural convection between horizontal concentric cylinders filled with a porous layer with internal heat generation. *Int. J. Heat Mass Transfer* **27**, 337–349. [6.17]

Vasseur, P., Robillard, L. and Anochiravani, I. 1988 Natural convection in an inclined rectangular porous cavity with uniform heat flux from the side. *Wärme-Stoffübertrag.* **22**, 69–77. [7.8]

Vasseur, P., Satish, M. G. and Robillard, L. 1987 Natural convection in a thin, inclined, porous layer exposed to a constant heat flux. *Int. J. Heat Mass Transfer* **30**, 537–550. [7.8]

Vasseur, P., Wang, C. H. and Sen, M. 1989 The Brinkman model for natural convection in a shallow porous cavity with uniform heat flux. *Numer. Heat Transfer* **15**, 221–242. [6.19.1.2, 7.6.2]

Vasseur, P., Wang, C. H. and Sen, M. 1990 Natural convection in an inclined rectangular porous slot: the Brinkman-extended Darcy model. *ASME J. Heat Transfer* **112**, 507–511. [7.8]

Vaszi, A. Z., Elliot, L., and Ingham, D. B. 2003 Conjugate free convection from vertical fins embedded in a porous medium. *Numer. Heat Transfer A* **44**, 743–770. [5.1.5]

Vaszi, A. Z., Elliot, L., Ingham, D. B. and Pop, I. 2001a Conjugate free convection above a cooled finite horizontal flat plate embedded in a porous medium. *Int. Comm. Heat Mass Transfer* **28**, 703–712. [5.2]

Vaszi, A. Z., Elliot, L., Ingham, D. B. and Pop, I. 2002a Conjugate free convection above a heated finite horizontal flat plate embedded in a porous medium. *Int. J. Heat Mass Transfer* **45**, 2777–2795. [5.2]

Vaszi, A. Z., Elliot, L., Ingham, D. B. and Pop, I. 2002b Conjugate free convection from a vertical plate fin embedded in a porous medium. *Heat Transfer 2002, Proc. 12th Int. Heat Transfer Conf.*, Elsevier, vol. 2, pp. 827–832. [5.12.1]

Vaszi, A. Z., Elliot, L., Ingham, D. B. and Pop, I. 2004 Conjugate free convection from a vertical plate fin with a rounded tip embedded in a porous medium. *Int. J. Heat Mass Transfer* **47**, 2785–2794. [5.12.1]

Vaszi, A. Z., Ingham, D. B., Lesnic, D., Munslow, D. and Pop, I. 2001b Conjugate free convection from a slightly inclined plate embedded in a porous medium. *Zeit. Angew. Math. Mech.* **81**, 465–479. [5.3]

Vaszi, A., Elliot, L. and Ingham, D. B. 2004 A study of conjugate natural convection from vertical fins at various shapes embedded in a porous medium. In *Applications of Porous Media (ICAPM 2004)*, (eds. A. H. Reis and A. F. Miguel), Évora, Portugal, pp. 96–106. [5.1.5]

Vedha-Nayagam, M., Jain, P. and Fairweather, G. 1987 The effect of surface mass transfer on buoyancy induced flow in a variable-porosity medium adjacent to a horizontal heated plate. *Int. Comm. Heat Mass Transfer* **14**, 495–506. [5.2]

Veinberg, A. K. 1967 Permeability, electrical conductivity, dielectric constant and thermal conductivity of a medium with spherical and ellipsoidal inclusions. *Soviet Phys. Dokl.* **11**, 593–595. [10.1.6]

Venkataraman, P. and Rao, P. R. M. 2000 Validation of Forchheimer's law for flow through porous media with converging boundaries. *J. Hydr. Engng.* **126**, 63–71. [1.5.2]

Vighnesan, N. V., Ray, S. N. and Soundalgekar, M. V. 2001 Oscillating plate temperature effects on mixed convection flow past a semi-infinite vertical porous plate. *Defence Sci. J.* **51**, 415–418. [8.1.1]

Vigo, T. L. and Bruno, J. S., 1987 Temperature-adaptable textiles containing durably bound polyethylene glycols. *Textile Res. J.* **57**, 427–429. [10.5]

Viljoen, H. J., Gatica, J. E. and Hlavacek, V. 1990 Bifurcation analysis of chemically driven convection. *Chem. Engng. Sci.* **45**, 503–517. [6.19.2]

Vincourt, M. C. 1989a Competition between two directions of convective rolls in a horizontal porous layer, non-uniformly heated. *Mechanics* **16**, 19–24. [6.15.1]

Vincourt, M. C. 1989b Influence of a heterogeneity on the selection of convective patterns in a porous layer. *Int. J. Engng. Sci.* **27**, 377–392. [6.15.1]

Viskanta, R. 2005 Combustion and heat transfer in inert porous media. *Handbook of Porous Media* (ed. K. Vafai), 2nd ed., Taylor and Francis, New York, pp. 607–644. [3.3]

Vitanov, N. K. 2000 Upper bounds on the heat transport in a porous layer. *Physica D* **136**, 322–339. [6.3]

Vorontsov, S.S., Gorin, A. V., Nakoyakov, V. Ye., Khoruzhenko, A.G. and Chupin, V.M. 1991 Natural convection in a Hele-Shaw cell. *Int. J. Heat Mass Transfer* **34**, 703–709. [2.5]

Vynnycky, M. and Kimura, S. 1994 Conjugate free convection due to a vertical plate in a porous medium. *Int. J. Heat Mass Transfer* **37**, 229–236. [5.1.5]

Vynnycky, M. and Kimura, S. 1995 Transient conjugate free convection due to a vertical plate in a porous medium. *Int. J. Heat Mass Transfer* **38**, 219–231. [5.1.5]

Vynnycky, M. and Pop, I. 1997 Mixed convection due to a finite horizontal flat plate embedded in a porous medium. *Journal Fluid Mech.* **351**, 359–378. [8.1.2]

Waite, M. W. and Amin, M. R. 1999 Numerical investigation of two-phase fluid flow and heat transfer in porous media heated from the side. *Numer. Heat Transfer A* **35**, 271–290. [11.9.3]

Wakao, N. and Kaguei, S. 1982 *Heat and Mass Transfer in Packed Beds*, Gordon and Breach, New York. [4.]

Wakao, N., Kaguei, S. and Funazkri, T. 1979 Effect of fluid dispersion coefficients on particle-to-fluid heat transfer coefficients in packed beds. *Chem . Engng. Sci.* **34**, 325–336. [2.2.2]

Wakao, N., Tanaka, K. and Nagai, H. 1976 Measurement of particle-to-gas mass transfer coefficients from chromatographic adsorption experiments. *Chem. Engng. Sci.* **31**, 1109–1113. [2.2.2]

Walker, K. L. and Homsy, G. M. 1977 A note on convective instabilities in Boussinesq fluids and porous media. *ASME J. Heat Transfer* **99**, 338–339. [6.6]

Walker, K. L. and Homsy, G. M. 1978 Convection in a porous cavity. *J. Fluid Mech.* **87**, 449–474. [7.1.3]

Wang, B. X. and Du, J. H. 1993 Forced convection heat transfer in a vertical annulus filled with porous media. *Int. J. Heat Mass Transfer* **36**, 4207–4213. [4.9]

Wang, B. X. and Zhang, X. 1990 Natural convection in liquid-saturated porous media between concentric inclined cylinders. *Int. J. Heat Mass Transfer* **33**, 827–833. [7.8]

Wang, C. and Tu, C. 1989 Boundary-layer flow and heat transfer of non-Newtonian fluids in porous media. *Int. J. Heat Fluid Flow* **10**, 160–165. [4.16.3]

Wang, C. Y. 1994 Thermal convective instability of a horizontal saturated porous layer with a segment of inhomogeneity. *Appl. Sci. Res.*, **52**, 147–160. [6.13.4]

Wang, C. Y. 1997 Onset of natural convection in a sector-shaped box containing a fluid-saturated porous medium. *Physics Fluids* **9**, 3570–3571. [6.15.1]

Wang, C. Y. 1998a Modeling multiphase flow and transport in porous media. *Transport Phenomena in Porous Media* (eds. D. B. Ingham and I. Pop). Elsevier, Oxford, pp. 383–410. [10.3.1]

Wang, C. Y. 1998b Onset of natural convection in a fluid-saturated porous medium inside a cylindrical enclosure bottom heated by constant flux. *Int. Comm. Heat Mass Transfer* **25**, 593–598. [6.16.1]

Wang, C. Y. 1999a Onset of convection in a fluid saturated porous layer overlying a solid layer which is heated by a constant flux. *ASME J. Heat Transfer* **121**, 1094–1097. [6.19.1]

Wang, C. Y. 1999b Onset of convection in a fluid-saturated rectangular box, bottom heated by constant flux. *Phys. Fluids* **11**, 1673–1675. [6.15.1]

Wang, C. Y. 1999c Thermo-convective stability of a fluid-saturated porous medium inside a cylindrical enclosure: Permeable top constant flux heating. *Mech. Res. Commun.* **26**, 603–608. [6.16.1]

Wang, C. Y. 2002 Convective stability in a rectangular box of fluid-saturated porous medium with constant pressure top and constant flux bottom heating. *Transport Porous Media* **46**, 37–42. [6.15.1]

Wang, C. Y. and Beckermann, C. 1993 A two-phase model of liquid-gas flow and heat transfer in capillary porous media.—II. Application to pressure-driven boiling flow adjacent to a vertical heated plate. *Int. J. Heat Mass Transfer* **36**, 2759–2768. [10.3.1]

Wang, C. Y. and Beckermann, C. 1995 Boundary layer analysis of buoyancy driven two-phase flow in capillary porous media. *ASME J. Heat Transfer* **117**, 1082–1087. [10.3.1, 10.4]

Wang, C. Y. and Cheng, P. 1996 A multiphase mixture model for multiphase, multicomponent transport in capillary porous media.—I. Model development. *Int. J. Heat Mass Transfer* **39**, 3607–3618. [3.5.5]

Wang, C. Y. and Cheng, P. 1997 Multiphase flow and heat transfer in porous media. *Adv. Heat Transfer* **30**, 93–196. [3.6]

Wang, C. Y. and Cheng, P. 1998 Multidimensional modeling of steam injection into porous media. *ASME J. Heat Transfer* **120**, 286–290. [3.6]

Wang, C. Y., Beckermann, C. and Fan, C. 1994a Transient natural convection and boiling in a porous layer heated from below. *Heat Transfer 1994*, Inst. Chem. Engrs, Rugby, vol 5, pp. 411–416. [10.3.1]

Wang, C. Y., Beckermann, C. and Fan, C. 1994b Numerical study of boiling and natural convection in capillary porous media using the two-phase mixture model. *Numer. Heat Transfer A* **26**, 375–398. [10.3.1]

Wang, C. Y., Wu, C. Z., Tu, C. J. and Fukusako, S. 1990 Freezing around a vertical cylinder immersed in porous media incorporating the natural convection effect. *Wärme-Stoffübertrag.* **26**, 7–15. [10.2.1.2]

Wang, C., Liao, S. J. and Zhu, J. M. 2003a An explicit solution for the combined heat and mass transfer by natural convection from a vertical wall in a non-Darcy porous medium. *Int. J. Heat Mass Transfer* **46**, 4813–4822. [9.2.1]

Wang, C., Liao, S. J. and Zhu, J. M. 2003b An explicit analytic solution for non-Darcy natural convection over horizontal plate with surface mass and thermal dispersion effects. *Acta Mech.* **165**, 139–150. [9.2.1]

Wang, C., Tu, C. and Zhang, X. 1990 Mixed convection of non-newtonian fluids from a vertical plate embedded in a porous medium. *Acta Mech. Sinica* **6**, 214–220. [8.1.1]

Wang, C., Zhu, J. M., Liao, S. J. and Pop, I. 2003c On the explicit analytic solution of Cheng-Chang equation. *Int. J. Heat Mass Transfer* **46**, 1855–1860. [5.2]

Wang, H. and Takle, E. 1995 Boundary-layer flow and turbulence near porous obstacles. *Boundary-Layer Meteor.* **74**, 73–88. [1.8]

Wang, M. and Bejan, A. 1987 Heat transfer correlation for Bénard convection in a fluid saturated porous layer. *Int. Comm. Heat Mass Transfer* **14**, 617–626. [6.9.2]

Wang, M. and Georgiadis, J. G. 1996 Conjugate forced convection in crossflow over a cylinder array with volumetric heating. *Int. J. Heat Mass Transfer* **39**, 1351–1361. [4.16.2]

Wang, M., Kassoy, D. R. and Weidman, P. D. 1987 Onset of convection in a vertical slab of saturated porous media between two impermeable conducting blocks. *Int. J. Heat Mass Transfer* **30**, 1331–1341. [6.15.2]

Wang, S. C., Chen, C. K., Yang, Y. T. and Yang, Y. T. 2002 Natural convection of non-Newtonian fluids through permeable axisymmetric and two-dimensional bodies in a porous medium. *Int. J. Heat Mass Transfer* **45**, 393–408. [5.9]

Wang, S. C., Yang, Y. T. and Chen, C. K. 2003d Effect of uniform suction on laminar filmwise condensation on a finite-size horizontal flat surface in a porous medium. *Int. J. Heat Mass Transfer* **46**, 4003–4011. [10.4]

Wang, W. and Sangani, A. S. 1997 Nusselt number for flow perpendicular to arrays of cylinders in the limit of small Reynolds and large Peclet numbers. *Phys. Fluids* **9**, 1529–1539. [4.16.2]

Wang, X., Thauvin, F. and Mohanty, K. K. 1999 Non-Darcy flow through anisotropic porous media. *Chem. Engng. Sci.* **54**, 1859–1869. [1.5.2]

Ward, J. C. 1964 Turbulent flow in porous media. *ASCE J. Hydraul. Div.* **90** (HY5), 1–12. [1.5.2, 10.1.6]

Weaver, J. A. and Viskanta, R. 1986 Freezing of liquid saturated porous media. *ASME J. Heat Transfer* **108**, 654–659. [10.2.2]

Webb, S. W., Francis, N. D., Dunn, S. D., Itamura, M. T. and James, D. L. 2003 Thermally induced natural convection effects in Yucca mountain drifts. *J. Contam. Hydrol.* **62**, 713–730. [11.8]

Weber, J. E. 1974 Convection in a porous medium with horizontal and vertical temperature gradients. *Int. J. Heat Mass Transfer* **17**, 241–248. [7.9]

Weber, J. E. 1975a Thermal convection in a tilted porous layer. *Int. J. Heat Mass Transfer* **18**, 474–475. [7.8]

Weber, J. E. 1975b The boundary-layer regime for convection in a vertical porous layer. *Int. J. Heat Mass Transfer* **18**, 569–573. [7.1.2, 7.1.3, 7.2, 7.6.1, 10.1.2]

Weber, M. and Kimmich, R. 2002 Rayleigh-Bénard percolation transition of thermal convection in porous media: Computational fluid dynamics, NMR velocity mapping, NMR temperature mapping. *Phys. Rev. E* **66**, article no. 056301. [6.9.1]

Weber, M., Klemm, A. and Kimmich, R. 2001 Rayleigh-Bénard percolation transition study of thermal convection in porous media: Numerical simulation and NMR experiments. *Phys. Rev. Lett.* **86**, 4302–4305. [6.9.1]

Wei, Q. 2004 Bounds on convective heat transfer in a rotating porous layer. *Mech. Res. Commun.* **31**, 269–276. [6.22]

Weidman, P. D. and Amberg, M. F. 1996 Similarity solutions for steady laminar convection along heated plates with variable oblique suction: Newtonian and Darcian fluid flow. *Quart. J. Mech. Appl. Math.* **49**, 373–403. [8.1.1]

Weidman, P. D. and Kassoy, D. R. 1986 Influence of side wall heat transfer on convection in a confined saturated porous medium. *Phys. Fluids* **29**, 349–355. [6.15.2]

Weinert, A. and Lage, J. L. 1994 Porous aluminum-alloy based cooling devices for electronics. SMU-MED-CPMA Inter. Rep. 1.01/94, Southern Methodist University, Dallas, TX. [1.5.3]

Weinitschke, H. J., Nandakumar, K. and Sankar, S. R. 1990 A bifurcation study of convective heat transfer in porous media. *Phys. Fluids A* **2**, 912–921. [6.17]

Weir, G.J. 1991 Geometric properties of two phase flow in geothermal reservoirs. *Transport in Porous Media* **6**, 501–517. [11.9.1]

Weir, G. J. 1994a The relative importance of convective and conductive effects in two-phase geothermal fields. *Transport in Porous Media* **16**, 289–295. [11.9.1]

Weir, G. J. 1994b Nonreacting chemical transport in two-phase reservoirs—factoring diffusive and wave properties. *Transport in Porous Media* **17**, 201–220. [11.9.1]

Weisman, C., Le Quéré, P. and Firdaouss, M. 1999 On the closed form solution of natural convection in a cavity partially filled with a porous medium. *C. R. Acad. Sci. Paris, IIB*, **327**, 235–240. [7.7]

Weiss, D. W., Stickler, L. A. and Stewart, W. E. 1991 The effect of water density extremum on heat transfer within a cylinder containing a heat-generating porous medium. *Int. Comm. Heat Mass Transfer*, **18**, 259–271. [6.17]

Wettlaufer, J. S., Worster, M. G. and Huppert, H. E. 1997 Natural convection during solidification of an alloy from above with application to the evolution of sea ice. *J. Fluid Mech.* **344**, 291–316. [10.2.3]

Whitaker, S. 1986 Flow in porous media J: A theoretical derivation of Darcy's law. *Transport in Porous Media* **1**, 3–25. [1.4.3]

Whitaker, S. 1996 The Forchheimer equation: a theoretical development. *Transport in Porous Media* **25**, 27–61. [1.5.2]

White, S. M. and Tien, C. L. 1987 Analysis of laminar film condensation in a porous medium. *Proceedings 1987 ASME JSME Thermal Engineering Joint Conference*, ASME, New York, vol. 2, pp. 401–406. [10.4]

Wilcock, W. S. D. 1998 Cellular convection models of mid-ocean ridge hydrothermal circulation and the temperatures of black smoker fluids. *J. Geophys. Res.* **103**, 2585–2596. [11.8]

Wilkes, K. F. 1995 Onset of natural convection in a horizontal porous medium with mixed thermal boundary conditions. *ASME J. Heat Transfer* **117**, 543–547. [6.2]

Wong, W. S., Rees, D. A. S. and Pop, I. 2004 Forced convection past a heated cylinder in a porous medium using a nonequilibrium model; finite Péclet number effects. *Int. J. Thermal Sci.* **43**, 213–220. [4.10]

Wooding, R. A. 1957 Steady state free thermal convection of liquid in a saturated permeable medium. *J. Fluid Mech.* **2**, 273–285. [1.5.1]

Wooding, R. A. 1959 The stability of a viscous liquid in a vertical tube containing porous material. *Proc. Roy. Soc. London Ser. A* **252**, 120-134. [6.16.1, 11.3]

Wooding, R. A. 1960 Rayleigh instability of a thermal boundary layer in flow through a porous medium. *J. Fluid Mech.* **9**, 183–192 [8.1.1]

Wooding, R. A. 1963 Convection in a saturated porous medium at large Reynolds number or Péclet number. *J. Fluid Mech.* **15**, 527–544. [5.10.1.1, 5.11.1]

Wooding, R. A. 1978 Large-scale geothermal field parameters and convection theory. *N. Z. J. Sci.* **27**, 219–228. [6.12, 6.13.3]

Wooding, R. A. 1985 Convective plumes in saturated porous media. *Convective Flows in Porous Media* (eds. R. A. Wooding and I. White) Dept. Sci. Indust. Res., Wellington, N. Z., pp. 167–178. [5.11.1]

Wooding, R. A. 2005 Variable-density unstable plumes in a porous medium, a salt lake problem. *Water Resources Res.*, under review. [1.3, 9.1.6.4]

Wooding, R. A., Tyler, S. W. and White, I. 1997a Convection in groundwater below an evaporating salt lake: 1. Onset of instability. *Water Resources Res.* **33**, 1199–1217. [6.10, 9.1.6.4]

Wooding, R. A., Tyler, S. W., White, I. and Anderson, P. A. 1997b Convection in groundwater below an evaporating salt lake: 2. Evolution of fingers or plumes. *Water Resources Res.* **33**, 1219–1228. [6.10, 9.1.6.4]

Woods, A. W. 1999 Liquid and vapor flow in superheated rock. *Ann. Rev. Fluid Mech.* **31**, 171–199. [11.9.3]

Worster, M. G. 1991 Natural convection in a mushy layer. *J. Fluid Mech.* **224**, 335–359. [10.2.3]

Worster, M. G. 1992 Instabilities of the liquid and mushy regions during solidification of alloys. *J. Fluid Mech.* **237**, 649–669. [10.2.3]

Worster, M. G. 1997 Convection in mushy layers. *Annu. Rev. Fluid Mech.* **29**, 91–122. [10.2.3]

Worster, M. G. 2000 Solidification of fluids. *Perspectives in Fluid Dynamics* (eds. G. K. Batchelor, H. K. Moffat and M. G. Worster), Cambridge University press, Cambridge, UK, pp. 393–446. [10.2.3]

Worster, M. G. and Kerr, R. C. 1994 The transient behavior of alloys solidified from below prior to the formation of chimneys. *J. Fluid Mech.* **269**, 23–44. [10.2.3]

Wright, S. D., Ingham, D. B. and Pop, I. 1996 On natural convection from a vertical plate with a prescribed surface heat flux in porous media. *Transport in Porous Media* **22**, 181–193. [5.1.1]

Wu, C. C. and Hwang, G. J. 1998 Flow and heat transfer characteristics inside packed and fluidized beds. *ASME J. Heat Transfer* **120**, 667–673. [4.6.5]

Wu, R. S., Cheng, K. C. and Craggs, A. 1979 Convective instabilities in porous media with maximum density and throughflow effects by finite-difference and finite-element methods. *Numer. Heat Transfer* **2**, 303–318. [6.10]

Xiong, M. and Kuznetsov, A. V. 2000 Forced convection in a Couette flow in a composite duct: An analysis of thermal dispersion and non-Darcian effects. *J. Porous Media* **3**, 245–255. [4.11]

Xu, H. 2004 An explicit analytic solution for free convection about a vertical plate embedded in a porous medium by means of homotopy analysis method. *Appl. Math. Comput.* **158**, 433–443. [5.1.1]

Xu, W. and Lowell, R. P. 1998 Oscillatory instability of one-dimensional two-phase hydrothermal flow in heterogeneous porous media. *J. Geophys. Res.* **103**, 20859–20868. [11.9.1]

Yamaguchi, Y., Asako, Y. and Nakamura, H. 1993 Three-dimensional natural convection in a vertical porous layer with a hexagonal core of negligible thickness. *Int. J. Heat Mass Transfer*, **36**, 3403–3406. [7.3.7]

Yamamoto, K. 1974 Natural convection about a heated sphere in a porous medium. *J. Phys. Soc. Japan* **37**, 1164–1166. [5.6.2]

Yan, B. and Pop, I. 1998 Unsteady forced convection heat transfer about a sphere in a porous medium. In *Mathematics of Heat Transfer* (G. E. Tupholme and A. S. Wood, eds.) Clarendon Press, Oxford, pp. 337–344. [4.6.4]

Yan, B., Pop, I. and Ingham, D. B. 1997 A numerical study of unsteady free convection from a sphere in a porous medium. *Int J. Heat Mass Transfer* **40**, 893–903. [5.6.1]

Yan, Y. H., Ochterbeck, J. M. and Peng, X. F. 1998 Numerical study of capillary rewetting in porous media. *Heat Transfer 1998, Proc. 11th IHTC*, **4**, 509–514. [3.6]

Yang, C. H., Rastrogi, S. K. and Poulikakos, D. 1993a Freezing of a water-saturated inclined packed bed of beads. *Int. J. Heat Mass Transfer* **36**, 3583–3592. [10.2.1.2]

Yang, C. H., Rastrogi, S. K. and Poulikakos, D. 1993b Solidification of a binary mixture saturating an inclined bed of packed spheres. *Int. J. Heat Fluid Flow* **14**, 268–278. [10.2.3]

Yang, J. H. and Lee, S. L. 1999 Effect of anisotropy on transport phenomena in anisotropic porous media. *Int. J. Heat Mass Transfer* **42**, 2673–2681. [4.16.2]

Yang, J. W. and Edwards, R. N. 2000 Predicted groundwater circulation in fractured and unfractured anisotropic porous media driven by nuclear fuel waste heat generation. *Canad. J. Earth Sci.* **37**, 1301–1308. [2.6]

Yang, Y. T. and Hwang, C. Z. 2003 Calculation of turbulent flow and heat transfer in a porous-baffled channel. *Int. J. Heat Mass Transfer* **46**, 771–780. [1.8]

Yang, Y. T. and Wang, S. J. 1996 Free convection heat transfer of non-Newtonian fluids over axisymmetric and two-dimensional bodies of arbitrary shape embedded in a fluid-saturated porous medium. *Int. J. Heat Mass Transfer* **39**, 203–210. [5.9]

Yee, C. K. and Lai, F. C. 2001 Effects of a porous manifold on thermal stratification in a liquid storage tank. *Solar Energy* **71**, 241–254. [8.3.1]

Yee, S. S. and Kamiuto, K. 2002 Effect of viscous dissipation on forced-convection heat transfer in cylindrical packed beds. *Int. J. Heat Mass Transfer* **45**, 461–464. [4.9]

Yen, Y. C. 1974 Effects of density inversion on free convective heat transfer in porous layer heated from below. *Int. J. Heat Mass Transfer* **17**, 1349–1356. [6.9.1, 6.9.2]

Yih, K. A. 1997 The effect of transpiration on coupled heat and mass transfer in mixed convection over a vertical plate embedded in a saturated porous medium. *Int. Comm. Heat Mass Transfer* **24**, 265–275. [9.6]

Yih, K. A. 1997 The effect of uniform lateral mass flux on free convection about a vertical cone embedded in a saturated porous medium. *Int. Comm. Heat Mass Transfer* **24**, 1195–1205. [5.8]

Yih, K. A. 1998a Heat source/sink effect on MHD mixed convection in stagnation flow on a vertical permeable plate in porous media. *Int. Comm. Heat Mass Transfer* **25**, 427–442. [8.1.4]

Yih, K. A. 1998b Uniform lateral mass flux effect on natural convection on non-Newtonian fluids over a cone in porous media. *Int. Comm. Heat Mass Transfer* **25**, 959–968. [5.8]

Yih, K. A. 1998c Coupled heat and mass transfer in mixed convection over a wedge with variable wall temperature and convection in porous media: The entire regime. *Int. Comm. Heat Mass Transfer* **25**, 1145–1158. [9.6]

Yih, K. A. 1998d The effect of uniform suction/blowing on heat transfer of magnetohydrodynamic Hiemenz flow through porous media. *Acta Mech.* **130**, 147–158. [4.16.2]

Yih, K. A. 1998e Blowing/suction effect on non-Darcy forced convection flow about a flat plate with variable wall temperature in porous media. *Acta Mech.* **131**, 255–265. [4.16.2]

Yih, K. A. 1998f Coupled heat and mass transfer in mixed convection over a vertical flat plate embedded in saturated porous media: PST/PSC or PHF/PMF. *Heat Mass Transfer* **34**, 55–61. [9.6]

Yih, K. A. 1998g Coupled heat and mass transfer in mixed convection about a vertical cylinder in a porous medium: the entire regime. *Mech. Res. Commun.* **25**, 623–630. [9.6]

Yih, K. A. 1999a Uniform transpiration effect on combined heat and mass transfer by natural convection over a cone in saturated porous media: uniform wall temperature, concentration or heat flux. *Int. J. Heat Mass Transfer* **42**, 3533–3537. [9.2.1]

Yih, K. A. 1999b Uniform transpiration effect of coupled heat and mass transfer, in mixed convection about inclined surfaces in porous media: the entire regime. *Acta Mech.* **132**, 229–240. [9.6]

Yih, K. A. 1999c Coupled heat and mass transfer in mixed convection over a VHF/VMF wedge in porous media—the entire regime. *Acta Mech.* **137**, 1–12. [9.6]

Yih, K. A. 1999d Coupled heat and mass transfer by free convection over a truncated cone in porous media: VWT-VWC or VHF-VMF. *Acta Mech.* **137**, 83–97. [9.2.1]

Yih, K. A. 1999e Coupled heat and mass transfer in mixed convection over a wedge with variable wall temperature and convection in porous media: The entire regime. *Int. Comm. Heat Mass Transfer* **26**, 259–267. [5.7]

Yih, K. A. 1999f Coupled heat and mass transfer by natural convection adjacent to a permeable horizontal cylinder in a saturated porous medium. *Int. Comm. Heat Mass Transfer* **26**, 431–440. [9.2.1]

Yih, K. A. 1999g Mixed convection about a cone in a porous medium—the entire regime. *Int. Comm. Heat Mass Transfer* **26**, 1041–1050. [8.1.4]

Yih, K. A. 1999h Uniform transpiration effect on coupled heat and mass transfer in mixed convection about a vertical cylinder in porous media: the entire regime. *J. Chinese Soc. Mech. Engrs*. **20**, 81–86. [9.6]

Yih, K. A. 1999i Blowing/suction effect on combined convection in stagnation flow over a vertical plate embedded in a porous medium. *Chinese J. Mech. A*. **15**, 41–45. [8.1.1]

Yih, K. A. 2000a Viscous and Joule heating effects on non-Darcy MHD natural convection flow over a permeable sphere in porous media with internal heat generation. *Int. Comm. Heat Mass Transfer* **27**, 591–600. [5.6.1]

Yih, K. A. 2000b Combined heat and mass transfer in mixed convection adjacent to a VWT/VWC or VHF/VMF cone in a porous medium: The entire regime. *J. Porous Media* **3**, 185–191. [9.6]

Yih, K. A. 2001a Radiation effect on mixed convection over an isothermal wedge in porous media: The entire regime. *Heat Transfer Engng*. **22**, 26–32. [8.1.4]

Yih, K. A. 2001b Radiation effect on mixed convection over an isothermal cone in porous media. *Heat Mass Transfer* **37**, 53–57. [8.1.4]

Yokoyama, Y. and Kulacki, F. A. 1996 Mixed convection in a horizontal duct with a sudden expansion and local duct heating. *ASME HTD*. **331**, 33–41. [8.2.2]

Yokoyama, Y., Kulacki, F. A. and Mahajan, R. L. 1999 Mixed convection in a horizontal porous duct with a sudden expansion and local heating from below. *ASME J. Heat Transfer* **121**, 653–661. [8.2.1]

Yoo, H and Viskanta, R. 1992 Effect of anisotropic permeability in the transport process during solidification of a binary mixture. *Int. J. Heat Mass Transfer* **35**, 2335–2346. [10.2.3]

Yoo, J. S. 2003 Thermal convection in a vertical porous slot with spatially periodic boundary conditions: low Ra flow. *Int. J. Heat Mass Transfer* **46**, 381–384. [7.1.6]

Yoo, J. S. and Schultz, W. W. 2003 Thermal convection in a horizontal porous layer with spatially periodic boundary temperature temperatures: small Ra flow. *Int. J. Heat Mass Transfer* **46**, 4747–4750. [6.14]

Yoon, D. Y., Choi, C. K. and Yoo, J. S. 1992 Analysis of thermal instability in a horizontal, porous layer heated from below. *Int. Chem. Engng*. **32**, 181–191. [6.11.3]

Yoon, D. Y., Kim, D. S. and Chang, K. C. 2004 The onset of oscillatory convection in a horizontal porous layer saturated with viscoelastic liquid. *Transport Porous Media* **55**, 275–284. [6.23]

Yoon, D. Y., Kim, D. S. and Choi, C. K. 1998 Convective instability in packed beds with internal heat sources and throughflow. *Korean J. Chem. Engng*. **15**, 341–344. [6.11.2]

Yoon, D. Y., Kim, M. C. and Choi, C. K. 2003 Oscillatory convection in a horizontal porous layer saturated with a viscoelastic fluid. *Korean J. Chem. Engng*. **20**, 27–31. [6.23]

Yoshino, M. and Inamura, T. 2003 Lattice Boltzmann simulations for flow and heat/mass transfer problems in a three-dimensional porous structure. *Int. J. Numer. Meth. Fluids* **43**, 183–198. [2.6]

You, H. I. and Song, J. S. 1999 Heat transfer analysis of non-Darcian flow in local thermal non-equilibrium. *J. Chinese Soc. Mech. Engrs*. **20**, 75–80. [4.10]

Young, R. 1993a Two-phase brine mixtures in the geothermal context and the polymer flood model. *Transport in Porous Media* **11**, 179–185. [11.9.1]

Young, R. 1993b Two-phase geothermal flows with conduction and the connection with Buckley-Leverett theory. *Transport in Porous Media* **12**, 261–278. [11.9.1]

Young, R. and Weir, G. 1994 Constant rate production of geothermal fluid from a two-phase vertical column. I: Theory. *Transport in Porous Media* **14**, 265–286. [11.9.1]

Young, R. M. 1996a Phase transitions in one-dimensional steady state hydrothermal flows. *J. Geophys. Res.* **101**, 18011–18022. [11.9.2]

Young, R. M. 1996b A basic model for vapour-dominated geothermal reservoirs. *Proceedings of the 18th NZ Geothermal Workshop*, University of Auckland, Auckland, New Zealand, 301–304. [11.9.2]

Young, R. M. 1998a Classification of one-dimensional steady-state two-phase geothermal flows including permeability variations: Part 1. Theory and special cases. *Int. J. Heat Mass Transfer* **41**, 3919–3935. [11.9.2]

Young, R. M. 1998b Classification of one-dimensional steady-state two-phase geothermal flows including permeability variations: Part 2. The general case. *Int. J. Heat Mass Transfer* **41**, 3937–3948. [11.9.2]

Young, T. J. and Vafai, K. 1998 Convective flow and heat transfer in a channel containing multiple heated obstacles. *Int. J. Heat Mass Transfer* **41**, 3279–3298. [4.11]

Young, T. J. and Vafai, K. 1999 Experimental and numerical investigation of forced convective characteristics of arrays of channel mounted obstacles. *ASME J. Heat Transfer* **121**, 34–42. [4.11]

Younsi, R., Harkati, A. and Kalache, D. 2001 Heat and mass transfer in composite fluid-porous layer: effect of permeability. *Arab J. Sci. Engng.* **26**, 145–155. [9.4]

Younsi, R., Harkati, A. and Kalache, D. 2002a Numerical simulation of double-diffusive natural convection in a porous cavity: opposing flow. *Arab J. Sci. Engng.* **27**, 181–194. [9.6]

Younsi, R., Harkati, A. and Kalache, D. 2002b Numerical simulation of thermal and concentration natural convection in a porous cavity in the presence of an opposing flow. *Fluid Dyn.* **37**, 854–864. [9.6]

Yu, B. 2004 Discussion: "A numerical study of thermal dispersion in porous media" and "Numerical determination of thermal dispersion coefficients using a periodic porous structure. *ASME J. Heat Transfer* **126**, 1060–1061. [2.2.2]

Yu B.M. and Cheng P. 2002a Fractal models for the effective thermal conductivity of bidispersed porous media. *J. Thermophys. Heat Transfer* **16**, 22–29. [4.16.4]

Yu B.M. and Cheng P. 2002b A fractal permeability model for bi-dispersed porous media. *Int. J. Heat Mass Transfer* **45**, 2983–2993. [4.16.4]

Yu, W. B., Ya, Y. M., Zhang, X. F., Zhang, S.J. and Xiao, J. Z. 2004 Laboratory investigation on coding effect of coarse rock layer and fine rock layer in permafrost regions. *Cold Regions Sci. Tech.* **38**, 31–42. [4.16.2]

Yu, W. P., Wang, B. X. and Shi, M. H. 1993 Modelling of heat and mass transfer in unsaturated wet porous medal with consideration of capillary hysteresis. *Int. J. Heat Mass Transfer* **36**, 3671–3676. [3.6]

Yu, W. S., Lin, H. T. and Lu, C. S. 1991 Universal formulations and comprehensive correlations for non-Darcy natural convection and mixed convection in porous media. *Int. J. Heat Mass Transfer* **34**, 2859–2868. [8.1.1, 8.1.2]

Yücel, A. 1990 Natural convection heat and mass transfer along a vertical cylinder in a porous medium. *Int. J. Heat Mass Transfer* **33**, 2265–2274. [9.4]

Yücel, A. 1993 Mixed convective heat and mass transfer along a vertical surface in a porous medium. *ASME HTD* **240**, 49–57. [9.6]

Zaturska, M. B. and Banks, W. H. H. 1987 On the spatial stability of free-convection flows in a saturated porous medium. *J. Engng. Math.* **21**, 41–46. [5.1.9.9]

Zebib, A. 1978 Onset of natural convection in a cylinder of water saturated porous media. *Phys. Fluids* **21**, 699–700. [6.16.1]

Zebib, A. and Kassoy, D. R. 1978 Three dimensional natural convection motion in a confined porous medium. *Phys. Fluids* **21**, 1–3. [6.15.1]

Zenkovskaya, S. M. 1992 Action of high-frequency vibration on filtration convection. *J. Appl. Mech. Tech. Phys.* **32**, 83–86. [6.24].

Zenkovskaya, S. M. and Rogovenko, T. N. 1999 Filtration convection in a high-frequency vibration field. *J. Appl. Mech. Tech. Phys.* **40**, 379–385. [6.24]

Zhang, B. L. and Zhao, Y. 2000 A numerical method for simulation of forced convection in a composite porous/fluid system. *Int. J. Heat Fluid Flow* **21**, 432–441. [4.11]

Zhang, H. Y. and Huang, X. Y. 2001 A two-equation analysis of convection in porous media. *Transport Porous Media* **44**, 305–324. [Correction in **46** (2002), 113–115.] [4.10]

Zhang, K., Liao, X. H. and Schubert, G. 2005 Pore water convection within carbonaceous chondrite parent bodies: Temperature-dependent viscosity and flow structure. *Phys. Fluids* **17**, Art. No. 086602. [6.17]

Zhang, X. 1993 Natural convection and heat transfer in a vertical cavity filled with an ice-water saturated porous medium. *Int. J. Heat Mass Transfer* **36**, 2881–2890. [10.1.7]

Zhang, X. L. and Kahawita, R. 1994 Ice water convection in an inclined rectangular cavity filled with a porous medium. *Wärme-Stoffübertrag.* **30**, 9–16. [7.8]

Zhang, X. and Nguyen, T. H. 1990 Development of convective flow during the melting of ice in a porous medium heated from above. *ASME HTD* **156**, 1–6. [10.1.7]

Zhang, X. L. and Nguyen, T. H. 1994 Numerical study of convection heat transfer during the melting of ice in a porous layer. *Numer. Heat Transfer A* **25**, 559–574. [10.1.7]

Zhang, X. L. and Nguyen, T. H. 1999 Solidification of a superheated fluid in a porous medium: effects of convection. *Int. J. Numer. Meth. Heat Fluid Flow* **9**, 72–91. [10.2.2]

Zhang, X. L., Nguyen, T. H. and Kawawita, R. 1997 Effects of anisotropy in permeability on two-phase flow and heat transfer in porous cavity. *Heat Mass Transfer* **32**, 167–174. [10.1.7]

Zhang, X., Hung, N. T. and Kawawita, R. 1993 Convective flow and heat transfer in an anisotropic porous layer with principal axes non-coincident with the gravity vector. *ASME HTD* **264**, 79–86.

Zhang, X., Nguyen, T. H. and Kahawita, R. 1991a Melting of ice in a porous medium heated from below. *Int. J. Heat Mass Transfer* **34**, 389–405. [10.1.7]

Zhang, Y., Khodadadi, J. M. and Shen, F. 1999 Pseudosteady-state natural convection inside spherical containers partially filled with a porous medium. *Int. J. Heat Mass Transfer* **42**, 2327–2336. [7.7]

Zhang, Y., Lu, N. and Ross, B. 1994 Convective instability of moist gas in a porous medium. *Int. J. Heat Mass Transfer*, **37**, 129–138. [6.7]

Zhang, Z. and Bejan, A. 1987 The horizontal spreading of thermal and chemical deposits in a porous medium. *Int. J. Heat Mass Transfer* **30**, 2289–2303. [9.2.3]

Zhang, Z., Bejan, A. and Lage, J. L 1991b Natural convection in a vertical enclosure with internal permeable screen. *ASME J. Heat Transfer* **113**, 377–383. [7.7]

Zhang, Z. J., Du, J. H. and Wang, B. X. 1999 Effect of viscous dissipation on forced-convection heat transfer in porous media. *J. Shangai Jiaotong Univ.* **33**, 979–982. [4.9]

Zhao, C. B., Hobbs, B. E. and Mühlhaus, H. B. 1998a Finite element modeling of temperature gradient driven rock alteration and mineralization in porous rock masses. *Comput. Meth. Appl. Math.* **165**, 175–187. [11.8]

Zhao, C. B., Hobbs, B. E. and Mühlhaus, H. B. 1999a Effects of medium thermoelasticity on high Rayleigh number steady-state heat transfer and mineralization in deformable fluid-saturated porous media heated from below. *Comput. Meth. Appl. Mech. Engng.* **173**, 41–54. [11.8]

Zhao, C. B., Hobbs, B. E. and Mühlhaus, H. B. 1999b Theoretical and numerical analyses of convective instability in porous media with upward throughflow. *Int. J. Numer. Anal. Methods Geomech.* **23**, 629–646. [6.10]

Zhao, C. B., Hobbs, B. E. and Ord, A. 2003a Effect of material anisotropy on the onset of convective flow in three-dimensional fluid-saturated faults. *Math. Geology* **35**, 141–154. [11.8]

Zhao, C. B., Hobbs, B. E., Baxter, K., Mühlhaus, H. B. and Ord, A. 1999c A numerical study of pore-fluid, thermal and mass flow in fluid-saturated porous rock basins. *Engng. Comput.* **16**, 202–214. [11.8]

Zhao, C. B., Hobbs, B. E., Mühlhaus, H. B. and Ord, A. 1999d Finite-element analysis of flow problems near geological lenses in hydrodynamic and hydrothermal systems. *Geophys. J. Inter.* **138**, 146–158. [11.8]

Zhao, C. B., Hobbs, B. E., Mühlhaus, H. B., Ord, A. and Lin, G. 2002 Analysis of steady-state heat transfer through mid-crustal vertical cracks with upward throughflow in hydrothermal systems. *Int. J. Numer. Anal. Meth. Geomech.* **26**, 1477–1491. [11.8]

Zhao, C. B., Hobbs, B. E., Mühlhaus, H. B., Ord, A. and Lin, G. 2000a Numerical modeling of double diffusion driven reactive flow transport in deformable fluid-saturated porous media with particular consideration of temperature-dependent chemical reaction rates. *Engng. Comput.* **17**, 367–385. [11.8]

Zhao, C. B., Hobbs, B. E., Mühlhaus, H. B., Ord, A. and Lin, G. 2001a Finite element modeling of three-dimensional convection problems in fluid-saturated porous media heated from below. *Comm. Numer. Methods Engng.* **17**, 101–114. [11.8]

Zhao, C. B., Hobbs, B. E., Mühlhaus, H. B., Ord, A. and Lin, G. 2003b Convective instability of 3–D fluid-saturated geological fault zones heated from below. *Geophys. J. Int.* **155**, 213–220. [11.8]

Zhao, C. B., Hobbs, B. E., Ord, A., Lin, G. and Mühlhaus, H. B. 2003c An equivalent algorithm for simulating thermal effects of magma intrusion problems in porous rocks. *Comput. Meth. Appl. Mech. Engng.* **192**, 3397–3408. [11.8]

Zhao, C. B., Hobbs, B. E., Ord, A., Peng, P. L., Mühlhaus, H. B. and Liu, L. M. 2004a Theoretical investigation of convective instability in inclined and fluid-saturated three-dimensional fault zones. *Tectonophysics* **387**, 47–64. [11.8]

Zhao, C. B., Hobbs, B. E., Ord, A., Peng, S. L., Mühlhaus, H. B. and Liu, L. M. 2005 Double diffusion-driven convective instability of three-dimensional fluid-saturated geological fault zones heated from below. *Math. Geology* **37**, 373–391. [11.8]

Zhao, C. B., Hobbs, B. E., Walshe, J. L., Mühlhaus, H. B., and Ord, A. 2001b Finite element modeling of fluid-rock interaction problems in pore-fluid saturated hydrothermal/sedimentary basins. *Comput. Meth. Appl. Mech. Engng.* **190**, 2277–2293. [11.8]

Zhao, C. B., Mühlhaus, H. B. and Hobbs, B. F. 1997 Finite element analysis of steady state natural convection problems in fluid-saturated porous media heated from below. *Int. J. Numer. Anal. Methods Geomech.* **21**, 863–881. [11.8]

Zhao, C. Y. and Lu, T. J. 2002 Analysis of microchannel heat sinks for electronic cooling. *Int. J. Heat Mass Transfer* **45**, 4857–4869. [4.10]

Zhao, C. Y., Kim, T., Lu, T. J. and Hodson, H. P. 2004b Thermal transport in high porosity cellular metal foams. *J. Thermophys. Heat Transfer* **18**, 309–317. [4.9]

Zhao, C., Mühlhaus, H. B. and Hobbs, B. F. 1998b Effects of geological inhomogeneity on high Rayleigh number steady state heat and mass transfer in fluid-saturated porous media heated from below. *Numer. Heat Transfer A* **33**, 415–431. [11.8]

Zhao, J. Z. and Chen, T. S. 2002 Inertia effects on non-parallel thermal instability of natural convection flow over horizontal and inclined plates in porous media. *Int. J. Heat Mass Transfer* **45**, 2265–2276. [5.4]

Zhao, J. Z. and Chen, T. S. 2003 Non-Darcy effects on non-parallel thermal instability of horizontal natural convection flow. *J. Thermophys. Heat Transfer* **17**, 150–158. [5.4]

Zhao, M., Robillard, L. and Prud'homme, M. 1996 Effect of weak rotation on natural convection in a horizontal porous cylinder. *Heat Mass Transfer* **31**, 403–409. [6.17]

Zhao, P. and Chen, C. F. 2001 Stability analysis of double-diffusive convection in superposed fluid and porous layers using a one-equation model. *Int. J. Heat Mass Transfer* **44**, 4625–4633. [9.4]

Zhao, T. S. and Liao, Q. 2000 On capillary-driven flow and phase-change heat transfer in a porous structure heated by a finned surface: measurements and modeling. *Int. J. Heat Mass Transfer* **43**, 1141–1155. [3.6]

Zhao, T. S. and Song, Y. J. 2001 Forced convection in a porous medium heated by a permeable wall perpendicular to the flow direction: analyses and measurements. *Int. J. Heat Mass Transfer* **44**, 1031–1037. [4.16.2]

Zhao, T. S., Cheng, P. and Wang, C. Y. 2000b Buoyancy-induced flows and phase-change heat transfer in a vertical capillary structure with symmetric heating. *Chem. Engng. Sci.* **55**, 2653–2661. [11.9.3]

Zhao, T. S., Liao, Q. and Cheng, P. 1999e Variations of buoyancy-induced mass flux from single-phase to two-phase flow in a vertical porous tube with constant heat flux. *ASME J. Heat Transfer* **121**, 646–652. [11.9.3]

Zhekamukhov, M. K. and Zhekamukhova, I. M. 2002 On convective instability of air in the snow cover. *J. Engng. Phys. Thermophys.* **75**, 849–858. [11.1]

Zheng, W., Robillard, L. and Vasseur, P. 2001 Convection in a square cavity filled with an anisotropic porous medium saturated with water near 4 degrees C. *Int. J. Heat Mass Transfer* **44**, 3463–3470. [7.3.2]

Zhou, M. J. and Lai, F. C. 2002 Aiding and opposing mixed convection from a cylinder in a saturated porous medium. *J. Porous Media* **5**, 103–111. [8.1.3]

Zhu, J. and Kuznetsov, A. V. 2005 Forced convection in a composite parallel plate channel: modeling the effect of interface roughness and turbulence using a $k-\varepsilon$ model. *Int. Comm. Heat Mass Transfer* **32**, 10–18. [1.8]

Zhu, N. and Vafai, K. 1996 The effects of liquid-vapor coupling and non-Darcian transport on asymmetrical disk-shaped heat pipes. *Int. J. Heat Mass Transfer* **39**, 2095–2113. [3.6]

Zhu, N. and Vafai, K. 1999 Analysis of cylindrical heat pipes incorporating the effects of liquid-vapor coupling and non-Darcian transport—a closed form solution. *Int. J. Heat Mass Transfer* **42**, 3405–3418. [11.9.2]

Zili, L. and Ben Nasrallah, S. 1999 Heat and mass transfer during drying in cylindrical packed beds. *Numer. Heat Transfer A* **36**, 210–228. [3.6]

Zili-Ghedira, L, Slimi, K. and Ben Nasrallah, S. 2003 Double diffusive natural convection in a cylinder filled with moist porous grains and exposed to a constant wall heat flux. *J. Porous Media* **6**, 123–136. [3.6]

Zimmerman, W., Painter, B. and Behringer, R. 1998 Pattern formation in an inhomogeneous environment. *Europ. Phys. J. B* **5**, 757–770. [11.2]

Zimmermann, W., Seeβelberg, M. and Petruccione, F. 1993 Effects of disorder in pattern formation. *Phys. Rev. E* **48**, 2699–2703. [6.9.1]

Zukauskas, A. 1987 Convective heat transfer in cross flow. *Handbook of Single-Phase Convective Heat Transfer* (eds. S. Kakac, R. K. Shah and W. Aung), Wiley, New York, Chapter 6. [4.15]

Index